Vascular Flora of Illinois

Robert H. Mohlenbrock

Vascular Flora of Illinois

A Field Guide

FOURTH EDITION

Southern Illinois University Press / *Carbondale*

First edition published as *Guide to the Vascular Flora of Illinois* in 1975. Second edition published as *Guide to the Vascular Flora of Illinois,* revised and enlarged edition, in 1986. Third edition published as *Vascular Flora of Illinois* in 2002.
Fourth edition 2014.
Printed in the United States of America

17 16 15 14 4 3 2 1

Library of Congress Cataloging-in-Publication Data

Mohlenbrock, Robert H., 1931–
Vascular flora of Illinois : a field guide / Robert H. Mohlenbrock — 4th ed.
p. cm.
"First edition published as Guide to the vascular flora of Illinois in 1975. Second edition published as Guide to the vascular flora of Illinois, revised and enlarged edition, in 1986. Third edition published as Vascular flora of Illinois in 2002."
Includes indexes.
ISBN 978-0-8093-3208-3 (pbk. : alk. paper) — ISBN 0-8093-3208-6 (pbk. : alk. paper) — ISBN 978-0-8093-3209-0 (ebook) — ISBN 0-8093-3209-4 (ebook)
1. Plants—Illinois—Identification. I. Title.
QK157.M646 2013
581.9773—dc23 2013005013

The paper used in this publication meets the minimum requirements of American National Standard for Information Sciences—Permanence of Paper for Printed Library Materials, ANSI Z39.48-1992. ♾

This book, my sixty-third, is dedicated to the following people who have played major roles in my career as botanist and teacher:

Miss E. Esther Smith, a dedicated and devoted biology teacher at Murphysboro (Illinois) Township High School, who introduced me to the wonderful world of plants and encouraged me to continue my study of plants

John W. Voigt, my master's advisor, who later coauthored two books with me, was a colleague of mine for thirty years at Southern Illinois University Carbondale, and was a cherished friend

Robert E. Woodson Jr. and Rolla M. Tryon Jr., of the Missouri Botanical Garden, who taught me the fundamentals of plant taxonomy and how to teach them while I pursued my doctorate at Washington University, St. Louis

Julian A. Steyermark, premier botanist, who shared with me an infectious enthusiasm for plants, which I have never lost, while he shared my office as a visiting professor in the spring of 1959

my parents, Robert H. and Elsie T. Mohlenbrock, who encouraged me to pursue my dreams at every stage of my life

my dear wife, Beverly Ann Kling Mohlenbrock, who has inspired me, assisted me, encouraged me, and been my constant companion for nearly sixty years

Contents

Preface

When Andre Michaux came to Illinois Territory for a few months during 1795, he apparently was the first person to collect plants in what is now Illinois. Among his collections were fifty-two species that were new to science. Since that time, countless botanists, both amateur and professional, have been contributing to the knowledge of the plants of this state.

The first flora of the state was prepared by Dr. Samuel B. Mead, a physician from Augusta, Illinois, when he wrote *Catalogue of Plants Growing Spontaneously in the State of Illinois, the Principal Part near Augusta, Hancock County,* in 1846. In 1857, Increase A. Lapham published *Catalogue of the Plants of the State of Illinois.* In 1876, Harry N. Patterson, a printer and botanist from Oquawka, Illinois, published *Catalogue of the Phaenogamous and Vascular Cryptogamous Plants of Illinois.*

Between 1876 and 1945, several people published plant lists and floras for various parts of Illinois, but the next statewide flora appeared in 1945, when Dr. G. Neville Jones, a professor of botany at the University of Illinois, published his first edition of *Flora of Illinois.* Jones followed with his second edition in 1950, and his third edition in 1963. My subsequent *Guide to the Vascular Flora of Illinois* was published in 1975, with a second revised and enlarged edition published in 1986 and a third edition published in 2002. The following chart shows the number of taxa (species and hybrids) known from Illinois between 1846 and this volume:

Flora	*Species and Hybrids*
Mead (1846)	900
Lapham (1857)	1,127
Patterson (1876)	1,542
Jones (1945)	2,125
Jones (1950)	2,200
Jones (1963)	2,400
Mohlenbrock (1975)	2,782
Mohlenbrock (1986)	2,954
Mohlenbrock (2002)	3,248
Mohlenbrock (2014)	3,598

The publication of each flora seems to stimulate interest among Illinois's plant collectors, so that each succeeding flora has a substantial number of additions.

With the rapid increase in Illinois's population and with the subsequent increase in disturbed areas, many new non-native species are added to the flora each year. I am including all of these additions to the flora based on the tradition that any plant should be included in a flora if it grows spontaneously (that is, not having been purposely planted) and persists for one year or more.

In addition, previously little visited patches of prairies, wetlands, and forested canyons have yielded several new native species. While there has been intense exploration for plants throughout the state, I also have been researching in herbaria in an attempt to verify previous reports of records of Illinois plants.

Many plants have not been seen in Illinois during the last fifty years, and several of these that are native may be extirpated from the state. Ebinger et al. (2009) discussed the status of some of these presumably extirpated native species. I have not indicated in this book whether a plant is extirpated or not. However, Ebinger et al. stated that some of the plants were misidentified originally, and that is true for a few of them. Other specimens could not be located, including a few that were in the herbarium at Southern Illinois University Carbondale (SIUC).

Several persons have alerted me to new additions to the Illinois flora. I am indebted to Gerould Wilhelm, Gordon Tucker, Scott Kobel, Robbie Sliwinski, Ken Klick, Michael Hayes, John Schwegman, Eric Ulaszek, and Jack Brooks for their cooperation.

I subtitled this book *A Field Guide* on purpose because it is intended to be used by persons in the field who study the obvious morphological features that the plants exhibit. I am aware of what is called the angiosperm group, which is using phytochemical techniques to try to figure the relationships of plants. This group has published three papers, in 1998, 2002, and 2009, reorganizing plant families based on phytochemical experiments. Their conclusions are revolutionary, and a number of botanists have followed their recommendations.

The question is whether phytochemical characters are more important than morphological characters. For a field botanist who needs to rely on morphological characters, the answer is a resounding "No."

While some of the phytochemical studies may come to some of the same conclusions as morphological studies, several unexpected conclusions have been proposed. For example, the Boraginaceae has been merged with the Hydrophyllaceae, overlooking the fact that the Boraginaceae has axile placentation and usually 2–4 ovules, while the Hydrophyllaceae has parietal placentation and numerous ovules.

It is also surprising to see that woody genera such as *Viburnum* and *Sambucus* have been removed from the Caprifoliaceae and merged with *Adoxa,* a small herb. All are now in the Adoxaceae, in the view of the phytochemists. Likewise, placing the small herbaceous *Parnassia* in the Celastraceae, which contains *Celastrus* (the bittersweets) and *Euonymus* (trees and shrubs), is equally unexpected. From a morphological standpoint, it is beyond my imagination to place *Mazus* (a group of small figworts), *Mimulus* (larger figworts), and *Phryma* (lopseeds) in the Phrymaceae.

The Scrophulariaceae has suffered the greatest destruction, with only *Verbascum* (with 5 petals forming an actinomorphic flower) and *Scrophularia* (with 4 petals forming a zygomorphic flower) remaining in the Scrophulariaceae.

While the phytochemists have data that may be interpreted as showing relationships among groups that we have not considered in the past, phytochemistry is only one of many tools and cannot replace the study of morphological features.

I have been studying plants in the field for more than sixty years, identifying them by the morphological features each plant exhibits. The phytochemical characters are not usable in the field and leave one with the distinct impression that the new system based on these characters is not consistent with morphological studies.

This book reflects, for the most part, the traditional relationships that have evolved from George Bentham to Engler and Prantl to John Hutchinson to Arthur Cronquist. Because the new proposed family arrangements do not necessarily have a definitive set of morphological characters, it is impossible to construct a workable key to families.

It is gratifying to look at plants, to know what they are, and to give them their correct names. To me, each plant is to be approached more as a friend, with a set of features I recognize, not as a group of mysterious interacting molecules.

Of the 3,598 species and hybrids of vascular plants known from Illinois to date, 1,156, or about 30 percent, are not native. In 2011, the Illinois Endangered Species Protection Board issued a list of the endangered and threatened plants of Illinois.

There has been an increase in the number of genera in Illinois in this book. This is due mostly to a reinterpretation of existing genera. Recent biosystematic techniques have begun to substantiate the genera that botanists such as John Kunkel Small, Per Axel Rydberg, and Edmund C. Greene proposed a century ago. During the last twenty years, while I have had the opportunity to spend most of my time in the field instead of splitting my time between field work and administrative and lecturing duties, I have made a concentrated effort to look into some of the more difficult genera in Illinois. I have paid particularly close attention in the field to plants in *Carex, Crataegus, Dichanthelium, Rubus, Rudbeckia, Solidago, Symphyotrichum,* and *Xanthium.* I am recognizing several species in these genera that many botanists do not believe should be given species status. However, having studied many of the types of *Rubus* named by Bailey, the Davises, and others, and many of the types of *Crataegus* named by Sargent, I believe many of the differences they observed are as important as some of the small differences exhibited by certain species of *Carex,* which few botanists seem to object to. Rationale for my treatments of *Rudbeckia, Solidago, Symphyotrichum,* and *Xanthium* will be explained in my forthcoming volumes on the Asteraceae in the Illustrated Flora of Illinois series.

Since the last edition in 2002, many studies have altered the status of a great number of species. Some of these changes I agree with; some I do not, based on my detailed observations as a field botanist.

The sequence of groups in this book is ferns, conifers, and flowering plants, with dicotyledons given before monocotyledons. Within each group, the families are arranged alphabetically, as are the genera within each family and the species within each genus.

For each taxon recognized in this book, there is given a common name if one is generally used in Illinois. This is followed by an indication of flowering time for flowering plants and of spore production time for ferns and their relatives. A habitat statement and a general comment on its distribution in Illinois are given for each taxon. For some other scientific names used previously for a taxon, those synonyms are printed in italics.

I would like to thank the curators of all the herbaria I have studied in for the privilege of utilizing their collections. I wish to thank the Illinois Nature Preserves Commission, which kindly let me use *The Natural Divisions of Illinois* in its entirety, and Mr. John Schwegman, who was the principal author. I would also like to thank Madison Preece who prepared the glossary and the indexes. I have taken the liberty of changing the scientific names of some of these plants to coincide with the nomenclature used in this book.

County Map of Illinois

Vascular Flora of Illinois

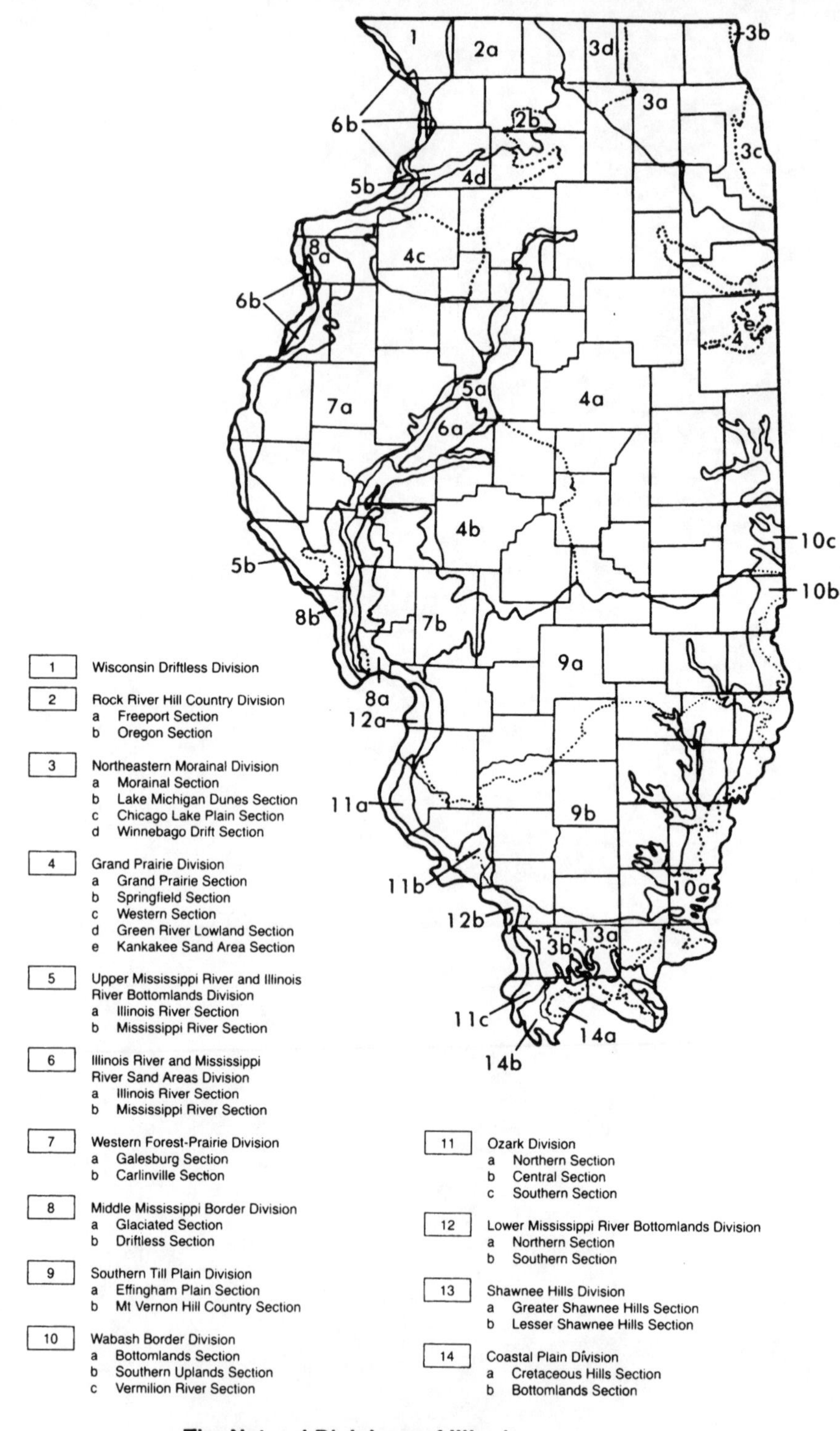

The Natural Divisions of Illinois

Courtesy of the Illinois Nature Preserves Commission

The Natural Divisions of Illinois

Introduction

The Natural Divisions of Illinois[1] is a classification of the natural environments and biotic communities of Illinois based on physiography, flora, and fauna. The general approach was first applied to Illinois by Vestal,[2] who developed a preliminary vegetation map of the state based mainly on topography, soil type, and natural vegetation.

The natural divisions of Illinois have been derived through an expansion of Vestal's approach. Factors considered in delimiting the 14 natural divisions are topography, soils, bedrock, glacial history, and the distribution of plants and animals. The divisions are divided into 33 sections based on differences that are of the same nature as those used to separate the divisions but deemed less significant. The role of each factor in developing this system is discussed in the text.

Topography

Glacial history has played an important role in shaping Illinois's topography. Except for relatively small driftless areas in northwestern, west-central, and extreme southern Illinois, the topography can be characterized as a more or less dissected plain of glacial till. Rugged topography is characteristic of the driftless areas and of much of the till plain along the major river valleys. Other areas of significant relief within the till plain are the youthful moraines of northeastern Illinois and areas of thin drift and rolling topography in the Rock River valley and southeastern Illinois. The remainder of this till plain is an arc of older, eroded Illinoian till south and west of the younger, less dissected Wisconsinan till.

Topography has influenced the biota of Illinois by controlling the diversity of habitats. In the glaciated regions, forest was restricted mainly to the rugged topography of stream valleys and moraines, while prairie occupied most of the level uplands and some of the broad floodplains. In general, the more rugged the topography, the greater the diversity of habitats. Youthful topography, such as the Wisconsinan till plain, is poorly drained, resulting in an abundance of aquatic habitats. Drainage has been improved by ditching in most prairie regions of the till plain.

1. Schwegman, J. E., G. D. Fell, M. Hutchison, G. Paulson, W. M. Shepherd, and J. White. 1973. Comprehensive Plan for the Illinois Nature Preserves System. Part II—The Natural Divisions of Illinois. Illinois Nature Preserves Commission, Springfield. 32 pages, plus map.
2. Vestal, A. G. 1931. A Preliminary Vegetation Map of Illinois. Transactions of the Illinois State Academy of Science 23: 204–17.

Soils

The principal soil features influencing the recognition of divisions and sections are sand areas, soils derived from glacial till, deep loess soils, and claypan soils. The description of soil conditions within the divisions is only generalized. Alluvial soils occur in every division, but they are not usually mentioned in the text.

Some of the most important features of soils are parent material, texture, and degree of development. The general terms "heavy soils," "light soils," and "sandy soils" refer to texture. Heavy soils are predominantly clay; moisture generally moves slowly through such soils, and aeration is poor. They often provide a more rigorous environment for plant growth. Light soils contain a mixture of clay, silt, and sand, are well aerated, and readily yield moisture to plants. Sandy soils are well aerated and have poor moisture-holding ability. They provide a severe environment for plants because of their droughtiness and susceptibility to wind erosion.

The parent material for most Illinois soils is loess, a wind-deposited silt. Loess deposits are deepest near the major river valleys and become thinner and of finer texture with increased distance. This mantle of loess has reduced the diversity of soil types in much of the state. Deep loess soils are generally less weathered and more fertile than those derived from thin loess, especially in the southern third of Illinois.

Soils derived primarily from glacial till are restricted to northeastern Illinois, where geographic position has prevented deep accumulations of loess. The texture of these soils ranges from gravel to clay. The diversity of soil parent material is partly responsible for the great variety of environments and varied biota of this region.

Sand derived from glacial outwash is another important parent material that occurs as terraces along major streams throughout Illinois. In some instances, the soils developed from sand support biota distinctive enough to warrant recognition in the natural divisions. Local sand areas are not listed because they are generally not considered to be of statewide significance.

Alluvial soils tend to be lighter in central and northern Illinois than in southern Illinois. They frequently support distinctive biota.

Peat and bedrock are parent materials that give rise to soils that often support unusual plants and animals. Although limited in area, they have contributed to the recognition of some divisions and sections.

The degree of soil development is a function of time and climate. Older soils tend to be more strongly developed and have many of their soluble minerals and clay particles leached from the surface layers. The same is true of soils developed in areas of higher precipitation, as in southern Illinois. A strongly developed soil may have a claypan formed by the accumulation of clay particles in the subsoil. Claypan restricts penetration by roots and movement of groundwater, thus sometimes determining the dominant vegetation over wide areas. For example, the strongly developed claypan soils of south-central Illinois determine to a great extent the distribution of post oak flatwoods. These soils are a primary factor in the recognition of the Southern Till Plain Division.

Younger soils are derived from relatively recent deposits of till, loess, or alluvium or exposure of bedrock. They are not strongly developed, are relatively rich in minerals, and lack claypans.

Bedrock

Bedrock differences are often reflected in the topography, especially in driftless areas. Bedrock also controls the plant life to some extent through the influence it has on the thin soils developed from it. Bedrock type can determine the presence of caves and other habitats.

The Ozark Division shows the effect of bedrock differences. The Northern Section (Monroe County) and the Southern Section (Union and Alexander counties) are both parts of the Salem Plateau of the Ozarks and were presumably uplifted at about the same time. The Southern Section is underlain by cherty limestone, which is resistant to erosion and has produced very rugged topography with surface drainage. Steep cherty slopes in this section support acid-soil plants such as shortleaf pine, azalea, and farkleberry. On the other hand, the Northern Section is underlain by relatively pure limestone that is less resistant to erosion and groundwater solution, producing gentler topography with some internal drainage through sinkholes and solution cavities. Natural sinkhole ponds are common. Most of the acid-soil plants are lacking, and the flora includes a distinctive assemblage of Ozark "limestone glade" species.

Natural Vegetation

The dominant species and some of the most frequent associates in the principal plant communities of each natural division are listed in the text, but no attempt is made to characterize the total vegetation of each division. Generally, the descriptions pertain to natural vegetation in Illinois about the time of settlement. At present, this vegetation has been completely destroyed by man over vast sections of the state.

The striking contrast between prairie and forest vegetation that existed at the time of settlement is of little consequence to the recognition of natural divisions, as both forest and prairie or prairielike communities occur in every division. However, the relative abundance of the prairie community was important in delimiting some divisions.

At the time of the settlement of Illinois, there were considerable areas termed "barrens" by land surveyors. Apparently the barrens developed into forest or were cleared for agriculture soon after settlement. In any event, they were destroyed before being thoroughly described. They probably occurred in all the natural divisions, but they are not included in the vegetation accounts because their species composition is unknown.

Flora

An analysis of the distribution of vascular plants in relation to physiographic provinces, bedrock types, soil types, topography, and drainage systems reveals a considerable number of distinctive species apparently restricted to certain provinces or natural features. These restricted distributions, especially for such dominant plants as trees, were important in delimiting the natural divisions. Many of these distinctive plants are listed in the descriptions of the divisions.

Fauna

The distribution of birds in Illinois is less adequately documented than that of other classes of vertebrates. Nor does their distribution correlate as well with the natural divisions recognized here, being more heavily influenced by climate and by artificial and transitory conditions. For these reasons, distinctive species of birds are mentioned in the descriptions of only a few natural divisions.

Examination of the distribution of non-avian vertebrates in Illinois reveals many correlations with physiographic features, vegetation, watersheds, and soil types. The present work indicates considerable correlation between the natural divisions and patterns of fish distribution and to a lesser extent with those of mammals.

The vertebrates that are considered distinctive of certain divisions and species that have influenced the location of divisional boundaries are mentioned in the descriptions of the divisions. There is no attempt to characterize the fauna of each division.

No attempt has been made to examine the distribution of invertebrates or their relation to the natural divisions. The plains scorpion is mentioned as a distinctive feature of the Ozark Division, but this is the only exception.

Climate

Climate has played a role in the delineation of some divisions and sections by its effects on the distribution of plants and animals, especially in extreme northern Illinois and southern Illinois. However, there is little direct correspondence between climatic regions and the natural divisions.

Illinois has a continental climate with hot humid summers and cold winters. This climate varies considerably from north to south in temperature and precipitation. The northern counties have lower mean annual temperatures, shorter growing seasons, and less precipitation than the southern counties.

The mean annual temperature ranges from about 47°F in the north to about 59°F in the south. The mean January temperature ranges from 22°F in the north to 36°F in the south, and the mean July temperature ranges from 73°F in the north to 80°F in the south. The average length of the growing season ranges from less than 160 days in northwestern Illinois to more than 200 days in the southern tip of the state.

Average annual precipitation increases from about 32 inches in the north to 47 inches in the south. Although the southern counties receive the most rainfall, much of this occurs in winter. Precipitation during the growing season is more nearly uniform throughout the state.

Wisconsin Driftless Division

The Wisconsin Driftless Division is part of an area extending from northwestern Illinois into Wisconsin, Iowa, and Minnesota that apparently escaped Pleistocene glaciation. This division is one of the most maturely developed land surfaces in Illinois and is characterized by rugged terrain that originally was mostly forested. It has the coldest climate in the state. It contains several distinctive plants of northern affinity and some species that may represent relicts of the pre-ice-age flora. The division contains lead deposits.

Bedrock

The Wisconsin Driftless Division is a maturely dissected upland of Ordovician and Silurian limestone, dolomite, and shale. Bedrock crops out along the major watercourses. Prominent "mounds" capped with the more resistant dolomite are common. A mineralized zone containing deposits of lead and zinc is an important feature. Caves are known in the dolomite.

Topography

The topography of the Wisconsin Driftless Division is one of rolling hills and great relief, particularly along interior stream canyons. High erosional remnants (including Charles Mound, the highest point in Illinois, with an elevation of 1,257 feet) are prominent features. There are loess-capped bluffs and palisades along the Mississippi River valley and ravines and bluffs throughout the division.

Soils

The soils of this division have developed from loess or, on steeper slopes, from loess on bedrock. The loess soils are derived from thick deposits and are weakly to moderately developed. The soils on bedrock are thin to moderately thick and well drained.

Plant Communities

Forest. The original vegetation of the Wisconsin Driftless Division was predominantly upland hardwood forest dominated by black oak (*Quercus velutina*) and white oak (*Quercus alba*) on dry sites and by sugar maple (*Acer saccharum*), basswood (*Tilia americana*), and red oak (*Quercus rubra*) on mesic sites. Floodplain forests dominated by silver maple (*Acer saccharinum*), American elm (*Ulmus americana*), and green ash (*Fraxinus lanceolata*) occupy alluvial soils of the stream valleys. Cliffs and cool, shaded slopes of this division often support white pine (*Pinus strobus*), Canada yew (*Taxus canadensis*), and white birch (*Betula papyrifera*).

Prairie. Dry prairie on the rolling uplands contained such species of the northern Great Plains as the plains buttercup (*Ranunculus rhomboideus*), pasque flower (*Anemone patens*), June grass (*Koeleria macrantha*), and Wilcox's panic grass (*Dichanthelium wilcoxianum*), along with the dominant little bluestem (*Schizachyrium scoparium*) and side-oats grama (*Bouteloua curtipendula*). Areas of mesic and wet prairie were infrequent. Frink's Prairie and Jules Prairie were two extensive prairies on the uplands. Loess hill prairies dominated by little bluestem and side-oats grama occur on the steep southwest-facing bluffs above the Mississippi River floodplain.

Aquatic Habitats

The main aquatic habitats of the Wisconsin Driftless Division are creeks and rivers. Springs are local, and intermittent streams are characteristic of the ravines.

Special Features

The Wisconsin Driftless Division contains several distinctive plants considered relicts of preglacial or interglacial floras, such as jeweled shooting star (*Dodecatheon amethystinum*), sullivantia (*Sullivantia sullivantii*), and cliff goldenrod (*Solidago sciaphila*). Cool, shaded ravines, cliffs, and river bluffs provided habitat for relict populations of some distinctive northern plants including woodland white violet (*Viola blanda*), bird's-eye primrose (*Primula *mistassinica*), American stickseed (*Hackelia americana*), and moschatel (*Adoxa moschatellina*). Some of these distinctive plants are restricted to the river bluffs of the Mississippi River valley, and others grow only in interior stream canyons.

Principal Natural Features

Forest: Dry upland, mesic upland, floodplain.
Prairie: Dry, mesic, wet.
Loess Hill Prairie
Bedrock: Outcrops of dolomite and shale, zinc and lead deposits, caves.
Topography: Mississippi River bluffs, interior stream canyons, ravines, ridges and "mounds," level to rolling upland.
Aquatic Habitats: Creeks, rivers.
Special Features: Northern and preglacial or interglacial plants.

Rock River Hill Country Division

The Rock River Hill Country Division is a region of rolling topography that is drained by the Rock River. It has a thin mantle of glacial till. Prairie formerly occupied the larger expanses of level uplands, but forest was equally abundant along the watercourses and in the more dissected uplands. Several distinctive plant species occur in this division. The Freeport and Oregon sections are recognized on the basis of bedrock types and resultant floral differences.

Glacial History

The Rock River Hill Country Division is thinly mantled with glacial drift from the Illinoian and early Wisconsinan stages of Pleistocene glaciation. The Pecatonica lobe of the Altonian substage of the Wisconsinan glacial stage extended westward in this division into Stephenson and Ogle counties.

Bedrock

The bedrock is primarily Ordovician and Silurian dolomite and limestone. Outcrops of the dolomite occur throughout the division, particularly along the streams. St. Peter sandstone underlies the Oregon Section and crops out frequently. There are caves in the dolomite near the Mississippi River.

Topography

The topography varies from level to rolling, with river valleys, and "dells" or bluffs along streams throughout the division. The Oregon Section is very rough, with bluffs, ridges, and ravines in the sandstone. The Pecatonica River meanders through a broad plain formed by sediments of glacial lake.

Soils

The soils have developed primarily from moderately thick loess and thin loess on bedrock. Soils developed from glacial outwash occur in the major river valleys. Small areas of soils developed from glacial till occur throughout the division.

Plant Communities

Forest. The forests are similar to those of the Wisconsin Driftless Division. They occurred on slopes and areas protected from prairie fires. The dry upland forests are dominated by black oak, white oak, bur oak (*Quercus macrocarpa*), and wild black cherry (*Prunus serotina*). Mesic sites support forests dominated by sugar maple, basswood, slippery elm (*Ulmus rubra*), and red oak. White pine, Canada yew, and yellow birch (*Betula lutea*) are occasionally found on cool, shaded bluffs. The floodplain forests are particularly well developed along the Pecatonica River and are dominated by silver maple, black willow (*Salix nigra*), cottonwood (*Populus deltoides*), American elm, and ashes (*Fraxinus* spp.). The Oregon Section is heavily forested with black oak, white oak, and bur oak.

Prairie. Prairies occupied much of the level to rolling uplands. The largest was known as Shannon Prairie. The dry upland prairies contained the floral elements of the northern Great Plains listed for the Wisconsin Driftless Division. Mesic prairies were the most common type and contained the species typical of the Grand Prairie, the dominant grasses being big bluestem (*Andropogon gerardii*), Indian grass (*Sorghastrum nutans*), and prairie dropseed (*Sporobolus heterolepis*). Wet prairies contained cord grass (*Spartina pectinata*), bluejoint grass (*Calamagrostis canadensis*), and big bluestem.

Marsh. Marshes in poorly drained parts of the prairies and along the streams and river floodplains are dominated by cattail (*Typha* sp.), bulrushes (*Scirpus* spp.), sedges (Cyperaceae), and occasionally common reed (*Phragmites australis*).

Aquatic Habitats

The major aquatic habitats of the Rock River Hill Country Division are rivers and creeks. Meander scar sloughs characterized by Pecatonica River floodplain. Springs and seepage areas occur locally, especially in the sandstone areas of the Oregon Section.

Freeport Section

The Freeport Section includes most of the Rock River Hill Country Division and is characterized by rolling hills and dolomite and limestone bedrock.

Principal Natural Features

Forest: Dry upland, mesic upland, floodplain.
Prairie: Dry, mesic, wet.
Marsh Bedrock: Outcrops of dolomite and limestone caves.
Topography: Stream canyons and bluffs, floodplain, rolling uplands, meander scars.
Aquatic Habitats: Rivers, creeks, sloughs.
Special Feature: Northern relict plants.

Oregon Section

The Oregon Section is distinguished from the rest of the division by its sandstone bedrock and the unique northern plants associated with it. These distinctive northern relicts include ground pine (*Dendrolycopodium dendroideum*), rusty woodsia (*Woodsia ilvensis*), oak fern (*Gymnocarpium dryopteris*), and American wintergreen (*Pyrola americana*).

Principal Natural Features

Forest: Dry upland, mesic upland, floodplain.
Prairie: Dry, mesic, wet.
Bedrock: Sandstone outcrops.

Topography: Rolling uplands, floodplain, ravines, bluffs.
Aquatic Habitats: Rivers, creeks.
Special Feature: Northern relict plants.

Northeastern Morainal Division

The Northeastern Morainal Division is the region of most recent glaciation in Illinois. Glacial landforms are common features and are responsible for the rough topography over most of the area. Lake-bed deposits and beach sands are also frequent features. Unlike most of Illinois, the soils of this division are derived from glacial drift rather than loess. Drainage is poorly developed, and many natural lakes are found. This division contains distinctive northern and eastern floral elements including the bog community. Several species of animals are known in Illinois only from this area.

The sections are recognized because of differences in topography, soil, glacial history, flora, and fauna.

Bedrock

The bedrock is primarily Ordovician and Silurian limestone and dolomite with some shale. The bedrock is deeply buried by glacial drift, but limestone crops out along some of the streams.

Glacial History

The Northeastern Morainal Division is covered with deep glacial drift from the Altonian and Woodfordian substages of the Wisconsinan glacial stage. Moraines, kames, eskers, and other glacial landforms occur throughout the division.

Topography

Moraines and morainic systems are dominant topographic features and account for the rough, hilly, and rolling terrain of most of the division. There are outwash plains at the fronts of major terminal moraines, such as the Marengo Ridge. The Chicago lake plain and ancient beach ridges are prominent features of the Chicago area and were formed by sediments of glacial Lake Chicago. Bluffs were formed along Lake Michigan north of Chicago during high-water stages of glacial Lake Chicago.

Sand dunes are present along Lake Michigan north of Waukegan and east of the Sugar River in Winnebago County. The Lake Michigan dunes are well developed and are associated with the beach-and-dunes association. Ridges and swales occur in the sand area north of Waukegan and in the Chicago lake plain.

Soils

The soils are derived primarily from glacial drift, lake bed sediments, beach deposits, and peat. They range from very poorly drained to well drained on the uplands. They are diverse in texture, ranging from gravel and sand to silty clay loams. The many different soils are responsible for the diversity of plant communities found in this division.

Plant Communities

Forest. Bur oak and white oak dominate the dry upland forests on moraines and other glacial landforms of this division. Many of the forests were typical oak openings but have been heavily grazed or have rapidly grown up to closed, dense forests due to the effective exclusion of fire. The mesic upland forests were dominated by sugar maple, basswood, red oak, and white ash (*Fraxinus americana*). Notable northern shrubs of these mesic forests are highbush cranberry (*Viburnum trilobum*) and red elder (*Sambucus racemosus* var. *pubens*). Beech (*Fagus grandifolia*) occurred in a few ravines along Lake Michigan. The dominant species of the floodplain forests are silver maple, green ash, and American elm. Some of the poorly drained upland forests are dominated by swamp white oak (*Quercus bicolor*). Black oak and Hill's oak (*Quercus ellipsoidalis*) occur on the sandy soils of the dunes and ridges, forming savannalike or scrub-oak forests. Jack pine (*Pinus banksiana*) and white pine once grew on sandy ridges in the Chicago lake plain. Tamarack (*Larix laricina*) occurs in some poorly drained depressions of the Valparaiso morainic system. Plants essentially restricted in Illinois to the forests of the Northeastern Morainal Division include maple-leaf viburnum (*Viburnum acerifolium*), round-lobed hepatica (*Hepatica nobilis* var. *obtusa*), wood reed (*Cinna latifolia*), large-leaved aster (*Eurybia macrophylla*), moccasin flower (*Cypripedium acaule*), and purple trillium (*Trillium erectum*).

Prairie. The presettlement vegetation of the Northeastern Morainal Division was about 60 percent prairie. Dry prairie on the gravel moraines and eroded bluffs was dominated by little bluestem and side-oats grama and contained several species typical of the western plains. The mesic prairie and wet prairie were dominated by prairie dropseed, big bluestem, Indian grass, switch grass (*Panicum virgatum*), cord grass, and bluejoint grass and contained many characteristic forbs. Extensive areas of sand prairie occurred in the sand areas of the Chicago lake plain, east of the Sugar River in Winnebago County, and in the Lake Michigan Dunes Section. Some distinctive plants of sandy prairie and sandy open woods are sweet-fern (*Comptonia peregrina*), speckled alder (*Alnus rugosa*), yellow fringed orchid (*Platanthera ciliaris*), fringed gentian (*Gentianopsis crinita*), and small fringed gentian (*Gentianopsis virgata*). Colic-root (*Aletris farinosa*), hardhack (*Spiraea tomentosa*), and lupine (*Lupinus perennis*) occur in sand prairies near Lake Michigan.

Fen. A fen is a type of wet prairie with an alkaline water source. Fens are associated with calcareous springs and seeps and also occur in swales and on low ground near lakes in areas of calcareous groundwater. They warrant recognition because of their distinctive species composition. Some of the notable plants are small white lady's slipper (*Cypripedium candidum*), grass-of-Parnassus (*Parnassia glauca*), meadow spikemoss (*Selaginella apoda*), Ohio goldenrod (*Oligoneuron ohioense*), Kalm's lobelia (*Lobelia kalmii*), shrubby cinquefoil (*Dasiphora fruticosa*), low calamint (*Clinopodium arkansanum*), and white camass (*Zigadenus glaucus*).

Marsh. Marshes are conspicuous plant communities of the Northeastern Morainal Division, common because of the poorly drained soils. Marshes are generally dominated by cattails and bulrushes, with common reed locally abundant.

Sedge meadow. Many of the marshes of this division grade into an open community where sedges dominate instead of cattails and bulrushes, forming sedge meadows. These are often associated with wet-shrub communities dominated by dogwoods (*Cornus* spp.), willows (*Salix* spp.) and, infrequently, speckled alder (*Alnus rugosa*).

Bog. True bogs are found in Illinois only in the Northeastern Morainal Division. These have formed in poorly drained depressions in the Valparaiso morainic system and contain many distinctive plants such as a pitcher plant (*Sarracenia purpurea*), sundew (*Drosera* sp.), cranberry (*Oxycoccus macrocarpus*), leatherleaf (*Chamaedaphne calyculata*), poison sumac (*Toxicodendron vernix*), winterberry (*Ilex verticillata*), and dwarf birch (*Betula pumila*). All stages of bog succession—young, mature, and old—are represented.

Aquatic Habitats

The Northeastern Morainal Division is poorly drained and has many aquatic habitats. It contains all of Illinois's glacial lakes and is the only division on Lake Michigan. The glacial lakes are generally of two types: those with a peat base and those with a sand or marl base. The biota of each type is different. Some species of water marigold (*Megalodonta beckii*) and two species of bladderwort (*Utricularia* spp.) are restricted to the glacial lakes of this division.

Distinctive Fauna

The pugnose shiner, blackchin shiner, and banded killifish are known in Illinois only from the glacial lakes of this division. The alewife, American smelt, lake chub, and ninespine stickleback are fishes that occur in Illinois only along the shore of Lake Michigan. Lake trout and lake whitefish are species restricted to the deeper waters of Lake Michigan. The common tern breeds in Illinois in only a few locations near Lake Michigan, and the piping plover did so until fairly recently. Breeding of the golden-winged warbler in Illinois is restricted to this division, where it has been known to hybridize with the blue-winged warbler. Breeding of the Nashville warbler in Illinois is reported only from this division. Except for casual occurrences, the large population of oldsquaws that winters on Illinois waters is restricted to Lake Michigan; and much the same is true for wintering white-winged, common, and surf scoters. The occurrence in Illinois of several species of migrant waterbirds such as the parasitic, pomarine, and long-tailed jaegers, the knot, and Bonaparte's and little gull is limited almost entirely to Lake Michigan or its shores. Other animals restricted to this division are the spotted turtle, blue-spotted salamander, and pigmy shrew.

Morainal Section

The Morainal Section of the Northeastern Morainal Division encompasses the moraines and morainic systems of the late advances of the Woodfordian substage of Wisconsinan glaciation. This section contains most of Illinois's glacial lakes as well as its true bogs. Glacial landforms are well represented.

Principal Natural Features

Forest: Dry upland, mesic upland, wet upland, floodplain, tamarack swamp.
Prairie: Dry, mesic, wet.
Fen
Marsh
Sedge Meadow
Bog: Youthful, mature, old.
Glacial Landforms: Moraines, kames, eskers, drumlins, kettle holes.

Topography: Rolling upland, ravines, lake bluffs.
Aquatic Habitats: Rivers, creeks, glacial lakes, sloughs.

Lake Michigan Dunes Section

This section is recognized because of the unique flora of its dunes and beaches. This flora includes some sand-binding plants that differ from those of the inland sand deposits, such as beach grass (*Ammophila breviligulata*), creeping juniper (*Juniperus horizontalis*), and bearberry (*Arctostaphylos uva-ursi*). Plant communities range from sand prairie, to marsh, to scrub-oak forest. The continuing vegetational succession from shifting sand to stabilized sand results in a wide variety of plant associations.

Principal Natural Features

Forest: Scrub-oak.
Sand Prairie: Dry, mesic, and wet.
Fen
Marsh
Topography: Beach, ridges and swales, dunes.
Aquatic Habitats: Creeks, Lake Michigan.

Chicago Lake Plain Section

The Chicago lake plain is a flat, poorly drained area of lake-bed sediments deposited by glacial Lake Chicago. It is recognized because of its special topography and physiographic history. Long ridges of shore-deposited sands are conspicuous topographic features. A few natural lakes occur near Calumet City. The original vegetation was mostly prairie and marsh, with scrub-oak forests on sandy ridges. Black gum (*Nyssa sylvatica*) and sassafras (*Sassafras albidum*) are found in some of the wet forests. Thismia (*Thismia americana*), one of the most unusual plants of the American flora, occurred only in the Chicago Lake Plain Section, near Lake Calumet, but is now considered extinct because this area has been converted to industry.

Principal Natural Features

Forest: Scrub-oak, mesic upland, floodplain.
Prairie: Dry, mesic, wet.
Fen
Marsh
Topography: Lake plain, ridges and swales.
Aquatic Habitats: Lakes, creeks, Lake Michigan.

Winnebago Drift Section

This section encompasses the Winnebago Formation of Altonian drift. This early Wisconsinan drift is better drained than that of the Morainal Section, having fewer marshes and no glacial lakes. The original vegetation was predominantly prairie, with oak openings, dry upland forest, and well-developed floodplain forests. Glacial outwash is extensive along many of the creeks and rivers. Wet prairie and marsh with large sedge meadows or "prairie bogs" occur in the sand area along Coon Creek. Dunes have formed along the east bank of the Sugar River and support sand prairie vegetation and dry upland forests of black oak and Hill's oak.

Distinctive features of this section are the extensive gravel hill prairies that once extended along the eroded east bluffs of the Rock River valley into Wisconsin. These prairies contained many western elements, including pasque flower, plains buttercup, and prairie smoke (*Geum triflorum*).

Principal Natural Features

Forest: Dry upland, mesic upland, floodplain.
Prairie: Dry, mesic, wet.
Sand Prairie: Dry, mesic, wet.
Marsh
Sedge Meadow
Bedrock: Outcrops of limestone and sandstone.
Topography: Dunes, outwash plains, river terraces, meander scars.
Aquatic Habitats: Rivers, creeks, sloughs.

Grand Prairie Division

The Grand Prairie Division is a vast plain formerly occupied primarily by tall-grass prairie. The soils were developed from recently deposited loess, lake-bed sediments, and outwash and are generally very fertile. Natural drainage was poor, resulting in many marshes and prairie potholes. Forest bordered the rivers, and there were occasional groves on moraines and other prominent glacial landforms. The sections of this division are differentiated on the basis of soils, topography, and glacial history.

At one time, bison grazed the prairies, and waterfowl in great numbers occupied the marshes and potholes. The steel plow brought about the rapid destruction of the vast Illinois prairies. Ditches and tile lines drained almost all of the marshes and potholes. The bison were gone by 1814. The abundant waterfowl were displaced. The giant Canada goose was extirpated as a breeding bird, and other characteristic species disappeared or became scarce. The prairie, once seemingly limitless, is now one of the rarest plant communities in Illinois, with only pitifully small and often degraded patches remaining.

Glacial History

The Grand Prairie is a rather level, poorly drained plain of glacial drift from the Illinoian and Wisconsinan stages of Pleistocene glaciation. Repeated advances and retreats of the Wisconsinan glaciers created a series of moraines and morainic systems of which the Shelbyville and Bloomington morainic systems are conspicuous.

Bedrock

The bedrock, deeply buried by glacial drift, crops out only along the larger rivers. Major outcrops of sandstone are found near Ottawa along the Illinois and Fox rivers. Dolomite crops out along the Kankakee River west of Kankakee.

Topography

The topography of the Grand Prairie Division is generally level to rolling, with the major stream valleys and the extensive systems of moraines providing the greatest relief. Large, flat expanses of lake-bed deposits are found in LaSalle, Kendall, Will, Grundy, Livingston, Ford, Iroquois, Kankakee, and Douglas counties. Extensive outwash plains and sand dunes are found in Kankakee and Iroquois counties and in the valleys of the Green River and lower Rock River. The major rivers have well-developed floodplains, and in many areas there are ravines in the bluffs.

Soils

The soils are relatively young and high in organic content, having developed from a thin to moderately thick layer of loess, glacial drift, or lake-bed sediments. Soils developed from sand, muck, and peat exist in the Kankakee Sand Area and Green River Lowland sections. Deep loess occurs along the Illinois and the lower Sangamon rivers.

Plant Communities

Forest. The forests of the Grand Prairie Division are generally associated with the stream valleys and moraines. On dry sites, the forests are dominated by white oak, black oak, and shagbark hickory (*Carya ovata*), with shingle oak (*Quercus imbricaria*) and bur oak frequent associates. On mesic sites, these species are replaced by basswood, sugar maple, slippery elm, American elm, hackberry (*Celtis occidentalis*), red oak, and white ash. Black walnut (*Juglans nigra*), bitternut hickory (*Carya cordiformis*) and, in the northern part, bigtooth aspen (*Populus grandidentata*) are common. The floodplain forests are of the silver maple–American elm–ash type. The prairie groves were influenced by recurrent fires and are generally of two types: one dominated by bur oak and the other dominated by American elm and hackberry. Sandy soils in the Green River Lowland and Kankakee Sand Area sections support scrub forests of black oak.

Prairie. The vast prairies of Illinois were once one of its most remarkable features. They contained several hundred species of grasses and forbs and were interspersed by numerous marshes and prairie potholes. At the time of settlement, wet and mesic prairie were the most widespread plant communities of the Grand Prairie Division. Mesic prairie was dominated by big bluestem, Indian grass, prairie dropseed, switch grass, and little bluestem and contained many characteristic prairie plants such as leadplant (*Amorpha canescens*), compass plant (*Silphium laciniatum*), prairie dock (*Silphium terebinthinaceum*), and rattlesnake master (*Eryngium yuccifolium*). The wet sites were dominated by cord grass, sedges, and bluejoint grass and supported such species as ironweed (*Vernonia fasciculata*), boneset (*Eupatorium perfoliatum*), swamp milkweed (*Asclepias incarnata*), and water hemlock (*Cicuta maculata*).

Dry upland prairie occurs mainly on steep slopes along the Illinois River, on loess bluffs along the lower Sangamon River, and on gravel moraines and kames. The dominant species of the dry prairies are little bluestem and sideoats grama. They commonly contain such forbs as scurf pea (*Psoralea tenuiflora*), pale beard-tongue (*Penstemon pallidus*), false boneset (*Brickellia eupatorioides*), cylindrical blazing star (*Liatris cylindracea*), and fringed puccoon (*Lithospermum incisum*).

Sand prairie occurs in the Kankakee Sand Area and Green River Lowland sections. Little bluestem, fall witch-grass (*Leptoloma cognatum*), and sand dropseed (*Sporobolus cryptandrus*) are the important grasses, and goat's rue (*Tephrosia virginiana*) and spotted monarda (*Monarda punctata*) are common forbs. June grass and porcupine grass (*Heterostipa spartea*) are common. The sand prairies contain several species of Great Plains affinity such as western ragweed (*Ambrosia psilostachya*), prickly-pear cactus (*Opuntia rafinesquii*), poppy mallow (*Callirhoe triangulata*), hairy grama (*Bouteloua hirsuta*), western sunflower (*Helianthus occidentalis*), silky aster (*Symphyotrichum sericeum*), and flax-leaved aster (*Ionactis linariifolius*).

Marsh. The formerly common marshes and prairie potholes of the Grand Prairie Division were dominated by bulrushes, sedges, bur-reeds (*Sparganium* spp.), cattails, and common reed and contained many species of aquatic and semiaquatic plants such as arrowhead (*Sagittaria* spp.), water plantain (*Alisma subcordatum*), pondweed (*Potamogeton* spp.), pickerelweed (*Pontederia cordata*), beggar-ticks (*Bidens* spp.), and water crowfoot (*Ranunculus* spp.).

Aquatic Habitats

The aquatic habitats of the Grand Prairie Division are rivers, creeks, and prairie potholes.

Distinctive Fauna

Distinctive animals of the Grand Prairie Division are Blanding's turtle, western smooth green snake, western fox snake, eastern plains garter snake, Kirtland's water snake, northern lined snake, Franklin's ground squirrel, and thirteen-lined ground squirrel. None of these animals are completely restricted to this division, but all are most abundant here. Several amphibians and reptiles common outside the limits of Wisconsinan glaciation are conspicuously absent from the Grand Prairie.

Grand Prairie Section

The Grand Prairie Section encompasses the area outside the Northeastern Morainal Division that was covered by the Woodfordian substage of the Wisconsinan stage of Pleistocene glaciation, excluding the outwash and sand areas. The Shelbyville and Bloomington morainic systems form the boundaries of this section. Mesic black-soil prairie, marshes, and prairie potholes in the young, poorly-drained drift are characteristic. Glacial landforms are common. The Kankakee mallow (*Iliamna remota*) and Chase's aster (*Eurybia chasei*) are endemic to Illinois and restricted to small areas of the Grand Prairie Section. The Kankakee mallow is found only on an island in the Kankakee River, and Chase aster is known only from three counties near Peoria.

Principal Natural Features

Forest: Floodplain, mesic upland, dry upland.
Prairie Grove
Prairie: Wet, mesic, dry.
Marsh
Bedrock: Sandstone and dolomite outcrops.
Topography: Level to rolling upland, floodplain, ravines, river bluffs, lake plains, glacial landforms.
Aquatic Habitats: Prairie potholes, rivers, creeks.
Special Features: Endemic plants.

Springfield Section

The Springfield Section is part of the Illinoian drift, and the drainage system is better developed than in the younger Wisconsinan drift of the Grand Prairie Section. This section was mostly prairie in presettlement times. Deep loess deposits that support dry hill prairie occur along the lower Sangamon River. Large tracts of floodplain forest exist in the valley of the lower Sangamon River and its tributaries.

Principal Natural Features

Forest: Floodplain, mesic upland, dry upland.
Prairie: Wet, mesic, dry.
Loess Hill Prairie
Marsh
Bedrock: Outcrops.
Topography: Level to rolling upland, floodplain, ravines, river bluffs.
Aquatic Habitats: River, creeks.

Western Section

The Western Section was predominantly prairie in presettlement times and therefore is in the Grand Prairie Division. It is part of the older, dissected Illinoian drift.

Principal Natural Features

Forest: Floodplain, mesic upland, dry upland.
Prairie: Wet, mesic, dry.
Marsh
Bedrock: Outcrops.
Topography: Level to rolling upland, floodplain, ravines.
Aquatic Habitats: Creeks.

Green River Lowland Section

The broad valley of the Green River and lower Rock River was formed by glacial meltwaters. Much glacial outwash was deposited, and sand flats and dunes developed. This section had extensive marshes and wet prairies. It has scrub-oak forests on the dry sandy ridges and floodplain forests along the rivers. Sand prairie occupied the sand flats and dunes. Most of this section has been disturbed by grazing, drainage, and cultivation. There are some active dunes where the stabilizing cover has been removed.

Principal Natural Features.

Forest: Scrub-oak, floodplain.
Prairie: Wet, mesic.
Sand Prairie: Wet, mesic, dry.
Marsh
Topography: Outwash plain, dunes.
Aquatic Habitats: Rivers.

Kankakee Sand Area Section

The sand of the Kankakee Sand Area Section was deposited by the Kankakee Flood during late Wisconsinan glaciation. Sand prairie and marsh were the predominant vegetation of this section before the land was drained for cultivation. Primrose violet (*Viola primulifolia*) and Carey's smartweed (*Persicaria careyi*) are restricted in Illinois to this section. Scrub-oak forests occur on drier sites. The clear, well-vegetated, sand-bottomed streams contain such unusual fishes as the weed shiner, iron color shiner, and least darter.

Principal Natural Features

Forest: Scrub-oak.
Sand Prairie: Wet, mesic, dry.
Marsh
Topography: Outwash plain, dunes.
Aquatic Habitats: Creeks.

Upper Mississippi River and Illinois River Bottomlands Division

The Upper Mississippi River and Illinois River Bottomlands Division encompasses the rivers and floodplains of the Mississippi River above its confluence with the Missouri River and the bottomlands and associated backwater lakes of the Illinois River and its major tributaries south of La Salle. It does not include the major sand deposits, which are in a separate division. Much of the division was originally forested, but prairie and marsh also occurred. The more sluggish nature of the Illinois River and its distinctive backwater lakes distinguishes the Illinois River Section from the Upper Mississippi River Section.

Bedrock

The bedrock of the two river valleys is deeply covered by alluvia deposits.

Topography

The bottomlands of the Mississippi River and the Illinois River are characterized by broad floodplains and gravel terraces formed by glacial floodwaters.

Soils

The soils are from recent alluvium and glacial outwash. They are poorly drained, are alkaline to slightly acidic, and vary from sandy to clayey. In general they are lighter than the alluvial soils of the Lower Mississippi River Bottomlands Division.

Plant Communities

Forest. The bottomland forests are generally dominated by silver maple, American elm, and green ash. Pin Oak (*Quercus palustris*) is the most important oak; pecan (*Carya illinoensis*), bur oak, sycamore (*Platanus occidentalis*), honey locust (*Gleditsia triacanthos*), hickories (*Carya* spp.), and black walnut are frequent. Black willow and river birch (*Betula nigra*) are common in the Mississippi River Section. A few southern lowland species, including water locust (*Gleditsia aquatica*), overcup oak (*Quercus lyrata*), sugarberry (*Celtis laevigata*), deciduous holly (*Ilex decidua*), and swamp privet (*Forestiera acuminata*) range into the southern part of this division.

Prairie. In presettlement times, mesic prairie and wet prairie occurred in the broad bottomlands. The species composition of these prairies was similar to that of the prairies of the Grand Prairie Division.

Marsh. Marshes containing the typical marsh species are important features throughout both sections.

Spring bogs. Spring-fed bogs with peat deposits are found on terraces along the Illinois River. These are unique to the Illinois River and the species composition differs from that of the bogs of the Northeastern Morainal Division. Distinctive plants include black ash (*Fraxinus nigra*), willows, poison sumac, and skunk cabbage (*Symplocarpus foetidus*).

Aquatic Habitats

Oxbow lakes occur in both the Illinois River valley and the Mississippi River valley. Backwater lakes are distinctive of the Illinois River valley, and the Illinois River is more sluggish than the Mississippi River. Springs are common in gravel terraces along the Illinois River. The bottomland forests are subject to prolonged periods of flooding.

Distinctive Fauna

The fish faunas of the Illinois River and the upper Mississippi River are similar although they differ somewhat from that of the silt-laden Mississippi River south of its confluence with the Missouri River.

Illinois River Section

The Illinois River Section is distinguished from the Mississippi River Section by its distinctive backwater lakes and differences in forest vegetation. The spring bogs along the river bluffs are a special feature.

Principal Natural Features

Forest: Bottomland.
Prairie: Wet, mesic.
Marsh
Topography: River floodplain, river terraces.
Aquatic Habitats: Backwater lakes, oxbow lakes, rivers.

Mississippi River Section

This section is composed of several distinct bottomlands along the Mississippi River, from Wisconsin to Calhoun County. Most of the prairies of this section have been drained for agriculture. Forests are still found inside the levees and on the river islands.

Principal Natural Features

Forest: Bottomland.
Prairie: Wet, mesic.
Marsh
Topography: River floodplain, river terraces.
Aquatic Habitats: Oxbow lakes, rivers.

Illinois River and Mississippi River Sand Areas Division

The Illinois River and Mississippi River Sand Areas Division encompasses the sand areas and dunes in the bottomlands of the Illinois and Mississippi rivers and includes the "perched dunes" atop the bluffs near Hanover in Jo Daviess County. Scrub-oak forest and dry sand prairie are the natural vegetation of this division. Several plant species found here are more typical of the short-grass prairies to the west of Illinois. Several "relict" western amphibians and reptiles are known only from these sand areas. The two sections are distinguished because of differences in flora and fauna.

Topography

The topography is generally one of level to rolling plains of sand deposited by glacial meltwaters and blown into widespread areas east of the rivers. In many areas the sand has migrated onto the bluffs and uplands east of the river terraces. In places, dunes 20 to 40 feet high have formed, and blowouts are common in unstabilized sand.

Soils

The soils are derived from sand and sandy material. Other soils in depressions surrounded by sand are also in this division. The soils are generally droughty and subject to wind erosion. Low areas are generally wet.

Plant Communities

Forest. The forests of this division are limited to scrubby stands dominated by black oak in the Mississippi River Section and black and blackjack oaks (*Quercus marilandica*) in the Illinois River Section.

Prairie. Sand prairie, composed of such species as little bluestem, June grass, Indian grass, and porcupine grass, is the major community of this division. The sand prairie habitats range from dry to wet and include such plants as goat's rue, spotted monarda, prickly-pear cactus, tubercled three-awned grass (*Aristida tuberculosa*), poppy mallow, and fall witch-grass. The dry sites have western floral elements, including sand love-grass (*Eragrostis trichodes*), hairy grama, and rarely bladderpod (*Physarea ludoviciana*) and Patterson's bindweed (*Stylisma pickeringii* var. *pattersonii*). The mesic and wet sites have prairie vegetation similar to that of the Grand Prairie Division. There are also various plant associations related to unstabilized sand. Long-leaved calamovilfa (*Calamovilfa longifolia*) is one of the principal sand binders.

Marsh. Marsh occurs in low, poorly drained areas.

Distinctive Fauna

Distinctive animals are the bull snake, plains hognose snake, Illinois mud turtle, and Illinois chorus frog. The Illinois chorus frog is restricted to the Illinois River Section. The white-tailed jackrabbit is found in Illinois only in the northern part of the Mississippi River Section. The lark sparrow breeds most commonly in Illinois in the sandy habitats of this division, and a northern outlier population of the summer tanager breeds in forests of the Illinois River Section.

Illinois River Section

The Illinois River Section is distinguished from the Mississippi River Section on the basis of floral and faunal differences.

Principal Natural Features

Forest: Scrub-oak.
Sand Prairie: Dry, mesic, wet.
Marsh
Topography: Dunes, blowouts, level to rolling plain.

Mississippi River Section

The Mississippi River Section has floral elements absent from the Illinois River Section, including rock spikemoss (*Selaginella rupestris*), rough-seeded rock-pink (*Phemeranthus rugospermum*), and beach-heath (*Hudsonia tomentosa*). Rock spikemoss and beach-heath form large mats that stabilize blowouts.

Principal Natural Features

Forest: Scrub-oak.
Sand Prairie: Dry, mesic, wet.
Marsh
Topography: Dunes, blowouts, level to rolling plain.

Western Forest-Prairie Division

The Western Forest-Prairie Division is a strongly dissected glacial till plain of Illinoian and Kansan age. At the time of settlement, forest was the predominant vegetation, but there was also considerable prairie on the level uplands. The prairie soils were developed from loess and are fertile. The two sections are geographically separated by the Illinois River valley and also have some faunal differences.

Glacial History

Most of the bedrock is covered by glacial drift from the Illinoian stage of Pleistocene glaciation. There is an area of older Kansan drift in the western parts of Pike and Adams counties.

Bedrock

Pennsylvanian and Mississippian bedrocks of limestone, sandstone, shale, and coal crop out frequently along the major streams.

Topography

The till plain is strongly dissected, with many ravines in the level to rolling uplands. Floodplains are developed along the major streams.

Soils

Most of the soils are fairly young, having developed from four to five feet of loess. The prairie soils are high in organic matter and are similar to those of the Grand Prairie Division. The forest soils are acidic and low in organic matter. Relatively small areas of droughty, fine-textured soils have developed in till on some steep slopes.

Plant Communities

Forest. The upland forests consist of an oak-hickory association with black oak, white oak, and several species of hickory as the dominants. Scattered sites of fine-textured soils support a post oak–black jack oak community. This forest community also occurred on the margins of the prairies, perhaps because of fires. The mesic forests contain white oak, red oak, basswood, sugar maple, and slippery elm. The floodplain forests are dominated by silver maple, American elm, ashes, and box elder (*Acer negundo*).

Prairie. In presettlement times, large prairies existed on the uplands of this division. Carthage Prairie, Hancock Prairie, and Bushnell Prairie were in the Galesburg Section; and String Prairie and Brown's Prairie were in the Carlinville Section. The prairie vegetation was similar to that of the Grand Prairie Division, but wet prairie was less frequent.

Marsh. Poorly drained areas with marsh vegetation are less frequent than in the Grand Prairie Division but are similar in composition.

Aquatic Habitats

The aquatic habitats of this well-drained division consist mainly of rivers and creeks.

Distinctive Fauna

There is some continuity of animal life between the Western Forest Prairie Division and the Southern Till Plain Division. The five-lined skink, the broad-headed skink, and the ornate box turtle are species that occur throughout the Southern Till Plain Division and range into this division but are absent from the adjoining Grand Prairie Division.

Galesburg Section

The Galesburg Section is the area north of the broad Illinois River valley and is distinguished from the Carlinville Section because of its separate location. There were about equal amounts of forest and prairie in this section at the time of settlement, with the forests primarily in the well-dissected areas along tributaries of the Illinois River.

Principal Natural Features

Forest: Dry upland, mesic upland, floodplain.
Prairie: Dry, mesic, wet.
Marsh
Bedrock: Outcrops.
Topography: Level to rolling uplands, ravines, floodplain.
Aquatic Habitats: Rivers, creeks.

Carlinville Section

The Carlinville Section is the area of well-dissected land southeast of the Illinois River valley. The original vegetation of this section was mostly forest, with only about 12 percent of the area prairie.

Principal Natural Features

Forest: Dry upland, mesic upland, floodplain.
Prairie: Dry, mesic, wet.
Marsh
Bedrock: Outcrops.
Topography: Level to rolling uplands, ravines, floodplain.
Aquatic Habitats: Creeks.

Middle Mississippi Border Division

The Middle Mississippi Border Division consists of a relatively narrow band of river bluffs and rugged terrain bordering the Mississippi River floodplain from Rock Island County to St. Clair County and the lower Illinois River floodplain. Forest is the predominant natural vegetation, but hill prairies are common on the west-facing bluffs. The soils were generally developed from very deep loess. Limestone cliffs are common features. This division is best distinguished from the river bluffs to the north and south of it by the absence of certain floral and faunal elements. The Driftless Section was never glaciated and is distinguished from the remainder of the division.

Bedrock

The bedrock of the Middle Mississippi Border Division consists of limestone and sandstone, with dolomite associated with the sandstone in Calhoun County. Outcrops and cliffs of limestone are common along the bluffs, and sandstone outcrops occur in Rock Island and Calhoun counties. Cretaceous gravels and clays occur in Pike and Adams counties. There are caves throughout the division in the limestone and dolomite, but they are most abundant in unglaciated Calhoun County.

Glacial History

Most of the division was glaciated during the Illinoian stage of Pleistocene glaciation. An area of older Kansan drift is in western Hancock, Adams, and Pike counties. Calhoun County and parts of Pike and Adams counties apparently escaped Pleistocene glaciation.

Topography

The Middle Mississippi Border Division is greatly dissected, particularly along the major streams where there are bluffs and ravines. Sinkhole plains are most common in the southern part of the division. The Driftless Section is higher and has more rugged topography than the Glaciated Section.

Soils

Most of the soils on the uplands have developed from deep, well-drained loess. Isolated areas of heavy soils are also present, especially in the southern part of the division.

Plant Communities

Forest. The vegetation of the Middle Mississippi Border Division is mostly mesic and dry forests associated with the dissected uplands. The forests of the dry sites are dominated by black oak and white oak. Sugar maple, basswood, red oak, hackberry, slippery elm, and black walnut are major components of the forests on mesic sites. Floodplain forests along the creeks contain silver maple, hickories, cottonwood, and sycamore. Post oak (*Quercus stellata*) is common on the heavy soils and near ridge tops.

Prairie. Prairies of the Middle Mississippi Border Division are limited to the steep slopes and ridges of deep loess atop the river bluffs. The prairies are dominated by little bluestem and side-oats grama, with purple prairie clover (*Dalea purpurea*) and flowering spurge (*Euphorbia corollata*) among the most frequent forbs. Scurf pea, a distinctive western plant, is also common. Stickleaf (*Mentzelia oligosperma*), another western plant, reaches its northeastern limits on the exposed limestone ledges of this division and of the Northern Section of the Ozark Division.

Aquatic Habitats

There are creeks throughout the division and sinkhole ponds in the Driftless Section.

Distinctive Fauna

The dark-sided salamander and western worm snake are restricted in Illinois to the Middle Mississippi Border Division. Forested glens of this division serve as major nighttime roosting places for wintering bald eagles.

Glaciated Section

The topography of this section has been modified by the Illinoian and Kansan stages of Pleistocene glaciation. Limestone underlies most of the Glaciated Section and frequently forms cliffs along the river bluffs.

Principal Natural Features

Forest: Dry upland, mesic upland, floodplain.
Loess Hill Prairie
Bedrock: Limestone outcrops, caves.
Topography: River bluffs, ravines, floodplain, sinkhole plain.
Aquatic Habitats: Creeks.
Special Feature: Roosting areas for wintering bald eagles.

Driftless Section

The Driftless Section apparently escaped Pleistocene glaciation. Its topography is rougher than that of the Glaciated Section, and it has many sinkholes and sinkhole ponds. Except for the jeweled shooting star, the Driftless Section is not known to harbor preglacial relict plants.

Principal Natural Features

Forest: Dry upland, mesic upland, floodplain.
Loess Hill Prairie
Bedrock: Limestone and sandstone outcrops, caves.
Topography: River bluffs, ravines, floodplain, sinkhole plain.
Aquatic Habitats: Creeks, sinkhole ponds.
Special Feature: Roosting areas for wintering bald eagles.

Southern Till Plain Division

The Southern Till Plain Division encompasses most of the area of dissected Illinoian glacial till plain south of the Shelbyville Moraine and the Sangamon River and Macoupin Creek watersheds. Both forest and prairie were present at the time of settlement. The soils are relatively poor because of their high clay content and the frequent occurrence of a "claypan" subsoil. Post oak flatwood forest is characteristic of the division. The two sections are distinguished because of topographic differences.

Bedrock

The bedrock of the Southern Till Plain Division consists of sandstone, limestone, coal, and shale, which commonly crop out of the eastern and southeastern parts of the division. Bedrock lies near the surface of the Mt. Vernon Hill Country Section.

Glacial History

The Illinoian stage of Pleistocene glaciation reached the southernmost limit of North American continental glaciation just beyond the limits of this division. The Southern Till Plain Division is entirely covered by Illinoian till. Glacial landforms are common only in the northwestern part of the division.

Topography

The glacial till of the Southern Till Plain Division becomes thinner from north to south. The bedrock of the Mt. Vernon Hill Country Section is near the surface, accounting for its hilly and rolling topography. The Effingham Plain Section is a nearly level to dissected till plain. There are broad floodplains along the major streams and there are ravines in the bluffs along the stream valleys.

Soils

The soils on the uplands are light colored and strongly developed, with poor internal drainage. They have developed from thin loess and till under both forest and prairie vegetation. Fragipan and claypan layers are characteristic of the upland soils. Some of the prairie soils have a high sodium content and are known locally as "alkaline slicks."

Plant Communities

Forest. The level, poorly drained, heavy soils on the uplands of the Southern Till Plain Division support a characteristic flatwoods forest of post oak, swamp white oak, blackjack oak, and pin oak. Forests on the uplands, where drainage is slightly improved, include black oak, shingle oak, mockernut hickory (*Carya tomentosa*), and shagbark hickory in the post oak community. The forests on the slopes along stream valleys are dominated by white oak, shingle oak, and black oak on the drier southern and western exposures, with hickories, white ash, basswood, sugar maple, wild black cherry, slippery elm, and black walnut with the oaks on the more mesic sites. Forests in the broad floodplains of the Kaskaskia and Big Muddy rivers are dominated by silver maple, willows, sycamore, and American elm near the rivers, with pin oak, white oak, hickories, ashes, hackberry, and honey locust on the heavier soils farther from the rivers. Pin oak occasionally grows in nearly pure stands over large areas of the floodplain. The floodplain forests of the smaller streams have a higher percentage of oaks than the floodplain forests of central and northern Illinois. Pin oak and shingle oak are dominant, with white oak, red oak, hickories, black walnut, river birch, and cottonwood occasional associates. Shumard oak (*Quercus shumardii*) and sweet gum (*Liquidambar styraciflua*) grow in the floodplain of the Big Muddy River but are not generally abundant.

Prairie. At the time of settlement, about 40 percent of the uplands of the Southern Till Plain Division supported prairie vegetation. Most of the prairie was of the mesic tall-grass type characteristic of the Grand Prairie Division. Twelve Mile Prairie and Looking Glass Prairie were two large expanses of prairie. Mesic prairies extended along the west side of the division almost to the limit of glaciation but were rare in the southeastern part. Wet prairie was not common but did occur in parts of the Kaskaskia River floodplain. It is not known whether the alkaline slicks of this division supported a unique prairie flora.

Marsh. Marshes were associated with the stream floodplains of this division.

Aquatic Habitats

The aquatic habitats of the Southern Till Plain Division consist of rivers, creeks, and oxbow lakes.

Distinctive Fauna

The northern crayfish frog, northern fence lizard, ground skink, five-lined skink, and broad-headed skink are common in the Southern Till Plain Division but are rare or absent from the Grand Prairie Division.

Effingham Plain Section

The Effingham Plain Section is a relatively flat plain drained by the Kaskaskia River. It originally was mostly prairie. Post oak flatwoods are characteristic of the uplands of this section. A few flocks of the greater prairie chicken remain within this section. Sanctuaries are being established to maintain the population by providing nesting habitat.

Principal Natural Features

Forest: Upland flatwoods, dry upland, mesic upland, and floodplain.
Prairie: Wet, mesic, dry.
Marsh
Bedrock: Outcrops.
Topography: Level to rolling upland, dissected till plain, ravines, floodplain.
Aquatic Habitats: Creeks, rivers, oxbow lakes.
Special Features: Greater prairie chicken population, alkaline slicks.

Mt. Vernon Hill Country Section

The Mt. Vernon Hill Country Section is distinguished from the Effingham Plain Section by its rolling, hilly topography. In presettlement times, upland forests covered most of this section. The striped shiner and the stoneroller, two fishes of nearly statewide distribution, are absent from most of the Mt. Vernon Hill Country Section. The broad bottomlands of the major rivers that drain the eastern part of this section are considered to be part of the Wabash Border Division.

Principal Natural Features

Forest: Upland flatwoods, dry upland, mesic upland, floodplain.
Prairie: Wet, mesic, dry.
Marsh
Bedrock: Outcrops.
Topography: Rolling till plain, dissected till plain, ravines, floodplain.
Aquatic Habitats: Creeks, rivers, oxbow lakes.

Wabash Border Division

The Wabash Border Division includes the bottomlands of the Wabash River and its major tributaries, the loess-covered uplands bordering the Wabash River, and the forests of the Vermilion River, Little Vermilion River, and Crab

Apple Creek. This is a region of lowland oak forests containing beech, tulip tree (*Liriodendron tulipifera*), and other trees typical of the forests to the east of Illinois. The Wabash River drainage contains several distinctive fishes. The sections are distinguished by differences in topography, glacial history, and flora and fauna.

Bedrock

The bedrock of the Wabash Border Division consists of limestone, sandstone, coal, and shale. Small outcrops of bedrock occur along some of the streams throughout the division. Bedrock outcrops along the river bluffs are few and never form the towering cliffs characteristic of the western border of Illinois.

Glacial History

All but a small area in the southern part of the Wabash Border Division was subject to Pleistocene glaciation. The Vermilion River dissects younger Wisconsinan drift, whereas the remainder of the division is the older Illinoian till plain. There are areas of outwash sand along the lower Wabash River.

Topography

The topography of the Wabash Border Division is relatively gentle. The Southern Uplands Section is a low, eroded till plain, with bluffs above the bottomlands. The Vermilion River Section has more rugged topography resulting from erosion of its streams into Wisconsinan drift. The rivers have broad floodplains formed by glacial lakes, which include terrace deposits and many meander scars.

Soils

The soils in the Bottomlands Section range from floodplain soils to terrace soils. The soils of the Southern Uplands Section have developed from moderately deep loess deposits. Soils of the Vermilion River Section are derived from thin loess over loamy till. Despite the diversity of soils, this division is united by the continuity of its forest vegetation and its fish and amphibian faunas.

Plant Communities

Forest. The forests of the bottomlands are developed either on recent alluvium or on Pleistocene terraces, which are generally better drained. Bottomland forests in the floodplains of the major rivers are dominated by pin oak, overcup oak, swamp white oak, swamp chestnut oak (*Quercus michauxii*), bur oak, cherrybark oak (*Quercus pagoda*), and Shumard oak. Commonly associated with the oaks are sweet gum, hackberry, American elm, kingnut hickory (*Carya laciniosa*), silver maple, and pecan. The forests on the terraces contain a mixture of the tree species found on the floodplains and those found on the uplands. Shumard oak, bur oak, and sweet gum are the commonest, with pin oak and swamp white oak in poorly drained areas. The best-drained terraces contain shagbark hickory and tulip tree. Other characteristic trees are kingnut hickory, Kentucky coffee tree (*Gymnocladus dioica*), and hackberry. Along the rivers, black willows, cottonwood, sycamore, and silver maple predominate. Sloughs near the

Wabash River and Saline River contain bald cypress (*Taxodium distichum*) and some other southern swamp species.

The mesic forests of the Southern Uplands Section and the Vermilion River Section are dominated by white oak, red oak, and sugar maple and frequently contain beech and tulip tree. Forests on drier sites are dominated by black oak and hickories. Forests in the floodplains of the Southern Uplands Section and Vermilion River Section contain silver maple, cottonwood, willows, sycamore, and American elm.

Prairie. At the time of settlement, some of the Southern Uplands Section supported mesic prairie, while mesic and wet prairie occurred in the Bottomlands Section. The prairies of this division were similar to those of the Grand Prairie Division.

Marsh. The poorly drained bottomlands contain large areas of marsh associated with the sloughs and meander scars. These are dominated by cord grass and river bulrush (*Scirpus fluviatilis*).

Aquatic Habitats

The poorly drained Bottomlands Section of the Wabash Border Division contains many aquatic habitats, including rivers, oxbow lakes, and sloughs. The Southern Uplands and Vermilion River sections are better drained, and the aquatic habitats consist of creeks and rivers.

Distinctive Fauna

Fishes limited in Illinois to the Wabash Border Division are the river chub, river redhorse, mountain madtom, and greenside darter. The northern bigeye chub and bluebreast darter are known in Illinois only from the Vermilion River system, and the harlequin darter is known only from the Embarras River. The red-backed salamander is essentially restricted in Illinois to this division.

Bottomlands Section

The Bottomlands Section of the Wabash Border Division encompasses the bottomland forests, sloughs, marshes, and oxbow lakes in the floodplains of the Wabash River, the Ohio River, and their major tributaries. Bottomland forests are the predominant vegetation with wet prairie and marsh associated with the sloughs.

Principal Natural Features

Forest: Floodplain, terrace, swamp.
Prairie: Mesic, wet.
Marsh
Topography: River floodplain, terrace deposits, meander scars.
Aquatic Habitats: Oxbow lakes, sloughs, rivers.

Southern Uplands Section

The Southern Uplands Section contains the dry and mesic upland forests on the deep loess bluffs along the Wabash River. The upland forests of white oak, sugar maple, beech, and sweet gum are the predominant plant community. Some sandstone ravines support an unusual combination of plant species that includes some relict northern species.

Principal Natural Features

Forest: Dry upland, mesic upland, floodplain.
Prairie: Mesic, dry.
Bedrock: Outcrops.
Topography: Dissected till plain, river bluffs, ravines.
Aquatic Habitats: Creeks.
Special Feature: Relict northern plants.

Vermilion River Section

The Vermilion River Section is characterized by rugged topography and the beech-maple forests in the ravines along the Vermilion River and its tributaries. The beech-maple forests represent an important climax deciduous forest type of the northeastern United States, which is found in Illinois only in the extreme eastern and southern portions.

Principal Natural Features

Forest: Dry upland, mesic upland, floodplain.
Prairie: Dry, mesic, wet.
Topography: Dissected till plain, ravines, floodplain.
Aquatic Habitats: Rivers, creeks.

Ozark Division

The Ozark Division consists of the Illinois part of the Salem Plateau of the Ozark uplift from Northern Monroe County southward and includes the glaciated sandstone ravines in Randolph County. The area is mostly forested, but many hill prairies occur in the Northern Section. The division contains many Ozarkian, southern, and southwestern plants and animals that are rare or absent elsewhere in Illinois. The sections are based on differences in bedrock, topography, flora, and fauna.

Bedrock

Most of the Ozark Division is part of the Salem Plateau of the Ozark uplift. The northern part of the division is underlain by relatively pure limestone, while the southern part is underlain by cherty limestone that is more resistant to erosion. Sandstone underlies the Central Section of the division. Bedrock crops out in all sections. Caves and sinkholes are numerous in the limestone of the Northern Section and less so in the Southern Section.

Glacial History

Part of the Northern Section and all of the Central Section of the Ozark Division were glaciated during the Illinoian stage of Pleistocene glaciation. The Southern Section is driftless.

Topography

The topography of the Ozark Division is that of a maturely dissected plateau with steep bluffs along the Mississippi River. There are ravines and stream canyons throughout the division, especially in the sandstone of the Central Section. The Northern Section has a well-developed sinkhole plain topography.

Soils

Most of the soils of the Ozark Division are derived from deep loess, but thin soils occur over the bedrock outcrops along the river bluffs and in interior ravines. Some of the soils of the Southern Section are derived from bedrock and are acidic.

Plant Communities

Forest. At the time of settlement, the Ozark Division was almost entirely forested. The forests of the Northern Section consist in part of red oak, sugar maple, basswood, and Ohio buckeye (*Aesculus glabra*) on the mesic sites, with white oak, black oak, and hickories on the ridge tops. Beech and tulip tree occur in the mesic forests of the Central Section. The forests of the Southern Section contain a rich assemblage of tree species including cucumber tree (*Magnolia acuminata*), black gum, butternut (*Juglans cinerea*), black walnut, and bitternut hickory in addition to red oak, sugar maple, basswood, white oak, black oak, Ohio buckeye, beech, and tulip tree. Yellow-wood (*Cladrastis kentukea*) occurs in the Southern Section in Alexander County. A mixed association of white oak, red oak, sycamore, American elm, river birch, wild black cherry, and cottonwood is found along the stream floodplains throughout the division. Stands of shortleaf pine (*Pinus echinata*) are found in the Southern Section and as far north as Piney Creek in Randolph County.

Prairie. Loess hill prairies are common on the river bluffs in the Northern Section of the Ozark Division but are rare in the Southern Section and absent from the Central Section. These prairies have a similar species composition to that of loess hill prairies of the Middle Mississippi Border Division, with little bluestem and side-oats grama dominating; but they have several species of plants that are restricted to this division in Illinois.

Aquatic Habitats

The Ozark Division has few aquatic habitats. Ponds occur in some of the sinkholes. Springs occur at cave entrances and at the bases of some of the bluffs. Creeks are the commonest aquatic feature of the division.

Distinctive Fauna

The plains scorpion, eastern narrow-mouthed toad, and eastern coachwhip are restricted in Illinois to the Northern Section of the Ozark Division, while the spring cavefish, northern blacktail shiner, and scarlet snake are known only from the Southern Section. The northern flatheaded snake occurs in both the Northern and Southern sections but is found nowhere else in the state.

Northern Section

The Northern Section is distinguished from the other sections of the Ozark Division by its limestone bedrock, numerous caves and sinkholes, unique plant and animal species, and forest composition. Plant species unique to this section are the reticulate-seeded spurge (*Euphorbia spathulata*), stiff bedstraw (*Galium virgatum*), Missouri black-eyed Susan (*Rudbeckia missouriensis*) and small heliotrope (*Heliotropium tenellum*). These plants grow in hill prairies or on exposed limestone ledges.

Principal Natural Features

Forest: Dry upland, mesic upland, floodplain.
Loess Hill Prairie
Bedrock: Sinkholes, caves, limestone outcrops.
Topography: Sinkhole plain, ravines, river bluffs, floodplain.
Aquatic Habitats: Sinkhole ponds, creeks, springs.
Special Features: Ozark "limestone glade" plants, distinctive reptiles and amphibians.

Central Section

The Central Section of the Ozark Division is distinguished because of its sandstone bedrock, forest composition, and distinctive flora. Distinctive plants of this section include Harvey's buttercup (*Ranunculus harveyi*), large-flowered rock-pink (*Phemeranthus calycinum*), and Bradley's spleenwort (*Asplenium bradleyi*). These species apparently entered Illinois from the Missouri Ozarks after Illinoian glaciation.

Principal Natural Features

Forest: Dry upland, mesic upland, floodplain.
Bedrock: Sandstone outcrops.
Topography: Hills, ravines, stream canyons, floodplain.
Aquatic Habitats: Creeks.
Special Feature: Ozarkian floral element.

Southern Section

The Southern Section of the Ozark Division is distinguished by its bedrock, topography, glacial history, unique fauna, forest composition, and distinctive southern and Ozarkian flora. Black spleenwort (*Asplenium resiliens*), shortleaf pine, azalea (*Rhododendron prinophyllum*), and big leaf snowbell bush (*Styrax grandifolia*) are part of the distinctive floral element.

Principal Natural Features

Forest: Dry upland, mesic upland, floodplain.
Loess Hill Prairie
Bedrock: Outcrops.
Topography: Steep ravines, river bluffs, floodplain.
Aquatic Habitats: Creeks, springs, sinkhole ponds.
Special Features: Ozarkian floral element, spring cavefish.

Lower Mississippi River Bottomlands Division

The Lower Mississippi River Bottomlands Division includes the Mississippi River and its floodplain from Alton to the Thebes Gorge. The Mississippi River is muddy here due to the silt load brought in by the Missouri River. Its fish fauna contains a distinctive assemblage of silt-tolerant plains species.

The Northern Section (the American Bottom) originally contained prairies, marshes, and forest. The Southern Section was densely forested. The forests of this division contain a greater number of tree species than the forests of the Upper Mississippi River, including some southern lowland species.

Topography

The broad bottomlands of the Lower Mississippi River Bottomlands Division were formed by glacial floodwaters. Since the retreat of the glaciers the river has meandered through this broad floodplain, and many meander scars and oxbow lakes remain.

Soils

The soils are generally fine textured, with areas of both sandy, well-drained soils and clay soils with poor internal drainage. The soils of this division have developed from alluvium.

Plant Communities

Forest. Except for areas of wet prairie and marsh in the Northern Section, the division was entirely forested in presettlement times. The bottomland forests on the light soils include silver maple, ashes, American elm, honey locust, sugarberry, and pecan. Beech, basswood, and red buckeye (*Aesculus pavia*) grew as associated species on the loamy soils in the Southern Section; but most of these soils, being better drained and fertile, have been cleared for agriculture. Bottomland forests on heavy soils of the Southern Section are dominated by pin oak, overcup oak, Shumard oak, and cherrybark oak in association with kingnut hickory, sugarberry, and sweet gum. Bottomland swamps in the Southern Section are dominated by pumpkin ash (*Fraxinus profunda*), swamp cottonwood (*Populus heterophylla*), Drummond's red maple (*Acer drummondii*), and water locust. Bald cypress grows in the Southern Section as far north as the southern edge of Union County. Tupelo (*Nyssa aquatica*) and some other coastal plain species are absent.

Prairie. There were relatively large areas of wet and mesic prairie in the Northern Section of the Lower Mississippi River Bottomlands Division, but most have been drained for agriculture.

Marsh. The Northern Section contained large marshes dominated by river bulrush, cattail, lotus (*Nelumbo lutea*), and pickerelweed. Some of the marshes remain, even though much of the area is organized into drainage districts.

Aquatic Habitats

The silt-laden Mississippi River below the Missouri River provides aquatic habitats somewhat different from those of the upper Mississippi River. Oxbow lakes and sloughs are common features of the bottomlands. The spring-fed swamps of northwestern Union County provide a unique habitat for several species of fish.

Distinctive Fauna

The herpetofauna of the Lower Mississippi River Bottomlands Division is similar to that of the bottomlands of the Coastal Plain Division. It includes the western cottonmouth, green water snake, green tree frog, western bird-voiced tree frog, and mole salamander. The fish fauna of this division includes several species not found elsewhere in Illinois. The bantam sunfish is found only in the spring-fed swamps in northwestern Union County; and the Alabama shad, plains minnow, sturgeon chub, flathead chub, and sicklefin chub are found in the Mississippi River. The banded pigmy sunfish is known only from the spring-fed swamps of this division and from the bottomland swamps of the Coastal Plain Division.

Northern Section

The Northern Section is distinguished by its forest composition, the presence of wet prairies and marshes, and the absence of the coastal plain trees of the Southern Section. The bottomlands of this section near St. Louis are called the "American Bottoms."

Principal Natural Features

Forest: Bottomland.
Prairie: Wet, mesic.
Marsh
Topography: River floodplain, meander scars.
Aquatic Habitats: Oxbow lakes, Mississippi River.

Southern Section

The bottomland forests of the Southern Section contain a greater number of tree species, including some bottomland swamp species typical of the coastal plain. The composition of the forests in this section varies with the soils.

Principal Natural Features

Forest: Bottomland on heavy soils, bottomland on light soils, bottomland swamp.
Topography: River floodplain, meander scars.
Aquatic Habitats: Oxbow lakes, Mississippi River.
Special Feature: Spring-fed swamps.

Shawnee Hills Division

The Shawnee Hills extend across the southern tip of the state from Fountain Bluff on the Mississippi River to the Shawneetown Hills near the mouth of the Wabash River. This unglaciated hill country is characterized by a high east-west escarpment of sandstone cliffs forming the Greater Shawnee Hills and a series of lower hills underlain by limestone and sandstone known as the Lesser Shawnee Hills. Originally, this division was mostly forested, and considerable forest remains to the present time. There are a number of distinctive plant species restricted to this division of Illinois.

Bedrock

The Greater Shawnee Hills form a band along the northern edge of the division and consist of massive Pennsylvanian sandstone strata that dip northward toward the Illinois Basin. The range of hills averages 10 miles wide and borders the Lesser Shawnee Hills to the south. The Lesser Shawnee Hills are underlain by Mississippian limestone and sandstone, and sinkholes and caves are locally common features. Mineralized faults containing fluorspar and zinc, silver, and other metals exist in the eastern part of the Shawnee Hills Division. Iron deposits are found in Hardin County. There is a dome containing an igneous rock core in western Hardin

County, and outcrops of igneous rock occur in the Lesser Shawnee Hills Section.

Topography

The topography of the Shawnee Hills Division is very rugged, with many bluffs and ravines. The north slopes of the Greater Shawnee Hills Section are relatively gentle, but the south slopes consist of many escarpments, cliffs, and overhanging bluffs. Streams have eroded canyons in the sandstone. The Lesser Shawnee Hills average about 200 feet lower than the Greater Shawnee Hills. The Lesser Shawnee Hills have local areas of sinkhole topography.

Soils

The soils are derived mainly from loess. Narrow bands of moderately developed deep loess soils occur along the Mississippi River in Jackson County and along the Ohio River in eastern Hardin County; however, most of the soils are derived from thinner loess and are strongly developed. Claypan and fragipan layers are frequent.

Plant Communities

Forest. At the time of settlement, most of the Shawnee Hills Division supported forest, and considerable land remains timbered. Most of the upland forests are dominated by white oak, black oak, and shagbark hickory with post oak, blackjack oak, scarlet oak (*Quercus coccinea*), and pignut hickory (*Carya glabra*) on dry sites. Deep mesic ravines have a forest community of red oak, beech, tulip tree, bitternut hickory, sugar maple, and white ash with black walnut, butternut, Ohio buckeye, and basswood occasional. The floodplain forests also contain sycamore, Kentucky coffee tree, sugarberry, and honey locust.

Prairie. The Lesser Shawnee Hills Section contains limestone glades that support a dry prairie vegetation like that of hill prairies in western Illinois but with the addition of such southern plants as wild blue sage (*Salvia pitcheri*) and heart-leaved tragia (*Tragia cordata*).

Aquatic Habitats

The Shawnee Hills Division has numerous clear rocky streams and creeks. Sinkhole ponds are found in the Lesser Shawnee Hills Section.

Distinctive Fauna

The streams support several distinctive fishes including black-spotted topminnow, spottail darter, and stripetail darter. The latter species is restricted to extreme southeastern Illinois.

Greater Shawnee Hills Section

The Greater Shawnee Hills Section is distinguished by its sandstone bedrock, topography, and distinctive plants. Filmy fern (*Vandenboschia boschiana*), Virginia saxifrage (*Micranthes virginiensis*), small-flowered rock-pink (*Phemeranthus parviflorum*), thread-leaved evening primrose (*Oenothera linifolia*), synandra (*Synandra hispidula*), French's shooting star (*Dodecatheon frenchii*), and small-flowered alumroot (*Heuchera parviflora* var. *rugelii*) are some of the distinctive plants in this section. French's shooting star and small-flowered alumroot are abundant in their restricted habitats and may have persisted since preglacial times. Deep ravines and sandstone ledges along the larger streams support relict northern plants such as club-mosses (*Lycopodium* spp.), sphagnum (*Sphagnum* sp.), and barren straw-berry (*Waldsteinia fragarioides*). Except for synandra, these distinctive plants are absent from a seemingly suitable habitat in the glaciated area of southwestern Williamson County.

Principal Natural Features

Forest: Dry upland, mesic ravine, floodplain.
Bedrock: Outcrops, faults.
Topography: Bluffs, ravines, stream canyons, floodplain, over hanging cliffs.
Aquatic Habitats: Creeks.
Special Feature: Northern relict plants, preglacial relict plants.

Lesser Shawnee Hills Section

The Lesser Shawnee Hills Section is distinguished by its limestone bedrock and sinkhole topography. The fluorspar deposits near Cave-in-Rock and Rosiclare in Hardin County are world famous. Caves are common features of the limestone bluffs. Distinctive plants of this section are wild mock-orange (*Philadelphus verrucosus*) and great chickweed (*Stellaria pubera*).

Principal Natural Features

Forest: Dry upland, mesic ravine, floodplain.
Limestone Glades
Bedrock: Sinkholes, caves, sandstone outcrops, limestone outcrops, igneous outcrops, mineralized faults.
Topography: Sinkhole plain, interior bluffs, river bluffs, ravines, floodplain.
Aquatic Habitats: Sinkhole ponds, creeks.
Special Feature: Fluorspar deposits.

Coastal Plain Division

The Coastal Plain Division is a region of swampy forested bottomlands and low clay and gravel hills that is the northernmost extension of the Gulf Coastal Plain Province in North America. Bald cypress–tupelo gum swamps are a unique feature of this division in Illinois, as are many southern animals and plants found within them. The division encompasses the bottom-lands of the Cache, Ohio, and Mississippi rivers and hills capped by Creta-ceous and Tertiary sand, gravel, and clay. It has a relatively mild climate, the warmest in the state. The two sections distinguish between upland and low-land environments.

Glacial History

The Coastal Plain Division was never subjected to Pleistocene glaciation, but it has been influenced by glacial floodwaters. Glacial Lake Cache was formed in the valleys of the Cache and Ohio rivers during a late stage of gla-ciation. Pleistocene deposits form the terraces in the bottomlands of the rivers.

Bedrock

The Cretaceous Hills Section is composed of unconsolidated sediments of Cretaceous and Tertiary sands, gravels, and clays. The bedrock of the Bottomlands Section is deeply buried by alluvium.

Topography

The Bottomlands Section of the Coastal Plain Division consists of the broad floodplain at the confluence of the Ohio and Mississippi rivers, the broad floodplain of the Cache River, and the terraces and meander scars in the floodplains. The topography of the Cretaceous Hills Section is steep to rolling.

Soils

The soils of the uplands are derived from relatively thin loess and in a few places from gravel. Some areas in the eastern part of the Cretaceous Hills Section have gravel exposures. The soils of the Bottomlands Section range from recent alluvium to older terrace soils. The terrace soils tend to be hardpan clays; but areas of sand occur, especially in the western part of the section. The alluvial soils are generally heavy except near the rivers.

Plant Communities

Forest. The presettlement vegetation of the Coastal Plain Division was mostly forest. The upland forests of the Cretaceous Hills Section are similar to those of the Shawnee Hills with black oak, white oak, red oak, cherrybark oak, black gum, tulip tree, shagbark hickory, and pignut hickory the common trees. In the extreme western end of these hills beech and cucumber tree are also found. A large stand of native chestnuts (*Castanea dentata*) formerly grew near Olmsted in Pulaski County. Mesic ravines along the Ohio River contain southeastern floral elements that include the silverbell (*Halesia carolina*).

The bottomland forests consist in part of Shumard oak, cherrybark oak, swamp white oak, swamp chestnut oak, pin oak, overcup oak, kingnut hickory, shagbark hickory, bitternut hickory, ashes, sweet gum, black gum, honey locust, sugarberry, pecan, wild black cherry, and catalpa (*Catalpa speciosa*). Beech, tulip tree, and cucumber tree grow on the better-drained bottomland soils. Pin oak is the commonest tree on the heavier terrace soils, along with post oak and willow oak (*Quercus phellos*). Silver maple and American elm grow along the streams.

The bottomland swamps of this division contain an association of bald cypress, tupelo gum, swamp cottonwood, Drummond's red maple, water locust, pumpkin ash, and overcup oak. Water hickory (*Carya aquatica*) and planer-tree (*Planera aquatica*) are occasional on better-drained soils.

Prairie. The Cretaceous Hills Section contained small dry prairies on the uplands of the eastern part and mesic prairies in some of the broad creek bottoms. The few persisting remnants indicate that little bluestem was the dominant species of the dry upland prairies and that big bluestem, Indian grass, and gama grass (*Tripsacum dactyloides*) were the dominant species of the mesic prairies. Prairie dropseed and cord grass are absent from the prairies of the Cretaceous Hills Section.

Southern seep spring bog. Seep springs in the eastern end of the Cretaceous Hills Section are very acidic. They are generally dominated by sedges, royal fern (*Osmunda spectabilis*), lady fern (*Athyrium filix-femina* var. *asple-*

nioides), and cinnamon fern (*Osmunda cinnamomea*) and may contain large areas of sphagnum moss. These seep springs contain a distinctive southeastern flora that includes netted chain fern (*Woodwardia areolata*), screw-stem (*Bartonia paniculata*), and incomperta sedge (*Carex atlantica*).

Aquatic Habitats

The aquatic habitats of the bottomlands of the Coastal Plain Division include rivers, creeks, oxbow lakes, and sloughs. The Cretaceous Hills have creeks and seep springs.

Distinctive Fauna

The herpetofauna of the swampy lowlands of the Coastal Plain Division is similar to that of the Lower Mississippi River Bottomlands Division and includes the western cottonmouth, green water snake, green tree frog, western bird-voiced tree frog, and mole salamander. The range of the dusky salamander in Illinois is essentially restricted to the Cretaceous Hills Section.

Cretaceous Hills Section

The Cretaceous Hills Section encompasses the rolling hills of unconsolidated Cretaceous and Tertiary sands, gravels, and clays. Found in these hills are fossil beds from the Cretaceous period.

Principal Natural Features

Forest: Dry upland, mesic upland.
Prairie: Dry, mesic.
Southern Seep Spring Bog
Bedrock: Cretaceous and Tertiary sands, gravels and clays, Cretaceous fossil beds.
Topography: Steep to rolling hills.
Aquatic Habitats: Creeks, seep springs.

Bottomlands Section

The Bottomlands Section encompasses the bottomland forests, oxbow lakes, sloughs, and rivers of the Coastal Plain Division. This section includes the remnants of the once vast bald cypress and tupelo gum swamps along the rivers.

Principal Natural Features

Forest: Floodplain, terrace, swamp.
Topography: Terrace ridges, floodplain, meander scars.
Aquatic Habitats: Backwater swamps, sloughs, oxbow lakes, creeks, rivers.

General Key to Groups of Illinois Vascular Plants

1. Plants reproducing by spores, either ferns, quillworts, clubmosses, or horsetails.... .... Group I, p. 38
1. Plants reproducing by seeds.
   2. Plants cacti or succulents or pitcherlike or completely nongreen or with leaves merely represented by scales .... Group II, p. 41
   2. Plants not succulent nor pitcherlike, green, with leaves not scalelike.
      3. Plants climbing or twining .... Group III, p. 42
      3. Plants erect, ascending, prostrate, or trailing, not climbing or twining.
         4. Plants aquatic, living in water the entire year .... Group IV, p. 43
         4. Plants terrestrial, at least during most of the year.
            5. Plants with latex or colored sap ....Group V, p. 46
            5. Plants with clear sap.
               6. Some part of plant prickly or spiny .... Group VI, p. 46
               6. Plants without prickles or spines.
                  7. Plants woody (excluding woody vines that are in Group III).
                     8. Leaves less than 3 mm wide, usually needlelike .... Group VII, p. 47
                     8. Leaves more than 3 mm wide, not needlelike.
                        9. Leaves opposite or whorled .... Group VIII, p. 48
                        9. Leaves alternate.
                           10. Leaves simple, entire, toothed, or lobed ....Group IX, p. 49
                           10. Leaves compound ....Group X, p. 52
                  7. Plants herbaceous.
                     11. Plants monocotyledonous (germinating with a single cotyledon, or seed leaf), usually with elongated, parallel-veined leaves; flower parts usually in 3s ....Group XI, p. 53
                     11. Plants dicotyledonous (germinating with a pair of cotyledons, or seed leaves), usually with net-veined leaves; flower parts usually in 2s, 4s, or 5s.
                        12. Leaves all basal ....Group XII, p. 58
                        12. Leaves on the stem (basal leaves present sometimes as well).
                           13. Leaves, or some of them, whorled......Group XIII, p. 59
                           13. Leaves opposite or alternate.
                              14. At least some of the leaves opposite.
                                 15. Leaves simple.
                                    16. Leaves entire.........................Group XIV, p. 60
                                    16. Leaves toothed or lobed ....Group XV, p. 62
                                 15. Leaves compound..................... Group XVI, p. 63
                              14. Leaves alternate.
                                 17. Leaves simple.
                                    18. Leaves entire....................... Group XVII, p. 63
                                    18. Leaves toothed or lobed ... Group XVIII, p. 65
                                 17. Leaves compound.
                                    19. Leaves trifoliolate or ternately compound .... Group XIX, p. 67

19. Leaves pinnately or palmately compound.
20. Leaves pinnately compound .... Group XX, p. 67
20. Leaves palmately compound .... Group XXI, p. 68

Group I
Plants reproducing by spores, either ferns, quillworts, clubmosses, or horsetails.

1. Stem longitudinally striate throughout, conspicuously jointed, with sheaths at each joint, the sheaths bearing short teeth .... 6. Equisetaceae, p. 74
1. Stem neither longitudinally striate nor jointed, without tooth-bearing sheaths.
2. Leaves very slender, grasslike, needlelike, or scalelike.
3. Leaves long, grasslike; plant with bilobed corm at base ... 8. Isoetaceae, p. 75
3. Leaves shorter, scalelike or soft needlelike; basal corm absent.
4. Leaves with a small, membranous structure (ligule) near axil .... 16. Selaginellaceae, p. 81
4. Leaves lacking a ligule .... 9. Lycopodiaceae, p. 76
2. Leaves broader, usually divided, if simple and undivided, then at least 4 mm broad.
5. True aquatic ferns, rooted or floating in water.
6. Plants rooted; leaves with 4 pinnae (like a 4-leaf clover) .... 10. Marsileaceae, p. 77
6. Plants floating; leaves bilobed .... 2. Azollaceae, p. 70
5. Terrestrial ferns, at least under normal environmental conditions.
7. Plants known only from the gametophyte stage .... 7. Hymenophyllaceae, p. 75
7. Plants known from the sporophyte stage.
8. Leaves simple and entire.
9. Leaves heart-shaped at base, tapering to a long tip; midvein conspicuous .... 1. Aspleniaceae, p. 69
9. Leaves tapering to base, obtuse or short-acute at tip; midvein absent or inconspicuous .... *Ophioglossum* in 12. Ophioglossaceae, p. 78
8. Leaves simple and pinnatifid to compound.
10. Leaves pinnatifid (if only the lowest pair or pairs of pinnae are divided to the rachis, see second number 10).
11. Veins of leaf united to form a network.
12. Pinnae scalloped along the margin .... *Onoclea* in 11. Onocleaceae, p. 77
12. Pinnae finely serrulate along the margin .... *Woodwardia* in 3. Blechnaceae, p. 71
11. Veins of leaf free.
13. Blades translucent, 1 cell layer thick, glabrous .... 7. Hymenophyllaceae, p. 75
13. Blades more opaque, several cell layers thick, glabrous or pubescent.
14. Blades deciduous, thin, triangular, the lowest (and largest) pinnae projecting downward .... *Phegopteris* in 17. Thelypteridaceae, p. 81
14. Blades evergreen, subcoriaceous, elongated, the lowest pinnae not projecting downward.
15. Sinuses between pinnules extending nearly to the winged rachis .... 14. Polypodiaceae, p. 79
15. Sinuses between pinnules very shallow .... 1. Aspleniaceae, p. 69
10. Leaves compound with at least the lowest pair of pinnae divided all the way to the rachis.

16. Petiole forked at tip; ultimate segments toothed along upper margin, entire along lower margin .... *Adiantum* in 15. Pteridaceae, p. 80

16. Petiole not forked at tip; ultimate segments not with the above pattern on the margins.

17. Upper part of leaves pedately divided, lower part pinnately divided .... *Pteris* in 15. Pteridaceae, p. 80

17. Leaves not divided as above.

18. Blades once-pinnate.

19. Pinnae with an auricle on 1 margin.

20. Margin of pinnae spiny-toothed; petioles densely scaly .... *Polystichum* in 5. Dryopteridaceae, p. 71

20. Margin of pinnae not spiny-toothed; petioles not densely scaly .... 1. Aspleniaceae, p. 69

19. Pinnae without auricles on the margin.

21. Pinnae entire, several times longer than broad .... *Athyrium* in 5. Dryopteridaceae, p. 71

21. Pinnae toothed or lobed, at most only twice as long as broad, usually shorter .... 1. Aspleniaceae, p. 69

18. Blades pinnate-pinnatifid, bipinnate, bipinnate-pinnatifid, or tripinnate.

22. Leaves covered with a white, powdery substance .... *Argyrochosma* in 15. Pteridaceae, p. 80

22. Leaves not covered with a white, powdery substance.

23. Rachis smooth, without scales or hairs or glands.

24. Blades ternate, appearing broadly triangular.

25. Ultimate pinnules cut into linear divisions.... .... 12. Ophioglossaceae, p. 78

25. Ultimate pinnules round-lobed or unlobed.

26. Ultimate pinnules toothed .... 12. Ophioglossaceae, p. 78

26. Ultimate pinnules entire.

27. Blades delicate, less than 15 cm long .... *Gymnocarpium* in .... 5. Dryopteridaceae, p. 71

27. Blades more firm, well over 15 cm long .... *Pteridium* in 4. Dennstaedtiaceae, p. 71

24. Blades elongated, not ternate.

28. Veins of blade forming a single row of closed areoles on either side of midvein....... .... *Woodwardia* in 3. Blechnaceae, p. 71

28. All veins free.

29. Lower ½ of rachis brown, the upper ½ green .... 1. Aspleniaceae, p. 69

29. Rachis all 1 color, or at least not as above.

30. Rachis purple-brown .... *Pellaea* in 15. Pteridaceae, p. 80

30. Rachis brown or green.

31. Blades bipinnate-pinnatifid to tri-pinnate ... 5. Dryopteridaceae, p. 71

31. Blades pinnate-pinnatifid or bipinnate.

32. Pinnae tapering to a long-pointed apex........... *Athyrium* in .... 5. Dryopteridaceae, p. 71

32. Pinnae not tapering to a long-pointed apex.
33. Blades glabrous.
34. Ultimate pinnules finely toothed ....13. Osmundaceae, p. 79
34. Ultimate pinnules not finely toothed.
35. Blades delicate; plants less than 25 cm tall; plants of limestone cliffs ....*Cryptogramma* in 15. Pteridaceae, p. 80
35. Blades more firm; plants at least 50 cm tall; plants in woods or lowlands or on sandstone.
36. Blades pinnate-pinnatifid .... 13. Osmundceae, p. 79
36. Blades bipinnate... ......*Dryopteris* in 5. Dryopteridceae, p. 71
33. Blades pubescent.
37. Fronds up to 75 cm long. .... *Thelypteris* in 17. Thelypteridaceae, p. 81
37. Fronds generally always well over 75 cm long....... 13. Osmundaceae,p. 79
23. Rachis hairy, scaly, or glandular.
38. Rachis both hairy and scaly.
39. Blades triangular, the rachis winged nearly to base ....*Phegopteris* in 17. Thelypteridaceae, p. 81
39. Blades elongated, the rachis winged only near apex or unwinged .... *Woodsia* in 5. Dryopteridaceae, p. 71
38. Rachis hairy or scaly but not both.
40. Rachis glandular.
41. Petiole and rachis hairy............................. ....4. Dennstaedtiaceae, p. 71
41. Petiole and rachis scaly............................. .... *Woodsia* in 5. Dryopteridaceae, p. 71
40. Rachis eglandular.
42. Rachis hairy.
43. Rachis purple-brown .... 15. Pteridaceae, p. 80
43. Rachis not purple-brown.
44. Blades ternate, appearing triangular.
45. Ultimate pinnules glabrous....... ....12. Ophioglossaceae, p. 78
45. Ultimate pinnules hairy .... *Pellaea* in 15. Pteridaceae, p. 80
44. Blades elongated, not ternate.

46. Pinnae with a tuft of brown wool in the axils ....13. Osmundaceae, p. 79
46. Pinnae without a tuft of brown wool in the axils.
47. Lower pair of pinnae gradually reduced in size .... 11. Onocleaceae, p. 77
47. Lower pair of pinnae not much smaller than those above .... *Dryopteris* in ..... 5. Dryopteridaceae, p. 71
42. Rachis scaly.
48. Blade pinnate-pinnatifid or bipinnate.
49. Blade subcoriaceous, evergreen.... .... *Dryopteris* in .... 5. Dryopteridaceae, p. 71
49. Blade thinner, deciduous.
50. Blade distinctly triangular; all but the lowest pair of pinnae connected by a winged rachis................... *Phegopteris* in .... 17. Thelypteridaceae, p. 81
50. Blade not triangular; most of the pinnae stalked...................... .... *Athyrium* in .... 5. Dryopteridaceae, p. 71
48. Blade bipinnate-pinnatifid or tripinnate .... 5. Dryopteridaceae, p. 71

Group II

Plants reproducing by seeds, either cacti, succulent, or pitcherlike, or completely non-green, or with leaves merely represented by scales.

1. Plants green.
2. Plants succulent.
3. Plants without leaves, prickly ....46. Cactaceae, p. 175
3. Plants usually with leaves or the leaves scalelike and the stem succulent, not prickly.
4. Petals absent.
5. Plants with scarious perianth .... 27. Amaranthaceae, p. 87
5. Plants without a scarious perianth.............56. Chenopodiaceae, p. 194
4. Petals present.
6. Petals 4 .... *Cakile* in 42. Brassicaceae, p. 162
6. Petals 5.
7. Sepals 5; stamens 5 ....62. Crassulaceae, p. 206
7. Sepals 2; stamens 4–40............................. 129. Portulacaceae, p. 294
2. Plants not succulent.
8. Leaves modified into pitchers .... 142. Sarraceniaceae, p. 341
8. Leaves not modified into pitchers but reduced to scales.
9. Low bushy shrubs or small trees.
10. Low bushy shrubs; leaves tomentose.. .... *Hudsonia* in 57. Cistaceae, p. 200
10. Small trees; leaves glabrous........................... 151. Tamaricaceae, p. 357
9. Delicate herbs...................................... *Bartonia* in 75. Gentianaceae, p. 239
1. Plants non-green.
11. Stems twining or creeping, orange or yellow ....64. Cuscutaceae, p.208
11. Stems upright or ascending, neither orange nor yellow.
12. Plants less than 1.5 cm tall; stamens 3 ....196. Thismiaceae, p. 488
12. Plants well over 1.5 cm tall; stamens 1, 4, 8, or 10.

13. Flowers zygomorphic; stamens 1 or 4.
14. Stamen 1; leaves parallel-veined .... 187. Orchidaceae, p. 428
14. Stamens 4; leaves net-veined .... 113. Orobanchaceae, p. 279
13. Flowers actinomorphic; stamens 8 or 10 .... 104. Monotropaceae, p. 271

Group III
Plants climbing or twining.

1. Plants woody.
2. Leaves simple.
3. Leaves whorled .... 173. Dioscoreaceae, p. 414
3. Leaves opposite or alternate.
4. Leaves opposite.
5. Leaves entire .... *Lonicera* in 52. Caprifoliaceae, p. 180
5. Leaves serrate .... *Euonymus* in 54. Celastraceae, p. 193
4. Leaves alternate.
6. Leaves evergreen .... *Hedera* in 33. Araliaceae, p. 100
6. Leaves deciduous.
7. Some or all the leaves lobed.
8. Plants without tendrils, spiny .... *Lycium* in 147. Solanaceae, p. 352
8. Plants with tendrils, not spiny.
9. Leaves coarsely toothed .... *Vitis* in 160. Vitaceae, p. 366
9. Leaves entire .... 100. Menispermaceae, p. 270
7. Leaves unlobed.
10. Leaves peltate .... *Menispermum* in 100. Menispermaceae, p. 270
10. Leaves not peltate.
11. Leaves toothed.
12. Tendrils present .... 160. Vitaceae, p. 366
12. Tendrils absent .... *Celastrus* in 54. Celastraceae, p. 193
11. Leaves entire.
13. Leaves woolly, more than 10 cm long .... *Aristolochia* in 34. Aristolochiaceae, p. 101
13. Leaves glabrous or nearly so, usually less than 10 cm long.
14. Plants often with spines or stiff bristles; flowers unisexual; perianth parts 6; stamens 6; fruit globose .... 194. Smilacaceae, p. 486
14. Plants without spines or stiff bristles; flowers perfect; sepals 5; petals 5; stamens 5; fruit ellipsoid .... *Berchemia* in 134. Rhamnaceae, p. 306
2. Leaves compound.
15. Leaves opposite.
16. Leaflets 7–11; tendrils absent .... *Campsis* in 40. Bignoniaceae, p. 158
16. Leaflets 2; tendrils present .... *Bignonia* in 40. Bignoniaceae, p. 158
15. Leaves alternate.
17. Leaves bipinnate or tripinnate .... *Ampelopsis* in 160. Vitaceae, p. 366
17. Leaves once-pinnate, palmate, or trifoliolate.
18. Leaves palmately compound.
19. Tendrils present .... *Parthenocissus* in 160. Vitaceae, p. 366
19. Tendrils absent .... 89. Lardizabalaceae, p. 263
18. Leaves once-pinnate or trifoliolate.
20. Leaves pinnate, with 9 or more leaflets .... 72. Fabaceae, p. 218
20. Leaves trifoliolate .... *Toxicodendron* in 28. Anacardiaceae, p. 89
1. Plants not woody.
21. Leaves opposite or whorled.
22. Leaves compound .... *Clematis* in 132. Ranunculaceae, p. 298
22. Leaves simple.
23. Leaves unlobed.
24. Latex present.

25. Ovaries 2; fruits in pairs .... *Trachelospermum* in 31. Apocynaceae, p. 99
25. Ovary 1; fruit usually solitary .... 35. Asclepiadaceae, p. 102
24. Latex absent.
26. Leaves whorled .... 173. Dioscoreaceae, p. 414
26. Leaves opposite .... *Cynanchum* in 35. Asclepiadaceae, p. 102
23. Leaves 3- to 7-lobed .... *Humulus* in 51. Cannabinaceae, p. 180
21. Leaves alternate.
27. Leaves compound.
28. Leaves ternate or biternate; tendrils present; stamens 8 .... *Cardiospermum* in 140. Sapindaceae, p. 341
28. Leaves once-pinnate or trifoliolate; tendrils absent; stamens 6 or 10.
29. Sepals 2; petals 4; stamens 6.......... *Adlumia* in 74. Fumariaceae, p. 238
29. Sepals 5; petals 5; stamens 10 .... 72. Fabaceae, p. 218
27. Leaves simple, although they may be deeply lobed.
30. Tendrils present.
31. Leaves lobed and/or toothed; petals 5.
32. Fringed corona present .... 117. Passifloraceae, p. 282
32. Fringed corona absent .... 63. Cucurbitaceae, p. 207
31. Leaves unlobed and entire; petals 0 or perianth parts 6.
33. Petals 0; stamens 8; flowers perfect .... *Brunnichia* in 128. Polygonaceae, p. 287
33. Perianth parts 6; stamens 6; flowers unisexual .... 194. Smilacaceae, p. 486
30. Tendrils absent.
34. Flowers unisexual.
35. Leaves serrate .... *Tragia* in 71. Euphorbiaceae, p. 213
35. Leaves entire .... 173. Dioscoreaceae, p. 414
34. Flowers perfect.
36. Some part of plant prickly .... 128. Polygonaceae, p. 287
36. Plants not prickly.
37. Flowers crowded into a head, all sharing the same receptacle... .... *Mikania* in 36. Asteraceae, p. 104
37. Flowers not crowded into a head, not sharing the same receptacle.
38. Leaves entire .... 59. Convolvulaceae, p. 201
38. Some or all the leaves lobed or pinnatifid.
39. Leaves with a pair of small lobes at base .... *Solanum* in 147. Solanaceae, p. 352
39. Leaves several-lobed or pinnatifid .... 59. Convolvulaceae, p. 201

Group IV

Plants aquatic, living in water the entire year.

1. Plants monocotyledonous (germinating with a single cotyledon, or seed leaf), usually with elongated, parallel-veined leaves; flower parts often in 3s.
2. Free-floating aquatics; leaves thick, in a tight rosette .... *Pistia* in 167. Araceae, p. 372
2. Plants not free floating or, if free floating, without thick leaves in a tight rosette.
3. Flowers crowded together on a spadix .... 162. Acoraceae, p. 368
3. Flowers not crowded together on a spadix (in *Ruppia,* 2 flowers are borne on a spadixlike structure).
4. Plants thalloid, floating on water .... 181. Lemnaceae, p. 424
4. Plants with roots, stems, and leaves.
5. Perianth absent, or reduced to very minute scales or bristles.
6. Plants erect; inflorescence terminal, spicate, thick; leaves long, linear, strap-shaped .... 199.Typhaceae, p. 489
6. Plants not erect, free-floating or sometimes rooted in bottom mud; inflorescence axillary or terminal and slenderly spicate; leaves not very long, linear, or strap-shaped.

7. Leaves alternate; stamens 2 or 4; inflorescence spicate and usually terminal, or with flowers borne in pairs in a spadixlike structure.
   8. Stamens 2; flowers on a short spadixlike structure, concealed within the leaf sheath; fruit stipitate, drupelike ....191. Ruppiaceae, p. 485.
   Stamens 4; flowers in a spike or head; fruit sessile, appearing as an achene upon drying ....190. Potamogetonaceae, p. 483
7. Leaves opposite; stamen 1; inflorescence not spicate, axillary.
   9. Carpel 1; fruit beakless.......... 185. Najadaceae, p. 427
   9. Carpels 2–4; fruit beaked..........200. Zannichelliaceae, p. 490

5. Perianth present, composed of either calyx or corolla or both.
   10. Pistils simple, more than 1, separate or slightly coherent at base.
      11. Calyx and corolla undifferentiated (similar in color and texture).
         12. Inflorescence a spikelike raceme, without bracts; leaves all basal ....180. Juncaginaceae, p. 424
         12. Inflorescence loosely racemose, bracteate; leaves basal and also alternate .... 193. Scheuchzeriaceae, p. 486
      11. Calyx and corolla differentiated in color and texture.
         13. Inflorescence umbellate; pistils 6, coherent at base; fruit a follicle ....169. Butomaceae, p. 373
         13. Inflorescence not umbellate; pistils 10 or more, free to base; fruit an achene .... 164. Alismataceae, p. 369
   10. Pistil 1, compound.
      14. Ovary superior.
         15. Calyx and corolla differentiated in color and texture ....200. Xyridaceae, p. 490
         15. Calyx and corolla undifferentiated (similar in color and texture).
            16. Flowers unisexual .... 195. Sparganiaceae, p. 488
            16. Flowers perfect.
               17. Perianth scarious .... 179. Juncaceae, p. 419
               17. Perianth petaloid.......... 189. Pontederiaceae, p. 482
      14. Ovary inferior.
         18. Leaves whorled..........176. Hydrocharitaceae, p. 416
         18. Leaves basal, or cauline and alternate.
            19. Stamens 2, or 6–12, never 3; flowers unisexual; styles not petaloid..........176. Hydrocharitaceae, p. 416
            19. Stamens 3; flowers bisexual; styles petaloid.......... ....178. Iridaceae, p. 416

1. Plants dicotyledonous (germinating with a pair of cotyledons, or seed leaves), usually with net-veined leaves; flower parts often in 4s or 5s.
   20. Some of the leaves deeply divided or compound.
      21. Flowers borne together in a solitary, yellow head, each flower sharing a common receptacle.......... *Megaladonta* in 36. Asteraceae, p. 104
      21. Flowers not crowded, each with its own receptacle.
         22. Leaves trifoliolate.......... *Menyanthes* in 101. Menyanthaceae, p. 270
         22. Leaves 5- to 9-parted or pinnatisect.
            23. Leaves 5- to 9-parted .... *Nasturtium* in 42. Brassicaceae, p. 162
            23. Leaves pinnatisect.
               24. Underwater structures bearing small bladders; flowers zygomorphic .... 91. Lentibulariaceae, p. 263
               24. Underwater structures not bearing bladders; flowers actinomorphic.
                  25. Some or all the flowers unisexual.
                     26. Calyx (or involucre) 8- to 12-cleft; stamens 10–20; ovary unlobed ....55. Ceratophyllaceae, p. 194
                     26. Calyx 3- or 4-cleft; stamens 3–8; ovary 2- to 4-lobed.......... ....78. Haloragidaceae, p. 243

25. All flowers perfect.
27. Pistils 2–several, free, or pistil 1 but deeply 2- to 4-lobed.
28. Sepals 3; petals 3; stamens 3–6 .... 45. Cabombaceae, p. 175
28. Sepals 4–5; petals 0, 4, or 5; stamens 3, 4, 8, or numerous.
29. Sepals 5; petals 5; stamens numerous .... *Ranunculus* in 132. Ranunculaceae, p. 298
29. Sepals 4; petals 0 or 4; stamens 3, 4, or 8 .... 78. Haloragidaceae, p. 243
27. Pistil 1, unlobed.
30. Sepals 5; petals 5, united at base; stamens 5 .... *Hottonia* in 130. Primulaceae, p. 294
30. Sepals 4; petals 4, free; stamens 6 .... *Neobeckia* in 42. Brassicaceae, p. 162
20. Leaves simple, none deeply divided.
31. Leaves with conspicuous sheathing stipules (ocreae) at base .... 128. Polygonaceae, p. 287
31. Leaves without sheathing stipules at base.
32. Flowers unisexual; petals absent.
33. Calyx absent; fruit heart-shaped .... 48. Callitrichaceae, p. 177
33. Sepals 3; fruit ellipsoid....... *Proserpinaca* in 78. Haloragidaceae, p. 243
32. Flowers perfect; perianth, or at least the calyx, present.
34. Leaves peltate.
35. Sepals, petals, and stamens each 5 .... *Hydrocotyle* in 30. Apiaceae, p. 91
35. Sepals, petals, and stamens some number other than 5.
36. Sepals 3–4; petals 3–4; stamens 12–18; leaves up to 6 cm across .... *Brasenia* in 45. Cabombaceae, p. 175
36. Sepals and petals indistinguishable, together totaling more than 8 segments; stamens more than 20; leaves more than 6 cm across.................................... 107. Nelumbonaceae, p. 272
34. Leaves not peltate.
37. Leaves opposite or whorled.
38. Stamen 1; leaves linear .... 82. Hippuridaceae, p. 244
38. Stamens 3–5; leaves not linear.
39. Sepals and petals usually 3 each .... *Elatine* in 69. Elatinaceae, p. 210
39. Sepals and petals (if present) usually 4 or 5 each.
40. Corolla absent ....*Didiplis* in 96. Lythraceae, p. 265
40. Corolla present.
41. Stamens 2; leaves lanceolate .... *Justicia* in 23. Acanthaceae, p. 83
41. Stamens 4; leaves more or less orbicular .... *Bacopa* in 145. Scrophulariaceae, p. 342
37. Leaves alternate or basal.
42. Petals absent (sepals appearing petaloid in *Caltha*).
43. Flowers white; pistil 1 ....143. Saururaceae, p. 341
43. Flowers yellow; pistils 4–12 .... *Caltha* in 132. Ranunculaceae, p. 298
42. Petals present (what appears to be petals in *Caltha* are actually sepals).
44. Petals united........... *Nymphoides* in 75. Gentianaceae, p. 239
44. Petals free.
45. Flowers borne in umbels; stamens 5 .... *Hydrocotyle* in 30. Apiaceae, p. 91
45. Flowers not borne in umbels; stamens 8–numerous.
46. Petals 4–6; stamens 8–11 .... *Ludwigia* in 112. Onagraceae, p. 275
46. Petals more than 6; stamens more than 12 .... 109. Nymphaeaceae, p. 273

Group V
Plants with latex or colored sap.

1. Plants woody.
  2. Leaves palmately lobed....................24. Aceraceae, p. 85
  Leaves 2-lobed, 3-lobed, or unlobed, regularly toothed or entire.
    3. Leaves entire.................... *Maclura* in 105. Moraceae, p. 271
    3. Leaves toothed .... *Morus* in 105. Moraceae, p. 271
1. Plants not woody.
  4. Many ray flowers crowded into heads, each sharing the same receptacle; basal leaves different from cauline leaves ....36. Asteraceae, p. 104
  4. Flowers not in heads sharing the same receptacle.
    5. Sap milky (latex).
      6. Leaves opposite or whorled.
        7. Leaves serrate....................71. Euphorbiaceae, p. 213
        7. Leaves entire.
          8. Petals absent....................71. Euphorbiaceae, p. 213
          8. Petals present.
            9. Pistils 2; fruit in pairs....................31. Apocynaceae, p. 99
            9. Pistil 1; fruit solitary....................35. Asclepiadaceae, p. 102
      6. Leaves alternate.
        10. Leaves toothed....................50. Campanulaceae, p. 177
        10. Leaves entire.
          11. Pistils 2; fruit in pairs....................31. Apocynaceae, p. 99
          11. Pistil 1; fruit solitary ....35. Asclepiadaceae, p. 102
    5. Sap colored ....115. Papaveraceae, p. 280

Group VI
Some part of plant prickly or spiny.

1. Plants woody.
  2. Latex present.................... *Maclura* in 105. Moraceae, p. 271
  2. Latex absent.
    3. Leaves compound.
      4. Leaves twice-compound.
        5. Spines 2–10 cm long.................... *Gleditsia* in 47. Caesalpiniaceae, p. 176
        5. Spines up to 5 mm long.
          6. Leaflets 1 cm long or longer; stamens 5; fruit fleshy .... *Aralia* in 33. Araliaceae, p. 100
          6. Leaflets less than 1 cm long; stamens numerous; fruit dry, prickly.................... *Schrankia* in 102. Mimosaceae, p. 270
      4. Leaves once-compound or trifoliolate.
        7. Leaves trifoliolate.
          8. Petiole winged; fruit a sour orange.................... *Poncirus* in 137. Rutaceae, p. 335
          8. Petiole unwinged; fruit not a sour orange.
            9. Fruit a legume; flowers zygomorphic .... *Ononis* in 72. Fabaceae, p. 218
            9. Fruit a rose hip, blackberry, or raspberry; flowers actinomorphic.
              10. Base of petiole dilated.................... *Rosa* in 135. Rosaceae, p. 308
              10. Base of petiole not dilated .... *Rubus* in 135. Rosaceae, p. 308
        7. Leaves once-pinnate, with 5 or more leaflets.
          11. Leaflets more than 5.
            12. Stamens more than 10 .... *Rosa* in 135. Rosaceae, p. 308
            12. Stamens 3, 5, or 10.
              13. Stamens 5; pistils 3–5.................... *Zanthoxylum* in 137. Rutaceae, p. 335
              13. Stamens 10 (occasionally fewer in *Gleditsia*); pistil 1.
                14. Spines in pairs at point of leaf attachment .... *Robinia* in 72. Fabaceae, p. 218

14. Spines scattered on trunk and on branches, not in pairs .. .. *Gleditsia* in 47. Caesalpiniaceae, p. 176

11. Leaflets 5.

15. Leaves pinnately compound......... *Rosa* in 135. Rosaceae, p. 308

15. Leaves palmately compound...... *Rubus* in 135. Rosaceae, p. 308

3. Leaves simple.

16. Leaves with spinulose teeth, evergreen..... *Ilex* in 32. Aquifoliaceae, p. 100

16. Leaves without spinulose teeth, usually deciduous.

17. Stamens 15 or more .... 135. Rosaceae, p. 308

17. Stamens 5 or 6.

18. Sepals, petals, and stamens each 6.. .... *Berberis* in 38. Berberidaceae, p. 156

18. Sepals, petals, and stamens each 5.

19. Plants woody vines.................. *Lycium* in 147. Solanaceae, p. 352

19. Plants shrubs or small trees.

20. Leaves usually lobed..................... 77. Grossulariaceae, p. 242

20. Leaves not lobed.

21. Hypanthium present; petaloid staminodia absent............ .... 134. Rhamnaceae, p. 306

21. Hypanthium absent; petaloid staminodia 5........................ .... 141. Sapotaceae, p. 341

1. Plants not woody.

22. Flowers crowded into heads, each sharing the same receptacle.......................... ....36. Asteraceae, p. 104

22. Flowers not crowded into heads and not sharing the same receptacle.

23. Leaves opposite, sometimes nearly perfoliate; flowers borne in prickly heads ....65. Dipsacaceae, p.209

23. Leaves alternate or basal, never perfoliate; flowers not in prickly heads.

24. Plants with colored sap .... *Argemone* in 115. Papaveraceae, p. 280

24. Plants with clear sap.

25. Petals absent (sepals petaloid in Polygonaceae).

26. Leaves spine-tipped .... *Salsola* in 56. Chenopodiaceae, p. 194

26. Leaves not spine-tipped.

27. Leaves with a sheathing stipule (ocrea) at base .... 128. Polygonaceae, p. 287

27. Leaves without a sheathing stipule at base................................. ....27. Amaranthaceae, p. 87

25. Petals present.

28. Leaves compound .... 33. Araliaceae, p. 100

28. Leaves simple.

29. Leaves, or some of them, lobed .... 147. Solanaceae, p. 352

29. Leaves not lobed.

30. Leaves with spinulose teeth; flowers in spherical heads........ .... *Eryngium* in 30. Apiaceae, p. 91

30. Leaves without spinulose teeth; flower solitary .... *Hydrolea* in 84. Hydrophyllaceae, p. 245

Group VII

Plants woody; leaves less than 3 mm wide, usually needlelike.

1. Some or all the leaves scalelike, overlapping, up to 0.4 mm long.

2. Plants producing flowers.

3. Leaves downy; flowers yellow .... *Hudsonia* in 57. Cistaceae, p. 200

3. Leaves glabrous; flowers pink.................................. 151. Tamaricaceae, p. 357

2. Plants not producing flowers... 18. Cupressaceae, p. 81

1. Some or all the leaves needlelike or awl-shaped.

4. Needlelike leaves borne in fascicles of 2 or more, at least 2 cm long................... ....20. Pinaceae, p. 82

4. Needlelike leaves borne singly, up to 2 cm long.

5. All or most of the needlelike leaves 1.2 cm long or longer, usually flattened or with a keel; seeds either borne in a fleshy scarlet cup or in a subglobose fruit with shield-shaped scales.
    6. Shrubs with evergreen leaves; fruit borne in a fleshy scarlet cup............... .... 21. Taxaceae, p. 88
    6. Trees with deciduous leaves; seeds borne in a subglobose fruit with shield-shaped scales.. 22. Taxodiaceae, p. 83
5. None of the needlelike leaves more than 1.2 cm long, usually awl-shaped; seeds borne in a fleshy purple "berry"........................ 18. Cupressaceae, p. 81

Group VIII

Plants woody (excluding vines); leaves opposite or whorled.

1. Leaves simple.
  2. Some or all the leaves whorled.
    3. Trees; flowers zygomorphic.
      4. Fertile stamens 4; fruit an ovoid capsule; flowers basically purple............. .... *Paulownia* in 118. Paulowniaceae, p. 282
      4. Fertile stamens 2; fruits an elongated capsule; flowers basically white .... *Catalpa* in 40. Bignoniaceae, p. 158
    3. Shrubs; flowers actinomorphic.
      5. Flowers white; petals 4, united; stamens 4; ovary inferior; flowers and fruits in globose clusters.............. *Cephalanthus* in 136. Rubiaceae, p. 331
      5. Flowers pink; petals 5, free; stamens 10; ovary superior; flowers and fruits in axillary whorls.......................... *Decodon* in 96. Lythraceae, p. 265
  2. Leaves opposite, not whorled.
    6. Leaves entire.
      7. Plants parasitic; petals absent; stamens 3................. 159. Viscaceae, p. 366
      7. Plants not parasitic; petals present; stamens 4 to numerous.
        8. Sepals and petals similar and indistinguishable, numerous.................. .... 49. Calycanthaceae, p. 177
        8. Sepals and petals distinct and dissimilar, 4 or 5 each.
          9. Sepals 5; petals 5, united.
            10. Trees; stamens 4; ovary superior.. .... *Paulownia* in 118. Paulowniaceae, p. 282
            10. Small trees or shrubs; stamens 5; ovary inferior.
              11. Flowers more or less zygomorphic... .... *Lonicera* in 52. Caprifoliaceae, p. 180
              11. Flowers actinomorphic ....*Symphoricarpos* in 52. Caprifoliaceae, p. 180
          9. Sepals 4; petals 4, free (united in *Cephalanthus* and Oleaceae).
            12. Flowers in globose heads .... *Cephalanthus* in 136. Rubiaceae, p. 331
            12. Flowers not in globose heads.
              13. Petals united at base; stamens 2............ 111. Oleaceae, p. 273
              13. Petals free; stamens 4 or 8.
                14. Stamens 4; leaves not with scales...................................... .... 60. Cornaceae, p. 204
                14. Stamens 8; leaves with scales .... *Shepherdia* in 68. Elaeagnaceae, p. 210
    6. Leaves toothed or lobed.
      15. Leaves lobed.
        16. Petals small and inconspicuous or absent; fruit a samara..................... .... 24. Aceraceae p. 85
        16. Petals white, conspicuous; fruit fleshy.. .... *Viburnum* in 52. Caprifoliaceae, p. 180
      15. Leaves toothed.
        17. Sepals 4; petals 4 or absent.
          18. Petals absent; sepals yellow......... *Forestiera* in 111. Oleaceae, p. 273
          18. Petals present; sepals green.

19. Stamens numerous; hypanthium present .... *Rhodotypos* in 135. Rosaceae, p. 308
19. Stamens 2, 4, 5, 8, or 10; hypanthium absent.
20. Petals united; stamens 2 or 5.
21. Stamens 2 .... *Forsythia* in 111. Oleaeae, p. 273
21. Stamens 5 .... *Buddleja* in 43. Buddlejaceae, p. 175
20. Petals free; stamens 4, 5, 8, or 10.
22. Some flowers sterile; stamens 8 or 10 .... 83. Hydrangeaceae, p. 245
22. All flowers fertile; stamens 4 or 5 .... *Euonymus* in 54. Celastraceae, p. 193
17. Sepals 5; petals 5.
23. Petals free; stamens 10 or more .... 121. Philadelphaceae, p. 282
23. Petals united; stamens 5 .... 52. Caprifoliaceae, p. 180
1. Leaves compound.
24. Leaves palmately compound .... 81. Hippocastanaceae, p. 244
24. Leaves pinnately compound or trifoliolate.
25. Leaflets 3, finely toothed .... 148. Staphyleaceae, p. 356
25. Leaflets 5–13 (if 3 in *Acer negundo,* not finely toothed).
26. Leaflets aromatic; petals 5–8; stamens 5 or 6 .... *Phellodendron* in 137. Rutaceae, p. 335
26. Leaflets not aromatic; petals always 5 or absent; stamens 2 or 5.
27. Stamens 2; sepals 4 .... *Fraxinus* in 111. Oleaceae, p. 273
27. Stamens 5; sepals 5.
28. Petals white; fruit a berry ..... *Sambucus* in 52. Caprifoliaceae, p. 180
28. Petals yellow or green or absent; fruit a samara .... 24. Aceraceae. p. 85

Group IX

Plants woody (excluding vines); leaves alternate, simple, entire, toothed, or lobed.

1. Leaves entire.
2. Leaves scalelike.
3. Sepals 5; petals 5; stamens 10–30 .... *Hudsonia* in 57. Cistaceae, p. 200
3. Sepals 4; petals 4; stamens 4 .... 151. Tamaricaceae, p. 357
2. Leaves broader, not scalelike.
4. Latex present; fruit large and globose .... *Maclura* in 105. Moraceae, p. 271
4. Latex absent; fruit not large and globose.
5. Fruit an acorn .... *Quercus* in 73. Fagaceae, p. 234
5. Fruit not an acorn.
6. Petals absent.
7. Sepals present.
8. Sepals 6; stamens 9 .... 90. Lauraceae, p. 263
8. Sepals 4 or 5; stamens 5, 8, or 10.
9. Sepals 4; stamens 8 .... *Dirca* in 152. Thymelaeaceae, p. 357
9. Sepals 5; stamens 5 or 10.
10. Leaves 3-veined from the base; stamens 5 .... *Celtis* in 154. Ulmaceae, p. 357
10. Leaves not 3-veined from the base; stamens 10 .... 110. Nyssaceae, p. 273
7. Sepals absent.
11. Leaves glandular, aromatic .... 106. Myricaceae, p. 272
11. Leaves not glandular, not aromatic .... *Salix* in 138. Salicaceae, p. 335
6. Petals present.
12. Leaves cordate; flowers zygomorphic; fruit a legume .... *Cercis* in 47. Caesalpiniaceae, p. 176
12. Leaves not cordate; flowers actinomorphic; fruit not a legume.
13. Leaves with scales on lower surface .... *Elaeagnus* in 68. Elaeagnaceae, p. 210

13. Leaves without scales on lower surface.
14. Petals free.
15. Petals 6; leaves more than 6 cm long.
16. Petals maroon; leaves with an abrupt tip; bud scales absent .... 29. Annonaceae, p. 91
16. Petals white; leaves without an abrupt tip; bud scales present .... *Magnolia* in 97. Magnoliaceae, p. 266
15. Petals 4 or 5; leaves less than 6 cm long.
17. Ovary superior; hypanthium absent.
18. Sepals absent; petals yellow .... *Nemopanthus* in 32. Aquifoliaceae, p. 100
18. Sepals present; petals usually not yellow .... 70. Ericaceae, p. 211
17. Ovary inferior; hypanthium present.
19. Stamens numerous .... 135. Rosaceae, p. 308
19. Stamens 4–10.
20. Sepals 4; petals 4; stamens 4; veins strongly curved .... 60. Cornaceae, p. 204
20. Sepals 4–5; petals 4–5; stamens 4, 5, or 10; veins not strongly curved.................................... .... 70. Ericaceae, p. 211
14. Petals united.
21. Flowers unisexual; sepals 4; petals 4; stamens 4–16......... ....67. Ebenaceae, p. 191
21. Flowers perfect; sepals 5; petals 5; stamens 5 or 10.
22. Thorns often present; stamens 5................................... .... *Bumelia* in 141. Sapotaceae, p. 310
22. Thorns never present; stamens usually 10................... .... 70. Ericaceae, p. 211

1. Leaves toothed or lobed.
23. Leaves toothed.
24. Stems prickly or spiny.
25. Sepals 6; petals 6; stamens 6 .... *Berberis* in 38. Berberidaceae, p. 156
25. Sepals 5; petals 5; stamens 5–numerous... .... *Crataegus* in 135. Rosaceae, p. 308
24. Stems neither prickly nor spiny.
26. Leaves with spinulose teeth .... *Ilex* in 32. Aquifoliaceae, p. 100
26. Leaves without spinulose teeth.
27. Stems trailing; leaves evergreen, crowded at the tip of the branches... ....44. Buxaceae, p. 175
27. Stems upright; leaves usually deciduous, not crowded at the tip of the branches.
28. Plants with latex .... 105. Moraceae, p. 271
28. Plants without latex.
29. Flowers unisexual.
30. Petals present.
31. Margins of leaves strongly scalloped; leaves 4 cm long or longer; fruit dry.. .... *Hamamelis* in 79. Hamamelidaceae, p. 243
31. Margins of leaves not strongly scalloped; leaves less than 4 cm long; fruit fleshy .... *Ilex* in 32. Aquifoliaceae, p. 100
30. Petals absent.
32. Leaves aromatic, glandular .... 106. Myricaceae, p. 272
32. Leaves not aromatic, not glandular.
33. Fruit an acorn or enclosed in a spiny bur .... 73. Fagaceae, p. 234
33. Fruit neither an acorn nor enclosed in a spiny bur.
34. Sepals 5; fruit fleshy or a samara............................ .... 154. Ulmaceae, p. 357

34. Sepals absent (at least in the staminate flowers); fruit dry but not a samara.
35. Carpels 2; seeds with a tuft of hair .... 138. Salicaceae, p. 335
35. Carpel 1; seeds without a tuft of hair.
36. Staminate flowers without a calyx; pistillate flowers not in catkins ....61. Corylaceae, p. 205
36. Staminate flowers with a calyx; pistillate flowers in catkins........ 39. Betulaceae, p. 157
29. Flowers perfect.
37. Petals absent..................................... 134. Rhamnaceae, p. 306
37. Petals present.
38. Petals 4.
39. Flowers yellow; petals free; leaves strongly scalloped .... *Hamamelis* in 79. Hamamelidaceae, p. 243
39. Flowers not yellow; petals united; leaves not strongly scalloped.
40. Stamens 4, 5, or 10................... 70. Ericaceae, p. 211
40. Stamens 8 .... *Halesia* in 150. Styracaceae, p. 356
38. Petals 5.
41. Petals united.
42. Pubescence stellate .... *Styrax* in 150. Styracaceae, p. 356
42. Pubescence not stellate, or absent .... 70. Ericaceae, p. 211
41. Petals free.
43. Stamens 5.
44. Flowers white, in racemes ....86. Iteaceae, p. 248
44. Flowers not white, not in racemes .... 134. Rhamnaceae, p. 306
43. Stamens 15 or more.
45. Fruit attached to a paddlelike structure...153. Tiliaceae, p. 357
45. Fruit not attached to a paddlelike structure ....135. Rosaceae, p. 308
23. Leaves lobed.
46. Leaves bilobed.. 19. Ginkgoaceae, p. 82
46. Leaves not bilobed.
47. Leaves star-shaped................. *Liquidambar* in 79. Hamamelidaceae, p. 243
47. Leaves not star-shaped.
48. Leaves 4-lobed; petals more than 2 cm long, orange-based .... *Liriodendron* in 97. Magnoliaceae, p. 266
48. Leaves not 4-lobed; petals (if present) less than 2 cm long, not orange-based.
49. Flowers unisexual.
50. Petals present; fruit dry, globose............125. Platanaceae, p. 284
50. Petals absent; fruit an acorn or a nut, not globose.
51. Leaves glandular, aromatic; fruit a nut .... *Comptonia* in 106. Myricaceae, p. 272
51. Leaves not glandular, not aromatic; fruit an acorn... *Quercus* in 73. Fagaceae, p. 234
49. Flowers perfect.
52. Petals absent; leaves aromatic... *Sassafras* in 90. Lauraceae, p. 263
52. Petals present; leaves not aromatic.
53. Stamens 5; leaves usually palmately lobed .... 77. Grossulariceae, p. 242
53. Stamens more than 5; leaves usually not palmately lobed.... ....135. Rosaceae, p. 308

Group X
Plants woody (excluding vines); leaves alternate, compound.

1. Leaves palmately compound with at least 5 leaflets.
  2. Plants prickly; flowers white.
    3. Stamens and pistils numerous; fruit a blackberry or raspberry .... *Rubus* in 135. Rosaceae, p. 308
    3. Stamens and pistils 5; fruit not a blackberry or raspberry .... *Acanthopanax* in 33. Araliaceae, p. 100
  2. Plants not prickly; flowers yellow .... *Dasiphora* in 135. Rosaceae, p. 308
1. Leaves pinnately compound or trifoliolate.
  4. Leaves trifoliolate.
    5. Plants prickly or spiny.
      6. Petioles winged; fruit a sour orange........ *Poncirus* in 137. Rutaceae, p. 335
      6. Petioles not winged; fruit not a sour orange.
        7. Flowers actinomorphic; stamens numerous; fruit fleshy.. 135. Rosaceae, p. 308
        7. Flowers zygomorphic; stamens 10; fruit a legume .... *Ononis* in 72. Fabaceae, p. 218
    5. Plants neither prickly nor spiny.
      8. Stamens 5.
        9. Fruit fleshy; styles 3; leaflets not punctate ...... 28. Anacardiaceae, p. 89
        9. Fruit dry, winged; style 1; leaflets punctate.. *Ptelea* in 137. Rutaceae, p. 335
      8. Stamens 20 or more .... *Sibbaldiopsis* in 135. Rosaceae, p. 308
  4. Leaves pinnately compound with 5 or more leaflets.
    10. Plants prickly or spiny.
      11. Leaves bipinnately or tripinnately compound.
        12. Leaflets serrate; fruit fleshy.................... *Aralia* in 33. Araliaceae, p. 100
        12. Leaflets entire or nearly so; fruit a legume.. *Gleditsia* in 47. Caesalpiniaceae, p. 176
      11. Leaves once pinnately compound.
        13. Shrubs; flowers pink or rose; stamens and pistils numerous .... *Rosa* in 135. Rosaceae, p. 308
        13. Trees; flowers white or greenish; stamens 5 or 10; pistils 1–8.
          14. Flowers white, zygomorphic; stamens 10; pistil 1; fruit a legume.... .... *Robinia* in 72. Fabaceae, p. 218
          14. Flowers greenish, actinomorphic; stamens 5; pistils 3–8; fruit a follicle..................................... *Zanthoxylum* in 137. Rutaceae, p. 335
    10. Plants neither prickly nor spiny.
      15. Leaves bipinnately or tripinnately compound.
        16. Leaflets at least 1 cm broad.
          17. Flowers perfect; stamens 5; fruit fleshy .... *Aralia* in 33. Araliaceae, p. 100
          17. Flowers unisexual; stamens 10; fruit a legume .... *Gymnocladus* in 47. Caesalpiniaceae, p. 176
        16. Leaflets up to 5 mm broad.............. *Albizia* in 102. Mimosaceae, p. 270
      15. Leaves once pinnately compound.
        18. Flowers, or some of them, unisexual.
          19. Stamen 1; all flowers unisexual; pistil 1; fruit a drupe with a hard nut inside .... 87. Juglandaceae, p. 248
          19. Stamens 3, 5, or 10; some flowers usually perfect; pistils 2–5, or 1 and deeply lobed; fruit a samara ....146. Simaroubaceae, p. 352
        18. All flowers perfect.
          20. Petals 4; stamens 8; fruit an inflated bladder.. *Koelreuteria* in 140. Sapindaceae, p. 341
          20. Petals 5; stamens 1, 5, 9, 10, 15, or more; fruit not an inflated bladder.
            21. Flowers zygomorphic; stamens 1, 9, or 10; fruit a legume........... .... 72. Fabaceae, p. 218

21. Flowers actinomorphic; stamens 5 or 15 or more; fruit not a legume.
    22. Stamens 5; pistil 1; flowers yellowish .... 28. Anacardiaceae, p. 89
    22. Stamens 15 or more; pistils 2 or more; flowers white .... 135. Rosaceae, p. 308

Group XI

Plants herbaceous, monocotyledonous, usually with elongated parallel-veined leaves; flower parts usually in 3s.

1. Plants climbing or twining (if erect, then usually with a few weak tendrils from the upper axils); leaves net-veined; flowers unisexual.
   2. Inflorescence umbellate; ovary superior; fruit a berry.. 194. Smilacaceae, p. 486
   2. Inflorescence glomerulate or paniculate; ovary inferior; fruit a capsule .... 173. Dioscoreaceae, p. 414
1. Plants erect or floating in water (tendrils never present); leaves mostly parallel-veined; flowers bisexual or unisexual; if floating in water, then either thalloid or with fleshy leaves forming a basal rosette.
   3. Plants with 1 or 2 whorls of leaves.
      4. Flowers actinomorphic; ovary superior; stamens 6.
         5. Plants never more than 50 cm tall; flowers usually borne singly .... *Trillium* and *Medeola* in 198. Trilliaceae, p. 488
         5. Plants more than 50 cm tall; flowers usually more than 1 .... *Lilium* in 182. Liliaceae, p. 426
      4. Flowers zygomorphic; ovary inferior; stamen 1 .... *Isotria* in 187. Orchidaceae, p. 428
   3. Plants with leaves alternate, opposite, basal, or absent, but not in whorls.
      6. Plants free floating.
         7. Leaves thick, more than 1 cm wide, forming a basal rosette .... *Pistia* in 167. Araceae, p. 372
         7. Plants thalloid; leaves less than 1 cm wide, not forming basal rosettes.. 181. Lemnaceae, p. 424
      6. Plants not free floating.
         8. Flowers crowded together in a spadix.
            9. Spathe at least partially surrounding the spadix .... 167. Araceae, p. 372
            9. Spathe absent, the spadix borne naked .... 162. Acoraceae, p. 368
         8. Flowers not crowded into a spadix (in *Ruppia,* 2 flowers are borne on a spadixlike structure).
            10. Perianth absent or reduced to very minute scales (lodicules) or bristles.
                11. Each flower subtended by 1 or more scales; plants generally not true aquatics.
                    12. Leaves 2-ranked; sheaths usually open; stems usually hollow with solid nodes, often terete; anthers attached above the base .... 188. Poaceae, p. 435
                    12. Leaves (when present) 3-ranked; sheaths closed; stems solid with soft nodes, often 3-angled; anthers attached at the base .... 172. Cyperaceae, p. 375
                11. Flowers not subtended by individual scales; plants mostly aquatic.
                    13. Plants erect; inflorescence terminal, spicate, thick; leaves very long, linear, strap-shaped.......................... 199. Typhaceae, p. 489
                    13. Plants not erect, free-floating or sometimes rooted in bottom mud; inflorescence axillary or terminal and slenderly spicate; leaves not very long, usually not linear, usually not strap-shaped.

14. Leaves alternate; stamens 2 or 4; inflorescence spicate and usually terminal or with flowers borne in pairs in a spadix-like structure.

15. Stamens 2; flowers on a short spadixlike structure, concealed within the leaf sheath; fruit stipitate, drupelike .... 191. Ruppiaceae, p. 485

15. Stamens 4; flowers in a spike or head; fruit sessile, appearing as an achene upon drying .... 190. Potamogetonaceae, p. 483

14. Leaves opposite; stamen 1; inflorescence not spicate, axillary.

16. Carpel 1; fruit beakless .... 185. Najadaceae, p. 427

16. Carpels 2–4; fruit beaked .... 201. Zannichelliaceae, p. 490

10. Perianth present, composed of either calyx or corolla or both (plants with perianth reduced to minute scales or bristles should be sought under the first 10).

17. Pistils simple, more than 1, separate or slightly coherent at base.

18. Calyx and corolla differentiated in color and texture.

19. Inflorescence umbellate; pistils 6, coherent at base; fruit a follicle .... 169. Butomaceae, p. 373

19. Inflorescence not umbellate; pistils 10 or more, free to base; fruit an achene .... 164. Alismataceae, p. 369

18. Calyx and corolla not differentiated in color and texture.

20. Carpels usually 6, the axis between them slender .... 180. Juncaginaceae, p. 424

20. Carpels 3, the axis between them broadly 3-winged .... 193.Scheuchzeriaceae, p. 486

17. Pistil 1, compound.

21. Ovary superior.

22. Calyx and corolla differentiated in color and texture.

23. Flowers crowded together in a dense head; leaves basal .... 200. Xyridaceae, p. 490

23. Flowers borne in cymes or umbels; leaves cauline .... 171.Commelinaceae, p. 374

22. Calyx and corolla undifferentiated in color and texture.

24. Flowers unisexual.

25. Leaves reduced to minute scales in the axils of which are filiform branchlets (commonly mistaken for leaves) .... 168. Asparagaceae, p. 373

25. Leaves not reduced to minute scales.

26. Leaves net-veined; flowers in umbels .... 194. Smilacaceae, p. 486

26. Leaves parallel-veined; flowers in globose clusters, racemes, or panicles.

27. Perianth small, greenish; flowers aggregated in dense globose clusters; stamens 5 .... 195. Sparganiaceae, p. 488

27. Perianth usually conspicuous, greenish, yellowish, white, or bronze-purple; flowers in racemes or panicles; stamens 6 .... 184. Melanthiaceae. p. 427

24. Flowers bisexual.

28. Perianth scarious .... 179. Juncaceae, p. 419

28. Perianth petaloid.

29. Stamens 3 .... 189. Pontederiaceae, p. 482
29. Stamens 6 (or 4).
30. Stamens of different sizes .... 189. Pontederiaceae, p. 482
30. Stamens all alike.
31. Leaves evergreen, rigid; stems woody .... *Yucca* in 163. Agavaceae, p. 368
31. Leaves deciduous (evergreen in Liriope), mostly not rigid; stems herbaceous.
32. Leaves evergreen; flowers pinkish, 4–6 mm across .... *Liriope* in 186. Nartheciaceae, p. 428
32. Leaves deciduous; flowers not pinkish or, if pinkish, not 4–6 mm across.
33. Flowers borne in umbels.
34. Leaves in 2 whorls below the flowers ...... *Medeola* in 198. Trilliaceae, p.488
34. Leaves never whorled, usually all basal.
35. Bulbs with a strong odor of garlic or onion .... 165. Alliaceae, p. 370
35. Bulbs lacking an odor of garlic or onion.
36. Leaves linear, up to 5 mm wide .... *Nothoscordum* in 165. Alliaceae, p. 370
36. Leaves broadly elliptic to obovate, at least 3 cm wide .... *Clintonia* in 182. Liliaceae, p. 426
33. Flowers borne variously, but not in umbels.
37. Perianth parts 4; stamens 4; styles 2-lobed; leaves 1–3, ovate, cordate ... . *Maianthemum* in 192. Ruscaceae, p. 486
37. Perianth parts 6; stamens 6; style 1, simple or 3-cleft, or styles 3, distinct; leaves 1–several, narrower, if ovate, not cordate.
38. Flower solitary, 1 per plant, borne on a leafless scape; leaves basal.
39. Flower nodding; leaves usually speckled .... Erythronium in 182. Liliaceae, p. 426
39. Flower erect; leaves usually not speckled .... Tulipa in 182. Liliaceae, p. 426
38. Flowers numerous or, if borne singly, then the stem leafy.

40. Flowers axillary from the cauline leaves.

41. Flowers yellow, at first terminal, at length appearing axillary; style deeply 3-cleft; fruit a capsule.................................*Uvularia* in 170. Colchicaceae, p. 374

41. Flowers white, axillary from the beginning; style unbranched; fruit a berry........................*Polygonatum* in 192. Ruscaceae, p. 486

40. Flowers terminal.

42. Flower borne singly.

43. Style deeply 3-cleft; flowers at most 4 cm long, basically yellowish; plants rhizomatous, to 50 cm tall. .....*Uvularia* in 170. Colchicaceae, p. 374

43. Style undivided; flowers at least 5 cm long, basically orange; plants bulbous, almost always over 50 cm tall.............................*Lilium* in 182. Liliaceae p. 426

42. Flowers in spikes, racemes, panicles, or irregular clusters.

44. Flowers at least 5 cm long irregularly arranged (rarely 4 cm long in *Hosta*).

45. Plants bulbous; stems leafy..........................Lilium in 182. Liliaceae, p. 426

45. Plants with fleshy roots, tubers, or rhizomes; stems nearly leafless.

46. Flowers yellow or orange; leaves linear, to 2 cm wide .... 174. Hemerocallaceae, p. 415

46. Flowers lilac or whitish; leaves lanceolate, at least 3 cm wide .... *Hosta* in 182. Liliaceae, p. 426

44. Flowers less than 5 cm long, arranged in spikes, racemes, or panicles.

47. Leaves cauline, elliptic to ovate, not grasslike; style 1.

48. Perianth parts united into a tube.

49. Perianth campanulate; flowers white; fruit a berry .... *Convallaria* in 192. Ruscaceae, p. 486
49. Perianth rotate; flowers usually blue; fruit a capsule .... *Chionodoxa* in 175. Hyacinthaceae, p. 415
48. Perianth parts free, except at base... *Smilacina* in 192. Ruscaceae, p. 486
47. Leaves basal or, if cauline, then long and narrow (grasslike); styles 1 or 3.
50. Perianth scaly .... *Aletris* in 175. Hyacinthaceae, p. 415
50. Perianth not scaly.
51. Style 1; plants bulbous with a racemose inflorescence .... 175. Hyacinthaceae, p. 415
51. Styles 3; plants rhizomatous or, if bulbous, with a paniculate inflorescence.
52. Inflorescense glutinous; flowers perfect .... *Triantha* in 197. Tofieldiaceae, p. 488
52. Inflorescence not glutinous; flowers unisexual .... 184. Melanthiaceae, p. 427

21. Ovary inferior.
53. Plants growing in water.
54. Leaves whorled ....176. Hydrocharitaceae, p. 416
54. Leaves basal, or cauline and alternate.
55. Stamens 2 or 6–12, never 3; flowers unisexual; styles not petaloid ....176. Hydrocharitaceae, p. 416
55. Stamens 3; flowers bisexual; styles petaloid ....178. Iridaceae, p. 416
53. Plants growing on land.
56. Flowers zygomorphic; stamens 1 or 2 or composed of only a single anther half.
57. Leaves pinnately veined; stamen reduced to ½ an anther..183. Maranthaceae, p. 427
57. Leaves parallel-veined; stamens 1 or 2.....187. Orchidaceae, p. 428
56. Flowers actinomorphic or nearly so; stamens 3 or 6.
58. Stamens 3; styles sometimes petaloid ....178. Iridaceae, p. 416
58. Stamens 6; styles not petaloid.

59. Leaves reduced to scales; plants lacking chlorophyll, at most 3 cm tall.......196. Thismiaceae, p. 488
59. Leaves blade-bearing; plants with chlorophyll, well over 3 cm tall.
60. Flowers greenish yellow; at least some of the leaves more than 2 cm broad, fleshy.................................... *Manfreda* in 163. Agavaceae, p. 368
60. Flowers not greenish yellow; none of the leaves as much as 2 cm broad, not fleshy.
61. Flowers with a corona or, if without a corona, the flowers red...166. Amaryllidaceae, p. 372
61. Flowers without a corona, yellow...177. Hypoxidaceae, p. 416

Group XII
Plants herbaceous, dicotyledonous; leaves all basal.

1. Flowers crowded in heads, each flower sharing the same receptacle..36. Asteraceae, p. 104
1. Flowers not crowded in heads, not sharing the same receptacle.
2. Leaves compound.
3. Leaves ternately decompound (fertile stem with a pair of opposite leaves in *Adoxa*).
4. Sepals 5; petals 5; stamens 5..................................... 33. Araliaceae, p. 100
4. Sepals 4 or more, often petal-like; petals absent; stamens 8 or more.
5. Sepals 4; stamens 8 (sterile leaves all basal; flowering stems with a pair of opposite leaves).. 25. Adoxaceae, p. 87
5. Sepals (4–) 5 or more; stamens numerous....................................... 132. Ranunculaceae, p. 298
3. Leaves not ternately decompound.
6. Leaflets 3; petals 5.
7. Leaflets notched at tip; fruit a follicle; vegetative parts acrid to taste.......114. Oxalidaceae, p. 280
7. Leaflets not notched at tip; fruit fleshy; vegetative parts not acrid to taste ....135. Rosaceae, p. 308
6. Leaflets usually 5 or more; petals 4........................ 42. Brassicaceae, p. 162
2. Leaves simple.
8. Leaves, or some of them, lobed.
9. Leaves peltate.
10. Leaves up to 2 cm across.................. *Hydrocotyle* in 30. Apiaceae, p. 91
10. Leaves at least 10 cm across.................................. *Podophyllum* in 38. Berberidaceae, p. 156
9. Leaves not peltate.
11. Leaves deeply 2-lobed.
12. Sepals 2; petals 8 or 16; stamens numerous; sap red.................. *Sanguinaria* in 115. Papaveraceae, p. 280
12. Sepals 4; petals 8; stamens 8; sap clear.................................*Jeffersonia* in 38. Berberidaceae, p. 156
11. Leaves several-lobed.
13. Flowers in umbels; plants usually aquatic.................................*Hydrocotyle* in 30. Apiaceae, p. 91
13. Flowers not in umbels; plants not usually aquatic.
14. Leaves 3-lobed; sepals 5–12, petal-like; petals absent.................. *Hepatica* in 132. Ranunculaceae, p. 298
14. Leaves not 3-lobed; sepals 4–5, not petal-like; petals 4–16.
15. Sap red; petals 8 or 16.................................. *Sanguinaria* in 115. Papaveraceae, p. 280
15. Sap clear; petals 4–5.

16. Flowers zygomorphic; petals united; stamens 2 .... *Salvia* in 88. Lamiaceae, p. 250
16. Flowers actinomorphic; petals free; stamens 5.
17. Ovary superior; leaves irregularly lobed; flowers purple, at least 1 cm long........... 158. Violaceae, p. 363
17. Ovary inferior or half-inferior; leaves round-lobed; flowers not purple, up to 5 mm long .... 144. Saxifragaceae, p. 341
8. Leaves not lobed.
18. Leaves toothed.
19. Petals united, at least at base .... 130. Primulaceae, p. 294
19. Petals free.
20. Stamens 10 .... 131. Pyrolaceae, p. 297
20. Stamens 5 .... 158. Violaceae, p. 363
18. Leaves entire.
21. Leaves 2, cordate; sepals 3, maroon; petals absent; stamens 12 .... *Asarum* in 34. Aristolochiaceae, p. 101
21. Leaves usually more than 2, usually not cordate; sepals 4–5, usually green; petals 4–5; stamens usually not 12.
22. Leaves with sticky hairs .... 66. Droseraceae, p. 209
22. Leaves without sticky hairs.
23. Sepals 4; petals 4.
24. Petals white or yellow, free; stamens 6; ovary superior .... 42. Brassicaceae, p. 162
24. Petals translucent, united; stamens 2–4; ovary inferior .... 124. Plantaginaceae, p. 283
23. Sepals 5; petals 5.
25. Leaves linear; sepals spurred; pistils numerous .... *Myosurus* in 132. Ranunculaceae, p. 298
25. Leaves not linear; sepals not spurred; pistil 1.
26. Petals united, at least at base...... 130. Primulaceae, p. 294
26. Petals free.
27. Stamens 5............................ 116. Parnassiaceae, p. 282
27. Stamens 10............................... 131. Pyrolaceae, p. 297

Group XIII
Plants herbaceous, dicotyledonous; leaves whorled.

1. Flowers crowded in heads, each sharing the same receptacle .... 36. Asteraceae, p. 104
1. Flowers not crowded in heads, at least not sharing the same receptacle.
2. Leaves compound.
3. Stems inflated; leaves below water pinnatisect .... *Hottonia* in 130. Primulaceae, p. 294
3. Stems firm; plants not aquatic.
4. Leaves divided into 3–5 serrate leaflets ...... *Panax* in 33. Araliaceae, p. 100
4. Leaves cleft, each division cleft again, or leaves divided into 3 round-lobed segments.
5. Leaves cleft, each division cleft again.
6. Sepals 4, green; petals 4; stamens 6 .... *Dentaria* in 42. Brassicaceae, p. 162
6. Sepals 4–20, petal-like; petals absent; stamens numerous .... *Anemone* in 132. Ranunculaceae, p. 298
5. Leaves divided into 3 round-lobed segments .... *Anemone* in 132. Ranunculaceae, p. 298
2. Leaves simple.
7. Latex present.
8. Petals 5; ovary not 3-lobed .... 35. Asclepiadaceae, p. 102
8. Petals absent; ovary 3-lobed .... 71. Euphorbiaceae, p. 213

7. Latex absent.
  9. All leaves lying flat on the ground.................... 103. Molluginaceae, p. 271
  9. Leaves on more or less sprawling or on upright stems, not lying flat on the ground.
    10. Leaves toothed.
      11. Leaves evergreen; flowers actinomorphic.. *Chimaphila* in 131. Pyrolaceae, p. 287
      11. Leaves deciduous; flowers zygomorphic.. *Veronicastrum* in 145. Scrophulariaceae, p. 342
    10. Leaves entire.
      12. Petals absent; stamens 3 .... *Polycnemum* in 56. Chenopodiaceae, p. 194
      12. Petals present; stamens 4 or more.
        13. Flowers zygomorphic; petals 3; stamens 8..127. Polygalaceae, p. 286
        13. Flowers actinomorphic; petals 4–5 (rarely 3 in some species of *Galium*); stamens 3, 4, 5, or 10.
          14. Petals 4.
            15. Petals free; flowers subtended by 4 large, white bracts........................60. Cornaceae, p. 204
            15. Petals united; flowers not subtended by 4 large, white bracts.
              16. Plants up to 2 m tall; leaves more than 6 cm broad.......................... *Frasera* in 75. Gentianaceae, p. 239
              16. Plants less than 2 m tall; leaves less than 1 cm broad.
                17. Sepals 2; leaves pale green; ovary superior .... *Obolaria* in 75. Gentianaceae, p. 239
                17. Sepals 4; leaves bright green; ovary inferior ....136. Rubiaceae, p. 331
          14. Petals 3 or 5.
            18. Petals free ....53. Caryophyllaceae, p. 186
            18. Petals united ....136. Rubiaceae, p. 331

Group XIV

Plants herbaceous, dicotyledonous; leaves simple, opposite, entire.

1. Flowers crowded into heads, each sharing the same receptacle........................36. Asteraceae, p. 104
1. Flowers not crowded into heads, at least not sharing the same receptacle.
  2. Latex present.
    3. Petals absent...71. Euphorbiaceae, p. 213
    3. Petals present.
      4. Pistils 2; fruits borne in pairs...................................31. Apocynaceae, p. 99
      4. Pistil 1; fruits borne singly..................................35. Asclepiadaceae, p. 102
  2. Latex absent.
    5. Plants succulent.. *Sesuvium* in 26. Aizoaceae, p. 87
    5. Plants not succulent.
      6. Petals absent.
        7. Stamens numerous; pistils numerous.. *Clematis* in 132. Ranunculaceae, p. 298
        7. Stamens 1, 3–5, 8, or 10; pistil 1.
          8. Stamen 1; dwarf plant; fruits heart-shaped.. 48. Callitrichaceae, p. 177
          8. Stamens 3–5, 8, or 10; plants not dwarf; fruits not heart-shaped.
            9. Stamens 8 or 10 ....53. Caryophyllaceae, p. 186
            9. Stamens 3–5.
              10. Sepals 4; stamens 4.
                11. Plants prostrate.
                  12. Ovary inferior...... *Ludwigia* in 112. Onagraceae, p. 275
                  12. Ovary superior .... *Didiplis* in 96. Lythraceae, p. 265

11. Plants upright ....96. Lythraceae, p. 265

10. Sepals 3 or 5; stamens 3 or 5.

13. Calyx tubular and showy, petaloid .... 108. Nyctaginaceae, p. 272

13. Calyx not tubular nor showy, not petaloid.

14. Stipules present ....53. Caryophyllaceae, p. 186

14. Stipules absent.

15. Flowers with scarious bracts and scarious sepals .. .. 27. Amaranthaceae, p. 87

15. Flowers without scarious bracts and scarious sepals ....56. Chenopodiaceae, p. 194

6. Petals present.

16. Petals free.

17. Stamens 8–numerous.

18. Stamens numerous; leaves often punctate .... 85. Hypericaceae, p. 246

18. Stamens 8–12.

19. Stamens 8.

20. Leaves with 3–5 main veins .... 99. Melastomaceae, p. 270

20. Leaves with 1 main vein ....*Epilobium* in 112. Onagraceae, p. 275

19. Stamens 9–12.

21. Stamens 9; flowers maroon or pink ....*Triadenum* in 85. Hypericaceae, p. 246

21. Stamens 10–12; flowers purple or white.

22. Stipules present ....53. Caryophyllaceae, p. 186

22. Stipules absent ....96. Lythraceae, p. 265

17. Stamens 4–5.

23. Flowers large, more than 3 cm long; plants glandular-hairy; fruits with long, curved projections ....119. Pedaliaceae, p. 282

23. Flowers up to 2 cm long; plants not glandular-hairy; fruits without long, curved projections.

24. Sepals 4; petals 4 ....*Rotala* in 96. Lythraceae, p. 265

24. Sepals 2 or 5; petals 5.

25. Sepals 2................ *Claytonia* in 129. Portulacaceae, p. 294

25. Sepals 5.

26. Stipules present ....53. Caryophyllaceae, p. 186

26. Stipules absent ....93. Linaceae, p. 264

16. Petals united.

27. Petals 4.

28. Ovary superior.

29. Ovary notched at apex; flowers actinomorphic...................... .... *Polypremum* in 43. Buddlejaceae, p. 175

29. Ovary 2- or 4-lobed; flowers zygomorphic............................. ....88. Lamiaceae, p. 250

28. Ovary inferior.

30. Petals translucent; stipules absent.. .... 124. Plantaginaceae, p. 283

30. Petals not translucent; stipules present.................................. ....136. Rubiaceae, p. 331

27. Petals 5.

31. Ovary inferior.

32. Sepals absent.................................156. Valerianaceae, p. 360

32. Sepals 5 .... *Triosteum* in 52. Caprifoliaceae, p. 180

31. Ovary superior.

33. Stamens 5; flowers actinomorphic.

34. Pistils 2; fruits borne in pairs; flowers red on the outside, yellow on the inside... ....*Spigelia* in 95. Loganiaceae, p. 265

34. Pistil 1; fruit borne singly; flowers not red on the outside and yellow on the inside.
35. Ovary 1-locular.
36. Corolla tubular ....75. Gentianaceae, p. 239
36. Corolla rotate .... 130. Primulaceae, p. 294
35. Ovary 2- or 3-locular.
37. Ovary 2-locular ....145. Scrophulariaceae, p. 342
37. Ovary 3-locular .... 126. Polemoniaceae, p. 284
33. Stamens 2 or 4; flowers zygomorphic.
38. Ovary 4-parted; fruit separating into 4 nutlets.................. ....88. Lamiaceae, p. 250
38. Ovary not 4-parted; fruit not separating into 4 nutlets.
39. Flowers nearly actinomorphic....................................... .... 23. Acanthaceae, p. 83
39. Flowers zygomorphic.....145. Scrophulariaceae, p. 342

Group XV

Plants herbaceous, dicotyledonous; leaves simple and opposite, toothed or lobed.

1. Leaves toothed.
2. Flowers crowded into heads, each sharing the same receptacle ....36. Asteraceae, p. 104
2. Flowers not crowded into heads, at least not sharing the same receptacle.
3. Plants trailing; leaves evergreen .... *Linnaea* in 136. Rubiaceae, p. 331
3. Plants upright; leaves not evergreen.
4. Stems and flower-heads prickly ....65. Dipsacaceae, p. 209
4. Stems and flower-heads not prickly.
5. Petals absent.
6. Flower parts scarious ....*Atriplex* in 27. Amaranthaceae, p. 87
6. Flower parts not scarious .... 155. Urticaceae, p. 359
5. Petals present.
7. Petals 2 or 4.
8. Petals 2; sepals 2; ovary inferior .... *Circaea* in 112. Onagraceae, p. 275
8. Petals 4; sepals 4; ovary superior.
9. Flowers zygomorphic; stamens 2 or 4.................................... ....88. Lamiaceae, p. 250
9. Flowers actinomorphic; stamens 6 .... *Lunaria* in 42. Brassicaceae, p. 162
7. Petals 5.
10. Petals free; leaves glandular-toothed; flowers actinomorphic..... ....*Bergia* in 69. Elatinaceae, p. 210
10. Petals united; leaves not glandular-toothed; flowers mostly zygomorphic.
11. Calyx absent; ovary inferior...........156. Valerianaceae, p. 360
11. Calyx present; ovary superior.
12. Stamens 5; teeth of calyx with a hooked tip; fruit appressed to rachis .... 122. Phrymaceae, p. 283
12. Stamens 2 or 4; teeth of calyx not hooked at tip; fruit not appressed to rachis.
13. Ovary 4-parted.
14. Ovary deeply 4-lobed, the style arising from between the lobes ....88. Lamiaceae, p. 250
14. Ovary shallowly 4-lobed or, if more deeply lobed, the style arising from the top of the ovary.............. .... 157. Verbenaceae, p. 361
13. Ovary single, not 4-parted.
15. Each cell of the ovary with one ovule...................... .... 157. Verbenaceae, p. 361

15. Each cell of the ovary with several ovules .... 145. Scrophulariaceae, p. 342

1. Leaves lobed.
   16. Flowers crowded into heads, each sharing the same receptacle .... 36. Asteraceae, p. 104
   16. Flowers not crowded into heads, at least not sharing the same receptacle.
      17. Plants prickly .... *Dipsacus* in 65. Dipsacaceae, p. 209
      17. Plants not prickly.
         18. Flowers in dense, hemispherical heads .... *Knautia* in 65. Dipsacaceae, p. 209
         18. Flowers not in dense, hemispherical heads.
            19. Basal leaves present in addition to the cauline ones.
               20. Basal leaves ternately compound .... 25. Adoxaceae, p. 87
               20. Basal leaves palmately lobed .... *Mitella* in 144. Saxifragaceae, p. 341
            19. Basal leaves not present.
               21. Petals absent (sepals petaloid); stamens and pistils numerous........ .... *Clematis* in 132. Ranunculaceae, p. 298
               21. Petals present; stamens 2 or 4; pistil 1.
                  22. Ovary 4-parted; fruit separating into 4 nutlets .... 88. Lamiaceae, p. 250
                  22. Ovary not 4-parted; fruit solitary .... 145. Scrophulariaceae, p. 342

Group XVI
Plants herbaceous, dicotyledonous; leaves compound, opposite.

1. Leaves pinnately compound.
   2. Flowers crowded into heads, each sharing the same receptacle .... 36. Asteraceae, p. 104
   2. Flowers not crowded into heads, never sharing the same receptacle.
      3. Petals free; ovary superior.
         4. Stamens 5 .... 33. Araliaceae, p. 100
         4. Stamens 10 or 15 .... 161. Zygophyllaceae, p. 368
      3. Petals united; ovary inferior .... 156. Valerianaceae, p. 360
1. Leaves palmately compound or trifoliolate.
   5. Leaves trifoliolate; flowers perfect.
      6. Petals 4; stamens 6; pistil 1 .... *Dentaria* in 42. Brassicaceae, p. 162
      6. Petals absent (sepals petaloid); stamens and pistils numerous....................... .... *Clematis* in 132. Ranunculaceae, p. 298
   5. Leaves with several leaflets; flowers unisexual .... *Cannabis* in 51. Cannabinaceae, p. 180

Group XVII
Plants herbaceous, dicotyledonous; leaves simple, alternate, entire.

1. Flowers crowded into heads, each sharing the same receptacle................................ .... 36. Asteraceae, p. 104
1. Flowers not crowded into heads, at least not sharing the same receptacle.
   2. Latex present.
      3. Petals 5.
         4. Flowers zygomorphic; corolla split nearly to base...................................... .... *Lobelia* in 50. Campanulaceae, p. 177
         4. Flowers actinomorphic; corolla not split nearly to base.
            5. Pistils 2; fruit borne in pairs............ *Amsonia* in 31. Apocynaceae, p. 99
            5. Pistil 1; fruit not borne in pairs .... 50. Campanulaceae, p. 177
      3. Petals absent .... 71. Euphorbiaceae, p. 213
   2. Latex absent.
      6. Flowers unisexual.
         7. Perianth and bracts scarious .... 27. Amaranthaceae, p. 87
         7. Perianth and bracts not scarious.

8. Stamens 4 .... *Parietaria* in 155. Urticaceae, p. 359
8. Stamen 1 .... 71. Euphorbiaceae, p. 213
6. Flowers perfect.
9. Petals absent.
10. Flowers zygomorphic; stamens 6 .... *Aristolochia* in 34. Aristolochiaceae, p. 101
10. Flowers actinomorphic; stamens other than 6.
11. Perianth and bracts scarious .... *Celosia* in 27. Amaranthaceae, p. 87
11. Perianth and bracts not scarious.
12. Leaves with sheaths (ocreae) at base .... 128. Polygonaceae, p. 287
12. Leaves without sheaths at base.
13. Sepals absent; stamens 6–8, bright white; pistils 3–4 .... 143. Saururaceae, p. 341
13. Sepals present; stamens usually not 6–8, not bright white; pistil 1.
14. Ovary inferior .... 139. Santalaceae, p. 340
14. Ovary superior.
15. Sepals 8 ...... *Thymelaea* in 152. Thymelaeaceae, p. 357
15. Sepals 1, 3, or 5.
16. Flowers showy, because of petaloid sepals, or solitary.
17. Leaves more than 6 cm long; flowers numerous in racemes, white; pistil 1; fruit a berry .... 123. Phytolaccaceae, p. 283
17. Leaves less than 6 cm long; flower solitary, green; pistils numerous; fruit a follicle .... *Helleborus* in 132. Ranunculaceae, p. 298
16. Flowers not showy, not solitary .... 56. Chenopodiaceae, p. 194
9. Petals present.
18. Flowers zygomorphic.
19. Petals united; stamens 2 or 4 .... 145. Scrophulariaceae, p. 342
19. Petals free; stamens 5, 8, or 10.
20. Stamens 10; fruit a legume ..... *Crotalaria* in 72. Fabaceae, p. 218
20. Stamens 5 or 8; fruit not a legume.
21. Petals 3; stamens 8 .... 127. Polygalaceae, p. 286
21. Petals 5; stamens 5 .... 158. Violaceae, p. 363
18. Flowers actinomorphic.
22. Leaves perfoliate or strongly clasping.
23. Flowers in umbels, white; petals free .... *Bupleurum* in 30. Apiaceae, p. 91
23. Flower solitary in the axils of the leaves, blue; petals united .... *Triodanis* in 50. Campanulaceae, p. 177
22. Leaves neither perfoliate nor strongly clasping.
24. Petals united.
25. Plants more than 2 m tall, densely woolly .... *Verbascum* in 145. Scrophulariaceae, p. 342
25. Plants less than 2 m tall, not densely woolly.
26. Petals 4; stamens 4 .... *Anagallis* in 130. Primulaceae, p. 294
26. Petals 5; stamens 5–15.
27. Flowers with a corona; seeds with a tuft of hairs .... 35. Asclepiadaceae, p. 102
27. Flowers without a corona; seeds without a tuft of hairs.
28. Ovary distinctly 4-lobed .... 41. Boragniaceae, p. 159
28. Ovary not distinctly 4-lobed.

29. Flowers in a scorpioid cyme .... 80. Heliotropaceae p. 244
29. Flowers not in a scorpioid cyme.
30. Ovary inferior .... *Samolus* in 130. Primulaceae, p. 294
30. Ovary superior.
31. Stigmas 3 ...... 126. Polemoniaceae, p. 284
31. Stigmas 1 or 2.
32. Stigmas 2 .... 59. Convolvulaceae, p. 201
32. Stigma 1.
33. Stamens opposite the petals .... 130. Primulaceae, p. 294
33. Stamens alternate with the petals .... 59. Convolvulaceae, p. 201
24. Petals free.
34. Petals 5.
35. Stamens united into a tube .... 93. Linaceae, p. 264
35. Stamens free.
36. Sepals all the same size; stamens 10 .... *Ludwigia* in 112. Onagraceae, p. 275
36. Sepals of 2 different sizes; stamens 5–15 .... 57. Cistaceae, p. 200
34. Petals 4.
37. Stamens 6 .... 42. Brassicaceae, p. 162
37. Stamens 8.
38. Pistil 1 .... 112. Onagaraceae, p. 275
38. Pistils 3 .... 133. Resedaceae, p. 306

Group XVIII
Plants herbaceous, dicotyledonous; leaves simple and alternate, toothed or lobed.

1. Flowers crowded into heads, each sharing the same receptacle .... 36. Asteraceae, p. 104
1. Flowers not crowded into heads, at least not sharing the same receptacle.
2. Leaves toothed.
3. Teeth spinulose; flowers crowded into white, globose heads .... *Eryngium* in 30. Apiaceae, p. 91
3. Teeth not spinulose; flowers not crowded into white, globose heads.
4. Petals absent.
5. Leaves with a sheathing stipule (ocrea) at base .... *Rumex* in 128. Polygonaceae, p. 287
5. Leaves without a sheathing stipule at base.
6. Sepals petaloid.
7. Leaves cordate; stamens numerous; pistils 4–12 .... *Caltha* in 132. Ranunculaceae, p. 298
7. Leaves not cordate; stamens 10; pistils 5 .... 120. Penthoraceae, p. 282
6. Sepals not petaloid, often green and obscure.
8. Sepals 3 .... *Proserpinaca* in 78. Haloragidaceae, p. 243
8. Sepals 4 or 5, or calyx 4-parted.
9. Stamens 3 or 8 .... 71. Euphorbiaceae, p. 213
9. Stamens 4 or 5.
10. Flowers unisexual; leaves strongly 3-veined at base.
11. Stamens 4 .... *Fatoua* in 105. Moraceae, p. 271
11. Stamens 5 .... 155. Urticaceae, p. 359
10. Flowers, or most of them, perfect; leaves not 3-veined at base .... 56. Chenopodiaceae, p. 194
4. Petals present.
12. Flowers zygomorphic.

13. Petals free, 1 of them spurred.
14. Petals 5; stems opaque .... 158. Violaceae, p. 363
14. Petals 2, each of them 2-lobed; stems translucent .... 37. Balsaminaceae, p. 156
13. Petals united, rarely spurred.
15. Corolla split; stamens 5 .... *Lobelia* in 50. Campanulaceae, p. 177
15. Corolla not split; stamens 2 or 4 .... 145. Scrophulariaceae, p. 342
12. Flowers actinomorphic.
16. Petals free.
17. Stamens and pistils attached to a central column; stamens numerous .... 98. Malvaceae, p. 267
17. Stamens and pistils not attached to a central column; stamens 5 or 6.
18. Pistils 2; stem with a single leaf (all other leaves basal) .... *Sullivantia* in 144. Saxifragaceae, p. 341
18. Pistil 1; stem usually with more than 1 leaf.
19. Ovary inferior .... 94. Loasaceae, p. 265
19. Ovary superior.
20. Sepals 4; petals 4; stamens 2 or 6 .... 42. Brassicaceae, p. 162
20. Sepals 5; petals 5; stamens 5.
21. Flowers in a head .... *Eryngium* in 30. Apiaceae, p. 91
21. Flowers not in a head...... 149. Sterculiaceae, p. 356
16. Petals united.
22. Anthers splitting lengthwise .... *Verbascum* in 145. Scrophulariaceae, p. 342
22. Anthers opening at the tip .... 147. Solanaceae, p. 352
2. Leaves lobed.
23. Petals absent.
24. Leaves with a sheathing stipule (ocrea) at base .... 128. Polygonaceae, p. 287
24. Leaves without a sheathing stipule at base.
25. Sepals usually petaloid; stamens numerous .... 132. Ranunculaceae, p. 298
25. Sepals inconspicuous, not petaloid; stamens 3–5.
26. Plants more than 2 m tall; leaves palmately lobed .... *Ricinus* in 71. Euphorbiaceae, p. 213
26. Plants up to 1.5 m tall; leaves pinnatifid.
27. Sepals 3; stamens 3 .... *Proserpinaca* in 78. Haloragidaceae, p. 243
27. Sepals 5; stamens 5 .... 56. Chenopodiaceae, p. 194
23. Petals present.
28. Petals free.
29. Stamens and pistils borne in a central column.... 98. Malvaceae, p. 267
29. Stamens and pistils not borne in a central column.
30. Flowers borne in a head .... *Eryngium* in 30. Apiaceae, p. 91
30. Flowers not borne in a head.
31. Petals apparently 10; stamens 5; ovary inferior .... 94. Loasaceae, p. 265
31. Petals 4–6; stamens 6–numerous; ovary superior.
32. Sepals 4; petals 4; stamens 6 (rarely 2); pistil 1 .... 42. Brassicaceae, p. 162
32. Sepals 5–6; petals 5–6; stamens numerous; pistils 3 .... 133. Resedaceae, p. 306
28. Petals united.
33. Flowers zygomorphic; stamens 4............ 145. Scrophulariaceae, p. 342
33. Flowers actinomorphic; stamens 5.
34. Stigmas 3 .... 126. Polemoniaceae, p. 284

34. Stigmas 2.
35. Styles 2, free .... 84. Hydrophyllaceae, p. 245
35. Style 1, becoming lobed at tip but always united below ....
.... 147. Solanaceae, p. 352

Group XIX
Plants herbaceous, dicotyledonous; leaves alternate, and trifoliolate or ternately compound.

1. Leaves trifoliolate.
2. Sepals 4; petals 4 .... 58. Cleomaceae, p. 201
2. Sepals 5; petals 5.
3. Flowers zygomorphic; fruit a legume or loment .... 72. Fabaceae, p. 218
3. Flowers actinomorphic; fruit neither a legume nor loment.
4. Stamens 10 or more.
5. Leaflets notched at tip; plants with acrid sap ....
.... 114. Oxalidaceae, p. 280
5. Leaflets not notched at tip; plants without acrid sap ....
.... 135. Rosaceae, p. 308
4. Stamens 5.
6. Petals free; flowers usually in umbels .... 30. Apiaceae, p. 91
6. Petals united; flowers not in umbels .... *Menyanthes* in
.... 101. Menyanthaceae, p. 270
1. Leaves ternately compound.
7. Stamens and pistils numerous .... 132. Ranunculaceae, p. 298
7. Stamens 5–6; pistil 1.
8. Sepals 6; petals reduced to 6 glands; stamens 6 ....
.... *Caulophyllum* in 38. Berberidaceae, p. 156
8. Sepals 5; petals 5; stamens 5 .... 30. Apiaceae, p. 91

Group XX
Plants herbaceous, dicotyledonous; leaves alternate, pinnately compound.

1. Flowers crowded into heads, each sharing the same receptacle ....
.... 36. Asteraceae, p. 104
1. Flowers not crowded into heads, at least not sharing the same receptacle.
2. Flowers strongly zygomorphic.
3. Sepals 5; petals 5; stamens 10 .... 72. Fabaceae, p. 218
3. Sepals 2; petals 4; stamens 6 .... 74. Fumariaceae, p. 238
2. Flowers actinomorphic (slightly zygomorphic in Caesalpiniaceae).
4. Leaves bipinnately or tripinnately compound.
5. Petals absent .... 132. Ranunculaceae, p. 298
5. Petals present.
6. Plants dioecious; pistils 3–5 .... *Aruncus* in 135. Rosaceae, p. 308
6. Flowers perfect; pistil 1.
7. Stamens 5; flowers usually in umbels .... 30. Apiaceae, p. 91
7. Stamens 8–numerous; flowers usually not in umbels.
8. Stamens numerous; flowers in spherical clusters ....
.... 102. Mimosaceae, p. 270
8. Stamens 8–10; flowers not in spherical clusters ....
.... *Ruta* in 137. Rutaceae, p. 335
4. Leaves once pinnately compound.
9. Plants with colored sap .... 115. Papaveraceae, p. 280
9. Plants without colored sap.
10. Petals absent (sepals may be petaloid).
11. Stamens numerous; pistils numerous ....
.... *Cimicifuga* in 132. Ranunculaceae, p. 298
11. Stamens 4; pistils 1–2 .... *Sanguisorba* in 135. Rosaceae, p. 308
10. Petals present.

12. Petals free.
13. Sepals 3; petals 3 .... 92. Limnanthaceae, p. 264
13. Sepals 4 or more; petals 4 or more.
14. Sepals 4; petals 4.
15. Stamens 6 (rarely 2); ovary superior .... 42. Brassicaceae, p. 162
15. Stamens 8; ovary inferior .... 112. Onagraceae, p. 275
14. Sepals 5 (–7); petals 5 (–7).
16. Stamens usually more than 10 (sometimes as few as 5 in *Agrimonia*); pistils 2–numerous....... 135. Rosaceae, p. 308
16. Stamens 5–10; pistil 1.
17. Petioles with a gland .... 47. Caesalpiniaceae, p. 176
17. Petioles without glands.
18. Flowers in umbels....................... 30. Apiaceae, p. 91
18. Flowers not in umbels .... *Erodium* in 76. Geraniaceae, p. 241
12. Petals united.
19. Stigmas 3.. 126. Polemoniaceae, p. 284
19. Stigmas 1 or 2.
20. Styles 2, free; fruit not fleshy....... 84. Hydrophyllaceae, p. 245
20. Styles seemingly 1; fruit fleshy .... 147. Solanaceae, p. 352

Group XXI
Plants herbaceous, dicotyledonous; leaves alternate, palmately compound.

1. Flowers crowded into heads, each sharing the same receptacle.. 36. Asteraceae, p. 104
1. Flowers not crowded into heads, at least not sharing the same receptacle.
2. Sepals 4; petals 4.
3. Flowers spurred; plants not viscid .... 132. Ranunculaceae, p. 298
3. Flowers not spurred; plants viscid .... 58. Cleomaceae, p. 201
2. Sepals 5; petals 5.
4. Stamens more than 10 (sometimes 10 in *Geum*); pistils numerous... 135. Rosaceae, p. 308
4. Stamens 5 or 10 (rarely 9); pistil 1.
5. Stamens 5; flowers usually in umbels .... *Cynosciadium* in 30. Apiaceae, p. 91
5. Stamens (9–) 10; flowers not in umbels.
6. Flowers zygomorphic; fruit a legume .... 72. Fabaceae, p. 218
6. Flowers actinomorphic; fruit not a legume .... 76. Geraniaceae, p. 241

Descriptive Flora

1. ASPLENIACEAE—SPLEENWORT FAMILY

1. **Asplenium** L.—Spleenwort

1. Leaves simple, unlobed (rarely a single pair of lobelike auricles at base) .... 9. *A. rhizophyllum*
1. Leaves pinnatifid, pinnate, or bipinnate-pinnatifid.
  2. Rachis green throughout.
    3. Petiole green throughout .... 10. *A. ruta-muraria*
    3. Petiole brown at base, sometimes throughout.
      4. Blades entirely pinnatifid or with merely the lowest pair of pinnae distinct; spores normal .... 6. *A. pinnatifidum*
      4. Blades pinnatifid above, pinnate below for at least two pairs of pinnae; spores abortive.
        5. Petiole brown throughout .... 3. *A. X gravesii*
        5. Petiole brown at base, green above .... 13. *A. X trudellii*
  2. Rachis partly or entirely brown or black.
    6. Rachis brown below for about ½ its length, green above.
      7. Blades bipinnate to bipinnate-pinnatifid; spores normal .... 1. *A. bradleyi*
      7. Blades once-pinnate below, pinnatifid above; spores abortive.
        8. Pinnae separated to remote; base of most of the pinnae cuneate; outline of frond linear .... 4. *A. X herb-wagneri*
        8. Pinnae approximate to overlapping; base of most of the pinnae truncate; outline of frond oblong-lanceolate.
          9. Basal pinnae 2–4, overlapping .... 11. *A. X shawneense*
          9. Basal pinnae 8 or more, separate .... 5. *A. X kentuckiense*
    6. Rachis brown or black throughout or at least ¾ its length.
      10. Rachis brown for ¾ its length, green above; blade pinnatifid for ½ its length, the tip narrowly caudate .... 2. *A. X ebenoides*
      10. Rachis brown or black throughout; blade pinnatifid only at apex, the tip not caudate.
        11. Pinnae auriculate.
          12. Pinnae pairs opposite (less commonly alternate); fertile and sterile leaves similar; auricle of pinna not overlapping rachis .... 8. *A. resiliens*
          12. Pinnae pairs alternate; fertile leaves more erect and taller than sterile leaves; auricle of pinna overlapping rachis .... 7. *A. platyneuron*
        11. Pinnae not auriculate .... 12. *A. trichomanes*

1. **Asplenium bradleyi** D. C. Eaton. Bradley's Spleenwort. June–Sept. Crevices of sandstone or cherty cliffs, rare; s. ⅙ of IL; also Randolph Co.

2. **Asplenium X ebenoides** R. R. Scott. Scott's Spleenwort. May–Sept. Rocky, sandstone woodlands, rare; s. ⅙ of IL. Considered to be the hybrid between *A. platyneuron* (L.) Oakes and *A. rhizophyllum* L.

3. **Asplenium X gravesii** Maxon. Graves's Spleenwort. May–Sept. Crevices of sandstone cliffs, very rare; Union and Williamson cos. Considered to be the hybrid between *A. bradleyi* D. C. Eaton and *A. pinnatifidum* Nutt.

4. **Asplenium X herb-wagneri** W. C. Taylor & Mohlenbr. Wagner's Spleenwort. May–Oct. Crevice of cherty cliff, very rare; Union Co. Considered to be the hybrid between *A. pinnatifidum* Nutt. and *A. trichomanes* L.

5. **Asplenium X kentuckiense** McCoy. Kentucky Spleenwort. May–Sept. Crevice of cherty cliff, very rare; Union Co. Considered to be the hybrid between *A. pinnatifidum* Nutt. and *A. platyneuron* (L.) Oakes.

6. **Asplenium pinnatifidum** Nutt. Pinnatifid Spleenwort. May–Sept. Crevices of sandstone cliffs, occasional; s. ⅙ of IL; also Cumberland, Fulton, and Wabash cos.

7. **Asplenium platyneuron** (L.) Oakes. Three taxa occur in Illinois:

a. Pinnae finely toothed and singly pinnate........... 7a. *A. platyneuron* var. *platyneuron*
a. Pinnae either coarsely toothed or doubly pinnate.
  b. Pinnae coarsely toothed.. 7b. *A. platyneuron* var. *incisum*
  b. Pinnae doubly pinnate..7c. *A. platyneuron* f. *hortonae*

7a. **Asplenium platyneuron** (L.) Oakes var. **platyneuron**. Ebony Spleenwort. May–Sept. Dry or moist woods, steep banks; common in the s. ½ of IL, becoming less abundant northward.

7b. **Asplenium platyneuron** (L.) Oakes var. **incisum** (Park) Robins. Ebony Spleenwort. May–Sept. Dry woods, not common; scattered in IL.

7c. **Asplenium platyneuron** (L.) Oakes var. **incisum** (Park) Robins. f. **hortonae** (Davenp.) L. B. Smith. Ebony Spleenwort. May–Sept. Dry woods, rare; Jackson Co.

8. **Asplenium resiliens** Kunze. Black Spleenwort. May–Sept. Crevices of limestone cliffs, rare; confined to the s. ⅙ of IL.

9. **Asplenium rhizophyllum** L. Walking Fern. May–Sept. Rocky woods, often on sandstone or limestone; scattered throughout the state. *Camptosorus rhizophyllus* (L.) Link.

10. **Asplenium ruta-muraria** L. var. **cryptolepis** (Fern.) Wherry. Wall-rue Spleenwort. May–Sept. Crevices of limestone cliffs, very rare and not seen in IL since the middle of the 1800s; s. IL, exact locality unknown. Our plant differs slightly from the European var. *ruta-muraria.*

11. **Asplenium X shawneense** (R. C. Moran) H. E. Ballard. Shawnee Spleenwort. Sori unknown. Known only from the type; crevice of sandstone cliff, Williamson Co.

12. **Asplenium trichomanes** L. Two subspecies may be distinguished:

a. Spores up to 30 mμ in diameter; scales of rhizome mostly more than 5 mm long .... 12a. *A. trichomanes* ssp. *trichomanes*
a. Spores 34–43 mμ in diameter; scales of rhizome mostly less than 5 mm long.......... ....12b. *A. trichomanes* ssp. *quadrivalens*

12a. **Asplenium trichomanes** L. ssp. **trichomanes.** Maidenhair Spleenwort. May–Sept. Crevices of sandstone and limestone cliffs; occasional in the s. ⅓ of IL; also Cumberland Co.

12b. **Asplenium trichomanes** L. ssp. **quadrivalens** D. E. Meyer. Maidenhair Spleenwort. May–Sept. Limestone rocks, very rare; Union Co.

13. **Asplenium X trudellii** Wherry. Trudell's Spleenwort. May–Sept. Crevice of sandstone cliff, very rare; Jackson Co. Considered to be the hybrid between *A. montanum* L. and *A. pinnatifidum* Nutt. *Asplenium montanum* is unknown from Illinois.

2. AZOLLACEAE—MOSQUITO FERN FAMILY

1. **Azolla** Lam.—Mosquito Fern

1. Plants dark green to reddish; stems 0.5–1.0 cm long; megaspores densely covered with tangled hairs .... 1. *A. caroliniana*
1. Plants green or blue-green to reddish; stems 1.0–1.5 cm long; megaspores sparsely covered with tangled hairs.. 2. *A. mexicana*

1. **Azolla caroliniana** Willd. Mosquito Fern. July–Oct. Stagnant or slow-moving water in ponds, lakes, and streams, floating on the water, very rare; Fulton and St. Clair cos. Not seen in IL for more than 60 years.

2. **Azolla mexicana** C. Presl. Mosquito Fern. July–Oct. Stagnant or slow-moving water in ponds, lakes, and streams, floating on the water; scattered throughout IL.

3. BLECHNACEAE—CHAIN-FERN FAMILY

1. **Woodwardia** Sm.—Chain-fern

1. Leaves dimorphic; sterile blades pinnatifid .... 1. *W. areolata*
1. Leaves all of one form; sterile blades once-pinnate............................. 2. *W. virginica*

1. **Woodwardia areolata** (L.) T. Moore. Netted Chain-fern. July–Sept. Springy marshes in lowland woods, sandstone cliffs, rare; s. ⅙ of IL. *Lorinseria areolata* (L.) Presl.

2. **Woodwardia virginica** (L.) Sm. Virginia Chain-fern. July–Sept. Bogs, very rare; Lake Co. Collected in IL in 1944 by Verne O. Graham. *Anchistea virginica* (L.) Presl.

4. DENNSTAEDTIACEAE—HAY-SCENTED FERN FAMILY

1. Sori distinct; leaves pubescent .... 1. *Dennstaedtia*
1. Sori continuous; leaves glabrous or nearly so ....2. *Pteridium*

1. **Dennstaedtia** Bernh.—Hay-scented Fern

1. **Dennstaedtia punctilobula** (Michx.) T. Moore. Hay-scented Fern. July–Oct. Moist, shaded sandstone ravines, rare; in the s. ⅙ of IL; also Ogle Co.

2. **Pteridium** Gled.—Bracken Fern

1. **Pteridium aquilinum** (L.) Kuhn. Two varieties occur in Illinois:

a. Margins of ultimate leaf segments pubescent, the terminal segments 5–8 mm broad ....1a. *P. aquilinum* var. *latiusculum*
a. Margins of ultimate leaf segments glabrous or nearly so, the terminal segments to 4.5 mm broad ....1b. *P. aquilinum* var. *pseudocaudatum*

1a. **Pteridium aquilinum** (L.) Kuhn var. **latiusculum** (Desv.) L. Underw. Bracken Fern. June–Sept. Open woods, fields, roadsides, sandy areas; common throughout IL.

1b. **Pteridium aquilinum** (L.) Kuhn var. **pseudocaudatum** (Clute) A. Heller. Bracken Fern. June–Sept. Open woods, rare; Hardin Co.

5. DRYOPTERIDACEAE—SHIELD FERN FAMILY

1. Leaves more or less ternate; indusium absent...............................6. *Gymnocarpium*
1. Leaves not ternate; indusium present.
  2. Leaves once-pinnate.
    3. Pinnae with an auricle on upper side near base, serrate........... 7. *Polystichum*
    3. Pinnae without a basal auricle, entire..4. *Diplazium*
  2. Leaves pinnate-pinnatifid to tripinnate.
    4. Fronds leathery, evergreen; indusium reniform, attached at the center of the sorus.. 5. *Dryopteris*
    4. Fronds not leathery, not evergreen; indusium various, attached laterally to the sorus or beneath the sorus.
      5. Fronds at least 50 cm long; indusium usually attached laterally to the sorus.
        6. Fronds pinnate-pinnatifid; indusium light brown at maturity, firm, parallel to the veins; sori straight or nearly so........................3. *Deparia*
        6. Fronds bipinnate to tripinnate; indusium dark brown at maturity, membranous, many crossing the veins; sori more or less curved ....1. *Athyrium*
      5. Fronds less than 50 cm long; indusium attached beneath the sorus (sometimes appearing lateral in *Cystopteris*).
        7. Stipes densely chaffy; indusium separating into shreds...... 8. *Woodsia*
        7. Stipes sparsely chaffy or glabrous; indusium hoodlike, not separating into shreds... 2. *Cystopteris*

1. **Athyrium** Roth—Lady Fern

1. **Athyrium filix-femina** (L.) Martens. Two subspecies occur in Illinois:

a. Fourth or fifth pair of pinnae from the base the largest; rachis glandular; indusium eglandular ....1a. *A. filix-femina* ssp. *angustum*

a. Second or third pair of pinnae from the base the largest; rachis eglandular; indusium glandular .... 1b. *A. filix-femina* ssp. *asplenioides*

1a. **Athyrium filix-femina** (L.) Martens ssp. **angustum** (Willd.) R.T. Clausen. Lady Fern. June–Sept. Moist, open woods, borders of swamps; occasional to common throughout the state. *Athyrium angustum* (Willd.) Presl; *A. filix-femina* var. *michauxii* (Spreng.) Farw.

1b. **Athyrium filix-femina** (L.) Martens ssp. **asplenioides** (Michx.) R.T. Clausen. Southern Lady Fern. June–Sept. Moist, often rocky woods; occasional in the s. ⅓ of IL. *Athyrium asplenioides* Michx.; *A. filix-femina* var. *asplenioides* (Michx.) Farw.

2. **Cystopteris** Bernh.—Fragile Fern

1. Some or all parts of the plant with gland-tipped hairs.
  2. Bulblets frequently present; glandular hairs common; blades long-attenuate at the apex .... 1. *C. bulbifera*
  2. Bulblets sparse and misshapen or absent; glandular hairs sparse; blades short-attenuate at the apex.
    3. Spores aborted, appearing dark and irregularly shaped ...... 3. *C. X illinoensis*
    3. Spores normal, appearing translucent and reniform to round.
      4. Spores 34–38 mμ long; plants primarily of sandstone cliffs .... 4. *C. laurentiana*
      4. Spores 25–35 mμ long; plants primarily of calcareous areas .... 6. *C. tennesseensis*
1. Glandular hairs not present.
  5. Stems pubescent with yellowish hairs; spores 28–34 mμ long .... 5. *C. protrusa*
  5. Stems glabrous; spores 39–60 mμ long.
    6. Margins of pinnae with rounded teeth; pinnae often curving toward the apex of the blade .... 7. *C. tenuis*
    6. Margins of pinnae with pointed teeth; pinnae not curving toward the apex of the blade .... 2. *C. fragilis*

1. **Cystopteris bulbifera** (L.) Bernh. Bladder Fern. June–Sept. Mostly limestone areas including limey springs and cliffs; occasional throughout the state.

2. **Cystopteris fragilis** (L.) Bernh. Fragile Fern. June–Sept. Moist woods, rare; Lake and McHenry cos.

3. **Cystopteris X illinoensis** R. C. Moran. Illinois Fragile Fern. Old quarry, in calcareous soil, very rare; known only from the type locality in Winnebago Co. Considered to be the hybrid between *C. bulbifera* (L.) Bernh. and *C. tenuis* (Michx.) Desv.

4. **Cystopteris laurentiana** (Weatherby) Blasdell. Laurentian Bladder Fern. Limestone cliff, very rare; Lee and Ogle cos.

5. **Cystopteris protrusa** (Weatherby) Blasdell. Fragile Fern. June–Sept. Moist woods; common throughout IL. *Cystopteris fragilis* (L.) Bernh. var. *protrusa* Weatherby.

6. **Cystopteris tennesseensis** Shaver. Fragile Fern. June–Sept. Limestone or sometimes sandstone areas; occasional throughout IL.

7. **Cystopteris tenuis** (Michx.) Desv. Fragile Fern. June–Sept. Sandstone cliffs; scattered in IL. *Cystopteris fragilis* (L.) Bernh. var. *mackayi* G. Lawson.

3. **Deparia** Hook. & Grev.—Silvery Spleenwort

1. **Deparia acrostichoides** (Sw.) M. Kato. Silvery Spleenwort. July–Sept. Rich woods; occasional to common throughout IL. *Athyrium thelypterioides* (Michx.) Desv.

4. **Diplazium** Sw.—Narrow-leaved Spleenwort

1. **Diplazium pycnocarpon** (Spreng.) M. Broun. Glade Fern. Aug.–Sept. Moist, shaded woods; occasional throughout IL. *Athyrium pycnocarpon* (Spreng.) Tidestr.

5. **Dryopteris** Adans.—Shield Fern

1. Fronds dimorphic, the outer sterile leaves smaller than the inner fertile leaves........ .... 5. *D. cristata*
1. Fronds all alike.

2. Teeth of leaves with short, spiny tips; blades mostly tripinnate.
  3. Indusium eglandular .... 2. *D. carthusiana*
  3. Indusium glandular.
    4. Leaves broadest near middle; spores abortive, shriveled..... 1. *D. X boottii*
    4. Leaves broadest near base; spores normal and plump or abortive and shriveled.
      5. Pinnae at right angles to rachis, abruptly narrowed to the long tip; rhizome suberect; spores normal, plump...................... 8. *D. intermedia*
      5. Pinnae ascending from the rachis, gradually narrowed to the short tip; rhizome short-creeping; spores abortive, shriveled.......................... ....11. *D. X triploidea*
2. Teeth of leaves without spiny tips; blades mostly bipinnate or bipinnatifid.
  6. Sori marginal ....9. *D. marginalis*
  6. Sori not marginal.
    7. Sori borne midway between midnerve and margin of leaf segment.
      8. Leaves with 10–15 pairs of pinnae .... 4. *D. X clintoniana*
      8. Leaves with 20 or more pairs of pinnae.
        9. Sori closer to midvein than to margin; spores normal, plump......... .... 6. *D. filix-mas*
        9. Sori closer to margin than to midvein; spores abortive, shriveled... ....10. *D. X neo-wherryi*
    7. Sori borne near midvein of leaf segment.
      10. Scales at base of stipe dark brown to black; fronds usually more than ½ as broad as long ....7. *D. goldiana*
      10. Scales at base of stipe pale brown; fronds usually less than ½ as broad as long ....3. *D. celsa*

1. **Dryopteris X boottii** (Tuckerm.) Underw. Boott's Woodfern. June–Aug. Moist, rocky woods, very rare; LaSalle Co. Considered to be the hybrid between *D. cristata* (L.) Gray and *D. intermedia* (Willd.) Gray.

2. **Dryopteris carthusiana** (Villars) H. P. Fuchs. Spinulose Woodfern. June–Aug. Moist, rocky woods, rarely in dry situations, sandy ridges; occasional throughout the state. *Dryopteris spinulosa* (O. F. Muell.) Watt.

3. **Dryopteris celsa** (Wm. Palmer) Knowlton, Palmer, & Pollard. Log Fern. June–Aug. Near swamp, very rare; Johnson Co.

4. **Dryopteris X clintoniana** (D. C. Eaton) Dowell. Clinton's Woodfern. June–Sept. Moist, rich woods, very rare; St. Clair Co. Not seen since the 1800s. Considered to be the hybrid between *D. cristata* (L.) Gray and *D. goldiana* (Hook.) Gray.

5. **Dryopteris cristata** (L.) Gray. Crested Fern. June–Sept. Low, moist woods, boggy ground; occasional in the n. ½ of the state.

6. **Dryopteris filix-mas** (L.) Schott. Male Fern. June–Sept. Steep northwest-facing slope of a ravine, very rare; Cook Co.

7. **Dryopteris goldiana** (Hook.) Gray. Goldie's Fern. June–Sept. Moist, shaded, often rocky woods, springy ground, not common; scattered throughout IL.

8. **Dryopteris intermedia** (Willd.) Gray. Common Woodfern. June–Aug. Moist, rocky ravines and woods; occasional in IL. *Dryopteris spinulosa* (O. F. Muell.) Watt. var. *intermedia* (Willd.) Underw.

9. **Dryopteris marginalis** (L.) Gray. Marginal Shield Fern. June–Oct. Rocky woods; occasional to common throughout IL.

10. **Dryopteris X neo-wherryi** W. H. Wagner. Wherry's New Fern. June–Aug. Rich, rocky woods, very rare; Pope Co. Considered to be the hybrid between *D. goldiana* (Hook.) Gray and *D. marginalis* (L.) Gray.

11. **Dryopteris X triploidea** Wherry. Woodfern. June–Sept. Moist, rocky woods, very rare; DuPage, Lake, Will, and Woodford cos. Considered to be the hybrid between *D. carthusiana* (Villars) H. P. Fuchs and *D. intermedia* (Willd.) Gray.

6. **Gymnocarpium** Newm.—Oak Fern

1. Blades, stipes, and rachises glabrous or nearly so .... 1. *G. dryopteris*
1. Blades, stipes, and rachises glandular-pubescent........................... 2. *G. robertianum*

1. **Gymnocarpium dryopteris** (L.) Newm. Oak Fern. June–Sept. Deep, moist woods, rare; n. ⅙ of IL; also St. Clair Co. *Dryopteris disjuncta* (Ledeb.) C. V. Morton.

2. **Gymnocarpium robertianum** (Hoffm.) Newm. Scented Oak Fern. June–Sept. Steep bluff, very rare; Carroll Co. *Dryopteris robertiana* (Hoffm.) C. Chr.

7. **Polystichum** Roth—Christmas Fern

1. **Polystichum acrostichoides** (Michx.) Schott. Christmas Fern. June–Oct. Woods, particularly in rocky soil; common throughout IL.

8. **Woodsia** R. Br.—Woodsia

1. Blades bipinnate-pinnatifid; leaves generally over 25 cm long; petiole stramineous, unjointed; scales of rhizome few .... 2. *W. obtusa*
1. Blades pinnate-pinnatifid; leaves generally less than 20 cm long; petiole brown, jointed below the middle; scales of rhizome numerous .... 1. *W. ilvensis*

1. **Woodsia ilvensis** (L.) R. Br. Rusty Woodsia. June–Oct. Sandstone cliffs, rare; LaSalle, Lee, and Ogle cos.

2. **Woodsia obtusa** (Spreng.) Torr. Common Woodsia. May–Oct. Dry or moist rocky woods, cliffs; occasional to common in IL.

6. EQUISETACEAE—HORSETAIL FAMILY

1. **Equisetum** L.—Horsetail; Scouring Rush

1. Aerial stems persisting 1 year or less; stomates on upper surface, or absent; cone apex rounded.
2. Aerial stems unbranched.
3. Aerial stems green.
4. Teeth of sheaths more than 11, often black or with narrow white margins .... 3. *E. fluviatile*
4. Teeth of sheaths fewer than 11, with prominent white margins .... 9. *E. palustre*
3. Aerial stems not green.
5. Teeth of sheaths reddish, papery .... 12. *E. sylvaticum*
5. Teeth of sheaths black or brown, firm.
6. Aerial stems persistent, becoming green, with stomates .... 10. *E. pratense*
6. Aerial stems dying back after spores are shed, without stomates .... 1. *E. arvense*
2. Aerial stems branched with regular whorls of branches.
7. First internode of each branch shorter than subtending stem sheath.
8. Teeth of sheaths reddish, papery .... 12. *E. sylvaticum*
8. Teeth of sheaths black or brown, firm .... 9. *E. palustre*
7. First internode of each branch equal to or longer than subtending stem sheath.
9. Teeth of sheaths reddish, papery .... 12. *E. sylvaticum*
9. Teeth of sheaths dark, firm.
10. Branch sheath teeth deltate; branches spreading .... 10. *E. pratense*
10. Branch sheath teeth attenuate; branches ascending.
11. Lowest whorl of branches with first internode longer than sheath; spores green .... 1. *E. arvense*
11. Lowest whorl of branches with first internode nearly equal to sheath; spores white .... 6. *E. X litorale*
1. Aerial stems persisting more than 1 year (except *E. laevigatum*); stomates sunken; cone apex pointed (except some *E. laevigatum*).
12. Apex of cone rounded; aerial stems annual .... 5. *E. laevigatum*
12. Apex of cone pointed; aerial stems perennial (except *E. laevigatum*).
13. Spores white.
14. Sheaths green; teeth of sheaths persistent .... 8. *E. X nelsonii*
14. Sheaths dark; teeth of sheaths persistent or deciduous.
15. Teeth of sheaths 14 or fewer, persistent .... 7. *E. X mackaii*

15. Teeth of sheaths more than 14, deciduous........................2. *E. X ferrissii*
13. Spores green.
16. Sheaths dark-girdled; teeth of sheaths 14 or more.................4. *E. hyemale*
16. Sheaths green; teeth of sheaths 3–32.
17. Teeth of sheaths 3; aerial stems crooked....................... 11. *E. scirpoides*
17. Teeth of sheaths 3–32; aerial stems straight.
18. Teeth of sheaths deciduous; cone often rounded at tip.....................
....5. *E. laevigatum*
18. Teeth of sheaths persistent; cone pointed at tip....13. *E. variegatum*

1. **Equisetum arvense** L. Common Horsetail. Apr.–June. Railroad embankments, roadsides, fields, shores, prairies, calcareous fens; common in IL; in every co.

2. **Equisetum X ferrissii** Clute. Intermediate Scouring Rush. May–Aug. Shores, banks, railroad embankments; common throughout IL. Considered to be a hybrid between *E. hyemale* L. and *E. laevigatum* A. Br.

3. **Equisetum fluviatile** L. Water Horsetail. May–Aug. In water of lakes, streams, and swamps, ditches, bogs; occasional in the n. ½ of IL; also Jackson Co.

4. **Equisetum hyemale** L. ssp. **affine** (Engelm.) Calder & Roy L. Taylor. Scouring Rush. May–Aug. Shores, banks, roadsides, railroad embankments, sand dunes, pastures, degraded prairies; in every co. *Equisetum hyemale* var. *affine* (Engelm.) A. A. Eaton.

5. **Equisetum laevigatum** A. Br. Smooth Scouring Rush. May–July. Shores, banks, roadsides, prairies; common throughout IL.

6. **Equisetum X litorale** Kuhl. Horsetail. June–Sept. Shores, banks, occasional in the n. ½ of IL. Considered to be a hybrid between *E. arvense* L. and *E. fluviatile* L.

7. **Equisetum X mackaii** (Newm.) Brichan. Horsetail. June–Sept. Lake shores, interdunal swales, pannes, very rare; Cook, Lake, and Mason cos. Considered to be a hybrid between *E. hyemale* L. and *E. variegatum* Schleich. *Equisetum X trachyodon* A. Br.

8. **Equisetum X nelsonii** (A. A. Eaton) J. H. Schaffner. Horsetail. June–Sept. Marly fens, interdunal swales, pannes; occasional in the n. ⅙ of IL. Considered to be a hybrid between *E. laevigatum* A. Br. and *E. variegatum* Schleich.

9. **Equisetum palustre** L. Marsh Horsetail. June–Sept. Banks of rivers and streams, very rare; Peoria, Tazewell, and Woodford cos.

10. **Equisetum pratense** Ehrh. Meadow Horsetail. May–July. North-facing wooded slopes, very rare; Carroll, Jo Daviess, and Ogle cos.

11. **Equisetum scirpoides** Michx. Dwarf Scouring Rush. Aug.–Sept. Wooded ravines; n. ⅙ of IL.

12. **Equisetum sylvaticum** L. Woodland Horsetail. July–Sept. Moist woods, very rare; Lee and Ogle cos.

13. **Equisetum variegatum** Schleich. Variegated Scouring Rush. July–Oct. Moist sand, fens, shores, lakes; n. ⅔ of IL; also Greene Co.

7. HYMENOPHYLLACEAE—FILMY FERN FAMILY

1. Plants known only from the filamentous gametophyte stage..........1. *Crepidomanes*
1. Sporophyte fronds flat...2. *Vandenboschia*

1. **Crepidomanes** C. Presl—Filmy Fern

1. **Crepidomanes intricatum** (Farrar) Ebihara & Weakley. Appalachian Filmy Fern. On shaded sandstone rocks; Pope Co. *Trichomanes intricatum* Farrar.

2. **Vandenboschia** Copel.—Filmy Fern

1. **Vandenboschia boschiana** (Sturm) Ebihara & K. Iwats. Filmy Fern. June–Sept. Beneath moist, overhanging sandstone cliffs, rare; s. ⅙ of IL. *Trichomanes boschianum* Sturm.

8. ISOETACEAE—QUILLWORT FAMILY

1. **Isoetes** L.—Quillwort

1. Sporangia punctate or striate; megaspores tuberculate; microspores spinulose or papillose.

2. Leaves mostly 8–15 cm long; ligule elongated, cordate; sporangia striate; megaspores 480–650 mμ in diameter; microspores 27–37 mμ in diameter, papillose .... 1. *I. butleri*
2. Leaves mostly 15–40 cm long; ligule subulate, triangular; sporangia punctate; megaspores 280–440 mμ in diameter; microspores 20–30 mμ in diameter, spinulose .... 3. *I. melanopoda*

1. Sporangia neither punctate nor striate; megaspores reticulate, with narrow ridges; microspores smooth or minutely roughened .... 2. *I. engelmannii*

1. **Isoetes butleri** Engelm. Butler's Quillwort. May–June. Dolomite prairies, very rare; DuPage and Will cos.

2. **Isoetes engelmannii** A. Br. Engelmann's Quillwort. May–Oct. Shallow water, very rare; St. Clair Co. Not seen since the latter part of the 1800s.

3. **Isoetes melanopoda** Gay & Dur. Black Quillwort. May–Sept. Shallow water of ponds and ditches, wet meadows, flat woods, shallow depressions on sandstone cliffs; occasional in IL. Plants with pale rather than black sporangia may be called f. *pallida* (Engelm.) Fern.

9. LYCOPODIACEAE—CLUBMOSS FAMILY

1. Sporangia axillary along the stem; gemmae frequently present in the axils of the upper leaves .... 3. *Huperzia*
1. Sporangia aggregated in terminal cones; gemmae absent.
   2. Leaves (sporophylls) present on peduncles .... 4. *Lycopodiella*
   2. Leaves (sporophylls) not present on peduncles, or peduncles absent.
      3. Leaves 4- to 5-ranked .... 2. *Diphasiastrum*
      3. Leaves 6-ranked or more.
         4. Leaves with hairlike tips; bracts coarsely jagged at tip; strobilus pedunculate .... 5. *Lycopodium*
         4. Leaves without hairlike tips; bracts entire or merely erose at tip; strobilus sessile .... 1. *Dendrolycopodium*

1. **Dendrolycopodium** A. Haines—Ground Pine

1. Leaves of main axis pale green, spreading, prickly .... 1. *D. dendroideum*
1. Leaves of main axis dark green, ascending, soft .... 2. *D. hickeyi*

1. **Dendrolycopodium dendroideum** (Michx.) A. Haines. Ground Pine. July–Sept. Wet woods, wet sandy detritus of St. Peter sandstone, rare; n. ½ of IL. *Lycopodium dendroideum* Michx.

2. **Dendrolycopodium hickeyi** (W. H. Wagner, Beitel, & R. C. Moran) A. Haines. Hickey's Ground Pine. One IL location, very rare. (G. Tucker, 2010.) *Lycopodium hickeyi* W. H. Wagner, Beitel, & R. C. Moran.

2. **Diphasiastrum** Holub—Ground Pine

1. Peduncles 3–10 cm long; leaves with appressed tips, 1–2 mm apart; sporophylls without scarious margins, ovate .... 1. *D. digitatum*
1. Peduncles 12–18 cm long; leaves with spreading tips, about 4 mm apart; sporophylls with scarious margins, orbicular .... 2. *D. X habereri*

1. **Diphasiastrum digitatum** (Dill.) Holub. Ground Pine. July–Sept. Sandstone ledges in Gallatin, Ogle, and Pope cos; adventive under conifers in much of IL. *Lycopodium digitatum* Dill.; *Lycopodium complanatum* L. var. *flabelliforme* Fern.

2. **Diphasiastrum X habereri** (House) Holub. Hybrid Ground Pine. July–Sept. Cook Co. Considered to be a hybrid between *D. digitatum* (Dill.) Holub and *D. trachycaulum*. *Diphasiastrum trachycaulum* is not known from IL.

3. **Huperzia** Bernh.—Clubmoss

1. Spores normal.
   2. Leaves oblanceolate, broadest near middle, serrulate or rarely entire; stomates present only on the lower surface (with 10X magnification); spores usually less than 35 mμ in diameter .... 2. *H. lucidula*

2. Leaves linear or lanceolate, broadest near base, entire; stomates present on both surfaces (with 10X magnification); spores usually more than 35 mμ in diameter.......... 3. *H. porophila*
1. Spores shriveled, abortive.......... 1. *H. X bartleyi*

1. **Huperzia X bartleyi** (Cusick) Kartesz & Gandhi. Bartley's Clubmoss. June–Sept. Shaded sandstone cliff; LaSalle Co. Considered to be a hybrid between *H. lucidula* (Michx.) Trev. and *H. porophila* (Lloyd & Underw.) Holub. It grows with *H. porophila* at its only Illinois location. Specimen of R. H. Mohlenbrock deposited at the Missouri Botanical Garden.

2. **Huperzia lucidula** (Michx.) Trev. Two varieties occur in Illinois:

a. Leaves serrulate.......... 2a. *H. lucidula* var. *lucidula*
a. Leaves entire.......... 2b. *H. lucidula* var. *tryonii*

2a. **Huperzia lucidula** (Michx.) Trev. var. **lucidula.** Shining Clubmoss. June–Sept. Shaded sandstone cliffs and ravines, boggy woods; scattered in IL. *Lycopodium lucidulum* (Michx.) Trev.

2b. **Huperzia lucidula** (Michx.) Trev. var. **tryonii** (Mohlenbr.) Mohlenbr. Tryon's Shining Clubmoss. June–Sept. Shaded sandstone cliff, very rare; known only from the type locality in Jackson Co.

3. **Huperzia porophila** (Lloyd & Underw.) Holub. Cliff Clubmoss. June–Sept. Moist, shaded sandstone cliffs, rare; in the n. ½ and s. ⅙ of IL. *Lycopodium porophilum* Lloyd & Underw.

4. **Lycopodiella** Holub—Bog Clubmoss

1. Fertile stems 8–35 cm long; leaves on peduncles spreading or appressed.......... 1. *L. appressa*
1. Fertile stems 3.5–14.0 cm long; leaves on peduncles all spreading.......... 2. *L. inundata*

1. **Lycopodiella appressa** (Chapm.) Cranfill. Appressed Bog Clubmoss. July–Sept. Wet woods, very rare; Pulaski Co., collected by J. Henderson near Perks. Not seen in more than 60 years. *Lycopodium appressum* (Chapm.) Lloyd & Underw.

2. **Lycopodiella inundata** (L.) Holub. Bog Clubmoss. July–Sept. Bogs, very rare; Cook, Lee, and Ogle cos. *Lycopodium inundatum* L.

5. **Lycopodium** L.—Clubmoss

1. Strobili 2–6 per peduncle; leaves of the upright branches spreading.......... 1. *L. clavatum*
1. Strobilus 1 per peduncle; leaves of the upright branches ascending.......... 2. *L. lagopus*

1. **Lycopodium clavatum** L. Common Clubmoss. July–Sept. North-facing slope of sandy soil, below outcrop of St. Peter sandstone, and seepage stream banks draining from loess bluff; also sand areas; rare; Carroll, DuPage, Kankakee, Ogle, and Rock Island cos.

2. **Lycopodium lagopus** (Laes.) Zins. One-cone Clubmoss. July–Sept. Woodland opening, very rare; Will Co. *Lycopodium clavatum* L. var. *megastachyon* Fern. & Bissell; *L. clavatum* L.

10. MARSILEACEAE—WATER-CLOVER FAMILY

1. **Marsilea** L.—Water-clover

1. **Marsilea quadrifolia** L. Water-clover. June–Dec. Introduced into lakes and ponds; scattered and uncommon in the s. ⅔ of IL.

11. ONOCLEACEAE—SENSITIVE FERN FAMILY

1. Sterile leaves pinnate to pinnate-pinnatifid, with free venation, usually more than 1 m long.......... 1. *Matteuccia*
1. Sterile leaves pinnatifid, with net venation, less than 1 m long.......... 2. *Onoclea*

1. **Matteuccia** Todaro—Ostrich Fern

1. **Matteuccia struthiopteris** (L.) Todaro var. **pensylvanica** (Willd.) C. V. Morton. Ostrich Fern. July–Sept. Rich, moist woods, floodplain woods; n. ½ of IL.

2. **Onoclea** L.—Sensitive Fern

1. **Onoclea sensibilis** L. Sensitive Fern. June–Oct. Moist woods, swampy woods, low, open ground, marshes, common; probably in every co. Plants with fronds partly fertile and partly sterile may be called f. *obtusilobata* (Schkuhr) Gilbert.

12. OPHIOGLOSSACEAE—ADDER'S-TONGUE FAMILY

1. Leaves pinnately compound or simple and pinnatifid, the venation free; sporangia stipitate.
  2. Leaf blades deltate, usually 5–25 cm long, usually sterile; plants at least 12 cm tall; leaf sheaths closed (except *Botrypus virginianus*).
    3. Sterile blades sessile below long stalk of fruiting panicle, deciduous .... 3. *Botrypus*
    3. Sterile blades on long petioles arising near base of plant, deciduous or persistent .... 4. *Sceptridium*
  2. Leaf blades oblong to linear, 2–4 cm long, all fertile; plants up to 10 cm tall; leaf sheaths open .... 1. *Botrychium*
1. Leaves simple, entire, the venation netlike; sporangia sessile .... 3. *Ophioglossum*

1. **Botrychium** Sw.—Moonwort

1. Blades linear .... 1. *B. campestre*
1. Blades lanceolate to ovate.
  2. Leaves distinctly petiolate .... 3. *B. simplex*
  2. Leaves sessile or short-petiolate .... 2. *B. matricariifolium*

1. **Botrychium campestre** W. H. Wagner & Farrar. Prairie Moonwort. May–June. Grassy area, very rare; Kane Co.

2. **Botrychium matricariifolium** (Doll.) A. Br. Daisy-leaved Grape Fern. June–July. Young, second growth, upland sugar maple woods, rare; Lee and Winnebago cos.

3. **Botrychium simplex** E. Hitchc. Least Moonwort. May–June. Woods and adjacent thickets, rare; Cook, Lee, and Winnebago cos. These plants are usually found under slippery elm.

2. **Botrypus** Michx.—Rattlesnake Fern

1. **Botrypus virginianus** (L.) Michx. Rattlesnake Fern. June–July. Dry or moist woods; in every co. *Botrychium virginianum* (L.) Sw.

3. **Ophioglossum** L.—Adder's-tongue

1. Leaf rounded or subacute at apex, the veins forming principal areoles without secondary areoles.
  2. Leaf deep green, shiny; spores 37–42 mμ in diameter .... 3. *O. vulgatum*
  2. Leaf pale green, dull; spores 50–54 mμ in diameter .... 2. *O. pusillum*
1. Leaf acute, apiculate, the veins forming principal areoles surrounding secondary areoles .... 1. O. engelmannii

4. **Sceptridium** Lyon—Grape Fern

1. Ultimate segments of the leaf obtuse or rounded; blade fleshy or coriaceous, not turning bronze in winter .... 3. S. multifidum
1. Ultimate segments of the leaf acute or subacute; blade subcoriaceous or membranous, turning bronze or remaining green in winter.
  2. Pinnules sharply serrate; blade membranous, remaining green in winter .... 1. *S. biternatum*
  2. Pinnules entire, crenulate, or lobed; blade subcoriaceaous, green or turning bronze in winter.
    3. Fronds turning bronze in winter, the segments of the sterile blade lanceolate to narrowly oblong .... 2. *S. dissectum*
    3. Fronds remaining green during the winter, the segments of the sterile blade broadly ovate and rounded .... 4. *S. oneidense*

1. **Sceptridium biternatum** (Sav.) Lyon. Southern Grape Fern. Sept.–Nov. Open oak-hickory woods; occasional in the s. ⅙ of IL; also Mason Co. *Botrychium biternatum* (Sav.) Underw.

2. **Sceptridium dissectum** (Spreng.) Lyon. Two intergrading varieties occur in Illinois:

a. Blade of leaf very finely dissected .... 2a. *S. dissectum* var. *dissectum*
a. Blade of leaf shallowly divided ....2b. *S. dissectum* var. *obliquum*

2a. **Sceptridium dissectum** (Spreng.) Lyon var. **dissectum.** Cut-leaved Grape Fern. Sept.–Nov. Open oak-hickory woods; scattered in IL. *Botrychium dissectum* Spreng.

2b. **Sceptridium dissectum** (Spreng.) Lyon var. **obliquum** (Muhl.) Mohlenbr., comb. nov. (basionym: *Botrychium obliquum* Muhl.). Grape Fern. Sept.–Nov. Open oak-hickory woods, pastures; throughout IL. *Botrychium obliquum* Muhl.; *Botrychium dissectum* Spreng. f. *obliquum* (Muhl.) Fern.; *Botrychium dissectum* Spreng. var. *obliquum* Muhl.) Clute.

3. **Sceptridium multifidum** (Gmel.) Nishida & Tagawa. Northern Grape Fern. Aug.–Sept. Rich woodlands; confined to the n. ⅙ of IL. *Botrychium multifidum* (Gmel.) Rupr.; *Botrychium multifidum* (Gmel.) Rupr. var. *intermedium* (D. C. Eaton) Fern.

4. **Sceptridium oneidense** (Gilb.) Holub. Grape Fern. Sept.–Nov. Woods, in sandy soil, very rare; Ogle Co. *Botrychium obliquum* Muhl. var. *oneidense* Gilb.; *Botrychium oneidense* (Gilb.) House; *Botrychium dissectum* Spreng. f. *oneidense* (Gilb.) Clute.

13. OSMUNDACEAE—ROYAL FERN FAMILY

1. **Osmunda** L.—Royal Fern

1. Leaves bipinnate, the pinnules serrulate; sporangia borne on upper ½ of leaf.......... ....3. *O. spectabilis*
1. Leaves always pinnate-pinnatifid, the pinnules entire; sporangia borne on a separate fertile leaf or centrally on the leaf.
  2. Sporangia borne on a separate fertile leaf; stipe densely covered with cinnamon-colored wool ....1. *O. cinnamomea*
  2. Sporangia produced centrally between sterile pinnae; stipe not densely covered with cinnamon-colored wool ....2. *O. claytoniana*

1. **Osmunda cinnamomea** L. Cinnamon Fern. Mar.–June. Swamps, swampy woods, bogs, ledges of sandstone cliffs; occasional in IL. *Osmundastrum cinnamomeum* (L.) C. Presl.

2. **Osmunda claytoniana** L. Interrupted Fern. Apr.–June. Moist, low woods, base of slopes in oak woods, prairie remnants, ledges of sandstone cliffs; occasional in IL. *Osmundastrum claytonianum* (L.) Tagawa.

3. **Osmunda spectabilis** Willd. Royal Fern. Apr.–June. Swamps, swampy woods, bogs, sandstone ledges; occasional throughout IL. *Osmunda regalis* L. var. *spectabilis* (Willd.) Gray.

14. POLYPODIACEAE—POLYPODY FAMILY

1. Lower surface of blades and stipes scaly, appearing pustular, the segments of the blade entire ....1. *Pleopeltis*
1. Blades and stipes scaleless, appearing smooth, the segments of the blade minutely toothed ....2. *Polypodium*

1. **Pleopeltis** Willd.—Resurrection Fern

1. **Pleopeltis polypodioides** (L.) E. G. Andrews & Windham. Gray Polypody; Resurrection Fern. June–Oct. Sandstone rocks, tree trunks and branches, not common; s. ¼ of IL; also Calhoun and Jersey cos. *Polypodium polypodioides* (L.) Watt var. *michauxianum* Weatherby.

2. **Polypodium** L.—Polypody

1. **Polypodium virginianum** L. Common Polypody. June–Oct. Sandstone rocks, usually under some shade, slopes of woods, dunes; occasional to rare in IL.

15. PTERIDACEAE—MAIDENHAIR FAMILY

1. Lower surface of leaves covered with a white, powdery substance.......................... .... 2. *Argyrochosma*
1. Lower surface of leaves some other color and not covered with a white powdery substance.
  2. Stipe branched at tip.. 1. *Adiantum*
  2. Stipe straight and unbranched at tip.
    3. Stipe and rachis purple-brown to nearly black.
      4. Leaves densely woolly, the ultimate pinnules less than 5 mm broad......... .... 3. *Cheilanthes*
      4. Leaves variously pubescent or glabrous but never woolly, the ultimate pinnules more than 5 mm broad... 5. *Pellaea*
    3. Stipe and rachis green.
      5. Stem very weak; plant rarely more than 15 cm long...... 4. *Cryptogramma*
      5. Stem firm; plant more than 15 cm long....................................... 6. *Pteretis*

1. **Adiantum** L.—Maidenhair Fern

1. **Adiantum pedatum** L. Maidenhair Fern. June–Sept. Moist, shaded woods; in every co.

2. **Argyrochosma** (J. Sm.) Windham

1. **Argyrochosma dealbata** (Pursh) Windham. Powdery Notholaena. July–Sept. Limestone cliff, very rare; Pike Co. *Notholaena dealbata* (Pursh) Windham.

3. **Cheilanthes** Sw.—Lip Fern

1. Leaves densely brown-woolly beneath, to 15 cm long, tripinnate, with 7–12 (–15) pairs of pinnae; on limestone... 1. *C. feei*
1. Leaves white-villous beneath, some usually well over 15 cm long, bipinnate-pinnatifid, with 12–20 pairs of pinnae; chiefly on sandstone .... 2. *C. lanosa*

1. **Cheilanthes feei** T. Moore. Baby Lip Fern; Resurrection Fern. June–Sept. Dry, exposed limestone cliffs; in cos. bordering the Mississippi River and Ohio River in the s. ½ of IL; also Adams, Calhoun, Carroll, Greene, Jersey, Jo Daviess, Ogle, Pike, and Scott cos.

2. **Cheilanthes lanosa** (Michx.) D. C. Eaton. Hairy Lip Fern. June–Sept. Dry, exposed cliffs, chiefly sandstone; local in the s. ⅓ of IL; also Calhoun, Cumberland, and Ogle cos.

4. **Cryptogramma** R. Br.—Cliffbrake Fern

1. **Cryptogramma stelleri** (S. G. Gmel.) Prantl. Slender Cliffbrake. June–Sept. Deep, shaded limestone ravines, very rare; confined to the n. ⅓ of IL.

5. **Pellaea** Link—Cliffbrake Fern

1. Petiole and rachis pubescent nearly throughout; fertile pinnules much narrower than the sterile ones.. 1. *P. atropurpurea*
1. Petiole and rachis glabrous or sparsely pubescent; fertile and sterile pinnae similar in shape.. 2. *P. glabella*

1. **Pellaea atropurpurea** (L.) Link. Purple Cliffbrake. June–Sept. On limestone cliffs, frequently under very dry conditions; in cos. bordering the Mississippi River in the s. ½ of IL; in cos. bordering the Ohio River; inland in Adams, Johnson, Ogle, Scott, and Williamson cos.

2. **Pellaea glabella** Kuhn. Smooth Cliffbrake. June–Sept. Limestone and dolomite outcrops; local in the n. cos., extending southward along the Mississippi and Ohio rivers; also LaSalle and Lee cos.

6. **Pteretis** L.—Brake Fern

1. **Pteretis multifida** Poir. Spider Brake. June–Sept. Native to e. Asia; on rocks of shaded masonry; Wabash Co. Not seen since the 1800s.

16. SELAGINELLACEAE—SPIKEMOSS FAMILY

1. **Selaginella** Beauv.—Spikemoss

1. Stems weak, herbaceous; leaves 4-ranked, flaccid, obtuse to acute.
  2. Smaller (dorsal) leaves long-attenuate, with the apex costate; megaspores shiny, laxly reticulate .... 2. *S. eclipes*
  2. Smaller (dorsal) leaves acute or, if long-attenuate, not costate; megaspores dull, closely reticulate .... 1. *S. apoda*
1. Stems wiry, evergreen; leaves spirally arranged, stiff, subulate-tipped .... 3. *S. rupestris*

1. **Selaginella apoda** (L.) Spring. Small Spikemoss. June–Oct. Moist shaded areas, grassy margins of streams; local in the s. ⅓ of IL.

2. **Selaginella eclipes** W. R. Buck. Small Spikemoss. June–Oct. Moist to wet, often calcareous habitats; n. ⅔ of IL; also Union Co.

3. **Selaginella rupestris** (L.) Spring. Rock Spikemoss. June–Sept. Dry, rocky areas, mostly on sandstone; nw. IL; also LaSalle, Pope, and Union cos.

17. THELYPTERIDACEAE—THELYPTERIS FAMILY

1. Rachis winged throughout or only the basal pair of pinnae winged; indusium none .... 2. *Phegopteris*
1. Rachis not winged, the pinnae separate from it nearly to apex of blade; indusium cordate or, if not cordate, only up to 0.3 mm long.
  2. Blades once-pinnate to pinnate-pinnatifid .... 3. *Thelypteris*
  2. Blades twice-pinnatifid .... 1. *Macrothelypteris*

1. **Macrothelypteris** (Ito) Ching

1. **Macrothelypteris torresiana** (Gaudichaud-Beaupre) Ching. Torres' Fern. June–Sept. Native to Asia and Africa; rich woods, very rare; Johnson Co.

2. **Phegopteris** (Presl) Fee—Beech Fern

1. Rachis winged only above the 2 basal pairs of pinnae, the wings not extending to the lowest pair of pinnae; blades pinnate-pinnatifid; rachis rather densely brown-scaly .... 1. *P. connectilis*
1. Rachis winged throughout; blades bipinnatifid; rachis sparsely white-scaly .... 2. *P. hexagonoptera*

1. **Phegopteris connectilis** (Michx.) Watt. Long Beech Fern. June–Aug. Moist shaded woods in sandstone areas, rare; in the n. ½ of IL. *Thelypteris phegopteris* (L.) Slosson; *Dryopteris phegopteris* (L.) C. Chr.

2. **Phegopteris hexagonoptera** (Michx.) Fee. Broad Beech Fern. June–Sept. Rich woods; scattered throughout IL. *Dryopteris hexagonoptera* C. Chr.

3. **Thelypteris** Schmidel—Marsh Fern

1. Lowest pinnae strongly reduced .... 1. *T. noveboracensis*
1. Lowest pinnae only slightly or not at all reduced .... 2. *T. palustris*

1. **Thelypteris noveboracensis** (L.) Nieuwl. New York Fern. June–Sept. Moist or rarely dry woods, rare; Kane, Kankakee, Monroe, Pope, and Wabash cos. *Dryopteris noveboracensis* (L.) Gray; *Parathelypteris noveboracensis* (L.) Ching.

2. **Thelypteris palustris** Schott var. **pubescens** (Laws.) Fern. Marsh Fern. June–Oct. Marshy ground, bogs, calcareous fens, swamps; common in n. IL, rarer southward. *Dryopteris thelypteris* (L.) Gray var. *pubescens* (Laws.) Nakai.

18. CUPRESSACEAE—CYPRESS FAMILY

1. Branches not flattened; leaves all subulate or some scalelike and subulate on the same plant; fruit a fleshy, berrylike cone .... 1. *Juniperus*
1. Branches flattened; leaves all scalelike; fruit a woody cone .... 2. *Thuja*

1. **Juniperus** L.—Juniper

1. All leaves subulate, borne in whorls of 3 .... 1. *J. communis*
1. Some leaves subulate, others scalelike, borne opposite.

2. Prostrate, trailing shrub; fruits more than 6 mm in diameter, 3- to 5-seeded .... 2. *J. horizontalis*
2. Upright tree; fruits 4–6 mm in diameter, 1- to 2-seeded .... 3. *J. virginiana*

1. **Juniperus** communis L. Two varieties occur in Illinois:

a. Central stem erect .... 1a. *J. communis* var. *communis*
a. All stems decumbent .... 1b. *J. communis* var. *depressa*

1a. **Juniperus communis** L. var. **communis.** Ground Juniper. Sand dunes, rare; Lake Co.

1b. **Juniperus communis** L. var. **depressa** Pursh. Ground Juniper. Sand dunes, rare; Cook and Lake cos; adventive in DuPage Co.

2. **Juniperus horizontalis** Moench. Trailing Juniper. Sand dunes, sandy prairies, rare; Cook and Lake cos.

3. **Juniperus virginiana** L. Red Cedar. Woods, cliffs, fields, dune slopes; common throughout IL. Plants with a pyramidal shape and strongly ascending branches may be known as var. *crebra* Fern. & Grisc.

2. **Thuja** L.—Arbor Vitae; White Cedar

1. **Thuja occidentalis** L. Arbor Vitae; White Cedar. Cliffs, bluffs, bogs, calcareous springy slopes, rare; Cook, Kane, Lake, LaSalle, and Will cos.

19. GINKGOACEAE—GINKGO FAMILY

1. **Ginkgo** L.—Ginkgo

1. **Ginkgo biloba** L. Native to China; apparently escaped from cultivation; DuPage Co. (Wilhelm, 2010.)

20. PINACEAE—PINE FAMILY

1. Needles spirally arranged, borne singly .... 2. *Picea*
1. Needles in clusters of 2–many.
2. Needles in clusters of 2–5; plants evergreen .... 3. *Pinus*
2. Needles in clusters of 20 or more; plants deciduous .... 1. *Larix*

1. **Larix** Mill.—Larch; Tamarack

1. Twigs glabrous; scales of cones glabrous, 10–20 in number; most of the leaves less than 2.5 cm long .... 2. *L. laricina*
1. Twigs pubescent; scales of cones pubescent, at least 30 in number; most of the leaves at least 2.5 cm long .... 1. *L. decidua*

1. **Larix decidua** (DuRoi) K. Koch. American Larch; Tamarack. Bogs, rare; Cook, DuPage, Lake, and McHenry cos.

2. **Larix laricina** Mill. European Larch. Native to Europe; rarely escaped from cultivation in a few ne. cos.

2. **Picea** Dietr.—Spruce

1. Branchlets glabrous; leaves dark green; cones more than 3 cm long.
2. Branches spreading to ascending; cones 3–6 cm long .... 1. *P. abies*
2. Branches pendulous; cones 5 cm long or longer .... 2. *P. glauca*
1. Branchlets pubescent; leaves pale blue-green; cones 2–3 cm long .... 3. *P. mariana*

1. **Picea abies** (L.) H. Karst. Norway Spruce. Native to Europe; rarely escaped from cultivation but apparently established in a woods in Jo Daviess Co; also DuPage Co.

2. **Picea glauca** (Moench) A. Voss. White Spruce. Native n. of IL; rarely escaped from cultivation; DuPage Co.

3. **Picea mariana** (Mill.) BSP. Black Spruce. Native n. and e. of IL; apparently adventive in a bog in Lake Co.

3. **Pinus** L.—Pine

1. Leaves 5 in a cluster.
2. Leaves up to 12.5 cm long, blue-green; cones up to 15 cm long......... 7. *P. strobus*
2. Leaves, or most of them, more than 12.5 cm long, gray-green; cones 15 cm long or longer .... 11. *P. wallichiana*

1. Leaves 2–3 in a cluster.
3. Leaves, or some of them, 3 in a cluster.
4. Leaves slender, at most 1.5 mm broad .... 2. *P. echinata*
4. Leaves stout, 1.5–3.0 mm broad.
5. Most of the leaves more than 7 cm long; scales of cone with a spine up to 3 mm long.
6. Leaves up to 15 cm long; cone almost as broad as long .... 6. *P. rigida*
6. Most of the leaves over 15 cm long; cone longer than broad .... 9. *P. taeda*
5. Most of the leaves less than 7 cm long; scales of cone with a spine 5–6 mm long .... 4. *P. pungens*
3. Leaves, or some of them, 2 in a cluster.
7. Leaves up to 7 cm long.
8. Cones without a spine.
9. Cones persistent; leaves up to 2.5 cm long .... 1. *P. banksiana*
9. Cones deciduous; leaves 2–6 cm long .... 8. *P. sylvestris*
8. Cones with a spine.
10. Scales of cone with a spine 2–4 mm long .... 10. *P. virginiana*
10. Scales of cone with a spine 5–6 mm long .... 4. *P. pungens*
7. Most of the leaves 7 cm long or longer.
11. None of the leaves over 17 cm long, dark green.
12. Scales of cone without a spiny tip....................................... 5. *P. resinosa*
12. Scales of cone spine-tipped.
13. Leaves thick; cones up to 8 cm long .... 3. *P. nigra*
13. Leaves slender; cones up to 5 cm long .... 2. *P. echinata*
11. Some of the leaves 17–25 cm long, light green........................... 9. *P. taeda*

1. **Pinus banksiana** Lamb. Jack Pine. Sandy soil, sometimes escaped from plantings; n. ¼ of IL.

2. **Pinus echinata** Mill. Shortleaf Pine. Rocky soil, rare in the native condition, where it is known from Randolph and Union cos.; frequently planted in plantations.

3. **Pinus nigra** Arn. Austrian Pine. Native to Europe; commonly planted but rarely escaped in IL.

4. **Pinus pungens** Lamb. Table Mountain Pine. Native e. and s. of IL; rarely encountered as an introduction in IL.

5. **Pinus resinosa** Ait. Red Pine. Dry woods, rare; LaSalle Co; adventive elsewhere.

6. **Pinus rigida** Mill. Pitch Pine. Native to the e. of IL; occasionally encountered as an introduction in IL.

7. **Pinus strobus** L. White Pine. Woods; local in the n. ¼ of IL.

8. **Pinus sylvestris** L. Scots Pine. Native to Europe; occasionally encountered as an introduction in IL.

9. **Pinus taeda** L. Loblolly Pine. Native primarily s. of IL; frequently planted in IL.

10. **Pinus virginiana** Mill. Virginia Pine. Dry soil; escaped from cultivation or planted; scattered in s. IL.

11. **Pinus wallichiana** A. B. Jacks. Himalayan White Pine. Native to Asia; infrequently planted in IL and rarely persisting except in Union Co.

21. TAXACEAE—YEW FAMILY

1. **Taxus** L.—Yew

1. **Taxus canadensis** Marsh. Canada Yew. Wooded hillsides, rare; n. ¼ of IL.

22. TAXODIACEAE—BALD CYPRESS FAMILY

1. **Taxodium** Rich.—Bald Cypress

1. **Taxodium distichum** (L.) Rich. Bald Cypress. Swamps; s. IL, extending n. along the Wabash River to Lawrence and Richland cos.; planted elsewhere.

23. ACANTHACEAE—ACANTHUS FAMILY

1. Corolla zygomorphic, 2-lipped, less than 2 cm long; stamens 2; sepals less than 10 mm long.
2. Petioles at least 1 cm long; flowers bracteate .... 1. *Dicliptera*

2. Petioles less than 1 cm long; flowers ebracteate .... 2. *Justicia*
1. Corolla nearly actinomorphic, not 2-lipped, at least 2.5 cm long; stamens 4; sepals at least 10 mm long .... 3. Ruellia

1. **Dicliptera** Juss.—Dicliptera

1. **Dicliptera brachiata** (Pursh) Spreng. Dicliptera. Aug.–Oct. Bottomland woods, along streams, rare; Massac and Pope cos.

2. **Justicia** L.—Water Willow

1. Leaves linear to lanceolate; flowers densely clustered in a headlike spike .... 1. *J. americana*
1. Leaves elliptic to oblong; flowers loosely scattered along the peduncle .... 2. *J. ovata*

1. **Justicia americana** (L.) Vahl. Water Willow. May–Oct. In and along streams, shallow water at the edge of lakes, occasional; throughout the state.

2. **Justicia ovata** (Walt.) Landau. Two varieties occur in Illinois:

a. Spike with flowers borne opposite .... 2a. *J. ovata* var. *ovata*
a. Spike with flowers secund .... 2b. *J. ovata* var. *lanceolata*

2a. **Justicia ovata** (Walt.) Landau var. **ovata.** Broad-leaved Water Willow. May–June. Swamps, bottomland woods, sometimes in shallow water, rare; Alexander, Massac, and Pulaski cos.

2b. **Justicia ovata** (Walt.) Landau var. **lanceolata** (Chapm.) R. W. Long. Narrow-leaved Water Willow. May–June. Bottomland woods, very rare; Alexander Co.

3. **Ruellia** L.—Wild Petunia

1. Calyx lobes not more than 1.5 mm broad; stems and leaves pubescent throughout.
   2. Leaves with petioles longer than 3 mm.
      3. Flowers sessile or subsessile in glomerules; capsules glabrous .... 1. *R. caroliniensis*
      3. Flowers pedunculate, solitary or in cymes; capsules puberulent .... 3. *R. pedunculata*
   2. Leaves sessile or with petioles at most 3 mm long .... 2. *R. humilis*
1. Calyx lobes at least 2 mm broad; stems and leaves glabrous or sparsely pubescent .... 4. *R. strepens*

1. **Ruellia caroliniensis** (J. F. Gmel.) Steud. var. **dentata** (Nees) Fern. Wild Petunia. May–Oct. Dry woods, not common; s. ¼ of IL.

2. **Ruellia humilis** Nutt. Two varieties occur in Illinois:

a. Corolla up to 4.5 cm long, the tube up to 2.5 cm long...... 2a. *R. humilis* var. *humilis*
a. Corolla 5 cm long or longer, the tube 3 cm long or longer .... 2b. *R. humilis* var. *longiflora*

2a. **Ruellia humilis** Nutt. var. **humilis.** Wild Petunia. May–Oct. Prairies, dry woods, bluffs, occasional; throughout IL. (Including var. *frondosa* Fern.)

2b. **Ruellia humilis** Nutt. var. **longiflora** (Gray) Fern. Wild Petunia. May–Oct. Prairies, dry woods, bluffs; scattered throughout IL; apparently less common than var. *humilis.* (Including var. *expansa* Fern.)

3. **Ruellia pedunculata** Torr. Wild Petunia. May–Sept. Rocky woods, occasional; s. ⅓ of IL.

4. **Ruellia strepens** L. Two forms are known from Illinois:

a. Corolla 3 cm long or longer, the tube open ....4a. *R. strepens* f. *strepens*
a. Corolla less than 3 cm long, the tube more or less closed.. ....4b. *R. strepens* f. *cleistantha*

4a. **Ruellia strepens** L. f. **strepens.** Smooth Ruellia. May–Oct. Low woods, mesic woods, upland woods, floodplains, along streams, occasional; s. ⅘ of IL; also DuPage, Jo Daviess, and Kendall cos.

4b. **Ruellia strepens** f. **cleistantha** (Gray) McCoy. Smooth Ruellia. May–Oct. Woods; scattered in s. ⅓ of IL. This is apparently the cleistogamous form of the species.

24. ACERACEAE—MAPLE FAMILY

1. **Acer** L.—Maple

1. Leaves compound .... 5. *A. negundo*
1. Leaves simple.
2. Leaves white, glaucous, or silver-white on the lower surface; sinuses between the major lobes of the leaves angular; margins of leaves crenate or serrate; flowers red or bright yellow, appearing before the leaves or, if flowers greenish yellow, borne in pendulous panicles and appearing with the leaves.
3. Leaves glaucous on the lower surface, the margins crenate; base of leaf distinctly cordate; lobes of leaves rounded; flowers yellow-green, in pendulous panicles, borne with the leaves .... 9. *A. pseudoplatanus*
3. Leaves white or silvery white on the lower surface, the margins serrate; base of leaf truncate; lobes of leaves acute; flowers red or bright yellow, appearing before the leaves.
4. Leaves cut more than ½ way to the middle, the central lobe tapering to its base .... 11. *A. saccharinum*
4. Leaves cut less than ½ way to the middle, the central lobe broadest at its base.
5. Samaras 3–4 cm long; petioles and lower surface of leaves white-tomentose .... 3. *A. drummondii*
5. Samaras less than 3 cm long; petioles and lower surface of leaves glabrous or pubescent, but not white-tomentose .... 10. *A. rubrum*
2. Leaves pale green, yellow-green, gray, or reddish on the lower surface; sinuses between major lobes of the leaf rounded; margins of leaves usually entire (doubly serrate in *A. ginnala*); flowers yellow or yellow-green, appearing with the leaves.
6. Latex present; samaras horizontally spreading .... 8. *A. platanoides*
6. Latex absent; samaras parallel or spreading but not horizontal.
7. Leaves divided more than ½ way to the middle .... 7. *A. palmatum*
7. Leaves divided less than ½ way to the middle, or to the middle.
8. Lobes of leaves obtuse .... 2. *A. campestre*
8. Lobes of leaves acute to acuminate.
9. Edges of leaves drooping.
10. Lower surface of leaves green or yellow-green, glabrous or moderately pubescent; sinuses between the major lobes of the leaves forming angles more than 90 degrees; axillary buds covered by stipules .... 6. *A. nigrum*
10. Lower surface of leaves gray or gray-green, densely pubescent; sinuses between the major lobes of the leaves forming angles less than 90 degrees; axillary buds not covered by stipules .... 12. *A. saccharum*
9. Edges of leaves flat.
11. Leaves 3-lobed, the margins doubly serrate; flowers in pendulous panicles .... 4. *A. ginnala*
11. Leaves 5-lobed or, if only 3-lobed, the margins entire; flowers in umbel-like corymbs.
12. Leaves usually green on both sides, glabrous to sparsely pubescent (densely so in one variety), the lobes more or less acute; blades usually at least 8 cm across at maturity; bark dark gray to black, scaly or furrowed .... 12. *A. saccharum*
12. Leaves pale green below, usually pubescent throughout below, the lobes more or less obtuse; blades usually less than 8 cm across at maturity; bark pale gray to whitish, smooth .... 1. *A. barbatum*

1. **Acer barbatum** Michx. Southern Sugar Maple. Apr. Low woods, rare; s. ⅙ of IL. *Acer floridanum* (Chapm.) Pax; *A. saccharum* Marsh. ssp. *floridanum* (Chapm.) Desmarais.

2. **Acer campestre** L. Hedge Maple. May–June. Native to Europe; rarely escaped from cultivation; Champaign, DuPage, Jackson, and Lake cos.

3. **Acer drummondii** H. & A. Swamp Red Maple; Drummond's Red Maple. Mar.–Apr. Swampy woods, often in standing water; s. ¼ of IL. *Acer rubrum* L. var. *drummondii* (H. & A.) Sarg.

4. **Acer ginnala** Maxim. Amur Maple. May–June. Native to Asia; infrequently escaped from cultivation.

5. **Acer negundo** L. Three varieties have been found in Illinois:

a. Twigs green, not glaucous, glabrous............................. 5a. *A. negundo* var. *negundo*
a. Twigs gray-purple and glaucous or twigs pubescent.
  b. Twigs gray-purple and glaucous, glabrous .... 5c. *A. negundo* var. *violaceum*
  b. Twigs green, pubescent... 5b. *A. negundo* var. *texanum*

5a. **Acer negundo** L. var. **negundo.** Box Elder. Apr.–May. Floodplain woods, along streams, low woods, disturbed areas, common; in every co.

5b. **Acer negundo** L. var. **texanum** Pax. Hairy Box Elder. Apr.–May. Low woods, apparently rare; s. ⅙ of IL.

5c. **Acer negundo** L. var. **violaceum** (Kirsch.) Jaeg. Purple-stemmed Box Elder. Apr.–May. Floodplain woods, along streams, low woods, disturbed areas, occasional; throughout the state.

6. **Acer nigrum** Michx. f. Two varieties are known from Illinois:

a. Most of the leaves 5-lobed .... 6a. *A. nigrum* var. *nigrum*
a. Most of the leaves 3-lobed .... 6b. *A. nigrum* var. *palmeri*

6a. **Acer nigrum** Michx. f. var. **nigrum.** Black Maple. May–June. Rich woods, along streams; occasional in the n. ¾ of IL; also Pope Co. *Acer saccharum* Marsh. var. *nigrum* (Michx. f.) Desmarais.

6b. **Acer nigrum** Michx. f. var. **palmeri** Sarg. Palmer's Black Maple. May–June. Woods, rare; Johnson Co.

7. **Acer palmatum** Thunb. Japanese Maple. May–June. Native to Asia; rarely escaped from cultivation; Adams, Champaign, DuPage, Jackson, and Wabash cos.

8. **Acer platanoides** L. Norway Maple. May–June. Native to Europe; occasionally planted and sometimes found escaped from cultivation or persisting around old homesites.

9. **Acer pseudoplatanus** L. Sycamore Maple. May–June. Native to Europe and Asia; rarely escaped from cultivation; Champaign, DeKalb, and McLean cos.

10. **Acer rubrum** L. Several variations occur in Illinois:

a. Leaves 3- or 5-lobed, the central lobe 4–8 cm long, acute to acuminate, the margins coarsely toothed.
  b. Leaves glabrous or sparsely pubescent on the lower surface...............................
  .... 10a. *A. rubrum* var. *rubrum* f. *rubrum*
  b. Leaves densely pubescent on the lower surface ....
  .... 10b. *A. rubrum* var. *rubrum* f. *tomentosum*
a. Leaves all 3-lobed, the central lobe up to 5 cm long, often shorter, acute, the margins shallowly toothed.. 10c. *A. rubrum* var. *trilobum*

10a. **Acer rubrum** L. var. **rubrum** f. **rubrum.** Red Maple. Mar.–Apr. Rocky woods, slopes, moist woods, swamps (in n. IL), bogs (in n. IL); occasional to common throughout the state. A hybrid with *A. saccharinum, Acer X freemanii* A. A. Murray, is known from the ne. cos.

10b. **Acer rubrum** L. var. **rubrum** f. **tomentosum** (Desf.) Dansereau. Hairy Red Maple. Mar.–Apr. Upland slopes, rare; Union Co.

10c. **Acer rubrum** L. var. **trilobum** K. Koch. Three-lobed Red Maple. Mar.–Apr. Woods, rare; Johnson, Pope, and Williamson cos.

11. **Acer saccharinum** L. Silver Maple. Feb.–Apr. Bottomland woods, floodplain forests, along streams, common; throughout the state.

12. **Acer saccharum** Marsh. Three varieties occur in Illinois:

a. Most of the leaves 5-lobed.
  b. Leaves glabrous or nearly so on the lower surface, green or pale green, the margins flat .... 12a. *A. saccharum* var. *saccharum*
  b. Leaves densely hairy on the lower surface, usually gray-green, the margins turned downward .... 12c. *A. saccharum* var. *schneckii*
a. Most of the leaves 3-lobed .... 12b. *A. saccharum* var. *rugelii*

12a. **Acer saccharum** Marsh. var. **saccharum.** Sugar Maple. Apr.–May. Rich woods, along streams, rocky glades, common; in every co.

12b. **Acer saccharum** Marsh. var. **rugelii** (Pax) Rehd. Three-lobed Sugar Maple. Apr.–May. Mesic woods, rare; Johnson and Pope cos.

12c. **Acer saccharum** Marsh. var. **schneckii** Rehd. Schneck's Sugar Maple. Apr.–May. Rich woods, rocky woods; occasional in the s. ⅙ of IL. *Acer saccharum* Marsh. ssp. *schneckii* (Rehd.) Desmarais.

25. ADOXACEAE—MOSCHATEL FAMILY

1. **Adoxa** L.—Moschatel

1. **Adoxa moschatellina** L. Moschatel. May–July. Moist cliffs, very rare; Jo Daviess Co.

26. AIZOACEAE—AIZOON FAMILY

1. **Sesuvium** L.—Sea Purslane

1. **Sesuvium maritimum** (Walt.) BSP. Sea Purslane. July. Native to the U.S. Coastal Plain; adventive in sand along the Mississippi River; Rock Island Co.

27. AMARANTHACEAE—PIGWEED FAMILY

1. Leaves alternate.
  2. Flowers unisexual; utricles 1-seeded .... 3. *Amaranthus*
  2. Flowers perfect; utricles with 3 or more seeds .... 4. *Celosia*
1. All or most of the leaves opposite.
  3. Flowers borne in inconspicuous axillary clusters .... 7. *Tidestromia*
  3. Flowers in terminal and axillary spikes, the spikes sometimes arranged in panicles.
    4. Flowers unisexual; plants dioecious; inflorescence an open panicle .... 6. *Iresine*
    4. Flowers perfect; inflorescence in heads or spikes.
      5. Sepals united into a tube .... 5. *Froelichia*
      5. Sepals free from each other.
        6. Inflorescence in short, dense spikes; sepals erect during fruiting .... 2. *Alternanthera*
        6. Inflorescence at maturity in elongated, open spikes; sepals reflexed during fruiting .... 1. *Achyranthes*

1. **Achyranthes** L.—Chaff Flower

1. **Achyranthes japonica** (Miq.) Nakai. Japanese Chaff Flower. July–Sept. Native to China and Japan; adventive along a river bank, rare; Massac Co.

2. **Alternanthera** Forsk.—Alligator Weed

1. **Alternanthera philoxeroides** Griseb. Alligator Weed. June–Aug. Introduced from the s. states; wet ground along rivers, very rare; Alexander Co.

3. **Amaranthus** L.—Amaranth; Pigweed

1. Stems with pairs of axillary spines at some or all the nodes .... 13. *A. spinosus*
1. Stems without spines.
  2. Plants prostrate.
    3. Plants monoecious; stamens 3; sepals 4–5 in the pistillate flowers .... 4. *A. blitoides*
    3. Plants dioecious; stamens 5; sepals 1–2, at least one of them rudimentary in the pistillate flowers .... 14. *A. tuberculatus*

2. Plants ascending to erect.
4. Plants monoecious.
5. Plants ascending, diffusely branched; flowers in small axillary clusters; stamens 2–3 .... 1. *A. albus*
5. Plants erect, branched or unbranched but rarely diffusely so; flowers in spikes or panicles of spikes; stamens 5 (3 in *A. powellii*).
6. Inflorescence robust, purple, dark red, or bright red; bracts not longer than the styles.
7. Inflorescence stiff, erect .... 8. *A. hypochondriacus*
7. Inflorescence lax, erect or pendulous.
8. Tepals of pistillate flowers oblong to lanceolate, acute; inflorescence usually dark red or purple .... 6. *A. cruentus*
8. Tepals of pistillate flowers spatulate to obovate, obtuse or emarginate; inflorescence usually bright red .... 5. *A. caudatus*
6. Inflorescence moderately large, green or silvery green; bracts longer than the styles.
9. Tepals of pistillate flowers obtuse or emarginate; plants usually densely pubescent .... 11. *A. retroflexus*
9. Tepals of pistillate flowers acute to acuminate to aristate; plants sparsely pubescent or glabrate.
10. Bracts 2–4 mm long; inflorescence lax, with spreading branches .... 7. *A. hybridus*
10. Bracts 4–7 mm long; inflorescence stiff, with erect branches .... 10. *A. powellii*
4. Plants dioecious.
11. Pistillate flowers with 5 well developed sepals.
12. Bracts 3–6 mm long, almost always longer than the sepals, usually awned .... 9. *A. palmeri*
12. Bracts 1.5–2.5 mm long, about ½ as long to sometimes almost as long as the sepals, rarely awned.
13. Fruits indehiscent; pistillate flowers borne in short, thick spikes .... 2. *A. ambigens*
13. Fruits dehiscent; pistillate flowers borne in long, slender spikes .... 3. *A. arenicola*
11. Pistillate flowers with 1–2 sepals, if 2, then 1 of them rudimentary.
14. Fruits regularly dehiscent, circumscissile; sepals of staminate flowers usually awn-tipped .... 12. *A. rudis*
14. Fruits bursting irregularly upon dehiscence, not circumscissile; sepals of staminate flowers acute, rarely awn-tipped .... 4. *A. tuberculatus*

1. **Amaranthus albus** L. Tumbleweed; White Amaranth. July–Sept. Fields, disturbed soil, common; throughout the state. *Amaranthus graecaezans* L., misapplied.

2. **Amaranthus ambigens** Standl. Water Hemp. July–Oct. Moist soil, very rare; Winnebago Co. Not seen in IL since the last part of the 1800s.

3. **Amaranthus arenicola** I. M. Johnst. Sandhills Amaranth. July–Oct. Disturbed sandy soil, rare; Cook, Crawford, DuPage, Fayette, Morgan, and Tazewell cos. *Amaranthus torreyi* (Gray) Benth., misapplied.

4. **Amaranthus blitoides** S. Wats. Prostrate Pigweed. July–Sept. Native to the w. U.S.; adventive in disturbed soil; common throughout the state. *Amaranthus graecizans* L., misapplied.

5. **Amaranthus caudatus** L. Love-lies-bleeding. July–Oct. Native to tropical America; rarely escaped from cultivation into disturbed soil; Champaign, Cook, DuPage, Fayette, Jackson, Kane, Lawrence, and Wabash cos.

6. **Amaranthus cruentus** L. Purple Amaranth. July–Oct. Native to the w. U.S.; adventive in disturbed soil; n. ½ of IL; also St. Clair and Wabash cos.

7. **Amaranthus hybridus** L. Green Pigweed. Aug.–Oct. Native to tropical America; naturalized in fields, disturbed soil; common throughout IL.

8. **Amaranthus hypochondriacus** L. Prince's Feather. July–Oct. Native to tropical America; rarely escaped from cultivation; Cook and DuPage cos.

9. **Amaranthus palmeri** S. Wats. Careless Weed. Aug.–Oct. Native to the w. U.S.; adventive in disturbed soil; scattered in IL.

10. **Amaranthus powellii** S. Wats. Tall Amaranth. July–Oct. Native to the sw. U.S.; adventive in cultivated fields, disturbed soil; occasional throughout the state.

11. **Amaranthus retroflexus** L. Rough Pigweed. Aug.–Oct. Native to tropical America; naturalized in fields, disturbed soil; common throughout the state.

12. **Amaranthus rudis** J. Sauer. Water Hemp. July–Oct. Moist, disturbed soil; scattered throughout the state. *Acnida altissima* (Nutt.) A. W. Wood; *Amaranthus tamariscinus* (Nutt.) Wood. Some botanists consider *A. rudis* to be the same as *A. tuberculatus.*

13. **Amaranthus spinosus** L. Spiny Pigweed. July–Oct. Native to tropical America; naturalized in disturbed soil; occasional throughout the state.

14. **Amaranthus tuberculatus** (Moq.) J. Sauer. Three varieties may be distinguished in Illinois:

a. Plants erect; leaves up to 15 cm long; pistillate flowers crowded into slender spikes .... 14a. *A. tuberculatus* var. *tuberculatus*
a. Plants prostrate to ascending; leaves usually no more than 8 cm long; pistillate flowers arranged in dense, many-flowered glomerules or loose, few-flowered glomerules.
  b. Plants entirely prostrate; pistillate flowers arranged in loose, few-flowered glomerules .... 14b. *A. tuberculatus* var. *prostratus*
  b. Plants prostrate to ascending; pistillate flowers aggregated into dense, many-flowered, globular glomerules .... 14c. *A. tuberculatus* var. *subnudus*

14a. **Amaranthus tuberculatus** (Moq.) Sauer var. **tuberculatus.** Water Hemp. Aug.–Oct. Moist soil, particularly stream banks, pond margins, and sandbars; also in moist, disturbed soils; scattered throughout the state.

14b. **Amaranthus tuberculatus** (Moq.) Sauer var. **prostratus** (Uline & Bray) Mohlenbr. Prostrate Water Hemp. Aug.–Oct. Disturbed, moist sandy soil; not common in IL. *Amaranthus altissima* L. var. *prostrata* (Uline & Bray) Fern.

14c. **Amaranthus tuberculatus** (Moq.) Sauer var. **subnudus** (S. Wats.) Mohlenbr. Water Hemp. Aug.–Oct. Moist sandy soil; scattered in IL. *Amaranthus altissima* L. var. *subnuda* (S. Wats.) Fern.

4. **Celosia** L.—Cockscomb

1. **Celosia argentea** L. Celosia; Cockscomb. July–Oct. Native to the tropics; rarely escaped from cultivation into disturbed soil; Coles, Peoria, and Vermilion cos.

5. **Froelichia** Moench—Cottonweed; Snake-cotton

1. Stems stout, usually more than 0.5 m tall; leaves 8–20 mm wide; calyx in fruit merely denticulate or entire .... 1. *F. floridana*
1. Stems slender, usually less than 0.5 m tall; leaves up to 9 mm wide; calyx in fruit spinose .... 2. *F. gracilis*

1. **Froelichia floridana** (Nutt.) Moq. var. **campestris** (Small) Fern. Cottonweed. June–Sept. Disturbed soil in sandy fields; occasional in the n. ⅔ of IL; also Jackson and Madison cos.

2. **Froelichia gracilis** (Hook.) Moq. Cottonweed. May–Sept. Sandy or gravelly soil, particularly along railroads; scattered throughout the state.

6. **Iresine** P. Br.—Bloodleaf

1. **Iresine rhizomatosa** Standl. Bloodleaf. Aug.–Oct. Wet woods, very rare; Champaign, Crawford, Massac, Pulaski, and Wabash cos.

7. **Tidestromia** Standl.—Tidestromia

1. **Tidestromia lanuginosa** (Nutt.) Standl. Tidestromia. July–Oct. Native to the w. U.S.; adventive in disturbed soil, very rare; Cook and St. Clair cos.

28. ANACARDIACEAE—SUMAC FAMILY

1. Leaves simple .... 1. *Cotinus*
1. Leaves compound.
  2. Leaves divided into 3 leaflets.

3. Leaves not poisonous to touch, very fragrant when crushed; middle leaflet sessile or nearly so; flowers yellowish, in dense clusters; drupes red .... 2. *Rhus*
3. Leaves poisonous to touch, not fragrant; middle leaflet with a petiolule at least 1 cm long; flowers greenish white, in loose clusters; drupes whitish .... 3. *Toxicodendron*

2. Leaves with 7 or more leaflets.
4. Leaves not poisonous to touch; flowers in terminal panicles; drupes red .... 2. *Rhus*
4. Leaves poisonous to touch; flowers in axillary panicles; drupes whitish .... 3. *Toxicodendron*

1. **Cotinus** Mill.—Smoke-tree

1. **Cotinus coggygria** Scop. European Smoke-tree. May–June. Introduced from Eurasia; rarely escaped from cultivation; Adams, Champaign, DuPage, and Jackson cos.

2. **Rhus** L.—Sumac

1. Leaflets 3, very fragrant when crushed; shrubs up to 2 m tall .... 1. *R. aromatica*
1. Leaflets 7 or more, not fragrant when crushed; shrubs or small trees usually more than 2 m tall.
2. Twigs glabrous.
3. Rachis winged between leaflets .... 2. *R. copallina*
3. Rachis unwinged .... 3. *R. glabra*
2. Twigs velvety-pubescent .... 4. *R. hirta*

1. **Rhus aromatica** Ait. Four varieties may be found in Illinois:

a. Flowers appearing before the leaves unfold; apex of middle leaflet acute.
b. Lower surface of leaflets glabrous or sparsely pubescent .... 1a. *R. aromatica* var. *aromatica*
b. Lower surface of leaflets densely pubescent, often velvety .... 1c. *R. aromatica* var. *illinoensis*
a. Flowers appearing after the leaves unfold; apex of middle leaflet obtuse.
c. Terminal leaflet up to 4 cm long; stems spreading to nearly prostrate .... 1b. *R. aromatica* var. *arenaria*
c. Terminal leaflet 2.5–6.0 cm long; stems ascending to erect .... 1d. *R. aromatica* var. *serotina*

1a. **Rhus aromatica** Ait. var. **aromatica.** Fragrant Sumac. Mar.–Apr. Dry woods, rocky woods, glades, occasional; throughout the state.

1b. **Rhus aromatica** Ait. var. **arenaria** (Greene) Fern. Dunes Sumac. Apr.–May. Sandy areas, dunes; n. ½ of IL.

1c. **Rhus aromatica** Ait. var. **illinoensis** (Greene) Rehd. Velvet-leaved Sumac. Mar.–Apr. Dry, rocky woods, very rare; Monroe Co.

1d. **Rhus aromatica** Ait. var. **serotina** (Greene) Rehd. Late Fragrant Sumac. Apr.–May. Rocky woods, very rare; Union Co.

2. **Rhus copallina** L. Dwarf Sumac; Winged Sumac; Shining Sumac. July–Sept. Woods, fields; common in the s. ¼ of IL, occasional to rare elsewhere. Some botanists segregate Illinois's plants as var. *latifolia* Engl.

3. **Rhus glabra** L. Smooth Sumac. June–July. Woods, fields, roadsides, common; in every co.

4. **Rhus hirta** L. Staghorn Sumac. June–July. Woods; occasional in the n. ½ of IL, uncommon and apparently adventive in the s. ½. *Rhus typhina* L.

3. **Toxicodendron** Mill.

1. Leaflets 3.
2. Terminal leaflet obtuse; most leaved clustered at tip of stem; drupes pubescent .... 1. *T. pubescens*
2. Terminal leaflet acute; leaves scattered along the stem; drupes usually glabrous.

3. Vines, or plants trailing on the ground, rarely suberect; aerial roots present .... 2. *T. radicans*
3. Plants erect; aerial roots absent .... 3. *T. rydbergii*
1. Leaflets 7 or more .... 4. *T. vernix*

1. **Toxicodendron pubescens** Mill. Poison Oak. May–July. Introduced at edge of a pine plantation; Pope Co. *Toxicodendron toxicarium* (Salisb.) Gillis; *Rhus toxicodendron* L.

2. **Toxicodendron radicans** (L.) Kuntze. Two varieties may be recognized in Illinois:

a. Leaves entire or shallowly toothed, subcoriaceous .... 2a. *T. radicans* var. *radicans*
a. Leaves coarsely toothed, thin or membranous .... 2b. *T. radicans* var. *negundo*

2a. **Toxicodendron radicans** (L.) Kuntze var. **radicans.** Poison Ivy. May–July. Fields, woods, bluffs, disturbed areas, common; throughout the state. More or less erect plants that have aerial roots are being referred to this taxon. *Rhus radicans* L.

2b. **Toxicodendron radicans** (L.) Kuntze var. **negundo** (Greene) Reveal. Poison Ivy. May–July. Fields, woods, bluffs, disturbed areas, common; throughout the state.

3. **Toxicodendron rydbergii** (Small) Greene. Western Poison Ivy. May–July. Fields, woods; scattered in IL. *Rhus radicans* L. var. *rydbergii* (Small) Rehd.

4. **Toxicodendron vernix** (L.) Kuntze. Poison Sumac. May–July. Bogs and marshes; occasional to rare in the ne. cos.; also Coles and Woodford cos. *Rhus vernix* L.

29. ANNONACEAE—CUSTARD APPLE FAMILY

1. **Asimina** Adans.—Pawpaw

1. **Asimina triloba** (L.) Dunal. Pawpaw. Apr.–May. Low woods, wooded slopes; common in the s. cos., becoming less common northward.

30. APIACEAE—CARROT FAMILY

1. Leaves all simple.
  2. Uppermost stem leaves perfoliate, entire .... 8. *Bupleurum*
  2. None of the leaves perfoliate; leaves toothed or lobed.
    3. Leaves reniform or orbicular; flowers in umbels .... 23. *Hydrocotyle*
    3. Leaves strap-shaped, ovate, or lanceolate; flowers in capitate clusters .... 19. *Eryngium*
1. Leaves, or some of them, compound.
  4. Leaves once compound, either pinnate or ternate.
    5. Leaves trifoliolate or ternate.
      6. Petiolules of leaflets all equal in length.
        7. Flowers white; fruits smooth.
          8. Lateral leaflets deeply divided .... 15. *Cryptotaenia*
          8. Lateral leaflets merely serrate .... 20. *Falcaria*
        7. Flowers greenish or yellow-green; fruit with hooked prickles .... 32. *Sanicula*
      6. Petiolule of middle leaflet longer than petiolules of lateral leaflets.
        9. Central flower of each ultimate umbel sessile; fruit more or less flat .... 39. *Zizia*
        9. Central flower of each ultimate umbel pedicellate; fruit terete .... 36. *Thaspium*
    5. Leaves palmately or pinnately divided, with 5 or more leaflets.
      10. Leaves palmately divided.
        11. Leaflets up to 7 mm wide, entire, cross-septate .... 16. *Cynosciadium*
        11. Leaflets more than 7 mm wide, serrate, not cross-septate.
          12. Leaflets coarsely toothed; flowers greenish or yellow-green; fruits with hooked prickles .... 32. *Sanicula*
          12. Leaflets finely toothed; flowers white; fruits smooth .... 20. *Falcaria*
      10. Leaves pinnately divided.
        13. Leaflets filiform, at most 1 mm wide .... 31. *Ptilimnium*
        13. Leaflets broader, always more than 1 mm wide.
          14. Bulblets present in some of the upper leaf axils .... 11. *Cicuta*
          14. Bulblets absent.

15. Margins of leaflets entire.

16. Leaflets cross-septate; bracts at base of inflorescence many. .... 24. *Limnosciadium*

16. Leaflets not cross-septate; bracts at base of inflorescence 0–2 .... 27. *Oxypolis*

15. Margins of leaflets serrate.

17. Stems pubescent throughout.

18. Leaflet divisions less than 1 cm wide........ 37. *Torilis*

18. Leaflet divisions more than 1 cm wide .... 22. *Heracleum*

17. Stems glabrous throughout.

19. Margins of leaflets with up to 8 teeth........ 27. *Oxypolis*

19. Margins of leaflets with more than 8 teeth.

20. Leaves linear to lanceolate, serrate or incised; flowers white; bracts present at base of umbels.

21. Leaflets regularly toothed........ 33. *Sium*

21. Leaflets unequally incised .... 7. *Berula*

20. Leaflets oblong to ovate to obovate, some lobed; flowers yellow; bracts absent........ 28. *Pastinaca*

4. Leaves more than once compound.

22. Ultimate leaflet segments filiform.

23. Petals yellow; involucre of bracts absent.

24. Annual with a strong odor........ 3. *Anethum*

24. Coarse perennial without a particularly strong odor........ 21. *Foeniculum*

23. Petals white; involucre well developed or reduced to 1 or 2 subulate bracts, rarely absent.

25. Fruits tuberculate........ 34. *Spermolepis*

25. Fruits smooth, although they may be ribbed.

26. Coarse biennial; fruits flat; bracts subulate, undivided, or absent .... 9. *Carum*

26. Annual; fruits not particularly flat; bracts cleft into filiform segments.

27. Fruits linear-oblong, 8–10 mm long; leaflet divisions capillary........ 38. *Trepocarpus*

27. Fruits oval to elliptic to suborbicular, 1.5–3.5 mm long; leaflet divisions filiform.

28. Umbels simple, opposite the leaf .... 6. *Apium*

28. Umbels compound, terminal or in the leaf axils .... 31. *Ptilimnium*

22. Ultimate leaflet segments broader than filiform, more than 1 mm wide.

29. Bulblets present in the axils of some of the upper leaves........ 11. *Cicuta*

29. Bulblets absent.

30. Leaflets entire........ 35. *Taenidia*

30. Leaflets serrate, lobed, or variously divided.

31. Stems hairy only at the nodes.

32. Flowers yellow or maroon; leaf divisions more than 5 mm wide .... 36. *Thaspium*

32. Flowers white; leaf divisions less than 5 mm wide .... 5. *Anthriscus*

31. Stems hairy, at least at base, or glabrous throughout.

33. Stems hairy, at least at base.

34. Involucre of bracts deeply pinnatifid .... 17. *Daucus*

34. Involucre of bracts absent or present but not deeply pinnatifid.

35. Leaflet divisions at least 1 cm wide.

36. Umbels with at least 25 rays; fruit as wide as long, not bristly .... 4. *Angelica*

36. Umbels with 1–6 rays; fruit longer than wide, bristly........ 25. *Osmorhiza*

35. Leaflet divisions less than 1 cm wide.
37. Flowers yellow; fruit with corky wings ....29. *Polytaenia*
37. Flowers white; fruit without corky wings.
38. Fruit glabrous; bracts at base of secondary umbels present.
39. Fruit 5–10 mm long; secondary bracts obtuse, usually glabrous.................... 10. *Chaerophyllum*
39. Fruit 5–6 mm long; secondary bracts acute, pubescent .... 5. *Anthriscus*
38. Fruit with prickles; bracts at base of secondary umbels absent.
40. Leaf sheaths of upper leaves pubescent... 5. *Anthriscus*
40. Leaf sheaths of upper leaves glabrous .... 36. *Torilis*
33. Stems glabrous.
41. Plants scapose, with all leaves basal .... 18. *Erigenia*
41. Plant with cauline leaves.
42. Ultimate leaflet divisions less than 4.5 mm wide.
43. Flowers not all alike, the outer ones in an umbel with at least 2 petals larger than the others .... 14. *Coriandrum*
43. Flowers all alike.
44. Main rays of umbel 1–4 .... 10. *Chaerophyllum*
44. Main rays of umbel 7 or more.
45. Upper cauline leaves ternate ....28. *Perideridia*
45. All cauline leaves pinnate.
46. Fruit flattened laterally, with slender ribs; biennial...9. *Carum*
46. Fruit more or less terete, with thickened ribs; annual .... 2. *Aethusa*
42. Ultimate leaflet divisions more than 5 mm wide.
47. Leaves 2- to 3-ternate.
48. Involucre of bracts absent.
49. Flowers yellow; some leaves often simple.. 38. *Zizia*
49. Flowers white; all leaves ternate.
50. Some leaflets over 2 cm wide .... 1. *Aegopodium*
50. None of the leaflets more than 1.5 cm wide ....12. *Conioselinum*
48. Involucre of bracts present.
51. Involucre of many filiform bracts; biennial.. 20. *Falcaria*
51. Involucre of few, sometimes foliaceous bracts; perennial..................................12. *Conioselinum*
47. Leaves 2- to 3-pinnate.
52. Flowers yellow; leaves all bipinnate; involucre never present; fruit 6–8 mm long ....29. *Polytaenia*
52. Flowers white; at least some of the leaves tripinnate; involucre usually present; fruit 2.0–4.5 mm long.
53. Fruits dorsally flattened, 4.0–5.5 mm long, the lateral wings forming a double border..12. *Conioselinum*
53. Fruits laterally flattened, 2–4 mm long, not with a double border.
54. Some of the ultimate leaflet divisions at least 10 mm wide.
55. Veins of leaves running only to the sinuses............................ 11. *Cicuta*
55. Veins of leaves running to the teeth ....25. *Oenanthe*
54. None of the ultimate leaflet divisions 10 mm wide.................... 13. *Conium*

1. **Aegopodium** L.—Goutweed

1. **Aegopodium podagraria** L. Goutweed. May–Aug. Native to Europe; infrequently cultivated and rarely escaped into shaded ground; Champaign, Cook, DeKalb, DuPage, Kendall, Lake, McHenry, and Sangamon cos.

2. **Aethusa** L.—Fool's Parsley

1. **Aethusa cynapium** L. Fool's Parsley. June–Aug. Native to Europe and Asia; rarely escaped from cultivation into disturbed soil; Cook, DuPage, Fayette, Henry, Kane, Kendall, and Winnebago cos.

3. **Anethum** L.—Dill

1. **Anethum graveolens** L. Dill. June–Aug. Native to Europe and Asia; occasionally cultivated and escaped; scattered throughout the state.

4. **Angelica** L.—Angelica

1. Uppermost leaf sheaths inflated to at least 2 cm broad ....1. *A. atropurpurea*
1. Uppermost leaf sheaths slender, not more than 1 cm broad..............2. *A. venenosa*

1. **Angelica atropurpurea** L. Great Angelica. May–Aug. Woodlands, calcareous fens, thickets, calcareous marshes, occasional; in the n. ½ of IL.
2. **Angelica venenosa** (Greenway) Fern. Wood Angelica. May–July. Prairies, rich woods, rocky woods, not common; confined to the s. ¼ of IL.

5. **Anthriscus** Bernh.—Chervil

1. Leaves and stems glabrous or soft hairy; fruit glabrous.
  2. Stems pubescent at the nodes; rays of umbels puberulent; fruit linear, the beak about ⅓ as long as the body.. 2. *A. cerefolium*
  2. Stems pubescent only at base; rays of umbels glabrous; fruit lanceolate, the beak about ⅙ as long as the body...3. *A. sylvestris*
1. Leaves and stems hispidulous; fruit hispid, with hooked hairs .... 1. *A. caucalis*

1. **Anthriscus caucalis** M. Bieb. Chervil. June–July. Native to Europe. Disturbed soil; Cook, DuPage, and Sangamon cos.
2. **Anthriscus cerefolium** (L.) Hoffm. Chervil. May–Aug. Native to Europe; rarely adventive in IL; DuPage Co.
3. **Anthriscus sylvestris** (L.) Hoffm. Chervil. May–Aug. Native to Europe; adventive in IL and apparently spreading; DeKalb, DuPage, Kane, Lake, and Ogle cos.

6. **Apium** L.—Marsh Parsley

1. **Apium leptophyllum** (Pers.) F. Muell. Marsh Parsley. Sept. Native to the tropics; rarely adventive in IL; Effingham and Fayette cos.; *Cyclospermum leptophyllum* (Pers.) Sprague ex Britt. & P. Wilson.

7. **Berula** Hoffm.—Water Parsnip

1. **Berula erecta** (Huds.) Coville. Low Water Parsnip. Marshes, ponds, wet ditches, low woods, not common; confined to the n. ½ of IL. *Berula pusilla* (Nutt.) Fern.

8. **Bupleurum** L.—Thoroughwax

1. **Bupleurum rotundifolium** L. Thoroughwax. May–July. Native to Europe; rarely escaped from cultivation; St. Clair Co.

9. **Carum** L.—Caraway

1. **Carum carvi** L. Caraway. May–July. Native to Europe; occasionally cultivated but rarely escaped into grassy areas.

10. **Chaerophyllum** L.—Chervil

1. Pedicel of fruit the same diameter throughout; fruits broadest at the middle............ ....1. *C. procumbems*
1. Pedicel of fruit wider at the top than at the bottom; fruits broadest below the middle .... 2. *C. tainturieri*

1. **Chaerophyllum procumbens** (L.) Crantz. Two varieties occur in Illinois:

a. Fruits glabrous .... 1a. *C. procumbens* var. *procumbens*
a. Fruits puberulent .... 1b. *C. procumbens* var. *shortii*

1a. **Chaerophyllum procumbens** (L.) Crantz var. **procumbens.** Wild Chervil. Apr.–June. Rich woods, along streams, alluvial floodplains, along railroads and highways, common; throughout the state.

1b. **Chaerophyllum procumbens** (L.) Crantz var. **shortii** Torr. & Gray. Wild Chervil. Apr.–June. Alluvial woods, not common; scattered in s. IL.

2. **Chaerophyllum tainturieri** Hook. Wild Chervil. Mar.–May. Fields, roadsides, disturbed soil, not common; mostly in the s. ⅓ of IL.

11. **Cicuta** L.—Water Hemlock

1. Upper leaves bearing bulblets in some of their axils; leaflets linear .... 1. *C. bulbifera*
1. None of the leaves bearing bulblets; leaflets narrowly lanceolate or broader .... 2. *C. maculata*

1. **Cicuta bulbifera** L. Bulblet Water Hemlock. July–Sept. Calcareous marshes, swamps, not common; n. ½ of IL; also Union and Washington cos.

2. **Cicuta maculata** L. Water Hemlock. May–Sept. Marshes, wet prairies, wet ditches, moist woods, common; throughout IL. This plant is poisonous if eaten.

12. **Conioselinum** Hoffm.—Hemlock Parsley

1. **Conioselinum chinense** (L.) BSP. Hemlock Parsley. July–Sept. Shaded, springy ground, often calcareous, very rare; Cook, Jo Daviess, and Kane cos. Native to the U.S. despite the specific epithet.

13. **Conium** L.—Poison Hemlock

1. **Conium maculatum** L. Poison Hemlock. May–Aug. Native to Europe; naturalized in disturbed soil, fields, thickets, and along roads, occasional; throughout the state. This plant is poisonous if eaten.

14. **Coriandrum** L.—Coriander

1. **Coriandrum sativum** L. Coriander. May–June. Native to Europe; occasionally cultivated but rarely escaped in IL; Champaign, Cook, Jackson, and Will cos.

15. **Cryptotaenia** DC.—Honewort

1. **Cryptotaenia canadensis** (L.) DC. Honewort. May–Aug. Rocky woods, upland woods, low ground, floodplains, common; throughout the state.

16. **Cynosciadium** DC.

1. **Cynosciadium digitatum** DC. May–June. Swampy woods, very rare; Jackson Co.

17. **Daucus** L.—Carrot

1. Primary involucral bracts bipinnatifid .... 1. *D. carota*
1. Primary involucral bracts once-pinnate .... 2. *D. pusilla*

1. **Daucus carota** L. Wild Carrot; Queen Anne's Lace. May–Oct. Native to Europe; common throughout the state; in every co. Plants with flower clusters lacking the purple center may be called f. *epurpuratus* Farw. Plants with all flowers rosy-pink may be called f. *roseus* Millsp.

2. **Daucus pusillus** Michx. Small Wild Carrot. Apr.–June. Woods and in a cemetery, rare; Jackson and Perry cos.

18. **Erigenia** Nutt.—Harbinger-of-Spring

1. **Erigenia bulbosa** (Michx.) Nutt. Harbinger-of-Spring; Pepper-and-Salt. Feb.–Apr. Rich mesic woods, occasional; throughout the state except for the nw. cos.

19. **Eryngium** L.—Eryngo

1. Stems erect; flowering heads over 5 mm in diameter.
  2. Leaves with parallel veins, prickly .... 3. *E. yuccifolium*
  2. Leaves with net veins, not prickly .... 1. *E. planum*
1. Stems prostrate; flowering heads less than 5 mm in diameter .... 2. *E. prostratum*

1. **Eryngium planum** L. Sea Holly. Native to Europe and Asia; rarely escaped from gardens; Cook Co.

2. **Eryngium prostratum** Nutt. Spreading Eryngo. May–Nov. Edge of ponds, wet ground, rare; Alexander, Pope, and Williamson cos.

3. **Eryngium yuccifolium** Michx. Rattlesnake Master. July–Aug. Prairies, openings in woods, occasional; throughout the state.

20. **Falcaria** Bernh.—Sickleweed

1. **Falcaria vulgaris** Bernh. Sickleweed. July–Sept. Native to Europe; rarely adventive in IL; Schuyler Co.

21. **Foeniculum** Mill.—Fennel

1. **Foeniculum vulgare** Mill. Fennel. May–Sept. Native to Europe; occasionally cultivated but rarely escaped; Champaign, Cook, and Madison cos.

22. **Heracleum** L.—Cow Parsnip

1. Leaflets up to 130 cm wide; inflorescence of 50 or more rays; umbels to 50 cm wide .... 1. *H. mantegazzianum*
1. Leaflets up to 40 cm wide; inflorescence of 15–30 rays; umbels up to 20 cm wide .... 2. *H. maximum*

1. **Heracleum mantegazzianum** Sommier & Levier. Giant Hogweed. June–July. Native to Asia. Disturbed soil: Cook, DuPage, and Lake cos.

2. **Heracleum maximum** Bartr. Cow Parsnip. May–Aug. Low woods, rich woods, disturbed woods; occasional in the n. ⅔ of IL; rare elsewhere. *Heracleum lanatum* Michx.

23. **Hydrocotyle** L.—Water Pennywort

1. Leaves distinctly wider than high, not peltate .... 1. *H. ranunculoides*
1. Leaves orbicular, as wide as high, peltate .... 2. *H. umbellata*

1. **Hydrocotyle ranunculoides** L. f. Water Pennywort. Native to the s. U.S.; introduced into wet areas, detention pond, rare; Clinton, Lake, Lawrence, Union, and Whiteside cos.

2. **Hydrocotyle umbellata** L. Water Pennywort. Native to the s. U.S.; introduced along a wet roadside ditch, very rare; Williamson Co.

24. **Limnosciadium** Math. & Constance

1. **Limnosciadium pinnatum** (DC.) Math. & Constance. May–June. Native to the w. U.S.; rarely adventive in IL; Champaign, Jackson, and St. Clair cos.

25. **Oenanthe** L.

1. **Oenanthe javanica** (Blume) DC. May–June. Native to the south Pacific; wet, shaded soil; Champaign and DuPage cos. *Oenanthe sarmentosa* C. Presl, misapplied.

26. **Osmorhiza** Raf.—Sweet Cicely

1. Styles 2 mm long or longer, exceeding the petals; roots strongly anise-scented .... 2. *O. longistylis*
1. Styles up to 1.5 mm long, shorter than the petals; roots faintly anise-scented .... 1. *O. claytonii*

1. **Osmorhiza claytonii** (Michx.) C. B. Clarke. Sweet Cicely. Apr.–June. Rich woods, occasional; throughout the state.

2. **Osmorhiza longistylis** (Torr.) DC. Two varieties occur in Illinois:

a. Stems sparsely pubescent to nearly glabrous..........2a. *O. longistylis* var. *longistylis*
a. Stems densely villous..2b. *O. longistylis* var. *villicaulis*

2a. **Osmorhiza longistylis** (Torr.) DC. var. **longistylis.** Anise-root. Apr.–June. Rich woods, occasional; throughout the state.

2b. **Osmorhiza longistylis** (Torr.) DC. var. **villicaulis** Fern. Anise-root. Apr.–June. Rich woods; less common than the preceding but scattered throughout IL.

27. **Oxypolis** Raf.—Cowbane

1. **Oxypolis rigidior** (L.) Raf. Two varieties have been found in Illinois:
a. Leaflets lanceolate to oblong, at least 4 mm broad ....1a. *O. rigidior* var. *rigidior*
a. Leaflets linear, up to 4 mm broad....................................1b. *O. rigidior* var. *ambigua*

1a. **Oxypolis rigidior** (L.) Raf. var. **rigidior.** Cowbane. July–Sept. Wet prairies, calcareous fens, marshes; occasional throughout the state.

1b. **Oxypolis rigidior** (L.) Raf. var. **ambigua** (Nutt.) Robins. Cowbane. July–Sept. Wet ground, rare; confined to the s. ½ of IL.

28. **Pastinaca** L.—Parsnip

1. **Pastinaca sativa** L. Parsnip. May–Oct. Native to Europe; naturalized in waste ground and along roads and in pastures; common in the n. ¾ of IL, occasional in the s. ¼, but in every co. Persons susceptible to this species will form blisters when the plant comes in contact with their skin.

29. **Perideridia** Reichenb.

1. **Perideridia americana** (Nutt.) Reichenb. Thicket Parsley; Eastern Yampah. Apr.–July. Floodplains, savannas, thickets, rocky woods; occasional in the n. ½ of IL, becoming rare in the s. ½.

30. **Polytaenia** DC.—Prairie Parsley

1. **Polytaenia nuttallii** DC. Prairie Parsley. Apr.–June. Prairies, rocky woods, occasional; throughout the state.

31. **Ptilimnium** Raf.—Mock Bishop's-weed

1. Main leaves alternate or opposite; fruits about 1.5 mm long................ 1. *P. costatum*
1. Main leaves verticillate; fruits 2–4 mm long ....2. *P. nuttallii*

1. **Ptilimnium costatum** (Ell.) Raf. Mock Bishop's-weed. July–Sept. Swampy ground, rare; s. ½ of IL.

2. **Ptilimnium nuttallii** (DC.) Britt. Mock Bishop's-weed. June–Aug. Swampy ground, rare; Jackson, Pulaski, Randolph, St. Clair, and Union cos.

32. **Sanicula** L.—Black Snakeroot

1. Styles much longer than the bristles of the fruit; plants with rhizomes.
   2. Fruits sessile, at least 5 mm long; sepals of staminate flowers firm, 1–2 mm long ....2. *S. marilandica*
   2. Fruits stipitate, up to 4 mm long (excluding the stipe); sepals of staminate flowers soft, up to 1 mm long...3. *S. odorata*
1. Styles shorter than or merely equaling the bristles of the fruit; plants without rhizomes.
   3. Roots thick and cordlike; leaves thick..4. *S. smallii*
   3. Roots slender and fibrous; leaves thin.
      4. Pedicels of staminate flowers less than twice as long as the flowers; fruits stipitate, 3–5 mm long... 1. *S. canadensis*
      4. Pedicels of staminate flowers more than twice as long as the flowers; fruits sessile, at least 5.5 mm long ....5. *S. trifoliata*

1. **Sanicula canadensis** L. Two varieties occur in Illinois:

a. Leaves less than 8 cm long and less than 4 cm wide ....
....1a. *S. canadensis* var. *canadensis*
a. Leaves at least 8 cm long and at least 4 cm wide ....1b. *S. canadensis* var. *grandis*

1a. **Sanicula canadensis** L. var. **canadensis.** Canada Black Snakeroot. May–Aug. Woods, floodplains, north-facing bluffs, occasional to common; throughout IL.

1b. **Sanicula canadensis** L. var. **grandis** Fern. Large-leaved Snakeroot. May–Aug. Woods, not common; scattered in IL.

2. **Sanicula marilandica** L. Black Snakeroot. May–July. Wooded slopes; occasional in the n. ⅔ of IL, rare or absent elsewhere.

3. **Sanicula odorata** Pryer & Phillippe. Common Black Snakeroot. May–June. Woods, occasional to common; throughout the state. *Sanicula gregaria* Bickn.

4. **Sanicula smallii** Bickn. Southern Black Snakeroot. Rich woods, very rare; Hardin Co.

5. **Sanicula trifoliata** Bickn. Beaked Black Snakeroot. June–July. Woods, not common; mostly confined to the n. ½ of IL; also Pope Co.

33. **Sium** L.—Water Parsnip

1. **Sium suave** Walt. Water Parsnip. July–Sept. Marshes, ponds, wet ditches, low woods, swamps, occasional; throughout IL. Leaves that are submerged are very highly dissected.

34. **Spermolepis** Raf.—Scaleseed

1. Fruits bristly .... 1. *S. echinata*
1. Fruits warty .... 2. *S. inermis*

1. **Spermolepis echinata** (Nutt.) A. A. Heller. Bristly Scaleseed. Apr.–June. Dry soil, rare; confined to the s. ¼ of IL.

2. **Spermolepis inermis** (Nutt.) Math. & Constance. Smooth Scaleseed. May–June. Introduced from the w. U.S.; sandy soil; occasional in the nw. ¼ of IL; also Cook Co.

35. **Taenidia** Drude—Yellow Pimpernel

1. **Taenidia integerrima** (L.) Drude. Yellow Pimpernel. May–July. Prairies, dry, often rocky woods, wooded banks, occasional; throughout the state.

36. **Thaspium** Nutt.—Meadow Parsnip

1. Stems pubescent at the nodes; leaflets ciliate along the margins; basal leaves 2- to 3-ternate .... 1. *T. barbinode*
1. Stems glabrous; leaf margins not ciliate; basal leaves toothed or only once-ternate, often simple .... 2. *T. trifoliatum*

1. **Thaspium barbinode** (Michx.) Nutt. Hairy Meadow Parsnip. Apr.–June. Moist, rich woods, usually near streams; occasional in the n. ½ of the state, rare or absent elsewhere.

2. **Thaspium trifoliatum** (L.) Gray. Two varieties occur in Illinois:

a. Flowers purple .... 2a. *T. trifoliatum* var. *trifoliatum*
a. Flowers yellow .... 2b. *T. trifoliatum* var. *flavum*

2a. **Thaspium trifoliatum** (L.) Gray var. **trifoliatum.** Meadow Parsnip. Apr.–June. Prairies, rocky woods, thickets; scattered throughout the state.

2b. **Thaspium trifoliatum** (L.) Gray var. **flavum** S. F. Blake. Yellow Meadow Parsnip. Apr.–June. Prairies, rocky woods, thickets; occasional throughout the state.

37. **Torilis** Adans.—Hedge Parsley

1. Bracts of involucre 0 or 1 .... 1. *T. arvensis*
1. Bracts of involucre several .... 2. *T. japonica*

1. **Torilis arvensis** (Huds.) Link. Hedge Parsley. June–Aug. Native to Europe; adventive in disturbed areas, common; throughout IL.

2. **Torilis japonica** (Houtt.) DC. Hedge Parsley. June–Aug. Native to Europe and Asia; naturalized in disturbed areas; occasional throughout the state.

38. **Trepocarpus** Nutt.

1. **Trepocarpus aethusae** Nutt. May–June. Swampy woods, very rare; only in the s. ⅙ of the state.

39. **Zizia** Koch—Golden Alexanders

1. All of the basal leaves simple .... 1. *Z. aptera*
1. All or nearly all of the basal leaves ternately compound .... 2. *Z. aurea*

1. **Zizia aptera** (Gray) Fern. Heart-leaved Meadow Parsnip. Apr.–June. Dry prairies, rocky woods, not common; n. ⅕ of IL; also Hardin Co.

2. **Zizia aurea** (L.) Koch. Golden Alexanders. Apr.–June. Moist woods, floodplains, prairies, savannas; occasional in the n. ¾ of IL, less common in the s. ¼.

31. APOCYNACEAE—DOGBANE FAMILY

1. Leaves alternate .... 1. *Amsonia*
1. Leaves opposite.
  2. Leaves deciduous; flowers in cymes; corolla funnelform or campanulate.
    3. Plants erect; corolla campanulate, white or pink .... 2. *Apocynum*
    3. Plants twining; corolla funnelform, greenish yellow .... 3. *Trachelospermum*
  2. Leaves evergreen; flower solitary in the axils of the leaves; corolla salverform .... 4. *Vinca*

1. **Amsonia** Walt.—Blue Star

1. **Amsonia tabernaemontana** Walt. Two varieties occur in Illinois:

a. Leaves green on the lower surface .... 1a. *A. tabernaemontana* var. *tabernaemontana*
a. Leaves glaucous on the lower surface .... 1b. *A. tabernaemontana* var. *salicifolia*

1a. **Amsonia tabernaemontana** Walt. var. **tabernaemontana.** Blue Star. Apr.–June. Rocky woods, thickets, low, moist ground; occasional in the s. ¾ of IL.

1b. **Amsonia tabernaemontana** Walt. var. **salicifolia** (Pursh) Woodson. Blue Star. May–June. Rocky woods; occasional in the s. ½ of IL.

2. **Apocynum** L.—Dogbane; Indian Hemp

1. Corolla 4–10 mm long, pink or pink-tinged, rarely white; seeds 2.5–4.0 mm long.
  2. Flowers pendulous; corolla lobes recurved; seeds 2.5–3.0 mm long .... 1. *A. androsaemifolium*
  2. Flowers ascending; corolla lobes not recurved; seeds 3–4 mm long .... 3. *A. X floribundum*
1. Corolla 2–4 mm long, white or greenish white; seeds 4–6 mm long.
  3. Leaves petiolate; corolla greenish white; follicles over 10 cm long; coma of seed more than 2 cm long .... 2. *A. cannabinum*
  3. Leaves sessile or nearly so; corolla white; follicles up to 10 cm long; coma of seed up to 2 cm long .... 4. *A. sibiricum*

1. **Apocynum androsaemifolium** L. Two varieties occur in Illinois:

a. Leaves and stems glabrous or sparsely pubescent .... 1a. *A. androsaemifolium* var. *androsaemifolium*
a. Leaves and stems with short, gray pubescence .... 1b. *A. androsaemifolium* var. *incanum*

1a. **Apocynum androsaemifolium** L. var. **androsaemifolium.** Spreading Dogbane. May–July. Woods, prairies; occasional to common throughout the state.

1b. **Apocynum androsaemifolium** L. var. **incanum** A. DC. Gray Spreading Dogbane. May–July. Woods, prairies; scattered in the n. ½ of IL.

2. **Apocynum cannabinum** L. Indian Hemp; Common Dogbane. May–Aug. Prairies, fields, rocky woods, disturbed soil, common; in every co.

3. **Apocynum X floribundum** Greene. Intermediate Dogbane. May–Aug. Woods; occasional in the n. ½ of IL; also Shelby and Union cos. Considered to be the hybrid between *A. androsaemifolium* L. and *A. cannabinum* L. *Apocynum X medium* Greene.

4. **Apocynum sibiricum** Jacq. Two varieties of this species are known from Illinois:

a. Leaves rounded at the sessile base .... 4a. *A. sibiricum* var. *sibiricum*
a. Leaves cordate at the clasping base .... 4b. *A. sibiricum* var. *cordigerum*

4a. **Apocynum sibiricum** Jacq. var. **sibiricum.** Indian Hemp. June–Aug. Prairies, fields, rocky woods, common; throughout the state.

4b. **Apocynum sibiricum** Jacq. var. **cordigerum** (Greene) Fern. Heart-leaved Indian Hemp. June–Aug. Moist fields, not common; scattered in IL.

3. **Trachelospermum** Lemaire—Climbing Dogbane

1. **Trachelospermum difforme** (Walt.) Gray. Climbing Dogbane. May–July. Swamps, wet woods; confined to the s. ⅙ of IL; also Champaign Co.

4. **Vinca** L.—Periwinkle

1. Leaves over 3 cm long; calyx lobes ciliate; flowers over 3 cm broad.........1. *V. major*
1. Leaves up to 3 cm long; calyx lobes glabrous; flowers to 3 cm broad......2. *V. minor*

1. **Vinca major** L. Large Periwinkle. Apr.–May. Native to Europe; rarely escaped from cultivation; DuPage, Marion, and Pope cos.

2. **Vinca minor** L. Common Periwinkle. Apr.–May. Native to Europe; occasionally escaped from cultivation.

32. AQUIFOLIACEAE—HOLLY FAMILY

1. Petals white, oblong, united at base; stamens attached to base of the petals; leaves toothed, not mucronate ....1. *Ilex*
1. Petals yellow, linear, free; stamens free; leaves entire, mucronate ....2. *Nemopanthus*

1. **Ilex** L.—Holly

1. Leaves coriaceous, evergreen, spinulose-toothed ....2. *I. opaca*
1. Leaves not coriaceous, deciduous, not spinulose-toothed.
  2. Teeth of calyx acute, glabrous; nutlets ribbed; petals usually 4–5, at least in the pistillate flowers; apex of leaves obtuse.. 1. *I. decidua*
  2. Teeth of calyx obtuse, ciliate; nutlets not ribbed; petals usually 6, at least in the pistillate flowers; apex of leaves more or less acute to acuminate........................ ....3. *I. verticillata*

1. **Ilex decidua** Walt. Swamp Holly; Deciduous Holly; Possum Haw. Apr.–May. Swamps, wooded slopes, and bluffs; occasional in the s. cos., becoming rare northward; absent in the upper ⅖ of IL. A form with orange berries may be known as f. *aurantiaca* Mohlenbr. & Ozment.

2. **Ilex opaca** Sol. American Holly. May–June. Rocky, wooded slope, very rare; Union Co.

3. **Ilex verticillata** (L.) Gray. Winterberry. May–July. Swamps, edge of streams, wooded slopes, and bluffs; occasional in the n. cos., rare in the s. cos.

2. **Nemopanthus** Raf.—Mountain Holly

1. **Nemopanthus mucronatus** (L.) Trelease. June–July. Mountain Holly. Wet soil, very rare; Cook Co.

33. ARALIACEAE—SPIKENARD FAMILY

1. Leaves simple, palmately lobed, evergreen ....3. *Hedera*
1. Leaves compound, deciduous.
  2. Leaves alternate or basal.
    3. Leaves pinnately divided ....2. *Aralia*
    3. Leaves palmately divided .... 1. *Acanthopanax*
  2. Leaves whorled, the leaflets palmate...4. *Panax*

1. **Acanthopanax** Miq.—Palmate Hercules' Club

1. **Acanthopanax sieboldianus** Mak. Palmate Hercules' Club. July–Aug. Native to Asia; escaped from cultivation; Champaign and DuPage cos.

2. **Aralia** L.—Spikenard

1. Shrubs or small trees.
  2. Plants with prickles on stems and/or leaves.
    3. Leaflets petiolulate; leaves often tripinnate .... 6. *A. spinosa*

3. Leaflets sessile or nearly so; leaves bipinnate .... 2. *A. elata*
2. Plants without prickles .... 1. *A. chinensis*
1. Herbs.
4. Stems bristly .... 3. *A. hispida*
4. Stems not bristly.
5. Leaflets 3–7 per leaf; peduncle appearing to arise directly from the ground... .... 4. *A. nudicaulis*
5. Leaflets 9–21 per leaf; peduncle arising from the aerial stem .... 5. *A. racemosa*

1. **Aralia chinensis** L. Chinese Angelica Tree. Apr. Native to Asia; rarely escaped from cultivation; Champaign, Coles, Cook, and DuPage cos.

2. **Aralia elata** Seem. Japanese Angelica Tree. July–Sept. Native to Asia; rarely escaped from cultivation; Cook and DuPage cos.

3. **Aralia hispida** Vent. Bristly Sarsaparilla. June–Aug. Sandy soil, rare; Cook and Lake cos.

4. **Aralia nudicaulis** L. Wild Sarsaparilla. May–June. Rich, wooded slopes, dune woods, rocky woods, bogs; occasional in the n. ½ of IL; also Macoupin Co.

5. **Aralia racemosa** L. American Spikenard. June–Aug. Rich woods, shaded woods, rocky woods; occasional throughout the state.

6. **Aralia spinosa** L. Hercules' Club; Devil's Walking-stick. July–Sept. Wooded slopes, occasional; confined to the s. ⅓ of IL; escaped in Champaign and McLean cos.

3. **Hedera** L.—English Ivy

1. **Hedera helix** L. English Ivy. June–Sept. Native to Europe; commonly planted but rarely escaping from cultivation; DuPage and Jackson cos. and undoubtedly elsewhere.

4. **Panax** L.—Ginseng

1. **Panax quinquefolius** L. Ginseng. June–July. Rich woods, rocky woods; occasional throughout IL.

34. ARISTOLOCHIACEAE—BIRTHWORT FAMILY

1. Leaves more than 2, cauline; stamens 6; calyx irregular.
2. High-climbing woody vine; lower surface of leaves white-woolly; calyx yellow, with a purple orifice, over 15 mm long; roots without odor of turpentine .... 1. *Aristolochia*
2. Erect herbs; lower surface of leaves pubescent, not white-woolly; calyx dark purple, to 15 mm long; roots with odor of turpentine .... 3. *Endodeca*
1. Leaves 2, basal; stamens 12; calyx regular .... 2. *Asarum*

1. **Aristolochia** L.—Birthwort

1. **Aristolochia tomentosa** Sims. Dutchman's Pipevine. May–June. Chiefly calcareous woods, rare; confined to the s. ½ of IL. *Isotrema tomentosa* (Sims) Huber.

2. **Asarum** L.—Wild Ginger

1. **Asarum canadense** L. Three varieties occur in Illinois:

a. Calyx lobes spreading, 1–2 cm long, with attenuate tips to 1.5 cm long.................... .... 1a. *A. canadense* var. *canadense*
a. Calyx lobes reflexed, 6–20 mm long, abruptly contracted to a cusp 2–4 mm long.
b. Calyx lobes 6–12 mm long .... 1c. *A. canadense* var. *reflexum*
b. Calyx lobes 12–20 mm long .... 1b. *A. canadense* var. *ambiguum*

1a. **Asarum canadense** L. var. **canadense.** Wild Ginger. Apr.–May. Rich woods, rare; Kankakee Co.

1b. **Asarum canadense** L. var. **ambiguum** (Bickn.) Farw. Wild Ginger. Apr.–May. Rich woods; restricted to the n. ½ of IL.

1c. **Asarum canadense** L. var. **reflexum** (Bickn.) Robins. Wild Ginger. Apr.–May. Rich woods, common; throughout IL.

3. **Endodeca** Raf.—Snakeroot

1. Leaves never more than 2 cm wide at mid-point .... 1. *E. hastata*
1. Some or all of the leaves more than 2 cm wide at mid-point .... 2. *E. serpentaria*

1. **Endodeca hastata** (Nutt.) Raf. Narrow-leaved Snakeroot. June–July. Swampy woods; s. ⅙ of IL. *Aristolochia hastata* Nutt.; *Aristolochia serpentaria* L. var. *hastata* (Nutt.) Duchartre.

2. **Endodeca serpentaria** (L.) Raf. Virginia Snakeroot. May–July. Rich woods; local in most of IL except apparently absent from the nw. cos. *Aristolochia serpentaria* L.

35. ASCLEPIADACEAE—MILKWEED FAMILY

1. Plants erect .... 2. *Asclepias*
1. Plants climbing or twining.
   2. Leaves rounded at base .... 4. *Vincetoxicum*
   2. Leaves cordate at base.
      3. Flowers whitish; plants glabrous .... 1. *Ampelamus*
      3. Flowers greenish yellow or reddish purple; plants pubescent........ 3. *Matelea*

1. **Ampelamus** Raf.—Sandvine

1. **Ampelamus albidus** (Nutt.) Britt. Bluevine. July–Sept. Moist woods, fields, thickets, roadsides; occasional to common in the s. ⅔ of IL, rare or absent elsewhere. *Cynanchum laeve* (Michx.) Pers.

2. **Asclepias** L.—Milkweed

1. Most or all the leaves alternate.
   2. Leaves glabrous.
      3. Corolla lobes spreading; flowers green and purple; leaves at least 1 cm wide .... 19. *A. viridis*
      3. Corolla lobes reflexed; flowers green and white; leaves up to 5 mm wide .... 12. *A. stenophylla*
   2. Leaves pubescent.
      4. Latex absent; flowers orange or yellow; hood with incurved horn .... 15. *A. tuberosa*
      4. Latex present; flowers green and white; hood without a horn .... 3. *A. hirtella*
1. Most or all the leaves opposite or whorled.
   5. Leaves filiform to narrowly linear, up to 3 mm wide .... 17. *A. verticillata*
   5. Leaves broader, more than 3 mm wide.
      6. Hood without a horn; corolla green, the hood green or purple.
         7. Umbel solitary; stems and leaves hirsute .... 5. *A. lanuginosa*
         7. Umbels several; stems and leaves glabrous or pilose .... 18. *A. viridiflora*
      6. Hood with a horn; corolla not green or, if green, the hood white.
         8. Leaves sessile.
            9. Leaves obtuse, oblong or elliptic, broadest at or near the middle, clasping .... 1. *A. amplexicaulis*
            9. Leaves subacute to acute, lanceolate to ovate, broadest at or near the base, not clasping .... 6. *A. meadii*
         8. Leaves petiolate.
            10. Leaves pubescent, at least on the lower surface.
               11. Veins of leaves ascending; hoods yellow .... 7. *A. ovalifolia*
               11. Veins of leaves horizontally spreading; hoods not yellow.
                  12. Flowers dark purple-red; follicles without soft spines .... 9. *A. purpurascens*
                  12. Flowers variously pale purple or greenish, or mixed with white; follicles usually with soft spines.
                     13. Stems and leaves puberulent; lobes of corolla 6–9 mm long; hoods 3–4 mm high .... 14. *A. syriaca*
                     13. Stems and leaves tomentose; lobes of corolla 9–13 mm long; hoods 10–15 mm high .... 11. *A. speciosa*

10. Leaves glabrous.
  14. Never more than 6 pairs of leaves on the stem.
    15. Some of the leaves usually whorled; corolla pale pink, the lobes 4–6 mm long .... 10. *A. quadrifolia*
    15. None of the leaves whorled; corolla white with a purple center, the lobes 8–12 mm long .... 16. *A. variegata*
  14. At least 7 pairs of leaves on the stem.
    16. Leaves broadly rounded or subcordate at base; lobes of corolla at least 1 cm long .... 13. *A. sullivantii*
    16. Leaves tapering to base; lobes of corolla up to 1 cm long.
      17. Lobes of corolla greenish, 6–8 mm long, the hoods white or pink .... 2. *A. exaltata*
      17. Lobes of corolla rose-purple to white, 3–5 mm long, the hoods rose-purple or white.
        18. Flowers white; follicles glabrous; seeds not comose .... 8. *A. perennis*
        18. Flowers rose-purple (rarely white); follicles puberulent; seeds comose .... 4. *A. incarnata*

1. **Asclepias amplexicaulis** Small. Sand Milkweed. May–July. Prairies, sandy areas; occasional in the n. ¾ of IL, rare elsewhere.

2. **Asclepias exaltata** L. Poke Milkweed. June–July. Woodland borders, uncommon; scattered in most parts of IL. *Asclepias phytolaccoides* Pursh.

3. **Asclepias hirtella** (Pennell) Woodson. Tall Green Milkweed. May–Aug. Prairies, fields; occasional throughout the state. *Acerates hirtella* Pennell.

4. **Asclepias incarnata** L. Swamp Milkweed. June–Aug. Marshes, bogs, wet ditches, common; in every co.

5. **Asclepias lanuginosa** Nutt. Woolly Milkweed. May–June. Rocky or gravelly prairies, rare; confined to the n. ⅙ of IL. *Asclepias otarioides* Fourn.; *Acerates lanuginosa* (Nutt.) Dcne.

6. **Asclepias meadii** Torr. Mead's Milkweed. May–June. Mesic prairies, edge of sandstone bluffs, very rare; Cook, Ford, Hancock, Henderson, Peoria, Saline, and Tazewell cos.

7. **Asclepias ovalifolia** Dcne. Oval Milkweed. May–June. Prairies and dry woods, rare; Cook, Kankakee, Kendall, Lake, and McHenry cos.

8. **Asclepias perennis** L. White Swamp Milkweed. May–Sept. Wet woods, swamps; occasional in the s. ¼ of IL and along the Wabash River in se. IL; also Richland Co.

9. **Asclepias purpurascens** L. Purple Milkweed. May–July. Woodland borders, prairies; occasional throughout the state.

10. **Asclepias quadrifolia** Jacq. Whorled Milkweed. May–July. Rocky open woods; occasional in the s. ⅔ of IL; also Marshall, McDonough, and Rock Island cos.

11. **Asclepias speciosa** Torr. Showy Milkweed. June–Aug. Introduced from the w. U.S.; adventive along railroads; Cook, DeKalb, and Lake cos.

12. **Asclepias stenophylla** Gray. Narrow-leaved Green Milkweed. June–July. Dry upland woods, very rare; Adams, Calhoun, and Pike cos. *Acerates angustifolia* Dcne.

13. **Asclepias sullivantii** Engelm. Prairie Milkweed. June–July. Moist prairies; occasional in the n. ⅔ of IL, rare elsewhere.

14. **Asclepias syriaca** L. Two varieties may be distinguished in Illinois:

a. Follicles with spiny processes up to 3 mm long .... 14a. *A. syriaca* var. *syriaca*
a. Follicles with spiny processes 3–10 mm long .... 14b. *A. syriaca* var. *kansana*

14a. **Asclepias syriaca** L. var. **syriaca**. Common Milkweed. May–Aug. Fields, edge of prairies, waste ground, roadsides, common; in every co.

14b. **Asclepias syriaca** L. var. **kansana** (Vahl) Palmer & Steyerm. Common Milkweed. May–Aug. Fields, waste ground, roadsides; occasional and scattered in IL.

15. **Asclepias tuberosa** L. ssp. **interior** Woodson. Butterfly-weed. May–Sept. Woods, prairies, black oak savannas; common throughout the state. (Including f. *lutea* Clute.)

16. **Asclepias variegata** L. Variegated Milkweed. May–July. Rocky woods, occasional; restricted to the s. 1/4 of the state.

17. **Asclepias verticillata** L. Horsetail Milkweed. May–Sept. Dry rocky woods, prairies, fields; occasional to common throughout IL.

18. **Asclepias viridiflora** Raf. Green Milkweed. May–Aug. Prairies, sandy soil, gravelly soil, fields; occasional throughout the state. (Including var. *lanceolata* [Ives] Torr.) *Acerates viridiflora* (Raf.) Eaton.

19. **Asclepias viridis** Walt. Green-flowered Milkweed. May–July. Prairies; occasional in the s. 1/3 of IL; also Logan Co.

3. **Matelea** Aubl.—Climbing Milkweed

1. Flowers greenish yellow; calyx and pedicels glabrous or nearly so; follicles smooth, angular; blooming usually in July and Aug. ....2. *M. gonocarpos*
1. Flowers rose, maroon, or rarely cream; calyx and pedicels pubescent; follicles muricate, rounded; blooming usually in May and June.
  2. Petals 3–6 mm wide, maroon ....1. *M. decipiens*
  2. Petals 1.5–2.5 mm wide, rose or rarely cream ....3. *M. obliqua*

1. **Matelea decipiens** (Alex.) Woodson. Climbing Milkweed. May–June. Low floodplain woods, very rare; Jackson, Saline, and Williamson cos. *Gonolobus decipiens* (Alex.) Perry.

2. **Matelea gonocarpos** (Walt.) Shinners. Climbing Milkweed. July–Aug. Rocky woods, rare; confined to the s. 1/6 of IL; also Wabash Co. *Gonolobus gonocarpos* (Walt.) Perry.

3. **Matelea obliqua** (Jacq.) Woodson. Climbing Milkweed. May–June. Rocky woods, very rare; Hardin, Massac, Pope, and Saline cos. *Gonolobus obliquus* (Jacq.) Schultes.

4. **Vincetoxicum** Medic.—Swallow-wort

1. **Vincetoxicum nigrum** (L.) Moench. Black Swallow-wort. June–Sept. Native to Europe; occasionally adventive in the n. 2/3 of IL. *Cynanchum nigrum* (L.) Pers.; *C. louiseae* Kartesz & Gandhi.

36. ASTERACEAE—ASTER FAMILY

1. Flowering heads with only ray flowers; latex present .... Group 1
1. Flowering heads with disc flowers, the ray flowers present or absent; latex absent.
  2. All the leaves, or at least those on the lower part of the stem, opposite or whorled .... Group 2
  2. None of the leaves opposite or whorled.
    3. Ray flowers present.
      4. Rays yellow or orange .... Group 3
      4. Rays blue, purple, pink, rose, or white, not yellow or orange .... Group 4
    3. Ray flowers absent.
      5. Leaves simple, entire, toothed, or shallowly lobed .... Group 5
      5. Leaves deeply pinnatifid or pinnately compound........................... Group 6

Group 1

All flowering heads radiate; latex present.

1. Flowering heads blue, purple, or pinkish.
  2. Cypselae beakless.
    3. Pappus a crown of scales in 2–3 series...28. *Cichorium*
    3. Pappus of capillary bristles in 1 series ....82. *Nabalus*
  2. Cypselae with a beak about 0.5 mm long up to a long filiform beak.
    4. Pappus of capillary bristles; heads several; peduncles subtended by bractlets.
      5. Beak of cypselae about 0.5 mm long; phyllaries in 1–2 series..81. *Mulgedium*
      5. Beak of cypselae usually 1–6 mm long; phyllaries in 1 series ....71. *Lactuca*

4. Pappus of plumose bristles; head solitary; peduncles not subtended by bractlets .... 111. *Tragopogon*
1. Flowering heads yellow, orange, cream, or whitish.
6. Flowering heads cream or whitish .... 82. *Nabalus*
6. Flowering heads yellow or orange.
7. Cypselae with a beak 0.5–12.0 mm long.
8. Cypselae with a stout beak about 0.5 mm long.
9. Pappus of capillary bristles; heads usually numerous; bractlets at base of peduncle up to 10 .... 71. *Lactuca*
9. Pappus of plumose bristles; heads 1-very few; bractlets at base of peduncle up to 20 .... 73. *Leontodon*
8. Cypselae with a filiform beak 1–12 mm long.
10. Receptacle paleate; pappus of 2 kinds, the outer of capillary bristles, the inner of plumose bristles .... 66. *Hypochaeris*
10. Receptacle epaleate; pappus not as above.
11. Outer pappus of scales, inner pappus of capillary bristles; flowering heads 1–few .... 95. *Pyrrhopappus*
11. Pappus of all uniform bristles; heads solitary or numerous.
12. Pappus of capillary bristles.
13. Stems scapose .... 108. *Taraxacum*
13. Stems with some cauline leaves .... 26. *Chondrilla*
12. Pappus of plumose bristles.
14. Flowering head solitary, 4–8 cm across .... 111. *Tragopogon*
14. Flowering heads numerous, up to 2 cm across.
15. Peduncles subtended by 5 foliaceous bractlets .... 62. *Helminotheca*
15. Peduncles subtended by numerous narrow bractlets .... 89. *Picris*
7. Cypselae beakless.
16. Leaves with spinescent teeth; pappus in 4 series .... 104. *Sonchus*
16. Leaves not spinescent; pappus in 1–2 series, or absent.
17. Pappus absent.
18. Flowering head usually solitary; peduncles not subtended by bractlets .... 100. *Serinia*
18. Flowering heads few to several; peduncles subtended by 4–5 bractlets .... 72. *Lapsana*
17. Pappus present.
19. Outer pappus of scales, inner pappus of bristles.
20. Phyllaries up to 35, in 2–5 series; cypselae with 10 ribs .... 83. *Nothocalais*
20. Phyllaries up to 18, in 1–2 series; cypselae with 10–20 ribs .... 70. *Krigia*
19. All pappus of bristles.
21. Pappus of plumose bristles .... 73. *Leontodon*
21. Pappus of capillary bristles.
22. None of the leaves pinnatifid .... 64. *Hieracium*
22. Some of the leaves pinnatifid .... 25. *Crepis*

Group 2

Flowering heads with disc flowers, the ray flowers present or absent; all the leaves, or at least those on the lower part of the stem, opposite or whorled; latex absent.

1. Ray flowers present.
2. Rays yellow or orange.
3. Pappus of numerous capillary bristles; plants creeping .... 21. *Calyptocarpus*

3. Pappus of awns, scales, 1–3 small bristles, or absent; plants upright (procumbent in *Sanvitalia*).
 4. Leaves simple, entire, serrate, shallowly lobed, or palmately lobed.
  5. Leaves shallowly palmately lobed; phyllaries in 1 series........................ .... 102. *Smallanthus*
  5. Leaves not palmately lobed; phyllaries in 2 or more series.
   6. Disk flowers sterile with poorly developed cypselae......................... ....101. *Silphium*
   6. Disk flowers fertile with well-developed cypselae.
    7. Ray flowers persistent on the cypselae ....61. *Heliopsis*
    7. Ray flowers deciduous from the cypselae
     8. Pappus absent or of 1–3 small bristles.
      9. Petioles absent of nearly so; cypselae 4–5 mm long, strongly angular; pappus completely absent ....56. *Guizotia*
      9. Petioles at least 3 mm long; cypselae 1.0–2.5 mm long, weakly angular; pappus absent or of 1–3 small bristles.... .... 2. *Acmella*
     8. Pappus of awns or scales.
      10. Cypselae flat.
       11. Stems winged.
        12. Cypselae wingless; all leaves opposite .... 114. *Verbesina*
        12. Cypselae winged; some of the leaves alternate ....4. *Actinomeris*
       11. Stems wingless......................................115. *Ximenesia*
      10. Cypselae angular.
        13. Cypselae 3-angled; pappus of cypselae of ray flowers with 3 awns....................................99. *Sanvitalia*
        13. Cypselae 2- or 4-angled; pappus of cypselae of ray flowers with 2 or 4 awns, or absent.
         14. Cypselae with 2 or 4 stout, barbed awns............ .... 17. *Bidens*
         14. Cypselae with 2 small, barbless awns, or pappus absent.
          15. Rays usually 8 per head; phyllaries in 2 series .... 32. *Coreopsis*
          15. Rays of various numbers, not all of them 8 per head; phyllaries in several series.
           16. Cypselae wingless.
            17. Ray flowers neutral....... 60. *Helianthus*
            17. Ray flowers fertile......... 114. *Verbesina*
           16. Cypselae winged .... 116. *Ximenesia*
 4. Leaves deeply pinnately lobed or 1- to 2-pinnate.
  18. Plants aquatic; submerged leaves different than emergent leaves........ ....78. *Megalodonta*
  18. Plants terrestrial; submerged leaves absent.
   19. Pappus of 20 scales, each with 10 bristles at tip; receptacle bristly ....9. *Dyssodia*
   19. Pappus of small scales, not bristle-tipped, with 2, 4, or absent awns; receptacle paleate.
    20. Phyllaries in 1 series; pappus of unequal small scales................ .... 106. *Tagetes*
    20. Phyllaries in 2 series; pappus of 2–8 awns or awns absent.
     21. Leaves once-divided.
      22. Pappus of 2 or 4 stout, barbed awns.................. 17. *Bidens*
      22. Pappus of 2 weak, barbless awns................. 32. *Coreopsis*
     21. Leaves at least twice-divided................................ 33. *Cosmos*

2. Rays white, pink, or rose.
23. Leaves pinnately divided or simple and lobed.
24. Some of the leaves pinnately divided; phyllaries in 2 series; pappus present.
25. Leaves and leaflets more than 1 cm wide .... 17. *Bidens*
25. Leaflets 1–2 mm wide.. 33. *Cosmos*
24. Leaves simple and lobed; phyllaries in 1 series; pappus absent .... 92. *Polymnia*
23. Leaves simple, unlobed.
26. Pappus of ray flowers consisting of 15–20 fimbriate scales........................ .... 52. *Galinsoga*
26. Pappus of ray flowers reduced to a small crown or absent.
27. Plants more than 50 cm tall; stems green .... 114. *Verbesina*
27. Plants less than 50 cm tall; stems mauve................................42. *Eclipta*
1. Ray flowers absent.
28. Flowers unisexual, green.
29. Pistillate involucres nutlike or burlike... 7. *Ambrosia*
29. Pistillate involucres not nutlike or burlike.
30. Heads in racemose spikes, bracteate .... 69. *Iva*
30. Heads in paniculate spikes, ebracteate ....36. *Cyclachaena*
28. Flowers perfect, not green.
31. Plants climbing; major phyllaries 4 per head, subtended by short outer phyllaries...80. *Mikania*
31. Plants erect; phyllaries not as above.
32. Some of the leaves divided into 3–7 leaflets or 2- to 3-pinnate.
33. Leaves once-pinnate... 17. *Bidens*
33. Leaves 2- to 3-pinnate ....110. *Thelesperma*
32. Leaves simple, sometimes lobed but not compound.
34. Pappus of 2 or 4 stout, barbed awns....................................... 17. *Bidens*
34. Pappus of capillary bristles or absent.
35. Pappus absent; receptacle paleate; stems square............................. ....79. *Melanthera*
35. Pappus of capillary bristles; receptacle epaleate; stems not square.
36. Some or all the leaves whorled; flowers purple or rose ....49. *Eutrochium*
36. Leaves opposite (if rarely whorled, the flowers white); flowers white, blue, or pink.
37. Receptacle conical; flowers blue....................30. *Conoclinium*
37. Receptacle flat; flowers pink or white.
38. Flowers pink ....50. *Fleischmannia*
38. Flowers white.
39. Phyllaries all of same length.......................5. *Ageratina*
39. Phyllaries of different lengths............... 46. *Eupatorium*

Group 3

Plants with both ray and disc flowers; rays yellow or orange; leaves alternate or basal; latex absent.

1. Leaves simple, entire, serrate, or shallowly lobed.
2. Most of the leaves basal.
3. Head solitary.
4. Pappus of 5–8 translucent scales....................................... 109. *Tetraneuris*
4. Pappus of capillary bristles ....113. *Tussilago*
3. Heads several... 85. *Packera*
2. Leaves mostly cauline.
5. Rays reflexed; disc long-columnar or globose.
6. Disc long-columnar; cauline leaves clasping........................... 38. *Dracopis*
6. Disc globose; cauline leaves not clasping ....59. *Helenium*

5. Rays spreading, rarely slightly reflexed; disc globose, conical, or flat.
7. Pappus entirely of capillary or barbellate bristles.
8. Leaves spinulose-dentate .... 93. *Prionopsis*
8. Leaves not spinulose-dentate.
9. Upper leaves clasping..67. *Inula*
9. None of the leaves clasping.
10. Inflorescence more or less a flat-topped corymb.
11. Leaves up to 5 mm wide, some of them glandular-punctate .... 48. *Euthamia*
11. Leaves usually more than 5 mm wide, not glandular-punctate..84. *Oligoneuron*
10. Inflorescence paniculate, thyrsoid, or in axillary clusters .... 103. *Solidago*
7. Pappus of a short crown, awns, or scales or, if capillary bristles present, scales are present in addition.
12. Receptacle bristly; pappus of 6–10 awned scales .... 51. *Gaillardia*
12. Receptacle epaleate or paleate; pappus not as above.
13. Pappus of the cypselae of the disc flowers scaly, with short teeth, a crown, or absent.
14. Outer pappus scaly, inner pappus bristly on the cypselae of the disc flowers.
15. Cypselae of ray flowers thick, of disc flowers flat ....63. *Heterotheca*
15. All cypselae flat... 27. *Chrysopsis*
14. Pappus of cypselae of disc flowers with short teeth, a crown, or absent.
16. Disc flowers sterile; phyllaries in 2–3 series.......101. *Silphium*
16. Disc flowers fertile; phyllaries in several series...................... ....97. *Rudbeckia*
13. Pappus of the disc flowers with 2–8 awns.
17. Phyllaries gummy ....55. *Grindelia*
17. Phyllaries not gummy.
18. Pappus of 2 awns, these sometimes deciduous.
19. Disc flowers sterile; phyllaries in 2–3 series ....101. *Silphium*
19. Disc flowers fertile; phyllaries in several series.
20. Cypselae wingless .... 60. *Helianthus*
20. Cypselae winged.
21. Stems unwinged ....116. *Ximenesia*
21. Stems winged....................................3. *Actinomeris*
18. Pappus of 5 or more awns, persistent.
22. Disc flowers sterile....................................8. *Amphiachyris*
22. Disc flowers fertile .... 57. *Gutierrezia*
1. Leaves simple and deeply divided or leaves compound.
23. Receptacle epaleate.
24. Rays reflexed; pappus reduced to a short crown or absent; phyllaries in 2 series .... 96. *Ratibida*
24. Rays spreading; pappus of barbellate bristles; phyllaries in 1 series.
25. Stems glandular-pubescent; rays 1–3 mm long ....99. *Senecio*
25. Stems eglandular; rays 6–10 mm long.
26. Leaves once-pinnatifid .... 86. *Packera*
26. Leaves 2- to 3-pinnatifid...99. *Senecio*
23. Receptacle paleate.
27. Receptacle flat; phyllaries in 2–3 series; disc flowers sterile ....101. *Silphium*
27. Receptacle conical; phyllaries in several series; disc flowers fertile.
28. Leaves 2- to 3-pinnate or -pinnatifid... 34. *Cota*
28. Leaves lobed or 1-pinnatifid..97. *Rudbeckia*

Group 4

Plants with both ray flowers and disc flowers; rays not orange or yellow; latex absent; leaves alternate or basal.

1. Leaves simple and entire, serrate, or shallowly lobed (sometimes with a basal pair of pinnae in *Tanacetum*).
2. Rays blue, purple, or pink.
3. Rays reflexed; disc conical .... 40. *Echinacea*
3. Rays spreading; disc flat or subglobose.
4. Pappus a double series of capillary bristles .... 68. *Ionactis*
4. Pappus a single series of capillary bristles, awns, scales, or absent.
5. Leaves mostly basal; flower heads 1–few .... 16. *Bellis*
5. Leaves mostly cauline; flower heads usually numerous.
6. Pappus of capillary bristles.
7. Rays usually more than 50 .... 45. *Erigeron*
7. Rays usually less than 50.
8. Basal leaves cordate; inflorescence a flat-topped corymb .... 47. *Eurybia*
8. Basal leaves tapering or rounded at the base or, if cordate, the inflorescence racemose or paniculate.
9. Basal leaves cordate .... 105. *Symphyotrichum*
9. Basal leaves tapering or rounded at base, or absent at flowering time.
10. Annuals .... 105. *Symphyotrichum*
10. Perennials.
11. Leaves sericeous on both surfaces .... 105. *Symphyotrichum*
11. Leaves glabrous or pubescent, but not sericeous.
12. Stems glandular, at least above; leaves oblong; phyllaries densely glandular .... 105. *Symphyotrichum*
12. Stems eglandular; leaves linear or lanceolate or elliptic; phyllaries eglandular.
13. Leaves coarsely toothed, all of them 2.5 cm wide or wider .... 15. *Aster*
13. Leaves entire or finely and sparsely toothed, many of them less than 2.5 cm wide .... 105. *Symphyotrichum*
6. Pappus of short awns or scales.
14. Pappus of 6–10 awned scales; rays purple; phyllaries in 2–3 series; receptacle setose .... 51. *Gaillardia*
14. Pappus of 2 awns and tiny bristles; rays pink; phyllaries in several series; receptacle epaleate .... 18. *Boltonia*
2. Rays white.
15. Pappus of capillary bristles.
16. Pappus in a double series.
17. Inflorescence flat-topped .... 37. *Doellingeria*
17. Inflorescence paniculate .... 68. *Ionactis*
16. Pappus in a single series.
18. Rays less than 5 mm long .... 31. *Conyza*
18. Rays more than 5 mm long.
19. Inflorescence flat-topped .... 84. *Oligoneuron*
19. Inflorescence paniculate or flowers few, not in flat-topped clusters.
20. Rays usually more than 50 .... 45. *Erigeron*
20. Rays usually less than 50 .... 105. *Symphyotrichum*
15. Pappus of 2–3 awns, a short crown, or absent.
21. Pappus absent; cypselae 3-angled .... 25. *Chamaemelum*
21. Pappus present; cypselae not 3-angled.

22. Pappus reduced to a short crown; leaves sometimes with a basal pair of pinnae .... 107. *Tanacetum*
22. Pappus of 2–3 awns; leaves without a basal pair of pinnae.
23. Stems winged .... 114. *Verbesina*
23. Stems unwinged.
24. Pappus of 2 awns and tiny bristles; plants glabrous; receptacle epaleate .... 18. *Boltonia*
24. Pappus of 2–3 awns, without bristles; plants pubescent; receptacle paleate .... 87. *Parthenium*
1. Leaves deeply pinnatifid to 1- to 3-pinnate.
25. Leaves merely deeply lobed or pinnatisect.
26. Pappus absent; receptacle epaleate .... 74. *Leucanthemum*
26. Pappus of 2–3 awns; receptacle paleate .... 87. *Parthenium*
25. Leaves 1- to 3-pinnate.
27. Pappus of capillary bristles; phyllaries more or less squarrose; flowers blue or purple .... 76. *Machaeranthera*
27. Pappus a low crown or absent; phyllaries not squarrose; flowers white.
28. Receptacle paleate; phyllaries in several series; pappus absent.
29. Plants aromatic; receptacle conical .... 11. *Anthemis*
29. Plants not aromatic; receptacle flat .... 1. *Achillea*
28. Receptacle epaleate; phyllaries in 2–3 series; pappus a low crown or absent.
30. Cypselae 3-ribbed; plants not aromatic .... 112. *Tripleurospermum*
30. Cypselae 5-ribbed; plants aromatic .... 77. *Matricaria*

Group 5

Flowering heads with only disc flowers; leaves alternate or basal and simple, entire, serrate, shallowly lobed, or pinnatifid (or with a basal pair of pinnae in *Tanacetum*); latex absent.

1. Leaves with spine-tipped teeth.
2. Flowers yellow .... 24. *Centaurea*
2. Flowers purple, pink, pale blue, or white.
3. Pappus of barbellate bristles; phyllaries in 2 series; stems winged .... 85. *Onopordum*
3. Pappus of simple bristles, plumose bristles, or a crown of scales; phyllaries in several series; stems unwinged.
4. Pappus a crown of scales; flowers pale blue .... 41. *Echinops*
4. Pappus of bristles; flowers purple, white, or pink.
5. Pappus of plumose bristles .... 29. *Cirsium*
5. Pappus of simple bristles .... 22. *Carduus*
1. Leaves without spine-tipped teeth.
6. Outer row of disc flowers appearing ligulate.
7. Outer phyllaries entire, never spine-tipped.
8. Phyllaries coriaceous, yellowish, without a hyaline margin .... 6 *Amberboa*
8. Phyllaries thin, green, with a hyaline margin .... 3. *Acroptilon*
7. Outer phyllaries fimbriate or laciniate, some of them sometimes spine-tipped.
9. Involucre 3–4 cm high; pappus 6–12 mm long .... 90. *Plectocephalus*
9. Involucre 1.0–1.5 cm high; pappus 3 mm long or less .... 24. *Centaurea*
6. Outer row of disc flowers tubular, not appearing ligulate.
10. Flowers greenish.
11. Phyllaries and fruits with hooked bristles, prickly or appearing prickly.
12. Flowers unisexual .... 115. *Xanthium*
12. Flowers perfect .... 12. *Arctium*
11. Phyllaries and fruits without hooked bristles, not prickly .... 14. *Artemisia*
10. Flowers white, purple, cream, yellow, orange, blue, pink, brownish, or rusty.

13. Phyllaries and cypselae with hooked bristles ....12. *Arctium*
13. Phyllaries and cypselae without hooked bristles.
14. Flowers orange; pappus absent or of short scales; some leaves clasping; receptacle paleate ....23. *Carthamus*
14. Flowers not orange; pappus of capillary bristles, a short crown, or 5–8 scales; leaves not clasping; receptacle epaleate.
15. Flowers yellowish, cream, brownish, or rusty.
16. Phyllaries in 1 series; flowers yellow ....99. *Senecio*
16. Phyllaries in several series; flowers cream, brownish, or rusty.
17. Pappus of plumose bristles ....20. *Brickellia*
17. Pappus of capillary bristles.
18. Heads leafy bracted .... 54. *Gnaphalium*
18. Heads not leafy bracted................ 94. *Pseudognaphalium*
15. Flowers pink, purple, or white.
19. Flowers pink or purple.
20. Flowers in glomerules, subtended by a 3-lobed bract ....43. *Elephantopus*
20. Flowers and bracts not as above.
21. Phyllaries in 1 series, subtended by calyxlike bracts; leaves basal ....88. *Petasites*
21. Phyllaries in several series; leaves not basal.
22. Pappus of plumose or barbellate bristles......75. *Liatris*
22. Pappus of simple bristles.
23. Pappus in a double series....................115. *Vernonia*
23. Pappus in a single series.
24. Capillary bristles united at base ....53. *Gamochaeta*
24. Capillary bristles not united at base................... .... 91. *Pluchea*
19. Flowers white.
25. Phyllaries in 1 series.
26. Calyxlike bracts at base of phyllaries; leaves basal ....88. *Petasites*
26. Calyxlike bracts absent; leaves cauline.
27. Phyllaries 5; flowers 5 per head .... 13. *Arnoglossum*
27. Phyllaries more than 5; flowers more than 5 per head ....44. *Erechtites*
25. Phyllaries in 2–several series.
28. Some of the leaves hastate .... 58. *Hasteola*
28. None of the leaves hastate.
29. Pappus a short crown............................107. *Tanacetum*
29. Pappus of capillary bristles.
30. Phyllaries in 2 series, not scarious........................... .... 19. *Brachyactis*
30. Phyllaries in several series, scarious.
31. Most of the leaves at base of plant ....10. *Antennaria*
31. Most of the leaves cauline .... 9. *Anaphalis*

Group 6

Flowering heads with only disc flowers; leaves alternate or basal, deeply pinnatifid to pinnately compound; latex absent.

1. Flowers greenish.
2. Pistillate involucre with 2 flowers and 2 beaks; receptacle paleate...... 7. *Ambrosia*
2. Involucres not as above; receptacle epaleate or woolly.....................11. *Artemisia*
1. Flowers yellow or greenish yellow.
3. Pappus of capillary bristles; phyllaries in 1 series ....99. *Senecio*

3. Pappus of 12–20 hyaline scales, of short awns, or absent; phyllaries in 2–4 series.
    4. Pappus of 12–20 hyaline scales .... 65. *Hymenopappus*
    4. Pappus a short crown or absent.
        5. Receptacle conic; plants aromatic ....77. *Matricaria*
        5. Receptacle flat; plants not aromatic ....107. *Tanacetum*

1. **Achillea** L.—Yarrow; Milfoil

1. **Achillea millefolium** L. Two varieties may be distinguished in Illinois:

a. Leaves and stems arachnoid to glabrescent; corymb flat-topped ....1a. *A. millefolium* var. *millefolium*
a. Leaves and stems densely woolly; corymb round-topped.. ....1b. *A. millefolium* var. *lanulosum*

1a. **Achillea millefolium** L. var. **millefolium.** Common Yarrow; Milfoil. May–Oct. Native to Europe; naturalized in fields and other disturbed soil, common; in every co. Rose-purple–flowered specimens may be called f. *rosea* E. L. Rand. & Redfield.

1b. **Achillea millefolium** L. var. **lanulosum** (Nutt.) Piper. Western Yarrow; Western Milfoil. June–Oct. Adventive from the w. U.S.; fields and disturbed soil; occasional throughout IL. *Achillea lanulosa* Nutt.

2. **Acmella** Rich. in Pers.—Spot-flower

1. **Acmella repens** (Walt.) Rich. in Pers. Spot-flower. June–Sept. Moist soil, very rare; Johnson Co. *Acmella oppositifolia* (Lam.) R. K. Jansen var. *repens* (Walt.) R. K. Jansen; *Spilanthes americana* (Mutis) Hieronymus var. *repens* (Walt.) A. H. Moore.

3. **Acroptilon** Cass. in F. Cuv.—Knapweed

1. **Acroptilon repens** (L.) DC. Russian Knapweed. May–Sept. Native to the Caucasus region; adventive in disturbed soil, rare; DeKalb, Williamson, and Winnebago cos. *Centaurea repens* L.; *Rhaponticum repens* (L.) Hidalgo.

4. **Actinomeris** Nutt.—Wingstem

1. Stems 2-winged; phyllaries deflexed; flowering heads up to 50; ray flowers 2–10; cypselae spreading in all directions, forming a globose fruiting structure................. ....1. *A. alternifolia*
1. Stems 4-winged; phyllaries erect; flowering heads up to 10; rays flowers 8–13; cypselae not spreading in all directions, not forming a globose fruiting structure ....2. *A. helianthoides*

1. **Actinomeris alternifolia** (L.) Benth. Yellow Ironweed. Aug.–Oct. Moist soil in open woods; occasional throughout IL. *Verbesina alternifolia* (L.) Benth.

2. **Actinomeris helianthoides** (Michx.) Nutt. Yellow Crownbeard; Yellow Wingstem. May–July. Open woods, prairies; common in the s. ⅗ of the state, apparently absent elsewhere. *Verbesina helianthoides* Michx.

5. **Ageratina** Spach—Snakeroot

1. **Ageratina altissima** (L.) R. M. King & H. Robins. White Snakeroot. July–Sept. Woods, common; throughout IL. *Eupatorium rugosum* Houtt. This species was the cause of a pioneer ailment called milk sickness, which occurred when milk was drunk from cows that had eaten of this plant.

6. **Amberboa** (Pers.) Less.—Sweet Sultan

1. **Amberboa moschata** (L.) DC. Sweet Sultan. July–Sept. Native to Asia; rarely escaped from cultivation; disturbed soil; Champaign Co. *Centaurea moschata* L.

7. **Ambrosia** L.—Ragweed

1. All leaves alternate, bipinnatifid .... 6. *A. tomentosa*
1. At least the lowest leaves, and sometimes all the leaves, opposite.

2. Leaves not deeply divided, usually with a pair of teeth near base........................ ....2. *A. bidentata*
2. Most leaves deeply divided.
   3. Leaves either palmately 3- to 5-lobed or entire, but never pinnatifid.............. ....7. *A. trifida*
   3. Some or all the leaves pinnatifid to bipinnatifid.
      4. All leaves pinnatifid to bipinnatifid.
         5. Leaves petiolate, glabrous or nearly so above, involucre with 5–7 spines..1. *A. artemisiifolia*
         5. Leaves sessile or subsessile, scabrous above; involucre unarmed or with blunt tubercles.
            6. Involucre with pointed spines 1 mm long; stems slightly scabrous. ....4. *A. X intergradiens*
            6. Involucre with blunt tubercles 0.5–1.0 mm long; stems harshly scabrous ....5. *A. psilostachya*
      4. Some leaves palmately 3- to 5-lobed, others pinnatifid to bipinnatifid ....3. *A. X helenae*

1. **Ambrosia artemisiifolia** L. Common Ragweed. Aug–Oct. Fields and disturbed areas, very common; in every co. *Ambrosia artemisiifolia* L. var. *elatior* (L.) Desc.

2. **Ambrosia bidentata** Michx. Toothed Ragweed. July–Oct. Fields and disturbed areas; occasional to common in the s. 3/5 of IL.

3. **Ambrosia X helenae** Rouleau. Hybrid Ragweed. July–Oct. Disturbed soil, very rare; Champaign Co. Considered to be a hybrid between *A. artemisiifolia* L. and *A. trifida* L. and appearing to be intermediate between them.

4. **Ambrosia X intergradiens** W. H. Wagner. Intermediate Ragweed. July–Oct. Disturbed soil; rare in IL. This is the hybrid between *A. artemisiifolia* L. and *A. psilostachya* DC.

5. **Ambrosia psilostachya** DC. Western Ragweed. July–Oct. Dry sandy soil in fields and disturbed areas; occasional in the n. 3/4 of IL; also St. Clair and Williamson cos. *Ambrosia psilostachya* DC. var. *coronopifolia* (Torr. & Gray) Farw.

6. **Ambrosia tomentosa** Nutt. False Ragweed. July–Oct. Native to the w. U.S.; naturalized in disturbed areas, rare; Champaign, LaSalle, Mason, and McHenry cos. The common name refers to the placement of this species in a different genus, *Franseria,* at one time so that it was not considered to be a true ragweed. *Franseria discolor* Nutt.

7. **Ambrosia trifida** L. Two varieties occur in Illinois:

a. Petioles of at least the upper leaves more or less wing-margined; ribs of fruits ending in short spines ....7a. *A. trifida* var. *trifida*
a. Petioles wingless; ribs of fruits ending in blunt to almost obsolete tubercles ....7b. *A. trifida* var. *texana*

7a. **Ambrosia trifida** L. var. **trifida.** Giant Ragweed. July–Oct. Fields and disturbed areas, floodplains, common; in every co. The original habitat of this species was in river floodplains. An unnamed hybrid between this species and *A. bidentata* is known from Clay, Jersey, St. Clair, Sangamon, and Vermilion cos.

7b. **Ambrosia trifida** L. var. **texana** Scheele. Southern Giant Ragweed. July–Oct. Native to the sw. U.S.; adventive in disturbed areas; occasional in s. IL.

8. **Amphiachyris** Nutt.—Broomweed

1. **Amphiachyris dracunculoides** (DC.) Nutt. Prairie Broomweed. July–Oct. Native to the w. U.S.; adventive in disturbed soil; Jefferson and Tazewell cos. *Gutierrezia dracunculoides* (DC.) Blake.

9. **Anaphalis** DC.—Everlasting

1. **Anaphalis margaritacea** (L.) DC. Pearly Everlasting. July–Oct. Introduced from the n. U.S.; woods, very rare; Cook Co. Not seen in IL in more than 50 years. *Gnaphalium margaritaceum* L.

10. **Antennaria** Gaertn.—Pussy-toes

1. Basal leaves prominently 1-nerved or obscurely 3-nerved, usually less than 1.5 cm wide.
  2. Basal leaves obovate, abruptly contracted below the middle to a petiolelike base; stolons short, bearing several leaves .... 1. *A. howellii*
  2. Basal leaves cuneate-spatulate, gradually tapering to the sessile base; stolons long, with very few leaves .... 2. *A. neglecta*
1. Basal leaves prominently 3- to 5-nerved, usually 1.5 cm wide or wider.
  3. Heads several.
    4. Involucre of pistillate plants 7–10 mm high .... 3. *A. parlinii*
    4. Involucre of pistillate plants 5–7 mm high .... 4. *A. plantaginifolia*
  3. Head solitary .... 5. *A. solitaria*

1. **Antennaria howellii** Greene. Pussy-toes; Cat's-foot. May–June. Fields, open woods, rare; DeKalb, Henry, Kane, and Kendall cos. *Antennaria neodioica* Greene; *A. neglecta* Greene var. *neodioica* (Greene) Cronq.; *A. neglecta* Greene var. *attenuata* (Fern.) Cronq.; *A. howellii* Greene ssp. *neodioica* (Greene) R. J. Bayer.

2. **Antennaria neglecta** Greene. Pussy-toes; Cat's-foot. Apr.–May. Fields, prairie remnants, open woods; common in the n. ½ of IL, occasional in the s. ½.

3. **Antennaria parlinii** Fern. Two subspecies occur in Illinois:

a. Upper surface of leaves glabrous or nearly so from the beginning; upper part of stem usually with purple glands .... 3a. *A. parlinii* ssp. *parlinii*
a. Upper surface of leaves arachnoid at first, tardily glabrate in age; upper part of stem without purple glands .... 3b. *A. parlinii* ssp. *fallax*

3a. **Antennaria parlinii** Fern. ssp. **parlinii.** Pussy-toes. Apr.–June. Open woods, fields, pastures; occasional throughout IL. *Antennaria plantaginifolia* (L.) Hook. var. *arnoglossa* (Greene) Cronq.; *A. plantaginifolia* (L.) Hook. var. *parlinii* (Fern.) Cronq.

3b. **Antennaria parlinii** Fern. ssp. **fallax** (Greene) R. J. Bayer & Stebbins. Pussy-toes. May–June. Fields, open woods; common in the n. ½ of IL, occasional in the s. ½.

4. **Antennaria plantaginifolia** (L.) Hook. Pussy-toes. Apr.–May. Woods; occasional to common; scattered in IL. (Including var. *petiolata* [Fern.] Heller.)

5. **Antennaria solitaria** Rydb. One-headed Pussy-toes. Apr.–May. Woods, very rare; Hardin Co.

11. **Anthemis** L.—Dog-fennel; Chamomile

(See also Cota.)

1. Cypselae 4-angled, not glandular-tuberculate; plants odorless; ray flowers pistillate .... 1. *A. arvensis*
1. Cypselae subterete, glandular-tuberculate; plants with a foul odor; ray flowers sterile .... 2. *A. cotula*

1. **Anthemis arvensis** L. var. **agrestis** (Wallr.) DC. Corn Chamomile. May–Aug. Native to Europe; naturalized in waste ground, fields, roadsides; occasional throughout IL.

2. **Anthemis cotula** L. Dog-fennel; Mayweed. May–Nov. Native to Europe; naturalized in waste ground, particularly barnyards; common in the n. ¾ of IL, less common southward.

12. **Arctium** L.—Burdock

1. Heads more or less corymbose, long-peduncled; petioles strongly angled, solid or hollow.
  2. Petioles solid; involucre glabrous or nearly so; heads 3–4 cm across .... 1. *A. lappa*
  2. Petioles hollow; involucre tomentose; heads 1.5–2.5 cm across .... 3. *A. tomentosa*
1. Heads more or less racemose, sessile or short-peduncled; petioles scarcely angled, hollow .... 2. *A. minus*

1. **Arctium lappa** L. Great Burdock. July–Oct. Native to Europe and Asia; naturalized in disturbed soil; occasional to common in the n. ½ of IL; also Clark and Hardin cos.

2. **Arctium minus** (Hill) Bernh. Common Burdock. June–Sept. Native to Europe and Asia; naturalized in disturbed soil; common throughout IL.

3. **Arctium tomentosum** Mill. Cotton Burdock; Woolly Burdock. July–Oct. Native to Europe and Asia; naturalized in disturbed soil; occasional in the n. ½ of IL.

13. **Arnoglossum** Raf.—Indian Plantain

1. Lower leaves reniform to deltate-ovate, as wide as long, lobed or coarsely angular-dentate.
  2. Leaves glaucous on the lower surface; lower leaves deltate-ovate; stem not conspicuously grooved .... 1. *A. atriplicifolium*
  2. Leaves green on the lower surface; lower leaves reniform; stem conspicuously grooved .... 3. *A. reniforme*
1. Lower leaves lance-ovate, longer than wide, entire, shallowly dentate, or crenate .... 2. *A. plantagineum*

1. **Arnoglossum atriplicifolium** (L.) H. Robins. Indian Plantain. June–Oct. Moist woods, dry woods, prairies, dune slopes, occasional; throughout the state. *Cacalia atriplicifolia* L.

2. **Arnoglossum plantagineum** Raf. Prairie Indian Plantain. June–Aug. Prairies, springy ground in marshes and bogs, not common; widely scattered in IL. *Cacalia plantaginea* (Raf.) Shinners.

3. **Arnoglossum reniforme** (Hook.) H. Robins. Great Indian Plantain. July–Sept. Woods, not common; scattered in most of IL except the ne. cos. *Cacalia muhlenbergii* (Sch.-Bip.) Fern.; *Arnoglossum muhlenbergii* (Sch.-Bip.) Fern.

14. **Artemisia** L.—Wormwood

1. Leaves white-tomentose or densely pubescent, at least on the lower surface.
  2. Receptacle woolly.
    3. Divisions of leaves oblong to lanceolate, 1.5–5.0 mm wide .... 2. *A. absinthium*
    3. Divisions of leaves linear to filiform, 1 mm wide or less .... 8. *A. frigida*
  2. Receptacle glabrous.
    4. Subshrubby perennials .... 10. *A. pontica*
    4. Annual or biennial herbs.
      5. Leaves lanceolate to linear.
        6. Leaves regularly serrate .... 11. *A. serrata*
        6. Leaves entire or irregularly few-toothed .... 9. *A. ludoviciana*
      5. Leaves pinnatifid.
        7. Leaves green and glabrous above, densely white-tomentose below .... 12. *A. vulgaris*
        7. Leaves pubescent below, but not densely white-tomentose .... 7. *A. forwoodii*
1. Leaves glabrous or nearly so.
  8. Leaves or their divisions linear to filiform, entire or irregularly lobed, the lower 3- to 5-lobed or pinnatifid but not sharply toothed.
    9. Phyllaries glabrous.
      10. Leaves mostly entire, occasionally irregularly lobed or the lower 3- to 5-lobed .... 6. *A. dracunculus*
      10. Leaves mostly pinnatifid.
        11. Stems glabrous .... 5. *A. caudata*
        11. Stems gray-pubescent .... 7. *A. forwoodii*
    9. Phyllaries pubescent .... 1. *A. abrotanum*
  8. Leaves or their divisions lanceolate, sharply toothed.
    12. Inflorescence dense, spicate .... 4. *A. biennis*
    12. Inflorescence loose, paniculate .... 2. *A. annua*

1. **Artemisia abrotanum** L. Southern Wormwood. Aug.–Sept. Native to Europe, Asia, and Africa; rarely persisting after cultivation in disturbed soil; Cook, Lake, and LaSalle cos.

2. **Artemisia absinthium** L. Common Wormwood; Absinthe. July–Sept. Native to Europe; occasionally escaped from cultivation into disturbed soil; scattered in IL.

3. **Artemisia annua** L. Annual Wormwood; Sweet Wormwood. Aug.–Oct. Native to Europe and Asia; naturalized in disturbed soil; occasional throughout IL.

4. **Artemisia biennis** Willd. Biennial Wormwood. Aug.–Oct. Native to the nw. U.S.; naturalized in disturbed soil; occasional throughout IL.

5. **Artemisia caudata** Michx. Sand Wormwood; Beach Wormwood. July–Oct. Sandy soil; occasional in the n. ½ of IL; also Jersey Co. *Artemisia campestris* L. ssp. *caudata* (Michx.) Hall. & Clem.

6. **Artemisia dracunculus** L. False Tarragon; Wild Tarragon. July–Sept. Prairies; rare in the n. ½ of IL. *Artemisia glauca* Pall. ex Willd.; *A. dracunculoides* Pursh.

7. **Artemisia forwoodii** S. Wats. Forwood's Wormwood; Gray Beach Wormwood. Aug.–Oct. Sandy dunes; Cook and Lake cos. *A. caudata* Michx. var. *calvens* Lunell.

8. **Artemisia frigida** Willd. Prairie Sagebrush; Fringed Sage. July–Sept. Native to the w. U.S.; adventive along a railroad; Cook Co.

9. **Artemisia ludoviciana** Nutt. Two varieties occur in Illinois:

a. Pubescence of upper surface of leaves early deciduous, the leaves becoming bright green .... 9a. *A. ludoviciana* var. *ludoviciana*

a. Pubescence of upper surface of leaves persistent, the leaves whitish green .... 9b. *A. ludoviciana* var. *gnaphalodes*

9a. **Artemisia ludoviciana** Nutt. var. **ludoviciana.** Western Mugwort. July–Sept. Introduced from the w. U.S.; naturalized in disturbed soil; occasional in the n. ⅗ of IL.

9b. **Artemisia ludoviciana** Nutt. var. **gnaphalodes** (Nutt.) Torr. & Gray. White Sage; Prairie Sage. July–Sept. Disturbed soil; occasional in the n. ⅗ of IL; also Jackson and St. Clair cos. *Artemisia gnaphalodes* Nutt.

10. **Artemisia pontica** L. Roman Wormwood. July–Oct. Native to Europe; adventive in disturbed areas; Cass, Lake, Morgan, and Winnebago cos.

11. **Artemisia serrata** Nutt. Serrate-leaved Sage; Serrate-leaved Mugwort; Saw-toothed Sagebrush. Aug.–Sept. Moist ground; occasional in the n. ½ of IL.

12. **Artemisia vulgaris** L. Three varieties occur in Illinois:

a. Leaves cleft at least halfway to the midvein, the teeth either acuminate or absent.

  b. Lobes of leaves with acuminate teeth .... 12a. *A. vulgaris* var. *vulgaris*

  b. Lobes of leaves entire .... 12b. *A. vulgaris* var. *glabra*

a. Leaves cleft only about ¼ the way to the midvein, with obtuse to subacute teeth .... 12c. *A. vulgaris* var. *latiloba*

12a. **Artemisia vulgaris** L. var. **vulgaris.** Common Mugwort. July–Oct. Native to Europe; adventive in disturbed areas; scattered in the n. ½ of IL.

12b. **Artemisia vulgaris** L. var. **glabra** Ledeb. Mugwort. July–Oct. Native to Europe; scattered in disturbed areas.

12c. **Artemisia vulgaris** L. var. **latiloba** Ledeb. Mugwort. July–Oct. Native to Europe; scattered in disturbed areas.

15. **Aster** L.—Aster

(See also Brachyactis, Doellingeria, Euryba, Ionactic, Oligoneuron, and Symphyotrichum.)

1. **Aster tataricus** L. f. Tartarian Aster. Sept.–Oct. Native to Asia; rarely escaped from cultivation; scattered in the c. cos. of IL.

16. **Bellis** L.—English Daisy

1. **Bellis perennis** L. English Daisy. Apr.–June. Native to Europe; rarely escaped from cultivation; disturbed soil; Champaign and McLean cos.

17. **Bidens** L.—Beggar's-tick; Sticktight

1. Rays white .... 1. *B. alba*
1. Rays yellow, or absent.
  2. Leaves simple and unlobed or pinnately 3-lobed, never completely pinnately compound.
    3. All leaves pinnately 3-lobed .... 12. *B. tripartita*
    3. Some or all leaves unlobed or undivided.
      4. Leaves sessile or connate; heads nodding in fruit.
        5. Rays up to 15 mm long; outer cypselae 3–6 mm long; inner cypselae 4–8 mm long .... 4. *B. cernua*
        5. Rays (10–) 15–25 mm long; outer cypselae 6–8 mm long; inner cypselae 8–10 mm long .... 9. *B. laevis*
      4. Leaves petiolate or winged to base; heads erect in fruit.
        6. Cypselae usually with 4 awns and tuberculate on midvein; lobes of disc flowers 5 .... 6. *B. connata*
        6. Cypselae usually with 3 awns and smooth on midvein; lobes of disc flowers 4 .... 5. *B. comosa*
  2. Most or all the leaves pinnately compound.
    7. Leaves bipinnate; cypselae linear, at least 8 times longer than broad .... 3. *B. bipinnata*
    7. Leaves once-pinnate; cypselae ovate to oblong, much less than 8 times longer than broad.
      8. Rays longer than the involucre; leaflets 3–7, deeply cleft or coarsely serrate.
        9. Cypselae 2 ½–4 times as long as wide, with ciliate margins; leaflets commonly 7 .... 11. *B. trichosperma*
        9. Cypselae up to 2 ½ times as long as wide, with scabrous, rarely ciliate margins; leaflets commonly 3 or 5.
          10. Outer phyllaries of the involucre 8–12 .... 2. *B. aristosa*
          10. Outer phyllaries of the involucre more than 12 .... 10. *B. polylepis*
      8. Rays shorter than the involucre or absent; leaflets 3 or 5, uncleft.
        11. Outer phyllaries of the involucre 2–5, not ciliate .... 7. *B. discoidea*
        11. Outer phyllaries of the involucre 6 or more, ciliate.
          12. Outer phyllaries of the involucre usually 6–10 .... 8. *B. frondosa*
          12. Outer phyllaries of the involucre more than 10 .... 13. *B. vulgata*

1. **Bidens alba** (L.) DC. White Bidens. July–Aug. Wet roadside ditch, very rare; Massac Co.

2. **Bidens aristosa** (Michx.) Britt. Three varieties may be distinguished in Illinois:

a. Awns present on the cypsela.
  b. Awns upwardly barbed .... 2a. *B. aristosa* var. *aristosa*
  b. Awns downwardly barbed .... 2b. *B. aristosa* var. *fritcheyi*
a. Awns absent or nearly so .... 2c. *B. aristosa* var. *mutica*

2a. **Bidens aristosa** (Michx.) Britt. var. **aristosa.** Swamp Marigold. Aug.–Oct. Wet ground; common in the s. ⅔ of IL, less common elsewhere.

2b. **Bidens aristosa** (Michx.) Britt. var. **fritcheyi** Fern. Swamp Marigold. Aug.–Oct. Wet ground; occasional and scattered throughout IL.

2c. **Bidens aristosa** (Michx.) Britt. var. **mutica** Gray. Swamp Marigold. Aug.–Oct. Wet ground; rare but scattered throughout the state.

3. **Bidens bipinnata** L. Spanish Needles. July–Oct. Disturbed soil, open woods; occasional in the s. ¾ of IL, rare or absent elsewhere.

4. **Bidens cernua** L. Nodding Bur Marigold; Sticktight. June–Oct. Wet ground, including bogs; occasional throughout IL. (Including var. *elliptica* Wieg. and var. *integra* Wieg.)

5. **Bidens comosa** (Gray) Wieg. Swamp Beggar's-tick. Sept.–Oct. Wet ground; common throughout IL.

6. **Bidens connata** Muhl. ex Willd. Swamp Beggar's-tick; Purple-stemmed Beggar's-tick. Aug.–Oct. Wet ground; occasional throughout IL. (Including var. *petiolata* [Nutt.] Farw.)

7. **Bidens discoidea** (Torr. & Gray) Benth. Swamp Beggar's-tick; Sticktight. Aug.–Oct. Swamps, on submerged logs; occasional in the s. ⅔ of IL; also DuPage Co.

8. **Bidens frondosa** L. Common Beggar's-tick; Sticktight. June–Oct. Marshes, swamps, wet prairies, wet disturbed soil; common throughout IL.

9. **Bidens laevis** (L.) BSP. Large-flowered Beggar's-tick. Aug.–Sept. Wet soil, very rare; Union Co.

10. **Bidens polylepis** S. F. Blake. Bur Marigold; Sticktight. Aug.–Oct. Marshes, swamps, wet ditches; occasional to common throughout IL. *Bidens aristosa* (Michx.) Britt. var. *retrorsa* (Sherff) Wunderlin.

11. **Bidens trichosperma** (Michx.) Britt. Two varieties occur in Illinois:

a. Cypselae more than 6 mm long; leaflets lance-linear, most of them more than 12 mm wide .... 11a. *B. trichosperma* var. *trichosperma*

a. Cypselae up to 6 mm long; leaflets linear, up to 12 mm wide .... 11b. *B. trichosperma* var. *tenuiloba*

11a. **Bidens trichosperma** (Michx.) Britt. var. **trichosperma.** Tall Swamp Marigold; Sticktight. June–Oct. Wet ground; scattered throughout IL. *Coreopsis trichosperma* Michx.; *Bidens coronata* (L.) Britt.

11b. **Bidens trichosperma** (Michx.) Britt. var. **tenuiloba** (Gray) Britt. Tall Swamp Marigold; Sticktight. June–Oct. Wet ground; common in the n. ⅔ of IL, occasional elsewhere. *Bidens coronata* (L.) Britt. var. *tenuiloba* (Gray) Sherff.

12. **Bidens tripartita** L. Three-parted Beggar's-tick. July–Oct. Wet ground; scattered throughout the state.

13. **Bidens vulgata** Greene. Tall Beggar's-tick; Sticktight. Aug.–Oct. Moist, usually disturbed soil; occasional throughout IL. (Including f. *puberula* [Wieg.] Fern.)

18. **Boltonia** L'Her.—False Aster

1. Leaves linear to oblanceolate, 5–20 mm wide; disc 6–10 mm wide.

2. Leaves not decurrent .... 1. *B. asteroides*

2. Leaves decurrent .... 2. *B. decurrens*

1. Leaves linear, 1–5 mm wide; disc 3–6 mm wide .... 3. *B. diffusa*

1. **Boltonia asteroides** (L.) L'Her. Two varieties are known from Illinois:

a. Phyllaries obovate to spatulate, with membranous margins 2.5–6.0 mm wide; rays 8–10 mm long .... 1a. *B. asteroides* var. *latisquama*

a. Phyllaries narrowly oblanceolate to oblanceolate, with membranous margins 1.0–2.5 mm wide; rays 10–15 mm long .... 1b. *B. asteroides* var. *recognita*

1a. **Boltonia asteroides** (L.) L'Her. var. **latisquama** (Gray) Cronq. White Doll's-daisy; False Aster. May–Oct. Moist ground, very rare; Cook Co. *Boltonia latisquama* Gray.

1b. **Boltonia asteroides** (L.) L'Her. var. **recognita** (Fern. & Grisc.) Cronq. White Doll's-daisy; False Aster. July–Oct. Moist ground, marshes, prairies, sometimes in standing water; common throughout IL. *Boltonia latisquama* Gray var. *recognita* Fern. & Grisc.; *Boltonia latisquama* Gray var. *microcephala* Fern. & Grisc.; *Boltonia recognita* (Fern. & Grisc.) G. N. Jones.

2. **Boltonia decurrens** (Torr. & Gray) A. Wood. Clasping-leaf Doll's-daisy; Decurrent False Aster. July–Oct. Moist ground, usually along the Illinois River, very rare. *Boltonia asteroides* (L.) L'Her. var. *decurrens* (Torr. & Gray) Engelm. ex Gray; *B. latisquama* Gray var. *decurrens* (Torr. & Gray) Fern. & Grisc. Reported as an adventive from Lake Co.

3. **Boltonia diffusa** Ell. Two varieties occur in Illinois:

a. Phyllaries subulate; peduncles filiform; plants with stolons ....3a. *B. diffusa* var. *diffusa*
a. Phyllaries linear-oblong; peduncles thickened; plants without stolons ....3b. *B. diffusa* var. *interior*

3a. **Boltonia diffusa** Ell. var. **diffusa.** Small-headed Doll's-daisy; Narrow-leaved False Aster. July–Sept. Moist ground, very rare; Alexander Co.

3b. **Boltonia diffusa** Ell. var. **interior** Fern & Grisc. Narrow-leaved False Aster. July–Sept. Moist or dry open ground; occasional in the s. ½ of IL. *Boltonia interior* (Fern. & Grisc.) G. N. Jones.

19. **Brachyactis** Ledeb.—Rayless Aster

1. **Brachyactis ciliata** (Ledeb.) Ledeb. Rayless Aster. July–Oct. Native to the w. U.S.; adventive in disturbed soil, particularly along highways where much salt has been applied during the winter; ne. cos.; also Henry Co. *Aster brachyactis* Blake.

20. **Brickellia** Ell.—False Boneset

1. **Brickellia eupatorioides** (L.) Shinners. Two varieties occur in Illinois:

a. Outer phyllaries with greatly prolonged tips, much longer than the inner phyllaries .... 1a. *B. eupatorioides* var. *corymbulosa*
a. Outer phyllaries not greatly prolonged, longer than or shorter than the inner phyllaries .... 1b. *B. eupatorioides* var. *texana*

1a. **Brickellia eupatorioides** (L.) Shinners var. **corymbulosa** (Torr. & Gray) Shinners. False Boneset. Aug.–Oct. Dry prairies, black oak savannas; occasional to common in the n. ¾ of IL, much less common and giving way to the next in the sw. cos. *Kuhnia eupatorioides* L. var. *corymbulosa* Torr. & Gray.

1b. **Brickellia eupatorioides** (L.) Shinners var. **texana** (Shinners) Shinners. False Boneset. July–Oct. Prairies, dry open woods, not common; apparently confined to the sw. cos. *Kuhnia eupatorioides* L. var. *texana* Shinners.

21. **Calyptocarpus** Less.

1. **Calyptocarpus vialis** Less. Carpet Daisy. July–Aug. Native to tropical America; rarely adventive and persisting in IL; DeKalb Co.

22. **Carduus** L.—Musk Thistle

1. Heads 1–several, erect.
  2. Heads 18–25 cm in diameter; leaf surface glabrous or with septate hairs .... 1. *C. acanthoides*
  2. Heads up to 18 cm in diameter; leaf surface densely hairy, with some nonseptate hairs .... 2. *C. crispus*
1. Head solitary, nodding .... 3. *C. nutans*

1. **Carduus acanthoides** L. Plumeless Thistle. May–Oct. Native to Europe and Asia; naturalized in pastures and other disturbed soil; n. ⅙ of IL.

2. **Carduus crispus** L. Curled Thistle. May–Oct. Native to Europe and Asia; disturbed soil; rare in IL.

3. **Carduus nutans** L. Musk Thistle; Nodding Thistle. May–Nov. Native to Europe; naturalized along railroads and highways and in pastures; occasional throughout IL. (Including var. *leiolepis* [Petrovic] Arenes.)

23. **Carthamus** L.—Safflower

1. **Carthamus tinctorius** L. Safflower. July–Aug. Native to Europe; rarely adventive in IL; Cook, DuPage, and Kendall cos.

24. **Centaurea** L.—Star Thistle; Knapweed

(See also Acroptilon, Amberboa, and Plectocephalus.)

1. Phyllaries spine-tipped.
  2. Corolla yellow.
    3. Central spine of phyllaries 10–25 mm long .... 12. *C. solstitialis*
    3. Central spine of phyllaries up to 10 mm long.
      4. Central spine of phyllaries divided, about 5 mm long..........1. *C. benedicta*
      4. Central spine of phyllaries undivided, 5–10 mm long......... 6. *C. melitensis*
  2. Corolla white, pink, or purple.
    5. Central spine of phyllaries 1.5–3.0 mm long .... 4. *C. diffusa*
    5. Central spine of phyllaries 10–25 mm long .... 2. *C. calcitrapa*
1. Phyllaries not spine-tipped.
  6. Plants annual; cypselae 4–5 mm long...3. *C. cyanus*
  6. Plants perennial; cypselae up to 4 mm long, or 5–6 mm long.
    7. Corolla 25–45 mm long; cypselae 5–6 mm long..........................8. *C. montana*
    7. Corolla 15–18 (–25) mm long; cypselae 2.5–3.5 mm long.
      8. Inner phyllaries pectinately fringed along the margins.
        9. Corolla up to 25 mm long; cypselae 3.0–3.5 mm long ...... 13. *C. stoebe*
        9. Corolla 15–18 mm long; cypselae 2.5–3.0 mm long.
          10. Heads with only disc flowers; pappus black ....9. *C. nigra*
          10. Heads with ray flowers; pappus not black.
            11. Involucre as wide as high; phyllaries brown ....7. *C. X monctonii*
            11. Involucre longer than wide; phyllaries blackish ....10. *C. nigrescens*
      8. Inner phyllaries dentate or shallowly lobed along the margins.
        12. Corolla 15–18 mm long; cypselae 2.5–3.0 mm long ....5. *C. jacea*
        12. Corolla 20–25 mm long; cypselae 3–4 mm long ....11. *C phrygia*

1. **Centaurea benedicta** (L.) L. Blessed Thistle. May–Sept. Native to Europe; rarely escaped from cultivation; Champaign Co. *Cnicus benedictus* L.

2. **Centaurea biebersteinii** DC. Spotted Knapweed. July–Oct. Native to Europe; adventive along railroads, pastures, other disturbed soil; occasional to common in the n. ½ of IL, uncommon elsewhere. *Centaurea maculosa* Lam.

3. **Centaurea cyanus** L. Bachelor's Buttons; Corn-flower. July–Sept. Native to Europe; occasionally escaped from cultivation into disturbed soil; scattered throughout IL.

4. **Centaurea diffusa** Lam. Spreading Star Thistle. July–Sept. Native to Europe; naturalized in disturbed soil; occasional in the n. ½ of IL.

5. **Centaurea jacea** L. Brown Knapweed. July–Oct. Native to Europe; naturalized in disturbed soil; occasional in the n. ½ of IL.

6. **Centaurea melitensis** L. Maltese Star Thistle. July–Sept. Native to Europe, Asia, and Africa; escaped from cultivation into disturbed soil; Menard Co.

7. **Centaurea X monctonii** C. E. Britt. Meadow Knapweed. July–Sept. Native to Europe; disturbed soil; rare in IL. (Keim, 2011.)

8. **Centaurea montana** L. Mountain Corn-flower. July–Sept. Native to Europe; disturbed soil; Cook Co.

9. **Centaurea nigra** L. Black Knapweed. July–Sept. Native to Europe; naturalized in disturbed soil; n. ⅙ of IL.

10. **Centaurea nigrescens** Willd. Tyrolean Knapweed. July–Sept. Native to Europe; naturalized in disturbed soil; scattered in IL. *Centaurea dubia* Suter; *C. vochinensis* Bernh. ex Reich.

11. **Centaurea phrygia** L. Wig Knapweed. July–Sept. Native to Europe; disturbed soil; DuPage Co.

12. **Centaurea solstitialis** L. Yellow Star Thistle. July–Sept. Native to Europe; rarely naturalized in IL; Jackson Co.

13. **Centaurea stoebe** L. ssp. micranthos (S. G. Gmel. ex Gugler) Hayek. Spotted Knapweed. July–Oct. Native to Europe; adventive along railroads, pastures, other

disturbed soil; occasional to common in the n. ½ of IL, uncommon elsewhere. *Centaurea maculosa* Lam.; *Centaurea biebersteinii* DC., misapplied.

25. **Chamaemelum** Mill.—Chamomile

1. **Chamaemelum nobile** (L.) All. Garden Chamomile; Western Chamomile; Low Chamomile. June–Sept. Native to Europe; rarely escaping from cultivation; scattered in IL. *Anthemis nobilis* L.

26. **Chondrilla** L.—Skeletonweed

1. **Chondrilla juncea** L. Rush Skeletonweed. July–Sept. Native to Europe, Asia, and n. Africa; adventive in disturbed soil; rare in IL.

27. **Chrysopsis** (Nutt.) Ell.—Golden Aster

1. Leaves with few to several teeth; plants with rhizomes ....1. *C. camporum*
1. Leaves entire or nearly so; plants without rhizomes................................2. *C. villosa*

1. **Chrysopsis camporum** Greene. Prairie Golden Aster. June–Sept. Sandy soil; occasional throughout IL. *Heterotheca camporum* (Greene) Shinners; *Heterotheca villosa* (Pursh) Nutt. var. *camporum* (Greene) Wunderlin.

2. **Chryopsis villosa** (Pursh) Nutt. Hairy Prairie Golden Aster. June–Sept. Sandy soil; scattered in IL. *Heterotheca villosa* (Pursh) Shinners.

28. **Cichorium** L.—Chicory

1. **Cichorium intybus** L. Chicory. June–Nov. Native to Europe; naturalized along roads and in other disturbed soil; common throughout IL. Rose-flowered forms may be known as f. *roseum* Newm.; white-flowered forms may be known as f. *album* Newm.

29. **Cirsium** Mill.—Thistle

1. Involucre up to 1.5 (–2.0) cm high; plants with creeping rootstocks; many of the heads unisexual.. 2. *C. arvense*
1. Involucre more than 1.5 cm high; plants without creeping rootstocks; heads all bisexual.
  2. Phyllaries sticky, not spine-tipped, or minutely spine-tipped ....6. *C. muticum*
  2. Phyllaries not sticky, at least the outer ones spine-tipped.
    3. Leaves decurrent, forming a winged stem................................. 10. *C. vulgare*
    3. Leaves not decurrent; stem unwinged.
      4. Stems densely white-tomentose, even at maturity; leaves sparsely tomentose on the upper surface, at least when young.
        5. Flowers white; all cauline leaves decurrent; plants without horizontal runners ....7. *C. pitcheri*
        5. Flowers purple (rarely white); usually not all cauline leaves decurrent; plants with horizontal runners.
          6. Cypselae 3–5 mm long; tube of corolla up to 15 mm long; plants with shallow horizontal runners; none of the cauline leaves clasping ....5. *C. flodmanii*
          6. Cypselae 6–7 mm long; tube of corolla up to 28 mm long; plants with deep horizontal runners; some of the cauline leaves clasping. .... 9. *C. undulatum*
      4. Stems glabrous or pubescent but not white-tomentose; leaves never tomentose on the upper surface.
        7. Leaves densely white-tomentose on the lower surface.
          8. Involucre 1.5–2.5 cm high; cypselae 3–4 mm long .... 3. *C. carolinianum*
          8. Involucre 2.5–3.5 cm high; cypselae 4–6 mm long.
            9. Leaves deeply pinnatifid .... 4. *C. discolor*
            9. Leaves toothed or shallowly pinnatifid............... 1. *C. altissimum*
        7. Leaves gray-hairy beneath, becoming glabrate, never densely white-tomentose .... 8. *C. pumilum*

1. **Cirsium altissimum** (L.) Spreng. Tall Thistle. Aug.–Sept. Woods, disturbed soil; occasional to common throughout IL.

2. **Cirsium arvense** (L.) Scop. Two varieties have been found in Illinois:

a. Leaves merely toothed or shallowly lobed, weakly prickly .... 2a. *C. arvense* var. *arvense*
a. Leaves deeply pinnatifid, strongly prickly...................... 2b. *C. arvense* var. *horridum*

2a. **Cirsium arvense** (L.) Scop. var. **arvense.** Canada Thistle. May–Oct. Native to Europe; adventive in disturbed soil; common throughout IL. (Including var. *integrifolium* Wimm. & Grab. and var. *vestitum* Wimm. & Grab.)

2b. **Cirsium arvense** (L.) Scop. var. **horridum** Wimm. & Grab. Prickly Canada Thistle. May–Oct. Native to Europe; adventive in disturbed soil; common throughout IL.

3. **Cirsium carolinianum** (Walt.) Fern. & Schub. Carolina Thistle. June–July. Dry, open woods; extreme se. cos. of IL.

4. **Cirsium discolor** (Muhl. ex Willd.) Spreng. Pasture Thistle; Field Thistle. Aug.–Oct. Fields, open woods, degraded prairies, pastures, roadsides; common throughout IL. Plants with white flower heads may be known as f. *albiflorum* (Britt.) House.

5. **Cirsium flodmanii** (Rydb.) Arthur. Prairie Thistle. June–Sept. Prairies, very rare; Cook Co.

6. **Cirsium muticum** Michx. Swamp Thistle; Fen Thistle. Aug.–Oct. Calcareous fens; occasional in the n. ½ of IL, extending s. to Wabash Co.

7. **Cirsium pitcheri** (Torr.) Torr. & Gray. Dune Thistle. June–July. Sand dunes near Lake Michigan, very rare; Cook and Lake cos.

8. **Cirsium pumilum** (Nutt.) Spreng. var. **hillii** (Canby) C. Boivin. Hill's Thistle. June–Aug. Dry prairies; n. ⅗ of IL but becoming rare. *Cirsium hillii* (Canby) Fern.

9. **Cirsium undulatum** (Nutt.) Spreng. Wavy-leaved Thistle. July–Aug. Native to the w. U.S.; found in 1904 in Will Co. by Homer Skeele.

10. **Cirsium vulgare** (Savi) Tenore. Bull Thistle. June–Sept. Native to Europe; naturalized in fields and disturbed soil; common throughout IL.

30. **Conoclinium** DC.—Mistflower

1. **Conoclinium coelestinum** (L.) DC. Mistflower. July–Oct. Moist ground; common in the s. ⅔ of IL; rare northward where it may be introduced. White-flowered and reddish purple-flowered plants rarely occur. *Eupatorium coelestinum* L.

31. **Conyza** Less.

1. Stems simple or nearly so up to the inflorescence; rays white; plants up to 3.5 m tall .... 1. *C. canadensis*
1. Stems diffusely branched from near the base; rays purplish; plants rarely more than 25 cm tall .... 2. *C. ramosissima*

1. **Conyza canadensis** (L.) Cronq. Horseweed; Mare's-tail. May–Nov. Disturbed soil, very common; in every co. *Erigeron canadensis* L.

2. **Conyza ramosissima** Cronq. Dwarf Fleabane. June–Sept. Disturbed soil; occasional throughout IL. *Erigeron divaricatus* Michx.

32. **Coreopsis** L.—Tickseed

1. Leaves undivided or with 1–2 short lateral lobes.
  2. Leaves mostly basal, linear to oblanceolate to obovate.
    3. Stems glabrous or nearly so; leaves glabrous; nodes on aerial stems 1–5, the internodes up to 5 cm long .... 4. *C. lanceolata*
    3. Stems spreading-villous; leaves pubescent; nodes on aerial stems more than 5, the internodes 6 cm long or longer .... 2. *C. crassifolia*
  2. Leaves developed to middle of stem or higher, ovate to elliptic-lanceolate .... 6. *C. pubescens*
1. Leaves 3- to 5-lobed or -divided.
  4. Leaves sessile, deeply 3-lobed to or below the middle .... 5. *C. palmata*
  4. Leaves usually petiolate, divided into 3–5 segments.
    5. Ligules of ray flowers reddish brown at base or throughout; disc flowers reddish brown.

6. Cypselae linear, wingless; leaf segments linear to linear-lanceolate............ ....7. *C. tinctoria*
6. Cypselae obovate, cartilaginous-margined; leaf segments lanceolate to orbicular.................... 1. *C. basalis*

5. Ligules of ray flowers yellow; disc flowers yellow or reddish brown.
7. Leaf segments elliptic-lanceolate; cypselae 5–7 mm long; disc flowers yellow or reddish brown ....8. *C. tripteris*
7. Leaf segments linear-filiform to linear-lanceolate; cypselae 1–4 mm long; disc flowers yellow ....3. *C. grandiflora*

1. **Coreopsis basalis** (Otto & A. Dietr.) S. F. Blake. June–Sept. Coreopsis. Native to the se. U.S.; rarely escaped from cultivation into disturbed soil; Lake Co.

2. **Coreopsis crassifolia** Ait. Hairy Sand Coreopsis. Apr.–Aug. Sandy soil, dry prairies; scattered in IL. *Coreopsis lanceolata* L. var. *villosa* Michx.

3. **Coreopsis grandiflora** Hogg ex Sweet. Two varieties occur in Illinois:

a. Leaflets linear-lanceolate ....3a. *C. grandiflora* var. *grandiflora*
a. Leaflets linear-filiform ....3b. *C. grandiflora* var. *harveyana*

3a. **Coreopsis grandiflora** Hogg ex Sweet var. **grandiflora.** Large-flowered Coreopsis. June–Aug. Native to the se. U.S.; adventive in disturbed soil; less common than the following but scattered in IL.

3b. **Coreopsis grandiflora** Hogg ex Sweet var. **harveyana** (Gray) Sherff. Large-flowered Coreopsis. June–Aug. Native to the se. U.S.; adventive in disturbed soil; scattered throughout IL.

4. **Coreopsis lanceolata** L. Sand Coreopsis. May–Aug. Sandy soil, rocky soil, dry prairies; scattered throughout IL.

5. **Coreopsis palmata** Nutt. Prairie Coreopsis. June–Aug. Prairies, open woods, black oak savannas; common in the n. ¾ of IL, occasional elsewhere.

6. **Coreopsis pubescens** Ell. Star Tickseed. June–Sept. Dry woods, not common; apparently only in the s. ¼ of IL.

7. **Coreopsis tinctoria** Nutt. Golden Coreopsis. July–Sept. Native to the w. U.S.; escaped from gardens into disturbed soil; scattered throughout IL.

8. **Coreopsis tripteris** L. Three varieties have been found in Illinois:

a. All leaves compound.
b. Lower surface of leaves and outer phyllaries glabrous or nearly so.................... ....8a. *C. tripteris* var. *tripteris*
b. Lower surface of leaves and outer phyllaries densely pubescent.................... ....8b. *C. tripteris* var. *deamii*

a. Most or all the leaves simple ....8c. *C. tripteris* var. *intercedens*

8a. **Coreopsis tripteris** L. var. **tripteris.** Tall Tickseed; Tall Coreopsis. July–Oct. Dry woods, prairies, sandy woods; common throughout IL.

8b. **Coreopsis tripteris** L. var. **deamii** Standl. Tall Tickseed; Tall Coreopsis. July–Oct. Dry woods, prairies; scattered in IL.

8c. **Coreopsis tripteris** L. var. **intercedens** Sherff. Tall Tickseed; Tall Coreopsis. July–Oct. Dry woods, prairies; scattered in IL.

33. **Cosmos** Cav.—Cosmos

1. Rays pink, red, or white; leaf segments linear to lanceolate.................... 1. *C. bipinnatus*
1. Rays orange or golden yellow; leaf segments lanceolate to elliptic.................... ....2. *C. sulphureus*

1. **Cosmos bipinnatus** Cav. Common Cosmos. Aug.–Oct. Native to Mexico; adventive in disturbed areas; scattered in IL.

2. **Cosmos sulphureus** Cav. Yellow Cosmos. Aug.–Oct. Native to Mexico; adventive in disturbed areas; Will Co.

34. **Cota** J. Gay ex Gussone—Yellow Chamomile

1. **Cota tinctoria** (L.) J. Gay ex Gussone. Yellow Chamomile. June–Sept. Native to Europe; rarely escaped from cultivation; Cook, DuPage, Jo Daviess, Kendall, and Winnebago cos. *Anthemis tinctoria* L.

35. **Crepis** L.—Hawksbeard

1. Cypselae pale brown or tawny; inner phyllaries glabrous.
  2. Involucre 5–8 mm high; cypselae 1.5–2.5 mm long .... 1. *C. capillaris*
  2. Involucre 8–12 mm high; cypselae 4–6 mm long .... 2. *C. pulchra*
1. Cypselae dark purple-brown; inner phyllaries pubescent on the inner surface ....
.... 3. *C. tectorum*

1. **Crepis capillaris** (L.) Wallr. Smooth Hawksbeard. June–July. Native to Europe; adventive in disturbed soil; Cook, DuPage, and Lake cos.

2. **Crepis pulchra** L. Pretty Hawksbeard. May–June. Native to Europe and Asia; adventive in disturbed soil; Alexander and Jackson cos.

3. **Crepis tectorum** L. Narrow-leaved Hawksbeard. May–Aug. Native to Europe; adventive in disturbed soil; ne. cos; also Carroll and Macon cos.

36. **Cyclachaena** Fres.—Burweed

1. **Cyclachaena xanthifolia** (Nutt.) Fres. Burweed Marsh Elder. July–Sept. Moist disturbed areas; occasional in the n. ½ of IL. *Iva xanthifolia* Nutt.

37. **Doellingeria** Nees

1. Phyllaries pubescent; leaves pubescent on both surfaces; ray flowers 4–7 (–12) per head; disk flowers 8–15 (–30) per head .... 1. *D. pubens*
1. Phyllaries glabrous; leaves more or less glabrous on both surfaces; ray flowers 7–14 per head; disk flowers 30–40 per head .... 2. *D. umbellata*

1. **Doellingeria pubens** (Gray) Rydb. Hairy Flat-topped Aster. July–Oct. Calcareous fens, not common; n. ⅙ of IL. *Aster pubentior* Cronq.; *Aster umbellatus* Mill. var. *pubens* Gray; *Doellingeria umbellata* (Mill.) Nees ssp. *pubens* (Gray) A. Love & D. Love.

2. **Doellingeria umbellata** (Mill.) Nees. Flat-topped Aster. July–Oct. Calcareous fens, bogs, low ground; occasional in the n. ½ of IL, absent elsewhere. *Aster umbellatus* Mill.

38. **Dracopis** Cass.

1. **Dracopis amplexicaulis** (Vahl) Cass. Clasping Coneflower. July–Sept. Native to the s. and w. U.S.; rarely adventive in IL; Cook, DuPage, Greene, and Jackson cos. *Rudbeckia amplexicaulis* Vahl.

39. **Dyssodia** Cav.—Fetid Marigold

1. **Dyssodia papposa** (Vent.) A. Hitchc. Fetid Marigold. Sept.–Oct. Native to the s. and w. U.S.; adventive in disturbed soil and increasing rapidly in the ne. cos; scattered in IL.

40. **Echinacea** Moench—Purple Coneflower

1. Leaves lanceolate to linear, attenuate to base, all entire.
  2. Ray flowers 15–40 mm long... 1. *E. angustifolia*
  2. Ray flowers 40–90 mm long.
    3. Pollen white; rays 3–4 mm wide; cypselae glabrous....................... 2. *E. pallida*
    3. Pollen yellow; rays 4–7 mm wide; cypselae pubescent .... 4. *E. simulata*
1. Leaves ovate, rounded at base, serrate or dentate, the upper frequently entire........
.... 3. *E. purpurea*

1. **Echinacea angustifolia** DC. Narrow-leaved Purple Coneflower. June–Sept. Prairies, very rare; scattered in w. cos.

2. **Echinacea pallida** (Nutt.) Nutt. Pale Coneflower. May–Aug. Dry prairies, open woods; occasional throughout IL.

3. **Echinacea purpurea** (L.) Moench. Purple Coneflower; Broad-leaved Coneflower. July–Sept. Prairies, open woods, wooded floodplains; occasional throughout IL.

4. **Echinacea simulata** McGregor. Wavy-leaf Purple Coneflower. July–Aug. Open woods, dry prairies; not common but scattered in IL.

41. **Echinops** L.—Globe Thistle

1. **Echinops sphaerocephalus** L. Globe Thistle. July–Sept. Native to Europe; escaped into disturbed soil; ne. cos.; also Coles Co.

42. **Eclipta** L.

1. **Eclipta prostrata** (L.) L. Yerba de Tajo. July–Oct. Muddy shores, along streams, wet ditches, moist disturbed areas; common in the s. ¾ of IL, less common in the n. ¼ of IL, where it may be adventive. *Eclipta alba* (L.) Hassk.

43. **Elephantopus** L.—Elephant's-foot

1. **Elephantopus carolinianus** Raeusch. Carolina Elephant's-foot. Aug.–Sept. Dry woods; occasional in the s. ⅓ of IL.

44. **Erechtites** Raf.—Fireweed

1. **Erechtites hieracifolia** (L.) Raf. Three varieties occur in Illinois:

a. Most or all the leaves sessile or clasping at base, not reduced upwards .... 1a. *E. hieracifolia* var. *hieracifolia*
a. Some or all the leaves petiolate, not clasping, sometimes reduced upwards.
  b. Leaves rapidly reduced upwards....................... 1b. *E. hieracifolia* var. *intermedia*
  b. Leaves not rapidly reduced upwards.................... 1c. *E. hieracifolia* var. *praealta*

1a. **Erechtites hieracifolia** (L.) Raf. var. **hieracifolia.** Fireweed. Aug.–Oct. Moist woods, bogs, disturbed soil, recently burned areas; common throughout IL. *Erigeron hieracifolia* L.

1b. **Erectites hieracifolia** (L.) Raf. var. **intermedia** Fern. Fireweed. Aug.–Oct. Moist woods, bogs, disturbed soil; scattered throughout IL.

1c. **Erechtites hieracifolia** (L.) Raf. var. **praealta** (Raf.) Fern. Fireweed. Aug.–Oct. Moist woods, bogs, disturbed soil; scattered in IL.

45. **Erigeron** L.—Fleabane

1. Leaves broadly rounded at the sessile or clasping base.
  2. Heads few, 2.5–3.5 cm across; rays 50–100, about 1 mm wide ....3. *E. pulchellus*
  2. Heads several, 1.5–2.0 cm across; rays 150–200, about 0.5 mm wide .... 2. *E. philadelphicus*
1. Leaves tapering to an often petiolate base, never broadly sessile or clasping.
  3. Plants 0.3–1.5 m tall or taller; blades of lower leaves 2.5–15.0 cm long; pappus of ray flowers absent or simple.
    4. Cauline leaves many; basal leaves ovate, coarsely dentate; at least the middle of the stem with long, spreading hairs..............................1. *E. annuus*
    4. Cauline leaves few; basal leaves spatulate, entire or nearly so; middle part of stem with short, appressed hairs .... 4. *E. strigosus*
  3. Plants up to 0.4 m tall; blades of lower leaves 1–3 cm long; pappus of all flowers double ....5. *E. tenuis*

1. **Erigeron annuus** (L.) Pers. Daisy Fleabane; Annual Fleabane; White-top. June–Nov. Woods, pastures, fields, disturbed areas, common; in every co.

2. **Erigeron philadelphicus** L. Philadelphia Fleabane; Marsh Fleabane; Daisy Fleabane. Mar.–July. Wet meadows, prairies, fields, along rivers and streams, disturbed areas; common throughout IL. (Including f. *angustatus* Vict. & Rousseau and f. *scaturicola* [Fern.] Cronq.)

3. **Erigeron pulchellus** Michx. Robin's Plantain. Apr.–June. Open woods; occasional throughout IL. *Erigeron bellidifolius* Muhl. ex Willd.

4. **Erigeron strigosus** Muhl. Daisy Fleabane; White-top. May–Aug. Open woods, fields, dry prairies, disturbed areas; common throughout IL. Plants with the involucre 2–3 mm high may be called var. *beyrichii* (Fisch. & Mey.) Torr. & Gray.

5. **Erigeron tenuis** Torr. & Gray. Slender Fleabane. May–June. Native to the s. U.S.; adventive along a roadside; Union Co.

46. **Eupatorium** L.—Thoroughwort

(See also Conoclinium, Eutrochium, Fleischmannia, and Ageratina.)

1. Leaves pinnatifid, the segments filiform.. 2. *E. capillifolium*
1. Leaves simple, never filiform.
  2. Leaves distinctly petiolate...6. *E. serotinum*

2. Leaves sessile, subsessile, or perfoliate.
   3. Leaves perfoliate .... 4. *E. perfoliatum*
   3. Leaves sessile or subsessile.
      4. Leaves rounded or subcordate at the base .... 7. *E. sessilifolium*
      4. Leaves tapering to the base.
         5. Some of the leaves whorled, narrowly lanceolate, usually more than 6 times longer than wide .... 3. *E. hyssopifolium*
         5. Leaves not whorled, lanceolate to elliptic, less than 6 times longer than wide.
            6. Leaves serrate throughout .... 5. *E. X polyneuron*
            6. Leaves serrate only in the lower ⅔ of the leaf .... 1. *E. altissimum*

1. **Eupatorium altissimum** L. Tall Boneset; Tall Thoroughwort. Aug.–Oct. Woods, fields, pastures, prairies, disturbed soil; common throughout IL.

2. **Eupatorium capillifolium** (Lam.) Small. Dogfennel. Aug.–Sept. Old field, very rare; Alexander Co.

3. **Eupatorium hyssopifolium** L. Hyssop-leaved Boneset. Aug.–Oct. Dry open areas, rare; Pope Co.

4. **Eupatorium perfoliatum** L. Perfoliate Boneset; Common Boneset. July–Oct. Marshes, bogs, other wet ground; common throughout IL. Specimens with some of the leaves whorled may be known as f. *trifolium* Fassett; specimens with purple flower heads may be known as f. *purpureum* Britt.; specimens with truncate leaves may be known as f. *truncatum* (Muhl.) Fassett.

5. **Eupatorium X polyneuron** (F. J. Herm.) Wunderlin. Hybrid Eupatorium. Wet ground; scattered but not common in IL. Considered to be a hybrid between *E. perfoliatum* and *E. serotinum.*

6. **Eupatorium serotinum** Michx. Late Boneset. July–Oct. Moist open woods, pastures, disturbed areas; common throughout IL.

7. **Eupatorium sessilifolium** L. var. **brittonianum** Porter. Upland Boneset. Aug.–Oct. Woods; occasional in the s. ½ of IL, uncommon or rare elsewhere.

47. **Eurybia** (Cass.) Cass. in F. Cuv.—Bigleaf Aster

1. Petioles winged.
   2. Leaves more or less glabrous beneath, rather thin, with fragile petioles; branches of the inflorescence compact, not widely forking; cordate base of leaves strongly overlapping; heads about 2.5 cm across; involucre 8–10 mm high, narrowly campanulate; ray flowers about 8, 10–12 mm long; disc flowers about 25, red-brown; pappus bristles red-brown, 4–5 mm long .... 1. *E. chasei*
   2. Leaves spreading-hirsute beneath, firm, with rigid petioles; branches of the inflorescence widely forking; cordate base of leaves not overlapping; heads about up to 2 cm across; involucre 6–8 mm high, campanulate; ray flowers 9–20, 10–18 mm long; disc flowers up to 50, yellow; pappus bristles yellowish, 6–7 mm long .... 2. *E. furcata*
1. Petioles unwinged.
   3. Branches of inflorescence and phyllaries glandular; rays white but often tinged with purple .... 3. *E. macrophylla*
   3. Branches of inflorescence and phyllaries eglandular; rays white .... 4. *E. schreberi*

1. **Eurybia chasei** (G. N. Jones) Mohlenbr., comb. nov. (basionym: *Aster chasei* G. N. Jones). Chase's Aster. Aug.–Sept. Wooded slopes, very rare and restricted to Marshall, Peoria, and Tazewell cos.

2. **Eurybia furcata** (E. L. Burgess) G. L. Nesom. Forked Aster. Aug.–Oct. Wooded bluffs and slopes, shaded calcareous areas; uncommon in the ne. corner of IL; also Tazewell Co. *Aster furcatus* E. L. Burgess.

3. **Eurybia macrophylla** (L.) Cass. in F. Cuv. Bigleaf Aster. July–Oct. Dry open woods, slopes of wooded dunes, swampy forests; Boone, Cook, DuPage, and Lake cos. *Aster macrophyllus* L.

4. **Eurybia schreberi** (Nees) Nees. Schreber's Aster. July–Sept. Woods, rare; scattered in the n. ⅓ of IL. *Aster schreberi* Nees.

48. **Euthamia** Nutt.—Grass-leaved Goldenrod

1. Leaves up to 3 (–8) mm wide, usually 1-nerved, with numerous obvious glandular dots.
2. Involucre 3.0–4.7 mm high; leaves never more than 3 mm wide; disc flowers 3–22 .... 1. *E. caroliniana*
2. Involucre 4.5–6.0 mm high; leaves up to 8 mm wide; disc flowers up to 9 .... 3. *E. gymnospermoides*
1. Leaves 3–12 mm wide, some or all of them 3- or 5-nerved, with a few obscure glandular dots, or eglandular.
3. Stems and leaves often densely hirtellous; inner phyllaries oblong; ray flowers up to 35 per head; disc flowers up to 13 per head, the corolla 2.5–3.3 mm long .... 2. *E. graminifolia*
3. Stems and leaves glabrous; inner phyllaries linear; ray flowers up to 14 per head; disc flowers up to 6 per head, the corolla 3.3–4.4 mm long .... 4. *E. leptocephala*

1. **Euthamia caroliniana** (L.) Greene ex Porter & Britt. Carolina Grass-leaved Goldenrod. Aug.–Sept. Sandy soil, very rare; Cook and Lake cos. *Solidago remota* Greene; *Euthamia remota* (Greene) Friesn.; *Solidago graminifolia* (L.) Salisb. var. *remota* (Greene) S. K. Harris.

2. **Euthamia graminifolia** (L.) Nutt. Two varieties occur in Illinois:

a. Leaves and stems glabrous or nearly so .... 2a. *E. graminifolia* var. *graminifolia*
a. Leaves and stems hirtellous .... 2b. *E. graminifolia* var. *nuttallii*

2a. **Euthamia graminifolia** (L.) Nutt. var. **graminifolia** Grass-leaved Goldenrod. Aug.–Oct. Moist ground, fields, sandy shores; common throughout IL. *Solidago graminifolia* (L.) Salisb.

2b. **Euthamia graminifolia** (L.) Nutt. var. **nuttallii** (Greene) W. Stone. Grass-leaved Goldenrod. Aug.–Oct. Moist ground; scattered throughout IL. *Euthamia graminifolia* (L.) Salisb. var. *nuttallii* (Greene) Fern.; *Solidago graminifolia* (L.) Salisb. var. *nuttallii* (Greene) Fern.; *Euthamia nuttallii* Greene.

3. **Euthamia gymnospermoides** Greene. Viscid Grass-leaved Goldenrod. Aug.–Oct. Prairies, sandy soil; occasional in n. and w. IL. *Solidago gymnospermoides* (Greene) Fern.

4. **Euthamia leptocephala** (Torr. & Gray) Greene ex Porter & Britt. Mississippi Valley Grass-leaved Goldenrod. Aug.–Oct. Sandy woods, rare; s. 1/6 of IL. *Solidago leptocephala* Torr. & Gray.

49. **Eutrochium** Raf.—Joe-pye-weed

1. Heads 5- to 7-flowered; stems glaucous; inflorescence convex.
2. Stems hollow; leaves often more than 4 in a whorl .... 1. *E. fistulosum*
2. Stems pithy; leaves usually 4 in a whorl .... 3. *E. purpureum*
1. Heads 9- to 22-flowered; stems not glaucous; inflorescence more or less flat-topped .... 2. *E. maculatum*

1. **Eutrochium fistulosum** (Barratt) E. E. Lamont. Hollow Joe-pye-weed; Trumpetweed. July–Sept. Low, wet ground; occasional in the s. 1/2 of IL, less common or rare elsewhere. *Eupatorium fistulosum* Barratt; *Eupatoriadelphus fistulosus* (Barratt) R. M. King & H. Robins.

2. **Eutrochium maculatum** (L.) E. E. Lamont. Spotted Joe-pye-weed. June–Oct. Marshes, calcareous fens; common in n. IL, rare in s. IL. *Eupatorium maculatum* L.; *Eupatoriadelphus maculatus* (L.) R. M. King & H. Robins.

3. **Eutrochium purpureum** (L.) E. E. Lamont. Two varieties have been found in Illinois:

a. Leaves glabrous or nearly so on the lower surface .... 3a. *E. purpureus* var. *purpureus*
a. Leaves uniformly short-pubescent on the lower surface .... 3b. *E. purpureus* var. *holzingeri*

3a. **Eutrochium purpureum** (L.) E. E. Lamont var. **purpureum**. Purple Joe-pye-weed. June–Oct. Woods; common in the n. ¾ of IL, occasional elsewhere. *Eupatorium purpureum* L.; *Eupatoriadelphus purpureus* (L.) R. M. King & H. Robins.

3b. **Eutrochium purpureum** (L.) E. E. Lamont var. **holzingeri** (Rydb.) E. E. Lamont. Purple Joe-pye-weed. June–Oct. Woods; occasional in n. IL. *Eupatorium purpureum* L.; *Eupatoriadelphus purpureus* (L.) R. M. King & H. Robins. var. *holzingeri* (Rydb.) R. M. King & H. Robins.

50. **Fleischmannia** Sch.-Bip.

1. **Fleischmannia incarnata** (Walt.) R. M. King & H. Robins. Pink Thoroughwort. Aug.–Oct. Wet woods, swamps, rare; s. ⅙ of IL. *Eupatorium incarnatum* Walt.

51. **Gaillardia** Foug.—Gaillardia; Blanket-flower

1. Ray and disc flowers yellow (ray flowers sometimes purple near base); perennials.
  2. Heads up to 4 cm across; stems puberulent; some of the leaves clasping at the base .... 1. *G. aestivalis*
  2. Heads at least 5 cm across; stems hirsute; none of the leaves clasping at the base .... 2. *G. aristata*
1. Ray flowers yellow and purple; disc flowers purple; annuals .... 3. *G. pulchella*

1. **Gaillardia aestivalis** (Walt.) Rock var. **flavovirens** (Mohr) Cronq. Gaillardia. July–Oct. Prairies, very rare; Alexander Co. Not seen in IL since 1874. There has been conjecture that the specimen on which this report is made was not actually collected in IL.

2. **Gaillardia aristata** Pursh. Gaillardia. June–Aug. Native to the w. U.S.; adventive along a railroad; Jackson Co.

3. **Gaillardia pulchella** Foug. Blanket-flower. June–July. Native to the w. U.S.; rarely escaped from cultivation; scattered in IL.

52. **Galinsoga** Ruiz & Pavon—Peruvian Daisy

1. Pappus of disc flowers conspicuously fringed, not tapering to awn tips; pappus of ray flowers absent or minute; outer phyllaries 2–4 .... 1. *G. parviflora*
1. Pappus of disc flowers slightly fringed, tapering to awn tips; pappus of ray flowers well-developed; outer phyllaries 1–2 .... 2. *G. quadrisulcata*

1. **Galinsoga parviflora** Cav. Lesser Peruvian Daisy. June–Oct. Native to tropical America; naturalized in disturbed soil; occasional in the n. ½ of IL; also Jackson Co.

2. **Galinsoga quadrisulcata** Ruiz & Pavon. Peruvian Daisy. June–Nov. Native to tropical America; adventive in disturbed soil; occasional throughout IL. *Galinsoga ciliata* (Raf.) S. F. Blake.

53. **Gamochaeta** Wedd.—Cudweed

1. **Gamochaeta purpurea** (L.) Cabrera. Early Cudweed. May–July. Fields, open woods, occasional throughout IL, except for the ne. cos. *Gnaphalium purpureum* L.

54. **Gnaphalium** L.—Cudweed

(See also Gamochaeta and Pseudognaphalium.)

1. **Gnaphalium uliginosum** L. Low Cudweed. June–Sept. Native to Europe; rarely adventive in pastures; Cook and Lake cos.

55. **Grindelia** Willd.—Gumweed

1. Annuals; leaves serrulate to crenulate; phyllaries squarrose; cypselae 2–3 mm long .... 3. *G. squarrosa*
1. Perennials; leaves with bristle-tipped teeth; phyllaries not squarrose; cypselae 4–6 (–7) mm long.
  2. Pappus of 4 setiferous awns 4–8 mm long, equaling or longer than the disc corollas .... 1. *G. lanceolata*
  2. Pappus of 2–6 setiferous awns or scales 2–5 mm long, usually shorter than the disc corollas .... 2. *G. perennis*

1. **Grindelia lanceolata** Nutt. Narrow-leaved Gumweed; Spiny-toothed Gumweed. June–Sept. Sandy roadsides, very rare; Alexander Co.

2. **Grindelia perennis** A. Nelson. Perennial Gum-plant. July–Sept. Native to the s. and sw. U.S.; adventive in disturbed sandy soil; Alexander Co. *Grindelia squarrosa* (Pursh) Dunal var. *quasiperennis* Lunell; *Grindelia hirsutula* Hook. & Arn., misapplied.

3. **Grindelia squarrosa** (Pursh) Dunal. Two varieties occur in Illinois:

a. Upper and middle leaves 2–4 times longer than wide, ovate to oblong .... 3a *G. squarrosa* var. *squarrosa*
a. Upper and middle leaves 5–8 times longer than wide, linear-oblong to oblanceolate .... 3b. *G. squarrosa* var. *serrulata*

3a. **Grindelia squarrosa** (Pursh) Dunal. var. **squarrosa.** Broad-leaved Gumweed; Tarweed. July–Sept, Native to the w. U.S.; naturalized in fields and disturbed soil; occasional in the n. ½ of IL, rare elsewhere.

3b. **Grindelia squarrosa** (Pursh) Dunal. var. **serrulata** (Rydb.) Steyerm. Broad-leaved Gum-plant; Tarweed. July–Sept. Native to the w. U.S.; adventive and scattered in the n. ½ of IL, rare elsewhere. *Grindelia serrulata* Rydb.

56. **Guizotia** Cass. in F. Cuv.—Guizotia

1. **Guizotia abyssinica** (L. f.) Cass. in F. Cuv. Guizotia. July–Aug. Native to Africa; rarely adventive in disturbed soils; one collection from IL, probably now extirpated. (Baagoe, 1974.)

57. **Gutierrezia** Lagasca—Snakeweed

1. **Gutierrezia texana** (DC.) Torr. & Gray. Snakeweed. July–Oct. Native to the sw. U.S.; rarely adventive in disturbed soil; Madison and St. Clair cos.

58. **Hasteola** Raf.—Sweet Indian Plantain

1. **Hasteola suaveolens** (L.) Raf. Sweet Indian Plantain. July–Sept. Wet ground, calcareous fens; not common in n. IL, rare in s. IL. *Cacalia suaveolens* L.

59. **Helenium** L.—Sneezeweed

1. Leaves linear to linear-filiform, not decurrent .... 1. *H. amarum*
1. Leaves linear-lanceolate to ovate, decurrent.
  2. Disc depressed-globose, yellow .... 2. *H. autumnale*
  2. Disc globose, purplish .... 3. *H. flexuosum*

1. **Helenium amarum** (Raf.) Rock. Bitterweed. Aug.–Oct. Fields and disturbed areas; occasional in s. IL, extending n. to Champaign and Pike cos., where probably introduced. *Helenium tenuifolium* Nutt.

2. **Helenium autumnale** L. Three varieties have been found in Illinois:

a. Leaves up to 7 times longer than wide.
  b. Rays, or most of them, more than 13 mm long .... 2a. *H. autumnale* var. *autumnale*
  b. Rays up to 13 mm long .... 2c. *H. autumnale* var. *parviflorum*
a. Leaves more than 7 times longer than wide .... 2b. *H. autumnale* var. *canaliculatum*

2a. **Helenium autumnale** L. var. **autumnale.** Yellow Sneezeweed. July–Nov. Wet meadows, marshes, wet ditches, calcareous fens; common throughout the state.

2b. **Helenium autumnale** L. var. **canaliculatum** (Lam.) Torr. & Gray. Yellow Sneezeweed. Aug.–Oct. Wet ground; scattered in IL.

2c. **Helenium autumnale** L. var. **parviflorum** (Nutt.) Fern. Yellow Sneezeweed. Aug.–Oct. Wet ground; not common but scattered in IL.

3. **Helenium flexuosum** Raf. Purple Sneezeweed. June–Sept. Fields, sandy meadows; occasional in the s. ½ of IL, rare in n. IL, where it is probably introduced. (Including *H. X polyphyllum* Small.)

60. **Helianthus** L.—Sunflower

1. Leaves with conspicuously ciliate margins; rays less than 1 cm long .....4. *H. ciliaris*
1. Leaves without conspicuously ciliate margins; rays normally more than 1 cm long.
  2. Disc red or purple.

3. Leaves linear, or rhombic and broadest near the middle.
    4. Stems and leaves glabrous, usually glaucous ....20. *H. salicifolius*
    4. Stems and leaves strigose, pilose, or hispid, usually not glaucous.
        5. Stems and leaves strigose or pilose ....1. *H. angustifolius*
        5. Stems and leaves hispid ....23. *H. subrhomboideus*
3. Leaves lanceolate to ovate, broadest below the middle.
    6. Receptacle flat or nearly so.
        7. Phyllaries ovate, abruptly contracted above; receptacular bracts not bearded at apex ....2. *H. annuus*
        7. Phyllaries lanceolate, tapering to the tip; receptacular bracts bearded at apex ....19. *H. petiolaris*
    6. Receptacle convex or low conic.
        8. Leaves tapering to short, thick petioles, or sessile .....18. *H. pauciflorus*
        8. Leaves abruptly slender petiolate ....21. *H. silphioides*

2. Disc yellow.
    9. Stems scapose, or with 3–5 pairs of cauline leaves smaller than the basal leaves...17. *H. occidentalis*
    9. Stems leafy to the inflorescence.
        10. Stems glabrous or nearly so below the inflorescence, often glaucous.
            11. Heads 1.5–3.0 cm across; disc 0.4–1.0 cm across; rays 1.0–1.5 cm long ....15. *H. microcephalus*
            11. Heads 4–9 cm across; disc 1.0–2.5 cm across; rays 2–4 cm long.
                12. Leaves thin, membranous ....6. *H. decapetalus*
                12. Leaves thick, firm.
                    13. Most or all the leaves alternate ....10. *H. grosseserratus*
                    13. Most leaves opposite except sometimes for the very uppermost.
                        14. Leaves sessile or on petioles up to 5 mm long...7. *H. divaricatus*
                        14. Leaves petiolate, the petioles more than 5 mm long ....22. *H. strumosus*
        10. Stems pubescent or scabrous throughout.
            15. Leaves gray-pubescent on both surfaces.
                16. Leaves densely gray-pubescent on both surfaces; cypselae villous, at least near tip; leaves lance-ovate to lanceolate, up to 18 cm long.
                    17. Ray flowers 20–22 per head; heads 2.0–2.5 cm across; leaves lance-ovate to lanceolate .... 16. *H. mollis*
                    17. Ray flowers 17–19 per head; heads 1.5–2.0 cm across; leaves lanceolate ....5. *H. X cinereus*
                16. Leaves sparsely gray-pubescent on both surfaces; cypselae usually glabrous; leaves lanceolate, over 20 cm long...3. *H. X brevifolius*
            15. Leaves not gray-pubescent.
                18. Petioles 2–8 cm long.
                    19. Phyllaries acute at tip ....12. *H. X laetiflorus*
                    19. Phyllaries acuminate to long-attenuate at tip...24. *H. tuberosus*
                18. Petioles up to 2 cm long or absent.
                    20. All leaves opposite except for the upper 2 or 3.
                        21. Main side veins of leaves joining midvein at base of blade; leaves broadest at base ....11. *H. hirsutus*
                        21. Main side veins of leaves joining midvein about 1 cm above base of blade; leaves broadest near middle ....8. *H. X doronicoides*
                    20. Leaves opposite on lower ½ of stem, alternate on upper ½.
                        22. Stems with spreading hairs.............................9. *H. giganteus*
                        22. Stems with appressed hairs or merely scabrous.

23. Stems with appressed hairs.................. 14. *H. maximilianii*
23. Stems merely scabrous.......................... 13. *H. X luxurians*

1. **Helianthus angustifolius** L. Narrow-leaved Sunflower. Aug.–Oct. Moist ground, rare; Massac, Pope, Pulaski, and White cos.

2. **Helianthus annuus** L. Common Sunflower; Garden Sunflower. July–Nov. Native to the w. U.S.; adventive into fields and along roads; occasional to common throughout IL.

3. **Helianthus X brevifolius** E. E. Wats. Short-leaved Sunflower. July–Sept. Prairies, open areas; scattered in c. and n. IL. This is the hybrid between *H. mollis* Lam. and *H. grosseserratus* M. Martens.

4. **Helianthus ciliaris** DC. Ciliate Sunflower. Aug.–Sept. Native to the w. U.S.; adventive in disturbed soil; St. Clair Co.

5. **Helianthus X cinereus** Torr. & Gray. Gray Sunflower. July–Sept. Prairies; scattered in the n. ⅔ of IL. This is the hybrid between *H. mollis* Lam. and *H. occidentalis* Riddell.

6. **Helianthus decapetalus** L. Pale Sunflower; Thin-leaved Sunflower. July–Oct. Dry woods, savannas; occasional throughout IL.

7. **Helianthus divaricatus** L. Woodland Sunflower. July–Oct. Open woods, black oak savannas, rocky glades; common throughout IL.

8. **Helianthus X doronicoides** Lam. Sunflower. July–Oct. Dry woods, rare; Sangamon Co. Considered to be a hybrid between *H. giganteus* L. and *H. mollis* Lam.

9. **Helianthus giganteus** L. Tall Sunflower. Aug.–Oct. Calcareous prairies and fens, rare; Cook, Kane, Kankakee, McHenry, Tazewell, and Winnebago cos.

10. **Helianthus grosseserratus** Martens. Sawtooth Sunflower. July–Oct. Prairies, edge of woods, wet ditches, wet meadows; common throughout IL.

11. **Helianthus hirsutus** Raf. Three varieties have been found in Illinois:

a. Some or all the leaves more than 2 cm wide.
  b. Stems with long-spreading hairs .... 11a. *H. hirsutus* var. *hirsutus*
  b. Stems with short, stiff hairs ....11c. *H. hirsutus* var. *trachyphyllus*
a. Leaves up to 2 cm wide .... 11b. *H. hirsutus* var. *stenophyllus*

11a. **Helianthus hirsutus** Raf. var. **hirsutus.** Hispid Sunflower. July–Sept. Open woods, fields; scattered throughout IL.

11b. **Helianthus hirsutus** Raf. var. **stenophyllus** Torr. & Gray. Narrow-leaved Hispid Sunflower. Aug.–Sept. Open woods; scattered in IL.

11c. **Helianthus hirsutus** Raf. var. **trachyphyllus** Torr. & Gray. Hispid Sunflower. July–Sept. Open woods, fields; occasional throughout IL.

12. **Helianthus X laetiflorus** Pers. Hybrid Sunflower. Aug.–Sept. Roadsides; scattered in IL. Considered to be a hybrid between *H. subrhomboideus* Rydb. and *H. tuberosus* L.

13. **Helianthus X luxurians** E. E. Wats. Hybrid Sunflower. Aug.–Sept. Roadsides, occasional. Considered to be a hybrid between *H. giganteus* L. and *H. grosseserratus* Martens.

14. **Helianthus maximilianii** Schrad. Maximilian's Sunflower. July–Aug. Native to the w. U.S.; naturalized in prairies and along roads; occasional in IL. *Helianthus dalyi* Britt.

15. **Helianthus microcephalus** Torr. & Gray. Small-headed Sunflower. Aug.–Oct. Dry open woods; occasional in s. IL, rare elsewhere.

16. **Helianthus mollis** Lam. Downy Sunflower. July–Sept. Prairies; common in most of IL. (Including var. *cordatus* S. Wats.)

17. **Helianthus occidentalis** Riddell. Western Sunflower. July–Oct. Sandy prairies, hill prairies; common in the n. ½ of IL, extending s. to St. Clair Co.

18. **Helianthus pauciflorus** Nutt. Prairie Sunflower. July–Oct. Dry prairies; occasional throughout IL. *Helianthus rigidus* (Cass.) Desf.

19. **Helianthus petiolaris** Nutt. Petioled Sunflower. June–Nov. Native to the w. U.S.; adventive in sandy soil in fields, along roads, and along railroads; occasional throughout IL.

20. **Helianthus salicifolius** A. Dietr. Willow-leaved Sunflower. Aug.–Oct. Native to the w. U.S.; adventive in disturbed soil; Cook Co.

21. **Helianthus silphioides** Nutt. Silphium Sunflower. July–Oct. Prairies, rare; Alexander and St. Clair cos.

22. **Helianthus strumosus** L. Pale-leaved Sunflower. July–Oct. Open woods; occasional to common throughout IL.

23. **Helianthus subrhomboideus** Rydb. Prairie Sunflower. July–Oct. Dry prairies; scattered in the n. ½ of IL. *Helianthus pauciflorus* Nutt. var. *subrhomboideus* (Rydb.) Cronq.

24. **Helianthus tuberosus** L. Two varieties occur in Illinois:

a. Upper and middle leaves alternate, moderately and inconspicuously pubescent on the lower surface with mostly appressed hairs .... 24a. *H. tuberosus* var. *tuberosus*

a. All leaves opposite except sometimes for the uppermost 2–3, densely pubescent on the lower surface with loose or spreading hairs .... 24b. *H. tuberosus* var. *subcanescens*

24a. **Helianthus tuberosus** L. var. **tuberosus.** Jerusalem Artichoke. Aug.–Oct. Moist ground; common throughout IL.

24b. **Helianthus tuberosus** L. var. **subcanescens** Gray. Jerusalem Artichoke. Aug.–Oct. Moist ground; occasional throughout IL.

61. **Heliopsis** Pers.—Ox-eye Sunflower

1. **Heliopsis helianthoides** (L.) Sweet. Two varieties occur in Illinois:

a. Leaves glabrous or scarcely scabrous on the upper surface .... 1a. *H. helianthoides* var. *helianthoides*

a. Leaves harshly scabrous on the upper surface .... 1b. *H. helianthoides* var. *scabra*

1a. **Heliopsis helianthoides** (L.) Sweet var. **helianthoides.** Ox-eye Sunflower. June–Oct. Open woods, prairies; common throughout IL.

1b. **Heliopsis helianthoides** (L.) Sweet var. **scabra** (Dunal) Fern. Ox-eye Sunflower. June–Oct. Open woods, prairies; occasional throughout IL.

62. **Helminotheca** Zinn.—Ox-tongue

1. **Helminotheca echioides** (L.) Holub. Bristly Ox-tongue. July–Sept. Native to Europe; rarely adventive in disturbed soil; Hancock Co. *Picris echioides* L.

63. **Heterotheca** Cass.—Golden Aster

1. **Heterotheca subaxillaris** (Lam.) Britt. & Rusby. Camphorweed; Golden Aster. July–Oct. Sandy soil in Alexander Co. where native; disturbed soil in n. IL, where introduced. *Heterotheca latifolia* Buckl.

64. **Hieracium** L.—Hawkweed

1. Plants with well-developed clusters of basal leaves at flowering time.
  2. Basal leaves rounded to cordate at base; pappus white .... 5. *H. murorum*
  2. Basal leaves tapering to base; pappus sordid.
    3. Plants without stolons .... 6. *H. piloselloides*
    3. Plants with stolons.
      4. Flowers red-orange .... 1. *H. aurantiacum*
      4. Flowers yellow .... 2. *H. caespitosum*
1. Plants without well-developed clusters of basal leaves at flowering time.
  5. Leaves more than 24 per stem .... 8. *H. umbellatum*
  5. Leaves less than 24 per stem.
    6. At least some of the pubescence more than 1 cm long .... 4. *H. longipilum*
    6. None of the pubescence 1 cm long.
      7. Heads with more than 40 flowers; cypselae truncate at summit .... 7. *H. scabrum*
      7. Heads with less than 40 flowers cypselae tapering to summit .... 3. *H. gronovii*

1. **Hieracium aurantiacum** L. Orange Hawkweed; Devil's Paint Brush. July–Sept. Native to Europe; adventive in grassy areas and sandy fields; occasional in the ne. cos.

2. **Hieracium caespitosum** Dumort. King Devil; Field Hawkweed. May–Aug. Native to Europe; adventive in disturbed areas in the ne. cos. *Hieracium pratense* L.

3. **Hieracium gronovii** L. Hairy Hawkweed. June–Oct. Dry open woods, black oak savannas; occasional in IL, but rare or absent in the nw. cos.

4. **Hieracium longipilum** Torr. ex Hook. Long-bearded Hawkweed. July–Sept. Fields, prairies, open woods; occasional throughout IL.

5. **Hieracium murorum** L. Golden Lungwort. June–Aug. Native to Europe; adventive in disturbed soil; Sangamon Co.

6. **Hieracium piloselloides** Villars. Glaucous King Devil. May–June. Native to Europe; adventive in disturbed soil; Lake Co. *Hieracium florentinum* All.

7. **Hieracium scabrum** Michx. Two varieties have been found in Illinois:

a. Pubescence on the stem up to 3 mm long .... 7a. *H. scabrum* var. *scabrum*
a. Pubescence on the stem more than 3 mm long .... 7b. *H. scabrum* var. *intonsum*

7a. **Hieracium scabrum** Michx. var. **scabrum.** Rough Hawkweed. Aug.–Sept. Black oak savannas, fields, dry woods; occasional throughout IL. A hybrid between *H. scabrum* and *H. kalmii* L., called *Hieracium X fassetti,* has been reported from IL.

7b. **Hieracium scabrum** Michx. var. **intonsum** Fern. & St. John. Rough Hawkweed. Aug.–Sept. Black oak savannas, fields, dry woods; occasional throughout IL.

8. **Hieracium umbellatum** L. Canada Hawkweed. Aug.–Sept. Dry woods, sand barrens; occasional in the n. 1/6 of IL, absent elsewhere. *Hieracium kalmii* L., misapplied; *Hieracium canadense* Michx. var. *fasciculatum* (Pursh) Fern.

65. **Hymenopappus** L'Her.

1. **Hymenopappus scabiosaeus** L'Her. Old Plainsman. May–June. Sand prairies, rare; Cass, Iroquois, Kankakee, and Mason cos.

66. **Hypochaeris** L.—Cat's-ear

1. Leaves glabrous or pubescent only on the midrib; outer cypselae beakless, the inner beaked; annuals .... 1. *H. glabra*
1. Leaves hispid; all cypselae beaked; perennials .... 2. *H. radicata*

1. **Hypochaeris glabra** L. Smooth Cat's-ear. May–Aug. Native to Europe and Asia; adventive in disturbed soil; Jackson Co.

2. **Hypochaeris radicata** L. Rough Cat's-ear. May–Aug. Native to Europe; adventive in disturbed soil; Champaign, Cook, DuPage, Kane, McHenry, and St. Clair cos.

67. **Inula** L.—Elecampane

1. **Inula helenium** L. Elecampane. July–Aug. Native to Europe; escaped from cultivation and adventive in fields, pastures, roadsides, and open woods; not common but scattered in IL.

68. **Ionactis** Greene

1. **Ionactis linariifolius** (L.) Greene. Flax-leaved Aster; Stiff Aster. July–Oct. Black oak savannas, sandy prairies, sandy barrens; occasional in the n. 1/2 of IL, extending s. to St. Clair Co. *Aster linariifolius* L.

69. **Iva** L.—Marsh Elder

1. **Iva annua** L. Marsh Elder; Sumpweed. Aug.–Oct. Moist ground; common in the s. 1/3 of IL, less common to rare elsewhere and possibly adventive in the ne. cos. *Iva ciliata* Willd.; *I. caudata* Small.

70. **Krigia** Schreb.—Dwarf Dandelion

(See also Serinia.)

1. Plants scapose, or at least with most of the leaves at or near the base of the plant.
  2. Head solitary; involucre 9–14 mm high; inner pappus with 15–20 bristles .... 2. *K. dandelion*
  2. Heads 2 or more; involucre 4–7 mm high; pappus with 5 bristles .... 3. *K. virginica*

1. Plants leafy to the summit .... 1. *K. biflora*

1. **Krigia biflora** (Walt.) Blake. False Dandelion. May–Sept. Open woods, prairies; common throughout IL. (Including ssp. *glandulifera* [Fern.] Johnson & Iltis.)

2. **Krigia dandelion** (L.) Nutt. Dwarf Dandelion; Potato Dandelion. Apr.–June. Open woods, exposed bluff tops; occasional in the s. ⅓ of IL.

3. **Krigia virginica** (L.) Willd. Dwarf Dandelion. Apr.–Aug. Sandy soil; occasional in the n. and w. cos., rare elsewhere.

71. **Lactuca** L.—Lettuce

(See also Mulgedium.)

1. Leaves broadly ovate .... 8. *L. sativa*
1. Leaves elliptic to linear-lanceolate, often deeply pinnatifid, usually not ovate.
    2. Leaves prickly on the midvein beneath.
        3. Involucre 15–22 mm high; cypselae black, 1- to 3-nerved on each face .... 5. *L. ludoviciana*
        3. Involucre 8–15 mm high; cypselae yellow-gray, several-nerved on each face. .... 9. *L. serriola*
    2. Leaves not prickly on the midvein beneath or, if with a few prickles, the leaves never more than 1 cm wide.
        4. Pappus brown or gray .... 1. *L. biennis*
        4. Pappus white or cream.
            5. Involucre 15–22 mm high.
                6. Flowers reddish or salmon-colored; leaves without a prickly margin .... 4. *L. hirsuta*
                6. Flowers yellow or blue; leaves often with a prickly margin .... 5. *L. ludoviciana*
            5. Involucre up to 15 mm high (rarely to 18 mm in *L. saligna* which has all leaves less than 1 cm wide).
                7. None of the cauline leaves more than 1 cm wide; beak of cypsela twice as long as the body .... 7. *L. saligna*
                7. Some or all the leaves more than 1 cm wide; beak of cypsela nearly absent to almost as long as the body.
                    8. Flowers yellow; beak of cypsela at least ½ as long as to equaling the body .... 2. *L. canadensis*
                    8. Flowers blue; beak of cypsela nearly absent or less than ½ as long as the body.
                        9. Some of the cypselae beakless or nearly so; pappus white; terminal lobe of most leaves broadly triangular .... 3. *L. floridana*
                        9. Cypselae with a beak ¼–½ as long as the body; pappus cream; terminal lobe not broadly triangular .... 6. *L. X morssii*

1. **Lactuca biennis** (Moench) Fern. Two forms occur in Illinois:

a. Some or all of the leaves lobed .... 1a. *L. biennis* f. *biennis*
a. None of the leaves lobed .... 1b. *L. biennis* f. *integrifolia*

1a. **Lactuca biennis** (Moench) Fern. f. **biennis.** Tall Blue Lettuce. Aug.–Sept. Open woods, floodplain woods, disturbed woods; occasional throughout the state.

1b. **Lactuca biennis** (Moench) Fern. f. **integrifolia** (Torr. & Gray) Fern. Tall Blue Lettuce. Aug.–Sept. Open woods, floodplain woods, disturbed woods; scattered in IL.

2. **Lactuca canadensis** L. Four varieties have been found in Illinois:

a. All but the lowermost leaves unlobed.
    b. Cauline leaves entire or nearly so .... 2a. *L. canadensis* var. *canadensis*
    b. Cauline leaves regularly denticulate .... 2d. *L. canadensis* var. *obovata*
a. All but sometimes the uppermost leaves pinnatifid.
    c. Lobes of leaves linear, falcate .... 2c. *L. canadensis* var. *longifolia*
    c. Lobes of leaves broadly obovate, straight or falcate .... 2b. *L. canadensis* var. *latifolia*

2a. **Lactuca canadensis** L. var. **canadensis.** Wild Lettuce. June–Sept. Dry woods, pastures, prairies, disturbed soil; occasional throughout IL.

2b. **Lactuca canadensis** L. var. **latifolia** Ktze. Wild Lettuce. June–Sept. Dry woods, pastures, prairies, disturbed soil; common throughout IL.

2c. **Lactuca canadensis** L. var. **longifolia** (Michx.) Farw. Wild Lettuce. June–Sept. Dry woods, pastures, prairies, disturbed soil; occasional throughout IL.

2d. **Lactuca canadensis** L. var. **obovata** Wieg. Wild Lettuce. June–Sept. Dry woods, pastures, prairies, disturbed soil; occasional throughout IL.

3. **Lactuca floridana** (L.) Gaertn. Two varieties occur in Illinois:

a. All except sometimes the uppermost leaves lobed ......3a. *L. floridana* var. *floridana*
a. All leaves unlobed, denticulate ....3b. *L. floridana* var. *villosa*

3a. **Lactuca floridana** (L.) Gaertn. var. **floridana.** Woodland Lettuce. July–Sept. Woods; common throughout IL.

3b. **Lactuca floridana** (L.) Gaertn. var. **villosa** (Jacq.) Cronq. Woodland Lettuce. July–Sept. Woods; occasional in IL.

4. **Lactuca hirsuta** Muhl. var. **sanguinea** (Bigel.) Fern. Hairy Lettuce. July–Sept. Dry woods, rare; in the s. 1/6 of IL.

5. **Lactuca ludoviciana** (Nutt.) DC. Western Wild Lettuce; Prairie Lettuce. July–Sept. Dry prairies, rare; in the n. 1/2 of IL, extending s. to Monroe Co. Both blue- and yellow-flowered plants may be found. An unnamed hybrid between *L. ludoviciana* (Nutt.) DC. and *L. canadensis* L. is known from Cook Co.

6. **Lactuca X morssii** Robins. Hybrid Lettuce. July–Aug. Woods, very rare. This is the reputed hybrid between *L. biennis* (Moench) Fern. and *L. canadensis* L.

7. **Lactuca saligna** L. Willow Lettuce. July–Oct. Native to Europe; adventive in disturbed areas; occasional to common throughout IL.

8. **Lactuca sativa** L. Garden Lettuce. July–Sept. Native to Europe and Asia; escaped from cultivation but not naturalized; disturbed soil; occasional but not persistent in IL.

9. **Lactuca serriola** L. Prickly Lettuce. July–Nov. Native to Europe; adventive in disturbed soil; common throughout IL. (Including var. *integrata* [Gren. & Godr.] Farw.) *Lactuca scariola* L.

72. **Lapsana** L.—Nipplewort

1. **Lapsana communis** L. Nipplewort. June–Sept. Native to Europe; adventive in disturbed soil; occasional in the n. 1/3 of IL.

73. **Leontodon** L.—Hawkbit

1. Pappus of all flowers with a single row of plumose bristles ....1. *L. autumnalis*
1. Pappus of 2 types, inner double, with plumose and setiform bristles, outer reduced to a short irregular crown .... 2. *L. saxatilis*

1. **Leontodon autumnalis** L. Fall Dandelion. June–Oct. Native to Europe and Asia; adventive in disturbed soil, rare; Champaign and Christian cos. *Oporinia autumnalis* (L.) D. Don.

2. **Leontodon saxatilis** L. Hawkbit. June–Sept. Native to Europe; adventive in disturbed soil, rare; Cook and DuPage cos. *Leontodon leysseri* (Wallr.) G. Beck; *Leontodon taraxicoides* (Villars) Marat.

74. **Leucanthemum** Mill.—Ox-eye Daisy

1. **Leucanthemum vulgare** Lam. Ox-eye Daisy. May–Aug. Native to Europe and Asia; adventive in fields and disturbed soil; common throughout IL. *Chrysanthemum leucanthemum* L.

75. **Liatris** Schreb.—Blazing-star

1. Heads few to numerous, each with more than 15 flowers.
  2. Phyllaries closely appressed; stems glabrous or nearly so ....2. *L. cylindracea*
  2. Phyllaries loose or squarrose, rarely somewhat appressed; stems usually pubescent throughout or at least densely so in the inflorescence.
    3. Pappus plumose; corolla lobes hairy on the inner surface.

4. Phyllaries loosely spreading; pubescence more or less appressed .... 10. *L. squarrosa*
4. Phyllaries squarrose; pubescence spreading .... 3. *L. hirsuta*
3. Pappus merely barbellate; corolla lobes usually glabrous within, although the corolla tube is hairy within from the middle to the base.
5. Flowers 25–80 per head.
6. Margins of phyllaries entire or nearly so........................... 8. *L. scariosa*
6. Margins of phyllaries erose or laciniate .... 3. *L. ligulistylis*
5. Flowers 15–25 per head.
7. Middle phyllaries with irregularly lacerate, scarious margins.. 1. *L. aspera*
7. Middle phyllaries with entire or merely erose, scarious margins.
8. Phyllaries and leaves with short, spreading hairs........... 6. *L. scabra*
8. Phyllaries and leaves glabrous or nearly so .... 10. *L. squarrulosa*
1. Heads more than 20, each with fewer than 15 flowers.
9. Phyllaries spreading to reflexed; stems pubescent .... 5. *L. pycnostachya*
9. Phyllaries appressed; stems glabrous or nearly so.
10. Phyllaries obtuse to subacute; involucre 7–10 (–11) mm high; corolla tube glabrous within; pappus merely barbellate .... 8. *L. spicata*
10. Phyllaries, or some of them, acuminate and mucronate; involucre (10–) 11–18 mm high; corolla tube pilose within; pappus plumose .... 4. *L. punctata*

1. **Liatris aspera** Michx. Two varieties occur in Illinois:

a. Stems and leaves short-hairy.. 1a. *L. aspera* var. *aspera*
a. Lower part of the stems and the leaves glabrous or nearly so .... 1b. *L. aspera* var. *intermedia*

1a. **Liatris aspera** Michx. var. **aspera.** Rough Blazing-star. July–Nov. Prairies, black oak savannas; occasional to common throughout IL. A hybrid between this plant and *L. spicata* (L.) Willd. is known as *Liatris X steelei* Gaiser. It is known from Lake Co. White-flowered plants may be called f. *benkei* (Macb.) Fern.

1b. **Liatris aspera** Michx. var. **intermedia** (Lunell) Gaiser. Blazing-star. July–Nov. Prairies; black oak savannas; occasional throughout IL.

2. **Liatris cylindracea** Michx. Cylindrical Blazing-star. July–Oct. Dry prairies, sand flats; occasional in the n. ¾ of IL, rare southward. This species hybridizes with *L. aspera* Michx. to form *Liatris X gladewitzii* (Farw.) Shinners. Known from Lake Co.

3. **Liatris hirsuta** Rydb. Hirsute Blazing-star. June–Sept. Dry woods; scattered in sw. IL. *Liatris squarrosa* (L.) Michx. var. *hirsuta* (Rydb.) Gaiser.

4. **Liatris punctata** Hook. var. **nebraskana** Gaiser. Dotted Blazing-star. Aug.–Oct. Native to the w. U.S.; adventive along a railroad; DuPage Co. Not seen in IL since 1925.

5. **Liatris pycnostachya** Michx. Prairie Blazing-star. Sept.–Oct. Prairies, marly fens; common throughout IL. This plant hybridizes with *L. squarrosa* (L.) Michx. to form *Liatris X ridgwayi* Standl. The hybrid known from Richland Co.

6. **Liatris scabra** (Greene) K. Schum. Blazing-star. Sept.–Oct. Prairies, open woods; occasional in the s. ⅖ of IL.

7. **Liatris scariosa** (L.) Willd. var. nieuwlandii (Lunell) E. G. Voss. Savanna Blazing-star. July–Sept. Savannas; occasional in c. IL; also Cook, DuPage, Lake, and Will cos. *Liatris ligulistylis* (A. Nels.) K. Schum.; *L. X nieuwlandii* (Lunell) Gaiser.

8. **Liatris spicata** (L.) Willd. Marsh Blazing-star. July–Sept. Prairies, sand flats, wet meadows; scattered throughout IL. White-flowered plants may be known as f. *albiflora* Britt.

9. **Liatris squarrosa** (L.) Michx. Blazing-star. June–Sept. Dry open woods, prairies; occasional in s. IL. White-flowered plants may be known as f. *alba* Evers & Thieret.

10. **Liatris squarrulosa** Michx. Blazing-star. Aug.–Oct. Dry woods, glades, rare; Alexander and Union cos.

76. **Machaeranthera** Nees—Tadoka Daisy

1. **Machaeranthera tanacetifolia** (Kunth in HBK.) Nees. Tadoka Daisy. Native to the sw. U.S.; adventive in disturbed soil; Jackson Co. *Aster tanacetifolia* Kunth.

77. **Matricaria** L.—Wild Chamomile

1. Heads rayless; disc corollas 4-lobed; plants pineapple-scented when crushed .... 2. *M. discoidea*
1. Heads with evident white rays; disc corollas 5-lobed; plants not pineapple-scented when crushed .... 1. *M. chamomilla*

1. **Matricaria chamomilla** L. German Chamomile. May–Oct. Native to Europe and Asia; adventive in disturbed soil; occasional in the s. ½ of IL, n. to Cook, Knox, and Woodford cos. *Matricaria recutita* L.

2. **Matricaria discoidea** DC. Pineapple-weed. Apr.–Sept. Native to the w. U.S.; adventive in disturbed areas; common throughout IL. *Matricaria matricarioides* (Less.) Porter.

78. **Megalodonta** Greene—Water Marigold

1. **Megalodonta beckii** (Torr.) Greene. Water Marigold. June–Aug. Ponds and deep, clear lakes, rare; Cook, Lake, and St. Clair cos. *Bidens beckii* Torr.

79. **Melanthera** Rohr

1. **Melanthera nivea** (L.) Small. Snow Squarestem. June–Oct. Floodplain woods, very rare; Massac and Pulaski cos. *Melanthera hastata* Michx.

80. **Mikania** Willd.—Climbing Hempweed

1. **Mikania scandens** (L.) Willd. Climbing Hempweed. July–Oct. Low woods, swamps, banks of streams, not common; s. IL; also Kankakee Co.

81. Mulgedium Cass. in F. Cuv.—Blue Lettuce

1. **Mulgedium pulchellum** (Pursh) G. Don in R. Sweet. Showy Blue Lettuce. July–Aug. Native to the w. U.S.; adventive in disturbed soil; ne. cos. *Lactuca pulchella* Pursh; *Lactuca tatarica* L. var. *pulchellum* (Pursh) Breitung.

82. **Nabalus** Cass.—Rattlesnake Root; White Lettuce

1. Phyllaries glabrous.
 2. Phyllaries 7–10; flowers 8–15 per head; pappus red-brown .... 1. *N. albus*
 2. Phyllaries 4–6; flowers 5–6 per head; pappus straw-colored or cinnamon-brown .... 2. *N. altissimus*
1. Phyllaries pubescent.
 3. Leaves petiolate; inflorescence corymbose-paniculate .... 4. *N. crepidineus*
 3. Leaves sessile or nearly so, except the basal; inflorescence narrow and elongate.
  4. Stems and leaves glabrous and usually glaucous; flowers purplish .... 5. *N. racemosus*
  4. Stems and leaves pubescent, not glaucous; flowers cream-colored .... 3. *N. asper*

1. **Nabalus albus** (L.) Hook. White Lettuce. Aug.–Nov. Woods, wooded dune slopes; occasional in the n. ⅔ of IL; also Pulaski and Union cos. *Prenanthes alba* L.

2. *Nabalus altissimus* (L.) Hook. Two varieties occur in Illinois:

a. Pappus straw-colored .... 2a. *N. altissimus* var. *altissimus*
a. Pappus cinnamon-colored .... 2b. *N. altissimus* var. *cinnamomea*

2a. **Nabalus altissimus** (L.) Hook. var. **altissimus.** Tall White Lettuce. Aug.–Oct. Woods; occasional in e. and s. IL; also Cook, Lake, and Will cos. *Prenanthes altissimus* L. var. *altissima.*

2b. **Nabalus altissimus** (L.) Hook. var. **cinnamomea** (Fern.) Mohlenbr., comb. nov. (basionym: *Prenanthes altissima* L. var. *cinnamomea* Fern.). Tall White Lettuce. Aug.–Oct. Woods; occasional in e. and s. IL; also Cook and Lake cos. *Prenanthes altissima* L. var. *cinnamomea* Fern.

3. **Nabalus asper** (Michx.) Torr. & Gray. Rough White Lettuce. Aug.–Oct. Dry prairies; occasional in the n. ⅔ of IL, less common southward. *Prenanthes asper* Michx.

4. **Nabalus crepidineus** (Michx.) DC. Great White Lettuce. Aug.–Oct. Mesic woods,

floodplain woods, banks of streams; occasional throughout IL. *Prenanthes crepidinea* Michx.

5. **Nabalus racemosus** (Michx.) Hook. Two varieties occur in Illinois:

a. Phyllaries 7–10; flowers 9–16 per head.................. 5a. *N. racemosus* var. *racemosus*
a. Phyllaries 10–14; flowers 17–26 per head............... 5b. *N. racemosus* var. *multiflorus*

5a. **Nabalus racemosus** (Michx.) Hook. var. **racemosus.** Rattlesnake Root. July–Oct. Prairies, moist soil; occasional in the n. ¾ of IL, less common southward. *Prenanthes racemosa* Michx. var. *racemosa.*

5b. **Nabalus racemosus** (Michx.) Hook. var. **multiflorus** (Cronq.) Mohlenbr., comb. nov. (basionym: *Prenanthes racemosa* Michx. var. *multiflora* (Cronq.) Cronq.). Rattlesnake Root. July–Oct. Prairies, moist soil; occasional in the n. ⅓ of IL. *Prenanthes racemosa* Michx. var. *multiflora* (Cronq.) Cronq.

83. **Nothocalais** (Gray) Greene—Prairie Dandelion

1. **Nothocalais cuspidata** (Pursh) Sch.-Bip. Prairie Dandelion. Apr.–June. Prairies, very rare; n. ½ of IL. *Agoseris cuspidata* (Pursh) D. Dietr.; *Microseris cuspidata* (Pursh) Sch.-Bip.

84. **Oligoneuron** Small—Flat-topped Goldenrod

1. Basal leaves ovate to elliptic, usually very scabrous.
2. Largest basal leaves up to 20 cm long; inner phyllaries lanceolate to oblong, glabrous or nearly so; ray flowers 4–7 (–12) per head; disc flowers 8–15 (–30) per head; cypselae glabrous .... 6. *O. rigidum*
2. Largest basal leaves up to 12 cm long; inner phyllaries linear, strigillose throughout; ray flowers 7–14 per head; disc flowers 30–40 per head; cypselae strigose throughout .... 2. *O. canescens*
1. Basal leaves linear to lanceolate, glabrous, except sometimes near apex.
3. Inflorescence puberulent; leaves conduplicate .... 5. *O. riddellii*
3. Inflorescence glabrous; leaves flat.
4. Ray flowers white .... 1. *O. album*
4. Ray flowers yellow.
5. Leaves usually serrate above the middle; rays up to 5 mm long .... 4. *O. ohiense*
5. Leaves entire or nearly so throughout; rays 5–8 mm long.......................... .... 3. *O. X lutescens*

1. **Oligoneuron album** (Nutt.) G. L. Nesom. Stiff Aster. June–Oct. Prairies, sandy soil; occasional in the n. ½ of IL, apparently absent elsewhere. *Aster ptarmicoides* (Nees) Torr. & Gray; *Solidago ptarmicoides* (Nees) Boivin.

2. **Oligoneuron canescens** Rydb. Stiff Goldenrod. July–Oct. Prairies; confined to the extreme ne. cos. of IL. *Solidago rigida* L. var. *humilis* Porter in T. C. Porter & J. M. Coult.; *Oligoneuron rigidum* (L.) Small var. *humile* (Porter) G. L. Nesom.

3. **Oligoneuron X lutescens** (Lindl.) G. L. Nesom. Yellow Stiff Aster. Aug.–Sept. Dry prairies, rare; Cook Co. *Aster X lutescens* (Lindl.) Torr. & Gray; *Solidago X lutescens* (Lindl.) Boivin. Considered to be a hybrid between *O. album* (Nutt.) G. L. Nesom and *O. riddellii* (Frank) Rydb.

4. **Oligoneuron ohiense** (Riddell) Small. Ohio Goldenrod. June–Oct. Calcareous fens, low sand flats, other moist soil; occasional in ne. IL, extending s. to Peoria and Woodford cos. *Solidago ohiensis* Riddell.

5. **Oligoneuron riddellii** (Frank) Rydb. Riddell's Goldenrod. Aug.–Nov. Prairies, calcareous fens, other moist ground; occasional in the n. ½ of IL, rare in the s. ½. *Solidago riddellii* Frank.

6. **Oligoneuron rigidum** (L.) Small. Two varieties occur in Illinois.

a. Outer phyllaries strigillose; leaves and stems hispid or strigose; branches of inflorescence open... 6a. *O. rigidum* var. *rigidum*
a. Outer phyllaries glabrous; leaves and stems glabrous or sparsely hirsute; branches of inflorescence contracted.......................... 6b. *O. rigidum* var. *glabratum*

6a. **Oligoneuron rigidum** (L.) Small var. **rigidum.** Stiff Goldenrod. July–Oct. Prairies; common throughout IL. except for the extreme s. tip of the state. *Solidago rigida* L.

6b. **Oligoneuron rigidum** (L.) Small var. **glabratum** (E. L. Braun) G. L. Nesom. Smooth Stiff Goldenrod. Aug.–Oct. Prairies, very rare; Jackson Co. *Solidago rigida* L. var. *glabratum* E. L. Braun.

85. **Onopordum** L.—Scotch Thistle

1. **Onopordum acanthium** L. Scotch Thistle; Cotton Thistle. June–Aug. Native to Europe; rarely adventive in disturbed soil; Champaign and Cook cos.

86. **Packera** A. Love & D. Love—Groundsel

1. Leaves chiefly basal, the basal ones crenulate, dentate, or entire, the cauline ones (if present) sometimes pinnatifid.
  2. Basal leaves with winged petioles .... 3. *P. obovata*
  2. Basal leaves without winged petioles.
    3. Basal leaves cordate or subcordate at the base.
      4. Basal leaves deeply cordate.. 1. *P. aurea*
      4. Basal leaves shallowly subcordate.................................... 6. *P. pseudaurea*
    3. Basal leaves tapering to the base, neither cordate nor subcordate.
      5. Leaves and stems, especially at the nodes, floccose-tomentose at maturity; peduncles tomentose.. 5. *P. plattensis*
      5. Leaves and stems glabrous or nearly so at maturity; peduncles glabrous or nearly so... 4. *P. paupercula*
1. Stems leafy to the summit; leaves all pinnatifid or coarsely sinuate-dentate .... 2. *P. glabella*

1. **Packera aurea** (L.) A. Love & D. Love. Golden Ragwort; Squaw-weed. Apr.–June. Wet ground; occasional throughout IL. *Senecio aureus* L.

2. **Packera glabella** (Poir.) C. Jeffrey. Butterweed. Apr.–June. Moist shady ground, along streams, floodplain woods, wet meadows; common in the s. ½ of IL, extending north to Peoria Co.; also DuPage and Will cos., where it is adventive. *Senecio glabellus* Poir.

3. **Packera obovata** (Muhl. ex Willd.) W. A. Weber & A. Love. Round-leaved Groundsel. Apr.–June. Rich woods, rocky outcrops, rare; scattered in the s. ¾ of IL. *Senecio obovatus* Muhl. ex Willd.

4. **Packera paupercula** (Michx.) A. Love & D. Love. Balsam Ragwort; Northern Ragwort. May–June. Wet prairies, moist sand flats, sedge meadows, open woods; common in the n. ½ of IL, rare in the s. ½. *Senecio pauperculus* Michx.

5. **Packera plattensis** (Nutt.) W. A. Weber & A. Love. Prairie Groundsel; Prairie Ragwort. May–June. Dry prairies, woods, fields; common in n. and w. IL, rare or absent elsewhere. *Senecio plattensis* Nutt.

6. **Packera pseudaurea** (Rydb.) W. A. Weber & A. Love var. **semicordata** (Mack. & Bush) Trock & T. M. Barkley. Western Heart-leaved Groundsel. Apr.–June. Wet ground, very rare; Cook and Lake cos. *Senecio pseudaureus* Rydb.

87. **Parthenium** L.—Feverfew

1. Leaves pinnately dissected.. 2. *P. hysterophorus*
1. Leaves serrate or dentate.
  2. Stems and leaves variously glabrous or variously pubescent but not hispid....... .... 3. *P. integrifolium*
  2. Stems and leaves hispid .... 1. *P. hispidum*

1. **Parthenium hispidum** Raf. Hispid Feverfew. June–Sept. Prairies, dry woods; scattered in IL. *Parthenium integrifolium* L. var. *hispidum* (Raf.) Mears.

2. **Parthenium hysterophorus** L. Santa Maria. Aug.–Oct. Native to tropical America; rarely adventive in disturbed areas; Cook Co. Not seen in IL since 1890.

3. **Parthenium integrifolium** L. American Feverfew; Wild Quinine. June–Sept. Prairies, dry woods, glades; common throughout IL.

88. **Petasites** Mill.- Sweet Coltsfoot

1. **Petasites hybridus** (L.) Gaertn., Mey., & Scherb. Sweet Coltsfoot. Apr.–May. Native to Europe; adventive in a wooded ravine; Lake Co.

89. **Picris** L.—Bitterweed

(See also Helminotheca.)

1. **Picris hieracioides** L. Cat's-ear. July–Sept. Native to Europe and Asia; rarely adventive in disturbed soil; Menard Co.

90. **Plectocephalus** D. Don in R. Sweet—Basket-flower.

1. **Plectocephalus americanus** (Nutt.) D. Don in R. Sweet. American Basket-flower. July–Sept. Native to the sw. U.S.; rarely escaped from cultivation; Lawrence and Wabash cos. *Centaurea americana* Nutt.

91. **Pluchea** Cass.—Marsh Fleabane; Stinkweed

1. Leaves not succulent, petiolate; phyllaries glabrous .... 1. *P. camphorata*
1. Leaves succulent, sessile or nearly so; phyllaries pubescent on the outer face......... .... 2. *P. odorata*

1. **Pluchea camphorata** (L.) DC. Camphor-weed; Stinkweed. July–Oct. Swamps and sloughs, marshes; occasional in s. IL, extending n. to DeWitt Co.
2. **Pluchea odorata** L. var. **succulenta** (Fern.) Cronq. Salt Marsh Fleabane. Aug.–Sept. Native to the e. coast of the U.S.; adventive in a ditch in Cook Co.

92. **Polymnia** L.—Leaf-cup

(See also Smallanthus.)

1. **Polymnia canadensis** L. Leaf-cup. June–Nov. Moist or dry woods, shaded limy springy slopes; common throughout IL. (Including f. *radiata* [Gray] Fassett.)

93. **Prionopsis** Nutt.—Prionopsis

1. **Prionopsis ciliata** (Nutt.) Nutt. Prionopsis. Aug.–Sept. Sandy soil along a roadside, very rare; Pulaski Co. *Haplopappus ciliata* (Nutt.) DC.; *Grindelia ciliata* Nutt.

94. **Pseudognaphalium** Kirpich—Cudweed

1. Leaves decurrent .... 1. *P. macounii*
1. Leaves not decurrent .... 2. *P. obtusifolium*

1. **Pseudognaphalium macounii** (Greene) Kartesz. Western Cudweed; Clammy Cudweed. July–Sept. Sandy soil, very rare; Clark Co. *Gnaphalium macounii* Greene; *Gnaphalium viscosum* HBK.; *Pseudognaphalium viscosum* (HBK.) W. A. Weber.
2. **Pseudognaphalium obtusifolium** (L.) Hilliard & Burtt. Sweet Everlasting; Cat's-foot; Old Field Balsam. July–Oct. Fields, pastures, prairies, open woods; common throughout IL; in every co. *Gnaphalium obtusifolium* L.

95. **Pyrrhopappus** DC.—False Dandelion

1. **Pyrrhopappus carolinianus** (Walt.) DC. Pale False Dandelion. May–June. Dry woods, prairies, roadsides; occasional to common in the s. ½ of IL.

96. **Ratibida** Raf.—Prairie Coneflower

1. Disk columnar, equaling or exceeding the rays, 2–4 times as long as wide .... 1. *R. columnifera*
1. Disk ellipsoid, shorter than the rays, 1–1 ½ times as long as wide........................... .... 2. *R. pinnata*

1. **Ratibida columnifera** (Nutt.) Wooton & Standl. Long-headed Coneflower. June–Nov. Native to the w. U.S.; adventive along railroads; scattered in the n. ⅔ of IL.
2. **Ratibida pinnata** (Vent.) Barnh. Drooping Coneflower; Yellow Coneflower; Gray-headed Coneflower. July–Oct. Prairies; common in the n. ¾ of IL, occasional elsewhere.

97. **Rudbeckia** L.—Coneflower; Black-eyed Susan

1. Disc greenish yellow; leaves deeply 5- to 9-cleft; plants usually glabrous .... 6. *R. laciniata*

1. Disc brown or purple; leaves unlobed or 3- (or 5-) lobed; plants usually pubescent.
  2. Lower cauline leaves 3- (or 5-) lobed or -parted.
    3. Plants with basal leafy offshoots; palea of receptacle glabrous; leaves rough-hairy on the lower surface .... 14. *R. triloba*
    3. Plants without basal leafy offshoots; palea of receptacle pubescent toward apex; leaves downy-pubescent on the lower surface ...... 11. *R. subtomentosa*
  2. Lower cauline leaves toothed or entire, not lobed or parted.
    4. Pappus absent; style branches elongate, acute.
      5. Heads up to 1.5 cm across; leaves nearly uniform; annual ...... 1. *R. bicolor*
      5. Heads more than 1.5 cm across; lower leaves larger than upper leaves; perennial.
        6. Basal leaves ovate to broadly elliptic, 4–7 cm wide; cauline leaves ovate to occasionally lanceolate, coarsely toothed, the lowest ones 3–6 cm wide .... 5. *R. hirta*
        6. Basal leaves oblanceolate to lanceolate, 1–4 (–5) cm wide; cauline leaves linear-lanceolate to oblanceolate, entire or finely serrate, the lowest ones up to 3 cm wide .... 9. *R. serotina*
    4. Pappus present, consisting of a minute, short crown, commonly toothed on the angles; style branches short, obtuse.
      7. Lobes of disc corollas reflexed; rays 4 cm long or longer .... 4. *R. grandiflora*
      7. Lobes of disc corollas ascending; rays 1–4 cm long.
        8. Cauline leaves entire, linear-spatulate .... 7. *R. missouriensis*
        8. Cauline leaves toothed, elliptic to ovate.
          9. Basal leaves about three times longer than wide; rays 5–20 (–25) mm long.
            10. Basal leaves 2.0–4.5 cm wide; stems villous-hirsute; paleae ciliate .... 3. *R. fulgida*
            10. Basal leaves 1–2 cm wide; stems glabrous or strigose; paleae ciliate only at the tip .... 13. *R. tenax*
          9. Basal leaves not more than twice as long as wide; rays 15–40 mm long.
            11. Upper cauline leaves noticeably smaller than lower cauline leaves.
              12. Rays 25–40 mm long; paleae eciliate .... 12. *R. sullivantii*
              12. Rays 15–25 mm long; paleae ciliate .... 8. *R. palustris*
            11. Upper cauline leaves not noticeably smaller than lower cauline leaves.
              13. Ray flowers 8–12 per head; basal leaves coarsely dentate..... .... 15. *R. umbrosa*
              13. Ray flowers 12–20 per head; basal leaves crenate or entire.
                14. Stems densely villous-hirsute; basal leaves up to 3.5 cm wide, coarsely crenate; phyllaries densely hairy; rays 15–25 mm long .... 2. *R. deamii*
                14. Stems glabrous or sparsely villous; basal leaves up to 6.5 mm wide, finely crenate to entire; phyllaries glabrous or sparsely hairy; rays 20–24 mm long ...... 10. *R. speciosa*

1. **Rudbeckia bicolor** Nutt. Small-headed Black-eyed Susan. June–Sept. Wet areas along streams and in woods, rare; Jackson and Union cos.

2. **Rudbeckia deamii** S. F. Blake. Deam's Coneflower. Aug.–Sept. Along streams, rare; se. cos. of IL. *Rudbeckia fulgida* Ait. var. *deamii* (S. F. Blake) Perdue.

3. **Rudbeckia fulgida** Ait. Orange Coneflower. July–Oct. Woods; scattered in the s. ½ of IL.

4. **Rudbeckia grandiflora** (Sweet) C. C. Gmel. ex DC. Large Black-eyed Susan. Aug. Native to the w. U.S.; adventive in a degraded prairie; DuPage Co.

5. **Rudbeckia hirta** L. Black-eyed Susan. June–Oct. Prairies, pastures, black oak savannas, fields; occasional in the e. ½ of the state.

6. **Rudbeckia laciniata** L. Goldenglow. July–Nov. Floodplain woods, along streams,

calcareous springy places; occasional to common throughout IL. Pubescent plants sometimes occur.

7. **Rudbeckia missouriensis** Engelm. ex Boynt. & Beadle. Missouri Coneflower. July–Oct. Prairies, glades, not common; Hardin, Monroe, and Randolph cos.

8. **Rudbeckia palustris** Eggert ex Boynt. & Beadle. Marsh Coneflower. July–Sept. Wet woods and fields; Jackson and Union cos. *Rudbeckia fulgida* Ait. var. *palustris* (Eggert ex Boynt. & Beadle) Perdue.

9. **Rudbeckia serotina** Nutt. Two varieties occur in Illinois:

a. Rays 1.0–3.5 cm long .... 9a. *R. serotina* var. *serotina*
a. Rays 3.5–5.0 mm long .... 9b. *R. serotina* var. *lanceolata*

9a. **Rudbeckia serotina** Nutt. var. **serotina** Black-eyed Susan. May–Oct. Prairies, pastures, open woods, roadsides; common throughout the state. *Rudbeckia hirta* L. var. *pulcherrima* Farw.; *Rudbeckia hirta* L. f. *pulcherrima* (Farw.) Fern. & Schub.

9b. **Rudbeckia serotina** Nutt. var. **lanceolata** (Bisch.) Fern. & Schub. Large-flowered Black-eyed Susan. July–Sept. Cherty woods, very rare; Union Co.

10. **Rudbeckia speciosa** Wenderoth. Showy Coneflower. Aug.–Oct. Moist woods; occasional in IL. *Rudbeckia fulgida* Ait. var. *speciosa* (Wenderoth) Perdue.

11. **Rudbeckia subtomentosa** Pursh. Fragrant Coneflower; Sweet Black-eyed Susan. July–Sept. Open woods, prairies, thickets; occasional throughout the state.

12. **Rudbeckia sullivantii** Boynt. & Beadle. Sullivant's Coneflower. July–Oct. Wet woods, wooded swamps, calcareous fens; scattered in IL but not common. *Rudbeckia fulgida* Ait. var. *sullivantii* (Boynt. & Beadle) Cronq.

13. **Rudbeckia tenax** Boynt. & Beadle. Narrow-leaved Black-eyed Susan. July–Oct. Woods; s. ⅙ of the state.

14. **Rudbeckia triloba** L. Two varieties occur in Illinois:

a. Some of the leaves 3-parted or 3-lobed; phyllaries lanceolate to lance-ovate, strigose to hispid .... 14a. *triloba* var. *triloba*
a. Some of the leaves 5- to 7-parted or 5- to 7-lobed; phyllaries linear to linear-lanceolate, spreading-villous .... 14b. *R. triloba* var. *beadleyi*

14a. **Rudbeckia triloba** L. var. **triloba.** Brown-eyed Susan. June–Oct. Woods, fields, stream banks; common throughout IL.

14b. **Rudbeckia triloba** L. var. **beadleyi** (Small) Fern. Five-lobed Brown-eyed Susan. June–Oct. Woods, fields; scattered but not common in IL.

15. **Rudbeckia umbrosa** Boynt. & Beadle. Shady Coneflower. Aug.–Oct. Floodplain forests, rare; s. ½ of IL, mostly along the Mississippi and Ohio rivers. *Rudbeckia fulgida* Ait. var. *umbrosa* (Boynt. & Beadle) Cronq.

98. **Sanvitalia** Lam.—Creeping Zinnia

1. **Sanvitalia procumbens** Lam. Creeping Zinnia. July. Native to Mexico; rarely escaped from cultivation; Cook and Jackson cos.

99. **Senecio** L.—Groundsel; Ragwort

1. Phyllaries black-tipped; rays absent .... 1. *S. vulgaris*
1. Phyllaries not black-tipped; rays 1–10 mm long.
  2. Stems glandular-hairy; rays 1–3 mm long .... 2. *S. viscosus*
  2. Stems glabrous or sometimes floccose-tomentose or arachnoid, not glandular; rays 6–10 mm long .... 1. *S. jacobaea*

1. **Senecio jacobaea** L. Stinking Willie; Tansy Ragwort. July–Aug. Native to Europe; rarely adventive along a railroad; DuPage Co.

2. **Senecio viscosus** L. Sticky Groundsel. Native to Europe; adventive in disturbed soil; Cook and Jackson cos.

3. **Senecio vulgaris** L. Common Groundsel. Apr.–Dec. Native to Europe; adventive in disturbed soil; occasional in the n. ½ of IL.

100. **Serinia** Raf.—Dwarf Dandelion.

1. **Serinia cespitosa** Raf. Dwarf Dandelion. May–July. Moist ground, sandy soil; occasional in the s. ½ of IL. *Krigia cespitosa* (Raf.) Chambers; *Krigia oppositifolia* Raf.

101. **Silphium** L.—Rosinweed

1. Leaves connate-perfoliate; stems conspicuously 4-angled .... 4. *S. perfoliatum*
1. Leaves not connate-perfoliate; stems nearly round or only slightly angled.
   2. Leaves deeply pinnatifid.
      3. Phyllaries appressed, obtuse to acute at the apex .... 3. *S. laciniatum*
      3. Phyllaries often reflexed, acuminate to caudate at the apex .... 5. *S. pinnatifidum*
   2. Leaves entire or serrate.
      4. Leaves all at base of plant .... 7. *S. terebinthinaceum*
      4. Leaves on stem.
         5. Stems, leaves, and phyllaries glabrous or nearly so .... 6. *S. speciosum*
         5. Stems, leaves, and phyllaries hispid, or at least scabrous.
            6. Cauline leaves opposite; basal leaves withered at flowering time .... 2. *S. integrifolium*
            6. Cauline leaves usually in whorls of 3; basal leaves persistent at flowering time .... 1. *S. asteriscus*

1. **Silphium asteriscus** L. var. **trifoliatum** (L.) J. A. Clevinger. Whorled Rosinweed. July–Sept. Dry soil, very rare; Hardin Co. *Silphium trifoliatum* L.

2. **Silphium integrifolium** Michx. Three varieties have been found in Illinois:

a. Phyllaries eglandular .... 2a. *S. integrifolium* var. *integrifolium*
a. Phyllaries glandular.
   b. Lower surface of leaves velvety-pubescent.......... 2b. *S. integrifolium* var. *deamii*
   b. Lower surface of leaves glabrous to hirsute but not velvety-pubescent .... 2c. *S. integrifolium* var. *neglectum*

2a. **Silphium integrifolium** Michx. var. **integrifolium**. Rosinweed. July–Aug. Prairies; occasional to common throughout IL.

2b. **Silphium integrifolium** Michx. var. **deamii** L. M. Perry. Rosinweed. Prairies; common throughout IL.

2c. **Silphium integrifolium** Michx. var. **neglectum** Settle & Fisher. Rosinweed. July–Aug. Prairies; occasional throughout IL.

3. **Silphium laciniatum** L. Two varieties occur in Illinois:

a. Phyllaries eglandular .... 3a. *S. laciniatum* var. *laciniatum*
a. Phyllaries glandular .... 3b. *S. laciniatum* var. *robinsonii*

3a. **Silphium laciniatum** L. var. **laciniatum**. Compass-plant. June–Sept. Prairies; common throughout IL, although less common in the s. cos.

3b. **Silphium laciniatum** L. var. **robinsonii** L. M. Perry. Compass-plant. June–Sept. Prairies; occasional in the n. ¾ of IL.

4. **Silphium perfoliatum** L. Cup-plant. July–Oct. Floodplain woods, along streams, wet ground; common throughout IL.

5. **Silphium pinnatifidum** Ell. Pinnatifid Dock. June–Sept. Dry woods, very rare; Hardin Co. *Silphium terebinthinaceum* Jacq. var. *pinnatifidum* (Ell.) Gray.

6. **Silphium speciosum** Nutt. Rosinweed. July–Sept. Native to the w. U.S.; adventive along a railroad; Lake Co. *Silphium integrifolium* Michx. var. *laeve* Torr. & Gray.

7. **Silphium terebinthinaceum** Jacq. Two varieties occur in Illinois:

a. Upper leaf surface harshly scabrous .... 7a. *S. terebinthinaceum* var. *terebinthinaceum*
a. Upper leaf surface completely glabrous .... 7b. *S. terebinthinaceum* var. *lucy-brauniae*

7a. **Silphium terebinthinaceum** Jacq. var. **terebinthinaceum**. Prairie Dock. June–Sept. Prairies, glades; common throughout IL.

7b. **Silphium terebinthinaceum** Jacq. var. **lucy-brauniae** Steyerm. Prairie Dock. June–Sept. Prairies, rare; Cook and Jackson cos.

102. **Smallanthus** Mack.—Bear's-foot

1. **Smallanthus uvedalius** (L.) Mack. Bear's-foot. July–Sept. Rich woods; occasional in s. IL, extending n. to St. Clair Co.; also Vermilion Co. *Polymnia uvedalia* L.

103. **Solidago** L.—Goldenrod

1. Flower heads in axillary clusters and also sometimes forming a thyrse above; flower heads not secund.
2. Stems glabrous, at least below the inflorescence.
3. Flowering heads only in axillary clusters; leaves subtending all but the uppermost flower clusters longer than the clusters.
4. Stems strongly glaucous, terete, often arching, not zigzag; leaves lanceolate to elliptic, to 3 cm wide .... 8. *S. caesia*
4. Stems not glaucous, often with ridges, usually erect, often zigzag; leaves broadly ovate to broadly lanceolate, to 10 cm wide .... 12. *S. flexicaulis*
3. Uppermost flowering heads forming a thyrse; the lowest cluster of flower heads subtended by leaves shorter than the clusters.
5. Basal leaves withered at flowering time; cauline leaves less than 2.5 cm wide .... 23. *S. rigidiuscula*
5. Basal leaves persistent at flowering time; cauline leaves usually 2.5 cm wide or wider.
6. Cypselae pubescent, 2.5–3.0 mm long .... 25. *S. sciaphila*
6. Cypselae glabrous, up to 2.5 (–2.6) mm long.
7. Lowest leaves at least seven times longer than wide; disc flowers 9–15; petioles with a sheathing base .... 21. *S. purshii*
7. Lowest leaves less than seven times longer than wide; disc flowers 4–10; petioles without a sheathing base .... 27. *S. speciosa*
2. Stems pubescent below the inflorescence, although sometimes glabrous at base.
8. Basal leaves persistent at flowering time, larger than the middle and upper leaves.
9. Rays white .... 5. *S. bicolor*
9. Rays yellow .... 15. *S. hispida*
8. Basal leaves absent at flowering time; before anthesis, the basal leaves are smaller than the middle and upper leaves.
10. Leaves thin, serrate; phyllaries appressed; cypselae 2.0–2.7 mm long .... 7. *S. buckleyi*
10. Leaves thick and firm, sparsely serrate to entire; outer phyllaries squarrose; cypselae 3–4 mm long.
11. Leaves and phyllaries glutinous; leaves less than 3 cm wide; stems without spreading hairs .... 2. *S. angusta*
11. Leaves and phyllaries not glutinous; some or all the leaves more than 3 cm wide; stems with spreading hairs .... 20. *S. petiolaris*
1. Flower heads in terminal pyramidal panicles, usually with spreading branches; flower heads usually secund.
12. Stems pubescent.
13. Basal and lower cauline leaves cordate; pappus up to 1 mm long, shorter than the cypsela .... 28. *S. sphacelata*
13. None of the leaves cordate; pappus more than 1 mm long, as long as or longer than the cypsela.
14. Leaves prominently 3-veined above the base, the basal absent at flowering time.
15. Leaves ovate to elliptic; cypselae 1.5–2.5 mm long.
16. Leaves rigid, elliptic; all except the uppermost leaves petiolate, scabrous .... 22. *S. radula*
16. Leaves not rigid, broadly ovate to broadly elliptic; middle and upper leaves sessile, softly pubescent .... 11. *S. drummondii*
15. Leaves lanceolate to oblanceolate; cypselae 0.8–1.5 mm long.
17. Involucre 3.0–4.5 mm high; ray flowers 10–16, 3–4 mm long; disc flowers 3–7, 3.0–3.5 mm long .... 1. *S. altissima*
17. Involucre 2–3 mm high; ray flowers 6–12, 2–3 mm long; disc flowers 2–5, 2.3–2.7 mm high .... 9. *S. canadensis*
14. Leaves 1-veined, the basal usually present at flowering time (except in *S. aspera* and *S. rugosa*).
18. Basal leaves absent at flowering time.

19. Stems densely hispid; leaves scabrous, thick and firm, strongly rugose, the teeth more or less blunt .... 4. *S. aspera*

19. Stems spreading-villous; leaves villous, thin, scarcely rugose, the teeth sharp-pointed .... 24. *S. rugosa*

18. Basal leaves present at flowering time.

20. Lower leaves broadly ovate to obovate to broadly elliptic, not gray-puberulent.

21. Stems strongly angled and sometimes slightly winged; leaves harshly scabrous on the upper surface..................... 19. *S. patula*

21. Stems terete or nearly so, unwinged; leaves glabrous or slightly scabrous on the upper surface.

22. Leaves rather thin, glabrous or slightly scabrous above, sparsely pubescent or glabrous below; cypselae short-hispid .... 3. *S. arguta*

22. Leaves thick, glabrous above, densely pubescent below; cypselae glabrous or sparsely pubescent .... 6. *S. boottii*

20. Lower leaves oblanceolate, gray-puberulent.

23. Involucre 3.0–4.5 mm high; cypselae strigose; lower leaves less than seven times longer than wide .... 18. *S. nemoralis*

23. Involucre 4.5–6.0 (–6.5) mm high; cypselae sericeous; lower leaves more than seven times longer than wide .... 10. *S. decemflora*

12. Stems glabrous, at least below the inflorescence.

24. Branches of the inflorescence glabrous.

25. Leaves distinctly 3-nerved above the base; cypselae glabrous or sparsely hispid.. 14. *S. glaberrima*

25. Leaves 1-nerved (sometimes obscurely 3-nerved); cypselae puberulent throughout... 17. *S. juncea*

24. Branches of the inflorescence pubescent.

26. Leaves somewhat fleshy, usually entire........................26. *S. sempervirens*

26. Leaves not fleshy, some or all of them serrate.

27. Basal leaves absent at flowering time.

28. Leaves 3-veined above the base; stems usually strongly glaucous .... 13. *S. gigantea*

28. Leaves 1-veined; stems not glaucous.

29. Lower leaves larger than the middle and upper leaves .... 17. *S. juncea*

29. Lower leaves smaller than the middle and upper leaves............ ....30. *S. ulmifolia*

27. Basal leaves present at flowering time.

30. Stems strongly angular, sometimes narrowly winged; leaves harshly scabrous above.. 19. *S. patula*

30. Stems terete, unwinged; leaves glabrous or spreading-pubescent, not harshly scabrous above.

31. Leaves sharply serrate; lower leaves with winged petioles; upper and middle leaves more than 2 cm wide.........3. *S. arguta*

31. Leaves sparsely serrate to often entire; petioles of lower leaves unwinged; upper and middle leaves less than 2 cm wide.

32. Leaves sometimes slightly sweet-scented; heads not secund; involucre 5–6 mm high .... 16. *S. jejunifolia*

32. Leaves not sweet-scented; heads secund; involucre 3–5 mm high .... 29. *S. uliginosa*

1. **Solidago altissima** L. Tall Goldenrod. July–Oct. Mesic woods, dry woods, prairies, old fields, pastures, fens, roadsides; common throughout IL; in every co. *Solidago procera* Ait.; *Solidago canadensis* L. var. *gilvocanescens* Rydb.; *Solidago altissima* L. var. *gilvocanescens* (Rydb.) Semple.

2. **Solidago angusta** (Torr. & Gray) Gray. Sticky Goldenrod. June–Oct. Rocky woods, bluff tops; confined to the s. ⅙ of IL. *Solidago petiolaris* Ait. var. *angusta* (Torr. & Gray) Gray; *Solidago petiolaris* Ait. var. *wardii* (Britt.) Fern.

3. **Solidago arguta** Ait. Two varieties occur in Illinois:

a. Leaves and stems pubescent ....3a. *S. arguta* var. *arguta*
a. Leaves and stems glabrous..3b. *S. arguta* var. *caroliniana*

3a. **Solidago arguta** Ait. var. **arguta.** Sharp-toothed Goldenrod. Aug.–Sept. Dry forests, rocky slopes; known only from Jackson and Union cos.

3b. **Solidago arguta** Ait. var. **caroliniana** Gray. Smooth Sharp-toothed Goldenrod. Aug.–Sept. Dry slopes, very rare; Union Co.

4. **Solidago aspera** Ait. Rough Wrinkle-leaved Goldenrod. Aug.–Sept. Wet woods, mesic woods, dry woods, roadsides; scattered in IL but more common in the s. ½ of the state.

5. **Solidago bicolor** L. White Goldenrod; Silverrod. Aug.–Oct. Dry open woods, very rare; Alexander, Jackson, and Union cos.

6. **Solidago boottii** Hook. Boott's Goldenrod. Rocky slopes beneath *Pinus echinata,* very rare; Union Co. One specimen from Union Co., with inner phyllaries only 0.5 mm wide, cypselae 1.3–1.5 mm long, and pappus bristles up to 2.5 mm long, is a hybrid between *Solidago arguta* Ait. and *Solidago ulmifolia* L. This hybrid may be called *Solidago X neurolepis* Fern.

7. **Solidago buckleyi** Torr. & Gray. Buckley's Goldenrod. Aug.–Oct. Bluff tops, dry forests, rocky woods; occasional in the s. ⅙ of IL.

8. **Solidago caesia** L. Wreath Goldenrod; Blue-stem Goldenrod. Aug.–Oct. Mesic woods, bottomland forests, upland woods, along rivers and streams; occasional throughout IL.

9. **Solidago canadensis** L. Two varieties occur in Illinois:

a. Stems densely pubescent throughout ....9a. *S. canadensis* var. *canadensis*
a. Stems moderately to densely pubescent above, glabrous near base........................ .... 9b. *S. canadensis* var. *hargeri*

9a. **Solidago canadensis** L. var. **canadensis.** Canada Goldenrod; Old Field Goldenrod. July–Oct. Prairies, dry woods, savannas, old fields, open disturbed areas, along rivers and streams, roadsides, pastures; common throughout IL.

9b. **Solidago canadensis** L. var. **hargeri** Fern. Smooth-based Canada Goldenrod. July–Oct. Dry woods, old fields, pastures; scattered in IL.

10. **Solidago decemflora** DC. Gray Goldenrod. Aug.–Oct. Fields; scattered in the n. ¼ of IL. *Solidago nemoralis* Ait. var. *longipetiolata* (Mack. & Bush) Palmer & Steyerm.

11. **Solidago drummondii** Torr. & Gray. Ozark Goldenrod; Drummond's Goldenrod. Aug.–Nov. Crevices of limestone bluffs; most cos. along the Mississippi River from Pike Co. southward; also Hardin and Pulaski cos.

12. **Solidago flexicaulis** L. Broadleaf Goldenrod; Zigzag Goldenrod. Aug.–Oct. Rich woods, bottomland forests, calcareous springy places; occasional throughout IL. *Solidago latifolia* L.

13. **Solidago gigantea** Ait. Two varieties occur in Illinois:

a. Leaves pubescent on 1 or both surfaces...................... 13a. *S. gigantea* var. *gigantea*
a. Leaves glabrous or nearly so .... 13b. *S. gigantea* var. *leiophylla*

13a. **Solidago gigantea** Ait. var. **gigantea**. Smooth Goldenrod. Aug.–Oct. Mesic woods, wet woods, along rivers and streams, calcareous fens, marshes, wet ground; common throughout IL; in every co.

13b. **Solidago gigantea** Ait. var. **leiophylla** Fern. Aug.–Oct. Smooth Goldenrod. Wet soil; scattered in IL.

14. **Solidago glaberrima** M. Martens. Eastern Missouri Goldenrod. July–Sept. Woods, prairies; occasional throughout the state. *Solidago missouriensis* Nutt., misapplied; *Solidago moritura* Steele.

15. **Solidago hispida** Muhl. ex Willd. Hispid Goldenrod. Aug.–Oct. Dry open woods, rare; Jackson, Randolph, and Union cos.

16. **Solidago jejunifolia** Steele. Few-leaved Goldenrod. Aug.–Sept. Wet, sandy soil, very rare; Lake and McHenry cos. *Solidago speciosa* Nutt. var. *jejunifolia* (Steele) Cronq. The leaves have a faintly sweet odor early in the morning.

17. **Solidago juncea** Ait. Two forms occur in Illinois:

a. Branches of panicle glabrous .... 17a. *S. juncea* f. *juncea*
a. Branches of panicle hirtellous .... 17b. *S. juncea* f. *scabrella*

17a. **Solidago juncea** Ait. f. **juncea.** Early Goldenrod. June–Sept. Prairies, glades, dry woods, mesic woods, roadsides, open disturbed areas; scattered in IL; probably in every co.

17b. **Solidago juncea** Ait. f. **scabrella** (Torr. & Gray) Fern. Early Goldenrod. June–Sept. Old fields, rare; scattered in IL.

18. **Solidago nemoralis** Ait. Gray Goldenrod; Field Goldenrod. July–Nov. Old fields, pastures, prairies, dry woods; scattered throughout IL; probably in every co.

19. **Solidago patula** Muhl. ex Willd. Swamp Goldenrod; Rough-leaved Goldenrod. Aug.–Oct. Swamps, calcareous fens, bogs; scattered in the state, but absent from the nw. and extreme s. cos.

20. **Solidago petiolaris** Ait. Downy Goldenrod. June–Oct. Bluff tops, dry woods; confined to the s. ⅙ of IL.

21. **Solidago purshii** Porter. Pursh's Bog Goldenrod. Aug.–Sept. Acid bogs, rare; Kane, Lake, and McHenry cos.

22. **Solidago radula** Nutt. Western Rough Goldenrod. Aug.–Oct. Mesic woods, dry woods, savannas, prairies, roadsides; confined to the w. ½ of IL, extending n. to Henderson Co. Specimens with broader and shorter phyllaries are called var. *laeta* (Greene) Fern. Specimens with very narrow phyllaries are called var. *stenolepis* Fern. Both varieties have been found in Illinois.

23. **Solidago rigidiuscula** (Torr. & Gray) Porter. Aug.–Oct. Prairie Goldenrod. Prairies, dry woods; scattered throughout the state.

24. **Solidago rugosa** Mill. Wrinkle-leaved Goldenrod. Aug.–Oct. Bottomland woods, mesic forests, calcareous fens, roadsides; scattered in the s. ½ of IL. *Solidago altissima* L. var. *rugosa* (Mill.) Torr.; *Solidago rugosa* Mill. var. *celtidifolia* (Small) Fern.

25. **Solidago sciaphila** Steele. Shadowy Goldenrod. Aug.–Sept. Sandstone and limestone cliffs, very rare; Carroll, Jo Daviess, LaSalle, and Ogle cos.

26. **Solidago sempervirens** L. Two varieties have been found in Illinois:

a. Leaves not ciliolate, some of them more than 3 cm wide ....
.... 26a. *S. sempervirens* var. *sempervirens*
a. Leaves ciliolate, up to 3 cm wide .... 26b. *S. sempervirens* var. *mexicana*

26a. **Solidago sempervirens** L. var. **sempervirens.** Seaside Goldenrod. Sept.–Oct. Native to e. and s. U.S.; adventive in the ne. cos. along roadsides where salt has been applied heavily during the winter season.

26b. **Solidago sempervirens** L. var. **mexicana** (L.) Fern. Seaside Goldenrod. Sept.–Oct. Native to e. and s. U.S.; adventive in the ne. cos. along roadsides where salt has been applied heavily during the winter season.

27. **Solidago speciosa** Nutt. Showy Goldenrod. Aug.–Oct. Dry woods, prairies, black oak savannas; scattered throughout IL, but less common in the s. tip of the state.

28. **Solidago sphacelata** Raf. Heart-leaved Goldenrod. Aug.–Sept. River bluffs, very rare; Hardin and Pope cos.

29. **Solidago uliginosa** Nutt. Bog Goldenrod. Aug.–Sept. Acid bogs, calcareous fens; occasional in the n. ½ of IL. *Solidago uniligulata* Nutt.

30. **Solidago ulmifolia** Muhl. ex Willd. Elm-leaved Goldenrod. July–Nov. Dry woods; common throughout IL; in every co.

104. **Sonchus** L.—Sow Thistle

1. Heads 3–5 cm across; flowers bright yellow.
    2. Phyllaries and peduncles stipitate-glandular .... 1. *S. arvensis*
    2. Phyllaries and peduncles eglandular or with sessile glands .... 4. *S. uliginosus*
1. Heads 1.0–2.5 cm across; flowers pale yellow.
    3. Basal auricles of leaves rounded; cypselae longitudinally ribbed, otherwise smooth .... 2. *S. asper*
    3. Basal auricles of leaves acute; cypselae longitudinally ribbed and papillate........ .... 3. *S. oleraceus*

1. **Sonchus arvensis** L. Field SowThistle. July–Sept. Native to Europe; adventive in fields and disturbed soils; occasional in the n. ½ of IL.

2. **Sonchus asper** (L.) Hill. Spiny SowThistle. June–Oct. Native to Europe; adventive in fields and disturbed areas; occasional throughout IL.

3. **Sonchus oleraceus** L. Common SowThistle. July–Oct. Native to Europe; adventive in fields and disturbed soils; common throughout IL.

4. **Sonchus uliginosus** Bieb. Smooth SowThistle. July–Sept. Native to Europe; adventive in fields and disturbed soil; abundant in the n. ½ of IL, extending southward to Clay Co. *Sonchus arvensis* L. var. *glabrescens* Gunther, Grab., & Wimmer.

105. **Symphyotrichum** Nees—Aster

1. Basal or lower leaves cordate or subcordate, on long petioles.
  2. Cauline leaves clasping .... 34. *S. undulatum*
  2. Cauline leaves sessile or short-petiolate, not clasping.
    3. Leaves entire or nearly so.
      4. Leaves glabrous or nearly so above .... 30. *S. shortii*
      4. Leaves scabrous above.
        5. Phyllaries reflexed; leaves soft-hairy below .... 2. *S. anomalum*
        5. Phyllaries appressed; leaves rough-hairy below .... 19. *S. oolentangiense*
    3. Leaves, or most of them, regularly serrate.
      6. Petioles unwinged .... 7. *S. cordifolium*
      6. Petioles winged.
        7. Peduncles and branches of the inflorescence with few or no bracts; heads relatively few, less than 50; some or all the rays at least 10 mm long .... 4. *S. ciliolatum*
        7. Peduncles and branches of the inflorescence with many bracts; heads relatively many, usually more than 50; rays up to 8 (–10) mm long.
          8. Stems evenly and densely pilose or hirsute throughout.
            9. Stems pilose; leaves firm, not becoming brittle with age; heads on short, ascending branches, the peduncles with a few scattered bracts; involucre campanulate, 4–7 mm high; rays bright blue or lavender; cypselae glabrous .... 8. *S. drummondii*
            9. Stems hirsute; leaves membranous, becoming brittle with age; heads on long, spreading branches, the peduncles with many crowded bracts; involucre turbinate to hemispheric, 3.8–5.2 mm high; rays bluish white; cypselae pubescent .... 32. *S. texanum*
          8. Stems glabrous or sparsely pubescent along decurrent lines on upper part .... 35. *S. urophyllum*

1. Basal or lower leaves neither cordate nor subcordate, sessile or petiolate.
  10. Stem leaves clasping or frequently auriculate at base.
    11. Leaves sericeous on both surfaces .... 29. *S. sericeum*
    11. Leaves glabrous or variously pubescent, but not sericeous.
      12. Leaves coarsely serrate .... 26. *S. prenanthoides*
      12. Leaves entire or finely and irregularly serrate.
        13. Stems glaucous.
          14. Leaves strongly clasping .... 13. *S. laeve*
          14. Leaves scarcely clasping .... 5. *S. concinnum*
        13. Stems green.
          15. Phyllaries glandular.
            16. At least the upper part of the stems strongly hirsute and often glandular.
              17. Leaves strongly auriculate-clasping .... 16. *S. novae-angliae*
              17. Leaves scarcely auriculate-clasping .... 17. *S. oblongifolium*
            16. Stems short-hairy, never glandular.
              18. Leaves thick and firm, up to 3 ½ times as long as wide.
                19. Involucre campanulate, 5.5–7.5 mm long; phyllaries in 4–5 series, squarrose, acute to acuminate at the apex, sparsely strigillose, densely stipitate-glandular, the middle phyllaries 1.0–1.2 mm wide .... 21. *S. patens*

19. Involucre turbinate, 8–12 mm long; phyllaries in 5–7 series, appressed, obtuse at the apex, densely strigillose, sparsely stipitate-glandular or eglandular, the middle phyllaries 1.2–1.7 mm wide .... 22. *S. patentissimum*

18. Leaves thin and membranous, more than 3 ½ times longer than wide.. 23. *S. phlogifolium*

15. Phyllaries eglandular.

20. None of the leaves more than 5 cm long; phyllaries uniformly pubescent .... 1. *S. X amethystinum*

20. Some or all the leaves at least 5 cm long; phyllaries glabrous or nearly so.

21. Plants very slender, the stems at most 2.5 mm in diameter; phyllaries acute; rays white or pale lavender .....3. *S. boreale*

21. Plants stout, the stems more than 2.5 mm in diameter; phyllaries long-attenuate; rays blue (sometimes white in *S. firmum*).

22. Stems coarsely hispid; rhizomes short .....27. *S. puniceum*

22. Stems glabrous or nearly so; rhizomes elongated............ .... 12. *S. firmum*

10. Leaves neither clasping nor auriculate at base.

23. Annuals; plants along highways heavily salted in winter.

24. Rays 4–7 mm long, 1.0–1.3 mm wide, in 1 series; disc flowers 25–50 .... 7. *S. divaricatum*

24. Rays 1.5–4.0 mm long, 0.2–0.6 mm wide, in 1–3 series; disc flowers up to 20... 31. *S. subulatum*

23. Perennials; plants usually not typically along highways heavily salted in winter.

25. Leaves sericeous on both surfaces...................................... 29. *S. sericeum*

25. Leaves glabrous or pubescent but not sericeous.

26. Stems glandular, at least near the top; leaves oblong; phyllaries densely glandular...17. *S. oblongifolium*

26. Stems eglandular; leaves linear or lanceolate or elliptic, not oblong; phyllaries eglandular.

27. Phyllaries uniformly pubescent on back.

28. Heads numerous; involucre 3–5 mm high; rays 8–20 ....10. *S. ericoides*

28. Heads solitary to few; involucre 5–7 mm high; rays 20–35 ....11. *S. falcatum*

27. Phyllaries glabrous or at most ciliate, never uniformly pubescent on back.

29. Phyllaries subulate-tipped, inrolled and more or less twisted at apex.

30. Some or all the leaves 1 cm wide or wider; rays 16–35, 5–10 mm long .... 24. *S. pilosum*

30. Leaves up to 6 mm wide; rays 12–16, 2–5 mm long................ ....21. *S. parviceps*

29. Phyllaries obtuse to acute, not subulate-tipped nor inrolled nor twisted at apex.

31. Involucre up to 4 mm high.

32. Bracteal leaves tiny, less than 1 cm long ....28. *S. racemosum*

32. Bracteal leaves larger, some of them at least 1.5 cm long .... 14. *S. lanceolatum*

31. Involucre 4–12 mm high.

33. Involucre 7–12 mm high, the phyllaries usually more than 1 mm wide (rarely so in *S. praealtum*); rays usually blue.

34. Rays 6–15; leaves conspicuously reticulate-veined below .... 25. *S. praealtum*

34. Rays 15–30; leaves not conspicuously reticulate-veined below.

35. Phyllaries obtuse; plants not glaucous; some of the involucres usually more than 9 mm high ....33. *S. turbinellum*

35. Phyllaries acute; plants more or less glaucous; none of the involucres more than 9 mm high ....5. *S. concinnum*

33. Involucre to 7 mm high, the phyllaries up to 1 mm wide; rays usually white (blue in *S. concinnum, S. praealtum,* and sometimes *S. boreale*).

36. Leaves conspicuously reticulate-veined below............ .... 25. *S. praealtum*

36. Leaves not conspicuously reticulate-veined below.

37. Leaves pubescent beneath, at least on the veins.

38. Leaves soft-hairy throughout on the lower surface..18. *S. ontarionis*

38. Leaves pubescent only on the midvein on the lower surface...........................15. *S. lateriflorum*

37. Leaves more or less glabrous.

39. Heads 1–several, solitary at the tips of stiff, ascending pedicels .... 9. *S. dumosum*

39. Heads few to many, on short peduncles.

40. Rays 30–50; most of the leaves less than 6 mm wide; heads few...............3. *S. boreale*

40. Rays usually less than 30; most of the leaves more than 6 mm wide; heads several to numerous.

41. Flowers blue; involucre 5–7 mm high ....5. *S. concinnum*

41. Flowers usually white; involucre 3–4 mm high ....14. *S. lanceolatum*

1. **Symphyotrichum X amethystinum** (Nutt.) G. L. Nesom. Amethyst Aster. June–Oct. Moist ground; scattered in the n. ½ of IL. *Aster amethystinum* Nutt. Considered to be a hybrid between *S. ericoides* (L.) G. L. Nesom and *S. novae-angliae* (L.) G. L. Nesom.

2. **Symphyotrichum anomalum** (Engelm. ex Torr. & Gray) G. L. Nesom. Manyray Aster. Sept.–Oct. Prairies, dry woods; occasional in sw. IL, n. to Peoria and Woodford cos. *Aster anomalus* Engelm. ex Torr. & Gray. Specimens with white rays have been called f. *albiflora,* but this form has not been transferred to *Symphyotrichum.*

3. **Symphyotrichum boreale** (Torr. & Gray) A. Love & D. Love. Northern Bog Aster; Rush Aster. Aug.–Oct. Calcareous fens, bogs; occasional in ne. IL, s. to Peoria Co. *Aster borealis* (Torr. & Gray) Prov.; *Aster junciformis* Rydb.

4. **Symphyotrichum ciliolatum** (Lindl.) A. Love & D. Love. Fringed Aster. Aug.–Sept. Woods, very rare; Cook Co. *Aster ciliolatus* Lindl.

5. **Symphyotrichum concinnum** (Willd.) Mohlenbr., comb. nov. (basionym: *Aster concinnus* Willd.). Narrow-leaved Smooth Aster. Sept.–Oct. Rich woods, rare; Hardin, Pope, and Saline cos. *Aster concinnus* Willd.; *Symphyotrichum laeve* (L.) G. L. Nesom var. *concinnum* (Willd.) G. L. Nesom.

6. **Symphyotrichum cordifolium** (L.) G. L. Nesom. Heart-leaved Aster. Aug.–Oct. Dry woods; occasional to common throughout IL. (Including var. *moratus* Shinners.) *Aster cordifolius* L.

7. **Symphyotrichum divaricatum** (Nutt.) G. L. Nesom. Southern Annual Saltmarsh Aster. Aug. Native to the s. and se. U.S.; adventive along highways where salt is applied heavily during the winter season; Cook and DuPage cos; also shallow marsh in sand deposits, Alexander Co. *Tripolium divaricatum* Nutt.; *Symphyotrichum subulatum* (Michx.) G. L. Nesom var. *ligulatum* (Shinners) S. D. Sundberg; *Aster exilis* Ell., *nomen confusum.*

8. **Symphyotrichum drummondii** (Lindl.) G. L. Nesom. Drummond's Aster. Aug.–Oct. Moist forests, dry open woods, prairies, fields, pastures, glades, edge of cliffs;

common throughout IL. *Aster drummondii* Lindl.; *Aster sagittifolius* Willd. var. *drummondii* (Lindl.) Shinners.

9. **Symphyotrichum dumosum** (L.) G. L. Nesom. Rice-button Aster; Bushy Aster. Aug.–Oct. Moist woods, calcareous fens, sand flats, prairies, along roads; locally common and scattered in IL. *Aster dumosus* L.(Including var. *coridifolius* [Michx.] Torr. & Gray and var. *strictior* Torr. & Gray.)

10. **Symphyotrichum ericoides** (L.) G. L. Nesom. Two varieties occur in Illinois:

a. Stems with appressed or ascending short hairs .... 10a. *S. ericoides* var. *ericoides*
a. Stems with spreading or slightly reflexed hairs..... 10b. *A. ericoides* var. *prostratum*

10a. **Symphyotrichum ericoides** (L.) G. L. Nesom var. **ericoides.** Heath Aster. July–Oct. Dry or moist prairies, dry disturbed soil; occasional to common throughout IL. Blue-flowered plants may be known as f. *caeruleus* (Benke) S. F. Blake; rose-flowered plants may be known as f. *gransii* Benke. These two forms have not been transferred to *Symphyotrichum. Aster ericoides* L.

10b. **Symphyotrichum ericoides** (L.) G. L. Nesom var. **prostratum** (Kuntze) G. L. Nesom. Heath Aster. July–Oct. Dry or moist prairies; common throughout IL. *Aster ericoides* L. var. *prostratus* Kuntze; *Aster exiguus* (Fern.) Rydb.

11. **Symphyotrichum falcatum** (Lindl.) G. L. Nesom var. **commutatum** (Torr. & Gray) G. L. Nesom. White Prairie Aster. July–Oct. Prairies, rare; Cook, DuPage, and Kane cos. *Aster commutatus* (Torr. & Gray) Gray.

12. **Symphyotrichum firmum** (Nees) G. L. Nesom. Glossy-leaved Aster; Swamp Aster. Aug.–Oct. Wet prairies, calcareous fens; occasional in the n. ½ of IL; also Fayette Co. *Aster firmus* Nees; *Aster puniceus* L. var. *firmus* (Nees) Torr. & Gray; *A. lucidulus* (Gray) Wieg.

13. **Symphyotrichum laeve** (L.) A. Love & D. Love. Smooth Aster. July–Oct. Prairies, glades, rock ledges, roadsides, occasionally in calcareous fens; common throughout IL. *Aster laevis* L. Rare white-flowered plants may be called f. *beckwithiae* House. This form has not been transferred to *Symphyotrichum.*

14. **Symphyotrichum lanceolatum** (Willd.) G. L. Nesom. Four varieties occur in Illinois:

a. Stems more or less glabrous.
  b. Involucre 3.5–5.5 mm high; heads not crowded.
    c. Leaves linear to narrowly lanceolate, the uppermost much reduced .... 14a. *S. lanceolatum* var. *lanceolatum*
    c. Leaves broadly oblanceolate, the uppermost not much reduced .... 14d *S. lanceolatum* var. *latifolium*
  b. Involucre 3–4 mm high; heads crowded ....14c. *S. lanceolatum* var. *interior*
a. Stems densely tomentose.................................14b. *S. lanceolatum* var. *hirsuticaule*

14a. **Symphyotrichum lanceolatum** (Willd.) G. L. Nesom var. **lanceolatum** White Panicled Aster. Aug.–Oct. Wet meadows, wet ditches, moist ground; common throughout IL. *Aster simplex* Willd.; *Aster lanceolatus* Willd. var. *simplex* (Willd.) A. G. Jones.

14b. **Symphyotrichum lanceolatum** (Willd.) G. L. Nesom var. **hirsuticaule** (Semple & Chmielewski) G. L. Nesom. Hairy White Panicled Aster. Aug.–Oct. Wet soil, very rare in s. IL. *Aster lanceolatus* Willd. var. *hirsuticaulis* Semple & Chmielewski.

14c. **Symphyotrichum lanceolatum** (Willd.) G. L. Nesom var. **interior** (Wieg.) G. L. Nesom. Small-flowered White Panicled Aster. Aug.–Oct. Moist ground; scattered in the s. ½ of IL. *Aster simplex* Willd. var. *interior* (Wieg.) Cronq.; *Aster lanceolatus* Willd. ssp. *interior* (Wieg.) A. G. Jones.

14d. **Symphyotrichum lanceolatum** (Willd.) G. L. Nesom var. **latifolium** (Semple & Chmielewski) G. L. Nesom. Broad-leaved White Panicled Aster. Aug.–Oct. Moist soil; scattered in IL. *Aster lanceolatus* Willd. var. *latifolius* Semple & Chmielewski.

15. **Symphyotrichum lateriflorum** (L.) G. L. Nesom, A. Love, & D. Love. One-sided Aster; Calico Aster. Aug.–Oct. Floodplain woods, calcareous fens, sloughs, along rivers and streams, around ponds and lakes; common throughout IL. *Aster lateriflorus* L. Plants with widely spreading branches may be called var. *horizontale* (Desf.) G. L Nesom.

16. **Symphyotrichum novae-angliae** (L.) G. L. Nesom. New England Aster. July–Oct. Mesic prairies, wet meadows, calcareous fens, pastures; common in the n. ⅔ of IL, less common in the s. ⅓. *Aster novae-angliae* L. White-flowered plants may be called var. *geneseensis* House; rose-flowered plants may be called f. *rosarius* House. Neither of these taxa has been transferred to *Symphyotrichum.*

17. **Symphyotrichum oblongifolium** (Nutt.) G. L. Nesom. Oblong-leaved Aster; Aromatic Aster. Aug.–Oct. Dry open woods, calcareous hill prairies, limestone barrens; uncommon to rare throughout most of IL, apparently absent from the e. cent. cos. *Aster oblongifolius* Nutt. Rose-flowered plants have be called f. *roseoligulatus* (Benke) Shinners. This form has not been transferred to *Symphyotrichum.*

18. **Symphyotrichum ontarionis** (Wieg.) G. L. Nesom. Ontario Aster. Aug.–Oct. Floodplain woods, river terraces; common throughout IL. *Aster ontarionis* Wieg.

19. **Symphyotrichum oolentangiense** (Riddell) G. L. Nesom. Two varieties may be distinguished in Illinois:

a. Stems scabrous and puberulent .... 19a. *S. oolentangiense* var. *oolentangiense*
a. Stems glabrous or nearly so .... 19b. *S. oolentangiense* var. *laevicaule*

19a. **Symphyotrichum oolentangiense** (Riddell) G. L. Nesom var. **oolentangiense.** Azure Aster. Aug.–Nov. Prairies, glades, black oak savannas; occasional throughout IL. *Aster azureus* Lindl.; *Aster oolentangiensis* Riddell. White-flowered plants rarely occur.

19b. **Symphyotrichum oolentangiense** (Riddell) G. L. Nesom var. **laevicaule** (Fern.) Mohlenbr., comb. nov. (basionym: *Aster azureus* Lindl. f. *laevicaulis* Fern.). Aug.–Nov. Prairies, woods, not common; in the s. ½ of IL. *Aster azureus* Lindl. f. *laevicaulis* Fern.; *Aster oolentangiensis* Riddell var. *laevicaulis* (Fern.) A. G. Jones.

20. **Symphyotrichum parviceps** (E. S. Burgess) G. L. Nesom. Small White Aster. Aug.–Oct. Prairies, open woods, roadsides; occasional in the n. ⅗ of IL. *Aster parviceps* (E. S. Burgess) Mack. & Bush; *Aster ericoides* L. var. *parviceps* E. S. Burgess.

21. **Symphyotrichum patens** (Ait.) G. L. Nesom. Spreading Aster. Aug.–Oct. Open woods; occasional in the s. ⅓ of IL; *Aster patens* Ait.

22. **Symphyotrichum patentissimum** (Lindl. ex DC.) Mohlenbr., comb. nov. (basionym: *Aster patentissimus* Lindl. ex DC.). Western Spreading Aster. Aug.–Oct. Dry woods, rare; sw. IL. *Aster patentissimus* Lindl.; *Symphyotrichum patens* (Ait.) G. L. Nesom var. *patentisimum* (Lindl. ex DC.) G. L. Nesom.

23. **Symphyotrichum phlogifolium** (Muhl. ex Willd.) G. L. Nesom. Thin-leaved Purple Aster. Aug.–Sept. Woods, rare; Hamilton Co. *Aster patens* Ait. var. *phlogifolius* (Muhl. ex Willd.) Nees; *Aster phlogifolius* Muhl. ex Willd.

24. **Symphyotrichum pilosum** (Willd.) G. L. Nesom. Two varieties occur in Illinois.

a. Stems and leaves pubescent; branches of the inflorescence rather long ....
.... 24a. *S. pilosum* var. *pilosum*
a. Stems and leaves glabrous; branches of the inflorescence short...............................
.... 24b. *S. pilosum* var. *pringlei*

24a. **Symphyotrichum pilosum**(Willd.) G. L. Nesom var. **pilosum.** Hairy White Aster. Aug.–Nov. Old fields, pastures, prairies, woods, roadsides; common throughout IL. *Aster pilosum* Willd.

24b. **Symphyotrichum pilosum** (Willd.) G. L. Nesom var. **pringlei** (Gray) G. L. Nesom. Pringle's Aster. Aug.–Oct. Dry sandy soil, very rare; Lake Co. *Aster pringlei* Gray.

25. **Symphyotrichum praealtum** (Poir.) G. L. Nesom. Three varieties occur in Illinois:

a. Leaves rather thick, firm, lanceolate to elliptic.
  b. Leaves more or less smooth above .... 25a. *S. praealtum* var. *praealtum*
  b. Leaves harshly scabrous above............................ 25c. *S. praealtum* var. *subasper*
a. Leaves thin, linear to linear-lanceolate .... 25b. *S. praealtum* var. *angustior*

25a. **Symphyotrichum praealtum** (Poir.) G. L. Nesom var. **praealtum.** Net-veined Aster. Aug.–Oct. Wet meadows, calcareous fens, moist soil; occasional to common in the n. ¾ of IL. *Aster praealtus* Poir.

25b. **Symphyotrichum praealtum** (Poir.) G. L. Nesom var. **angustior** (Wieg.) G. L.

Nesom. Willow-leaved Aster. Aug.–Oct. Wet meadows, moist soil; occasional in the n. ¾ of IL. *Aster praealtus* Poir. var. *angustior* Wieg.

25c. **Symphyotrichum praealtum** (Poir.) G. L. Nesom var. **subasper** (Lindl.) G. L. Nesom. Veiny Aster. Aug.–Oct. Moist ground; occasional in the n. ½ of IL. *Aster praealtus* Poir. var. *subasper* Lindl.

26. **Symphyotrichum prenanthoides** (Muhl. ex Willd.) G. L. Nesom. Crookedstem Aster. Aug.–Oct. Moist ground, north-facing slopes, swamps; rare in the n. ⅖ of IL; also Jackson Co. *Aster prenanthoides* Muhl. ex Willd.

27. **Symphyotrichum puniceum** (L.) A. Love & D. Love. Swamp Aster; Purplestem Aster. Aug.–Oct. Calcareous fens, moist soil; occasional in the n. ¾ of IL. *Aster puniceus* L.

28. **Symphyotrichum racemosum** (L.) G. L. Nesom var. **subdumosum** (Wieg.) G. L. Nesom. Small White Oldfield Aster. Aug.–Oct. Moist open ground; occasional or common throughout IL. *Aster racemosus* L.; *Aster vimineus* Lam. var. *subdumosus* Wieg.

29. **Symphyotrichum sericeum** (Vent.) G. L. Nesom. Silky Aster. Sept.–Oct. Prairies, sand barrens; occasional in the n. ½ of IL, rare or absent elsewhere. *Aster sericeus* Vent.

30. **Symphyotrichum shortii** (Lindl.) G. L. Nesom. Short's Aster. Aug.–Oct. Rocky woods, mesic woods; occasional to common throughout IL. *Aster shortii* Lindl. Rose-flowered plants have been called f. *gronemannii* Benke. This form has not been transferred to *Symphyotrichum.*

31. **Symphyotrichum subulatum** (Michx.) G. L. Nesom. Expressway Aster. Aug.–Oct. Native to the se. U.S.; adventive along highways, particularly where salt has been applied heavily during the winter, rapidly spreading in ne. IL. *Aster subulatus* Michx.

32. **Symphyotrichum texanum** (E. S. Burgess) Semple. Texas Aster. Aug.–Oct. Crevice of limestone cliff, very rare; Monroe Co. *Aster texanus* E. S. Burgess.

33. **Symphyotrichum turbinellum** (Lindl.) G. L. Nesom. Top-shaped Aster; Prairie Aster. Sept.–Oct. Prairies, dry woods; occasional in the s. ⅗ of IL. *Aster turbinellus* Lindl.

34. **Symphyotrichum undulatum** (L.) G. L. Nesom. Wavyleaf Aster. Aug.–Oct. Dry woods; scattered in the extreme se. cos. *Aster undulatus* L.

35. **Symphyotrichum urophyllum** (Lindl. ex DC.) G. L. Nesom. Arrow-leaved Aster. Aug.–Nov. Dry woods, mesic woods, prairies, pastures, old fields; common throughout IL. *Aster urophyllus* Lindl. ex DC.; *Aster sagittifolius* Willd.

106. **Tagetes** L.—Marigold

1. **Tagetes erecta** L. French Marigold. June–Oct. Native to Mexico; rarely escaped from cultivation; Alexander and DuPage cos. *Tagetes patula* L.

107. **Tanacetum** L.—Tansy; Golden Buttons

1. Leaves 1- to 3-pinnately lobed, with at least 3 pairs of lobes, glabrous, the lobes entire or dentate.
  2. Leaves 1- to 2-pinnately lobed, with 3–5 pairs of lobes, puberulent beneath; pappus none or with a crown of scales 0.1–0.2 mm long .... 2. *T. parthenium*
  2. Leaves 2- to 3-pinnately lobed, with 4 or more pairs of lobes, glabrous or sparsely pubescent beneath; pappus a crown of scales up to 0.5 mm long .... 3. *T. vulgare*
1. Leaves not pinnately lobed, or with a small pair of lobes near the base, the lobes more or less crenate .... 1. *T. balsamita*

1. **Tanacetum balsamita** L. Costmary; Mint Geranium. Native to Asia; escaped from gardens; scattered in IL. *Chrysanthemum balsamita* L.; *Balsamita major* Desf.

2. **Tanacetum parthenium** (L.) Sch.-Bip. Feverfew. June–Sept. Native to Europe and n. Africa; sparingly escaped from cultivation into disturbed soil; occasional in the n. ½ of IL. *Chrysanthemum parthenium* (L.) Bernh.

3. **Tanacetum vulgare** L. Tansy; Golden Buttons. July–Oct. Native to Europe and Asia; escaped from cultivation; adventive in fields and disturbed soil; occasional throughout IL.

108. **Taraxacum** F. H. Wiggers—Dandelion

1. Cypselae red to reddish brown; terminal lobe of leaves not noticeably larger than the other lobes.. 1. *T. erythrospermum*
1. Cypselae greenish brown to straw-colored; terminal lobe of leaves noticeably larger than the other lobes.. 2. *T. officinale*

1. **Taraxacum erythrospermum** Andrz. ex Besser. Red-seeded Dandelion. Feb.–Dec. Native to Europe; adventive in disturbed soil; occasional throughout IL. *Taraxacum laevigatum* (Willd.) DC.

2. **Taraxacum officinale** F. H. Wiggers. Common Dandelion. Feb.–Dec. Native to Europe; adventive in fields, lawns, and disturbed soils; very common throughout IL; in every co.

109. **Tetraneuris** Greene

1. **Tetraneuris herbacea** Greene. Lakeside Daisy; Four-nerved Starflower. Apr.–May. Dry gravelly banks and fields, very rare; Mason and Will cos. *Actinea herbacea* (Greene) B. L. Robins.; *Hymenoxys herbacea* (Greene) Cronq.; *Hymenoxys acaulis* (Pursh) K. K. Parker var. *glabra* (Gray) K. K. Parker.

110. **Thelesperma** Less.

1. **Thelesperma megapotamicum** (Spreng.) Kuntze. Green Thread. June–Oct. Native to the w. U.S.; adventive in disturbed soil; Kane Co. *Thelesperma gracile* (Torr.) Gray.

111. **Tragopogon** L.—Goat's-beard

1. Flowers purple.. 2. *T. porrifolius*
1. Flowers yellow.
    2. Phyllaries longer than the flowers; peduncles enlarged below head .... 1. *T. dubius*
    2. Phyllaries equaling or shorter than the flowers; peduncles not or scarcely enlarged below head.. 3. *T. pratensis*

1. **Tragopogon dubius** Scop. Sand Goat's-beard. May–Sept. Native to Europe; adventive in fields and disturbed soil; common throughout IL. *Tragopogon major* Jacq.

2. **Tragopogon porrifolius** L. Salsify; Vegetable Oyster. May–June. Native to Europe; adventive in fields and disturbed soil; occasional throughout IL. This species hybridizes with *T. pratensis* L. to form *Tragopogon X neohybridus,* which is known from Kane Co.

3. **Tragopogon pratensis** L. Common Goat's-beard. May–Nov. Native to Europe; adventive in fields and disturbed soil; common throughout IL.

112. **Tripleurospermum** Sch.-Bip.—Mayweed

1. **Tripleurospermum inodorum** (L.) Sch.-Bip. Scentless Chamomile; Mayweed. June–Sept. Native to Europe; adventive in disturbed soil, rare; Boone, DuPage, Grundy, Kane, and Lake cos. *Matricaria inodora* L.; *Matricaria perforata* Merat; *Matricaria maritima* L.; *Matricaria maritima* L. var. *agrestis* (Knaf) Wilmot; *Tripleurospermum perforatum* (Merat) Lainz.

113. **Tussilago** L.—Coltsfoot

1. **Tussilago farfara** L. Coltsfoot. Apr.–May. Native to Europe and Asia; adventive in a ravine; Lake Co.

114. **Verbesina** L.—Crownbeard; Wingstem

1. Most leaves opposite; rays yellow, 10–20 mm long; phyllaries in 2 series; disc flowers yellow; all cypselae flat.. 1. *V. occidentalis*
1. Most leaves alternate; rays white, 3–7 mm long; phyllaries in 1 series; disc flowers white; cypselae flat from the ray flowers, 3-sided from the disc flowers .... 2. *V. virginica*

1. **Verbesina occidentalis** (L.) Walt. Opposite-leaved Wingstem. Aug.–Oct. Bottomland forest, very rare; s. IL. *Siegesbeckia occidentalis* L.

2. **Verbesina virginica** L. Frostweed; Tickweed. July–Aug. Dry open woods, not common; extreme se. cos.

115. **Vernonia** Schreb.—Ironweed

1. Tips of phyllaries long-filiform; heads 12–20 mm across .... 1. *V. arkansana*
1. Tips of phyllaries obtuse to abruptly acuminate; heads 4–12 mm across.
    2. Leaves tomentose to tomentellous beneath.
        3. Tips of phyllaries obtuse to mucronate, erect or slightly spreading .... 5. *V. missurica*
        3. Tips of phyllaries abruptly acuminate, recurving or more or less appressed .... 2. *V. baldwinii*
    2. Leaves glabrous or scabrous-hirtellous beneath (sometimes tomentellous on the veins).
        4. Leaves glabrous, conspicuously punctate beneath; outer pappus of short capillary bristles.. 3. *V. fasciculata*
        4. Leaves scabrous-hirtellous, scarcely punctate beneath; outer pappus of scalelike bristles .... 4. *V. gigantea*

1. **Vernonia arkansana** DC. Ozark Ironweed. Aug.–Sept. Low open woods, prairies, very rare; Champaign Co. The original site has been destroyed in IL. *Vernonia crinita* Raf.

2. **Vernonia baldwinii** Torr. Two varieties occur in Illinois:

a. Phyllaries recurving at tip.. 2a. *V. baldwinii* var. *baldwinii*
a. Phyllaries more or less appressed.................................... 2b. *V. baldwinii* var. *interior*

2a. **Vernonia baldwinii** Torr. var. **baldwinii.** Baldwin's Ironweed; Western Ironweed. July–Sept. Prairies and open ground; occasional in the s. 2/3 of IL; also DuPage, Kane, and Lake cos.

2b. Vernonia baldwinii Torr. var. **interior** (Small) Schreb. Western Ironweed. July–Sept. Prairies, open ground; scattered in the s. 1/2 of IL.

3. **Vernonia fasciculata** Michx. Common Ironweed. July–Oct. Wet prairies, moist soil; common in the n. 1/2 of IL, occasional in the s. 1/2.

4. **Vernonia gigantea** (Walt.) Branner & Coville. Two varieties occur in Illinois:

a. Lower surface of leaves glabrous or with appressed hairs .... 4a. *V. gigantea* var. *gigantea*
a. Lower surface of leaves with spreading or erect hairs .... 4b. *V. gigantea* var. *taeniotricha*

4a. **Vernonia gigantea** (Walt.) Branner & Coville var. **gigantea.** Smooth Tall Ironweed. July–Oct. Low woods, open ground; occasional throughout IL. *Vernonia altissima* Nutt.

4b. **Vernonia gigantea** (Walt.) Branner & Coville var. **taeniotricha** S. F. Blake. Tall Ironweed. July–Oct. Low woods, moist soil; common in the s. 3/4 of IL, uncommon in the n. 1/4. *Vernonia altissima* Nutt. var. *taeniotricha* S. F. Blake. A hybrid between this taxon and *Vernonia missurica* Raf., called *Vernonia X illinoensis* Gl., is known from Champaign Co.

5. **Vernonia missurica** Raf. Missouri Ironweed. July–Sept. Low, open woods, prairies; common throughout IL. Rose-flowered forms may be called f. *carnea* Standl.; white-flowered forms may be called f. *swinkii* Steyerm.

116. **Xanthium** L.—Cocklebur

1. Leaves lanceolate, tapering to base, with 3-parted axillary spines..... 8. *X. spinosum*
1. Leaves ovate, cordate or truncate at base, without axillary spines.
    2. Burs 30–40 mm long, with more than 200 prickles in view at one time .... 7. *X. speciosum*
    2. Burs up to 30 mm long, with up to 200 prickles in view at one time.
        3. Burs and prickles glabrous or puberulent, but not hispid, pilose, or villous-hirsute; beaks of burs straight or nearly so.
            4. Burs 20–25 mm long, oblongoid to ellipsoid, more than twice as long as thick .... 2. *X. chinense*
            4. Burs up to 20 mm long, ovoid, up to twice as long as thick.

5. Leaves more or less thin; burs 10–18 mm thick; prickles 50–60 in view at one time, 4–6 mm long .... 3. *X. globosum*
5. Leaves thick; burs 6–9 mm thick; prickles 100–200 in view at one time, 2.0–3.5 mm long .... 1. *X. chasei*
3. Burs and prickles hispid, pilose, or villous-hirsute; beaks of burs usually uncinate.
6. Burs hispid, with 50–80 prickles in view at one time; leaves thin .... 4. *X. inflexum*
6. Burs pilose or villous-hirsute, with 100 or more prickles in view at one time; leaves thick.
7. Burs villous-hirsute, 6–8 mm thick .... 5. *X. italicum*
7. Burs pilose, 8–12 mm thick .... 6. *X. pensylvanicum*

1. **Xanthium chasei** Fern. Chase's Cocklebur. Sept.–Oct. Wet fields, along rivers; known only from central IL, usually near the Illinois River.

2. **Xanthium chinense** Mill. Mexican Cocklebur. Aug.–Oct. Wet fields and along roads; scattered and occasional throughout the state.

3. **Xanthium globosum** Shull. Subspherical Cocklebur. Sept.–Oct. Low wet fields; apparently restricted to the s. ¼ of IL.

4. **Xanthium inflexum** Mack. & Bush. Missouri Cocklebur. Sept.–Oct. Wet bottomlands; confined to the s. tip of IL.

5. **Xanthium italicum** Moretti. Common Cocklebur. Sept.–Oct. Low ground; common throughout the state.

6. **Xanthium pensylvanicum** Wallr. Pennsylvania Cocklebur. Sept.–Oct. Low ground in fields; scattered throughout the state.

7. **Xanthium speciosum** Kearney. Great Clotbur. Sept.–Oct. Low ground, mostly along roads; scattered in IL, but not common.

8. **Xanthium spinosum** L. Spiny Cocklebur. Aug.–Oct. Native to tropical America; adventive in disturbed areas; Alexander, Cook, and Pulaski cos.

117. **Ximenesia** Cav.—Golden Crownbeard

1. **Ximenesia encelioides** Cav. Golden Crownbeard. June–Aug. Native to the sw. U.S. and Mexico; rarely in disturbed soil; Champaign, Madison, and St. Clair cos. *Verbesina encelioides* (Cav.) Benth. & Hook. f. (Including var. *exauriculata* Rydb.)

37. BALSAMINACEAE—JEWELWEED FAMILY

1. **Impatiens** L.—Touch-me-not; Jewelweed

1. Spurred sepal and fruit glabrous; flowers in racemes.
2. Flowers yellow-orange; spur curved forward .... 2. *I. capensis*
2. Flowers lemon-yellow; spur spreading at right angles .... 3. *I. pallida*
1. Spurred sepal and fruit villous; flowers solitary or in pairs .... 1. *I. balsamina*

1. **Impatiens balsamina** L. Balsam. June–Aug. Native to Asia; rarely escaped from cultivation; Cook and Pope cos.

2. **Impatiens capensis** Meerb. Spotted Touch-me-not; Orange Touch-me-not; Orange Jewelweed. June–Oct. Wet soil; common throughout IL. *Impatiens biflora* Walt.

3. **Impatiens pallida** Nutt. Pale Touch-me-not; Yellow Touch-me-not; Yellow Jewelweed. June–Oct. Wet soil; occasional to common throughout IL.

38. BERBERIDACEAE—BARBERRY FAMILY

1. Prickly shrubs; fruit a red berry .... 1. *Berberis*
1. Unarmed herbs; fruit a capsule or a blue or yellow berry.
2. Flower solitary on each plant, white; leaves simple.
3. Leaves 7- to 9-lobed; sepals usually 6; stamens twice as many as the petals; fruit a berry .... 4. *Podophyllum*
3. Leaves 2-lobed; sepals usually 4; stamens the same number as the petals; fruit a capsule .... 3. *Jeffersonia*
2. Flowers in panicles, yellow-green; leaves ternately compound .... 2. *Caulophyllum*

1. **Berberis** L.—Barberry

1. Leaves toothed; flowers in racemes; prickles mostly forked; berries fleshy.
  2. Leaves with up to 20 teeth on the margin; branchlets brown; petals notched; berries ovoid .... 1. *B. canadensis*
  2. Leaves with 25 or more teeth on the margin; branchlets gray; petals not notched; berries ellipsoid .... 3. *B. vulgaris*
1. Leaves entire; flowers solitary or in clusters of 2–4; prickles unbranched; berries dry .... 2. *B. thunbergii*

1. **Berberis canadensis** Mill. American Barberry. May. Dry woods, sandstone cliffs, very rare; Jackson and Tazewell cos.

2. **Berberis thunbergii** DC. Japanese Barberry. Apr.–May. Native to Asia; spreading from cultivation; occasional to common in IL.

3. **Berberis vulgaris** L. Common Barberry; European Barberry. May–June. Native to Europe; escaped from cultivation; scattered in IL.

2. **Caulophyllum** Michx.—Blue Cohosh

1. **Caulophyllum thalictroides** (L.) Michx. Blue Cohosh. Apr.–May. Mesic woods; occasional to common throughout IL.

3. **Jeffersonia** Bart.—Twin-leaf

1. **Jeffersonia diphylla** (L.) Pers. Twin-leaf. Apr.–May. Mesic woods; scattered throughout IL. although rare in the s. ⅓ of the state.

4. **Podophyllum** L.—Mayapple

1. **Podophyllum peltatum** L. Mayapple. Apr.–June. Woods, very common; in every co. Plants with flowers developing 2–8 fruits have been called f. *polycarpum* Clute; plants with red fruits may be known as f. *deamii* Raymond; plants with orange fruits are called f. *biltmoreanum* Steyerm.

39. BETULACEAE—BIRCH FAMILY

1. Nuts borne in woody "cones"; pistillate scales 5-lobed; stamens 4; winter buds stalked .... 1. *Alnus*
1. Nuts borne in elongated catkins; pistillate scales 3-lobed; stamens 2; winter buds sessile .... 2. *Betula*

1. **Alnus** Mill.—Alder

1. Tree with a single trunk; leaves dentate, with up to 7 pairs of lateral veins .... 1. *A. glutinosa*
1. Small trees or shrubs with several stems; leaves serrate or serrulate, with 8 or more pairs of lateral veins.
  2. Leaves ovate to elliptic, green or whitened beneath.
    3. Lower surface of leaves green .... 2. *A. incana*
    3. Lower surface of leaves whitened .... 3. *A. rugosa*
  2. Leaves obovate, green beneath .... 4. *A. serrulata*

1. **Alnus glutinosa** (L.) Gaertn. Black Alder. Mar.–May. Native to Europe; naturalized along rivers in the n. ⅙ of IL; scattered throughout IL.

2. **Alnus incana** (L.) Muenchh. Mar.–May. Native to Europe; moist thickets; scattered in the ne. cos. of IL.

3. **Alnus rugosa** (DuRoi) Spreng. Mar.–May. Speckled Alder. Native to Europe. Moist thickets; restricted to a few extreme n. cos. *Alnus incana* var. *americana* Regel.

4. **Alnus serrulata** (Ait.) Willd. Smooth Alder; Brookside Alder. Apr.–May. Along rocky streams; occasional in the s. ⅓ of IL. (Including var. *noveboracensis* [Britt.] Fern.)

2. **Betula** L.—Birch

1. Leaves long-acuminate to caudate, broadest at base.
  2. Leaves long-caudate, with 18–47 teeth on each side .... 5. *B. populifolia*

2. Leaves acuminate to short-caudate, with 9–28 teeth on each side........................ ....4. *B. pendula*

1. Leaves acute, broadest above base.
    3. Leaves with 8 or more pairs of lateral veins.
        4. Bark yellowish or silvery gray, peeling off in thin layers; leaves green beneath........................1. *B. alleghaniensis*
        4. Bark brownish to pinkish, peeling off in shaggy pieces; leaves pale beneath ....2. *B. nigra*
    3. Leaves with up to 7 (–8) pairs of lateral veins.
        5. Bark peeling off in thin layers; small trees........................3. *B. papyrifera*
        5. Bark close, not peeling off in thin layers; shrubs.
            6. Leaves acute, some or all over 3 cm long.
                7. Bark dark brown; fruiting catkins less than 1 cm thick ....8. *B. X sandbergii*
                7. Bark gray; fruiting catkins at least 1 cm thick........................7. *B. X purpusii*
            6. Leaves obtuse to subacute, not more than 3 cm long........................6. *B. pumila*

1. **Betula alleghaniensis** Britt. Yellow Birch. Apr.–June. Boggy woods, rare; n. ⅙ of IL. *Betula lutea* Michx. f.

2. **Betula nigra** L. River Birch. Apr.–May. Along rivers and streams, floodplain woods; occasional in the s. ½ of IL, becoming less common northward.

3. **Betula papyrifera** Marsh. Paper Birch. Apr.–June. Wooded ravines, low dune ridges, rare; n. ⅙ of IL.

4. **Betula pendula** Roth. European Weeping Birch. Apr.–May. Native to Europe; rarely escaped from cultivation into low, sandy woods; scattered in the n. ½ of IL.

5. **Betula populifolia** Marsh. Gray Birch. Apr.–June. Native n. and e. of IL; thickets, rare; DuPage and Winnebago cos.

6. **Betula pumila** L. Three varieties occur in Illinois:

a. Twigs puberulent, at least when young; lower surface of leaves pubescent to nearly glabrous.
    b. Twigs and lower surface of leaves eglandular........................6a. *B. pumila* var. *pumila*
    b. Twigs and lower surface of leaves glandular........................6c. *B. pumila* var. *glandulifera*

a. Twigs and lower surface of leaves glabrous........................6b. *B. pumila* var. *glabra*

6a. **Betula pumila** L. var. **pumila.** Dwarf Birch. May. Bogs; confined to the extreme n. cos.

6b. **Betula pumila** L. var. **glabra** Regel. May. Bogs, rare; Lake Co.

6c. **Betula pumila** L. var. **glandulifera** Regel. May. Bogs, rare; Lake and Winnebago cos.

7. **Betula X purpusii** Schneid. Purpus' Birch. May–June. Boggy woods, rare; Lake Co. Considered to be a hybrid between *B. alleghaniensis* Britt. and *B. pumila* L. var. *glandulifera* Regel.

8. **Betula X sandbergii** Britt. Sandberg's Birch. May–June. Boggy woods, rare; Kane, Lake, and McHenry cos. Considered to be a hybrid between *B. alleghaniensis* Britt. and *B. pumila* L. var. *glandulifera* Regel.

40. BIGNONIACEAE—BIGNONIA FAMILY

1. Trees; leaves simple, usually whorled........................3. *Catalpa*
1. Vines; leaves compound, opposite.
    2. Leaflets 7–11; tendrils absent........................2. *Campsis*
    2. Leaflets 2; tendrils present........................1. *Bignonia*

1. **Bignonia** L.—Cross-vine

1. **Bignonia capreolata** L. Cross-vine. Apr.–June. Low woods, wooded slopes; confined to the s. ⅓ of IL. *Anisostichus capreolatus* (L.) Bureau.

2. **Campsis** Lour.—Trumpet Creeper

1. **Campsis radicans** (L.) Seem. Trumpet Creeper; Trumpet-vine. June–Aug. Roadsides, fields, edges of woods, swampy woods; native and common in the s. ½ of IL, adventive and rarer in the n. ½.

3. **Catalpa** Scop.—Catalpa

1. Lowest corolla lobe not notched; fruit 5–10 mm in diameter; crushed leaves with an unpleasant odor .... 1. *C. bignonioides*
1. Lowest corolla lobed notched; fruit 12 mm in diameter or wider; crushed leaves without an unpleasant odor .... 2. *C. speciosa*

1. **Catalpa bignonioides** Walt. Catalpa. May–June. Native to the se. U.S.; escaped along roads, railroads, and streams; occasional throughout the state.

2. **Catalpa speciosa** Warder. Catalpa. May–June. Low woods; native in the s. ¼ of IL, escaped from cultivation in disturbed areas in other parts of IL.

41. BORAGINACEAE—BORAGE FAMILY

1. Stems and leaves glabrous .... 12. *Mertensia*
1. Stems and leaves pubescent.
  2. Stamens exserted.
    3. Flowers blue .... 8. *Echium*
    3. Flowers not blue .... 1. *Amsinckia*
  2. Stamens included.
    4. Fruits echinate or prickly.
      5. Corolla 8–12 mm across; fruit 8–12 mm across .... 7. *Cynoglossum*
      5. Corolla 2–3 mm across; fruit 3–5 mm across.
        6. Racemes bracted throughout; pedicels not deflexed .... 10. *Lappula*
        6. Racemes bracted only at the base; pedicels deflexed .... 9. *Hackelia*
    4. Fruits smooth, wrinkled, rugose, or reticulate but not echinate or prickly.
      7. Corolla rotate .... 4. *Borago*
      7. Corolla funnelform, salverform, or tubular.
        8. Leaves decurrent by wings along the stem .... 16. *Symphytum*
        8. Leaves not decurrent.
          9. Corolla blue.
            10. Flower solitary in the leaf axils or at the end of branchlets .... 3. *Asperugo*
            10. Flowers in terminal racemes or cymes.
              11. Flowers not subtended by bracts; calyx tube not hispid nor bristly .... 5. *Brunnera*
              11. Flowers subtended by bracts; calyx tube hispid or bristly .... 2. *Anchusa*
          9. Corolla greenish white, white, orange, or yellow.
            12. All flowers subtended by bracts.
              13. Leaves without lateral veins .... 6. *Buglossoides*
              13. Leaves with distinct lateral veins.
                14. Stems pubescent but not spreading-hirsute; corolla lobes obtuse .... 11. *Lithospermum*
                14. Stems spreading-hirsute; corolla lobes acute .... 14. *Onosmodium*
            12. None or only a few of the flowers subtended by bracts.
              15. Corolla yellow; leaves lanceolate to ovate-lanceolate .... 1. *Amsinckia*
              15. Corolla white; leaves linear to oblong.
                16. Corolla with yellow appendages in the throat .... 13. *Myosotis*
                16. Corolla without yellow appendages in the throat .... 15. *Plagiobothrys*

1. **Amsinckia** Lehm.—Fiddleneck

1. Nutlets 1–2 mm long; leaves finely toothed or entire .... 4. *A. spectabilis*
1. Nutlets 2–4 mm long; leaves more or less entire.
  2. Corolla more or less enclosed by hairy bulges at the tip of the tube; stamens and styles included within the corolla .... 2. *A. lycopsoides*
  2. Corolla open; stamens and styles exserted from the corolla.

3. Corolla 7–11 mm long, 4–6 mm wide at summit, more or less orange .... 1. *A. intermedia*
3. Corolla 4–7 mm long, 2–3 mm wide at summit, pale yellow.. 3. *A. menziesii*

1. **Amsinckia intermedia** Fisch. & C. A. Meyer. Yellow Burweed. June–July. Native to California; rarely escaped from cultivation; DeKalb Co.
2. **Amsinckia lycopsoides** (Lehm.) Lehm. Tarweed. June–July. Native to the w. U.S.; adventive in disturbed areas; Cass Co.
3. **Amsinckia menziesii** (Lehm.) A. Nels. & J. F. Macbr. Menzies' Fiddleneck. June–July. Native to the w. U.S.; rarely adventive in IL; DuPage Co.
4. **Amsinckia spectabilis** Fisch. & Mey. Fiddleneck. June–July. Native to the w. U.S.; adventive along railroads; Champaign and DeKalb cos.

2. **Anchusa** L.—Alkanet

1. Bracts narrowly lanceolate; lobes of calyx linear-lanceolate; corolla 12–20 mm across .... 1. *A. azurea*
1. Bracts oblong to ovate; lobes of calyx lanceolate; corolla 5–10 mm across .... 2. *A. officinalis*

1. **Anchusa azurea** Mill. Blue Bugloss. May–Sept. Native to Europe; escaped from gardens in disturbed soil; w. IL.
2. **Anchusa officinalis** L. Common Alkanet. May–Oct. Native to Europe; adventive in waste ground; Jackson and Kane cos.

3. **Asperugo** L.—Madwort

1. **Asperugo procumbens** L. Madwort. May–Aug. Native to Europe and Asia; adventive in waste ground; Kane Co. Not seen in IL since 1973.

4. **Borago** L.—Borage

1. **Borago officinalis** L. Borage. July–Oct. Native to Europe; rarely escaped from cultivation; scattered in IL.

5. **Brunnera** C. Steven—Bugloss

1. **Brunnera macrophylla** (Adams) I. M. Johnst. Bugloss. June–July. Native to Siberia; rarely escaped from cultivation; DuPage Co.

6. **Buglossoides** Moench.—Gromwell

1. **Buglossoides arvense** (L.) I. M. Johnston. Corn Gromwell. Apr.–July. Native to Europe; adventive along roads and in fields and other disturbed areas; common throughout IL. *Lithospermum arvense* L.

7. **Cynoglossum** L.—Hound's-tongue

1. Corolla reddish purple; stem softly hairy; leaves borne all along the stem .... 1. *C. officinale*
1. Corolla pale blue to white; stem rough-hairy; leaves only on the lower ½ of the stem .... 2. *C. virginianum*

1. **Cynoglossum officinale** L. Common Hound's-tongue. May–July. Native to Europe and Asia; adventive in pastures, fields, roadsides; occasional to common in the n. ⅔ of IL, rare elsewhere.
2. **Cynoglossum virginianum** L. Wild Comfrey. Apr.–June. Rich woods, wooded slopes; occasional in the s. ⅓ of IL, rare or absent elsewhere.

8. **Echium** L.—Viper's Bugloss

1. **Echium vulgare** L. Viper's Bugloss. May–Oct. Native to Europe; adventive in disturbed areas, particularly along railroads; occasional in the n. ½ of IL, rare elsewhere.

9. **Hackelia** Opiz—Stickseed

1. Fruits pyramidal; basal leaves cuneate at base; cauline leaves linear to oblong .... 1. *H. deflexa*

1. Fruits globose; basal leaves cordate at base; cauline leaves ovate-oblong .... 2. *H. virginiana*

1. **Hackelia deflexa** (Wahlenb.) Opiz var. **americana** (Gray) Fern. & I. M. Johnston. American Stickseed. June–July. Rocky slopes, rare; Carroll, Jo Daviess, and Winnebago cos. *Hackelia americana* (Gray) Fern.

2. **Hackelia virginiana** (L.) I. M. Johnston. Stickseed. June–Sept. Moist or dry woods; occasional throughout IL.

10. **Lappula** Moench—Stickseed

1. Fruits with a single row of prickles .... 1. *L. occidentalis*
1. Fruits with a double row of prickles.................................... 2. *L. squarrosa*

1. **Lappula occidentalis** S. Wats. Western Stickseed; Western Beggar's Lice. May–Sept. Native to the w. U.S.; occasionally adventive in disturbed areas in the n. ⅙ of IL. *Lappula redowskii* (Hornem.) Greene var. *occidentalis* (S. Wats.) Rydb.

2. **Lappula squarrosa** (Retz.) Dumort. Two-ray Stickseed; Beggar's Lice. May–Sept. Native to Europe; adventive in disturbed areas; occasional in the n. ½ of IL; also Coles, Jersey, and St. Clair cos. *Lappula echinata* Gilib.

11. **Lithospermum** L.—Gromwell

1. Corolla white, greenish white, or pale yellow, up to 8 mm long.
   2. Most or all the cauline leaves 2 cm wide or wider, ovate .... 4. *L. latifolium*
   2. Most or all the cauline leaves up to 2 cm wide, lanceolate.............. 5. *L. officinale*
1. Corolla bright yellow or orange, at least 15 mm long.
   3. Leaves linear, acute, most of them not wider than 5 mm; corolla tube more than twice as long as the calyx, the lobes toothed .... 3. *L. incisum*
   3. Leaves linear-lanceolate to oblong, more or less obtuse, most or all of them more than 5 mm wide; corolla tube less than twice as long as the calyx, the lobes entire.
      4. Lobes of calyx during flowering up to 6 mm long, lengthening to 8 mm long during fruiting; stems softly pubescent; corolla tube glabrous within at the base .... 1. *L. canescens*
      4. Lobes of calyx during flowering up to 11 mm long, lengthening to 15 mm long during fruiting; stems somewhat rough-pubescent; corolla tube pubescent within at the base .... 2. *L. croceum*

1. **Lithospermum canescens** (Michx.) Lehm. Hoary Puccoon. Apr.–July. Prairies; occasional to common throughout the state.

2. **Lithospermum croceum** Fern. Hairy Puccoon. Apr.–June. Sandy prairies, sandy woods; occasional to common in the n. ½ of IL, less common in the s. ½. *Lithospermum caroliniense* (Walt.) MacM.; *L. caroliniense* (Walt.) Macm. var. *croceum* (Fern.) Cronq.

3. **Lithospermum incisum** Lehm. Yellow Puccoon; Fringed Puccoon. Apr.–June. Prairies; occasional in the n. ½ of IL, less common in the s. ½.

4. **Lithospermum latifolium** Michx. American Gromwell. Apr.–June. Rich woods; occasional throughout the state.

5. **Lithospermum officinale** L. Common Gromwell; European Gromwell. June–Aug. Native to Europe; rarely adventive in IL; Lake and Will cos.

12. **Mertensia** Roth—Bluebells

1. **Mertensia virginica** (L.) Pers. Bluebells; Virginia Cowslip. Mar.–June. Mesic woods, wooded floodplains; occasional throughout the state. Rose-flowered specimens may be called f. *rosea* Steyerm. White-flowered specimens also occur occasionally.

13. **Myosotis** L.—Forget-me-not; Scorpion-grass

1. Corolla white, 1–2 mm across; lobes of calyx 2-lipped.
   2. Fruiting pedicels spreading, the lowest 20 mm or more apart .... 1. *M. macrosperma*
   2. Fruiting pedicels more or less erect, the lowest to 20 mm apart......... 5. *M. verna*

1. Corolla blue, sometimes with a yellow center, 2–9 mm across; lobes of calyx equal or nearly so.
  3. Corolla 5–9 mm across, blue with a yellow center.
    4. Calyx with appressed hairs, the lobes much shorter than the tube .... 2. *M. scorpioides*
    4. Calyx with hooked hairs, the lobes about as long as the tube .... 4. *M. sylvatica*
  3. Corolla 2–4 mm across, blue but usually without a yellow center .... 3. *M. stricta*

1. **Myosotis macrosperma** Engelm. Big-fruited Scorpion-grass. Apr.–May. Dry woods, moist woods, fields; occasional in the s. ⅓ of IL; also Fayette and Jersey cos.

2. **Myosotis scorpioides** L. Common Forget-me-not. Apr.–Oct. Native to Europe and Asia; frequently cultivated; occasionally escaped into waste ground, often along streams; scattered in IL.

3. **Myosotis stricta** Link. Small-flowered Forget-me-not. Apr.–Aug. Native to Europe and Asia; infrequently adventive in the n. ¾ of IL. The report of *M. arvensis* (L.) Hill in *Guide to the Vascular Flora of Illinois* (1986) should be referred here.

4. **Myosotis sylvatica** Hoffm. Garden Forget-me-not. June–July. Native to Europe; commonly cultivated but rarely escaped; Cook and Jackson cos.

5. **Myosotis verna** Nutt. Small Scorpion-grass. Apr.–June. Dry woods, disturbed sandy soil, moist woods, fields; occasional to common throughout IL.

14. **Onosmodium** Michx.—Marbleseed

1. Nutlets constricted at base .... 1. *O. molle*
1. Nutlets not constricted at base .... 2. *O. occidentale*

1. **Onosmodium molle** Michx. Marbleseed. May–July. Rocky prairies, rocky woods, often in calcareous areas, not common; scattered throughout IL. *Onosmodium molle* Michx. var. *hispidissimum* (Mack.) Cronq.; *O. hispidissimum* Mack.

2. **Onosmodium occidentale** Mack. Marbleseed. May–July. Rocky prairies, rocky woods, not common; confined to the w. ½ of IL. *Onosmodium molle* Michx. var. *occidentale* (Mack.) Cochrane.

15. **Plagiobothrys** Fisch. & C. A. Mey.

1. Corolla 5–10 mm across .... 1. *P. hirtus*
1. Corolla 1.0–1.5 mm across .... 2. *P. scouleri*

1. **Plagiobothrys hirtus** (Greene) I. M. Johnston var. **figuratus** (Piper) I. M. Johnston. Bristly Plagiobothrys. June–Aug. Native to the w. U.S.; rarely adventive in IL; St. Clair Co.

2. **Plagiobothrys scouleri** (Hook. & Arn.) I. M. Johnston var. **penicillatus** (Greene) Cronq. Meadow Plagiobothrys. June–Aug. Native to the w. U.S.; adventive in DuPage Co.

16. **Symphytum** L.—Comfrey

1. **Symphytum officinale** L. Common Comfrey. May–Sept. Native to Europe; occasionally escaped from cultivation into waste ground; scattered in IL.

42. BRASSICACEAE—MUSTARD FAMILY

1. Petals white, greenish, pink, purple, lavender, or absent .... Group 1
1. Petals yellow or orange .... Group 2

Group 1
Petals white, greenish, pink, purple, lavender, or absent.

1. Cauline leaves opposite or whorled.
  2. Cauline leaves trifoliolate or palmately divided .... 16. *Dentaria*
  2. Cauline leaves undivided .... 28. *Lunaria*
1. Cauline leaves alternate or absent.
  3. Petals pink or purple.

4. Some of the cauline leaves pinnately divided.
5. Some of the leaves auriculate or clasping at the base; plants glabrous or nearly so; fruits 0.7–1.0 mm thick.
6. Hairs, if present, unbranched.. 24. *Iodanthus*
6. At least some of the hairs forked or stellate ....3. *Arabidopsis*
5. None of the leaves auriculate or clasping at the base; plants hispid; fruits more than 1 mm thick..37. *Raphanus*
4. None of the cauline leaves pinnately divided.
7. Plants succulent, with thick, coriaceous leaves..............................9. *Cakile*
7. Plants not succulent, the leaves neither thick nor coriaceous.
8. Plants glabrous or pubescent only near base.
9. Petals 7–12 mm long; hairs, if present, unbranched..... 26. *Iodanthus*
9. Petals 2.0–3.5 mm long; at least some of the hairs with 2–4 rays..... .... 8. *Boechera*
8. Plants pubescent nearly throughout.
10. Pubescence of glandular hairs....................................14. *Chorispora*
10. Pubescence eglandular.
11. Petals up to 15 mm long; cauline leaves fewer than 10.
12. Pubescence of stellate hairs .... 29. *Malcolmia*
12. Pubescence of unbranched hairs ....12. *Cardamine*
11. Petals usually 15 mm long or longer; cauline leaves more than 10.
13. Leaves entire; pubescence of stellate hairs.......30. *Mattthiola*
13. Leaves serrate; pubescence of unbranched or forked hairs, not stellate... 23. *Hesperis*
3. Petals white, greenish, or absent.
14. Some of the leaves sessile and/or auriculate-clasping.
15. Plants more or less aquatic; all leaves pinnately compound ....32. *Nasturtium*
15. Plants terrestrial; none or some but not all of the leaves pinnately compound.
16. Fruit at least 5 times longer than broad.
17. Stems glabrous or, if pubescent, with unbranched hairs.
18. Lowest leaves undivided..13. *Cardamine*
18. Lowest leaves pinnately divided .... 8. *Boechera*
17. At least some of the hairs forked.
19. Some of the leaves lyrate-pinnatifid ....3. *Arabidopsis*
19. None of the leaves lyrate-pinnatifid .... 4. *Arabis*
16. Fruit up to 2 times longer than broad.
20. Plants pubescent.
21. Pedicels densely pubescent; fruits neither triangular nor subglobose .... 26. *Lepidium*
21. Pedicels glabrous or nearly so; fruits triangular or subglobose.
22. Fruits triangular; basal leaves present at flowering time........ ....12. *Capsella*
22. Fruits subglobose; basal leaves absent at flowering time...... .... 26. *Lepidium*
20. Plants glabrous.
23. Upper cauline leaves perfoliate-clasping, suborbicular .... 26. *Lepidium*
23. Upper cauline leaves not clasping or, if clasping, lanceolate to lance-ovate.
24. Fruits flat, winged.
25. Cauline leaves sagittate, serrate; fruit to 15 mm long; seed coat reticulate...42. *Thlaspi*
25. Cauline leaves clasping, more or less entire; fruit to 6 mm long; seed coat smooth .... 31. *Microthlaspi*
24. Fruits inflated, unwinged .... 26. *Lepidium*
14. None of the leaves auriculate and/or clasping.

26. Plants aquatic; submerged leaves filiform-dissected ....33. *Neobeckia*
26. Plants terrestrial or, if aquatic, none of the leaves filiform-dissected.
27. Some or all the cauline leaves deeply lobed or divided or cauline leaves absent.
28. At least the lowest leaves 2- to 3-pinnate or -pinnatifid.
29. Fruit suborbicular; upper cauline leaves simple or once-pinnate .... 26. *Lepidium*
29. Fruit narrowly oblong to linear; all leaves 2- to 3-pinnate or pinnatifid... 17. *Descurainia*
28. Lowest leaves once-pinnate or -pinnatifid.
30. Petals at least 6 mm long.
31. Leaves pinnately compound .... 13. *Cardamine*
31. Leaves pinnatifid.
32. Upper member of fruit shorter than lower member.......... .... 20. *Eruca*
32. Upper member of fruit longer than lower member........... .... 37. *Raphanus*
30. Petals up to 5 mm long.
33. Fruit wrinkled, didymous..................................... 26. *Lepidium*
33. Fruit neither wrinkled nor didymous.
34. Plants aquatic; petals 4–5 mm long............ 32. *Nasturtium*
34. Plants terrestrial; petals up to 4 mm long.
35. Seeds narrowly winged; fruit 1.5–2.0 mm wide; stems pubescent near base .... 35. *Planodes*
35. Seeds unwinged; fruit up to 1 mm wide; stems usually glabrous near base .... 13. *Cardamine*
27. All or nearly all cauline leaves without deep lobes or pinnately divided or leaves all basal.
36. At least the lowest leaves petiolate, rounded to cordate at base.
37. Fruit up to 6 mm long; cauline leaves crenate to shallowly pinnatifid .... 5. *Armoracia*
37. Fruit more than 6 mm long; cauline leaves denticulate, sinuate, or nearly entire.
38. Plants with an onion scent; petals up to 7 mm long; fruiting pedicels up to 5 mm long .... 1. *Alliaria*
38. Plants without an onion scent; petals more than 7 mm long; fruiting pedicels more than 5 mm long.
39. Cauline leaves fewer than 10, entire to repand .... 13. *Cardamine*
39. Cauline leaves more than 10, serrulate to denticulate....... .... 23. *Hesperis*
36. Leaves tapering to the sessile or subsessile base.
40. Fruit more than twice as long as broad.
41. Fruit filiform, subterete, up to 1 mm wide........ 3. *Arabidopsis*
41. Fruit linear, flat, more than 1 mm wide.
42. Plants scapose, without cauline leaves............... 19. *Draba*
42. Plants with cauline leaves.
43. None of the cauline leaves auriculate............ 19. *Draba*
43. At least some of the cauline leaves auriculate.
44. Hairs, if present, unbranched .... 8. *Boechera*
44. Hairs forked .... 4. *Arabis*
40. Fruit up to twice longer than broad.
45. Plants glabrous or with unbranched hairs .... 26 *Lepidium*
45. Plants pubescent with forked or stellate hairs.
46. Fruit strongly ascending to appressed to the rachis .... 7. *Berteroa*
46. Fruit spreading.
47. Fruit elliptic, on pedicels more than twice as long; pubescence with forked hairs .... 27. *Lobularia*

47. Fruit orbicular, on pedicels less than twice as long; pubescence stellate .... 2. *Alyssum*

Group 2

Petals yellow or orange or cream.

1. All leaves unlobed.
  2. Plants glabrous.
    3. Cauline leaves clasping.
      4. Cauline leaves linear-lanceolate, less than 1 cm wide .... 25. *Isatis*
      4. Cauline leaves elliptic to ovate, more than 1 cm wide.
        5. Leaves entire, obtuse .... 15. *Conringia*
        5. Leaves toothed or sinuate, acute .... 9. *Brassica*
    3. Cauline leaves not clasping.
      6. Basal rosette of leaves present at flowering time .... 4. *Turritis*
      6. Basal rosette of leaves absent at flowering time .... 39. *Rorippa*
  2. Plants pubescent.
    7. Cauline leaves, or some of them, clasping.
      8. Basal rosette of leaves present at flowering time; fruit more than 1 cm long .... 4. *Turritis*
      8. Basal rosette of leaves absent at flowering time; fruit less than 1 cm long.
        9. Fruit several-seeded; stems glabrous or nearly so above .... 11. *Camelina*
        9. Fruit 1- to 2-seeded; stems pubescent throughout .... 34. *Neslia*
    7. None of the leaves clasping.
      10. Fruit at least 5 times longer than broad .... 22. *Erysimum*
      10. Fruit up to 3 times longer than broad.
        11. Petals 5–8 mm long .... 36. *Physaria*
        11. Petals up to 3 mm long.
          12. Fruit as long as broad .... 2. *Alyssum*
          12. Fruit longer than broad .... 19. *Draba*
1. At least some of the leaves lobed, pinnatifid, or pinnately compound.
  13. Some of the pedicels subtended by foliaceous bracts .... 21. *Erucastrum*
  13. None of the pedicels subtended by bracts.
    14. Petals to 5 (–6) mm long.
      15. Only the lowest leaves lobed or pinnatifid .... 4. *Arabis*
      15. All except sometimes the very uppermost leaves lobed or pinnatifid.
        16. Terminal lobe of leaf much larger than the others, obtuse .... 6. *Barbarea*
        16. Terminal lobe of leaf about the same size as the others, more or less acute.
          17. Some of the leaves 3-pinnate; glands sometimes present .... 17. *Descurainia*
          17. Leaves 1- to 2-pinnate or -pinnatifid; glands never present.
            18. Fruit linear, more than 5 times longer than broad .... 41. *Sisymbrium*
            18. Fruit oblong to nearly globose, usually less than 5 times longer than broad .... 39. *Rorippa*
    14. Petals more than 5 mm long.
      19. Most or all of the leaves basal .... 18. *Diplotaxis*
      19. Leaves on the stem as well as sometimes basal.
        20. Cauline leaves often glaucous, fetid when crushed .... 18. *Diplotaxis*
        20. Cauline leaves usually not glaucous, not fetid when crushed.
          21. Basal rosette of leaves present and green at flowering time .... 6. *Barbarea*
          21. Basal rosette of leaves absent or withered at flowering time.
            22. Uppermost leaves usually pinnatifid .... 20. *Eruca*
            22. Uppermost leaves not pinnatifid.
              23. Fruit up to 1 cm long, with a beak about as long as the fertile portion .... 38. *Rapistrum*

23. Fruit more than 1 cm long, the beak, if present, much shorter than the fertile portion.
24. Fruit bristly hispid ....40. *Sinapis*
24. Fruit not bristly hispid.
25. Petals with dark veins; fruit conspicuously torulose ....37. *Raphanus*
25. Petals without dark veins; fruit barely if at all torulose.
26. Fruit widely spreading, more or less without a beak, up to 1 mm broad; sepals up to 4 mm long... ....41. *Sisymbrium*
26. Fruit ascending, beaked, more than 1 mm broad; sepals more than 4 mm long.
27. Valves of fruit with 3–5 veins ....40. *Sinapis*
27. Valves of fruit with 1 vein...................9. *Brassica*

1. **Alliaria** Scop.—Garlic Mustard

1. **Alliaria petiolata** (Bieb.) Cavara & Grande. Garlic Mustard. May–June. Native to Europe; naturalized into woods and waste places; common in the n. ½ of IL, becoming less common in the s. cos., but spreading rapidly. *Alliaria officinalis* L.

2. **Alyssum** L.—Alyssum

1. **Alyssum alyssoides** (L.) L. Pale Alyssum. May–June. Native to Europe; naturalized along railroads and other waste places; occasional in the n. ⅔ of IL.

3. **Arabidopsis** Heynh.—Mouse-ear Cress

1. Biennials or perennials; petals 6–8 mm long; fruit flat, 0.8–1.0 mm wide.................. .... 1. *A. lyrata*
1. Annuals; petals 2–4 mm long; fruits terete, 0.5–0.8 mm wide ....2. *A. thaliana*

1. **Arabidopsis lyrata** (L.) O'Kane & Al-Shehbaz. Sand Cress. Apr.–Sept. Sand dunes, black oak savannas, dry gravelly prairies; in the n. ⅓ of IL. Common where the habitat is right. *Arabis lyrata* L.

2. **Arabidopsis thaliana** (L.) Heynh. Mouse-ear Cress. Mar.–June. Native to Europe; naturalized in waste areas; occasional in the s. ½ of IL, rare in the n. ½.

4. **Arabis** L.—Rock Cress

(See also Arabidopsis, Boechera, and Turritis.)

1. Plants with stolons; sepals 4.5–8.0 mm long; petals 10–20 mm long, 4–8 mm wide. ....1. *A. caucasica*
1. Plants without stolons; sepals 2.5–4.0 mm long; petals 3.5–5.5 mm long, 1–2 mm wide..2. *A. pycnocarpa*

1. **Arabis caucasica** Willd. Garden Rock Cress. May–June. Native to Asia; rarely escaped from cultivation; Macon Co.

2. **Arabis pycnocarpa** M. Hopkins. Three varieties occur in Illinois:

a. Stems pubescent all the way to the summit.
b. Stems hirsute ....2a. *A. pycnocarpa* var. *pycnocarpa*
b. Stems with appressed pubescence...............2b. *A. pycnocarpa* var. *adpressipilis*
a. Stems pubescent only on the lower ½ of the stem... ....2c. *A. pycnocarpa* var. *glabrata*

2a. **Arabis pycnocarpa** M. Hopkins var. **pycnocarpa**. Hairy Rock Cress. May–June. Limestone cliffs, woods; not common but scattered throughout IL. *Arabis hirsuta* (L.) Scop. var. *pycnocarpa* (M. Hopkins) Rollins.

2b. **Arabis pycnocarpa** M. Hopkins var. **adpressipilis** M. Hopkins. Hairy Rock Cress. May–June. Sandy soil, limestone woods; in the ne. cos.; also St. Clair Co. *Arabis hirsuta* (L.) Scop. var. *adpressipilis* (M. Hopkins) Rollins.

2c. **Arabis pycnocarpa** M. Hopkins var. **glabrata** (Torr. & Gray) Mohlenbr., comb. nov. (basionym: *Arabis hirsuta* (L.) Scop. var. *glabrata* Torr. & Gray). Rock Cress. June. Woods, very rare; Kankakee Co.

5. **Armoracia** Gaertn., Mey., & Scherb.

(See also Neobeckia.)

1. **Armoracia rusticana** (Lam.) Gaertn., Mey., & Scherb. Horseradish. May–July. Native to Europe; escaped from cultivation; occasional in IL. *Armoracia lapathifolia* Gilib.

6. **Barbarea** R. Br.—Winter Cress

1. All leaves pinnatifid; beak of fruit up to 1 mm long .... 1. *B. verna*
1. Uppermost leaves merely dentate; beak of fruit 1.5–3.0 mm long .... 2. *B. vulgaris*

1. **Barbarea verna** (Mill.) Aschers. Early Winter Cress. Apr.–June. Native to Europe; rarely naturalized in IL.
2. **Barbarea vulgaris** R. Br. Two varieties occur in Illinois:

a. Fruits strongly ascending .... 2a. *B. vulgaris* var. *vulgaris*
a. Fruits spreading .... 2b. *B. vulgaris* var. *arcuata*

2a. **Barbarea vulgaris** R. Br. var. **vulgaris.** Yellow Rocket. Apr.–June. Native to Europe; naturalized in waste ground; rare and scattered in IL.
2b. **Barbarea vulgaris** R. Br. var. **arcuata** (Opiz) Fries. Yellow Rocket. Apr.–June. Native to Europe; naturalized in waste ground; common throughout IL.

7. **Berteroa** DC.—Hoary Alyssum

1. **Berteroa incana** (L.) DC. Hoary Alyssum. May–Sept. Native to Europe; naturalized in fields, along roads, and in other disturbed areas; occasional, mostly in the n. cos.

8. **Boechera** Love & Love

1. None of the leaves sagittate nor auriculate at the base; fruit pendulous or horizontally spreading .... 1. *B. canadensis*
1. Some or all the cauline leaves sagittate or auriculate at the base; fruits spreading to ascending to erect (sometimes pendulous in *B. grahamii* and *B. canadensis*).
 2. Stems glabrous or only moderately pubescent at the base.
  3. Fruits less than 1 mm wide, straight, appressed, erect; seeds in 2 rows; stems usually 2–4; pubescence of forked hairs .... 6. *B. stricta*
  3. Fruits 1–4 mm wide, curved, pendulous to spreading to ascending, not erect and appressed; seeds in 1 row; stem 1; pubescence of simple hairs.
   4. Stems glabrous throughout .... 4. *B. laevigata*
   4. Stems pubescent at base.
    5. Fruits pendulous, on curved pedicels, secund, 2–4 mm wide; petals 3–5 mm long; none of the leaves lyrate-pinnatifid .... 1. *B. canadensis*
    5. Fruits ascending to spreading, on straight pedicels, not secund, 1–2 mm wide; petals 5–10 mm long; some of the basal leaves lyrate-pinnatifid .... 5. *B. missouriensis*
 2. Stems hairy throughout, often densely so at base.
  6. Pedicels curved; stem 1; seeds winged; petals 5.5–8.0 mm long; flowers sometimes lavender .... 3. *B. grahamii*
  6. Pedicels more or less straight; stems 2–4; seeds unwinged; petals 2.0–3.5 mm long; flowers white .... 2. *B. dentata*

1. **Boechera canadensis** (L.) Al-Shehbaz. Sicklepod. May–July. Mesic woods, dune woods, wooded slopes; occasional throughout IL. *Arabis canadensis* L.
2. **Boechera dentata** (Raf.) Al-Shehbaz & Zarucchi. Two varieties occur in Illinois:

a. Fruits stellate-pubescent .... 2a. *B. dentata* var. *dentata*
a. Fruits glabrous .... 2b. *B. dentata* var. *phalacrocarpa*

2a. **Boechera dentata** (Raf.) Al-Shehbaz var. **dentata.** Toothed Rock Cress. Apr.–June. Moist woods, floodplain forests; occasional in the n. ½ of IL, rare in the s. ½. *Arabis shortii* (Fern.) Gl.
2b. **Boechera dentata** (Raf.) Al-Shehbaz var. **phalacrocarpa** (M. Hopkins)

Mohlenbr., comb. nov. (basionym: *Arabis dentata* Raf. var. *phalacrocarpa* M. Hopkins). Toothed Rock Cress. May. Moist woods, rare; Lake Co. *Arabis shortii* (Fern.) Gl. var. *phalacrocarpa* (M. Hopkins) Steyerm.

3. **Boechera grahamii** (Lehmann) Windham & Al-Shehbaz. Purple Rock Cress. June–July. Rocky soil, very rare; Jo Daviess and Winnebago cos. *Arabis divaricarpa* A. Nelson, misapplied.

4. **Boechera laevigata** (Muhl. ex Willd.) Al-Shehbaz. Smooth Rock Cress; Smooth Bank Cress. Mar.–Aug. Moist woods, shaded banks; occasional to common throughout IL. *Arabis laevigata* (Muhl. ex Willd.) Poir.

5. **Boechera missouriensis** (Greene) Al-Shehbaz. Missouri Rock Cress. Mar.–June. Woods, rare; Union Co. *Arabis missouriensis* Greene.

6. **Boechera stricta** (Graham) Al-Shehbaz. Drummond's Rock Cress. May–July. Gravelly soil, very rare; n. ½ of IL. *Arabis drummondii* Gray.

9. **Brassica** L.—Mustard

(See also Synapis.)

1. None of the leaves auriculate and clasping at the base.
  2. Plants more or less hispid or hirsute, green; fruiting pedicels 3–7 mm long .... 3. *B. nigra*
  2. Plants glabrous or nearly so, glaucous; fruiting pedicels 7–10 mm long .... 1. *B. juncea*
1. Some of the cauline leaves auriculate and clasping at the base.
  3. Plants glaucous.
    4. Leaves not thick; petals 7–10 mm long; young leaves setose .... 5. *B. rapa*
    4. Leaves thick; petals 15 mm long or longer; young leaves glabrous .... 4. *B. oleracea*
  3. Plants green .... 2. *B. napus*

1. **Brassica juncea** (L.) Czern. Indian Mustard. Apr.–Oct. Native to Europe and Asia; naturalized in disturbed areas; occasional throughout IL.

2. **Brassica napus** L. Rape; Rutabaga. May–Sept. Native to Europe; rarely escaped from gardens in disturbed areas; scattered in IL.

3. **Brassica nigra** (L.) Koch. Black Mustard. Apr.–Oct. Native to Europe and Asia; naturalized in disturbed areas; occasional throughout IL.

4. **Brassica oleracea** L. Cabbage. Apr.–Sept. Native to Europe and Asia; rarely escaped from cultivation; Peoria Co.

5. **Brassica rapa** L. Field Mustard. Apr.–Sept. Native to Europe and Asia; naturalized in disturbed areas; occasional throughout IL.

10. **Cakile** Hill—Sea Rocket

1. **Cakile edentula** (Bigel.) Hook. ssp. **lacustris** (Fern.) Hulten. Sea Rocket. June–Nov. Sand beaches along Lake Michigan, rare; Cook and Lake cos.

11. **Camelina** Crantz—False Flax

1. Stems and leaves hirsutulous; fruits 4–5 (–7) mm long .... 1. *C. microcarpa*
1. Stems and leaves glabrous or with appressed pubescence; fruits 7–9 mm long ...... 2. *C. sativa*

1. **Camelina microcarpa** Andrz. Small-fruited False Flax. Apr.–Aug. Native to Europe; naturalized in disturbed areas; occasional throughout IL.

2. **Camelina sativa** (L.) Crantz. False Flax. Apr.–Aug. Native to Europe; naturalized in disturbed areas; scattered in IL.

12. **Capsella** Medic.—Shepherd's-purse

1. **Capsella bursa-pastoris** (L.) Medik. Shepherd's-purse. Jan.–Dec. Native to Europe; naturalized in disturbed areas; very common in IL; in every co.

13. **Cardamine** L.—Bitter Cress

1. Petals 7 mm long or longer; plants perennial.
  2. Some or all the leaves pinnate or pinnatifid .... 8. *C. pratensis*
  2. None of the leaves pinnate or pinnatifid.

3. Stems glabrous; sepals green .... 1. *C. bulbosa*
3. Stems hirsute; sepals purplish .... 2. *C. douglassii*
1. Petals up to 5 mm long; plants annual or biennial.
4. Leaves sagittate-auriculate at base .... 5. *C. impatiens*
4. None of the leaves sagittate-auriculate at base.
5. Petiole of cauline leaves ciliate at base.
6. Stems hirsute above; leaflets 4–20 mm wide; stamens 6...... 3. *C. flexuosa*
6. Stems glabrous; leaflets up to 4 mm wide; stamens 4............ 4. *C. hirsuta*
5. Petiole of cauline leaves not ciliate at base.
7. Terminal leaflet broader than the lateral leaflets; bases of leaflets decurrent along the rachis .... 7. *C. pensylvanica*
7. Terminal leaflet about the same width as the lateral leaflets; bases of leaflets not decurrent along the rachis .... 6. *C. parviflora*

1. **Cardamine bulbosa** (Muhl.) BSP. Spring Cress; Bulbous Cress. Mar.–June. Wet woods, along streams, calcareous fens; occasional to common throughout IL.

2. **Cardamine douglassii** (Torr.) Britt. Purple Spring Cress. Apr.–May. Low woods; occasional in the n. ⅔ of IL; also Johnson Co.

3. **Cardamine flexuosa** Withering. Cress. Mar.–June. Native to Europe and Asia; rarely adventive in gardens; Champaign and Coles cos.

4. **Cardamine hirsuta** L. Hairy Spring Cress. Feb.–June. Native to Europe and Asia; occasionally naturalized in disturbed areas throughout the state.

5. **Cardamine impatiens** L. Narrowleaf Bitter Cress. May–June. Native to Europe and Asia; disturbed wet soil; DuPage Co.

6. **Cardamine parviflora** L. var. **arenicola** (Britt.) O. E. Schultz. Small-flowered Bitter Cress. Mar.–July. Dry, open soil, wet ledges; occasional throughout IL.

7. **Cardamine pensylvanica** Willd. Bitter Cress. Mar.–Oct. Wet woods, wet fields, other moist areas; occasional throughout IL.

8. **Cardamine pratensis** L. var. **palustris** Wimm. & Grab. Cuckoo-flower. June–July. Wet ground, very rare; Lake, McHenry, and St. Clair cos.

14. **Chorispora** DC.—Blue Mustard

1. **Chorispora tenella** (Pallas) DC. Blue Mustard; Purple Rocket. Apr.–Oct. Native to Asia; adventive in disturbed areas, particularly along railroads; mostly in the ne. cos.

15. **Conringia** Link—Hare's-ear Mustard

1. **Conringia orientalis** (L.) Dumort. Hare's-ear Mustard. May–Aug. Native to Europe and Asia; naturalized in disturbed areas; occasional throughout IL.

16. **Dentaria** L.—Toothwort

1. Cauline leaves opposite, the segments broadly ovate .... 1. *D. diphylla*
1. Cauline leaves whorled, the segments narrowly oblong to lanceolate .... 2. *D. laciniata*

1. **Dentaria diphylla** Michx. Crinkleroot. May. Native n. and e. of IL; probably adventive in a mesic woods; Kane Co. *Cardamine diphylla* (Michx.) A. Wood.

2. **Dentaria laciniata** Muhl. Toothwort; Pennywort. Feb.–May. Woods, common; throughout the state. *Cardamine concatenata* (Michx.) O. Schwarz.

17. **Descurainia** Webb & Berthelot—Tansy Mustard

1. Fruits 1–2 mm thick; stems and leaves green, somewhat glandular-hairy .... 1. *D. pinnata*
1. Fruits up to 1 mm thick; stems and leaves grayish, not glandular-hairy .... 2. *D. sophia*

1. **Descurainia pinnata** (Walt.) Britt. ssp. **brachycarpa** (Richards.) Detling. Tansy Mustard. Apr.–June. Sandstone and limestone cliffs, where apparently native, and along railroads, where probably adventive; occasional to common throughout IL.

2. **Descurainia sophia** (L.) Prantl. Tansy Mustard; Flixweed. June–Aug. Native to Europe; adventive in disturbed soil, particularly barnyards and along railroads; scattered in IL.

18. **Diplotaxis** DC.—Rocket

1. Sepals 2.5–5.0 mm long; stems leafy only in lower ½; fruits sessile...... 1. *D. muralis*
1. Sepals 5–8 mm long; stems leafy throughout; fruits stipitate..............2. *D. tenuifolia*

1. **Diplotaxis muralis** (L.) DC. Wall Rocket. May–Oct. Native to Europe; naturalized in disturbed areas; occasional in the n. ½ of IL.

2. **Diplotaxis tenuifolia** (L.) DC. Sand Rocket. May–Sept. Native to Europe; rarely adventive in IL along a roadside; Cook Co.

19. **Draba** L.—Whitlow-grass

1. Petals notched ....5. *D. verna*
1. Petals entire.
  2. Leaves confined to the lower part of the stem, or all basal.
    3. Rachis and pedicels glabrous.
      4. Petals white; leaves entire; pedicels shorter than the fruits ....4. *D. reptans*
      4. Petals yellow; leaves usually dentate; pedicels as long as or longer than the fruits.. 3. *D. nemorosa*
    3. Rachis and pedicels pubescent ....2. *D. cuneifolia*
  2. Leaves all along the stem.
      5. Fruits up to 5 mm long, with up to 15 seeds............... 1. *D. brachycarpa*
      5. Fruits at least 6 mm long, with 20 or more seeds...........2. *D. cuneifolia*

1. **Draba brachycarpa** Nutt. Short-fruited Whitlow-grass. Mar.–May. Fields, lawns, prairies, woods, along railroads; common in the s. ⅔ of IL; also DeKalb Co., where it may be adventive.

2. **Draba cuneifolia** Nutt. Two varieties occur in Illinois:

a. Leaves basal and sometimes from the lowest nodes .... 2a. *D. cuneifolia* var. *cuneifolia*
a. Leaves cauline as well as basal... 2b. *D. cuneifolia* var. *foliosa*

2a. **Draba cuneifolia** var. **cuneifolia.** Wedge-leaved Whitlow-grass. Feb.–May. Limestone ledges, rare; Madison, Monroe, Randolph, and St. Clair cos.

2b. **Draba cuneifolia** var. **foliosa** Mohlenbr. Wedge-leaved Whitlow-grass. Mar. Limestone ledge, very rare; known only from the type locality in Monroe Co.

3. **Draba nemorosa** L. Whitlow-grass. Mar.–Apr. Dry soil, very rare; Jo Daviess Co.

4. **Draba reptans** (Lam.) Fern. Two varieties occur in Illinois:

a. Fruits glabrous...4a. *D. reptans* var. *reptans*
a. Fruits hispidulous ....4b. *D. reptans* var. *stellifera*

4a. **Draba reptans** (Lam.) Fern. var. **reptans.** Common Whitlow-grass. Feb.–May. Sand prairies, limestone ledges, hill prairies, dolomite glades; occasional in the n. cos., rare in the s. cos.

4b. **Draba reptans** (Lam.) Fern. var. **stellifera** (O. E. Schulz) Abrams. Common Whitlow-grass. Feb.–May. Prairies, glades; occasional in the n. cos. *Draba reptans* var. *micrantha* (Nutt.) Fern.

5. **Draba verna** L. Two subspecies occur in Illinois:

a. Seeds 40 or more per fruit; fruits more than twice as long as broad ....5a. *D. verna* ssp. *verna*
a. Seeds less than 40 per fruit; fruits never more than twice as long as broad ....5b. *D. verna* ssp. *praecox*

5a. **Draba verna** L. ssp. **verna.** Vernal Whitlow-grass. Feb.–May. Native to Europe and Asia; naturalized in waste areas, particularly in sandy soil; occasional in the s. cos., less common in the n. cos. *Eriophila verna* (L.) Chev.

5b. **Draba verna** L. ssp. **praecox** Stevens. Vernal Witlow-grass. Mar.–May. Native to Europe and Asia; naturalized in disturbed soil; Jackson Co. *Eriophila verna* (L.) Chev. ssp. *praecox* (Stevens) S. M. Walters.

20. **Eruca** Mill.—Garden Rocket

1. **Eruca vesicaria** (L.) Cav. ssp. **sativa** (Mill.) Thell. Garden Rocket. June–Sept. Native to Europe; rarely escaped from cultivation; Peoria Co. *Eruca sativa* Mill.

21. **Erucastrum** Presl—Rocket-weed

1. **Erucastrum gallicum** (Willd.) O. E. Schulz. Dog Mustard; Rocket-weed. Apr.–Oct. Native to Europe; naturalized in disturbed areas, particularly along railroads; occasional and scattered throughout IL.

22. **Erysimum** L.—Treacle Mustard

1. Petals well over 1 cm long.
  2. Basal leaves up to 2.5 mm wide; flowers yellow; fruits 4-angled, with 4 longitudinal stripes .... 1. *E. asperum*
  2. Basal leaves 3–25 mm wide; flowers usually orange, less frequently yellow; fruits usually not 4-angled, without 4 longitudinal stripes .... 2. *E. capitatum*
1. Petals up to 1 cm long.
  3. Petals up to 5 mm long .... 3. *E. cheiranthoides*
  3. Petals 5–10 mm long.
    4. Perennial; fruit up to 5 cm long; plants gray-green.
      5. Leaves denticulate, with some of the hairs on the upper surface 4-pronged .... 4. *E. hieracifolium*
      5. Leaves more or less entire, with none of the hairs on the upper surface 4-pronged .... 5. *E. inconspicuum*
    4. Annual; fruit more than 5 cm long; plants pale green .... 6. *E. repandum*

1. **Erysimum asperum** (Nutt.) DC. Rough Western Wallflower. May–July. Bluffs, very rare; w. IL.
2. **Erysimum capitatum** (Dougl.) Greene. Western Wallflower. May–July. Sandy bluffs, rare; w. cent. cos.; also Kendall, Lake, LaSalle, and Macon cos. *Erysimum arkansanum* Nutt.
3. **Erysimum cheiranthoides** L. Wormseed Mustard. May–Sept. Native to Europe and Asia; naturalized in disturbed soil in the n. ½ of IL; also Clark and St. Clair cos.
4. **Erysimum hieracifolium** L. Hawkweed Mustard. May–July. Native to Europe; naturalized along railroads and along roadsides; occasional in the n. ¼ of IL.
5. **Erysimum inconspicuum** (S. Wats.) MacM. Small Wormseed Mustard. May–July. Native to the w. U.S.; adventive in disturbed soil; occasional in the n. ½ of IL; also St. Clair and Washington cos.
6. **Erysimum repandum** L. Treacle Mustard. Apr.–June. Native to Europe and Asia; naturalized in disturbed soil; throughout the state.

23. **Hesperis** L.—Rocket

1. **Hesperis matronalis** L. Purple Rocket. May–Aug. Native to Europe and Asia; occasionally escaped from cultivation; scattered in IL.

24. **Iodanthus** Torr. & Gray—Purple Rocket

1. **Iodanthus pinnatifidus** (Michx.) Steud. Purple Rocket. May–July. Low woods, floodplain woods; occasional throughout the state, except for the nw. cos.

25. **Isatis** L.—Woad

1. **Isatis tinctoria** L. Dyer's Woad. May–June. Native to Europe; rarely escaped from cultivation; Cass, Champaign, and Cook cos.

26. **Lepidium** L.—Pepper-grass

1. At least some of the leaves auriculate, sagittate, or amplexicaul.
  2. Uppermost cauline leaves amplexicaul; plants glabrous or sparsely pubescent .... 6. *L. perfoliatum*
  2. Uppermost cauline leaves auriculate or sagittate; plants pubescent.
    3. Upper cauline leaves lobed or pinnatifid; fruits winged their entire length, 5–6 mm long; stamens 6 .... 8. *L. sativum*
    3. Upper cauline leaves entire, dentate, or serrulate; fruits unwinged or winged only near base, usually up to 4 mm long; stamens 2 (–4).
      4. Annual, with fibrous roots; fruits papillate, winged .... 1. *L. campestre*
      4. Perennial, with rhizomes; fruits smooth, unwinged .... 4. *L. draba*
1. None of the leaves auriculate, sagittate, or amplexicaul.
  5. Fruits rugose .... 3. *L. didymus*

5. Fruits smooth.
  6. Cauline leaves lance-ovate to ovate, some of them more than 1 cm wide; plants glaucous; perennials .... 5. *L. latifolium*
  6. Cauline leaves linear to linear-lanceolate, none of them as much as 1 cm wide; plants not glaucous; annuals or biennials.
    7. Petals absent.
      8. Plants with a fetid odor; basal leaves bipinnatifid .... 7. *L. ruderale*
      8. Plants without a fetid odor; basal leaves once-pinnatifid or coarsely toothed .... 2. *L. densiflorum*
    7. Petals about as long as the sepals .... 9. *L. virginicum*

1. **Lepidium campestre** (L.) R. Br. Field Pepper-grass; Field Cress. Apr.–June. Native to Europe; naturalized in disturbed areas, particularly pastures; occasional throughout IL.

2. **Lepidium densiflorum** Schrad. Small Pepper-grass; Small Pepper-cress. Apr.–Oct. Probably native to Europe and Asia; naturalized in disturbed areas; occasional throughout IL.

3. **Lepidium didymum** L. Wart Cress. June–Aug. Native to Europe; adventive in disturbed soil; Cook and Winnebago cos. *Coronopus didymus* (L.) Smith.

4. **Lepidium draba** L. Hoary Cress. Apr.–June. Native to Europe and Asia; naturalized in disturbed areas; occasional in the n. ½ of IL, rare in the s. ½. *Cardaria draba* (L.) Desv.

5. **Lepidium latifolium** L. Broad-leaved Pepper-cress. June–Aug. Native to Europe and Asia; naturalized along highways; ne. cos.; also Coles Co.

6. **Lepidium perfoliatum** L. Perfoliate Pepper-grass; Clasping Cress. Apr.–June. Native to Europe; naturalized in disturbed areas, particularly along roads and railroads, occasional; primarily in the n. cos.

7. **Lepidium ruderale** L. Stinking Pepper-grass; Stinking Pepper-cress. Apr.–June. Native to Europe; naturalized in disturbed soil; occasional in the n. ¼ of IL; also Mason Co.

8. **Lepidium sativum** L. Garden Pepper-grass; Garden Cress. May–June. Native to Europe; rarely escaped from cultivation; Cook, DuPage, and Kendall cos.

9. **Lepidium virginicum** L. Common Pepper-grass. Feb.–Dec. Disturbed soil, prairies, edge of woods, fields; common throughout IL; in every co.

27. **Lobularia** Desv.—Sweet Alyssum

1. **Lobularia maritima** (L.) Desv. Sweet Alyssum. June–Aug. Native to Europe; rarely escaped from cultivation; scattered in IL.

28. **Lunaria** L.

1. **Lunaria annua** L. Silver-dollar Plant. May–July. Native to Europe; occasionally planted but rarely escaped.

29. **Malcolmia** R. Br.

1. **Malcolmia africana** (L.) R. Br. Malcolmia. May. Native to Africa; rarely adventive in IL; along a railroad; Grundy Co.

30. **Matthiola** R. Br.—Stocks

1. **Matthiola incana** (L.) R. Br. Stocks. June–Aug. Native to Europe; occasionally planted but rarely escaped.

31. **Microthlaspi** F. K. Meyer

1. **Microthlaspi perfoliatum** (L.) F. K. Meyer. Perfoliate Penny Cress. June–July. Native to Europe; adventive in disturbed soil; not common; DuPage, Effingham, Hardin, Jackson, Shelby, and Williamson cos. *Thlaspi perfoliatum* L.

32. **Nasturtium** R. Br.—Water Cress

1. **Nasturtium officinale** R. Br. Water Cress. Apr.–Oct. Cool springs and branches; occasional throughout the state. (Including var. *silifolium* [Reichenb.] Koch.) *Rorippa nasturtium-aquaticum* (L.) Hayek.

33. **Neobeckia** Greene

1. **Neobeckia aquatica** Eaton. Lake Cress. May–Aug. Swamps and quiet streams; occasional throughout IL. *Armoracia aquatica* (Eaton) Wieg.; *A. lacustris* (Gray) Al-Shehbaz & V. Bates.

34. **Neslia** Desv.—Ball Mustard

1. **Neslia paniculata** (L.) Desv. Ball Mustard. May–Sept. Native to Europe; rarely adventive in IL; DuPage Co.

35. **Planodes** Greene

1. **Planodes virginica** (L.) Greene. Cress. Feb.–May. Open woods, waste ground; occasional to common. *Sibara virginica* (L.) Rollins; *Arabis virginica* L.

36. **Physaria** (Nutt. ex Torr. & Gray) Gray—Bladderpod

1. Annual; fruits glabrous or nearly so .... 1. *P. gracilis*
1. Perennial; fruits densely stellate-pubescent .... 2. *P. ludoviciana*

1. **Physaria gracilis** (Hook.) O'Kane & Al-Shehbaz. Slender Bladderpod. May–Aug. Native to the sw. U.S.; rarely adventive in IL. Cook Co. *Lesquerella gracilis* (Hook.) S. Wats.
2. **Physaria ludoviciana** (Nutt.) O'Kane & Al-Shehbaz. Silvery Bladderpod. May–Aug. Sandy soil, very rare; Mason Co. *Lesquerella ludoviciana* (Nutt.) S. Wats.

37. **Raphanus** L.—Radish

1. Petals yellowish or white; fruit up to 6 mm thick .... 1. *R. raphanistrum*
1. Petals pink or pale purple; fruit 6–10 mm thick .... 2. *R. sativus*

1. **Raphanus raphanistrum** L. Wild Radish. June–Aug. Native to Europe and Asia; occasionally adventive in disturbed soil; scattered in IL.
2. **Raphanus sativus** L. Radish. May–Sept. Native to Europe and Asia; commonly planted but rarely escaped from cultivation.

38. **Rapistrum** Crantz—Wild Rape

1. **Rapistrum rugosum** (L.) All. Wild Rape. May–July. Native to Europe; rarely adventive in disturbed soil; Champaign, DuPage, Fayette, and Kane cos.

39. **Rorippa** Scop.—Yellow Cress

(See also Nasturtium.)

1. Petals over 3 mm long, longer than the sepals; perennials.
  2. Some or all of the leaves lobed or pinnatifid.
    3. Leaves without auricles at base; seeds more than 1 mm long .... 7. *R. sylvestris*
    3. Leaves with auricles at base; seeds about 1 mm long.
      4. Stems erect, usually glabrous; cauline leaves entire to serrate; fruits straight, globose, 2.5–3.0 mm long .... 1. *R. austriaca*
      4. Stems decumbent or prostrate, usually hairy at base; at least some of the cauline leaves pinnatifid; fruits curved, narrowly oblong to linear, 5–15 mm long .... 6. *R. sinuata*
  2. None of the leaves lobed or pinnatifid, merely toothed .... 3. *R. indica*
1. Petals absent or up to 2.5 mm long, never longer than the sepals; annuals.
  5. Pedicels (1.5–) 3.0–8.0 mm long; fruits shorter than to up to 2 times longer than the pedicels.
    6. Stems erect; basal leaves in a rosette; petals 1.5–3.0 mm long .... 4. *R. palustris*
    6. Stems prostrate, decumbent, or ascending; basal leaves not in a rosette; petals 0.5–1.8 mm long .... 2. *R. curvipes*
  5. Pedicels up to 3 mm long; fruits 2–4 times longer than the pedicels.
    7. Petals absent or up to 0.5 mm long; seeds 150–200 per fruit .... 5. *R. sessiliflora*
    7. Petals 1–2 mm long; seeds 20–150 per fruit.

8. Petals about 1 mm long; seeds 20–70 per fruit.................... 8. *R. tenerrima*
8. Petals 1–2 mm long; seeds 100–150 per fruit............................... 9. *R. teres*

1. **Rorippa austriaca** (Crantz) Besser. Yellow Cress. May–July. Mud flats, ditches; occasional in IL.

2. **Rorippa curvipes** Greene. Yellow Cress. May–Sept. Muddy shores, wet fields, rare; Jackson and St. Clair cos.

3. **Rorippa indica** (L.) Hiern. Indian Yellow Cress. May–June. Native to Asia; rarely adventive in moist, disturbed areas; very rare; St. Clair Co.

4. **Rorippa palustris** (L.) Besser. Three varieties occur in Illinois:

a. Stems and leaves glabrous or nearly so.
b. Leaves pinnate or deeply pinnatifid, membranaceous .... 4a. *R. palustris* var. *palustris*
b. Leaves coarsely or shallowly toothed, firm..........4b. *R. palustris* var. *fernaldiana*
a. Stems and leaves hirsutulous...4c. *R. palustris* var. *hispida*

4a. **Rorippa palustris** (L.) Besser var. **palustris.** Marsh Yellow Cress. July–Sept. Mud and sand flats, rare; Gallatin, Jackson, and St. Clair cos. *Rorippa islandica* (Oeder) Borbas.

4b. **Rorippa palustris** (L.) Besser var. **fernaldiana** (Butters & Abbe) Stuckey. Marsh Yellow Cress. May–Nov. In ditches, along streams, other moist soil; common throughout IL. *Rorippa islandica* (Oeder) Borbas var. *fernaldiana* Butters & Abbe.

4c. **Rorippa palustris** (L.) Besser var. **hispida** (Desv.) Rydb. Hairy Marsh Yellow Cress. May–Oct. Wet ground; ne. ¼ of IL. *Rorippa islandica* (Oeder) Borbas var. *hispida* (Desv.) Butters & Abbe.

5. **Rorippa sessiliflora** (Nutt.) A. Hitchc. Sessile-flowered Yellow Cress. Apr.–Nov. Along rivers and streams; occasional to common in IL.

6. **Rorippa sinuata** (Nutt.) A. Hitchc. Spreading Yellow Cress. Apr.–July. Wet soil, particularly along rivers; occasional throughout IL.

7. **Rorippa sylvestris** (L.) Besser. Creeping Yellow Cress. May–Sept. Native to Europe; naturalized in wet areas; occasional throughout IL.

8. **Rorippa tenerrima** Greene. Slender Yellow Cress. May–Sept. Along rivers, rare; Jackson and St. Clair cos. *Rorippa truncata* (Jepson) Stuckey.

9. **Rorippa teres** (Michx.) Stuckey. Yellow Cress. May–June. Along Mississippi River, very rare; Jackson Co.

40. **Sisymbrium** L.—Hedge Mustard

1. Petals 5–6 mm long; fruits 2–10 cm long.
2. At least the upper leaves divided into threadlike divisions; fruits 5–10 cm long .... 1. *S. altissimum*
2. Leaves divided into triangular lobes; fruits 2–4 cm long.....................2. *S. loeselii*
1. Petals up to 3 mm long; fruits up to 2 cm long....................................3. *S. officinale*

1. **Sisymbrium altissimum** L. Tumble Mustard. May–Oct. Native to Europe and Asia; naturalized in disturbed soil, particularly along railroads; occasional throughout IL.

2. **Sisymbrium loeselii** L. Tall Hedge Mustard. May–Nov. Native to Asia; naturalized in disturbed soil, particularly in barnyards; mostly confined to the n. cos.

3. **Sisymbrium officinale** (L.) Scop. Two varieties occur in Illinois:

a. Fruits pubescent.. 3a. *S. officinale* var. *officinale*
a. Fruits glabrous or nearly so ....3b. *S. officinale* var. *leiocarpum*

3a. **Sisymbrium officinale** (L.) Scop. var. **officinale.** Hedge Mustard. May–Nov. Native to Europe and Asia; naturalized in disturbed soil; occasional to common throughout IL.

3b. **Sisymbrium officinale** (L.) Scop. var. **leiocarpum** DC. Hedge Mustard. May–Nov. Native to Europe and Asia; naturalized in disturbed soil; occasional to common throughout IL.

41. **Sinapis** L.—Mustard

1. Leaves all pinnatifid; fruits densely white-bristly..1. *S. alba*
1. At least the uppermost leaves merely toothed, not pinnatifid; fruits glabrous or nearly so..2. *S. arvensis*

1. **Sinapis alba** L. White Mustard. Apr.–July. Native to Europe and Asia; naturalized in disturbed soil; occasional throughout IL. *Brassica hirta* Moench; *B. alba* (L.) Rabenh.

2. **Sinapis arvensis** L. Charlock. Apr.–July. Native to Europe and Asia; naturalized in disturbed soil; occasional throughout IL. *Brassica kaber* (DC.) L. C. Wheeler var. *pinnatifida* (Stokes) L. C. Wheeler; *B. kaber* (DC.) L. C. Wheeler var. *schkuhriana* (Reichenb.) L. C. Wheeler; *B. kaber* (DC.) L. C. Wheeler.

42. **Thlaspi** L.—Penny Cress

1. **Thlaspi arvense** L. Penny Cress. Mar.–July. Native to Europe and Asia; naturalized into disturbed soil; occasional to common throughout IL.

43. **Turritis** L.—Tower Mustard

1. **Turritis glabra** L. Tower Mustard. May–July. Prairies, limestone woods, not common; throughout the state, but very rare in the s. cos. *Arabis glabra* (L.) Bernh.

43. BUDDLEJACEAE—BUDDLEJA FAMILY

1. Plants shrubby .... 1. *Buddleja*
1. Plants weak herbs .... 2. *Polypremum*

1. **Buddleja** L.—Butterfly Bush

1. **Buddleja davidii** Franch. Butterfly Bush. June–Oct. Native to Asia; rarely escaped from cultivation; Coles Co.

2. **Polypremum** L.—Polypremum

1. **Polypremum procumbens** L. Polypremum. June–Oct. Moist, sandy soil, very rare; Alexander Co.

44. BUXACEAE—BOXWOOD FAMILY

1. **Pachysandra** Michx.—Pachysandra

1. **Pachysandra terminalis** Sieb. & Zucc. Japanese Spurge. Native to Asia; occasionally planted as a ground cover, rarely escaped; Champaign, Cook, DuPage, Lake, and Kane cos.

45. CABOMBACEAE—WATERSHIELD FAMILY

1. Leaves uniform, reddish and usually gelatinous on the lower surface; stamens 12–18; carpels 4 or more .... 1. *Brasenia*
1. Leaves dimorphic, not reddish and not gelatinous on the lower surface; stamens 3–6; carpels 2–4 .... 2. *Cabomba*

1. **Brasenia** Schreb.—Watershield

1. **Brasenia schreberi** J. Gmel. Watershield. June–Sept. Ponds and quiet streams, rare; scattered in IL.

2. **Cabomba** Aubl.—Dimorphic Watershield

1. **Cabomba caroliniana** Gray. Carolina Watershield. May–Sept. Ponds; rare in the s. ⅔ of IL, absent elsewhere.

46. CACTACEAE—CACTUS FAMILY

1. Stems not jointed into segments; areoles bearing only spines .... 1. *Escobaria*
1. Stems jointed into segments, the segments cylindrical or flattened; areoles with both spines and glochidia .... 2. *Opuntia*

1. **Escobaria** Britt. & Rose

1. **Escobaria missouriensis** (Sweet) D. R. Hunt. Missouri Mammillaria. June–July. Among dry leaf litter near waterfall catchpool; Union Co. Apparently native. *Coryphantha missouriensis* (Sweet) Britt. & Rose.

2. **Opuntia** Mill.—Prickly Pear

1. Areoles with 3–9 spines; fruit dry, with spines and glochidia; stems not very flat, almost terete .... 1. *O. fragilis*

1. Areoles with 1–2 (–6) spines; fruit fleshy, without spines but with glochidia; stems very flat.
2. Joints of stem green; main roots fibrous; spines usually less than 3 cm long; margin of seed smooth, about 0.5 mm wide .... 2. *O. humifusa*
2. Joints of stem more or less glaucous; main roots tuberous-thickened; spines usually more than 3 cm long; margin of seed irregularly erose, about 1 mm wide .... 3. *O. macrorhiza*

1. **Opuntia fragilis** (Nutt.) Haw. Little Prickly Pear. June–July. Sandy soil, very rare; Jo Daviess Co.

2. **Opuntia humifusa** (Raf.) Raf. Common Prickly Pear. May–July. Dry soil, in sand, on exposed cliffs; scattered throughout IL. *Opuntia rafinesquii* Engelm.

3. **Opuntia macrorhiza** Engelm. Plains Prickly Pear. June–July. Edge of cliffs; confined to a few cliffs in the w. side of the state; also Grundy and Kane cos.

47. CAESALPINIACEAE—CAESALPINIA FAMILY

1. Trees.
2. Leaves simple; flowers pink .... 1. *Cercis*
2. Leaves once- or twice-pinnate; flowers whitish or greenish.
3. Some of the leaflets over 1.5 cm wide; flowers whitish; legume thick .... 4. *Gymnocladus*
3. None of the leaflets 1.5 cm wide; flowers greenish; legume thin .... 3. *Gleditsia*
1. Herbs.
4. Leaflets 4–18, at least 7 mm wide .... 5. *Senna*
4. Leaflets 20 or more, less than 7 mm wide .... 2. *Chamaechrista*

1. **Cercis** L.—Redbud

1. **Cercis canadensis** L. Redbud. Apr.–May. Woods, alluvial forests; common in the s. cos., becoming less abundant northward.

2. **Chamaechrista** Moench—Partridge Pea

1. Calyx 9–10 mm long; corolla 10–20 mm long; stamens 10; pedicels 10 mm long or longer; petiolar gland near base of petiole .... 1. *C. fasciculata*
1. Calyx 3–4 mm long; corolla less than 10 mm long; stamens 5; pedicels 2–4 mm long; petiolar gland near lowest pair of leaflets .... 2. *C. nictitans*

1. **Chamaechrista fasciculata** (Michx.) Greene. Two varieties occur in Illinois:

a. Stems with appressed pubescence .... 1a. *C. fasciculata* var. *fasciculata*
a. Stems spreading-hirsute .... 1b. *C. fasciculata* var. *robusta*

1a. **Chamaechrista fasciculata** (Michx.) Greene var. **fasciculata.** Partridge Pea. June–Oct. Open areas, prairies, fields; common throughout the state. *Cassia fasciculata* Michx.

1b. **Chamaechrista fasciculata** (Michx.) Greene var. **robusta** (Pollard) Pollard. Partridge Pea. June–Sept. Fields, rare; Jackson Co. *Cassia fasciculata* Michx. var. *robusta* (Pollard) Macbr.

2. **Chamaechrista nictitans** (L.) Moench. Sensitive Pea. July–Sept. Fields, roadsides, edge of woods; occasional in the s. ⅕ of IL, uncommon elsewhere. *Cassia nicticans* L.

3. **Gleditsia** L.—Honey Locust

1. Legume 15 cm long or longer, with sweet pulp between the several seeds .... 2. *G. triacanthos*
1. Legume to 5 cm long, with no pulp surrounding the 1–2 seeds .... 1. *G. aquatica*

1. **Gleditsia aquatica** Marsh. Water Locust. May–June. Swamps, rare; confined to s. IL, extending northward to Calhoun Co. in the w. and Lawrence Co. in the e.; also Henderson Co.

2. **Gleditsia triacanthos** L. Two forms occur in Illinois:

a. Trunk beset with thorns .... 2a. *G. triacanthos* f. *triacanthos*

a. Trunk without thorns.. 2b. *G. triacanthos* f. *inermis*

2a. **Gleditsia triacanthos** L. f. **triacanthos.** Honey Locust. May–June. Woods, along streams; common in the s. cos., becoming less common northward and possibly adventive in the n. cos.

2b. **Gleditsia triacanthos** L. f. **inermis** Schneid. Thornless Honey Locust. May–June. Woods; occasionally found with the typical form.

4. **Gymnocladus** Lam.—Coffee Tree

1. **Gymnocladus dioicus** (L.) K. Koch. Kentucky Coffee Tree. May–June. Mostly low woods, floodplains; occasional throughout the state.

5. **Senna** Mill.—Senna

1. Leaflets 4–6; petiolar gland midway between lowest pair of leaflets ....3. *S. obtusifolia*
1. Leaflets 8–18; petiolar gland near base of petiole.
  2. Leaflets obtuse; stems somewhat villous above .... 1. *S. hebecarpa*
  2. Leaflets acute; stems glabrous.
    3. Ovary and fruits pubescent; legume up to 10 cm long ....2. *S. marilandica*
    3. Ovary and fruits glabrous; legume 10–12 cm long.................4. *S. occidentalis*

1. **Senna hebecarpa** (Fern.) Irwin & Barneby. Wild Senna. July–Aug. Along streams, fens, open woods; occasional throughout IL. *Cassia hebecarpa* Fern.

2. **Senna marilandica** (L.) Link. Maryland Senna. July–Aug. Roadsides, thickets; occasional throughout IL. *Cassia marilandica* L.

3. **Senna obtusifolia** (L.) Irwin & Barneby. Sicklepod. July–Sept. Along streams and railroads, rare; Clinton, Coles, Cook, Jackson, Kankakee, Lake, Massac, Pulaski, Union, and Wabash cos. *Cassia obtusifolia* L.; *Cassia tora* L.

4. **Senna occidentalis** (L.) Link. Coffee Senna. Aug.–Sept. Native to the tropics; rarely adventive in IL; Cook Co. *Cassia occidentalis* L.

48. CALLITRICHACEAE—WATER STARWORT FAMILY

1. **Callitriche** L.—Water Starwort

1. Plants growing in water; flowers each with 2 bracts at base.
  2. Leaves all linear; fruit with a deep cleft................................1. *C. hermaphroditica*
  2. Floating leaves ovate to orbicular; fruit with a shallow cleft.
    3. Fruit as broad as long, rounded at base.................................2. *C. heterophylla*
    3. Fruit longer than broad, narrowed to base .... 4. *C. verna*
1. Plants growing on soil; flowers bractless ....3. *C. terrestris*

1. **Callitriche hermaphroditica** L. Linear-leaved Starwort. Apr.–Sept. Shallow water, very rare; Sangamon Co.

2. **Callitriche heterophylla** Pursh. Large Water Starwort. Apr.–Oct. Shallow water, occasional; scattered throughout IL.

3. **Callitriche terrestris** Raf. Terrestrial Starwort. Apr.–July. Moist soil; occasional in the s. ⅔ of IL, absent elsewhere.

4. **Callitriche verna** L. Common Water Starwort. Apr.–Oct. Shallow water, not common; scattered throughout IL. *Callitriche palustris* L.

49. CALYCANTHACEAE—CALYCANTHUS FAMILY

1. **Calycanthus** L.—Strawberry Shrub

1. **Calycanthus floridus** L. Strawberry Shrub. May–June. Native to the se. U.S.; rarely escaped from cultivation; Champaign, Henry, Jackson and Jersey cos.

50. CAMPANULACEAE—BELLFLOWER FAMILY

1. Flowers actinomorphic; corolla tube not split down 1 side; carpels 3–5.
  2. Leaves sessile, often clasping; flower solitary and sessile in the axils of the leaves... 4. *Triodanis*
  2. Leaves petiolate or at least not clasping; flowers in spikes or racemes or, if solitary, then pedicellate.
    3. Corolla rotate; capsule with pores nearly apical................2. *Campanulastrum*
    3. Corolla campanulate or funnelform; capsule with pores nearly basal .... 1. *Campanula*

1. Flowers zygomorphic; corolla tube split down 1 side; carpels 2 .... 3. *Lobelia*

1. **Campanula** L.—Bellflower

1. Cauline leaves linear to narrowly lanceolate (ovate, cordate leaves may be present at base of plant in *C. rotundifolia*).
  2. Stems glabrous or rarely closely puberulent; corolla 1.5 cm long or longer......... .... 4. *C. rotundifolia*
  2. Stems scabrous with retrorse hairs on the angles; corolla up to 1.2 cm long.
    3. Corolla white, 5–8 mm long; capsule up to 2 mm long .... 1. *C. aparinoides*
    3. Corolla bluish, 10–12 mm long; capsule 3–5 mm long............... 5. *C. uliginosa*
1. Cauline leaves oblong-lanceolate to ovate.
  4. Flowers short-pedicellate, drooping, borne in elongated, 1-sided racemes; lobes of calyx linear; capsule globose, nodding .... 3. *C. rapunculoides*
  4. Flowers sessile, erect, borne in glomerules; lobes of calyx lanceolate; capsule ovoid, erect .... 2. *C. glomerata*

1. **Campanula aparinoides** Pursh. Marsh Bellflower. June–Sept. Marshes, fens, extremely wet areas; occasional in the n. ½ of IL.

2. **Campanula glomerata** L. Clustered Bellflower. June–Aug. Native to Europe and Asia; rarely escaped from cultivation into disturbed areas.

3. **Campanula rapunculoides** L. European Bellflower. June–Sept. Native to Europe; adventive in waste areas; apparently confined to the n. ½ of IL.

4. **Campanula rotundifolia** L. Two varieties occur in Illinois:

a. Stems and cauline leaves glabrous .... 4a. *C. rotundifolia* var. *rotundifolia*
a. Stems and cauline leaves closely puberulent............ 4b. *C. rotundifolia* var. *velutina*

4a. **Campanula rotundifolia** L. var. **rotundifolia.** Bellflower. June–Sept. Woods, hill prairies, sandstone cliffs; occasional in the n. ½ of IL; also Jackson and Williamson cos. (Including var. *intercedens* [Witasek] Farw.)

4b. **Campanula rotundifolia** L. var. **velutina** A. DC. Hairy Bellflower. July–Sept. Crevices of cliffs, very rare; Jo Daviess Co.

5. **Campanula uliginosa** Rudb. Marsh Bellflower. June–Sept. Marshes, fens, very wet areas; occasional in the n. ½ of IL. *Campanula aparinoides* Pursh var. *grandiflora* Holz.

2. **Campanulastrum** Small—American Bellflower

1. **Campanulastrum americanum** (L.) Small. Two varieties occur in Illinois:

a. Lower leaves lanceolate, tapering to the base .... 1a. *C. americanum* var. *americanum*
a. Lower leaves ovate, abruptly contracted to a petiole .... 1b. *C. americanum* var. *illinoense*

1a. **Campanulastrum americanum** (L.) Small var. **americanum.** American Bellflower. June–Nov. Woods; common throughout IL; in every co. *Campanula americana* L.

1b. **Campanulastrum americanum** (L.) Small var. **illinoense** (Fresen.) Mohlenbr., comb. nov. (basionym: *Campanula illinoensis* Fresen.). American Bellflower. June–Nov. Woods; occasional throughout IL. *Campanula americana* L. var. *illinoense* (Fresen.) Farw.

3. **Lobelia** L.—Lobelia

1. Flowers 1.5 cm long or longer.
  2. Flowers red or deep rose (rarely white).
    3. Flowers red .... 2. *L. cardinalis*
    3. Flowers deep rose .... 7. *L. X speciosa*
  2. Flowers blue (rarely white).
    4. Auricles at base of calyx less than 2 mm long; stems densely puberulent throughout; flowers 1.5–2.5 cm long .... 5. *L. puberula*
    4. Auricles at base of calyx 2–5 mm long; stems glabrous or sparsely hirsute; flowers 2.0–3.3 cm long .... 6. *L. siphilitica*

1. Flowers up to 1.5 cm long.
   5. Cauline leaves linear to narrowly lanceolate, usually less than 3 mm wide; lower lip of corolla glabrous .... 4. *L. kalmii*
   5. Cauline leaves oblong, lanceolate, or obtuse, more than 3 mm wide; lower lip of corolla usually pubescent.
      6. Lower part of stems villous or hirsute; capsules inflated, completely enclosed by the calyx .... 3. *L. inflata*
      6. Stems glabrous or short-pubescent; capsules not inflated, partly exserted from the calyx.
         7. Leaves tapering to the base; lower part of stem short-pubescent .... 8. *L. spicata*
         7. Leaves rounded at the base; stems glabrous or with pubescence in lines .... 1. *L. appendiculata*

1. **Lobelia appendiculata** A. DC. Appendaged Lobelia. July–Aug. Very rare; Hancock Co. Collected by S. B. Mead in 1859.

2. **Lobelia cardinalis** L. Cardinal-flower. July–Oct. Wet ground; occasional to common throughout the state. White-flowered forms rarely occur.

3. **Lobelia inflata** L. Indian Tobacco. June–Oct. Woods, fields, disturbed areas, common; in every co.

4. **Lobelia kalmii** L. Bog Lobelia. July–Nov. Springy areas and dunes; occasional in the n. ⅓ of IL.

5. **Lobelia puberula** Michx. var. **simulans** Fern. Downy Lobelia. Aug.–Oct. Wet ground; occasional in the s. ⅓ of IL. Typical var. *puberula* occurs e. of IL.

6. **Lobelia siphilitica** L. Two varieties occur in Illinois:

a. Leaves strigose above, most of them more than 1.5 cm wide .... 6a. *L. siphilitica* var. *siphilitica*
a. Leaves glabrous above, less than 1.5 cm wide .... 6b. *L. siphilitica* var. *ludoviciana*

6a. **Lobelia siphilitica** L. var. **siphilitica.** Great Blue Lobelia. July–Oct. Wet ground, marshes, fens; occasional to common throughout IL.

6b. **Lobelia siphilitica** L. var. **ludoviciana** A. DC. Western Great Blue Lobelia. July–Oct. Wet ground, marshes, fens; scattered in IL.

7. **Lobelia X speciosa** Sweet. Hybrid Cardinal-flower. Sept. Low areas, very rare; Cass, Coles, DuPage, and Wabash cos. Considered to be a hybrid between *L. cardinalis* L. and *L. siphilitica* L.

8. **Lobelia spicata** Lam. Three varieties occur in Illinois:

a. Stems soft-hairy, at least near base.
   b. Appendages between calyx lobes up to 1 mm long, or absent; leaves mostly spreading .... 8a. *L. spicata* var. *spicata*
   b. Appendages between calyx lobes 1–5 mm long; leaves mostly ascending .... 8c. *L. spicata* var. *leptostachya*
a. Stems rough-hairy .... 8b. *L. spicata* var. *hirtella*

8a. **Lobelia spicata** Lam. var. **spicata**. Spiked Lobelia. July–Aug. Dry woods, prairies; occasional throughout the state.

8b. **Lobelia spicata** Lam. var. **hirtella** Gray. Hairy Spiked Lobelia. July–Aug. Dry woods, prairies; scattered in IL.

8c. **Lobelia spicata** Lam. var. **leptostachya** (A. DC.) Mack. & Bush. Spiked Lobelia. July–Aug. Dry soil, prairies; scattered in IL.

4. **Triodanis** Raf.—Venus' Looking-glass

1. Bracts suborbicular to ovate; seeds up to 0.7 mm long.
   2. Pores of capsules near apex .... 1. *T. biflora*
   2. Pores of capsules midway between apex and base .... 3. *T. perfoliata*
1. Bracts linear to lanceolate; seeds 0.7–1.0 mm long .... 2. *T. leptocarpa*

1. **Triodanis biflora** (R. & P.) Greene. Venus' Looking-glass. Apr.–June. Dry, often disturbed soil; confined to the s. ⅙ of IL. *Specularia biflora* R. & P.

2. **Triodanis leptocarpa** (Nutt.) Nieuwl. Slender-leaved Venus' Looking-glass. May–

July. Native to the w. U.S.; waste ground, along railroads, rare; Cass, DuPage, Livingston, Logan, Menard, and Will cos. *Specularia leptocarpa* (Nutt.) A. Gray.

3. **Triodanis perfoliata** (L.) Nieuwl. Common Venus' Looking-glass. Apr.–Aug. Dry, often disturbed, sandy soil, common; probably in every co. *Specularia perfoliata* (L.) A. DC.

51. CANNABINACEAE—CANNABIS FAMILY

1. Plants erect; leaves compound .... 1. *Cannabis*
1. Plants climbing or twining; leaves simple .... 2. *Humulus*

1. **Cannabis** L.—Hemp; Marijuana

1. **Cannabis sativa** L. Two varieties occur in Illinois:

a. Achenes 3.8 mm long or longer .... 1a. *C. sativa* var. *sativa*
a. Achenes less than 3.8 mm long .... 1b. *C. sativa* var. *spontanea*

1a. **Cannabis sativa** L. var. **sativa**. Hemp; Marijuana. June–Oct. Native to Asia; naturalized and escaped into waste areas; occasional throughout IL.

1b. **Cannabis sativa** L. var. **spontanea** Vavilov. Hemp. June–Oct. Native to Europe and Asia; naturalized and scattered in IL.

2. **Humulus** L.—Hops

1. Most of the leaves 5- to 7-lobed, without resinous glands beneath .... 1. *H. japonicus*
1. Most of the leaves 3-lobed or unlobed, with resinous glands beneath .... 2. *H. lupulus*

1. **Humulus japonicus** Sieb. & Zucc. Japanese Hops. Native to Asia; naturalized into disturbed areas; occasional in IL.

2. **Humulus lupulus** L. Three varieties occur in Illinois:

a. Plants sparsely pubescent and sparsely glandular on the lower surface .... 2a. *H. lupulus* var. *lupulus*
a. Plants moderately to copiously pubescent and glandular on the lower surface.
  b. Leaves usually 3-lobed, glabrous or nearly so between the veins on the lower surface .... 2b. *H. lupulus* var. *lupuloides*
  b. Leaves often unlobed, pubescent between the veins on the lower surface .... 2c. *H. lupulus* var. *pubescens*

2a. **Humulus lupulus** L. var. **lupulus.** Hops. July–Aug. Native to Europe; rarely escaped from cultivation; scattered in IL.

2b. **Humulus lupulus** L. var. **lupuloides** E. Small. Hops. July–Aug. Fencerows, thickets; occasional throughout IL.

2c. **Humulus lupulus** L. var. **pubescens** E. Small. Hops. July–Aug. Fencerows, thickets; occasional throughout IL.

52. CAPRIFOLIACEAE—HONEYSUCKLE FAMILY

1. Plants woody, either small trees, shrubs, or vines, the wood sometimes soft.
  2. Leaves pinnately compound .... 4. *Sambucus*
  2. Leaves simple.
    3. Mature leaves entire (leaves of vigorous shoots sometimes pinnatifid).
      4. Corolla tubular or funnelform, usually zygomorphic; berries several-seeded .... 3. *Lonicera*
      4. Corolla campanulate, actinomorphic; drupes 2-seeded .... 5. *Symphoricarpos*
    3. Leaves toothed or lobed.
      5. Corolla tubular; inflorescence 3-flowered; fruit a several-seeded capsule .... 1. *Diervilla*
      5. Corolla spreading; inflorescence almost always more than 3-flowered; fruit a 1-seeded drupe .... 7. *Viburnum*
1. Plants herbaceous and creeping, ascending, or erect.
  6. Stems creeping; leaves evergreen; flowers pedicellate; fruit a 1-seeded capsule .... 2. *Linnaea*

6. Stems ascending to erect; leaves deciduous; flowers sessile; fruit a 2- to 5-seeded drupe .... 6. *Triosteum*

1. **Diervilla** Mill.—Bush Honeysuckle

1. **Diervilla lonicera** Mill. Bush Honeysuckle; Dwarf Honeysuckle. May–Aug. Sandy woods, shaded woods; occasional in the n. ¼ of IL; also Brown, Coles, Piatt, and Pope cos.

2. **Linnaea** Gron.—Twin-flower

1. **Linnaea borealis** L. var. **longiflora** Torr. Twin-flower. June–Aug. Bogs, very rare; Cook Co. *Linnaea borealis* L. ssp. *americana* (Forbes) Hulten. Typical var. *borealis* does not occur in Illinois.

3. **Lonicera** L.—Honeysuckle

1. Some of the leaves connate.
  2. Corolla at least 3.5 cm long, bright red, the lobes more or less equal.
    3. Plants twining and climbing .... 16. *L. sempervirens*
    3. Plants erect, shrubby .... 5. *L. X heckrottii*
  2. Corolla up to 3 cm long, white, yellow, orange, purple, or brick-red, the lobes 2-lipped.
    4. Uppermost connate leaves broader than long, glaucous on the upper surface .... 14. *L. reticulata*
    4. Uppermost connate leaves longer than broad, not glaucous.
      5. Corolla orange-yellow; leaves gray on the lower surface .... 4. *L. flava*
      5. Corolla yellow, sometimes tinged with purple or brick-red; leaves whitened on the lower surface.
        6. Leaves without cilia, glabrous on the upper surface; tube of corolla glabrous or eglandular .... 3. *L. dioica*
        6. Leaves densely ciliate, strigose on the upper surface; tube of corolla glandular-pubescent .... 6. *L. hirsuta*
1. None of the leaves connate.
  7. Plants twining or climbing; corolla mostly 3 cm long or longer .... 7. *L. japonica*
  7. Plants erect or ascending; corolla up to 2 cm long.
    8. Peduncles much longer than the petioles.
      9. Branches with solid pith.
        10. Stems retrorsely bristly at the nodes .... 17. *L. standishii*
        10. Stems without retrorse bristles at the nodes .... 2. *L. canadensis*
      9. Branches hollow.
        11. Leaves averaging widest above the middle.
          12. Leaves averaging up to twice as long as wide.
            13. Bractlets and calyx lobes pubescent on the surface as well as on the margins .... 21. *L. xylosteum*
            13. Bractlets and calyx lobes merely ciliate along the margins .... 20. *L. X xylosteoides*
          12. Leaves more than twice as long as wide.
            14. Corolla tube up to 3 mm long .... 15. *L. ruprechtiana*
            14. Corolla tube 4–5 mm long or longer .... 11. *L. X muendeniensis*
        11. Leaves averaging widest at or below the middle.
          15. Plants glabrous throughout, or with an occasional gland on the calyx lobes and bractlets.
            16. Leaves widest below the middle .... 19. *L. tatarica*
            16. Leaves widest at the middle .... 13. *L. X notha*
          15. Plants sparsely to densely pubescent.
            17. Corolla averaging less than 12 mm long.
              18. Corolla tube up to 3 mm long .... 15. *L. ruprechtiana*
              18. Corolla tube 4–5 mm long.
                19. Bractlets much more than ½ to equaling the length of the ovary .... 9. *L. X minutiflora*
                19. Bractlets up to or scarcely exceeding ½ the length of the ovary .... 11. *L. X muendeniensis*

17. Corolla 12 mm long or longer.
20. Bractlets up to or scarcely exceeding ½ the length of the ovary.
21. Corolla up to 13 mm long; many of the leaves usually more or less acuminate .... 11. *L. X muendeniensis*
21. Corolla more than 13 mm long; leaves usually acute or obtuse .... 1. *L. X bella*
20. Bractlets much more than ½ to nearly equaling the length of the ovary.
22. Calyx lobes, bractlets, and bracts without stipitate glands .... 10. *L. morrowii*
22. Calyx lobes, bractlets, and bracts stipitate-glandular.
23. Bracts densely hirsute and glandular; margins of bractlets copiously ciliate and glandular .... 12. *L. X muscaviensis*
23. Bracts and margins of bractlets only sparsely hairy and glandular .... 9. *L. X minutiflora*
8. Peduncles shorter than the petioles.
24. Leaves acuminate at the tip .... 8. *L. maackii*
24. Leaves acute at the tip .... 18. *L. subsessilis*

1. **Lonicera X bella** Zabel. Showy Fly Honeysuckle. May–June. Woods, scattered in IL; considered to be the hybrid between *L. morrowii* Gray and *L. tatarica* L.

2. **Lonicera canadensis** Marsh. American Fly Honeysuckle. Apr.–May. Woods, very rare; Cook Co.

3. **Lonicera dioica** L. Two varieties occur in Illinois:

a. Leaves glabrous or nearly so .... 3a. *L. dioica* var. *dioica*
a. Leaves pubescent, at least on the lower surface .... 3b. *L. dioica* var. *glaucescens*

3a. **Lonicera dioica** L. var. **dioica.** Red Honeysuckle. May–June. Woods, not common; scattered in the n. ⅓ of IL.

3b. **Lonicera dioica** L. var. **glaucescens** (Rydb.) Butters. Red Honeysuckle. May–June. Rocky woods, very rare; Jackson Co.

4. **Lonicera flava** Sims. Yellow **Honeysuckle.** Apr.–May. Rocky woods, sandstone cliffs, rare; Jackson, Pope, and Randolph cos. *Lonicera flavescens* Small.

5. **Lonicera X heckrottii** Rehder. Gold Flame Honeysuckle. June–Sept. Escaped along highways in the ne. cos. Considered to be the hybrid between *L. americana* L. and *L. sempervirens* L.

6. **Lonicera hirsuta** Eat. Ciliate Honeysuckle. June–July. Native n. of IL; Cook Co.

7. **Lonicera japonica** Thunb. Two varieties occur in Illinois:

a. Corolla white on the outside; branchlets and leaves green.. .... 7a. *L. japonica* var. *japonica*
a. Corolla reddish on the outside; branchlets and leaves purplish .... 7b. *L. japonica* var. *chinensis*

7a. **Lonicera japonica** Thunb. var. **japonica.** Japanese Honeysuckle. May–Sept. Native to Asia; woods, thickets; very common in the s. ½ of IL, less common elsewhere. Juvenile leaves are sometimes deeply pinnatifid.

7b. **Lonicera japonica** Thunb. var. **chinensis** (P. W. Wats.) Baker. Chinese Honeysuckle. May–Sept. Native to Asia; fencerows; Pope and Williamson cos.

8. **Lonicera maackii** (Rupr.) Maxim. Amur Honeysuckle. May–June. Native to Asia; escaped into woodlands; throughout IL and becoming increasingly aggressive.

9. **Lonicera X minutiflora** Zabel. Small-flowered Fly Honeysuckle. May. Occasionally escaped in the ne. cos. Considered to be the hybrid between *L. morrowii* Gray and *L. X xylostioides* Tausch.

10. **Lonicera morrowii** Gray. Morrow's Honeysuckle. May–June. Native to Asia; scattered in the n. ½ of IL.

11. **Lonicera X muendeniensis** Rehd. Common Fly Honeysuckle. May. Occasional in the ne. cos. Considered to be the hybrid between *L. X bella* Zabel and *L. ruprechtiana* Regel.

12. **Lonicera X muscaviensis** Rehd. May. Occasional in the ne. cos. Considered to be the hybrid between *L. morrowii* Gray and *L. ruprechtiana* Regel.

13. **Lonicera X notha** Zabel. May. Occasional in the ne. cos. Considered to be the hybrid between *L. X bella* Zabel and *L. X muendeniensis* Rehd.

14. **Lonicera reticulata** Raf. Yellow Honeysuckle; Grape Honeysuckle. May–July. Wooded slopes, rocky banks; occasional to common in the n. 3/5 of IL. *Lonicera prolifera* (Kirchn.) Rehd.

15. **Lonicera ruprechtiana** Regel. Manchurian Honeysuckle. May. Native to Asia; adventive in the ne. cos.

16. **Lonicera sempervirens** L. Trumpet Honeysuckle. Apr.–June. Native to the s. U.S.; occasionally adventive in most of the state.

17. **Lonicera standishii** Jacques. Standish's Honeysuckle. Apr.–May. Native to Asia; escaped along a roadside in Jackson Co.

18. **Lonicera subsessilis** Rehd. Short-stalked Honeysuckle. May–June. Native to Asia; escaped in DuPage Co.

19. **Lonicera tatarica** L. Tartarian Honeysuckle. May–June. Native to Europe and Asia; escaped from cultivation and scattered in IL.

20. **Lonicera X xylosteoides** Tausch. May. Escaped from cultivation in DuPage, Grundy, and Lake cos. Considered to be the hybrid between *L. tatarica* L. and *L. xylosteum* L.

21. **Lonicera xylosteum** L. European Fly Honeysuckle. May. Native to Europe and Asia; escaped from cultivation and scattered in IL.

4. **Sambucus** L.—Elderberry

1. Cymes umbel-like, flat-topped; twigs with white pith; fruit dark purple .... 1. *S. nigra*

1. Cymes paniculate, ovoid; twigs with brown pith; fruit usually bright red .... 2. *S. racemosa*

1. **Sambucus nigra** L. Two varieties occur in Illinois:

a. Leaflets glabrous or hirtellous on the veins beneath .... 1a. *S. nigra* var. *nigra*

a. Leaflets puberulent throughout on the lower surface ...... 1b. *S. nigra* var. *submollis*

1a. **Sambucus nigra** L. var. **nigra**. Elderberry. June–July. Woods, along roads, common; in every co. *S. canadensis* L.

1b. **Sambucus nigra** L. var. **submollis** Rehd. Hairy Elderberry. June–July. Rocky woods; confined to the s. 1/6 of IL. *S. canadensis* L. var. *submollis* Rehd.

2. **Sambucus racemosa** L. ssp. **pubens** (Michx.) House. Red-berried Elder. Apr.–May. Mesic woods, swampy woods, not common; in the n. 1/4 of IL. *Sambucus pubens* Michx.

5. **Symphoricarpos** Duham.—Snowberry

1. Corolla 5–9 mm long, pink; drupes white or greenish white.

  2. Flowers pedicellate; style included within the corolla .... 1. *S. albus*

  2. Flowers sessile; style exserted from the corolla .... 2. *S. occidentalis*

1. Corolla 3–4 mm long, greenish and purplish; drupes coral-pink to purple .... 3. *S. orbiculatus*

1. **Symphoricarpos albus** (L.) Blake. Two varieties occur in Illinois:

a. Leaves more or less pilose on the lower surface; branchlets puberulent .... 1a. *S. albus* var. *albus*

a. Leaves and branchlets glabrous .... 1b. *S. albus* var. *laevigatus*

1a. **Symphoricarpos albus** (L.) Blake var. **albus** Snowberry. June. Edge of cliff, very rare; Kane Co.

1b. **Symphoricarpos albus** (L.) Blake var. **laevigatus** (Fern.) Blake. Snowberry. June–July. Native to the w. U.S.; escaped from cultivation in n. 1/2 of IL. *S. rivularis* Suksd.

2. **Symphoricapos occidentalis** Hook. Wolfberry. June–Aug. Dry open ground; occasional in the n. 1/2 of IL.

3. **Symphoricarpos orbiculatus** Moench. Coral-berry; Buckbrush; Indian Currant. July–Sept. Disturbed woods, thickets, pastures; occasional throughout the state.

6. **Triosteum** L.—Horse Gentian

1. Stems setose-hispid; calyx lobes glabrous except for the bristly margins .... 1. *T. angustifolium*
1. Stems glabrous or villous or glandular-pubescent; calyx lobes pubescent throughout.
  2. None or only up to 2 (–3) pairs of leaves connate at the base; style included within the corolla; corolla orange or red.
    3. Stems with some glandular pubescence .... 2. *T. aurantiacum*
    3. Stems without glandular pubescence .... 3. *T. illinoense*
  2. At least 3 or more pairs of leaves connate at the base; style exserted from the corolla; corolla yellow, green, or purplish .... 4. *T. perfoliatum*

1. **Triosteum angustifolium** L. Yellow-flowered Horse Gentian. Apr.–May. Low ground, rocky woods, not common; confined to the s. ½ of IL; also Pike Co.

2. **Triosteum aurantiacum** Bickn. Early Horse Gentian. Apr.–June. Dry woods, open woods, rich woods; occasional in the n. ¾ of IL; also St. Clair and Williamson cos.

3. **Triosteum illinoense** (Wieg.) Rydb. Illinois Horse Gentian. May–June. Dry woods, open woods, rich woods; occasional throughout the state. *Triosteum aurantiacum* Bickn. var. *illinoense* (Wieg.) Palmer & Steyerm.

4. **Triosteum perfoliatum** L. Late Horse Gentian. May–July. Dry woods, thickets, fencerows; occasional in the n. ⅔ of IL, less common in the s. ⅓.

7. **Viburnum** L.—Viburnum

1. Leaves more or less 3-lobed, palmately veined.
  2. Leaves dotted beneath; young branchlets more or less pilose; drupes purple-black at maturity.. 1. *V. acerifolium*
  2. Leaves not dotted beneath; young branchlets glabrous or nearly so; drupes red or orange.
    3. Petiole with concave glands; stipules slender-tipped...................... 7. *V. opulus*
    3. Petiole with convex glands; stipules broadened or thickened at the tip........... ....14. *V. trilobum*
1. Leaves not lobed, pinnately veined.
  4. Leaves evergreen, some of them usually denticulate, the lower surface yellow-tomentose.. 12. *V. rhytidophyllum*
  4. Leaves deciduous, regularly serrate or dentate, the lower surface not yellow-tomentose.
    5. Leaves finely or sharply serrate.
      6. Pubescence of stellate hairs; winter buds without scales ....4. *V. lantana*
      6. Pubescence not stellate or absent; winter buds with 1–2 pairs of outer scales.
        7. Petiole and midvein of lower surface of leaves rufous-pubescent......... .... 13. *V. rufidulum*
        7. Petiole and midvein of lower surface of leaves glabrous or without rufous pubescence.
          8. Leaves obtuse to acute at the apex; margins of the petioles neither undulate nor revolute.
            9. Lower surface leaves tomentellous throughout; branches in layers.. 8. *V. plicatum*
            9. Lower surface of leaves glabrous or pubescent but not tomentellous; branches not in layers........................... 9. *V. prunifolium*
          8. Leaves acuminate at the apex; margins of the petioles undulate and revolute... 5. *V. lentago*
    5. Leaves dentate.
      10. Leaves deeply cordate, with 40 or more teeth; bark peeling ......6. *V. molle*
      10. Leaves subcordate, rounded, or subcuneate at the base, with up to 40 (–44) teeth; bark not peeling.

11. Petioles with linear stipules, the leaves subtending the inflorescence with petioles up to 7 mm long................................10. *V. rafinesquianum*
11. Petioles without stipules, the leaves subtending the inflorescence usually with petioles longer than 7 mm.
12. Leaves pubescent over both surfaces; drupes red.....3. *V. dilatatum*
12. Leaves glabrous or somewhat pubescent on the lower surface, never pubescent throughout; drupes blue-black.
13. Branchlets and lower surface of leaves with pubescence............ ....2. *V. dentatum*
13. Branchlets and lower surface of leaves glabrous or nearly so .... 11. *V. recognitum*

1. **Viburnum acerifolium** L. Maple-leaved Arrowwood. May–July. Moist woods, mesic woods, dune slopes, swamps, occasional; confined to the ne. ¼ of IL; also Clark Co.

2. **Viburnum dentatum** L. Four varieties occur in Illinois:

a. Stipules and stipitate-glandular hairs absent.
b. Leaves not scabrous on the upper surface.............2a. *V. dentatum* var. *dentatum*
b. Leaves scabrous on the upper surface ....2d. *V. dentatum* var. *scabrellum*
a. Stipules usually present; stipitate-glandular hairs usually present.
c. Leaves stellate-pubescent beneath on the surface as well as on the veins........... ....2b. *V. dentatum* var. *deamii*
c. Leaves stellate-pubescent only on the veins beneath .... 2c. *V. dentatum* var. *indianense*

2a. **Viburnum dentatum** L. var. **dentatum**. Arrowwood. May–June. Thickets; escaped from cultivation; Cook Co.

2b. **Viburnum dentatum** L. var. **deamii** (Rehd.) Fern. Arrowwood. May–June. Woods, rare; Hardin, Jackson, and Pope cos.

2c. **Viburnum dentatum** L. var. **indianense** (Rehd.) Fern. Indiana Arrowwood. May–June. Woods, rare; se. IL.

2d. **Viburnum dentatum** L. var. **scabrellum** Torr. & Gray. Rough Arrowwood. May–June. Escaped from cultivation; Lake Co. *Viburnum scabrellum* (Torr. & Gray) Chapm.

3. **Viburnum dilatatum** Thunb. Viburnum. May–June. Native to Asia; rarely escaped from cultivation; Champaign and Jackson cos.

4. **Viburnum lantana** L. Wayfaring Tree. May–June. Native to Europe; rarely escaped from cultivation in the ne. cos.; also Adams Co.

5. **Viburnum lentago** L. Nannyberry. Apr.–June. Moist woods, along streams, fencerows; occasional in the n. ¼ of IL; also DeWitt, Jackson, Morgan, and Tazewell cos.

6. **Viburnum molle** Michx. Arrowwood. May–June. Rocky banks of streams, dry hillsides, rare; Adams, Brown, Marshall, Peoria, Pike, and Scott cos.

7. **Viburnum opulus** L. European High-bush Cranberry. May–July. Native to Europe; frequently cultivated and often escaped from cultivation and aggressive, particularly in the ne. cos.

8. **Viburnum plicatum** Thunb. var. **tomentosa** Miq. Shasta Viburnum. May–June. Native to Asia; escaped into a woods; Jackson Co.

9. **Viburnum prunifolium** L. Black Haw; Nannyberry. Apr.–June. Low woods, wooded slopes; occasional throughout IL.

10. **Viburnum rafinesquianum** Schultes. Two varieties occur in Illinois:

a. Leaves pubescent throughout................10a. *V. rafinesquianum* var. *rafinesquianum*
a. Leaves glabrous or pubescent only on the veins.. ....10b. *V. rafinesquianum* var. *affine*

10a. **Viburnum rafinesquianum** Schultes var. **rafinesquianum**. Downy Arrowwood. May–June. Wooded slopes, rocky woods; occasional in the n. ½ of IL, extending southward to Clark and Coles cos.

10b. **Viburnum rafinesquianum** Schultes var. **affine** (Bush) House. Smooth Arrowwood. May–June. Woods; scattered in the n. ½ of IL.

11. **Viburnum recognitum** Fern. Smooth Arrowwood. May–June. Woods, along streams; occasional in the s. ½ of IL; also Cook, DuPage, Henry, Kane, Kankakee, Lake, and LaSalle cos. *Viburnum dentatum* L. var. *lucidum* Aiton.

12. **Viburnum rhytidophyllum** Hemsl. Leather-leaved Viburnum. June. Native to Asia; rarely escaped from cultivation; Champaign, Jackson, and St. Clair cos.

13. **Viburnum rufidulum** Raf. Southern Black Haw; Rusty Nannyberry. Apr.–May. Rocky woods; occasional in the s. ¼ of IL, extending northward in cos. along the Mississippi River to Pike Co.; also Adams, Menard, and Schuyler cos.

14. **Viburnum trilobum** Marsh. High-bush Cranberry. June–July. Moist woods, rare; restricted to the n. ½ of IL; also Greene Co. *Viburnum opulus* L. var. *americanum* Aiton.

53. CARYOPHYLLACEAE—PINK FAMILY

1. Leaves with stipules, although sometimes the stipules reduced to thread-like structures less than 1 mm wide.
  2. Leaves whorled .... 17. *Spergula*
  2. Leaves opposite.
    3. Petals present; styles 3; fruit a capsule with several seeds ..... 18. *Spergularia*
    3. Petals absent; styles 2; fruit a 1-seeded utricle .... 11. *Paronychia*
1. Leaves without stipules.
  4. Fruit a 1-seeded utricle; petals absent.......... 15. *Scleranthus*
  4. Fruit a capsule with several seeds; petals present (absent in *Stellaria pallida, Sagina apetala,* and occasionally *Sagina procumbens, Silene antirrhina, Moenchia,* and on lateral branches of *Cerastium glomeratum*).
    5. Sepals free from each other; petals not long-clawed.
      6. Each petal 2-cleft, usually deeply so, or jagged toothed.
        7. Styles 3.
          8. Petals jagged-toothed at the apex .... 6. *Holosteum*
          8. Petals 2-cleft at the apex.
            9. Capsules curved, at least in the upper ½.......... 3. *Cerastium*
            9. Capsules straight .... 19. *Stellaria*
        7. Styles 5.
          10. Capsules curved, at least in the upper ½; styles opposite the sepals .... 3. *Cerastium*
          10. Capsules straight; styles alternate with the sepals ..... 10. *Myosoton*
      6. Each petal entire or slightly notched, or petals absent.
        11. Petals absent.
          12. Styles 3; stamens 1–3.......... 19. *Stellaria*
          12. Styles 4–5; stamens 4, 5, 8, or 10.
            13. Styles 4; stamens 4 or 8.
              14. Leaves linear to subulate; pedicels glandular-pubescent; sepals 1.5–2.0 mm long, glandular-pubescent.......... 13. *Sagina*
              14. Leaves linear to linear-lanceolate; pedicels glabrous; sepals 4–7 mm long, glabrous .... 9. *Moenchia*
            13. Styles 5; stamens 5 or 10.
              15. Plants usually matted, not viscid-pubescent; sepals 1–2 mm long.......... 13. *Sagina*
              15. Plants upright, viscid-pubescent; sepals 3.5–5.0 mm long .... 3. *Cerastium*
        11. Petals present.
          16. Styles 4; stamens 4 or 8; sepals 4; petals 4.......... 9. *Moenchia*
          16. Styles 5; stamens 5 or 10; sepals 5; petals 5.
            17. Leaves subulate, setaceous, filiform, or linear.......... 7. *Minuartia*
            17. Leaves oval, oblong, or ovate.
              18. Perennial with slender rhizomes; leaves obtuse to subacute; sepals obtuse, 2–3 mm long.......... 8. *Moehringia*
              18. Annual; leaves acuminate; sepals acuminate, 3–4 mm long .... 2. *Arenaria*
    5. Sepals united, the calyx tubular or urn-shaped; petals long-clawed.
      19. Styles 3–5; teeth of valves of fruit 3, 5, 6, or 10.
        20. Bracts 2–6.

21. Calyx lobes longer than the tube, usually at least 15 mm long .... 1. *Agrostemma*
21. Calyx lobes shorter than the tube, up to 13 mm long........16. *Silene*
20. Bracts absent.
22. Calyx up to 5 mm long, conspicuously nerved ....5. *Gypsophila*
22. Calyx at least 8 mm long, obscurely nerved.
23. Each petal with an awl-shaped scale at the base; flowers 1.8–2.5 cm wide, usually pink or white; calyx 20-nerved, tubular-cylindric .... 14. *Saponaria*
23. Each petal without a scale at the base; flowers less than 1.5 cm wide, red; calyx 5-nerved, ovoid................................. 20. *Vaccaria*
19. Styles 2; teeth of valves of fruit usually 4.
24. Veins or ribs of calyx 20–40 .... 4. *Dianthus*
24. Veins or ribs of calyx usually 5 ....12. *Petrorhagia*

1. **Agrostemma** L.—Corn Cockle

1. **Agrostemma githago** L. Corn Cockle. May–July. Native to Europe and Asia; naturalized in fields and other disturbed areas; occasional throughout the state.

2. **Arenaria** L.—Sandwort

(See also Minuartia and Moehringia.)

1. **Arenaria serpyllifolia** L. Thyme-leaved Sandwort. Apr.–Aug. Native to Europe and Asia; naturalized in disturbed sandy soil, often along roads and paths; scattered in IL.

3. **Cerastium** L.—Mouse-eared Chickweed

1. Styles 3; capsule 6-toothed at the apex...5. *C. dubium*
1. Styles 5; capsule 10-toothed at the apex.
2. Some or all bracts with scarious margins or at least the upper part of the bracts with scarious margins.
3. Petals 7–10 mm long, up to twice as long as the sepals; stems not viscid.
4. Petals 10 mm long or longer; sepals 5 mm long or longer .....1. *C. arvense*
4. Petals 7.5–9.0 mm long; sepals 3.5–6.0 mm long .... 11. *C. velutinum*
3. Petals less than 7 mm long, about as long as the sepals; stems more or less viscid.
5. Stamens 10; plants perennial, with basal offshoots....... 6. *C. fontanum*
5. Stamens 5; plants annual, without basal offshoots.
6. Uppermost bracts with narrow scarious margins, the lower bracts green throughout; petals deeply 2-cleft, with branched veins; capsules on erect pedicels.. 9. *C. pumilum*
6. Uppermost bracts as much as ½ composed of scarious margins; petals shallowly 2-cleft, with unbranched veins; capsules on deflexed pedicels... 10. *C. semidecandrum*
2. All bracts entirely green, not scarious.
7. Pedicels at least twice as long as the capsules.
8. Stamens 4 or 5 ....4. *C. diffusum*
8. Stamens 10.
9. Filaments glabrous; stems viscid throughout; sepals pubescent but not distinctly bearded at the apex.. 8. *C. nutans*
9. Filaments ciliate; stems viscid only near the apex; sepals long-bearded at the apex... 2. *C. brachypetalum*
7. Pedicels shorter than up to twice as long as the capsules.
10. Sepals acute; capsules nearly twice as long as the sepals.......................... ....7. *C. glomeratum*
10. Sepals obtuse to subacute; capsules about ¾ as long as the sepals .... 3. *C. brachypodum*

1. **Cerastium arvense** L. ssp. **strictum** Gaudin. Field Mouse-eared Chickweed. May–June. Rocky woods, prairies; occasional throughout the state.

2. **Cerastium brachypetalum** Pers. Mouse-eared Chickweed. Apr.–May. Native to Europe; adventive along roads; Fayette, Jackson, Montgomery, Pulaski, Shelby, Tazewell, and Union cos.

3. **Cerastium brachypodum** (Engelm.) B. L. Robins. Mar.–June. Mouse-eared Chickweed. Disturbed, usually moist soils, open woods, railroad ballast; occasional throughout IL. *Cerastium nutans* Raf. var. *brachypodum* Engelm.

4. **Cerastium diffusum** Pers. Mouse-eared Chickweed. Mar.–Apr. Native to Europe; rarely adventive in disturbed soil, particularly along roads; scattered in the cent. and s. parts of IL.

5. **Cerastium dubium** (Bast.) O. Swartz. Mouse-eared Chickweed. Apr.–May. Native to Europe; rarely adventive in the U.S.; Effingham, Fayette, and Shelby cos.

6. **Cerastium fontanum** Baum. Common Mouse-eared Chickweed. Mar.–Nov. Native to Europe and Asia; naturalized in disturbed soil, including lawns and gardens; common throughout IL. *Cerastium vulgatum* L.

7. **Cerastium glomeratum** Thuill. Clammy Mouse-eared Chickweed. Apr.–May. Native to Europe and Asia; naturalized in disturbed soil, including lawns; scattered throughout the state. *Cerastium viscosum* L.

8. **Cerastium nutans** Raf. Nodding Mouse-eared Chickweed. Mar.–June. Open, disturbed soil; throughout the state although apparently less common in the nw. cos.

9. **Cerastium pumilum** Curtis. Mouse-eared Chickweed. Apr.–June. Native to Europe; sparingly adventive in grassy areas; scattered in IL.

10. **Cerastium semidecandrum** L. Mouse-eared Chickweed. Mar.–June. Native to Europe; adventive in disturbed, usually sandy soil; ne. cos.; also Effingham and Marion cos.

11. **Cerastium velutinum** Raf. Large Field Mouse-eared Chickweed. May–June. Rocky woods, river banks, open woods; scattered throughout the state. *Cerastium arvense* L. var. *villosum* Holl. & Britt.

4. **Dianthus** L.—Pink; Carnation

1. Leaves elliptic to oblong, 9 mm wide or wider; flowers in dense heads .... 2. *D. barbatus*
1. Leaves linear, up to 9 mm wide; flowers not in dense heads.
  2. Annual or biennial; calyx densely pubescent; bracts about as long as the calyx .... 1. *D. armeria*
  2. Perennial; calyx glabrous or sparsely pubescent; bracts up to ½ as long as the calyx.
    3. Petals lacerate; leaves stiff; flowers strongly clove-scented.... 4. *D. plumarius*
    3. Petals sharply toothed; leaves lax; flowers not clove-scented .... 3. *D. deltoides*

1. **Dianthus armeria** L. Deptford Pink. May–Sept. Native to Europe; naturalized in fields, along roads, edge of woods; scattered throughout the state.

2. **Dianthus barbatus** L. Sweet William. June–Aug. Native to Europe; rarely escaped from cultivation; Jackson, DuPage, McLean, and Ogle cos.

3. **Dianthus deltoides** L. Maiden Pink. June–Aug. Native to Europe; seldom escaped from cultivation; Champaign, DuPage, Kane, Menard, Piatt, and Vermilion cos.

4. **Dianthus plumarius** L. Cottage Pink; Garden Pink. June. Native to Europe; rarely escaped from cultivation; Cook Co.

5. **Gypsophila** L.—Baby's Breath

1. Petals at least twice as long as the calyx; calyx 3–5 mm long.................. 1. *G. elegans*
1. Petals about as long as the calyx; calyx up to 3 (–4) mm long.
  2. Leaves cuneate, 1-nerved; calyx and pedicels eglandular; petals to up 4 mm long .... 2. *G. paniculata*
  2. Leaves subcordate, 3- to 5-nerved; calyx and pedicels glandular; petals 4–6 mm long .... 3. *G. scorzonerifolia*

1. **Gypsophila elegans** Bieb. Baby's Breath. May–June. Native to Europe and Asia; rarely escaped from cultivation; Champaign Co.

2. **Gypsophila paniculata** L. Baby's Breath. June–Aug. Native to Europe and Asia;

occasionally adventive in IL; Cook, Kane, Kankakee, LaSalle, Mason, Menard, and Winnebago cos.

3. **Gypsophila scorzonerifolia** Ser. Baby's Breath. June–Aug. Native to Europe; adventive in IL; Cook, Kane, Kankakee, and McHenry cos. *Gypsophila perfoliata* L., misapplied.

6. **Holosteum** L.—Jagged Chickweed

1. **Holosteum umbellatum** L. Jagged Chickweed. Apr.–May. Native to Europe; adventive in disturbed soil, particularly along roads; throughout IL. except for the extreme nw. cos.

7. **Minuartia** L.—Sandwort

1. Plants perennial, mat-forming; axillary leaves present; seeds 0.8–1.0 mm long........ .... 2. *M. michauxii*
1. Plants annual, not mat-forming; axillary leaves absent; seeds 0.5–0.8 mm long.
  2. Leaves up to 0.5 mm wide, dull; pedicels glabrous; sepals 1.5–4.0 mm long, obtuse at the apex; petals shallowly notched .... 1. *M. glabra*
  2. Leaves 0.5–1.5 mm wide, shiny; pedicels stipitate-glandular; sepals 4.0–5.5 mm long, acute at the apex; petals more deeply notched .... 3. *M. patula*

1. **Minuartia glabra** (Michx.) Mattfeld. Smooth Sandwort. June–July. Rocky soil; scattered in IL. *Arenaria glabra* Michx.

2. **Minuartia michauxii** (Fenzl) Farw. Stiff Sandwort. May–July. Sandy ridges near Lake Michigan, gravelly limestone glades; confined to the n. ⅓ of IL, where it is uncommon; also St. Clair Co. *Arenaria stricta* Michx.

3. **Minuartia patula** (Michx.) Mattfeld. Slender Sandwort. May–July. Wooded slopes; confined to a few ne. cos and St. Clair Co. *Arenaria patula* Michx.

8. **Moehringia** L.—Sandwort

1. **Moehringia lateriflora** (L.) Fenzl. Sandwort. Apr.–July. Moist or dry woods; prairie remnants; occasional in the n. ½ of IL; also St. Clair and Wabash cos. *Arenaria lateriflora* L.

9. **Moenchia** L.—Moenchia

1. **Moenchia erecta** (L.) P. Gaertn., Mey., & Scherb. Moenchia. May. Native to Europe; rarely adventive in the U.S.; disturbed ground in a field; Clay Co., collected by David Ketzner in 1992.

10. **Myosoton** Moench—Giant Chickweed

1. **Myosoton aquaticum** (L.) Moench. Giant Chickweed. May–Oct. Native to Europe; naturalized in moist soil along streams, sometimes in pastures; occasional in the n. ⅙ of IL; also Clark, Coles, Cumberland, and Sangamon cos. *Stellaria aquatica* (L.) Scop.

11. **Paronychia** Mill.—Forked Chickweed

1. Stems glabrous; sepals obtuse to subacute, 1.0–1.5 mm long........ 1. *P. canadensis*
1. Stems pubescent; sepals mucronulate, 2–3 mm long .... 2. *P. fastigiata*

1. **Paronychia canadensis** (L.) Wood. Slender Forked Chickweed. June–Oct. Dry woods, sandy soils, in shade of trees; occasional and scattered throughout IL.

2. **Paronychia fastigiata** (Raf.) Fern. Forked Chickweed. June–Oct. Dry, often sandy or rocky woods in much of the state but in grassy areas in the Chicago region; scattered in IL. (Including var. *paleacea* Fern.)

12. **Petrorhagia** (Ser.) Link—Saxifrage Pink

1. **Petrorhagia saxifraga** (L.) Link. Saxifrage Pink. June–July. Native to Europe; rarely adventive along roads; Champaign and Cook cos. *Tunica saxifraga* L.

13. **Sagina** L.—Pearlwort

1. Annuals without a persistent basal rosette; petals sometimes absent or usually 5 and often poorly developed.

- **2.** Petals usually present; sepals usually 5.
  - **3.** Pedicels without glandular-pubescence; seeds reddish brown, with slender ridges .... 2. *S. decumbens*
  - **3.** At least the upper part of the pedicels with glandular-pubescence; seeds dark brown, tuberculate .... 3. *S. japonica*
- **2.** Petals absent; sepals 4 .... 1. *S. apetala*

1. Perennials with a persistent basal rosette; petals usually 4, well-developed .... 4. *S. procumbens*

1. **Sagina apetala** Ard. Apetalous Pearlwort. Apr.–June. Native to Europe; adventive in disturbed soil; Union Co.

2. **Sagina decumbens** (Ell.) Torr & Gray. Annual Pearlwort. Apr.–June. Moist or dry, often sandy soil; disturbed moist soil; occasional in the s. ½ of IL; also Champaign, Coles, and Peoria cos.

3. **Sagina japonica** (Sw.) Ohwi. Japanese Pearlwort. Native to Asia; rarely adventive in the U.S.; Coles, Sangamon, and Williamson cos.

4. **Sagina procumbens** L. Perennial Pearlwort. June–Aug. Moist lawns, paths, and between bricks and patio blocks; apparently confined to the n. ¼ of IL.

14. **Saponaria** L.—Soapwort; Bouncing Bet

1. **Saponaria officinalis** L. Bouncing Bet. June–Sept. Native to Europe; naturalized in disturbed areas; common throughout the state.

15. **Scleranthus** L.—Knawel

1. **Scleranthus annuus** L. Knawel. June–Aug. Native to Europe and Asia; adventive into disturbed soil, particularly sandy areas; not common; Clay, Cumberland, Kankakee, Lake, McHenry, and Winnebago cos.

16. **Silene** L.—Catchfly

1. Styles 5.
- **2.** Each petal divided into four linear segments .... 9. *S. flos-cuculi*
- **2.** Each petal unlobed or 2-lobed.
  - **3.** Inflorescence a dense cyme; petals scarlet; stems hispid .... 3. *S. chalcedonica*
  - **3.** Inflorescence an open cyme; petals bright pink or white; stems softly pubescent or finely hirsute.
    - **4.** Flowers bisexual; leaves with silky pubescence .... 5. *S. coronaria*
    - **4.** Flowers unisexual; leaves hirsute or softly pubescent.
      - **5.** Petals white; leaves hirsute .... 11. *S. latifolia*
      - **5.** Petals bright pink; leaves softly pubescent .... 8. *S. dioica*

1. Styles 3.
- **6.** Petals deep red, crimson, or scarlet.
  - **7.** Stem with 15 or more pairs of leaves; petals entire or emarginate .... 14. *S. regia*
  - **7.** Stem with up to 8 pairs of leaves; petals bifid .... 16. *S. virginica*
- **6.** Petals white, pink, or purplish.
  - **8.** Leaves whorled .... 15. *S. stellata*
  - **8.** Leaves opposite.
    - **9.** Flowers up to 4 mm across or petals absent .... 1. *S. antirrhina*
    - **9.** Flowers 1 cm or more across.
      - **10.** Calyx with about 30 ribs .... 4. *S. conica*
      - **10.** Calyx with 5–10 ribs.
        - **11.** Stems glutinous below each node; flowers pink or purple .... 2. *S. armeria*
        - **11.** Stems not glutinous below each node; flowers white or only pink-based.
          - **12.** Stems viscid-pubescent or hirsute.
            - **13.** Flowers nodding, opening at dusk; viscid-pubescent annual .... 11. *S. noctiflora*
            - **13.** Flowers ascending, opening during the day; hirsute or pubescent biennial, but not viscid.

14. Calyx 9–14 mm long at anthesis; pubescence of calyx never glandular; lobes of calyx 2–4 mm long .... 7. *S. dichotoma*
14. Calyx up to 9 mm long at anthesis; pubescence of calyx sometimes glandular on the nerves; lobes of calyx 1.0–1.5 mm long .... 10. *S. gallica*
12. Stems glabrous.
15. Calyx glabrous, more or less inflated; petals 2-lobed.
16. Plants green; leaves acuminate; flower solitary in the upper axils .... 12. *S. nivea*
16. Plants glaucous; leaves acute; flowers in cymose panicles.
17. Calyx up to 15 mm long, scarcely inflated; vertical nerves of calyx sparsely or not at all connected by cross-veins .... 6. *S. cserei*
17. Calyx up to 20 mm long, inflated; vertical nerves of calyx regularly connected by cross-veins .... 18. *S. vulgaris*
15. Calyx pubescent, not inflated; petals fringed .... 14. *S. ovata*

1. **Silene antirrhina** L. Several variations occur in Illinois:

a. Stems erect, more or less unbranched, not spreading.
b. Glutinous areas present below some of the nodes.
c. Flowers with petals .... 1a. *S. antirrhina* var. *antirrhina*
c. Flowers apetalous .... 1b. *S. antirrhina* f. *apetala*
b. Glutinous areas absent on the stems .... 1c. *S. antirrhina* f. *deaneana*
a. Stems spreading, branched .... 1d. *S. antirrhina* var. *divaricata*

1a. **Silene antirrhina** L. var. **antirrhina.** Sleepy Catchfly. May–July. Disturbed soil, often along railroads; occasional to common throughout IL.

1b. **Silene antirrhina** L. f. **apetala** Farw. Sleepy Catchfly. May–July. Disturbed soil, often along railroads; occasional to common throughout IL.

1c. **Silene antirrhina** L. f. **deaneana** Fern. Sleepy Catchfly. May–July. Disturbed soil; occasional in IL.

1d. **Silene antirrhina** L. var. **divaricata** Robins. Sleepy Catchfly. May–July. Disturbed soil, rare; Winnebago Co.

2. **Silene armeria** L. Sweet William Catchfly. June–Aug. Native to Europe and Asia; occasionally escaped from cultivation; Champaign, Cook, DuPage, Kane, Lee, St. Clair, and Will cos.

3. **Silene chalcedonica** (L.) E. H. L. Krause in J. Sturm. Maltese Cross. June–Sept. Native to Russia and Siberia; rarely escaped from cultivation; DuPage, Lake, Marion, and Vermilion cos. *Lychnis chalcedonica* L.

4. **Silene conica** L. Striate Catchfly. June–July. Native to Europe and Asia; in cinders and sand along a railroad; Cass Co.

5. **Silene coronaria** (L.) E. H. L. Krause in J. Sturm. Mullein Pink. July–Sept. Native to Europe; occasionally escaped from cultivation; scattered in IL. *Lychnis coronaria* L.

6. **Silene cserei** Baumg. Glaucous Campion. May–Oct. Native to Europe; naturalized into waste ground, particularly along railroads; occasional to common in the n. ⅓ of IL, rarer elsewhere.

7. **Silene dichotoma** Ehrh. Forked Catchfly. June–Aug. Native to Europe and Asia; adventive into disturbed soil; scattered in the n. ½ of IL; also St. Clair Co.

8. **Silene dioica** (L.) Clairv. Red Campion. May–Oct. Native to Europe and Asia; adventive in waste ground, rare; scattered in the n. ⅔ of IL. *Lychnis dioica* L.

9. **Silene flos-cuculi** (L.) Clairville. Ragged Robin. June–July. Native to Europe; rarely escaped from gardens; DuPage Co. *Lychnis flos-cuculi* L.

10. **Silene gallica** L. Catchfly. Sept. Native to Europe and Asia; low ground; St. Clair Co.

11. **Silene latifolia** Poiret. White Campion; Evening Campion. Native to Europe and Asia; adventive in waste ground, particularly along railroads and in cultivated fields;

very common in the n. cos., becoming much rarer southward. *Lychnis alba* Mill.; *Silene pratensis* (Spreng.) Godron & Gren.

12. **Silene nivea** (Nutt.) Otth. Snowy Campion. June–Aug. Wooded ravines, calcareous fens, moist stream banks; uncommon in the n. ⅔ of IL.

13. **Silene noctiflora** L. Night-flowering Catchfly. June–Sept. Native to Europe; adventive into waste ground; occasional in the n. ⅗ of IL.

14. **Silene ovata** Pursh. Ovate-leaved Campion. Aug. Rich woods, very rare; Hardin Co.

15. **Silene regia** Sims. Royal Catchfly. July–Aug. Dry soil, usually along roads; prairies; rare; Clark, Cook, Lawrence, Madison, St. Clair, Wabash, White, Will, and Winnebago cos.

16. **Silene stellata** (L.) Ait. f. Two varieties occur in Illinois:

a. Stems glabrous or nearly so .... 16a. *S. stellata* var. *stellata*
a. Stems densely pubescent... 16b. *S. stellata* var. *scabrella*

16a. **Silene stellata** (L.) Ait. f. var. **stellata.** Starry Campion. June–Oct. Open woods, prairies; occasional throughout IL.

16b. **Silene stellata** (L.) Ait. f. var. **scabrella** (Nieuwl.) Palmer & Steyerm. Starry Campion. June–Oct. Open woods, prairies; scattered in IL.

17. **Silene virginica** L. Firepink. Apr.–July. Open woods, both rich and dry; scattered throughout IL. but not common.

18. **Silene vulgaris** (Moench) Garcke. Bladder Catchfly; Bladder Campion. May–Aug. Native to Europe and Asia; adventive in waste ground; occasional in the n. ½ of IL, less common elsewhere. *Silene cucubalus* Wibel.

17. **Spergula** L.—Corn Spurrey

1. **Spergula arvensis** L. Corn Spurrey. July–Sept. Native to Europe; adventive in disturbed soil, sometimes in grain fields, not common; Champaign, Cook and Marion cos.

18. **Spergularia** J. & C. Presl—Sand Spurrey

1. Seeds wingless; sepals up to 4.3 mm long; petals usually pink.
  2. Stamens 2–5; leaves obtuse to mucronate at the tip, more or less fleshy ....3. *S. salina*
  2. Stamens 6–10; leaves spinulose at the tip, scarcely fleshy .... 2. *S. maritima*
1. Seeds narrowly winged; sepals 4–6 mm long; petals white........................1. *S. media*

1. **Spergularia media** (L.) C. Presl. Sand Spurrey. July–Sept. Native to Europe; adventive along roads, particularly where salt has been used during the winter; ne. cos. of IL; also Marion and Peoria cos.

2. **Spergularia maritima** Chiov. Pink Sand Spurrey. July–Sept. Native to Europe; adventive in disturbed soil; Cook, Kane, and Kankakee cos. *Spergularia rubra* (L.) J. & C. Presl.

3. **Spergularia salina** J. Presl & C. Presl. Coastal Sand Spurrey. May–Sept. Native to Europe; adventive along highways that are heavily salted during the winter; confined to the ne. cos. of IL. *Spergularia marina* (L.) Griseb.

19. **Stellaria** L.—Chickweed

1. Median leaves ovate, usually perfoliate; petals shorter than the sepals, or absent; annuals from slender taproots.
  2. Petals 5; sepals 3–6 mm long; seeds 1.0–1.2 mm in diameter, usually dark reddish brown or purple-brown .... 4. *S. media*
  2. Petals absent; sepals 2–3 mm long; seeds 0.7–0.8 mm in diameter, light reddish brown.. 5. *S. pallida*
1. Median leaves linear, lanceolate, or elliptic or suborbicular and sessile; petals longer than the sepals; perennials from slender rhizomes.
  3. Stems puberulent in lines, at least in the upper ½ of the stem; leaves elliptic to suborbicular ....6. *S. pubera*
  3. Stems glabrous or essentially so; leaves lanceolate or linear.
    4. Leaves linear, up to 7 times longer than broad, completely glabrous; bracts without scarious margins or sometimes with a very narrow scarious margin .... 1. *S. crassifolia*

4. Leaves linear to lanceolate, at least some of them more than 8 times longer than broad, ciliate at base; bracts with obvious scarious margins.
5. Bracts and sepals ciliate; seeds roughened; sepals strongly nerved; inflorescence terminal .... 2. *S. graminea*
5. Bracts and sepals without cilia; seeds smooth; sepals nerveless or essentially so; inflorescence axillary .... 3. *S. longifolia*

1. **Stellaria crassifolia** Ehrh. Matted Chickweed. May. Low, springy places; McHenry Co.

2. **Stellaria graminea** L. Common Stitchwort. May–Aug. Native to Europe; naturalized in moist, disturbed areas; scattered throughout the state.

3. **Stellaria longifolia** Muhl. Long-leaved Stitchwort. May–July. Moist ground, including bogs and floodplains; scattered throughout IL, but rare in the s. cos.

4. **Stellaria media** (L.) Cyrillo. Common Chickweed. Feb.–Dec. Native to Europe and Asia; naturalized in waste ground; common throughout IL.

5. **Stellaria pallida** Pire. Apetalous Chickweed. Apr.–May. Native to Europe; adventive into cultivated areas; Fayette, Grundy, Kane, Kankakee, and Will cos.

6. **Stellaria pubera** Michx. Great Chickweed. Mar.–May. Wooded cliffs, rare; Cook, DuPage, Pope, and Will cos.

20. **Vaccaria** Moench—Cow Herb

1. **Vaccaria hispanica** (Mill.) Rausch. Cow Herb. June–Aug. Native to Europe; adventive in waste ground; scattered throughout IL. but uncommon in the s. ⅓ of IL. *Vaccaria pyramidata* Medic.; *Saponaria vaccaria* L.

54. CELASTRACEAE—BITTERSWEET FAMILY

1. Leaves alternate; most of the flowers unisexual .... 1. *Celastrus*
1. Leaves opposite; flowers perfect .... 2. *Euonymus*

1. **Celastrus** L.—Bittersweet

1. Flowers in axillary cymes; leaves suborbicular, crenate .... 1. *C. orbiculatus*
1. Flowers in terminal panicles or racemes; leaves ovate to ovate-oblong, finely serrate .... 2. *C. scandens*

1. **Celastrus orbiculatus** Thunb. Round-leaved Bittersweet; Japanese Bittersweet. May–June. Native to Asia; common and escaped from cultivation; scattered in IL, but becoming very invasive.

2. **Celastrus scandens** L. Bittersweet. May–June. Woods, thickets; occasional throughout IL.

2. **Euonymus** L.—Spindle Tree

1. Plants trailing or climbing, rooting at the nodes.
2. Leaves membranous, deciduous, obovate to oblong; sepals and petals each 5; capsules warty; plants trailing .... 10. *E. obovatus*
2. Leaves coriaceous, sometimes evergreen, ovate or elliptic; sepals and petals each 4; capsules smooth; plants climbing.
3. Leaves ovate, to 6 cm long; flowers in short-peduncled cymes .... 9. *E. kiautschovicus*
3. Leaves elliptic, to 4 cm long; flowers in long-peduncled cymes .... 7. *E. hederaceus*
1. Plants erect or ascending, not rooting at the nodes.
4. Stems winged .... 1. *E. alatus*
4. Stems unwinged.
5. Leaves sessile or on petioles up to 3 mm long; sepals and petals each 5; shrubs to 2.5 m tall .... 2. *E. americanus*
5. Leaves petiolate, the petioles mostly more than 3 mm long; sepals and petals usually each 4; shrubs usually more than 3 m tall.
6. Leaves pubescent on the lower surface; petals purple; plants deciduous .... 3. *E. atropurpureus*
6. Leaves glabrous on the lower surface except for sometimes along the veins; petals white to greenish; plants evergreen.
7. Petioles at least 12 mm long .... 4. *E. bungeanus*

7. Petioles 5–12 mm long.
 8. Fruits more or less lobed.
  9. Fruits purple or rose; anthers purple ....6. *E. hamiltonianus*
  9. Fruits yellow, tinged with pink, or pink; anthers yellow.............. ....5. *E. europaeus*
 8. Fruits globose, unlobed ....8. *E. japonicus*

1. **Euonymus alatus** (Thunb.) Sieb. Winged Euonymus. Apr.–June. Native to Asia; infrequently escaped from cultivation; scattered in IL.

2. **Euonymus americanus** L. Strawberry-bush. May–June. Rich woods, not common; confined to the extreme s. cos. and Vermilion and Wabash cos.

3. **Euonymus atropurpureus** Jacq. Wahoo. May–July. Woods, stream banks, shaded floodplains; occasional throughout IL.

4. **Euonymus bungeanus** Maxim. Chinese Spindle Tree. May–July. Native to Asia; rarely escaped from cultivation; Champaign, Cook, and DuPage cos.

5. **Euonymus europaeus** L. European Spindle Tree. May–June. Native to Europe; occasionally escaped from cultivation in the ne. cos.; also Champaign Co.

6. **Euonymus hederaceus** Champ. ex Benth. Climbing Euonymus; Wintercreeper. June–Aug. Native to Asia; infrequently escaped from cultivation. *Euonymus fortunei* (Turcz.) Hand.-Mazz.

7. **Euonymus hamiltonianus** Wall. Japanese Spindle Tree. May–June. Native to Asia; rarely escaped from cultivation; Cook and DuPage cos.

8. **Euonymus japonicus** L. Japanese Spindle Tree. July–Aug. Native to Asia; rarely escaped from cultivation; Jackson and Jersey cos.

9. **Euonymus kiautschovicus** Loes. Climbing Euonymus. July–Aug. Native to Asia; rarely escaped from cultivation; Champaign and Jackson cos.

10. **Euonymus obovatus** Nutt. Running Strawberry-bush. Apr.–June. Rich woods, not common; scattered in IL.

55. CERATOPHYLLACEAE—HORNWORT FAMILY

1. **Ceratophyllum** L.—Coontail

1. Achenes with 2 basal spines; ultimate leaf segments toothed on the margins.......... ....1. *C. demersum*
1. Achenes with several spines both lateral and basal; ultimate leaf segments not toothed on the margins ....2. *C. echinatum*

1. **Ceratophyllum demersum** L. Coontail. July–Sept. Quiet waters; common throughout IL.

2. **Ceratophyllum echinatum** Gray. Coontail. July–Sept. Quiet waters, not common; scattered in IL. *Ceratophyllum muricatum* Cham.

56. CHENOPODIACEAE—GOOSEFOOT FAMILY

Key to the Genera of Non-Flowering, Non-Fruiting, or Immature Plants

1. Stems very fleshy, jointed; leaves reduced to scales............................. 9. *Salicornia*
1. Stems not fleshy or only slightly fleshy, not jointed; leaves present.
 2. Leaves all opposite, crowded to appear fasciculate, thread-like to filiform ....8. *Polycnemum*
 2. Some or all the leaves alternate, never appearing fasciculate, linear or broader.
  3. Leaves spine-tipped .... 10. *Salsola*
  3. Leaves not spine-tipped.
   4. Stems and leaves and sometimes the calyx either glandular-pubescent, cobwebby hairy, white-mealy, scurfy-scaly, or silvery-scaly.
    5. Stems and leaves glandular-pubescent............................. 5. *Dysphania*
    5. Stems and leaves not glandular-pubescent.
     6. Stems and leaves and often the calyx white-mealy.
      7. Leaves not triangular or, if triangular, usually with coarse teeth ....2. *Chenopodium*
      7. Some of the leaves triangular, entire or sparsely toothed.......... ....7. *Monolepis*
     6. Stems and leaves not white-mealy.

8. Stems and leaves with cobwebby hairs............. 4. *Cycloloma*
8. Stems and leaves lacking cobwebby hairs but either scurfy-scaly or silvery-scaly .... 1. *Atriplex*
4. Stems and leaves neither glandular-pubescent, cobwebby hairy, white-mealy, scurfy-scaly, or silvery-scaly.
9. Some or all the leaves more than 5 mm wide.
10. Leaves and stems often short-pubescent or at least ciliate at base ....6. *Kochia*
10. Leaves and stems not pubescent or leaves not ciliate at base ....2. *Chenopodium*
9. None of the leaves more than 5 mm wide.
11. Leaves ciliate at base ....6. *Kochia*
11. Leaves not ciliate at base.
12. Leaves more or less terete, glaucous........................... 11. *Suaeda*
12. Leaves flat, not glaucous......................................3. *Corispermum*

Key to the Genera of Flowering and Fruiting Plants

1. Flowers sunken in the hollow of the uppermost joints of the very succulent stem; leaves reduced to scales.. 9. *Salicornia*
1. Flowers not sunken in the hollow of the uppermost joints of the stem; stems not very succulent nor jointed; leaves present.
2. All flowers unisexual; stems and leaves either scurfy-scaly or silvery-scaly .... 1. *Atriplex*
2. Some or all the flowers perfect; stems with various kinds of indument or glabrous but not scurfy-scaly or silvery-scaly.
3. Sepal 1 or calyx 3-cleft.
4. Sepal 1; styles not persistent on the fruit; leaves triangular to lanceolate, at least sparsely white-mealy...7. *Monolepis*
4. Calyx usually 3-cleft, rarely with only 1 or 2 sepals; styles persistent on the fruit; leaves linear, not white-mealy.............................3. *Corispermum*
3. Calyx 5-cleft.
5. Lobes of calyx tuberculate at tip..6. *Kochia*
5. Lobes of calyx not tuberculate at tip.
6. Seeds narrowly or broadly winged.
7. Seeds horizontal; wing of fruit more than 0.5 mm wide.
8. Fruits horizontally winged, flat; leaves not spine-tipped; bracts absent .... 4. *Cycloloma*
8. Fruits transversely winged, not flat; leaves spine-tipped; bracts present.. 10. *Salsola*
7. Seeds vertical; wing of fruit up to 0.5 mm wide......3. *Corispermum*
6. Seeds unwinged.
9. Stamens 3; styles 3; leaves opposite, filiform ....8. *Polycnemum*
9. Stamens 1–5, rarely 3; styles usually 2; leaves alternate.
10. Calyx lobes unequal; flowers perfect and pistillate on the same plant; plants not white-mealy .... 11. *Suaeda*
10. Calyx lobes equal; flowers all perfect; plants often white-mealy.
11. Leaves linear, 1-nerved; plants never white-mealy ....3. *Corispermum*
11. Leaves rarely linear, usually more than 1-nerved; plants sometimes white-mealy.
12. Plants with glandular hairs............................ 5. *Dysphania*
12. Plants without glandular hairs................ 2. *Chenopodium*

1. **Atriplex** L.—Orach; Saltbush; Spear Scale

1. All leaves alternate.
2. Plants silvery-pubescent; leaves entire or nearly so......................... 1. *A. argentea*
2. Plants hoary-mealy; leaves sinuate or dentate ....7. *A. rosea*
1. At least the lower leaves opposite.
3. Bracts subtending the fruits orbicular, thin, reticulate; pistillate flower with a calyx..3. *A. hortensis*

3. Bracts subtending the fruits deltate to rhombic, thick, smooth or tuberculate but scarcely reticulate; pistillate flower with no calyx.
4. Bracteoles subtending the fruit 5–12 mm long, smooth or sparsely tuberculate .... 2. *A. glabriuscula*
4. Bracteoles subtending the fruit up to 5 mm long, distinctly tuberculate (except in *A. patula*).
5. Lowest leaves linear, usually entire .... 4. *A. littoralis*
5. Lowest leaves lanceolate to ovate, usually hastate, entire, or dentate.
6. Leaves lance-hastate, entire or very sparsely toothed .... 5. *A. patula*
6. Leaves deltate-hastate to oval-hastate, dentate .... 6. *A. prostrata*

1. **Atriplex argentea** Nutt. Silver Orach. July–Oct. Native w. of the Mississippi River; adventive in disturbed soil; Cook, Hancock, and Menard cos.

2. **Atriplex glabriuscula** Edmonston. Smooth Orach. Aug. Native to the Atlantic coast; rarely adventive in IL. in roadside gravel; Kane Co.

3. **Atriplex hortensis** L. Garden Orache. Aug.–Sept. Native to Europe and Asia; seldom escaped from cultivation; Bond, DuPage, Lake, Marion, St. Clair, and Vermilion cos.

4. **Atriplex littoralis** L. Seaside Orach. July–Oct. Disturbed sandy soil; not common; ne. cos. *Atriplex patula* L. var. *littoralis* (L.) Gray.

5. **Atriplex patula** L. Spear Scale. July–Sept. Native to Europe, Asia, and N. Africa; adventive in disturbed soil, often in saline areas; scattered throughout IL.

6. **Atriplex prostrata** Moq. Common Orache; Spear Scale. July–Sept. Native to Europe; adventive in disturbed soil, often in saline areas; scattered throughout IL. *Atriplex hastata* L.; *A. patula* L. var. *hastata* (L.) Gray.

7. **Atriplex rosea** L. Red Orache. Aug.–Oct. Native to Europe and Asia; sparingly adventive in disturbed soil; Cook, Kane, and Will cos.

2. **Chenopodium** L.—Goosefoot

(See also Dysphania.)

1. Calyx completely covering fruit at maturity or barely reaching the tip in *C. standleyanum*.
2. All seeds vertical; calyx strawberry-red .... 5. *C. capitatum*
2. Seeds horizontal (some seeds vertical in *C. rubrum*); calyx not strawberry-red.
3. Pericarp free from seed or only very weakly adherent.
4. Leaves coarsely and regularly toothed or even hastate; some of the seeds vertical; leaves rhombic to ovate; calyx fleshy; seeds 0.8–1.0 mm wide, with sharp margins .... 17. *C. rubrum*
4. Leaves entire, sparsely low-serrate, or with a single rounded tooth on each side; all seeds horizontal; leaves lanceolate to oblong to ovate; calyx not fleshy; seeds 1.0–1.5 mm wide, with rounded margins.
5. Leaves more or less thick, lanceolate to elliptic, with 1 low, rounded tooth on each side; calyx sparsely white-mealy .... 16. *C. pratericola*
5. Leaves thin, lanceolate to ovate, entire or sparsely low-serrate; calyx usually densely white-mealy.
6. Calyx completely covering fruit; leaves obtuse to subacute at tip .... 6. *C. desiccatum*
6. Calyx barely covering top of fruit; leaves acute .... 19. *C. standleyanum*
3. Pericarp firmly adherent to the seed.
7. Leaves linear to lanceolate, entire or with a few low teeth; calyx glabrous or sparsely white-mealy.
8. Leaves linear; calyx sparsely white-mealy; seeds 1.2–1.5 mm wide .... 14. *C. pallescens*
8. Leaves lanceolate; calyx not white-mealy; seeds 1.0–1.2 mm wide .... 9. *C. lanceolatum*
7. Leaves rhombic to broadly ovate, regularly sinuate-dentate to 3-lobed; calyx densely white-mealy.
9. Leaves never 3-lobed, merely sinuate-dentate; plants fetid .... 2. *C. berlandieri*

9. At least some of the leaves 3-lobed and regularly toothed; plants not fetid.
    10. Calyx cleft nearly to the middle; leaves usually as broad as long; stems often glabrate and not white-mealy; seeds smooth, not puncticulate .... 13. *C. opulifolium*
    10. Calyx cleft nearly to the base; leaves usually up to 1 ½ times longer than broad; stems usually sparsely to densely white-mealy; seeds smooth, puncticulate, or reticulate.
        11. Seeds (1.3–) 1.5–2.3 mm broad, dull; leaves sparsely white-mealy on the lower surface; seeds reticulate.
            12. Seeds 1.5–2.3 mm broad; inflorescence pendulous at maturity; leaves ovate to rhombic..........................4. *C. bushianum*
            12. Seeds 1.3–1.7 mm broad; inflorescence erect at maturity; leaves lanceolate to lance-elliptic ....10. *C. macrocalycium*
        11. Seeds 1.3–1.5 mm broad, shiny; leaves densely white-mealy on the lower surface; seeds smooth or pitted .... 1. *C. album*

1. Calyx partially covering fruit at maturity.
    13. Some or all of the seeds vertical.
        14. Leaves triangular, thick; seeds 1.3–1.5 mm wide, the margins rounded.. 3. *C. bonus-henricus*
        14. Leaves rhombic to oblong to ovate, thin; seeds 0.5–1.0 mm wide, the margins acute.
            15. Seeds 0.5–0.6 mm wide; calyx glabrous, not fleshy; leaves densely white-mealy..8. *C. glaucum*
            15. Seeds 0.8–1.0 mm wide; calyx sometimes white-mealy, fleshy; leaves sparsely white-mealy on the lower surface............................17. *C. rubrum*
    13. All seeds horizontal.
        16. Seeds free or easily separated from the pericarp.
            17. Leaves up to 8 cm long, entire or shallowly toothed; calyx white-mealy; seeds 1.0–1.5 mm wide...19. *C. standleyanum*
            17. Leaves up to 20 cm long, usually with 4 large teeth on each side; calyx glabrous; seeds 1.8–2.0 mm wide.. 18. *C. simplex*
        16. Seeds firmly adherent to the pericarp.
            18. Leaves thick, entire; inflorescence often from base of the plant to the top ....15. *C. polyspermum*
            18. Leaves thin, toothed, rarely entire or with 1–2 pairs of teeth; inflorescence not at base of the plant.
                19. Calyx sparsely white-mealy; leaves usually glabrous or only sparsely white-mealy.
                    20. Leaves with irregular small teeth; seeds smooth.......20. *C. strictum*
                    20. Leaves with regular coarse teeth, rarely entire or with 1–2 pairs of teeth; seeds puncticulate or rugulate.
                        21. All except the uppermost leaves coarsely toothed; seeds puncticulate..11. *C. missouriense*
                        21. All leaves entire or with 1–2 pairs of teeth; seeds rugulate ....7. *C. foggii*
                19. Calyx glabrous; leaves densely white-mealy on the lower surface.
                    22. Seeds 0.9–1.0 mm wide, smooth, shiny, the margin rounded...21. *C. urbicum*
                    22. Seeds 1.2–1.5 mm wide, puncticulate, dull, the margin sharp ....12. *C. murale*

1. **Chenopodium album** L. Lamb's Quarters. May–Oct. Native to Europe and Asia; naturalized in disturbed soil, common; in every co.

2. **Chenopodium berlandieri** Moq. Stinking Lamb's Quarters. June–Oct. Disturbed soil; scattered in IL. *Chenopodium berlandieri* Moq. var. *zschackei* (Murr.) Murr.

3. **Chenopodium bonus-henricus** L. Good King Henry. July–Oct. Native to Europe; adventive in disturbed soil; Cook and DuPage cos.

4. **Chenopodium bushianum** Aellen. Goosefoot. Aug.–Oct. Disturbed soil; scat-

tered in IL but particularly in cos. that border the Mississippi and Illinois rivers. *Chenopodium berlandieri* Moq. var. *bushianum* (Aellen) Cronq.

5. **Chenopodium capitatum** (L.) Aschers. Strawberry Blite. May–Aug. Native to Europe and Asia; adventive in disturbed soil; scattered in IL.

6. **Chenopodium desiccatum** A. Nels. Entire-leaved Goosefoot. June–July. Sandy soil along rivers; black oak savannas; ne. cos. and Fayette, Iroquois, Jackson, Kankakee, and Mason cos.

7. **Chenopodium foggii** Wahl. Fogg's Goosefoot. Aug.–Oct. Fields, rare; Mason and Will cos.

8. **Chenopodium glaucum** L. Oak-leaved Goosefoot. July–Sept. Native to Europe and Asia; adventive in disturbed soil; scattered throughout IL but not common.

9. **Chenopodium lanceolatum** Muhl. Lance-leaved Lamb's Quarters. May–Oct. Disturbed soil; scattered throughout IL.

10. **Chenopodium macrocalycium** Aellen. Goosefoot. July–Sept. Along a road; Champaign Co. *Chenopodium berlandieri* Moq. var. *macrocalycium* (Aellen) Cronq.

11. **Chenopodium missouriense** Aellen. Missouri Goosefoot. Sept.–Oct. Fields, disturbed soil; scattered in IL.

12. **Chenopodium murale** L. Nettle-leaved Goosefoot. June–Nov. Native to Europe; adventive in disturbed soil; scattered but not common in IL.

13. **Chenopodium opulifolium** Schrad. Goosefoot. July–Aug. Native to Europe and Asia; rarely adventive in disturbed soil; Wabash and Winnebago cos.

14. **Chenopodium pallescens** Standl. Pale-leaved Goosefoot. June–Sept. Rocky ground; disturbed soil; confined to the n. ½ of IL. *Chenopodium leptophyllum* Nutt.

15. **Chenopodium polyspermum** L. Many-seeded Goosefoot. June–Oct. Native to Europe; adventive along a railroad; Jackson Co.; not seen since 1955.

16. **Chenopodium pratericola** Rydb. Narrow-leaved Goosefoot. June–Oct. Disturbed, often sandy soil; occasional throughout IL; some plants are apparently adventive.

17. **Chenopodium rubrum** L. Coast Blite. July–Oct. Native n., e., and w. of IL; adventive in disturbed soil; Cook and Peoria cos.

18. **Chenopodium simplex** (Torr.) Raf. Maple-leaved Goosefoot. June–Oct. Shaded ledges; rocky woods; occasional throughout the state. *Chenopodium gigantospermum* Aellen; *C. hybridum* L. var. *gigantospermum* (Aellen) Rouleau.

19. **Chenopodium standleyanum** Aellen. Goosefoot. June–Oct. Woodlands, roadsides; scattered throughout IL. *Chenopodium hybridum* L. var. *standleyanum* (Aellen) Fern.; *C. boscianum* Moq., misapplied.

20. **Chenopodium strictum** Roth var. **glaucophyllum** (Aellen) Wahl. Erect Goosefoot. July–Sept. Disturbed soil; Lake, Peoria, St. Clair, and Vermilion cos. *Chenopodium glaucophyllum* Aellen.

21. **Chenopodium urbicum** L. City Goosefoot. July–Oct. Native to Europe; adventive in disturbed soil; scattered but not common in all but the southernmost cos.

3. **Corispermum** L.—Bugseed

1. Fruits 1.8–3.0 mm long, wingless or nearly so, the wing less than 0.2 mm wide .... 4. *C. villosum*

1. Fruits 3.5–4.5 mm long, with a narrow, often translucent, the wing at least 0.3 mm wide.

  2. Inflorescence slender, not club-shaped, more or less interrupted, at least near base.

    3. Fruits broadest near the middle, never warty .... 2. *C. nitidum*

    3. Fruits broadest just above the middle, sometimes warty ..... 1. *C. americanum*

  2. Inflorescence broader, club-shaped, not interrupted near base .... 3. *C. pallasii*

1. **Corispermum americanum** (Nutt.) Nutt. American Hyssop-leaved Bugseed. July–Sept. Native to Europe and Asia; adventive in sandy soil, including sandy beaches along Lake Michigan; disturbed soil; Carroll, Cook, Lake, Macon, Mason, Menard, St. Clair, and Whiteside cos. *C. hyssopifolium* L. var. *americanum* Nutt.

2. **Corispermum nitidum** Kit. Bugseed. July–Sept. Native to Europe; adventive in sandy soil; Whiteside Co.

3. **Corispermum pallasii** Steven. Bugseed. July–Sept. Native to Europe; adventive in Lake Co.

4. **Corispermum villosum** Rydb. Hairy Bugseed. July–Sept. Native to Asia; adventive in sandy soil; St. Clair Co. C. *orientale* Lam. var. *emarginatum* (Rydb.) J. F. Macbr.

4. **Cycloloma** Moq.—Winged Pigweed

1. **Cycloloma atriplicifolium** (Spreng.) Coult. Winged Pigweed. July–Oct. Dry, sandy soil, often along rivers; scattered in IL.

5. **Dysphania** R. Brown

1. All seeds vertical; glands of calyx bright yellow, sessile ....4. *D. pumilio*
1. At least some of the seeds horizontal as well as vertical; glands of calyx not bright yellow, stipitate, or calyx merely puberulent.
  2. Leaves deeply sinuate-dentate to pinnatifid; calyx lobes obtuse, pubescent; calyx completely enclosing fruit; pericarp free or nearly so from the seed; seeds shiny, black.
    3. Inflorescence with leafy bracts; calyx keeled ....1. *D. ambrosioides*
    3. Inflorescence without leafy bracts; calyx without a keel.... 2. *D. anthelmintica*
  2. Leaves shallowly sinuate-dentate to entire; calyx lobes acute, with stipitate glands; calyx only partially enclosing fruit; pericarp firmly adherent to the seed; seeds dull, dark brown ....3. *D. botrys*

1. **Dysphania ambrosioides** (L.) Mosyakin & Clemants. Mexican Tea. May–Nov. Native to tropical America; adventive in waste ground; occasional to common in most of IL, although rare in the northernmost cos. *Chenopodium ambrosioides* L.

2. **Dysphania anthelmintica** (L.) Mosyakin & Clemants. Mexican Tea. May–Nov. Native to tropical America; adventive in waste ground; occasional in IL. *Chenopodium ambrosioides* L. var. *anthelmintica* (L.) Gray; *Chenopodium anthelmintica* L.

3. **Dysphania botrys** (L.) Mosyakin & Clemants. Jerusalem Oak. July–Oct. Native to Europe and Asia; naturalized in disturbed soil; occasional in IL. *Chenopodium botrys* L.

4. **Dysphania pumilio** (R. Brown) Mosyakin & Clemants. Aromatic Goosefoot. July–Aug. Native to Europe and Asia; adventive in disturbed soil; Coles and McDonough cos. *Chenopodium pumilio* R. Brown.

6. **Kochia** Roth—Kochia

1. **Kochia scoparia** (L.) Roth. Kochia; Cypress bush. July–Sept. Native to Asia; adventive in disturbed soil and old fields; scattered throughout the state. *Bassia hyssopifolia* L., misapplied.

7. **Monolepis** Schrad.—Poverty Weed

1. **Monolepis nuttalliana** (Roem. & Schultes) Greene. Poverty Weed. June–Oct. Native to the w. U.S.; adventive in disturbed dry, alkaline soil; Champaign, Grundy, Lee, McHenry, and Sangamon cos.

8. **Polycnemum** L.—Polycnemum

1. **Polycnemum majus** A. Br. Polycnemum. July–Sept. Native to Europe; rarely adventive along a railroad; Monroe Co.

9. **Salicornia** L.—Glasswort

1. **Salicornia rubra** A. Nels. Samphire; Glasswort. Aug.–Oct. Along the Little Calumet River, where it grew in soil contaminated by industrial discharge; Cook Co.; not seen in IL since 1948. *Salicornia europaea* L., misapplied; *S. europaea* L. var. *rubra* (A. Nels.) Breitung.

10. **Salsola** L.—Russian Thistle

1. Inflorescence densely spicate, with appressed bracts ....1. *S. collina*
1. Inflorescence loosely spicate, with spreading bracts...............................2. *S. tragus*

1. **Salsola collina** Pallas. Saltwort. July–Sept. Native to Europe and Asia; adventive in disturbed sandy soil; Bureau, Cass, Henry, Kankakee, Lee, Madison, and Whiteside cos.

2. **Salsola tragus** L. Russian Thistle. July–Sept. Native to Europe and Asia; adventive along railroads and on sand beaches; scattered throughout the state. *Salsola iberica* Sennen & Pav.; *S. pestifer* Nels.; *S. kali* L. var. *tenuifolia* Tausch.

11. **Suaeda** Forsk.—Sea Blite

1. **Suaeda calceoliformis** (Hook.) Moq. Sea Blite. Aug.–Oct. Native to the w. U.S.; adventive along roadsides where salt has been applied during the winter; confined to the ne. cos. *Suaeda depressa* (Pursh) S. Wats.

57. CISTACEAE—ROCKROSE FAMILY

1. At least some of the flowers with 5 yellow petals; capsules 1-locular; style present.
  2. Some of the flowers 1.5–2.5 cm across; style very short; leaves lanceolate to oblanceolate .... 1. *Helianthemum*
  2. None of the flowers as much as 1.5 cm across; style slender, elongated; leaves scalelike .... 2. *Hudsonia*
1. Flowers with 3 reddish petals; capsules partly 3-locular; style short or none .... 3. *Lechea*

1. **Helianthemum** Mill.—Rockrose

1. Terminal petaliferous flowers 2 or more; seeds not papillate .... 1. *H. bicknellii*
1. Terminal petaliferous flower solitary; seeds papillate .... 2. *H. canadense*

1. **Helianthemum bicknellii** Fern. Frostweed. June–July. Sandy woods, sandy prairies; occasional in the n. ½ of IL; also Madison, Richland, and St. Clair cos.

2. **Helianthemum canadense** (L.) Michx. Frostweed. May–July. Sandy woods, sandy prairies; occasional in the n. ½ of IL; also Effingham, Fayette, Madison, and St. Clair cos.

2. **Hudsonia** L.—False Heather

1. **Hudsonia tomentosa** Nutt. Two varieties occur in Illinois:

a. Flowers sessile or on very short bracteate pedicels .... 1a. *H. tomentosa* var. *tomentosa*
a. Flowers on leafless pedicels 1.5–6.0 mm long........ 1b. *H. tomentosa* var. *intermedia*

1a. **Hudsonia tomentosa** Nutt. var. **tomentosa.** Beach Heath. May–July. Sand dunes, hills, and blowouts, rare; Carroll, Fulton, Jo Daviess, Lee, and Whiteside cos.

1b. **Hudsonia tomentosa** Nutt. var. **intermedia** Peck. False Heather. May–July. Sandy soil, very rare; Jo Daviess Co.

3. **Lechea** L.—Pinweed

1. Outer 2 sepals conspicuously shorter than the inner 3.
  2. Leaves pubescent only on the veins on the lower surface.
    3. Seeds pale brown, partly covered by a grayish membrane; pedicels as long as or longer than the calyx .... 1. *L. intermedia*
    3. Seeds dark brown, smooth; pedicels mostly shorter than the calyx .... 4. *L. pulchella*
  2. Leaves pubescent throughout on the lower surface .... 5. *L. stricta*
1. Outer 2 sepals as long as or longer than the inner 3.
    4. Stems sparsely to moderately pubescent with ascending hairs; 3 inner sepals not appearing rough.
      5. Cauline leaves lanceolate to oblong, basal leaves elliptic-ovate; capsules about as long as the calyx .... 2. *L. minor*
      5. All leaves linear; capsules shorter than the calyx .... 6. *L. tenuifolia*
    4. Stems densely pubescent with spreading hairs; 3 inner sepals rough in appearance .... 3. *L. mucronata*

1. **Lechea intermedia** Britt. Savanna Pinweed. July–Oct. Sandy soil, rare; confined to the n. ⅙ of IL. *Lechea leggettii* Britt. & Hollick.

2. **Lechea minor** L. Small Pinweed. July–Sept. Sandy soil and cliffs, black oak savannas; occasional and sparsely distributed throughout IL.

3. **Lechea mucronata** Raf. Hairy Pinweed. July–Sept. Sandy soil and gravelly slopes; occasional in the n. ½ of IL, rare in the s. ½. *Lechea villosa* Ell.

4. **Lechea pulchella** Raf. Pretty Pinweed. July–Sept. Sandy soil, black oak savannas; restricted to the n. ½ of IL, except for Pope and Saline cos. (Including var. *moniliformis* Bickn.)

5. **Lechea stricta** Leggett. Bushy Pinweed. July–Sept. Sandy soil; sparsely scattered in the n. ½ of IL; also Monroe Co.

6. **Lechea tenuifolia** Michx. Two varieties occur in Illinois:

a. Leaves glabrous on the upper surface .... 6a. *L. tenuifolia* var. *tenuifolia*
a. Leaves pilose on the upper surface .... 6b. *L. tenuifolia* var. *occidentalis*

6a. **Lechea tenuifolia** Michx. var. **tenuifolia.** Narrow-leaved Pinweed. July–Sept. Sandy soil and cliffs, black oak savannas; occasional to common throughout IL.

6b. **Lechea tenuifolia** Michx. var. **occidentalis** Hodgdon. Pinweed. July–Sept. Dry sandy woods, very rare; Peoria Co.

58. CLEOMACEAE—SPIDER-FLOWER FAMILY

1. Stamens 8–32; fruits dehiscent only part way .... 3. *Polanisia*
1. Stamens 6; fruits dehiscent their entire length.
  2. Stipular spines present .... 4. *Tarenaya*
  2. Stipular spines absent.
    3. Petals 15–40 mm long; leaflets 5, 7, or 9; bracts ovate, cordate; seeds brown or pale green .... 1. *Cleroserrata*
    3. Petals 7–12 mm long; leaflets 3; bracts obovate, not cordate; seeds black .... 2. *Peritoma*

1. **Cleroserrata** H. H. Iltis—Bee Plant

1. **Cleroserrata speciosa** (Raf.) H. H. Iltis. Garden Spider-flower. June–Aug. Native to Europe; escaped from gardens; rare in IL. *Cleome speciosissima* Deppe ex Lindl.

2. **Peritoma** DC.—Bee Plant

1. **Peritoma serrulata** (Pursh) DC. Rocky Mountain Bee Plant. July–Sept. Native to the w. U.S.; occasional in disturbed soil; scattered in IL. *Cleome serrulata* Pursh.

3. **Polanisia** Raf.—Clammyweed

1. Leaflets obovate to lance-elliptic, all or most of them over 6 mm wide; fruits 5–10 mm wide.
  2. Largest petals 3.5–6.5 (–8.0) mm long; longest stamens 4–10 (–14) mm long, scarcely exceeding the petals .... 1. *P. dodecandra*
  2. Largest petals (7–) 8–16 mm long; longest stamens (9–) 12–30 mm long, much exceeding the petals .... 3. *P. trachysperma*
1. Leaflets linear, up to 4 mm wide; fruits 3–4 mm wide .... 2. *P. jamesii*

1. **Polanisia dodecandra** (L.) DC. Clammyweed. June–Sept. Gravelly soil, often along railroads, river banks; scattered throughout IL. *Polanisia graveolens* Raf.

2. **Polanisia jamesii** (Torr. & Gray) H. H. Iltis. James'-weed. June–Aug. Sandy soil, rare; confined to the n. ½ of IL. *Cristatella jamesii* Torr. & Gray.

3. **Polanisia trachysperma** Torr. & Gray. Clammyweed. June–Sept. Gravelly soil, sandy soil; scattered but not common throughout IL. *Polanisia dodecandra* (L.) DC. ssp. *trachysperma* (Torr. & Gray) H. H. Iltis.

4. **Tarenaya** Raf.—Spider-flower

1. Sepals and capsules glabrous; flowers pink or purple .... 1. *T. hassleriana*
1. Sepals and capsules glandular-pubescent; flowers usually white .... 2. *T. spinosa*

1. **Tarenaya hassleriana** (Chod.) H. H. Iltis. Pink-queen. June–Aug. Native to tropical America; rarely escaped from cultivation. *Cleome hassleriana* Chod.

2. **Tarenaya spinosa** (Jacq.) Raf. Spiny spider-flower. June–Sept. Native to tropical America; rarely escaped from cultivation; Jackson Co. *Cleome spinosa* Jacq.

59. CONVOLVULACEAE—MORNING-GLORY FAMILY

1. Plants prostrate, the leaves lying flat on the ground; flowers up to 3 mm wide .... 3. *Dichondra*

1. Plants upright or climbing, the leaves not lying flat on the ground; flowers more than 3 mm wide.
   2. Corolla rotate, usually blue; stigmas 4 .... 4. *Evolvulus*
   2. Corolla salverform, funnelform, or nearly campanulate; stigmas 1–2.
      3. Leaves linear, up to 3 (–6) mm wide; style deeply 2-cleft .... 7. *Stylisma*
      3. Leaves elliptic to oblong to ovate, usually at least 6 mm wide; style undivided although the stigmas sometimes lobed.
         4. Flowers scarlet .... 5. *Ipomoea*
         4. Flowers white, blue, or purple.
            5. Bracts large and foliaceous, sometimes concealing the calyx.
               6. Flowers in headlike clusters; calyx long-hirsute .... 6. *Jacquemontia*
               6. Flowers solitary or 2–4 in a group; calyx glabrous or pubescent, not long-hirsute .... 1. *Calystegia*
            5. Bracts small and never exceeding the calyx or absent.
               7. Sepals up to 8 mm long; stigmas 2, linear .... 2. *Convolvulus*
               7. Sepals 10 mm long or longer; stigmas 1 or, if 2, capitate .... 5. *Ipomoea*

1. **Calystegia** R. Br.—Bindweed

1. Flowers double, that is, with extra sets of petals .... 1. *C. hederacea*
1. Flowers single.
   2. Plants upright, either erect or ascending .... 4. *C. spithamaea*
   2. Plants trailing or climbing.
      3. Peduncles, or most of them, longer than the petioles; bracteoles usually not overlapping, acute, not saccate .... 2. *C. sepium*
      3. Peduncles, or most of them, shorter than the petioles; bracteoles overlapping, obtuse, saccate .... 3. *C. silvatica*

1. **Calystegia hederacea** Wallich. Japanese Bindweed. May–Sept. Native to Asia; rarely adventive in IL; Champaign, Cook, DuPage, Lake, Wabash, Washington, and Williamson cos. *Calystegia pubescens* Lindl.; *Convolvulus pellitus* Ledeb.

2. **Calystegia sepium** (L.) R. Br. Four subspecies occur in Illinois:

a. Bracteoles clearly differentiated from the sepals.
   b. Plants glabrous or sparsely pubescent; leaves mostly hastate.
      c. Leaves with a V-shaped sinus .... 2a. *C. sepium* ssp. *americanum*
      c. Leaves with a U-shaped sinus .... 2b. *C. sepium* ssp. *angulata*
   b. Plants densely soft-pubescent; leaves sagittate .... 2d. *C. sepium* ssp. *repens*
a. Bracteoles not clearly differentiated from the sepals but forming a continuous spiral with the sepals and gradually merging with them .... 2c. *C. sepium* ssp. *erratica*

2a. **Calystegia sepium** (L.) R. Br. ssp. **americanum** (Sims) Brummitt. American Bindweed. June–Aug. Moist soil, fields, disturbed areas; probably in every co. Convolvulus sepium L.

2b. **Calystegia sepium** (L.) R. Br. ssp. **angulata** Brummitt. Bindweed. June–Aug. Roadsides, fields; Cook and Woodford cos.

2c. **Calystegia sepium** (L.) R. Br. ssp. **erratica** Brummitt. Erratic Bindweed. May–Aug. Roadsides, fields; Richland Co.

2d. **Calystegia sepium** (L.) R. Br. ssp. **repens** (L.) Brummitt. Trailing Bindweed. July. Apparently adventive along a railroad; Lake Co. *Convolvulus sepium* L. var. *repens* (L.) Gray.

3. **Calystegia silvatica** (Kit.) Griseb. ssp. **fraterniflorus** (Mack. & Bush) Brummitt. Bindweed. June–Aug. Roadsides, fields; occasional in the s. ¾ of IL; also DeKalb Co. *Convolvulus sepium* L. var. *fraterniflorus* Mack. & Bush.

4. **Calystegia spithamaea** (L.) Pursh. Dwarf Bindweed. May–July. Open woods, prairies, sandy soil; occasional in the n. ⅘ of IL; also Hardin Co. *Convolvulus spithamaeus* L.

2. **Convolvulus** L.—Bindweed

1. Stems glabrous; leaves linear-oblong to ovate, glabrous or nearly so; sepals to 3 mm long; corolla to 2.5 cm wide .... 1. *C. arvensis*

1. Stems cinereous-pubescent; leaves oblong to elliptic, cinereous-pubescent; sepals to 8 mm long; corolla to 1.2 cm wide .... 2. *C. incanus*

1. **Convolvulus arvensis** L. Field Bindweed. May–Sept. Native to Europe; naturalized in disturbed areas; occasional to common throughout the state.

2. **Convolvulus incanus** Vahl. Ashy Bindweed. July–Sept. Native s. and w. of IL; adventive along a railroad; St. Clair Co. Not seen since the 1800s.

3. **Dichondra** J. R. Forst. & G. Forst.—Pony-foot

1. **Dichondra carolinensis** Michx. Pony-foot. May–June. Wet soil; Alexander Co., where it is apparently native. Specimen in the herbarium of the Missouri Botanical Garden.

4. **Evolvulus** L.—Ascending Morning-glory

1. **Evolvulus nuttallianus** Roemer & Schultes. Ozark Morning-glory. June–Aug. Native to the sw. U.S.; rarely adventive in disturbed soil; Kane Co; not seen in Illinois since 1976. *Evolvulus pilosus* Nutt.

5. **Ipomoea** L.—Morning-glory

1. Flowers scarlet; stamens and style exserted.
  2. Leaves cordate, entire or shallowly lobed .... 1. *I. coccinea*
  2. Leaves deeply pinnatifid .... 6. *I. quamoclit*
1. Flowers white, pink, purple, or blue; stamens and style included.
  3. Stems and petioles with purple setae .... 7. *I. setosa*
  3. Stems and petioles without purple setae.
    4. Perennials with glabrous or puberulent stems; sepals glabrous, obtuse to subacute; seeds pubescent .... 4. *I. pandurata*
    4. Annuals with pubescent stems; sepals pubescent, acute to acuminate; seeds glabrous or nearly so.
      5. Calyx lobes linear-lanceolate, with long-tapering tips; corolla essentially 3.0–4.5 cm long, sky blue (when fresh) .... 2. *I. hederacea*
      5. Calyx lobes oblong to lanceolate, acute to short-acuminate; corolla either less than 3 cm long or more than 4.5 cm long, not sky blue (when fresh).
        6. Corolla less than 3 cm long, essentially white; ovary and capsule 2-locular .... 3. *I. lacunosa*
        6. Corolla more than 4.5 cm long, usually not white; ovary and capsule 3-locular .... 5. *I. purpurea*

1. **Ipomoea coccinea** L. Red Morning-glory. July–Oct. Native to tropical America; occasionally naturalized in moist areas, particularly in the s. ⅔ of IL; also Cook Co.

2. **Ipomoea hederacea** (L.) Jacq. Ivy-leaved Morning-glory. June–Oct. Native to tropical America; naturalized in waste areas; occasional to common throughout the state. Plants with unlobed leaves sometimes occur.

3. **Ipomoea lacunosa** L. Small White Morning-glory. July–Oct. Fields and streams; occasional in the s. ⅘ of IL, apparently absent elsewhere. Purplish-tinged flowers sometimes occur.

4. **Ipomoea pandurata** (L.) G. F. W. Mey. Wild Sweet Potato Vine. June–Oct. Generally in disturbed areas, thickets, woods, fields; occasional to common in the s. ¾ of IL, less common elsewhere.

5. **Ipomoea purpurea** (L.) Roth. Common Morning-glory. July–Oct. Native to tropical America; naturalized in disturbed areas; occasional throughout the state.

6. **Ipomoea quamoclit** L. Cypress Vine. June–Sept. Native to tropical America; rarely escaped from cultivation; Pope Co.

7. **Ipomoea setosa** L. Brazilian Morning-glory. July–Aug. Native to Brazil; rarely escaped from cultivation; Johnson Co.

6. **Jacquemontia** Choisy

1. **Jacquemontia tamnifolia** (L.) Griseb. Jacquemontia. Aug.–Oct. Native to tropical America; adventive along a railroad; Grundy Co; not seen since 1972.

7. **Stylisma** Raf.

1. **Stylisma pickeringii** (Torr.) Gray var. **pattersonii** (Fern. & Schub.) Myint. Aug. Sandy prairies, very rare; Cass, Henderson, and Mason cos. *Breweria pickeringii* (Torr.) Gray var. *pattersonii* Fern. & Schub.

60. CORNACEAE—DOGWOOD FAMILY

1. **Cornus** L.—Dogwood

1. Flowers surrounded by 4 conspicuous, petal-like bracts; fruits red.
  2. Herbs to 30 cm tall; flowers pedicellate; drupes globose, soft .... 4. *C. canadensis*
  2. Trees to 12 m tall; flowers sessile; drupes ellipsoid, hard.................... 6. *C. florida*
1. Flowers not subtended by 4 conspicuous, petal-like bracts; fruits blue or white.
  3. Flowers yellow, appearing before the leaves expand .... 8. *C. mas*
  3. Flowers white or cream, appearing after the leaves expand.
    4. Leaves alternate.. 2. *C. alternifolia*
    4. Leaves opposite.
      5. Leaves with spreading or curled pubescence on the lower surface.
        6. Leaves broadly ovate to suborbicular, with 6–9 pairs of veins; drupes light blue.. 11. *C. rugosa*
        6. Leaves elliptic to lance-ovate, with 3–4 pairs of veins; drupes white.
          7. Leaves finely pubescent above; branchlets red; pith white.
            8. Stone of drupe round .... 12. *C. sericea*
            8. Stone of drupe flattened .... 1. *C. alba*
          7. Leaves scabrous above; branchlets gray; pith brown... 5. *C. drummondii*
      5. Leaves with appressed pubescence or glabrous on the lower surface.
        9. Leaves conspicuously whitish or glaucous beneath; drupes white (blue in *C. obliqua*).
          10. Young twigs glabrous or nearly so; drupes white.
            11. Twigs red; cymes flat-topped, broader than high.
              12. Stone of drupe round.. 12. *C. sericea*
              12. Stone of drupe flattened .... 1. *C. alba*
            11. Twigs gray; cymes elongated, as high as broad .... 10. *C. racemosa*
          10. Young twigs densely hairy; drupes blue .... 9. *C. obliqua*
        9. Leaves green or rufescent beneath; drupes blue.
          13. Pith white; leaves glabrous beneath.............................. 7. *C. foemina*
          13. Pith tawny; leaves with reddish pubescence beneath .... 3. *C. amomum*

1. **Cornus alba** L. European Red Osier. May–Sept. Native to Europe; adventive around old homesites; ne. cos; also Champaign Co.

2. **Cornus alternifolia** L. f. Alternate-leaved Dogwood; Pagoda Dogwood. May–June. Moist woods and slopes, along streams; not common in the n. ½ of IL, rarer in the s. ½.

3. **Cornus amomum** Mill. Swamp Dogwood. June–July. Moist thickets, swamps, not common; confined to the extreme s. tip of IL.

4. **Cornus canadensis** L. Bunchberry. May–June. Boggy woods, rare; Cook, Lake, LaSalle, McHenry, and Ogle cos.

5. **Cornus drummondii** C. A. Mey. Rough-leaved Dogwood. May–July. Rocky woods, prairies, low ground, fields, meadows; occasional to common in the s. ¾ of IL, rare elsewhere. *Cornus asperifolia* Michx., misapplied.

6. **Cornus florida** L. Flowering Dogwood. Apr.–June. Rocky woods, mesic woods, wooded slopes, low woods; occasional to common in the s. ½ of IL. Rare pink-flowered plants may be called f. *rubra* (Weston) Palmer & Steyerm.

7. **Cornus foemina** Mill. Stiff Dogwood; Swamp Dogwood. May–June. Swamps, low woods, occasional; restricted to the s. ½ of IL. *Cornus stricta* Lam.

8. **Cornus mas** L. Cornelian Cherry. Feb.–Mar. Native to Europe; rarely escaped from cultivation; Champaign, Jersey, and Peoria cos.

9. **Cornus obliqua** Raf. Pale Dogwood; Blue-fruited Dogwood; Silky Dogwood.

May–July. Swamps, low woods, wet prairies; occasional to common in most of IL. *Cornus amomum* Mill. var. *scheutzeana* (C. A. Mey.) Rickett.

10. **Cornus racemosa** Lam. Gray Dogwood. May–July. Moist woods, upland woods, prairies, roadsides; occasional to common throughout the state. *Cornus foemina* Mill., misapplied.

11. **Cornus rugosa** Lam. Round-leaved Dogwood. May–June. Shaded, rocky slopes, not common; confined to the n. ¼ of IL.

12. **Cornus sericea** L. Two varieties occur in Illinois:

a. Leaves glabrous or appressed-pubescent beneath .... 12a. *C. sericea* var. *sericea*
a. Leaves densely soft-pilose beneath .... 12b. *C. sericea* var. *baileyi*

12a. **Cornus sericea** L. var. **sericea.** Red-osier Dogwood. May–Sept. Marshes, calcareous fens; occasional in the n. ½ of IL, rare elsewhere. *Cornus stolonifera* Michx.

12b. **Cornus sericea** L. var. **baileyi** (Coult. & Evans) Mohlenbr., comb. nov. (basionym: *Cornus baileyi* Coult. & Evans). Bailey's Red-osier Dogwood. Dunes, rare; Cook and Lake cos. *Cornus stolonifera* Michx. var. *baileyi* (Coult. & Evans) Drescher.

61. CORYLACEAE—HAZELNUT FAMILY

1. Leaves broadly ovate, cordate, doubly serrate .... 2. *Corylus*
1. Leaves narrowly ovate or oval, not cordate, once serrate.
  2. Bracts inflated, enclosing the nut; bark rough; lateral nerves of leaves usually branched near margin .... 3. *Ostrya*
  2. Bracts foliaceous, subtending the nut; bark smooth; lateral nerves of leaves unbranched .... 1. *Carpinus*

1. **Carpinus** L.—Ironwood

1. **Carpinus caroliniana** Walt. Two varieties occur in Illinois:

a. Lower surface of leaves with dark glands .... 1b. *C. caroliniana* var. *virginiana*
a. Lower surface of leaves without glands .... 1a. *C. caroliniana* var. *caroliniana*

1a. **Carpinus caroliniana** Walt. var. **caroliniana**. Musclewood Tree; Blue Beech; Ironwood; American Hornbeam. Apr.–May. Moist woods, occasional; in the s. ⅙ of IL.

1b. **Carpinus caroliniana** Walt. var. **virginiana** (Marsh.) Fern. Musclewood Tree; Blue Beech; Ironwood; American Hornbeam. Apr.–May. Moist woods, common; probably in every co.

2. **Corylus** L.—Hazelnut

1. Leaves serrulate, densely hairy on the lower surface; bracts of the involucre free to the base, not bristly hairy, up to 3 cm long; petioles glandular-pubescent .... 1. *C. americana*
1. Leaves coarsely toothed as well as serrulate, usually hairy only along the veins of the lower surface; bracts of the involucre united at the base, bristly hairy, at least near tip, usually no more than 3 cm long; petioles not glandular-pubescent .... 2. *C. cornuta*

1. **Corylus americana** Walt. Hazelnut. Mar.–Apr. Thickets, dry, disturbed woods, common; in every co. (Including f. *missouriensis* [A. DC.] Fern.)

2. **Corylus cornuta** Marsh. Beaked Hazelnut. Apr.–May. Algific slopes, rare; Jo Daviess Co. *Corylus rostrata* Ait.

3. **Ostrya** Scop.—Hop Hornbeam

1. Ostrya virginiana (Mill.) K. Koch. Two varieties occur in Illinois:
a. Branchlets glabrous or sparsely pilose .... 1a. *O. virginiana* var. *virginiana*
a. Branchlets densely villous .... 1b. *O. virginiana* var. *lasia*

1a. **Ostrya virginiana** (Mill.) K. Koch var. **virginiana.** Hop Hornbeam; Ironwood. Apr.–May. Woodlands; in every co. (Including f. *glandulosa* [Spach] Macbr.)

1b. **Ostrya virginiana** (Mill.) K. Koch var. **lasia** Fern. Hop Hornbeam; Ironwood. Apr.–May. Woodlands, not common; confined to the s. ¼ of IL.

62. CRASSULACEAE—STONECROP FAMILY

1. Flowers yellow.
  2. Leaves entire; pistils united at base .... 3. *Sedum*
  2. Leaves crenate; pistils free to base .... 2. *Phedimus*
1. Flowers white, pink, or purplish.
  3. Leaves all opposite, entire near the base, denticulate above; pistils free to base .... 2. *Phedimus*
  3. Some or all the leaves alternate, either entire or distinctly serrate or denticulate; pistils united at base.
    4. Flowers white........ 3. *Sedum*
    4. Flowers pink or purplish.
      5. Leaves terete, linear to linear-spatulate; petals up to 8 mm long .... 3. *Sedum*
      5. Leaves flat, elliptic, ovate, or obovate; petals 8 mm long or longer........ 1. *Hylotelephium*

1. **Hylotelephium** H. Ohba—Orpine

1. Petals pink-tinged........ 1. *H. X erythrostictum*
1. Petals decidedly pink or purple.
  2. Flowers pink; petals more than twice as long as the sepals......2. *H. telephioides*
  2. Flowers purple; petals at most only twice as long as the sepals........ 3. *H. telephium*

1. **Hylotelephium X erythrostictum** (Miq.) H. Ohba. Garden Orpine. Sept. Native to Asia; rarely escaped from cultivation; DeKalb, DuPage and Lake cos. Considered to be a hybrid between *H. spectabile* (Boreau) H. Ohba and *Sedum viridescens* Nakai. *Sedum X alboroseum* Boreau.

2. **Hylotelephium telephioides** (Michx.) H. Ohba. American Orpine. Aug.–Sept. Sandstone cliffs, rare; in the s. ¼ of IL. *Sedum telephioides* Michx.

3. **Hylotelephium telephium** (L.) H. Ohba. Live-forever. Aug.–Sept. Native to Europe; occasionally escaped from cultivation. *Sedum purpureum* (L.) Schult.

2. **Phedimus** Raf.—False Stonecrop

1. Flowers yellow .... 1. *P. aizoon*
1. Flowers pink or purple........ 2. *P. spurius*

1. **Phedimus aizoon** (L.) 't Hart & U. Eggli. Orpine. June–July. Native to n. Asia; rarely escaped from cultivation; DuPage Co. *Sedum kamschaticum* (Fisch. & Mey.) Roem. & Schultes in J. J. Roem. Plants with larger clusters of flowers have been called var. *floriferum.* These have been found as escapes in DuPage Co. This latter variety has not been transferred to *Phedimus.*

2. **Phedimus spurius** (M. Bieb.) 't Hart. False Wild Stonecrop. June–July. Native to Europe and Asia; escaped from cultivation into cemeteries; Will Co. *Sedum spurium* M. Bieb.

3. **Sedum** L.—Stonecrop

1. Leaves terete.
  2. Flowers yellow.
    3. Leaves in six spiral rows........ 7. *S. sexangulare*
    3. Leaves not in six spiral rows.
      4. Leaves to 6 mm long; follicles spreading .... 1. *S. acre*
      4. Leaves about 10 mm long; follicles more or less erect..........5. *S. reflexum*
  2. Flowers pink or white.
    5. Creeping evergreen; flowers white........ 3. *S. album*
    5. Upright deciduous herb; flowers pink .... 4. *S. pulchellum*
1. Leaves flat.
  6. Flowers yellow .... 6. *S. sarmentosum*
  6. Flowers pink or rose or white.
    7. Flowers pink or rose; leaves not ternate .... 2. *S. alboroseum*
    7. Flowers white; leaves ternate........ 8. *S. ternatum*

1. **Sedum acre** L. Mossy Stonecrop; Yellow Stonecrop. May–Aug. Native to Europe; occasionally escaped from cultivation.

2. **Sedum alboroseum** L. Japanese Stonecrop. June–July. Native to Asia; escaped from cultivation in disturbed soil; DuPage Co.

3. **Sedum album** L. White Stonecrop. June–July. Native to Asia, Europe, and Africa; escaped from cultivation in disturbed soil; DuPage Co.

4. **Sedum pulchellum** Michx. Widow's-cross. May–July. Dry or moist sandstone; known only from the s. ⅙ of IL.

5. **Sedum reflexum** L. Rock Stonecrop. June–July. Native to Europe; rarely escaped from cultivation; Kane and St. Clair cos. *Sedum rupestre* L., misapplied.

6. **Sedum sarmentosum** Bunge. Yellow Stonecrop. May–July. Native to Asia; occasionally escaped from cultivation.

7. **Sedum sexangulare** L. Creeping Stonecrop. June–July. Native to Europe and Asia; escaped from gardens; DuPage Co.

8. **Sedum ternatum** Michx. Three-leaved Stonecrop. May–June. Limestone cliffs, moist wooded ravines, occasional; scattered throughout IL.

63. CUCURBITACEAE—GOURD FAMILY

1. Leaves pinnately cleft, often more than halfway to the middle.
  2. Corolla up to 3.5 cm across .... 1. *Citrullus*
  2. Corolla 4–8 cm across .... 6. *Luffa*
1. Leaves unlobed or shallowly lobed.
  3. Plants covered with soft, sticky hairs .... 5. *Lagenaria*
  3. Plants glabrous or pubescent but without soft, sticky hairs.
    4. Ovaries and fruits not spinulose.
      5. Corolla more than 5 cm long and more than 5 cm broad .... 3. *Cucurbita*
      5. Corolla up to 3.5 cm long and up to 3.5 cm broad.
        6. Corolla yellow.
          7. Corolla up to 2 cm across; leaves unlobed; staminate flower solitary .... 9. *Thladiantha*
          7. Corolla 2.5–3.5 cm across; leaves lobed; staminate flowers in clusters .... 2. *Cucumis*
        6. Corolla white or greenish .... 7. *Melothria*
    4. Ovaries and fruits spinulose.
      8. Plants pubescent; corolla 5-lobed; ovary 1-locular; fruit 1-seeded .... 8. *Sicyos*
      8. Plants glabrous or nearly so; corolla 6-lobed; ovary 2-locular; fruit 4-seeded .... 4. *Echinocystis*

1. **Citrullus** Neck—Melon

1. **Citrullus lanatus** (Thunb.) Matsumura & Nakai. Watermelon. Sept. Native to Africa; occasionally escaped from cultivation but seldom persisting; around dumps. *Citrullus vulgaris* Schrad.

2. **Cucumis** L.—Cucumber

1. Fruits glabrous, pubescent, or slightly warty .... 1. *C. melo*
1. Fruits muricate .... 2. *C. sativus*

1. **Cucumis melo** L. Muskmelon. Aug.–Oct. Native to Asia; occasionally escaped from cultivation into waste ground.

2. **Cucumis sativus** L. Cucumber. May–Sept. Native to Asia; rarely escaped from cultivation.

3. **Cucurbita** L.—Gourd

1. Leaves stiff, thick, often nearly twice as long as broad .... 1. *C. foetidissima*
1. Leaves flexible, thin, never twice as long as broad .... 2. *C. pepo*

1. **Cucurbita foetidissima** HBK. Missouri Gourd; Railroad Vine. May–Sept. Native to the w. U.S.; occasionally adventive along railroads.

2. **Cucurbita pepo** L. Two varieties have been found in Illinois:

a. Fruits usually 20 cm or more in diameter, the flesh edible......... 2a. *C. pepo* var. *pepo*
a. Fruits usually less than 20 cm in diameter, the flesh inedible..... 2b. *C. pepo* var. *ovifera*
2a. Cucurbita pepo L. var. **pepo.** Pumpkin. May–Sept. Native to tropical Am.; rarely adventive and persisting in IL.
2b. Cucurbita pepo L. var. **ovifera** (L.) Alef. Pear Gourd. May–Sept. Native to tropical America; moist soil; scattered in IL.

4. **Echinocystis** Torr. & Gray

1. **Echinocystis lobata** (Michx.) Torr. & Gray. Wild Balsam-apple. July–Oct. Moist soil, floodplains, thickets; occasional in the n. ¾ of IL, apparently absent in the s. ¼.

5. **Lagenaria** Seringe—Gourd

1. **Lagenaria siceraria** (Molina) Standl. Gourd. May–Sept. Native to the Eastern Hemisphere; rarely escaped from cultivation.

6. **Luffa** L.

1. **Luffa cylindrica** (L.) Roemer. Vegetable Sponge; Dishwater Gourd. June–Aug. Native to the Old World tropics; rarely escaped from cultivation.

7. **Melothria** L.

1. **Melothria pendula** L. Creeping Cucumber; Squirting Cucumber. June–Sept. Rocky soil, rare; Alexander, Jackson, Pope, and Union cos.

8. **Sicyos** L.—Bur Cucumber

1. **Sicyos angulatus** L. Bur Cucumber. July–Sept. Moist soil of fields and woods, floodplains; occasional and scattered throughout IL.

9. **Thladiantha** Bunge—Thladiantha

1. **Thladiantha dubia** Bunge. Thladiantha. July–Sept. Native to Asia; rarely adventive along fencerows; Cook, DuPage, and LaSalle cos.

64. CUSCUTACEAE—DODDER FAMILY

1. **Cuscuta** L.—Dodder

1. Sepals free to base.
2. Flowers pedicellate, borne in rather loose cymes or panicles; seeds 1.4–1.5 mm long .... 5. *C. cuspidata*
2. Flowers sessile, borne in dense glomerules; seeds 1.7 mm long or longer.
3. Bracts at base of sepals appressed; lobes of corolla obtuse; seeds 2.5–2.6 mm long .... 3. *C. compacta*
3. Bracts at base of sepals with recurved tips; lobes of corolla acute; seeds 1.7–1.8 mm long .... 6. *C. glomerata*
1. Sepals united below into a tube.
4. Most of the flowers with 4-lobed corollas.
5. Corolla lobes erect; flowers sessile or on pedicels up to 0.5 mm long.
6. Scales absent or reduced to minute teeth along the filaments; lobes of corolla acute, about as long as the corolla tube; seeds 1.3–1.4 mm long .... 10. *C. polygonorum*
6. Scales toothed from base to apex; lobes of corolla obtuse to subacute, shorter than the corolla tube; seeds 1.6–1.7 mm long ...... 2. *C. cephalanthi*
5. Corolla lobes inflexed; flowers on pedicels usually at least 1 mm long.. 4. *C. coryli*
4. Most of the flowers with 5-lobed corollas.
7. Lobes of corolla obtuse, erect or spreading................................ 7. *C. gronovii*
7. Lobes of corolla acute to acuminate, the tips inflexed.
8. Lobes of calyx obtuse; pedicels shorter than the flowers.
9. Lobes of corolla acuminate; scales about ½ as long as corolla tube; seeds 1.0–1.2 mm long .... 9. *C. pentagona*
9. Lobes of corolla acute; scales about as long as corolla tube; seeds 1.5–1.6 mm long .... 1. *C. campestris*

8. Lobes of calyx acute; pedicels as long as or longer than the flowers .... 8. *C. indecora*

1. **Cuscuta campestris** Yuncker. Dodder. June–Oct. Mostly disturbed areas; scattered but not common in IL. Mostly parasitic on *Polygonum, Convolvulus, Xanthium,* and *Brassica.*

2. **Cuscuta cephalanthi** Engelm. Dodder. Aug.–Oct. Low, wet areas; occasional and scattered throughout IL. Parasitic on *Cephalanthus, Justicia,* and many others.

3. **Cuscuta compacta** Juss. Compact Dodder. July–Oct. Moist areas; scattered throughout the state but not common. Parasitic on *Campsis, Cephalanthus,* and others.

4. **Cuscuta coryli** Engelm. Hazel Dodder. July–Oct. Moist or dry areas; scattered throughout IL. but apparently not common. Mostly parasitic on *Campsis, Corylus, Prunella,* and *Stachys.*

5. **Cuscuta cuspidata** Engelm. Dodder. July–Oct. Moist or dry areas; scattered throughout the state but apparently not common.

6. **Cuscuta glomerata** Choisy. Rope Dodder. July–Oct. Low, wet areas; occasional and scattered throughout IL. Parasitic mostly on members of the Asteraceae.

7. **Cuscuta gronovii** Willd. Two varieties occur in Illinois:

a. Calyx lobes shorter than the corolla; corolla cylindrical .... 7a. *C. gronovii* var. *gronovii*

a. Calyx lobes about as long as the corolla; corolla campanulate .... 7b. *C. gronovii* var. *latifolia*

7a. **Cuscuta gronovii** Willd. var. **gronovii.** Dodder. July–Oct. Moist areas; common throughout the state. Parasitic on many hosts.

7b. **Cuscuta gronovii** Willd. var. **latifolia** Engelm. Dodder. July–Oct. Low, moist areas, not common; scattered in IL.

8. **Cuscuta indecora** Choisy. Two varieties occur in Illinois:

a. Calyx lobes about ½ as long as corolla tube .... 8a. *C. indecora* var. *indecora*

a. Calyx lobes about as long as corolla tube .... 8b. *C. indecora* var. *neuropetala*

8a. **Cuscuta indecora** Choisy var. **indecora.** Dodder. July–Sept. Low areas; infrequent in the s. ⅔ of IL.

8b. **Cuscuta indecora** Choisy var. **neuropetala** (Engelm.) Hitchc. Dodder. July–Sept. Low areas; rare in the s. ½ of IL.

9. **Cuscuta pentagona** Engelm. Prairie Dodder. June–Oct. Dry, mostly disturbed areas; occasional to common throughout IL. Parasitic on many hosts.

10. **Cuscuta polygonorum** Engelm. Smartweed Dodder. July–Sept. Moist areas; occasional and scattered throughout IL. Parasitic on *Polygonum* and other hosts.

65. DIPSACACEAE—TEASEL FAMILY

1. Stems and leaves prickly .... 1. *Dipsacus*

1. Stems and leaves not prickly .... 2. *Knautia*

1. **Dipsacus** L.—Teasel

1. Leaves toothed, prickly on the margins .... 1. *D. fullonum*

1. Leaves pinnatifid or bipinnatifid, ciliate on the margins .... 2. *D. laciniatus*

1. **Dipsacus fullonum** L. Common Teasel. June–Oct. Native to Europe; adventive in disturbed areas, particularly along roads; scattered in IL and becoming more common. *Dipsacus sylvestris* Huds.

2. **Dipsacus laciniatus** L. Cut-leaved Teasel. July–Sept. Native to Europe; adventive in disturbed areas; scattered in IL.

2. **Knautia** L.—Bluebuttons

1. **Knautia arvensis** (L.) T. Coult. Bluebuttons. June–Aug. Native to Europe; rarely escaped from cultivation; Winnebago Co.

66. DROSERACEAE—SUNDEW FAMILY

1. **Drosera** L.—Sundew

1. Leaves spatulate; seeds obovoid, with a tight, papillate testa .... 1. *D. intermedia*

1. Leaves suborbicular; seeds fusiform, with a loose, striate testa ..... 2. *D. rotundifolia*

1. **Drosera intermedia** Hayne. Narrow-leaved Sundew. July–Sept. Bogs, rare; Cook, DuPage, Grundy, Iroquois, Kane, Kankakee, Lake, Lee, McHenry, Ogle, and Will cos.

2. **Drosera rotundifolia** L. Round-leaved Sundew. June–Sept. Bogs, rare; Cook, Kane, Lake, McHenry, and Ogle cos.

67. EBENACEAE—PERSIMMON FAMILY

1. **Diospyros** L.—Persimmon

1. **Diospyros virginiana** L. Four variations occur in Illinois:

a. Branchlets and leaves glabrous or nearly so.
  b. Fruit spherical, orange, black, or dark purple, maturing in June, July, or Oct.
    c. Fruit orange, maturing in Oct. .... 1a. *D. virginiana* var. *virginiana*
    c. Fruit black or dark purple, maturing in June and July .... 1b. *D. virginiana* f. *atra*
  b. Fruit depressed-globose, orange, maturing in June and July .... 1c. *D. virginiana* var. *platycarpa*
a. Branchlets and leaves pubescent........................ 1d. *D. virginiana* var. *pubescens*

1a. **Diospyros virginiana** L. var. **virginiana**. Persimmon. May–June. Dry woods, fields, roadsides, clearings, rarely in floodplain forests; occasional to common in the s. ⅔ of IL, apparently absent elsewhere. This variety fruits in October.

1b. **Diospyros virginiana** L. f. **atra** Sarg. Dark-fruited Persimmon. May. Dry woods, rare; Gallatin and Jackson cos. This form fruits in June and July.

1c. **Diospyros virginiana** L. var. **platycarpa** Sarg. Early-fruited Persimmon. May. Dry woods, rare; scattered in s. IL. This variety fruits in June and July.

1d. **Diospyros virginiana** L. var. **pubescens** (Pursh) Dippel. Hairy Persimmon. May–June. Dry woods, fields; scattered in s. IL. This variety fruits in Oct.

68. ELAEAGNACEAE—OLEASTER FAMILY

1. Leaves alternate; stamens 4; flowers bisexual .... 1. *Elaeagnus*
1. Leaves opposite; stamens 8; flowers unisexual .... 2. *Shepherdia*

1. **Elaeagnus** L.—Oleaster

1. Fruit yellow; branchlets and leaves with silvery scales .... 1. *E. angustifolia*
1. Fruit pink or red; branchlets and leaves with brown and silvery scales.
  2. Calyx tube about equal to calyx lobes; fruiting pedicels 1.5 cm long or longer .... 2. *E. multiflora*
  2. Calyx tube much longer than lobes; fruiting pedicels up to 1 cm long .... 3. *E. umbellata*

1. **Elaeagnus angustifolia** L. Russian Olive. May–June. Native to Europe and Asia; often escaped from cultivation.

2. **Elaeagnus multiflora** Thunb. Long-stalked Oleaster. May–June. Native to Asia; rarely adventive in ne. IL.

3. **Elaeagnus umbellata** Thunb. Autumn Olive. May–June. Native to Europe and Asia; commonly escaped from cultivation.

2. **Shepherdia** Nutt.—Buffalo-berry

1. **Shepherdia canadensis** (L.) Nutt. Buffalo-berry; Rabbit-berry. May–July. Shores of Lake Michigan, very rare; Cook and Lake cos.

69. ELATINACEAE—WATERWORT FAMILY

1. Sepals 5, acuminate; petals 5; plants pubescent .... 1. *Bergia*
1. Sepals 2–3, obtuse; petals 2–3; plants glabrous .... 2. *Elatine*

1. **Bergia** L.

1. **Bergia texana** (Hook.) Seub. Bergia. June–Sept. Wet soil, shallow marsh in sand deposits, very rare; Alexander and St. Clair cos.

2. **Elatine** L.—Waterwort

1. Petals 2; stamens 2; leaves obovate, not retuse at the apex .... 1. *E. brachysperma*
1. Petals 3; stamens 3; leaves oblong to spatulate, retuse at the apex .... 2. *E. triandra*

1. **Elatine brachysperma** Gray. Waterwort. July–Oct. Shallow water, muddy banks; Menard Co. *Elatine triandra* Schk. var. *brachysperma* (Gray) Fassett.

2. **Elatine triandra** Schk. Waterwort. July–Oct. Shallow water, muddy banks, very rare; Cass and Sangamon cos.

70. ERICACEAE—HEATH FAMILY

1. Leaves entire.
  2. Plants upright.
    3. Ovary inferior.
      4. Leaves with glandular dots; winter twigs predominantly black .... 6. *Gaylussacia*
      4. Leaves without glandular dots; winter twigs green, gray, or brown .... 10. *Vaccinium*
    3. Ovary superior.
      5. Leaves deciduous; anthers awnless; flowers over 1.5 cm long, bright pink or rose; capsules elongated .... 9. *Rhododendron*
      5. Leaves evergreen; anthers awned; flowers up to 1 cm long, pale pink or white; capsules depressed-globose .... 1. *Andromeda*
  2. Plants trailing.
    6. Fruit a capsule; flowers salverform; anthers awnless, splitting lengthwise; leaves cordate .... 4. *Epigaea*
    6. Fruit a drupe; flowers urn-shaped; anthers awned, opening by a terminal pore; leaves cuneate .... 2. *Arctostaphylos*
1. Leaves toothed.
  7. Ovary inferior.
    8. Corolla deeply 4-lobed; stems prostrate .... 7. *Oxycoccos*
    8. Corolla 5-lobed, urceolate; stems erect .... 10. *Vaccinium*
  7. Ovary superior.
    9. Tree to 15 m tall; flowers in terminal racemes; leaves deciduous .... 8. *Oxydendrum*
    9. Trailing or ascending shrub to 35 cm tall; flowers in axillary leafy racemes or solitary; leaves evergreen.
      10. Fruit dry; leaves scurfy beneath .... 3. *Chamaedaphne*
      10. Fruit fleshy; leaves not scurfy beneath .... 5. *Gaultheria*

1. **Andromeda** L.—Bog Rosemary

1. **Andromeda glaucophylla** Link. Bog Rosemary. May–June. Bogs, very rare; Lake and McHenry cos. *Andromeda polifolia* L. var. *glaucophylla* (Link) DC.

2. **Arctostaphylos** Adans.—Bearberry

1. **Arctostaphylos uva-ursi** (L.) Spreng. ssp. **coactilis** (Fern. & Macbr.) Love, Love, & Kapoor. Kinnikinnick; Bearberry. Apr.–June. Sand dunes and black oak savannas, rare; n. ¼ of IL; also Peoria and Tazewell cos. *Arctostaphylos uva-ursi* (L.) Spreng. var. *coactilis* Fern. & Macbr.

3. **Chamaedaphne** Moench—Leatherleaf

1. **Chamaedaphne calyculata** (L.) Moench var. **angustifolia** (Ait.) Rehder. Leatherleaf. May. Bogs, rare; Cook, Kane, Lake, McHenry, and Winnebago cos.

4. **Epigaea** L.—Trailing Arbutus

1. **Epigaea repens** L. Trailing Arbutus. Apr.–May. Woods, very rare; McHenry Co.; not seen since the 1800s. (Including var. *glabrifolia* Fern.)

5. **Gaultheria** L.—Wintergreen

1. **Gaultheria procumbens** L. Wintergreen; Checkerberry. June–Aug. Sandy soil of woods, bogs, edge of marshes, crests of ravines, rare; Cook, Lake, LaSalle, and Ogle cos.

6. **Gaylussacia** HBK.—Huckleberry

1. **Gaylussacia baccata** (Wang.) K. Koch. Black Huckleberry; Box Huckleberry. Apr.–June. Rocky woods and cliffs, black oak savannas; occasional throughout IL.

7. **Oxycoccos** Hedwig—Cranberry

1. Pedicels with a pair of bracts 2–4 mm long; leaves 6–15 mm long, 2–8 mm wide, scarcely if at all revolute, sometimes pale on the back but not strongly glaucous; corolla 8–10 mm long; berry 1–2 cm in diameter .... 1. *O. macrocarpon*
1. Pedicels with scalelike bracts up to 1.5 mm long; leaves 3–8 mm long, 1–3 mm wide, strongly revolute, strongly glaucous on the back; corolla 5–6 mm long; berry 5–6 mm in diameter .... 2. *O. microcarpon*

1. **Oxycoccos macrocarpon** (Ait.) Pers. Large Cranberry; American Cranberry. June–Aug. Bogs, rare; Cook, Lake, McHenry, and Will cos. *Vaccinium macrocarpon* Ait.

2. **Oxycoccos microcarpon** Turcz. Small Cranberry. May–July. Bogs, very rare: Lake Co. *Vaccinium oxycoccos* L.

8. **Oxydendrum** DC.—Sourwood

1. **Oxydendrum arboreum** (L.) DC. Sourwood. June–July. Native to the se. U.S.; naturalized in Adams, Christian, and Cook cos.

9. **Rhododendron** L.—Azalea

1. Bud scales glabrous; flowers not fragrant .... 2. *R. periclymenoides*
1. Bud scales pubescent; flowers fragrant.
  2. Pedicels usually eglandular; capsules usually densely pubescent........................ .... 3. *R. prinophyllum*
  2. Pedicels usually glandular; capsules sparsely pubescent.............. 1. *R. canescens*

1. **Rhododendron canescens** (Michx.) Sweet. Pink Azalea. May. Cherty slopes, rare; Union Co. *Azalea canescens* Michx.

2. **Rhododendron periclymenoides** (Michx.) Shinners. Pinkster-flower; Pink Azalea. May. Cherty slopes, rare; Union Co. *Azalea nudiflorum* (L.) Torr.

3. **Rhododendron prinophyllum** (Small) Millais. Pink Azalea; Rosebud Azalea. May. Cherty slopes, rare; Alexander, Jackson, and Union cos. *Azalea roseum* (Loisel.) Rehd.

10. **Vaccinium** L.—Blueberry

1. Stamens long-exserted from the corolla; leaves whitened beneath .... 7. *V. stamineum*
1. Stamens included within the corolla; leaves green or whitened beneath.
  2. Flowers in leafy, bracted racemes; anthers awned; pedicels jointed; berry dry ... . 2. *V. arboreum*
  2. Flowers solitary, in racemes or in glomerules but not conspicuously bracteate; anthers awnless; pedicels not jointed; berry juicy.
    3. Leaves with marginal spinulose teeth .... 1. *V. angustifolium*
    3. Leaves entire or serrulate but without spinulose teeth.
      4. Corolla white, sometimes tinged with pink; some or all the leaves over 5 cm long; tall shrubs over 1 m in height.
        5. Leaves glabrous, entire .... 3. *V. corymbosum*
        5. Leaves pubescent on the midvein below, serrulate .... 4. *V. fuscatum*
      4. Corolla green, greenish purple, or purple; leaves up to 5 cm long; low shrubs up to 1 m tall.
        6. Branchlets densely pubescent; berries strongly glaucous, sour .... 5. *V. myrtilloides*
        6. Branchlets glabrous or somewhat pubescent; berries faintly glaucous (strongly so in a few specimens), sweet........................... 6. *V. pallidum*

1. Vaccinium angustifolium Ait. Two varieties occur in Illinois:
a. Lower surface of leaves not glaucous........... 1a. *V. angustifolium* var. *angustifolium*
a. Lower surface of leaves glaucous...................... 1b. *V. angustifolium* var. *laevifolium*

1a. **Vaccinium angustifolium** Ait. var. **angustifolium.** Low-bush Blueberry. Apr.–Sept. Sandy, open woods, black oak savannas, slopes of dunes, bogs; confined to the n. ⅓ of IL.

1b. **Vaccinium angustifolium** Ait. var. **laevifolium** House. Low-bush Blueberry. Apr.–Sept. Black oak savannas, slopes of dunes, bogs; confined to the n. ⅓ of IL.

2. **Vaccinium arboreum** Marsh. Two varieties occur in Illinois:

a. Bracteal leaves much smaller than the foliage leaves .... 2a. *V. arboreum* var. *arboreum*
a. Bracteal leaves similar to foliage leaves in size..... 2b. *V. arboreum* var. *glaucescens*

2a. **Vaccinium arboreum** Marsh. var. **arboreum.** Farkleberry. May–June. Sandstone cliffs, locally abundant; restricted to the s. 3 tiers of cos., plus Jersey and Randolph cos.

2b. **Vaccinium arboreum** Marsh. var. **glaucescens** (Greene) Sarg. Farkleberry. May–June. Sandstone cliffs; occasional in the s. ⅙ of IL.

3. **Vaccinium corymbosum** L. High-bush Blueberry. May–June. Swamps, bogs, rare; Cook, Lake, LaSalle, McHenry, Stephenson, and Winnebago cos.

4. **Vaccinium fuscatum** Ait. Black High-bush Blueberry. May–June. Swampy ground, rare; Lake Co. *Vaccinium atrococcum* (Gray) Heller; *V. corymbosum* L. var. *atrococcum* Gray.

5. **Vaccinium myrtilloides** Michx. Canada Blueberry. May–June. Tamarack bogs, sandy or rocky slopes, rare; Lake, LaSalle, Lee, Ogle, and Winnebago cos.

6. **Vaccinium pallidum** Ait. Two varieties occur in Illinois:

a. Leaves more or less glabrous .... 6a. *V. pallidum* var. *pallidum*
a. Leaves densely pubescent .... 6b. *V. pallidum* var. *crinitum*

6a. **Vaccinium pallidum** Ait. var. **pallidum.** Low-bush Blueberry. Apr.–May. Sandstone cliffs, open sandy woods; occasional in the s. ⅙ of IL, becoming less common northward. *Vaccinium vacillans* Torr.

6b. **Vaccinium pallidum** Ait. var. **crinitum** (Fern.) Mohlenbr., comb. nov. (basionym: *Vaccinium vacillans* Torr. var. *crinitum* Fern). Low-bush Blueberry. Apr.–May. Sandstone cliffs; scattered in the s. ⅙ of IL but not common. *Vaccinium vacillans* Torr. var. *crinitum* Fern.

7. **Vaccinium stamineum** L. Deerberry. May–June. Edge of sandstone cliffs, very rare; Hardin and Pope cos.

71. EUPHORBIACEAE—SPURGE FAMILY

1. Leaves palmately 5- to 11-lobed; plants 3 m or more tall, glabrous .... 8. *Ricinus*
1. Leaves unlobed (if 3- to 5-lobed, then the plants pubescent); plants less than 1 m tall (up to 1.5 m in *Euphorbia lathyris*).
  2. Lower cauline leaves alternate, the upper opposite (*Euphorbia corollata*, with 5 white petal-like structures per flower, may be sought here)............ 7. *Poinsettia*
  2. Cauline leaves either all opposite, all alternate, or sometimes whorled.
    3. Leaves opposite along the stem; latex present.
      4. Leaves decussate, 4–10 cm long .... 5. *Euphorbia*
      4. Leaves not decussate, up to 3 cm long .... 2. *Chamaesyce*
    3. Leaves alternate along the stem; latex present or absent.
      5. Uppermost leaves subtending the inflorescence whorled; latex present.. 5. *Euphorbia*
      5. None of the leaves at the tip of the stem whorled; latex absent.
        6. Leaves sessile or nearly so; plants glabrous.................... 6. *Phyllanthus*
        6. Leaves petiolate; plants variously pubescent.
          7. Leaves entire.
            8. Some of the petioles at least ½ as long as the blades; capsule 2- to 3-celled (rarely 1-celled by abortion) .... 3. *Croton*
            8. Petioles much less than ½ as long as the blades; capsule 1-celled .... 4. *Crotonopsis*
          7. Leaves variously toothed.
            9. Plants twining; leaves cordate .... 9. *Tragia*
            9. Plants erect; leaves not cordate (except in *Acalypha ostryifolia*, a plant with a 3-celled ovary).
              10. Plants with stellate pubescence; leaves with 2 glands at base of blade; pistillate flowers not subtended by cleft bracts .... 3. *Croton*

10. Plants pubescent but not stellate; leaves without 2 glands at base of blade; pistillate flowers subtended by cleft bracts..... ....1. *Acalypha*

1. **Acalypha** L.—Three-seeded Mercury

1. Leaves cordate; fruit soft-echinate..3. *A. ostryifolia*
1. Leaves not cordate; fruit not soft-echinate.
   2. Some or all the petioles at least ⅓ as long as the blades.
      3. Bracts subtending pistillate flowers 5- to 9-lobed, often glandular; stems glabrous or with incurved hairs.
         4. Capsule 3-seeded; seeds 1.2–2.0 mm long ....4. *A. rhomboidea*
         4. Capsule 2-seeded; seeds 2.2–3.2 mm long .... 1. *A. deamii*
      3. Bracts subtending pistillate flowers 9- to 15-lobed, usually glandless; stems with spreading hairs...5. *A. virginica*
   2. Petioles at most only ¼ as long as the blades.................................2. *A. gracilens*

1. **Acalypha deamii** (Weatherby) Ahles. Large-seeded Mercury. July–Oct. Wooded river bottoms, rare; scattered in IL. *Acalypha rhomboidea* Raf. var. *deamii* Weatherby.

2. **Acalypha gracilens** Gray. Three varieties occur in Illinois:

a. Staminate spikes 5–15 mm long.
   b. Leaves oblong to oblong-lanceolate, the petioles more than 1/10 as long as the blades; capsules 3-seeded ....2a. *A. gracilens* var. *gracilens*
   b. Leaves linear to linear-lanceolate, the petioles usually not more than 1/10 as long as the blades; capsules 1-seeded .... 2c. *A. gracilens* var. *monococca*
a. Staminate spikes 3–4 cm long...2b. *A. gracilens* var. *fraseri*

2a. **Acalypha gracilens** Gray var. **gracilens.** Slender Three-seeded Mercury. June–Sept. Woods, fields, sandy soil, roadsides; occasional in the s. ⅘ of IL.

2b. **Acalypha gracilens** Gray var. **fraseri** (Muell.-Arg.) Weatherby. Slender Three-seeded Mercury. June–Sept. Woods, rare; confined to the s. ¼ of IL.

2c. **Acalypha gracilens** Gray var. **monococca** Engelm. & Gray. One-seeded Mercury. June–Sept. Fields, woods, not common; scattered in the s. ⅔ of IL.

3. **Acalypha ostryifolia** Riddell. Cordate Three-seeded Mercury. July–Oct. Riverbanks, fields, bluffs, roadsides; occasional in the s. ½ of IL.

4. **Acalypha rhomboidea** Raf. Three-seeded Mercury. July–Oct. Woods, fields, bluffs, roadsides, common; in every co. *Acalypha virginica* L. var. *rhomboidea* (Raf.) Cooperrider.

5. **Acalypha virginica** L. Three-seeded Mercury. July–Oct. Fields, roadsides, bluffs, woods; occasional to common in the s. ¾ of IL, rare elsewhere, where it may be adventive.

2. **Chamaesyce** S. F. Gray

1. Ovary and capsule pubescent.
   2. Ovary and capsule with spreading hairs ....7. *C. prostrata*
   2. Ovary and capsule with appressed hairs.
      3. Seeds with cross ridges; stems mostly reddish brown; style 0.3–0.5 mm long, cleft less than halfway to the base ....4. *C. maculata*
      3. Seeds smooth or minutely granular; stems mostly greenish; style about 0.7 mm long, cleft halfway to the base..................................... 3. *C. humistrata*
1. Ovary and capsule glabrous.
   4. Leaves entire.
      5. Leaves about as broad as long; capsule 1.0–1.2 mm in diameter; style 0.2 mm long; seeds about 1 mm long .... 8. *C. serpens*
      5. Leaves considerably longer than broad; capsule 2.0–3.5 mm in diameter; style 0.7–1.0 mm long; seeds 1.3–2.6 mm long.
         6. Glands conspicuously appendaged; seeds 1.3–1.6 mm long; capsule 2.0–2.5 mm in diameter...1. *C. geyeri*
         6. Glands scarcely appendaged; seeds 2.0–2.6 mm long; capsule 3.0–3.5 mm in diameter... 6. *C. polygonifolia*
   4. Leaves toothed.

7. Leaves toothed from tip to base.
   8. Stems usually prostrate or procumbent; capsules up to 1.9 mm long; seeds with sharp angles .... 10. *C. vermiculata*
   8. Stems usually ascending to erect; capsules 1.9 mm long or longer; seeds with rounded angles .... 5. *C. nutans*
7. Leaves toothed only at apex and at base.
   9. Leaves broadly oblong to ovate; seeds pitted and with short cross-ridges .... 9. *C. serpyllifolia*
   9. Leaves linear-oblong; seeds with 3–6 long cross-ridges .... 2. *C. glyptosperma*

1. **Chamaesyce geyeri** (Engelm.) Small. Geyer's Spurge. June–Sept. Sandy soil; confined to the nw. ¼ of IL; also Iroquois, Jackson, and Union cos. *Euphorbia geyeri* Engelm.

2. **Chamaesyce glyptosperma** (Engelm.) Small. Spurge. June–Oct. Sandy or gravelly soil; occasional in the n. ⅓ of IL; also Monroe Co. *Euphorbia glyptosperma* Engelm.

3. **Chamaesyce humistrata** (Engelm.) Small. Spreading Spurge. July–Sept. Sandy soil, riverbanks, disturbed soil; occasional throughout the state. *Euphorbia humistrata* Engelm.

4. **Chamaesyce maculata** (L.) Small. Spotted Spurge. July–Oct. Disturbed or cultivated habitats; common throughout IL; in every co. *Chamaesyce supina* (Raf.) Moldenke; *Euphorbia supina* Raf.

5. **Chamaesyce nutans** (Lag.) Small. Nodding Spurge. July–Oct. Disturbed areas, common; in every co. *Chamaesyce maculata* (L.) Small; *Euphorbia maculata* L.

6. **Chamaesyce polygonifolia** (L.) Small. Seaside Spurge. July–Sept. Dunes and beaches, rare; Cook, Fulton, and Lake cos. *Euphorbia polygonifolia* L.

7. **Chamaesyce prostrata** (Ait.) Small. Green Creeping Spurge. June–Sept. Native to tropical America; adventive in Cook, DuPage, and St. Clair cos. *Euphorbia chamaesyce* L.

8. **Chamaesyce serpens** (HBK.) Small. Round-leaved Spurge. June–Sept. Moist sandy soil; scattered throughout IL, but not common. *Euphorbia serpens* HBK.

9. **Chamaesyce serpyllifolia** (Pers.) Small. Thyme-leaved Spurge. July–Oct. Native to the w. U.S.; rarely adventive in sandy soil; Cook and DeKalb cos. *Euphorbia serpyllifolia* Pers.

10. **Chamaesyce vermiculata** (Raf.) House. Hairy Spurge. July–Sept. Introduced from the n. U.S.; rich disturbed soil, rare; Lake Co. *Euphorbia vermiculata* Raf.

3. **Croton** L.—Croton

1. Leaves toothed; calyx of staminate flowers 4-parted; rudimentary petals in pistillate flowers 5 .... 2. *C. glandulosus*
1. Leaves entire; calyx of staminate flowers 5-parted; petals in pistillate flowers absent.
   2. Calyx of pistillate flowers 7- to 12-parted; some or all of the leaves cordate .... 1. *C. capitatus*
   2. Calyx of pistillate flowers 5-parted; leaves tapering, rounded, or barely cordate at the base.
      3. Leaves ovate to oblong; stamens 3–8; plants monoecious; seeds pitted.
         4. Plants silvery, with stellate pubescence; stigmas 2-cleft; capsules 3–4 mm long, 1- to 2-celled; seeds spherical........................... 4. *C. monanthogynus*
         4. Plants white, tomentose; stigmas 3-cleft; capsules 5–7 mm long, 3-celled; seeds elongated .... 3. *C. lindheimerianus*
      3. Leaves linear to linear-oblong; stamens (8–) 10 (–12); plants dioecious; seeds reticulate... 5. *C. texensis*

1. **Croton capitatus** Michx. Capitate Croton; Woolly Croton; Hogwort. July–Sept. Sandy soil, limestone glades, pastures; occasional in the s. cos., less common in the n. cos., where it is probably adventive.

2. **Croton glandulosus** L. var. **septentrionalis** Muell.-Arg. Sand Croton. June–Oct. Sandy, often disturbed, soil; common in the s. cos., occasional in the n. cos., where it is probably adventive.

3. **Croton lindheimerianus** Scheele. Lindheimer's Croton. June–Oct. Native to the w. U.S.; adventive along a railroad; Madison Co.

4. **Croton monanthogynus** Michx. Croton. July–Oct. Dry fields, bluffs; occasional in the s. ½ of IL, much rarer in the n. ½.

5. **Croton texensis** (Klotzsch) Muell.-Arg. Texas Croton. June–Sept. Dry soil, very rare; Menard Co.

4. **Crotonopsis** Michx.—Rushfoil

1. Fruit without spines; leaves elliptic to lanceolate .... 1. *C. elliptica*
1. Fruit spiny at tip; leaves linear-lanceolate .... 2. *C. linearis*

1. **Crotonopsis elliptica** Willd. Rushfoil. July–Sept. Dry fields, dry woods, bluffs; occasional in the s. ½ of IL; also Kankakee and LaSalle cos. *Croton willdenowii* G. L. Webster.

2. **Crotonopsis linearis** Michx. Rushfoil. July–Sept. Sandy soil, rare; confined to the n. ½ of IL. *Croton michauxii* G. L. Webster.

5. **Euphorbia** L.—Spurge

(Also see Chamaesyce and Poinsettia.)

1. Leaves entire.
  2. Glands of the involucre with petal-like appendages.
    3. Bracts and uppermost leaves with white margins .... 8. *E. marginata*
    3. Bracts and leaves green or pale green throughout.
      4. Perennials; stipules absent; leaves on the stem alternate.
        5. Appendages of the cyathia 7–10 mm broad; some or all the pedicels more than 5 mm long .... 2. *E. corollata*
        5. Appendages of the cyathia 5–7 mm broad; pedicels 0.5–5.0 mm long .... 12. *E. pubentissima*
      4. Annuals; stipules glandular; some of the leaves on the stem opposite...... .... 6. *E. hexagona*
  2. Glands of the involucre without petal-like appendages.
    6. Leaves on the stem opposite and decussate .... 7. *E. lathyris*
    6. Leaves on the stem alternate.
      7. Cauline leaves linear to narrowly oblong-lanceolate, acute, sessile; seeds smooth; perennials with rhizomes.
        8. None of the leaves over 3 mm wide .... 3. *E. cyparissias*
        8. Some or all the leaves over 3 mm wide .... 4. *E. esula*
      7. Cauline leaves obovate, obtuse or retuse, short-petiolate; seeds pitted; annuals or biennials.
        9. Seeds 1.8–2.0 mm long, finely pitted throughout; capsules with rounded lobes .... 1. *E. commutata*
        9. Seeds 1.0–1.5 mm long, with 1–several rows of large pits; capsules with 2-keeled lobes .... 10. *E. peplus*
1. Leaves toothed (minutely serrulate in *E. platyphyllos*).
  10. Capsules smooth; seeds ovoid .... 5. *E. helioscopia*
  10. Capsules warty; seeds flattened or lenticular.
    11. Seeds dark brown, 1.7–2.0 mm long, obscurely reticulate; styles longer than the ovary; cyathium with 5 oblong glands.
      12. Leaves pubescent on the lower surface; rays of the terminal umbel usually 5; seeds about 2 mm long .... 11. *E. platyphyllos*
      12. Leaves glabrous; rays of the terminal umbel usually 3; seeds 1.7–2.0 mm .... 9. *E. obtusata*
    11. Seeds reddish brown, 1.3–1.5 mm long, distinctly reticulate; styles shorter than the ovary; cyathium with 4 lobes and a tuft of hairs ..... 13. *E. spathulata*

1. **Euphorbia commutata** Engelm. Wood Spurge. May–June. Wooded slopes, along streams, and in gravelly soils, rare; mostly in the n. ⅓ of IL; also Pope Co.

2. **Euphorbia corollata** L. Two varieties occur in Illinois:

a. Stem and lower surface of the leaves glabrous or nearly so .... 2a. *E. corollata* var. *corollata*

a. Stem and lower surface of the leaves softly hairy .... 2b. *E. corollata* var. *mollis*

2a. **Euphorbia corollata** L. var. **corollata.** Flowering Spurge. May–Oct. Prairies, woods, fields, roadsides, common; in every co.

2b. **Euphorbia corollata** L. var. **mollis** Millsp. Hairy Flowering Spurge. May–Oct. Prairies, woods, fields; occasional in the s. ¼ of IL.

3. **Euphorbia cyparissias** L. Cypress Spurge. Apr.–Sept. Native to Europe; naturalized along roadsides and in cemeteries; occasional throughout the state.

4. **Euphorbia esula** L. Leafy Spurge. May–Oct. Native to Europe; naturalized in fields and pastures; occasional in the n. ½ of IL.

5. **Euphorbia helioscopia** L. Wart Spurge. June–Oct. Native to Europe; adventive in waste areas; Cass, Cook, Hancock, and Menard cos.

6. **Euphorbia hexagona** Nutt. Spurge. July–Sept. Native to the n. and nw. U.S.; adventive in Mercer Co.

7. **Euphorbia lathyris** L. Caper Spurge. June–Sept. Native to Europe; rarely escaped from cultivation; Champaign and Jackson cos.

8. **Euphorbia marginata** Pursh. Snow-on-the-mountain. July–Oct. Native to the w. U.S.; occasionally adventive in IL.

9. **Euphorbia obtusata** Pursh. Blunt-leaved Spurge. May–July. Rich woods, wooded slopes; occasional throughout the state.

10. **Euphorbia peplus** L. Petty Spurge. June–Oct. Native to Europe; rarely adventive in IL; DuPage and Menard cos.

11. **Euphorbia platyphyllos** L. Broad-leaved Spurge. June–Aug. Native to Europe; adventive in disturbed soil; not common but scattered in IL.

12. **Euphorbia pubentissima** Michx. False Flowering Spurge. May–Oct. Sandy soil, rare; confined to the s. ¼ of IL. *Euphorbia corollata* L. var. *paniculata* Boiss.

13. **Euphorbia spathulata** Lam. Spurge. May–July. Limestone ledge at edge of hill prairie, rare; Monroe Co.

6. **Phyllanthus** L.—Leaf-flower

1. Leaves uniform, elliptic to oblong to obovate; capsules 1.6–2.0 mm long .... 1. *P. caroliniensis*

1. Leaves of 2 types, some of them scalelike, the others oblong; capsules 2.0–2.2 mm long.

2. Regular leaves hispidulous on the lower surface; flowers and fruits sessile or subsessile; ovaries verrucose; seeds transversely ribbed .... 3. *P. urinaria*

2. Regular leaves not hispidulous on the lower surface; flowers and fruits pedicellate; ovaries smooth; seeds with rounded processes......................... 2. *P. tenellus*

1. **Phyllanthus caroliniensis** Walt. Leaf-flower. June–Oct. Mostly sandy soil; occasional in the s. ½ of IL; also Cass, LaSalle, and Mason cos.

2. **Phyllanthus tenellus** Roxb. Mascarene Islands Leaf-flower. June–Aug. Native to Asia; escaped in greenhouses; Champaign Co.

3. **Phyllanthus urinaria** L. Chamber Bitter. June–Oct. Native to the tropics; rarely adventive in IL; Jackson Co.

7. **Poinsettia** Graham—Poinsettia

1. Floral bracts usually basally red; seeds not angular, sharply tuberculate; leaves glossy green .... 1. *P. cyathophora*

1. Floral bracts green or basally pale, never red; seeds angular, bluntly tuberculate; leaves dull green .... 2. *P. dentata*

1. **Pointsettia cyathophora** (Murr.) Kl. & Garcke. Two varieties occur in Illinois:

a. Leaves oval to lanceolate or even pandurate .... 1a. *P. cyathophora* var. *cyathophora*

a. Leaves linear to narrowly lanceolate.................. 1b. *P. cyathophora* var. *graminifolia*

1a. **Poinsettia cyathophora** (Murr.) Kl. & Garcke var. **cyathophora.** Wild Poinsettia. July–Oct. Fields, roadsides; occasional throughout IL. *Euphorbia cyathophora* Murr.; *Euphorbia heterophylla* L.

1b. **Poinsettia cyathophora** (Murr.) Kl. & Garcke var. **graminifolia** (Michx.) Mohlenbr. Narrow-leaved Wild Poinsettia. July–Oct. Moist soil along streams; Pope and Union cos.

2. **Poinsettia dentata** (Michx.) Kl. & Garcke. Two varieties occur in Illinois:

a. Leaves narrowly ovate to rhombic .... 2a. *P. dentata* var. *dentata*
a. Leaves linear to narrowly lanceolate........................ 2b. *P. dentata* var. *cuphosperma*

2a. **Poinsettia dentata** (Michx.) Kl. & Garcke var. **dentata.** Wild Poinsettia. June–Sept. Fields, roadsides, prairies; occasional to common throughout IL. *Euphorbia dentata* Michx.; *Euphorbia davidii* Subils.

2b. **Poinsettia dentata** (Michx.) Kl. & Garcke var. **cuphosperma** (Engelm.) Mohlenbr. Narrow-leaved Wild Poinsettia. June–Sept. Fields and roadsides; Cook and Hancock cos.

8. **Ricinus** L.—Castor Bean

1. **Ricinus communis** L. Castor Bean. June–Sept. Native to the tropics; occasionally escaped from cultivation.

9. **Tragia** L.—Noseburn

1. **Tragia cordata** Michx. Noseburn. July–Sept. Dry woods, bluffs, rare; Hardin, Johnson, and Pope cos.

72. FABACEAE—PEA FAMILY

1. Plants woody, including woody vines.
  2. Plants vines.
    3. Ovary and legume glabrous; standard reflexed near middle; seeds reniform .... 43. *Wisteria*
    3. Ovary and legume pubescent; standard reflexed at base; seeds lenticular..... ....33. *Rehsonia*
  2. Plants trees or shrubs.
    4. Leaflets 3 ....26. *Ononis*
    4. Leaflets 5 or more.
      5. Tip of rachis developed into a spine but soon deciduous........ 9. *Caragana*
      5. Tip of rachis not spinescent.
        6. Lower surface of leaflets glandular-punctate or canescent; flowers less than 1 cm long; shrubs to 7 m tall (herbaceous in *A. canescens*) ....1. *Amorpha*
        6. Lower surface of leaflets neither glandular-punctate nor canescent; flowers 1.4 cm long or longer; trees to 15 m tall.
          7. Stipular spines usually present; stems sometimes glandular or hispid.. 34. *Robinia*
          7. Stipular spines absent; stems glabrous ....10. *Cladrastis*
1. Plants herbaceous.
  8. Leaves simple...12. *Crotalaria*
  8. Leaves compound.
    9. Leaves palmately compound, not trifoliolate.
      10. Leaflets 7–11 ....22. *Lupinus*
      10. Leaflets 5.
        11. Calyx at least 8 mm long; plants silvery-silky ....28. *Pediomelum*
        11. Calyx up to 4 mm long; plants not silvery-silky .... 31. *Psoralidium*
    9. Leaves pinnately compound or trifoliolate.
      12. Leaves even-pinnate.
        13. Some or all of the leaf rachises ending in a tendril.
          14. Calyx lobes leaflike; lowermost stipules larger than the leaflets they subtend ....30. *Pisum*
          14. Calyx lobes not leaflike; none of the stipules larger than the leaflets they subtend.
            15. Wings and keel petal united; style with a terminal tuft of hairs .... 41. *Vicia*
            15. Wings and keel petal free from each other; style pubescent only along 1 side...19. *Lathyrus*
        13. None of the leaf rachises ending in a tendril.
          16. Plants trailing on the ground; leaflets 4 ....5. *Arachis*

16. Plants erect; leaflets more than 4 .... 36. *Sesbania*

12. Leaves odd-pinnate.

17. Leaflets 5 or more.

18. Climbing vines; flowers brownish purple or pink, in loose racemes .... 4. *Apios*

18. Upright herbs; flowers yellow, bright rose, red, or white, in umbels, spikes, or globose heads.

19. Flowers yellow or red, in umbels or globose heads; stamens 10.

20. Flowers densely crowded; calyx bladdery, enclosing the fruit .... 3. *Anthyllis*

20. Flowers loosely arranged; calyx not bladdery, not enclosing the legume .... 21. *Lotus*

19. Flowers bright rose, red, or white, in spikes or in globose heads; stamens 5 (if stamens 10, then flowers in globose heads).

21. Flowers red or yellow; stamens 10; calyx bladdery, enclosing the fruit .... 3. *Anthyllis*

21. Flowers bright rose, pink, or white; stamens 5; calyx not bladdery, not enclosing the fruit.

22. Plants glandular-viscid; fruit with hooked bristles .... 17. *Glycyrrhiza*

22. Plants not glandular viscid; fruit without hooked bristles.

23. Flowers in umbels .... 35. *Securigera*

23. Flowers in spikes or racemes.

24. Flowers in dense, elongated spikes, the rachis not exposed; fruits enclosed by the calyx, 1- to 2-seeded .... 13. *Dalea*

24. Flowers in racemes; fruits exserted, several-seeded (1-seeded in *Onobrychis*).

25. Inflorescence terminal; flowers cream, tipped with pink .... 39. *Tephrosia*

25. Inflorescence axillary; flowers pale pink, purple, cream, or yellow-green.

26. Fruit strongly reticulate and occasionally toothed along the margins; corolla pale pink .... 25. *Onobrychis*

26. Fruits neither strongly reticulate nor toothed; corolla yellow-green, cream, or purple .... 6. *Astragalus*

17. Leaflets 3.

27. Leaflets toothed.

28. Plants spiny .... 26. *Ononis*

28. Plants not spiny.

29. Flowers in elongated racemes .... 24. *Melilotus*

29. Flowers in heads or spikes.

30. Stipules entire; fruits usually straight, the petals mostly persistent .... 40. *Trifolium*

30. Stipules toothed (rarely entire); fruits coiled or twisted, the petals not persistent .... 23. *Medicago*

27. Leaflets entire.

31. Foliage with a few glandular dots .... 27. *Orbexilum*

31. Foliage without glandular dots.

32. Flowers 2 cm long or longer.

33. Flowers blue, white, or cream; plants upright .... 7. *Baptisia*

33. Flowers purple or violet; plants trailing or twining.

34. Flowers 2.0–2.5 cm long; stipels present; fruits 15 cm long or longer .... 8. *Canavalia*

34. Flowers 4–6 cm long; stipels absent; fruits less than 15 cm long .... 11. *Clitoria*

32. Flowers up to 1.6 cm long.
35. Leaflets without stipels.
36. Flowers yellow.
37. Flowers 1.3–1.6 cm long .... 7. *Baptisia*
37. Flowers less than 1 cm long.
38. Stipules united to petiole, forming a sheath which encircles the stem .... 38. *Stylosanthes*
38. Stipules free from petiole............. 20. *Lespedeza*
36. Flowers purple, violet, rose, or white with purple markings.
39. Uppermost leaves reduced to 1 leaflet; fruit a legume 1.5–2.5 cm long .... 21. *Lotus*
39. Uppermost leaves trifoliolate; fruit a loment less than 1 cm long or a legume opening by a lid.
40. Fruit a legume opening by a lid; flowers in a globose head.................................... 40. *Trifolium*
40. Fruit a loment; flowers variously arranged but not in a globose head.
41. Plants annual, with persistent, brown, glabrous stipules .... 18. *Kummerowia*
41. Plants perennial, with deciduous, greenish, usually pubescent stipules ..... 20. *Lespedeza*
35. Leaflets with stipels.
42. Calyx 2-lipped.
43. Keel petal spirally coiled.....................29. *Phaseolus*
43. Keel petal not spirally coiled.
44. Fruit a legume; plant a climbing vine................. .... 15. *Galactia*
44. Fruit a loment; plant not climbing ....14. *Desmodium*
42. Calyx teeth more or less equal, the calyx not 2-lipped.
45. Calyx 5-parted.
46. Fruit a 1-seeded, indehiscent loment .... 20. *Lespedeza*
46. Fruit a several-seeded, dehiscent legume.
47. Keel petal spirally coiled ....29. *Phaseolus*
47. Keel petal not spirally coiled.
48. Stems brown-villous, erect .... 16. *Glycine*
48. Stems glabrous or at least not brown-villous, twining ....42. *Vigna*
45. Calyx 4-parted.
49. Plants coarse; some of the leaflets more than 10 cm long ....32. *Pueraria*
49. Plants slender; leaflets never exceeding 10 cm in length.
50. Calyx subtended by 2 small bracts.
51. Style pubescent along 1 side .... 37. *Strophostyles*
51. Style glabrous .... 15. *Galactia*
50. Calyx not subtended by bracts...................... .... 2. *Amphicarpaea*

1. **Amorpha** L.—False Indigo

1. Some or all the leaflets 1 cm broad or broader, short-petiolulate; plants woody.
2. Leaflets and branchlets glabrous or nearly so; legume without resinous dots..... .... 3. *A. nitens*
2. Leaflets and branchlets pubescent; legume with resinous dots......2. *A. fruticosa*
1. None of the leaflets 1 cm broad, sessile or nearly so; plants herbaceous .... 1. *A. canescens*

1. **Amorpha canescens** Pursh. Leadplant. May–Aug. Prairies; common in the n. ⅔ of IL, less common elsewhere.

2. **Amorpha fruticosa** L. Three varieties occur in Illinois:

a. Leaflets mostly less than twice as long as broad ....2a. *A. fruticosa* var. *fruticosa*
a. Leaflets at least twice as long as broad.
  b. Leaflets grayish pubescent .... 2b. *A. fruticosa* var. *angustifolia*
  b. Leaflets softly tawny-villous................................2c. *A. fruticosa* var. *croceolanata*

2a. **Amorpha fruticosa** L. var. **fruticosa.** False Indigo. May–June. Moist soil; occasional throughout the state.

2b. **Amorpha fruticosa** L. var. **angustifolia** Pursh. False Indigo. May–June. Moist soil; occasional throughout the state. (Including var. *tennesseensis* [Shuttlew.] Palmer var. *oblongifolia* Palmer.) *Amorpha fragrans* Sweet.

2c. **Amorpha fruticosa** L. var. **croceolanata** (P. W. Wats.) Schneid. False Indigo. May–June. Low woods, rare; Pope Co. *Amorpha croceolanata* P. W. Wats.

3. **Amorpha nitens** Boynton. Smooth False Indigo. May–June. Woods along rivers and streams, very rare; confined to the s. ⅙ of IL.

2. **Amphicarpaea** Ell.—Hog Peanut

1. **Amphicarpaea bracteata** (L.) Fern. Two varieties occur in Illinois:

a. Stem with ascending white hairs; sides of legume glabrous......................................
....1a. *A. bracteata* var. *bracteata*
a. Stem with appressed tawny hairs; sides of legume pubescent..................................
....1b. *A. bracteata* var. *comosa*

1a. **Amphicarpaea bracteata** (L.) Fern. var. **bracteata.** Hog Peanut. Aug.–Sept. Woods and thickets; occasional to common throughout IL.

1b. **Amphicarpaea bracteata** (L.) Fern. var. **comosa** (L.) Fern. Hog Peanut. Aug.–Sept. Woods and thickets; occasional throughout IL. *Amphicarpaea pitcheri* Torr. & Gray.

3. **Anthyllis** L.—Lady's-finger

1. **Anthyllis vulneraria** L. Lady's-finger. July–Aug. Native to Europe; rarely adventive in IL; McHenry Co.

4. **Apios** Medic.—Groundnut

1. Flowers brownish purple; standard without a spongy knob at the apex...................
....1. *A. americana*
1. Flowers greenish white, tinged with pink; standard with a spongy knob at the apex
....2. *A. priceana*

1. **Apios americana** Medic. Groundnut. July–Sept. Thickets, prairies, moist soil; common throughout IL. (Including var. *turrigera* Fern.)

2. **Apios priceana** Robins. Price's Groundnut. July–Sept. Low woods, very rare; Union Co; not seen since 1948.

5. **Arachis** L.—Peanut

1. **Arachis hypogaea** L. Peanut. June–July. Native to S. America; rarely escaped from gardens; Cook Co. (Wilhelm, 2010.)

6. **Astragalus** L.—Milk Vetch

1. Calyx tube 2–6 mm long; fruit dry, dehiscent.
  2. Fruit with long hairs.
    3. Peduncles shorter than the raceme of flowers; petals ochroleucous, 12–16 mm long; fruits spreading to ascending ....3. *A. cicer*
    3. Peduncles longer than the raceme of flowers; petals purple, 15–22 mm long; fruits erect ....1. *A. agrestis*
  2. Fruit glabrous or nearly so.
    4. Fruits lunate, 2–3 cm long; corolla lilac, purple, or white, about 1 cm long; calyx tube 2–3 mm long..5. *A. distortus*

4. Fruits more or less straight, 1–2 cm long; corolla greenish yellow, 1.2–1.5 cm long; calyx tube 4.5–6.0 mm long .... 2. *A. canadensis*
1. Calyx tube 7–10 mm long; fruit fleshy, indehiscent.
5. Fruits glabrous .... 4. *A. crassicarpus*
5. Fruits pubescent............ 6. *A. tennesseensis*

1. **Astragalus agrestis** Doug. Field Milk Vetch. May–July. Native to the w. U.S.; adventive in Boone Co. *Astragalus goniatus* Nutt.

2. **Astragalus canadensis** L. Canada Milk Vetch. June–Aug. Prairies, thickets; occasional throughout IL. (Including var. *longilobus* Fassett.)

3. **Astragalus cicer** L. Chick-pea Milk Vetch. June–July. Native to Europe and Asia; rarely escaped from cultivation; Kane Co.

4. **Astragalus crassicarpus** Nutt. var. **trichocalyx** (Nutt.) Barneby. Large Ground Plum. Apr.–May. Rocky prairies, rare; Jersey, Macoupin, Madison, and St. Clair cos.; adventive in Will Co. *Astragalus mexicanus* A. DC. var. *trichocalyx* (Nutt.) Fern.; *A. trichocalyx* Nutt.

5. **Astragalus distortus** Torr. & Gray. Bent Milk Vetch. May–June. Rocky prairies, rare in w. IL; adventive in Cook Co.

6. **Astragalus tennesseensis** Gray. Ground Plum. May–June. Dry prairies, rare; confined to the n. ¼ of IL; also Mason and Tazewell cos.

7. **Baptisia** Vent.—Wild Indigo

1. Flowers yellow, 1.3–2.0 cm long.
2. Flowers bright yellow, 1.3–1.6 cm long .... 6. *B. tinctoria*
2. Flowers pale yellow, 1.5–2.0 cm long............ 4. *B. X deamii*
1. Flowers blue, white, or cream, 2 cm long or longer.
3. Flowers blue; stipules longer than the petioles.
4. Stipe of legume no longer than the calyx; petioles at least 5 mm long; keel petal up to 2.5 cm long .... 2. *B. australis*
4. Stipe of legume about twice as long as the calyx; petioles less than 5 mm long; keel petal 2.7–3.0 cm long............ 5. *B. minor*
3. Flowers white or cream; stipules shorter than the petioles.
5. Plants glabrous; stipules deciduous; legumes glabrous, up to 3 cm long .... 1. *B. alba*
5. Plants usually pubescent; stipules persistent; legumes pubescent, usually at least 4 cm long............ 3. *B. bracteata*

1. **Baptisia alba** (L.) Vent var. **macrophylla** (Larisey) Isely. White Wild Indigo. May–Aug. Prairies, woods, sandy alluvial plains, marshes; occasional to common throughout IL. *Baptisia lactea* (Raf.) Thieret; *B. leucantha* Torr. & Gray.

2. **Baptisia australis** (L.) R. Br. Blue Wild Indigo. May–June. Native s. and w. of IL; woods, very rare; Cook, DuPage, Johnson, Kane, Lake, and Union cos.

3. **Baptisia bracteata** Ell. Two varieties occur in Illinois:

a. Stems and leaves pubescent............ 3a. *B. bracteata* var. *bracteata*
a. Stems and leaves glabrous .... 3b. *B. bracteata* var. *glabrescens*

3a. **Baptisia bracteata** Ell. var. **bracteata**. Cream Wild Indigo. May–June. Moist prairies and open woods; occasional in the n. ⅘ of IL, rare elsewhere. *Baptisia leucophaea* Nutt.

3b. **Baptisia bracteata** Ell. var. **glabrescens** (Larisey) Isely. Cream Wild Indigo. May–June. Prairies; rare in w. cent. IL and s. IL. *Baptisia leucophaea* Nutt. var. *glabrescens* Larisey.

4. **Baptisia X deamii** Larisey. Deam's Wild Indigo. June–Aug. Moist sandy savannas, rare; Kankakee Co. Considered to be a hybrid between *B. bracteata* Ell. and *B. tinctoria* (Gl.) R. Br.

5. **Baptisia minor** Lehm. Lesser Blue Wild Indigo. June–July. Native to the sw. U.S.; adventive in Cook and DuPage cos. *Baptisia australis* (L.) R. Br. var. *minor* (Lehm.) Fern.

6. **Baptisia tinctoria** (Gl.) R. Br. var. **crebra** Fern. Yellow Wild Indigo. June–Aug. Sandy woods, very rare; Cook, DuPage, and Kankakee cos.

8. **Canavalia** Adans.—Jack Bean

1. **Canavalia ensiformis** (L.) DC. Jack Bean. June. Native to tropical America; rarely adventive in old fields; White Co.

9. **Caragana** Lam.—Pea Tree

1. **Caragana arborescens** Lam. Pea Tree. May. Native to Asia; escaped from cultivation and spreading somewhat; Champaign, DuPage, Menard, Perry, Sangamon, and Winnebago cos.

10. **Cladrastis** Raf.—Yellow-wood

1. **Cladrastis kentukea** (Dum.-Cours.) Rudd. Yellow-wood. May. Wooded slopes, very rare; Alexander and Gallatin cos. *Cladrastis lutea* (Michx. f.) K. Koch.

11. **Clitoria** L.—Butterfly Pea

1. **Clitoria mariana** L. Butterfly Pea. June–Aug. Dry woods, not common; confined to the s. ⅙ of IL.

12. **Crotalaria** L.—Rattlebox

1. Leaves acute at apex; flowers up to 1.5 cm long; legume less than 3 cm long .... 1. *C. sagittalis*
1. Leaves obtuse at apex; flowers over 1.5 cm long; legume 3 cm long or longer......... 2. *C. spectabilis*

1. **Crotalaria sagittalis** L. Rattlebox. June–Sept. Fields, edge of woods, sandy soil; occasional in the s. ½ of IL, becoming less common northward.
2. **Crotalaria spectabilis** Roth. Showy Rattlebox. Aug.–Oct. Native to the tropics; rarely adventive in IL; Alexander Co.

13. **Dalea** Juss.—Prairie Clover

1. Flowers pink or white; stamens 10; bracts about equaling the calyx.
  2. Annual; leaflets 19–35, linear-oblong .... 4. *D. leporina*
  2. Perennial; leaflets 5–13, linear .... 2. *D. enneandra*
1. Flowers purple, rose, fading to white, or white; stamens 5; bracts longer than the calyx.
  3. Leaflets 5–7 .... 3. *D. foliosa*
  3. Leaflets 19–27.
    4. Flowers white; calyx glabrous........ 1. *D. candida*
    4. Flowers purple or rose; calyx densely pubescent .... 5. *D. purpurea*

1. **Dalea candida** (Michx.) Willd. White Prairie Clover. June–Oct. Prairies; occasional throughout the state. *Petalostemum candidum* Michx.
2. **Dalea enneandra** Nutt. Sailpod Dalea. July–Aug. Native to the w. U.S.; naturalized in a degraded prairie; DuPage Co.
3. **Dalea foliosa** (Gray) Barneby. Leafy Prairie Clover. July–Sept. Dry soil above streams, shallow sand prairies; once known from a few cos. in the ne. ¼ of IL, now known only from Will Co. *Petalostemum foliosum* Gray.
4. **Dalea leporina** (Ait.) Bullock. Foxtail Dalea. July–Sept. Native to the w. U.S.; adventive in fields, roadsides, dry sandy banks; occasional in IL. *Dalea alopecuroides* Willd.
5. **Dalea purpurea** Vent. Purple Prairie Clover. June–Sept. Prairies; occasional throughout IL. *Petalostemum purpureum* (Vent.) Rydb.

14. **Desmodium** Desv.—Tick Trefoil

1. Flowers borne on a long, leafless peduncle .... 10. *D. nudiflorum*
1. Flowers borne on leafy stems.
  2. Leaves clustered in a whorl midway on the flowering stem....... 6. *D. glutinosum*
  2. Leaves scattered all along the flowering stem.
    3. Flowers white; stipe of fruit at least twice as long as the calyx .... 14. *D. pauciflorum*
    3. Flowers pink or purple; stipe of fruit not longer than the calyx.

4. Petioles up to 3 mm long or absent.................................17. *D. sessilifolium*
4. Petioles 4 mm long or longer.
 5. Stipules persistent, conspicuous.
  6. Plants lying on the ground; leaflets mostly suborbicular..16. *D. rotundifolium*
  6. Plants erect or ascending; leaflets variously shaped.
   7. Pubescence on lower surface of leaflets at least of some hooked hairs.
    8. Leaflets obscurely reticulate beneath; joints of fruit angular along the lower margin; rachis of inflorescence with spreading hairs..2. *D. canescens*
    8. Leaflets conspicuously reticulate beneath; joints of fruit rounded along both margins; rachis of inflorescence with some hooked hairs...7. *D. illinoense*
   7. Pubescence on lower surface of leaflets not of hooked hairs, or lower surface of leaflets glabrous......................4. *D. cuspidatum*
 5. Stipules deciduous, inconspicuous.
  9. Leaflets glaucous beneath...8. *D. laevigatum*
  9. Leaflets not glaucous beneath.
   10. Leaflets up to 3 cm long, up to 2 cm broad.
    11. Stems and upper surface of leaflets glabrous or nearly so; pedicels (8–) 10–20 mm long.....................9. *D. marilandicum*
    11. Stems and upper surface of leaflets pilose; pedicels 4–8 (–9) mm long.. 3. *D. ciliare*
   10. Leaflets, or most of them, longer than 3 cm, broader than 2 cm.
    12. Fruit with 1–2 segments; calyx 2–3 mm long...12. *D. obtusum*
    12. Fruit with 3 or more segments; calyx 3 mm long or longer.
     13. Most of the flowers 8–9 mm long.
      14. Leaflets velvety beneath, thick, coriaceous.. 18. *D. viridiflora*
      14. Leaflets pubescent beneath, but not velvety, thin, membranaceous...................................1. *D. canadense*
     13. Flowers less than 8 mm long.
      15. Leaflets velvety on the lower surface; stipules ovate-deltoid, at most about 2 ½ times longer than broad...11. *D. nuttallii*
      15. Leaflets variously pubescent to nearly glabrous on the lower surface but not velvety; stipules filiform or subulate, at least 3 times longer than broad.
       16. Leaflets rather thin, faintly reticulate; terminal leaflet at least 3 times longer than broad..13. *D. paniculatum*
       16. Leaflets rather thick, conspicuously or faintly reticulate; terminal leaflet less about 3 times longer than broad.
        17. Leaflets conspicuously reticulate..5. *D. glabellum*
        17. Leaflets faintly reticulate......... 15. *D. perplexum*

1. **Desmodium canadense** (L.) DC. Showy Tick Trefoil. July–Sept. Prairies; occasional throughout IL.

2. **Desmodium canescens** (L.) DC. Hoary Tick Trefoil. July–Aug. Dry woods; occasional in the s. ½ of IL, less common elsewhere.

3. **Desmodium ciliare** (Muhl. ex Willd.) DC. Hairy Tick Trefoil. Aug.–Oct. Dry woods; occasional in the s. ½ of IL.

4. **Desmodium cuspidatum** (Muhl.) Loud. Two varieties occur in Illinois:

a. Stems, leaflets, and bracts glabrous or nearly so.. 4a. *D. cuspidatum* var. *cuspidatum*

a. Stems, leaflets, and bracts pubescent..................4b. *D. cuspidatum* var. *longifolium*

4a. **Desmodium cuspidatum** (Muhl.) Loud. var. **cuspidatum.** Tick Trefoil. July–Sept. Dry woods; occasional throughout IL.

4b. **Desmodium cuspidatum** (Muhl.) Loud. var. **longifolium** (Torr. & Gray) B. G. Schub. Tick Trefoil. July–Sept. Dry woods; occasional throughout IL.

5. **Desmodium glabellum** (Michx.) DC. Smooth Tick Trefoil. July–Sept. Dry woods; occasional throughout IL except for the n. tier of cos. *Desmodium dillenii* Darl., in part.

6. **Desmodium glutinosum** (Muhl. ex Willd.) Wood. Pointed Tick Trefoil. June–Sept. Rich woods, oak savannas; occasional throughout IL. *Hylodesmum glutinosum* (Muhl. ex Willd.) Ohasi and R. R. Mill.

7. **Desmodium illinoense** Gray. Illinois Tick Trefoil. July–Aug. Dry prairies, roadsides; occasional throughout IL.

8. **Desmodium laevigatum** (Nutt.) DC. Glaucous Tick Trefoil. July–Sept. Dry woods, not common; apparently confined to the s. ¾ of IL.

9. **Desmodium marilandicum** (L.) DC. Small-leaved Tick Trefoil. July–Sept. Dry woods, not common; confined to the s. ⅔ of IL.

10. **Desmodium nudiflorum** (L.) DC. Bare-stemmed Tick Trefoil. July–Aug. Woods; occasional throughout IL except rare in the n. cos. *Hylodesmum nudiflorum* (L.) Ohasi and R. R. Mill.

11. **Desmodium nuttallii** (Schindl.) B. G. Schub. Nuttall's Tick Trefoil. Aug.–Sept. Dry woods, not common; confined to the s. ½ of IL.

12. **Desmodium obtusum** (Muhl. ex Willd.) DC. Stiff Tick Trefoil. July–Sept. Dry woods, not common; restricted to the s. ⅔ of IL; also Iroquois and Livingston cos. *Desmodium rigidum* (Ell.) DC., misapplied.

13. **Desmodium paniculatum** (L.) DC. Panicled Tick Trefoil. July–Sept. Dry woods; occasional to common throughout IL, except for the nw. cos.

14. **Desmodium pauciflorum** (Nutt.) DC. White-flowered Tick Trefoil. July–Sept. Woods; occasional in the s. ⅓ of IL, absent elsewhere. *Hylodesmum pauciflorum* (Nutt.) Ohasi and R. R. Mill.

15. **Desmodium perplexum** B. G. Schub. Tick Trefoil. July–Sept. Dry woods, not common; confined to the s. ½ of IL. *Desmodium dillenii* Darl., in part.

16. **Desmodium rotundifolium** (Michx.) DC. Round-leaved Tick Trefoil. Aug.–Sept. Dry woods, not common; mostly in the s. ½ of IL.

17. **Desmodium sessilifolium** (M. A. Curtis) Torr. & Gray. Sessile-leaved Tick Trefoil. July–Sept. Open woods, sand prairies; occasional throughout IL, except for the nw. cos.

18. **Desmodium viridiflorum** (L.) DC. Velvet Tick Trefoil. July–Aug. Woods, very rare; Alexander Co.

15. **Galactia** P. Br.—Milk Pea

1. Calyx 2-lipped..1. *G. mohlenbrockii*
1. Teeth of the calyx nearly equal, the calyx not 2-lipped........................ 2. *G. regularis*

1. **Galactia mohlenbrockii** Maxwell. Milk Pea. Aug.–Sept. Moist thickets, very rare; Massac Co. *Dioclea multiflora* (Torr. & Gray) C. Mohr. *Lackea multiflora* (Torr. & Gray) Fern.

2. **Galactia regularis** (L.) BSP. Two varieties occur in Illinois:

a. Leaflets glabrous or nearly so on the upper surface...
.... 2a. *G. regularis* var. *regularis*
a. Leaflets strigose-pilose on the upper surface...
.... 2b. *G. regularis* var. *mississippiensis*

2a. **Galactia regularis** (L.) BSP. var. **regularis.** Milk Pea. July–Sept. Rocky woods, not common; confined to the s. ⅙ of IL. *Galacatia volubilis* (L.) Britt.

2b. **Galactia regularis** (L.) BSP. var. **mississippiensis** Vail. Milk Pea. July–Sept. Rocky woods, occasional in the s. ⅙ of IL. *Galactia volubilis* (L.) Britt. var. *mississippiensis* Vail.

16. **Glycine** L.—Soybean

1. **Glycine max** (L.) Merr. Soybean. June–Aug. Native to Asia; frequently cultivated and occasionally escaped throughout the state.

17. **Glycyrrhiza** L.—Wild Licorice

1. **Glycyrrhiza lepidota** (Nutt.) Pursh. Wild Licorice. June–Aug. Native to the w. U.S.; adventive in disturbed soil in the n. ¾ of IL.

18. **Kummerowia** Schindl.

1. Petioles 4 mm long or longer; pubescence of stem pointing upward .... 1. *K. stipulacea*
1. Petioles 1–3 (–5) mm long; pubescence of stem pointing downward ...... 2. *K. striata*

1. **Kummerowia stipulacea** (Maxim.) Makino. Korean Bush Clover. Aug.–Oct. Native to Asia; roadsides and fields; common in the s. ⅔ of IL, rare or absent elsewhere. *Lespedeza stipulacea* Maxim.

2. **Kummerowia striata** (Thunb.) Schind. Japanese Bush Clover. Aug.–Oct. Native to Asia; roadsides and fields; common in the s. ⅔ of IL, rare or absent elsewhere. *Lespedeza striata* Thunb.

19. **Lathyrus** L.—Wild Pea

1. Leaflets 2.
  2. Corolla pink, purple, or white .... 7. *L. pratensis*
  2. Corolla bright yellow.
    3. Stems winged; plants not tuber-bearing.
      4. Perennial; peduncles 4- to 10-flowered ....2. *L. latifolius*
      4. Annual or biennial; peduncles 1- to 3-flowered.
        5. Flowers up to 1.5 cm long; flowers odorless.......................1. *L. hirsutus*
        5. Flowers 2 cm long or longer; flowers fragrant ....5. *L. odoratus*
    3. Stems unwinged; plants tuberous .... 8. *L. tuberosus*
1. Leaflets 4–12.
  6. Leaflets fleshy; stipules with 2 basal lobes.....................................3. *L. maritimus*
  6. Leaflets not fleshy; stipules with 1 basal lobe.
    7. Corolla yellowish white; petiolules glabrous..........................4. *L. ochroleucus*
    7. Corolla violet or purple; petiolules pubescent.
      8. Leaflets and stems glabrous ....6. *L. palustris*
      8. Leaflets and stems hirtellous ....9. *L. venosus*

1. **Lathyrus hirsutus** L. Vetchling. May–June. Native to Europe; adventive in disturbed soil; Alexander, Jackson, Perry, and Union cos.

2. **Lathyrus latifolius** L. Everlasting Pea. June–Sept. Native to Europe; occasionally escaped along roadsides from cultivation.

3. **Lathyrus maritimus** (L.) Bigel. Beach Pea. June–Aug. Beaches along Lake Michigan; Cook, Henry, and Lake cos. *Lathyrus japonicus* Willd. var. *glaber* (Ser.) Fern.; *L. maritimus* (L.) Bigel. var. *glaber* Ser.; *L. japonicus* Willd. var. *maritimus* (L.) Kartesz & Gandhi.

4. **Lathyrus ochroleucus** Hook. Pale Vetchling. May–July. Woods; rare in the n. ⅙ of IL; also Gallatin and St. Clair cos.

5. **Lathyrus odoratus** L. Sweet Pea. June–Sept. Native to Europe; escaped to roadsides; occasional in the s. ¼ of IL.

6. **Lathyrus palustris** L. Two varieties occur in Illinois:

a. Stems distinctly winged ....6a. *L. palustris* var. *palustris*
a. Stems unwinged ....6b. *L. palustris* var. *myrtifolius*

6a. **Lathyrus palustris** L. var. **palustris.** Marsh Vetchling. May–Sept. Moist open areas; occasional in the n. ½ of IL; also Massac, Pope, and Wabash cos.

6b. **Lathyrus palustris** L. var. **myrtifolius** (Muhl.) Gray. Vetchling. June–July. Moist soil; occasional in the n. ⅘ of IL. *Lathyrus myrtifolius* Muhl.

7. **Lathyrus pratensis** L. Yellow Vetchling. June–Aug. Native to Europe; rarely escaped from cultivation; Jackson Co.

8. **Lathyrus tuberosus** L. Tuberous Vetchling. June–Aug. Native to Europe; rarely adventive in IL; DuPage, Henderson, and Kane cos.

9. **Lathyrus venosus** Muhl. var. **intonsus** Butt. & St. John. Veiny Pea. May–June. Dry prairies, woods; occasional in the n. ¼ of IL.

20. **Lespedeza** Michx.—Bush Clover

(See also Kummerowia.)

1. Stems lying flat on the ground.
  2. Pubescence of stems spreading; flowers usually 6 or more in a cluster .... 14. *L. procumbens*
  2. Pubescence of stems appressed or absent; flowers usually up to 6 in a cluster .... 15. *L. repens*
1. Stems erect or ascending.
  3. Stems with spreading hairs or velvety-tomentose.
    4. Stems with spreading hairs.
      5. Peduncles longer than the subtending leaves.
        6. Flowers yellow-white.
          7. Leaflets obovate to oblong-ovate, the terminal leaflet 1–3 cm wide .... 7. *L. hirta*
          7. Leaflets narrowly oblong to linear, the terminal leaflet less than 1 cm wide .... 10. *L. X longifolia*
        6. Flowers purple.
          8. Leaflets more than twice longer than wide .... 11. *L. X manniana*
          8. Leaflets less than twice as long as wide .... 13. *L. X nuttallii*
      5. Peduncles shorter than the subtending leaves.
        9. Flowers creamy-white, with a purple spot at base .... 4. *L. capitata*
        9. Flowers purple or pink.
          10. Fruit 3–5 mm long, longer than the calyx; stipules apparently without veins.
            11. Leaflets elliptic to ovate-oblong, 1 ½–2 times longer than broad .... 17. *L. stuevei*
            11. Leaflets narrowly elliptic to linear, more than 2 times longer than broad .... 12. *L. X neglecta*
          10. Fruit 5 mm long or longer, as long as the calyx tube or a little shorter; stipules 3-nerved .... 16. *L. X simulata*
    4. Stems velvety-tomentose .... 3. *L. X brittonii*
  3. Stems with appressed hairs or glabrous.
    12. Flowers white or yellowish, with a purple spot at base.
      13. Leaflets linear, acute at the apex; calyx pilose .... 9. *L. leptostachya*
      13. Leaflets oblanceolate, truncate, retuse, obtuse, or subacute at the apex; calyx sericeous or nearly glabrous.
        14. Racemes sessile or nearly so, 1- to 4-flowered .... 5. *L. cuneata*
        14. Racemes long-pedunculate, more than 4-flowered .... 6. *L. daurica*
    12. Flowers purplish.
      15. Peduncles longer than the subtending leaves.
        16. Shrubs to 3 m tall; fruit 6–12 mm long.
          17. Calyx teeth shorter than the tube .... 2. *L. bicolor*
          17. Calyx teeth longer than the tube .... 18. *L. thunbergii*
        16. Herbs to 1 m tall; fruit up to 6 mm long .... 19. *L. violacea*
      15. Peduncles, or most of them, shorter than the subtending leaves.
        18. Fruit sharply pointed at apex .... 1. *L. X acuticarpa*
        18. Fruit obtuse or at most subacute at apex.
          19. Leaflets oval to elliptic .... 8. *L. intermedia*
          19. Leaflets linear to oblong .... 20. *L. virginica*

1. **Lespedeza X acuticarpa** Mack. & Bush. Sharp-fruited Bush Clover. July–Sept. Dry woods; rare where the parents both occur; Union Co. Presumed to be the hybrid between *L. violacea* (L.) Pers. and *L. virginica* (L.) Britt.

2. **Lespedeza bicolor** Turcz. Bicolor Lespedeza. Aug.–Oct. Native to Asia; rarely escaped from cultivation; Perry and Williamson cos.

3. **Lespedeza X brittonii** Bickn. Britton's Bush Clover. Aug.–Oct. Dry woods, usually with the parents; Union Co. Presumed to be the hybrid between *L. procumbens* Michx. and *L. virginica* (L.) Britt.

4. **Lespedeza capitata** Michx. Three varieties and a form occur in Illinois:

a. Leaflets oblong to elliptic.
   b. Leaflets shiny and silvery on the lower surface, grayish on the upper surface .... 4a. *L. capitata* var. *capitata*
   b. Leaflets not shiny on the lower surface, green on the upper surface .... 4d. *L. capitata* var. *vulgaris*

a. Leaflets lanceolate to broadly linear.
   c. Leaflets green on the upper surface.......... .... 4b. *L. capitata* var. *stenophylla* f. *stenophylla*
   c. Leaflets silvery on the lower surface.......... .... 4c. *L. capitata* var. *stenophylla* f. *argentea*

4a. **Lespedeza capitata** Michx. var. **capitata.** Round-headed Bush Clover. Aug.–Sept. Prairies, black oak savannas; occasional to common throughout the state. An unnamed hybrid with *L. leptostachya* is known from Lee Co.

4b. **Lespedeza capitata** Michx. var. **stenophylla** Bissell & Fern. f. **stenophylla.** Narrow-leaved Round-headed Bush Clover. Aug.–Sept. Prairies, black oak savannas; occasional in the n. ½ of IL.

4c. **Lespedeza capitata** Michx. var. **stenophylla** (Bissell & Fern.) f. **argentea** Fern. Silvery Round-headed Bush Clover. Aug.–Sept. Prairies, black oak savannas; occasional in the n. ½ of IL.

4d. **Lespedeza capitata** Michx. var. **vulgaris** Torr. & Gray. Round-headed Bush Clover. Aug.–Sept. Prairies, black oak savannas; occasional to common throughout IL.

5. **Lespedeza cuneata** (Dum.-Cours.) G. Don. Sericea Lespedeza. Sept.–Oct. Native to Asia; frequently planted in IL and often escaped in the s. ½ of the state; also Grundy and Kendall cos.

6. **Lespedeza daurica** (Laxm.) Schindl. Asian Bush Clover. July–Sept. Native to Asia; adventive in Perry Co.

7. **Lespedeza hirta** (L.) Hornem. Hairy Bush Clover. July–Sept. Dry woods, occasional in the extreme n. and s. cos., but virtually absent elsewhere.

8. **Lespedeza intermedia** (S. Wats.) Britt. Bush Clover. Aug.–Sept. Dry woods, prairies; occasional in the s. ½ of IL; also Cass and DuPage cos. *Lespedeza frutescens* (L.) Hornem.

9. **Lespedeza leptostachya** Engelm. Prairie Bush Clover. July–Sept. Prairies, very rare; confined to the n. ⅙ of IL.

10. **Lespedeza X longifolia** DC. Narrow-leaved Bush Clover. July–Aug. Open ground, very rare; Iroquois Co.; presumed hybrid between *L. hirta* L. and *L. capitata* Michx.

11. **Lespedeza X manniana** Mack. & Bush. Mann's Bush Clover. July–Sept. Woods; Union Co.; presumed to be the hybrid between *L. capitata* Michx. and *L. violacea* (L.) Pers.

12. **Lespedeza X neglecta** (Britt.) Mack. & Bush. Neglected Bush Clover. July–Sept. Woods, rare; presumed to be the hybrid between *L. stuevei* Nutt. and *L. virginica* (L.) Britt.

13. **Lespedeza X nuttallii** Darl. Nuttall's Bush Clover. July–Sept. Woods; rarely found with the parents; presumed to be the hybrid between *L. hirta* (L.) Hornem. and *L. intermedia* (S. Wats.) Britt.

14. **Lespedeza procumbens** Michx. Two varieties occur in Illinois:

a. Leaflets twice as long as broad .... 14a. *L. procumbens* var. *procumbens*
a. Leaflets 3–4 times longer than broad .... 14b. *L. procumbens* var. *elliptica*

14a. **Lespedeza procumbens** Michx. var. **procumbens.** Trailing Bush Clover. July–Sept. Dry woods; occasional in the s. ½ of the state; also LaSalle Co.

14b. **Lespedeza procumbens** Michx. var. **elliptica** Blake. Narrow-leaved Trailing Bush Clover. Dry woods; occasional in the s. ½ of IL.

15. **Lespedeza repens** (L.) Barton. Creeping Bush Clover. June–Sept. Dry woods; occasional in the s. ½ of the state; also Winnebago Co.

16. **Lespedeza X simulata** Mack. & Bush. Hybrid Bush Clover. July–Sept. Prairies, very rare; Crawford, Monroe, Ogle, and Pope cos.; presumed to be the hybrid between *L. capitata* Michx. and *L. virginica* (L.) Britt.

17. **Lespedeza stuevei** Nutt. Bush Clover. Aug.–Sept. Dry woods; occasional in the s. ½ of the state, rare or absent elsewhere.

18. **Lespedeza thunbergii** (DC.) Nakai. Tall Bush Clover. Aug.–Oct. Native to Asia; adventive in Coles, Jackson, Johnson, and Wayne cos.

19. **Lespedeza violacea** (L.) Pers. Violet Bush Clover. July–Sept. Dry woods; occasional in the s. ½ of IL, becoming less common northward.

20. **Lespedeza virginica** (L.) Britt. Slender Bush Clover. Aug.–Sept. Dry woods; occasional throughout the state but apparently more common in the s. cos.

21. **Lotus** L.—Bird's-foot Trefoil

1. Leaflets 5 .... 1. *L. corniculatus*
1. Leaflets 3 or the uppermost leaves reduced to 1 leaflet .... 2. *L. unifoliolatus*

1. **Lotus corniculatus** L. Bird's-foot Trefoil. May–Sept. Native to Europe; disturbed areas, particularly along roads; occasional to common throughout IL.

2. **Lotus unifoliolatus** Benth. Deer Vetch. June–Aug. Native to the w. U.S.; usually adventive along railroads; Cook, DuPage, Greene, Macoupin, and Peoria cos. *Hosackia americana* (Nutt.) Piper; *Lotus americanus* (Nutt.) Bisch.; *L. purshianus* (Benth.) Clements & E. Clements.

22. **Lupinus** L.—Lupine

1. Leaflets usually 8–11, each one 2.5–5.0 mm long, obtuse at the apex .... 1. *L. perennis*
1. Leaflets usually 12 or more, each one 6 cm long or longer, acute at the apex .... 2. *L. polyphyllus*

1. **Lupinus perennis** L. var. occidentalis S. Wats. Wild Lupine. May–June. Sandy woods; occasional in the n. ¼ of IL, absent elsewhere. Typical var. *perennis* occurs e. of IL.

2. **Lupinus polyphyllus** Lindl. Garden Lupine. May–June. Native to Europe and Asia; rarely escaped from gardens; DuPage Co.

23. **Medicago** L.—Medic

1. Flowers blue, purple, or yellow with purple blotches.
  2. Flowers blue or purple .... 6. *M. sativa*
  2. Flowers yellow with purple blotches .... 7. *M. X varia*
1. Flowers yellow, without purple blotches.
  3. Flowers 6–8 mm long.
    4. Plants perennial; fruits straight or falcate .... 2. *M. falcata*
    4. Plants annual; fruits coiled 2 times .... 7. *M. X varia*
  3. Flowers up to 6 mm long.
    5. Fruits spirally coiled, glabrous or spiny.
      6. Fruits glabrous; flowers about 3 mm long .... 5. *M. orbicularis*
      6. Fruits spiny; flowers 4–6 mm long.
        7. Each leaflet with a spot on its surface; spines of fruit curved .... 1. *M. arabica*
        7. Leaflets without a spot on the surfect; spines of fruit straight except for the uncinate tip .... 4. *M. minima*
    5. Fruits not spirally coiled, pubescent .... 3. *M. lupulina*

1. **Medicago arabica** (L.) Huds. Spotted Medic. May–Sept. Native to Europe; rarely adventive in IL; Jackson Co.

2. **Medicago falcata** L. Yellow Lucerne. May–Sept. Native to Europe; adventive in Cass, Kendall, McHenry, and Peoria cos. *Medicago sativa* L. var. *falcata* (L.) Arc.

3. **Medicago lupulina** L. Black Medic. May–Aug. Native to Europe; adventive in disturbed ground; in every co.

4. **Medicago minima** (L.) Desr. Least Medic. June–Aug. Native to Europe; adventive in Will Co.

5. **Medicago orbicularis** (L.) Bartal. Round Medic. May–July. Native to Europe; rarely adventive in disturbed ground; Jackson Co.

6. **Medicago sativa** L. Alfalfa. May–Oct. Native to Asia; commonly cultivated and often escaped; in every co.

7. **Medicago X varia** Martyn. Hybrid Medic. June–Aug. Native to Europe; occasion-

ally adventive in the n. cos.; presumed to be the hybrid between *M. falcata* L. and *M. sativa* L.

24. **Melilotus** Mill.—Sweet Clover

1. Flowers white .... 1. *M. albus*
1. Flowers yellow.
  2. Fruit 4.5–6.0 mm long, pubescent, obscurely reticulate .... 2. *M. altissimus*
  2. Fruit 2.5–3.5 mm long, glabrous or nearly so, prominently reticulate .... 3. *M. officinalis*

1. **Melilotus albus** Medic. White Sweet Clover. May–Nov. Native to Europe and Asia; naturalized in disturbed ground; in every co.
2. **Melilotus altissimus** Thuill. Tall Yellow Sweet Clover. June–Sept. Native to Europe; naturalized in disturbed ground; n. ½ of IL.
3. **Melilotus officinalis** (L.) Pallas. Yellow Sweet Clover. May–Nov. Native to Europe; naturalized in disturbed areas; in every co.

25. **Onobrychis** Gaertn.—Sainfoin

1. **Onobrychis viciifolia** Scop. Sainfoin. June–July. Native to Europe and Asia; rarely escaped from cultivation.

26. **Ononis** L.—Rest Harrow

1. **Ononis campestris** G. Koch. Rest Harrow. June–July. Native to Europe; rarely persisting from cultivation; DuPage Co. *Ononis spinosa* L., misapplied.

27. **Orbexilum** Rydb.

1. Some or all of the leaflets at least 2 cm broad; peduncles leafless .... 1. *O. onobrychis*
1. None of the leaflets 2 cm broad; peduncles leafy.
  2. Flowers less than 8 mm long; fruits transversely wrinkled, suborbicular .... 2. *O. pedunculatum*
  2. Flowers 8–10 mm long; fruits not transversely wrinkled, depressed-obovoid .... 3. *O. simplex*

1. **Orbexilum onobrychis** (Nutt.) Rydb. French Grass. June–Aug. Thickets, prairies, not common; in most sections of the state except for the nw. cos. *Psoralea onobrychis* Nutt.
2. **Orbexilum pedunculatum** (Miller) Rydb. var. **gracile** (Chapm.) Grimes. Sampson's Snakeroot. June–July. Dry woods, prairies; occasional in the s. ⅖ of IL, apparently absent elsewhere. *Psoralea psoralioides* (Walt.) Cory var. *eglandulosa* (Ell.) Freeman.
3. **Orbexilum simplex** (Nutt.) Rydb. Large-flowered Psoralea. May–June. Illinois habitat unknown although general habitat is woods and prairies; very rare; Clinton Co. This plant was collected by Buckley near Carlyle in the 1800s. *Psoralea simplex* Nutt.

28. **Pediomelum** Grimes

1. **Pediomelum argophyllum** (Pursh) Grimes. Silvery-leaved Scurf-pea. July. Native to the w. U.S.; adventive in DuPage Co. in a marsh. *Psoralea argophylla* Pursh.

29. **Phaseolus** L.—Kidney Bean

1. Flowers many in a raceme, the raceme longer than the subtending petiole; pedicels longer than the several small bracts .... 1. *P. polystachios*
1. Flowers few in a raceme, the raceme shorter than the subtending petiole; pedicels shorter than the pair of large bracts .... 2. *P. vulgaris*

1. **Phaseolus polystachios** (L.) BSP. Wild Kidney Bean. July–Sept. Dry woods; occasional in the s. ⅓ of IL; also Clark Co.
2. **Phaseolus vulgaris** L. Garden Bean. June–July. Native to Europe; rarely escaped from gardens; Cook Co.

30. **Pisum** L.—Garden Pea

1. **Pisum sativum** L. Garden Pea. May–June. Escaped from cultivation and scattered in IL.

31. **Psoralidium** Rydb.—Scurf-pea

1. **Psoralidium tenuiflorum** (Pursh) Rydb. Two varieties occur in Illinois:

a. Flowers 1–2 per node; corolla to 5 mm long ....1a. *P. tenuiflorum* var. *tenuiflorum*
a. Flowers 2–4 per node; corolla 6–7 mm long....... 1b. *P. tenuiflorum* var. *floribundum*

1a. **Psoralidium tenuiflorum** (Pursh) Rydb. var. **tenuiflorum.** Scurf-pea. June–Sept. Prairies; occasional in the n. ¾ of IL, absent elsewhere. *Psoralea tenuiflora* Pursh.

1b. **Psoralidium tenuiflorum** (Pursh) Rydb. var. **floribundum** Nutt. Scurf-pea. June–Sept. Prairies; occasional in the n. ¾ of IL, absent elsewhere. *Psoralea tenuiflora* Pursh var. *floribunda* (Nutt.) Rydb.; *P. floribunda* Nutt.

32. **Pueraria** DC.—Kudzu

1. **Pueraria montana** (Lour.) Merr. Kudzu-vine. Aug.–Sept. Native to Japan; occasionally cultivated and sometimes escaping in several parts of IL. *Pueraria lobata* (Willd.) Ohwi; *P. montana* (Lour.) Merr. var. *lobata* (Willd.) Maesen & S. Almeida.

33. **Rehsonia** Stritch—Wisteria

1. Standard petal up to 1.9 cm long, up to 1.8 cm wide ....1. *R. floribunda*
1. Standard petal more than 1.9 cm long, more than 1.8 cm wide............2. *R. sinensis*

1. **Rehsonia floribunda** (Willd.) Stritch. Japanese Wisteria. Apr.–June. Native to Japan; rarely escaped from cultivation. *Wisteria floribunda* (Willd.) DC.

2. **Rehsonia sinensis** (Sims) Stritch. Chinese Wisteria. Apr.–May. Native to Asia; rarely escaped from cultivation. *Wisteria sinensis* (Sims) Gl.

34. **Robinia** L.—Locust

1. Branchlets neither glandular-viscid nor bristly; flowers white ......2. *R. pseudoacacia*
1. Branchlets glandular-viscid or bristly; flowers pale pink, rose, or purple.
  2. Branchlets glandular-viscid ....3. *R. viscosa*
  2. Branchlets bristly ....1. *R. hispida*

1. **Robinia hispida** L. Bristly Locust. May–June. Native to the se. U.S.; occasionally escaped from cultivation, chiefly in the s. ½ of the state.

2. **Robinia pseudoacacia** L. May–June. Black Locust. Woods and thickets; native in extreme se. IL, commonly planted and escaped from cultivation elsewhere.

3. **Robinia viscosa** Vent. Clammy Locust. May–June. Native to the se. U.S.; rarely escaped from cultivation; Jo Daviess, Lee, Massac, and Stephenson cos.

35. **Securigera** Lassen—Crown Vetch

1. **Securigera varia** (L.) Lassen. Crown Vetch. May–Sept. Native to Europe; often planted and frequently escaped throughout the state. *Coronilla varia* L.

36. **Sesbania** Scop.—Sesbania

1. **Sesbania exaltata** (Raf.) Cory. Sesbania. July–Oct. Low ground, rare; Alexander, Jackson, Massac, and Pulaski cos. *Sesbania macrocarpa* Muhl.

37. **Strophostyles** Ell.—Wild Bean

1. Legume up to 3.5 cm long, with spreading pubescence; leaflets silky-pubescent on both surfaces; calyx tube densely pubescent, up to 1.5 mm long ....2. *S. leiosperma*
1. Legume 4 cm long or longer, with appressed pubescence; leaflets glabrous or sparsely pubescent, at least on the upper surface; calyx tube glabrous or sparsely pubescent, 1.5–2.5 mm long.
  2. Bracteoles as long as the calyx tube; larger leaflets up to 6 cm broad; seeds 6–12 mm long...1. *S. helvula*
  2. Bracteoles about ½ as long as the calyx tube; larger leaflets up to 2 cm broad; seeds 3–6 mm long ....3. *S. umbellata*

1. **Strophostyles helvula** (L.) Ell. Two varieties occur in Illinois:

a. Leaflets, or some of them, lobed .... 1a. *S. helvula* var. *helvula*
a. Leaflets unlobed .... 1b. *S. helvula* var. *missouriensis*

1a. **Strophostyles helvula** (L.) Ell. var. **helvula.** Wild Bean. June–Oct. Rocky woods, sandbars, roadsides, fields; occasional throughout the state.

1b. **Strophostyles helvula** (L.) Ell. var. **missouriensis** (S. Wats.) Britt. Wild Bean. June–Oct. Dry woods; occasional in cos. along the Mississippi River.

2. **Strophostyles leiosperma** (Torr. & Gray) Piper. Wild Bean. July–Sept. Prairies, dry woods, fields, roadsides; occasional in the s. ½ of the state, becoming less common northward.

3. **Strophostyles umbellata** (Muhl.) Britt. Wild Bean. July–Oct. Dry woods, stream banks; occasional in the s. ½ of the state, absent elsewhere.

38. **Stylosanthes** Sw.—Pencil-flower

1. **Stylosanthes biflora** (L.) BSP. Pencil-flower. June–Sept. Dry woods, fields; common in the s. ¼ of the state, rare or absent elsewhere. *Stylosanthes riparia* Kearney. (Including var. *hispidissima* [Michx.] Pollard & Ball.)

39. **Tephrosia** Pers.—Hoary Pea

1. **Tephrosia virginiana** (L.) Pers. Two varieties occur in Illinois:

a. Leaflets strigose above .... 1a. *T. virginiana* var. *virginiana*
a. Leaflets silky-sericeous above.................................... 1b. *T. virginiana* var. *holosericea*

1a. **Tephrosia virginiana** (L.) Pers. var. **virginiana**. Goat's-rue. May–Aug. Dry woods and bluffs; occasional throughout the state.

1b. **Tephrosia virginiana** (L.) Pers. var. **holosericea** (Nutt.) Torr. & Gray. Goat's-rue. May–Aug. Dry woods and bluffs; occasional throughout the state.

40. **Trifolium** L.—Clover

1. Flowers yellow.
  2. Petiolule of middle leaflet longer than the petiolules of lateral leaflets.
    3. Flowers 15 or more per head, the head 8–15 mm thick .... 3. *T. campestre*
    3. Flowers up to 15 per head, the head less than 8 mm thick........... 4. *T. dubium*
  2. Petiolule of middle leaflet as long as or shorter than the petiolules of lateral leaflets... 2. *T. aureum*
1. Flowers white, pink, purple, or scarlet.
  4. Stems pubescent.
    5. Flowers sessile or nearly so in the head.
      6. Flowering heads subglobose, about as broad as long.
        7. Heads subtended by a pair of leaves; corolla 12–15 mm long; heads usually 2–3 cm in diameter .... 8. *T. pratense*
        7. Heads not subtended by a pair of leaves; corolla 5–7 mm long; heads usually less than 2 cm in diameter.................................... 5. *T. fragiferum*
      6. Flowering heads elongated, cylindric, longer than broad.
        8. Flowers pale rose, the corolla shorter than the calyx; heads covered by silky hairs .... 1. *T. arvense*
        8. Flowers scarlet, the corolla longer than the calyx; heads not covered by silky hairs .... 7. *T. incarnatum*
    5. Flowers on pedicels 2 mm long or longer..................................... 9. *T. reflexum*
  4. Stems glabrous or nearly so.
    9. Flowers sessile or nearly so in the head.
      10. Heads not subtended by a pair of leaves; corolla 4–6 mm long.
        11. Stems repent; bracteoles as long as the calyx; corolla not resupinate .... 5. *T. fragiferum*
        11. Stems ascending; bracteoles shorter than the calyx; corolla resupinate.. 11. *T. resupinatum*
      10. Heads subtended by a pair of leaves; corolla 12–15 mm long.. 7. *T. pratense*

9. Flowers on pedicels 2 mm long or longer.
  12. Stems creeping, with basal runners or stolons.
    13. Calyx teeth shorter than the tube; peduncles leafless........ 10. *T. repens*
    13. Calyx teeth about twice as long as the tube; peduncles with a pair of leaves..12. *T. stololniferum*
  12. Stems ascending, without basal runners; calyx teeth as long as the tube or longer.
    14. Calyx up to 5 mm long, the teeth less than twice as long as the tube; stipules lanceolate..6. *T. hybridum*
    14. Calyx 6–8 mm long, the teeth at least twice as long as the tube; stipules ovate.. 9. *T. reflexum*

1. **Trifolium arvense** L. Rabbit-foot Clover. May–Oct. Native to Europe; occasionally adventive in disturbed ground in most regions of the state.

2. **Trifolium aureum** Pollich. Yellow Hop Clover. June–Sept. Native to Europe; adventive in disturbed ground; Cook, DuPage, Henry, Logan, Peoria, Union, and Winnebago cos. *Trifolium agrarium* L., misapplied.

3. **Trifolium campestre** Schreb. Low Hop Clover. May–Sept. Native to Europe; occasional to common throughout IL in disturbed ground. *Trifolium procumbens* L., misapplied.

4. **Trifolium dubium** Sibth. Little Hop Clover. May–Sept. Native to Europe; disturbed ground, not common; mostly in the s. ¼ of IL.

5. **Trifolium fragiferum** L. Strawberry Clover. June–Sept. Native to Europe; rarely adventive in disturbed ground; Cook and Jackson cos.

6. **Trifolium hybridum** L. Alsike Clover. May–Nov. Native to Europe; adventive in disturbed ground; in every co. (Including var. *elegans* [Savi] Boiss.)

7. **Trifolium incarnatum** L. Crimson Clover. May–July. Native to Europe; disturbed ground; occasionally adventive in the s. ¼ of the state; also Henry and DuPage cos.

8. **Trifolium pratense** L. Two varieties occur in Illinois:

a. Heads 1.2–3.0 cm in diameter..8a. *T. pratense* var. *pratense*
a. Heads 3–4 cm in diameter..8b. *T. pratense* var. *sativum*

8a. **Trifolium pratense** L. var. **pratense.** Red Clover. May–Nov. Native to Europe; naturalized in waste ground; in every co.

8b. **Trifolium pratense** L. var. **sativum** (Mill.) Schreb. Red Clover. May–Oct. Native to Europe; naturalized in waste ground in a few cos.

9. **Trifolium reflexum** L. Two varieties occur in Illinois:

a. Stems villous..9a. *T. reflexum* var. *reflexum*
a. Stems glabrous..9b. *T. reflexum* var. *glabrum*

9a. **Trifolium reflexum** L. var. **reflexum.** Buffalo Clover. May–June. Woods, not common; scattered in IL.

9b. **Trifolium reflexum** L. var. **glabrum** Lojacono. Buffalo Clover. May–June. Woods, not common; scattered in IL.

10. **Trifolium repens** L. White Clover. Jan.–Nov. Native to Europe; naturalized in disturbed ground; in every co.

11. **Trifolium resupinatum** L. Persian Clover. June–Sept. Native to Europe; rarely adventive in disturbed ground; Cook Co.

12. **Trifolium stoloniferum** Eat. Running Buffalo Clover. May–Aug. St. Clair Co.: Cahokia, collected by N. M. Glatfelter on June 1, 1890.

41. **Vicia** L.—Vetch

1. Peduncles as long as the leaflets or longer.
  2. Pubescence on stems and axis of inflorescence spreading................. 8. *V. villosa*
  2. Pubescence on stems and axis of inflorescence appressed or nearly absent.
    3. Flowers white, tinged with purple; racemes laxly flowered.
      4. Flowers 1–2 (–4) in a raceme; calyx lobes unequal; leaflets up to 10; fruit 4-seeded.. 7. *V. tetrasperma*
      4. Flowers 2–7 in a raceme; calyx lobes equal; leaflets 10 or more; fruit more than 4-seeded.. 3. *V. caroliniana*

3. Flowers purple; racemes densely flowered.
5. Flowers less than 10 per raceme; entire inflorescence shorter than the subtending leaves .... 1. *V. americana*
5. Flowers 10 or more per raceme; entire inflorescence longer than the subtending leaves.
6. Calyx gibbous at the base on the upper side .... 5. *V. dasycarpa*
6. Calyx merely rounded at the base on the upper side........ 4. *V. cracca*
1. Peduncles much shorter than the leaflets or nearly absent.
7. Stems more or less pubescent; leaflets all similar in shape; most or all the flowers 18 mm long or longer; legume torulose........ 6. *V. sativa*
7. Stems glabrous or nearly so; upper leaflets narrower than lower ones; all the flowers less than 18 mm long; legume flat .... 2. *V. angustifolia*

1. **Vicia americana** Muhl. American Vetch. May–Aug. Disturbed woods and disturbed prairies; occasional in the n. ⅓ of IL, absent elsewhere.

2. **Vicia angustifolia** Reich. Narrow-leaved Vetch. May–Oct. Native to Europe; disturbed soil; occasional in the extreme n. and extreme s. cos. *Vicia sativa* L. ssp. *nigra* (L.) Ehrh. (Including *V. sativa* L. var. *segetalis* [Thuill.] Ser.)

3. **Vicia caroliniana** Walt. Wood Vetch. Apr.–June. Dry, wooded slopes; occasional in the ne. corner of the state; also Adams Co.

4. **Vicia cracca** L. Cow Vetch. May–Aug. Native to Europe; disturbed soil, particularly roadsides and fields; occasional throughout the state.

5. **Vicia dasycarpa** Ten. Hairy-fruited Vetch. May–Sept. Native to Europe; scattered throughout the state in disturbed soil.

6. **Vicia sativa** L. Common Vetch. May–Sept. Native to Europe; roadsides and fields; scattered in IL.

7. **Vicia tetrasperma** (L.) Moench. Four-seeded Vetch. June–Sept. Native to Europe; rarely adventive in disturbed soil; DuPage Co.

8. **Vicia villosa** Roth. Winter Vetch. May–Oct. Native to Europe; fields and along roads; common throughout the state. *Vicia villosa* Roth ssp. *varia* (Host) Corbiere.

42. **Vigna** Savi—Cow Pea

1. **Vigna unguiculata** (L.) Walp. Cow Pea. July–Sept. Native to Asia; frequently cultivated and occasionally escaped in IL. *Vigna sinensis* (L.) Savi.

43. **Wisteria** Nutt.—American Wisteria

(See also Rehsonia.)

1. **Wisteria frutescens** (L.) Poir. American Wisteria. June–July. Swampy woods; generally confined to the s. ⅙ of IL but extending n. to Clark and Richland cos.; adventive in Peoria and Washington cos. *Wisteria macrostachya* Nutt., misapplied.

73. FAGACEAE—BEECH FAMILY

1. Apical bud solitary or absent; leaves sharply serrate; involucre persistent in fruit and spiny; nut compressed.
2. Apical bud solitary; staminate flowers in drooping, globose clusters; nut sharply 3-angled; bark gray, smooth........ 2. *Fagus*
2. Apical bud absent; staminate flowers in slender, ascending aments; nut subglobose, flattened on 1 or 2 sides; bark dark, at length roughened........ 1. *Castanea*
1. Apical buds 3 or more; leaves entire, lobed, dentate, or serrate; involucre persistent in fruit but not spiny; fruit an acorn, not compressed........ 3. *Quercus*

1. **Castanea** Mill.—Chestnut

1. Leaves glabrous; buds glabrous .... 1. *C. dentata*
1. Leaves pubescent; buds pubescent........ 2. *C. mollissima*

1. **Castanea dentata** (Marsh.) Borkh. Chestnut. June–July. Rocky woods; originally in the s. tip of the state, particularly Pulaski Co.; now nearly extinct except for a few plants scattered here and there and probably adventive.

2. **Castanea mollissima** Blume. Chinese Chestnut. June–July. Dry upland woods; scattered but rarely escaped in the e. U.S.; scattered in IL.

2. **Fagus** L.—Beech

1. **Fagus grandifolia** Ehrh. var. **caroliniana** (Loud.) Fern. & Rehd. Two forms occur in Illinois:

a. Leaves glabrous or nearly so on the lower surface ....1a. *F. grandifolia* var. *caroliniana* f. *caroliniana*
a. Leaves soft-pubescent on the lower surface..1b. *F. grandifolia* var. *caroliniana* f. *mollis*

1a. **Fagus grandifolia** Ehrh. var. **caroliniana** (Loud.) Fern. & Rehd. f. **caroliniana.** American Beech. Apr.–May. Rich woods; occasional in the s. ¼ of IL, extending along the extreme e. border of IL. to Vermilion Co.; also Cook and Lake cos. Typical var. *grandifolia* occurs n. and e. of IL.
1b. **Fagus grandifolia** Ehrh. var. **caroliniana** (Loud.) Fern. & Rehd. f. **mollis** Fern. & Rehd. Hairy American Beech. Apr.–May; Rich woods; occasional in the s. ¼ of IL.

3. **Quercus** L.—Oak

1. Leaves entire.
2. Leaves stellate-pubescent beneath; cup of acorn at least 1.5 cm broad ....7. *Q. imbricaria*
2. Leaves glabrous or nearly so; cup of acorn up to 1.2 cm broad...... 15. *Q. phellos*
1. Leaves toothed or lobed.
3. Leaves toothed.
4. Acorns on stalks at least 2 cm long; veins of leaves not reaching the tip of the teeth ....3. *Q. bicolor*
4. Acorns sessile or on stalks up to 1 cm long; veins of leaves reaching the tip of the teeth.
5. Leaves velvety-tomentose beneath; cup of acorn at least 2.5 cm broad ....11. *Q. michauxii*
5. Leaves glabrous or minutely pubescent but not velvety; cup of acorn less than 2.5 cm broad.
6. Leaves with pointed teeth; acorns up to 2 cm long.
7. Leaves with up to 13 sharp teeth on either margin; scales of acorn cup closely appressed...12. *Q. muhlenbergii*
7. Leaves with more than 13 sharp teeth on either margin; scales of acorn cup spreading ....1. *Q. acutissima*
6. Leaves with rounded teeth; acorns 2 cm long or longer....16. *Q. prinus*
3. Leaves lobed.
8. Lobes of leaves with bristle-tips.
9. Leaves permanently uniformly pubescent throughout on the lower surface.
10. Leaves much broader in the upper ½, with 3 (–5) broad, shallow lobes ....10. *Q. marilandica*
10. Leaves broader at or below the middle, with (3–) 5–11 narrower, deeper lobes.
11. Lower leaf surface grayish or yellowish; pubescence of stellate hairs; scales of buds reddish brown, glabrous or nearly so; terminal winter buds up to 6 mm long.
12. Terminal and usually 1 or 2 lateral lobes of leaf curved.............. ....6. *Q. falcata*
12. All lobes of leaf straight or nearly so ....13. *Q. pagoda*
11. Lower leaf surface brownish or reddish brown; pubescence not stellate; scales of buds grayish, tomentose; terminal winter buds about 6–12 mm long .... 21. *Q. velutina*
9. Leaves glabrous beneath, with tufts of axillary hairs, or irregularly pubescent.
13. Leaves much broader in the upper ½, with 3 (–5) broad, shallow lobes ....10. *Q. marilandica*
13. Leaves broader at or below the middle, with 5–11 narrower, deeper lobes.

14. Terminal winter buds 6–12 mm long, gray-tomentose; scales along the edge of the acorn cup not appressed; upper leaf surface with a pubescent midnerve .... 21. *Q. velutina*
14. Terminal winter buds up to 4 mm long, not gray-tomentose; scales along the edge of the acorn cup appressed; upper leaf surface with a glabrous midnerve.
 15. Leaves lobed less than halfway to middle; cup covering less than ¼ of acorn .... 17. *Q. rubra*
 15. Leaves lobed at least halfway to middle; cup covering at least ¼ of acorn.
  16. Acorn cup covering much more than ½ of the acorn.
   17. Acorn up to 1.5 cm long; cup not exceeding 1.5 cm in diameter .... 14. *Q. palustris*
   17. Acorn more than 1.5 cm long; cup more than 1.5 cm in diameter.
    18. Buds glabrous; leaves not long-tapering to the tip; acorn cup mostly 2 cm broad or broader .... 18. *Q. shumardii*
    18. Buds puberulent or at least ciliate; leaves tending to be long-tapering to the tip; acorn cup mostly 1.5–2.0 cm broad .... 20. *Q. texana*
  16. Acorn cup covering about ½ of the acorn.
   19. Acorn cup up to 1.5 cm in diameter, with pubescent scales... 5. *Q. ellipsoidalis*
   19. Acorn cup at least 1.5 cm in diameter, with glabrous scales.. 4. *Q. coccinea*

8. Lobes of leaves without bristle tips.
 20. Leaves completely glabrous beneath.. 2. *Q. alba*
 20. Leaves pubescent beneath, at least in the leaf axils.
  21. Leaves 3- to 5-lobed, the upper 3 with squarish tips, forming a cross; pubescence of stellate hairs; acorn up to 1 cm across, the cup unfringed and covering less than ½ the nut .... 19. *Q. stellata*
  21. Leaves 5- to 11-lobed, the upper 3 without squarish tips; pubescence not stellate; acorn 1 cm broad or broader, the cup either fringed or covering ½ to nearly all the nut.
   22. Scales of cup forming a fringe around the acorn; cup ½–⅘ covering the nut... 9. *Q. macrocarpa*
   22. Scales of cup not forming a fringe; cup nearly completely covering the nut .... 8. *Q. lyrata*

1. **Quercus acutissima** Carruth. Sawtooth Oak. Apr. Native to Europe; sometimes planted in IL but rarely escaped; disturbed woods; St. Clair Co.

2. **Quercus alba** L. White Oak. Apr.–May. Upland woods; common throughout the state; in every co. This is the state tree of Illinois. (Including f. *alexanderi* [Britt.] Trel.)

3. **Quercus bicolor** Willd. Swamp White Oak. Apr.–May. Low woods and swamps; occasional throughout the state.

4. **Quercus coccinea** Muench. Scarlet Oak. Apr.–May. Upland woods; occasional in the s. ⅓ of IL. (Including var. *tuberculata* Sarg.)

5. **Quercus ellipsoidalis** E. J. Hill. Hill's Oak. Apr.–May. Upland woods; occasional in the n. ⅓ of IL.

6. **Quercus falcata** Michx. Southern Red Oak. Apr.–May. Dry woods; occasional in the s. ⅓ of IL.

7. **Quercus imbricaria** Michx. Shingle Oak. Apr.–May. Moist or dry woods, edges of fields; occasional to common throughout IL.

8. **Quercus lyrata** Walt. Overcup Oak. Apr.–May. Bottomland woods; swampy woods; occasional in the s. ½ of IL, particularly along the Mississippi River.

9. **Quercus macrocarpa** Michx. Bur Oak. Apr.–May. Bottomland woods, savannas; occasional to common throughout the state; in every co.

10. **Quercus marilandica** Muench. Blackjack Oak. Apr.–May. Upland woods, bluff tops; occasional to common in the s. ¾ of IL, absent elsewhere.

11. **Quercus michauxii** Nutt. Swamp Chestnut Oak; Basket Oak. Apr.–May. Low woods, swamps; occasional in the s. ⅓ of IL; also Cass Co.

12. **Quercus muhlenbergii** Engelm. Chinquapin Oak; Yellow Chestnut Oak. Apr.–May. Moist or dry woods; occasional throughout the state. (Including f. *alexanderi* [Britt.] Trel.) *Quercus prinoides* Willd. var. *acuminata* (Michx.) Gl.

13. **Quercus pagoda** Raf. Cherrybark Oak. Apr.–May. Rich woods, bottomland woods; occasional in the s. ⅙ of IL. *Quercus pagodaefolia* (Ell.) Ashe; *Q. falcata* Michx. var. *pagodaefolia* Ell.

14. **Quercus palustris** Muench. Pin Oak. Apr.–May. Wet ground; occasional to common throughout IL.

15. **Quercus phellos** L. Willow Oak. Apr.–May. Swampy woods; Alexander, Johnson, Massac, Pulaski, and Union cos.

16. **Quercus prinus** L. Rock Chestnut Oak. Apr.–May. Rocky woods, rare; Alexander, Hardin, Saline, and Union cos. *Quercus montana* Willd.

17. **Quercus rubra** L. Red Oak. Apr.–May. Upland woods; common throughout the state. (Including var. *borealis* [Michx.] Farw.)

18. **Quercus shumardii** Buckl. Two varieties may be distinguished in Illinois:

a. Acorn cup saucer-shaped, enclosing the nut by ¼ or less.. 18a. *Q. shumardii* var. *shumardii*

a. Acorn cup turbinate, enclosing the nut by more than ¼.. 18b. *Q. shumardii* var. *schneckii*

18a. **Quercus shumardii** Buckl. var. **shumardii.** Shumard's Oak. Apr.–May. Low or alluvial woods; occasional in the s. ⅔ of IL; also McLean Co. (Including var. *stenocarpa* Laughlin.)

18b. **Quercus shumardii** Buckl. var. **schneckii** (Britt.) Sarg. Schneck's Oak. Apr.–May. Dry, rocky uplands; occasional in the s. ⅓ of IL.

19. **Quercus stellata** Wangh. Post Oak. Apr.–May. Upland woods, bluffs, flat woods; occasional to common in the s. ½ of IL; also McDonough Co.

20. **Quercus texana** Buckl. Nuttall's Oak. Apr.–May. Wet woods, very rare; Alexander, Massac, and Pulaski cos. *Quercus nuttallii* E. J. Palmer.

21. **Quercus velutina** Lam. Black Oak. Two forms occur in Illinois:

a. Young branchlets and lower surface of leaves glabrate at maturity, except in the leaf axils.. 21a. *Q. velutina* f. *velutina*

a. Branchlets and lower surface of leaves permanently pubescent.. 21b. *Q. velutina* f. *missouriensis*

21a. **Quercus velutina** Lam. f. **velutina.** Black Oak. Apr.–May. Upland woods; common throughout the state; in every co.

21b. **Quercus velutina** Lam. f. **missouriensis** (Sarg.) Trel. Hairy Black Oak. Apr.–May. Upland woods; Jackson, Saline, and Union cos.

Many oak hybrids have been recorded from Illinois. For many of the hybrids, there is little scientific evidence to support the hybrid status. Often it is the case that a tree looks like the possible hybrid between two species. I have made no effort to provide a key to the reported Illinois oak hybrids. A list of these hybrids that have been reported from Illinois is provided below:

Quercus X bebbiana C. Schneider (*alba X macrocarpa*)
Quercus X benderi Baenitz (*coccinea X rubra*)
Quercus X bushii Sarg. (*marilandica X velutina*)
Quercus X columnaris Laughlin (*palustris X rubra*)
Quercus X deamii Trel. (*macrocarpa X michauxii*)
Quercus X exacta Trel. (*imbricaria X palustris*)
Quercus X fernowii Trel. (*alba X stellata*)
Quercus X filialis Little (*phellos X velutina*)
Quercus X hawkinsiae Sudw. (*rubra X velutina*)
Quercus X hillii Trel. (*macrocarpa X muhlenbergii*)
Quercus X humidicola E. J. Palmer (*bicolor X lyrata*)

Quercus X jackiana C. Schneider (*alba X bicolor*)
Quercus X leana Nutt. (*imbricaria X velutina*)
Quercus X ludoviciana Sarg. (*falcata X phellos*)
Quercus X palaeolithicola Trel. (*coccinea X velutina*)
Quercus X palmeriana A. Camus (*falcata X imbricaria*)
Quercus X richteri Baenitz (*palustris X rubra*)
Quercus X runcinata (A. DC.) Engelm. (*imbricaria X rubra*)
Quercus X schochiana Dieck (*palustris X phellos*)
Quercus X schuettei Trel. (*bicolor X macrocarpa*)
Quercus X tridentata Engelm. (*imbricaria X marilandica*)

74. FUMARIACEAE—FUMITORY FAMILY

1. Corolla with 2 spurs.
  2. Leaves all arising at the base of the plant; plants erect.......................... 3. *Dicentra*
  2. Leaves cauline and alternate; plants climbing .... 1. *Adlumia*
1. Corolla with a single spur.
  3. Flowers yellow or pink; capsule elongated, several-seeded............... 2. *Corydalis*
  3. Flowers purplish, tipped with red; capsule nearly spherical, 1-seeded .... 4. *Fumaria*

1. **Adlumia** Raf.—Climbing Fumitory

1. **Adlumia fungosa** (Ait.) E. Greene. Climbing Fumitory. June–Sept. Escaped from cultivation; found in woods after a fire; Kankakee and Ogle cos.

2. **Corydalis** Medic.—Corydalis

1. Flowers pink with yellow tips... 7. *C. sempervirens*
1. Flowers wholly yellow.
  2. Spurred petal up to 10 mm long, the spur strongly incurved; pedicels 6–18 mm long .... 3. *C. flavula*
  2. Spurred petal 10–18 mm long, the spur straight or only slightly incurved; pedicels up to 6 mm long.
    3. Seeds 1.4–1.6 mm long; cleistogamous flowers often present; spurred petal 10–14 (–15) mm long.
      4. Capsules not torulose or only slightly torulose.....................5. *C. micrantha*
      4. Capsules strongly torulose...4. *C. halei*
    3. Seeds 1.8–2.1 mm long; cleistogamous flowers usually absent; spurred petal 14–18 mm long.
      5. Seeds smooth or slightly muricate; bracts 4–10 mm long.
        6. Capsules pendent to spreading, 18–24 mm long; leaves longer than the racemes...1. *C. aurea*
        6. Capsules erect, 12–20 mm long; leaves shorter than the racemes ....6. *C. montana*
      5. Seeds strongly muricate; bracts 10–17 mm long .... 2. *C. curvisiliqua*

1. **Corydalis aurea** Willd. Golden Corydalis. May–Aug. Rocky soil, rare; confined to the n. ½ of IL; also Jersey and Union cos.

2. **Corydalis curvisiliqua** Engelm. ssp. **grandibracteata** (Fedde) G. Ownbey. Large-bracted Corydalis. May–July. Woods, rare; Brown, Cass, Mercer, Morgan, and Pike cos.

3. **Corydalis flavula** (Raf.) DC. Pale Corydalis. Apr.–May. Moist woods, common in the s. cos., becoming less common northward.

4. **Corydalis halei** (Small) Fern. Hale's Corydalis. Apr.–May. Moist soil at base of cliffs, rare; Monroe and Pope cos.

5. **Corydalis micrantha** (Engelm.) Gray. Two subspecies occur in Illinois:

a. Racemes as long as or barely longer than the leaves; spur globose; capsules 10–15 mm long...5a. *C. micrantha* ssp. *micrantha*
a. Racemes much longer than the leaves; spur not globose; capsules 15–30 mm long .... 5b. *C. micrantha* ssp. *australis*

5a. **Corydalis micrantha** (Engelm.) Gray ssp. **micrantha**. Slender Corydalis. Apr.–July. Rocky woods, not common; scattered throughout the state.

5b. **Corydalis micrantha** (Engelm.) Gray ssp. **australis** (Chapm.) G. Ownbey. Corydalis. Apr.–May. Prairies, very rare; Grundy and Mercer cos. *Corydalis campestris* Buchholz & Palmer.

6. **Corydalis montana** Engelm. ex Gray. Western Corydalis. May–June. Wooded slopes, very rare; Jo Daviess Co. *Corydalis aurea* Willd. var. *occidentalis* Engelm. ex Gray; *Corydalis aurea* Willd. ssp. *occidentalis* (Engelm. ex Gray) G. B. Ownbey.

7. **Corydalis sempervirens** (L.) Pers. Pink Corydalis. May–Aug. Rocky woods, sandy soil after a fire, rare; Cook, LaSalle, Ogle, Stephenson, and Winnebago cos. *Capnoides sempervirens* (L.) Borkh.

3. **Dicentra** Bernh.—Bleeding-heart

1. Flowers white.
  2. Spurs of corolla spreading, subacute; flowers without an odor; plants from granular, white tubers .... 2. *D. cucullaria*
  2. Spurs of corolla not spreading, rounded; flowers with a sweet odor; plants from yellow, cormlike tubers .... 1. *D. canadensis*
1. Flowers purplish or reddish .... 3. *D. eximia*

1. **Dicentra canadensis** (Goldie) Walp. Squirrel Corn. Apr.–May. Rich woods, occasional; scattered throughout IL.

2. **Dicentra cucullaria** (L.) Bernh. Dutchman's-breeches. Mar.–May. Rich woods, common; throughout the state.

3. **Dicentra eximia** (Ker) Torr. Staggerweed; Bleeding-hearts. May–June. Native to the e. U.S.; occasionally cultivated but rarely escaped; Cook and Ogle cos.

4. **Fumaria** L.—Fumitory

1. **Fumaria officinalis** L. Fumitory. May–Aug. Native to Europe; adventive in disturbed soil, often after a fire; occasional in IL.

75. GENTIANACEAE—GENTIAN FAMILY

1. Leaves whorled; flowering stems 1 m or more tall; each lobe of the corolla with a large nectariferous gland .... 3. *Frasera*
1. Leaves opposite, alternate, or reduced to scales; flowering stems less than 1 m tall; corolla without large nectariferous glands.
  2. Leaves all reduced to scales .... 1. *Bartonia*
  2. Some or all of the leaves not reduced to scales.
    3. Lower leaves reduced to scales, upper leaves broader; calyx lobes 2 .... 7. *Obolaria*
    3. None of the leaves scalelike; calyx lobes 4–5.
      4. Lobes of the corolla longer than the tube; stem usually 4-angled .... 8. *Sabatia*
      4. Lobes of the corolla as long as or shorter than the tube; stem terete.
        5. Corolla red or pink, the tube up to 2 mm in diameter .... 2. *Centaurium*
        5. Corolla blue, purple, or white, the tube at least 5 mm in diameter.
          6. Corolla with petaloid outgrowths between the lobes .... 4. *Gentiana*
          6. Corolla lacking petaloid outgrowths between the lobes.
            7. Corolla entire at the apex; seeds smooth .... 5. *Gentianella*
            7. Corolla fringed or dentate at the apex; seeds papillose .... 6. *Gentianopsis*

1. **Bartonia** Muhl.—Screwstem

1. Most of the scale leaves alternate .... 1. *B. paniculata*
1. Most of the scale leaves opposite .... 2. *B. virginica*

1. **Bartonia paniculata** (Michx.) Muhl. Screwstem. Aug.–Sept. Shaded sandstone cliffs, very rare; Pope Co.

2. **Bartonia virginica** (L.) BSP. Yellow Bartonia. July–Oct. Bogs, shaded sandstone cliffs, rare; scattered in IL.

2. **Centaurium** Hill—Centaury

1. **Centaurium pulchellum** (Sw.) Druce. Showy Centaury. June–Oct. Native to Europe; in disturbed limey soil; ne. IL.

3. **Frasera** Walt.—Columbo

1. **Frasera caroliniensis** Walt. American Columbo. May–June. Rich woods, sandy woods; occasional in the s. ½ of IL; also Cook and DuPage cos. *Swertia caroliniensis* (Walt.) Ktze.

4. **Gentiana** L.—Gentian

1. Corolla white or yellowish white; margins of leaves glabrous .... 1. *G. alba*
1. Corolla blue or bluish purple; margins of leaves ciliolate.
  2. Anthers free; tube of corolla open at the top.
    3. Stems puberulent; leaves mostly 3–5 cm long; appendages between the petals 2-cleft .... 3. *G. puberulenta*
    3. Stems glabrous; leaves mostly 1.0–2.5 cm long; appendages between the petals several-cleft .... 5. *G. septemfida*
  2. Anthers coherent in a ring; tube of corolla nearly closed at the top.
    4. Lobes of the corolla smaller than the appendages between them .... 2. *G. andrewsii*
    4. Lobes of the corolla as long as or longer than the appendages between them .... 4. *G. saponaria*

1. **Gentiana alba** Muhl. Yellow Gentian. Aug.–Oct. Prairies, rich wooded slopes, not common; scattered throughout IL. *Gentiana flavida* Gray. A hybrid with *G. andrewsii*, called *Gentiana X pallidocyanea* J. Pringle, is known from Cook Co.

2. **Gentiana andrewsii** Griseb. There are two varieties in Illinois:

a. Lobes of corolla up to 1 mm long, acute, apiculate .... 2a. *G. andrewsii* var. *andrewsii*
a. Lobes of corolla up to 3 mm long, obtuse, mucronate .... 2b. *G. andrewsii* var. *dakotica*

2a. **Gentiana andrewsii** Griseb. var. **andrewsii.** Closed Gentian. Aug.–Oct. Moist woods, prairies, calcareous meadows; occasional in the n. ½ of the state, rare in the s. ½.

2b. **Gentiana andrewsii** Griseb. var. **dakotica** A. Nels. Western Closed Gentian. Aug.–Oct. Rich woods, rare; Brown, Kankakee, and Pope cos. *Gentiana clausa* Raf., misapplied.

3. **Gentiana puberulenta** J. Pringle. Downy Gentian. Aug.–Oct. Prairies; occasional in the n. ⅔ of the state, rare elsewhere.

4. **Gentiana saponaria** L. Soapwort Gentian. Aug.–Oct. Moist prairies, sandy woods, black oak savannas; occasional in the ne. cos.; also Gallatin, Pope, and Randolph cos.

5. **Gentiana septemfida** Pall. Gentian. July–Aug. Native to Asia; escaped into disturbed soil; Cook Co.

Two hybrid gentians have been reported from Illinois. They are *Gentiana X billingtonii* Farw. (*G. andrewsii X G. puberulenta*) and *Gentiana X curtisii* J. Pringle (*G. alba X G. puberulenta*).

5. **Gentianella** Small—Little Gentian

1. **Gentianella quinquefolia** (L.) Small ssp. **occidentalis** (Gray) J. M. Gillett. Stiff Gentian. Aug.–Oct. Meadows, prairies, calcareous woods; occasional in the n. ⅔ of IL; also Hardin Co. *Gentiana quinquefolia* L.

6. **Gentianopsis** Ma—Fringed Gentian

1. Uppermost leaves narrowly ovate to ovate, rounded at the base; fringe of corolla lobes uniformly at least 2 mm long .... 1. *G. crinita*
1. Uppermost leaves linear to linear-lanceolate, tapering to the base; fringe of corolla lobes 2–4 mm long at the edges but reduced to short teeth at the summit .... 2. *G. virgata*

1. **Gentianopsis crinita** (Froel.) Ma. Fringed Gentian. Aug.–Nov. Marshes; apparently introduced into sandy flats, not common; confined to the n. ¼ of IL. *Gentiana crinita* Froel.

2. **Gentianopsis virgata** (Raf.) Holub. Small Fringed Gentian. Aug.–Oct. Calcareous fens, not common; confined to the n. ¼ of IL. *Gentiana procera* Holm; *Gentianopsis procera* (Holm) Ma.

7. **Obolaria** L.—Pennywort

1. **Obolaria virginica** L. Pennywort. Apr.–May. Rich woods, rare; confined to the s. ¼ of IL; also Clark, Crawford, and Lawrence cos.

8. **Sabatia** Adans.—Rose Gentian

1. Tube of calyx 2–3 mm long; leaves at middle of stem clasping .... 1. *S. angulatus*
1. Tube of calyx 4–8 mm long; leaves at middle of stem not clasping .... 2. *S. campestris*

1. **Sabatia angulatus** (L.) Pursh. Rose Gentian; Marsh Pink; Rose Pink. June–Nov. Moist soil; occasional in the s. ⅔ of IL, rare or absent elsewhere. (Including f. *albiflora* [Raf.] House.)

2. **Sabatia campestris** Nutt. Prairie Rose Gentian. July–Sept. Prairies, very rare; scattered in s. cent. IL; also DuPage and Peoria cos.

76. GERANIACEAE—GERANIUM FAMILY

1. Leaves pinnatifid or pinnately compound .... 1. *Erodium*
1. Leaves palmately cleft or divided .... 2. *Geranium*

1. **Erodium** L'Her.—Storksbill

1. **Erodium cicutarium** (L.) L'Her. Storksbill; Pin Clover. Apr.–Oct. Native to Europe; naturalized in waste areas; not common but scattered in IL.

2. **Geranium** L.—Cranesbill; Wild Geranium

1. Petals more than 1 cm long.
  2. Peduncle 1-flowered; petals emarginate .... 9. *G. sanguineum*
  2. Peduncle 2- to several-flowered; petals rounded at tip.
    3. Stigmas yellow, white, or light brown .... 4. *G. maculatum*
    3. Stigmas red .... 11. *G. sylvaticum*
1. Petals less than 1 cm long.
  4. Leaves palmately compound .... 8. *G. robertianum*
  4. Leaves palmately lobed, often deeply so, but not compound.
    5. Sepals awnless; stamens 5.
      6. Fruits glabrous, rugose; beak of mature style column 2–5 mm long .... 5. *G. molle*
      6. Fruits pubescent, not rugose; beak of mature style column up to 1 mm long .... 7. *G. pusillum*
    5. Sepals awned or sharp-tipped; stamens 10.
      7. Peduncle 1-flowered .... 10. *G. sibiricum*
      7. Peduncle 2- to 7-flowered.
        8. Fruiting pedicels longer than the calyx.
          9. Stout rhizomes absent; beak of mature style 2.5–6.0 mm long .... 1. *G. bicknellii*
          9. Stout rhizomes present; beak of mature style less than 1 mm long .... 6. *G. nepalense*
        8. Fruiting pedicels shorter than the calyx.
          10. Carpels hirsute with spreading hairs; lobes of leaves acute .... 3. *G. dissectum*
          10. Carpels long-villous with ascending hairs; lobes of leaves obtuse .... 2. *G. carolinianum*

1. **Geranium bicknellii** Britt. Northern Cranesbill. June–Aug. Sandy woods, fields, rare; Cook, DuPage, and Lake cos.

2. **Geranium carolinianum** L. Wild Cranesbill. May–Aug. Woods, fields, roadsides; common in the s. ½ of IL, occasional in the n. ½. (Including var. *confertiflorum* Fern.)

3. **Geranium dissectum** L. Purple Cranesbill. June. Native to Europe; rarely escaped into the edge of a prairie; Lake Co.

4. **Geranium maculatum** L. Wild Geranium. Apr.–June. Rich woods, common; in every co.

5. **Geranium molle** L. Dove's-foot Cranesbill. May–Aug. Native to Europe and Asia; rarely adventive in disturbed soil.

6. **Geranium nepalense** Sweet var. **thunbergii** (Sieb. & Zucc.) Kudo. Thunberg's Geranium. May–June. Native to Asia; rarely escaped from cultivation; Kane Co. *Geranium thunbergii* Sieb. & Zucc.

7. **Geranium pusillum** L. Small Cranesbill. June–July. Native to Europe; scattered in waste ground in IL.

8. **Geranium robertianum** L. Herb Robert. July. Woods, fields, rare; Cook, DuPage, and Wabash cos.

9. **Geranium sanguineum** L. Geranium. May–July. Native to Europe and Asia; rarely adventive in waste ground; Morgan Co.

10. **Geranium sibiricum** L. Siberian Cranesbill. Aug.–Sept. Native to Europe and Asia; naturalized into waste ground, rare; Champaign, Ogle, Stephenson, and Winnebago cos.

11. **Geranium sylvaticum** L. Woodland Geranium. May–June. Native to boreal regions; rarely escaped from cultivation; Coles Co.

77. GROSSULARIACEAE—GOOSEBERRY FAMILY

1. **Ribes** L.—Currant; Gooseberry

1. Plants with spines or prickles.
  2. Berries prickly; lobes of the calyx shorter than the tube .... 3. *R. cynosbati*
  2. Berries lacking prickles; lobes of calyx as long as or longer than the tube.
    3. Flowers white to greenish white; berries at least 10–15 mm in diameter .... 5. *R. missouriense*
    3. Flowers greenish yellow or purplish; berries 8–10 (–12) mm in diameter........ .... 4. *R. hirtellum*
1. Plants lacking spines or prickles.
  4. Flowers golden yellow, very fragrant, clove-scented; tube of calyx more than twice as long as the lobes; leaves deeply 3-parted............................ 2. *R. aureum*
  4. Flowers yellow-green, yellow and white, or greenish purple, not fragrant; tube of calyx up to twice as long as the lobes; leaves shallowly 3-parted or unlobed.
    5. Leaves and fruits with yellow resinous dots; berries black.
      6. Flowers broadly campanulate; calyx 5–6 mm long, pubescent.................. .... 6. *R. nigrum*
      6. Flowers tubular-campanulate; calyx 8–10 mm long, glabrous.................... .... 1. *R. americanum*
    5. Leaves and fruits lacking resinous dots; berries red.
      7. Pedicels glandular; terminal leaf lobe with straight sides........... 8. *R. triste*
      7. Pedicels eglandular; terminal leaf lobe with curved sides .... 7. *R. rubrum*

1. **Ribes americanum** Mill. Wild Black Currant. Apr.–June. Moist woods, upland woods; common in the n. ½ of the state, absent elsewhere.

2. **Ribes aureum** Pursh var. **villosum** (Pursh) DC. Buffalo Currant. Apr.–June. Native to the w. U.S.; occasionally escaped from cultivation. *Ribes odoratum* Wendl. f.

3. **Ribes cynosbati** L. Prickly Gooseberry. Apr.–June. Moist, often rocky, woods; scattered in IL. (Including var. *glabratum* Fern.)

4. **Ribes hirtellum** Michx. Northern Gooseberry. Apr.–June. Fens, bogs, rare; confined to the n. ¼ of IL; also Menard Co.

5. **Ribes missouriense** Nutt. Missouri Gooseberry. Mar.–May. Woods; common in the n. ⅔ of IL, occasional elsewhere.

6. **Ribes nigrum** L. Black Currant. May–June. Native to Europe; rarely escaped from cultivation; Lake Co.

7. **Ribes rubrum** L. Red Currant. May–June. Native to Europe; occasionally escaped from cultivation. *Ribes sativum* (Rchb.) Syme.

8. **Ribes triste** Pallas. May–June. Swamp Red Currant. Seeps, very rare; Lake and McHenry cos.

78. HALORAGIDACEAE—WATER MILFOIL FAMILY

1. Leaves whorled or scattered, crowded; flowers 4-numerous; fruit 4-angled .... 1. *Myriophyllum*
1. Leaves alternate, rather remote; flowers 3-merous; fruit 3-angled .... 2. *Proserpinaca*

1. **Myriophyllum** L.—Water Milfoil

1. Bracts deeply pinnatifid or coarsely toothed.
  2. Bracts deeply pinnatifid; stamens 8 .... 6. *M. verticillatum*
  2. Bracts coarsely toothed; stamens 4 .... 4. *M. pinnatum*
1. Bracts entire or finely toothed.
  3. Bracts not exceeding the flower; stems white on drying; stamens 8.
    4. Leaf segments 12 or more per side .... 5. *M. spicatum*
    4. Leaf segments up to 12 per side .... 1. *M. exalbescens*
  3. Bracts exceeding the flower; stems not white on drying; stamens 4.
    5. All leaves verticillate; bracts linear to lanceolate, up to 1.5 mm wide .... 3. *M. hippuroides*
    5. Some leaves verticillate, other leaves alternate; bracts oblanceolate to elliptic, greater than 1.5 mm wide .... 2. *M. heterophyllum*

1. **Myriophyllum exalbescens** Fern. Spiked Water Milfoil. June–Sept. Quiet waters; occasional in IL. *Myriophyllum sibiricum* Komarov.

2. **Myriophyllum heterophyllum** Michx. Water Milfoil. June–Sept. Quiet waters; occasional throughout the state.

3. **Myriophyllum hippuroides** Nutt. Mare's-tail. June–Oct. Shallow water, very rare; Cook Co.

4. **Myriophyllum pinnatum** (Walt.) BSP. Rough Water Milfoil. June–Oct. Rooted in muddy shores or in shallow waters of ponds and lakes; occasional throughout the state.

5. **Myriophyllum spicatum** L. Amazon Water Milfoil. June–Sept. Native to the tropics; adventive in quiet waters; scattered in IL.

6. **Myriophyllum verticillatum** L. var. **pectinatum** Wallr. Whorled Water Milfoil. June–Sept. Shallow water, not common; scattered throughout the state.

2. **Proserpinaca** L.—Mermaid-weed

1. **Proserpinaca palustris** L. Two varieties occur in Illinois:

a. Fruits 4–6 mm wide, the sides concave and wing-angled .... 1a. *P. palustris* var. *palustris*
a. Fruits 2–4 mm wide, the sides flat or rounded and wingless .... 1b. *P. palustris* var. *crebra*

1a. **Proserpinaca palustris** L. var. **palustris.** Mermaid-weed. July–Oct. Shallow water, shores, roadside ditches; occasional throughout the state.

1b. **Proserpinaca palustris** L. var. **crebra** Fern. Small-fruited Mermaid-weed. July–Oct. Shallow water and shores; rare in the s. ¼ of IL.

79. HAMAMELIDACEAE—WITCH HAZEL FAMILY

1. Leaves oval to obovate to suborbicular; flowers with 4 yellow petals, opening from Sept. to Nov.; fruit an obovoid capsule .... 1. *Hamamelis*
1. Leaves star-shaped; flowers without petals, opening in Apr. and May; fruit a globose, echinate head .... 2. *Liquidambar*

1. **Hamamelis** L.—Witch Hazel

1. **Hamamelis virginiana** L. Witch Hazel. Sept.–Nov. Woods; occasional in the n. ½ of IL; also Richland, St. Clair, Wabash, and White cos.

2. **Liquidambar** L.—Sweet Gum

1. **Liquidambar styraciflua** L. Sweet Gum. Apr.–May. Low woods; occasional to common in the s. ⅓ of the state; rare or absent elsewhere; often planted as a street tree.

80. HELIOTROPACEAE—HELIOTROPE FAMILY

1. **Heliotropium** L.—Heliotrope

1. Plants glabrous throughout..1. *H. curassavicum*
1. Plants pubescent.
  2. Leaves ovate or oval, petiolate; flowers bractless.
    3. Flowers white; plants hoary-pubescent.................................2. *H. europaeum*
    3. Flowers blue; plants hirsute or hispid ....3. *H. indicum*
  2. Leaves linear, sessile or nearly so; flowers subtended by bracts..........................
  ....4. *H. tenellum*

1. **Heliotropium curassavicum** L. Seaside Heliotrope. May–Sept. Native to the s. U.S. and tropical America; probably adventive in IL; bottomlands, very rare; Monroe and St. Clair cos.

2. **Heliotropium europaeum** L. European Heliotrope. June–Sept. Native to Europe; rarely escaped from cultivation; Cook and Jackson cos.

3. **Heliotropium indicum** L. Indian Heliotrope. July–Nov. Native to Asia; moist, disturbed areas; occasional in the s. ½ of IL; also Menard Co.

4. **Heliotropium tenellum** (Nutt.) Torr. Slender Heliotrope. June–Aug. Limestone ledges, rare; Monroe and Randolph cos.

81. HIPPOCASTANACEAE—HORSE CHESTNUT FAMILY

1. **Aesculus** L.—Buckeye; Horse Chestnut

1. Petals red or pink, usually glandular along the margin; capsule without prickles or with very short processes.
  2. Petals red; capsule without prickles .... 5. *A. pavia*
  2. Petals pink; capsule without prickles or with a few very short processes ....
  .... 1. *A. X bushii*
1. Petals greenish yellow or white marked with red; capsule prickly or covered with scales.
  3. Petals all similar in size and shape; stamens exserted; capsule prickly.
    4. Petals greenish yellow; leaflets mostly 5..3. *A. glabra*
    4. Petals white marked with red; leaflets mostly 7 .... 4. *A. hippocastanum*
  3. 2 of the petals very unlike the others in shape and size; stamens included; capsule covered with scales .... 2. *A. flava*

1. **Aesculus X bushii** C. K. Schneid. Hybrid Buckeye. May–June. Rocky woods, rare; Jackson Co. Hybrid between *A. glabra* Willd. and *A. pavia* L.

2. **Aesculus flava** Soland. Sweet Buckeye. May–June. Rich woods, very rare; Gallatin Co. *Aesculus octandra* Marsh.

3. **Aesculus glabra** Willd. Two varieties occur in Illinois:

a. Bark brown; leaflets green on the lower surface...................3a. *A. glabra* var. *glabra*
a. Bark whitish; leaflets white on the lower surface....... 3b. *A. glabra* var. *leucodermis*

3a. **Aesculus glabra** Willd. var. **glabra.** Ohio Buckeye. Apr.–May. Rich woods; occasional throughout most of the state, except for most of the n. cos. (Including var. *sargentii* Rehder.)

3b. **Aesculus glabra** Willd. var. **leucodermis** Sarg. Chalky Buckeye. Apr.–May. Rich woods, very rare; Jackson Co.

4. **Aesculus hippocastanum** L. Horse Chestnut. May–June. Native to Europe and Asia; frequently planted as an ornamental, but rarely escaped.

5. **Aesculus pavia** L. Red Buckeye. Apr.–May. Rich woods, not common; confined to the s. ⅙ of IL; also Richland Co. *Aesculus discolor* Pursh.

82. HIPPURIDACEAE—MARE'S-TAIL FAMILY

1. **Hippuris** L.—Mare's-tail

1. **Hippuris vulgaris** L. Mare's-tail. June–Sept. Shores or shallow water, very rare; Kane, Lake, and McHenry cos.

83. HYDRANGEACEAE—HYDRANGEA FAMILY

1. **Hydrangea** L.—Hydrangea

1. Leaves glabrous or somewhat pubescent on either or both surfaces but not gray-tomentose nor with tuberculate hairs .... 1. *H. arborescens*
1. Leaves gray-tomentose, at least on the lower surface, with tuberculate hairs .... 2. *H. cinerea*

1. **Hydrangea arborescens** L. Wild Hydrangea. June–Aug. Woods; common in the s. ⅘ of IL; also DuPage Co. (Including var. *oblonga* Torr. & Gray.)

2. **Hydrangea cinerea** Small. Southern Wild Hydrangea. June–Aug. Rocky woods, rare; Hardin, Johnson, Pope, and Wabash cos. *Hydrangea arborescens* L. var. *discolor* Ser.; *H. arborescens* L. var. *deamii* St. John.

84. HYDROPHYLLACEAE—WATERLEAF FAMILY

1. Leaves entire, with axillary spines; styles 2, free .... 2. *Hydrolea*
1. Leaves toothed, pinnately lobed, or pinnately compound, without axillary spines; styles 2-cleft.
  2. Some or all of the cauline leaves opposite.
    3. Corolla 5–8 mm long; calyx with auricles between the lobes .... 1. *Ellisia*
    3. Corolla 3 mm long; calyx without auricles between the lobes......................... .... 4. *Nemophila*
  2. All of the cauline leaves alternate.
    4. Leaves palmately lobed .... 3. *Hydrophyllum*
    4. Leaves pinnately lobed, compound, or merely toothed.
      5. Lobes of corolla fimbriate .... 5. *Phacelia*
      5. Lobes of corolla entire.
        6. Flowers up to 5 mm across.. 5. *Phacelia*
        6. Flowers 1 cm across or wider.
          7. Branches of inflorescence glandular-pubescent............... 5. *Phacelia*
          7. Branches of inflorescence lacking glandular pubescence .... 3. *Hydrophyllum*

1. **Ellisia** L.

1. **Ellisia nyctelea** L. Aunt Lucy. Apr.–June. Wet ground; occasional in the n. ¾ of IL, rare elsewhere.

2. **Hydrolea** L.

1. **Hydrolea uniflora** Raf. Hydrolea. June–Sept. Swampy woods, wet ditches, rare; confined to the s. ⅙ of IL.

3. **Hydrophyllum** L.—Waterleaf

1. Leaves palmately lobed.
  2. Calyx with a reflexed appendage between 2 adjacent lobes; branches of inflorescence bristly hairy, with the hairs over 1 mm long .... 1. *H. appendiculatum*
  2. Calyx not appendaged between the lobes or with minute teeth; branches of inflorescence pubescent, with hairs up to 1 mm long............. 2. *H. canadense*
1. Leaves pinnately lobed or compound.
  3. Leaves 9- to 13-parted; stems hirsute .... 3. *H. macrophyllum*
  3. Leaves 3- to 7-parted; stems glabrous or sparsely pubescent............................... .... 4. *H. virginianum*

1. **Hydrophyllum appendiculatum** Michx. Great Waterleaf. Apr.–July. Rich woods; occasional throughout the state.

2. **Hydrophyllum canadense** L. Broad-leaved Waterleaf. May–July. Rich woods; occasional in the s. ¾ of IL; also LaSalle Co.

3. **Hydrophyllum macrophyllum** Nutt. Large-leaved Waterleaf. May–June. Rich woods; confined to the s. ¼ of IL.

4. **Hydrophyllum virginianum** L. Virginia Waterleaf. Apr.–July. Rich woods; occasional to common throughout the state.

4. **Nemophila** Nutt.

1. **Nemophila triloba** (Raf.) Thieret. Nemophila. Apr.–May. Rich woods, very rare; Massac Co. *Nemophila aphylla* (L.) Brummits.

5. **Phacelia** Juss.—Phacelia

1. Lobes of corolla fimbriate .... 4. *P. purshii*
1. Lobes of corolla not fimbriate.
  2. Corolla less than 5 mm across .... 5. *P. ranunculacea*
  2. Corolla 5–15 mm across.
    3. Corolla 10–15 mm across; filaments subtended by conspicuous scales; stamens exserted.
      4. Leaves pinnately 3- to 5-divided .... 1. *P. bipinnatifida*
      4. Leaves pinnatifid to once-divided .... 2. *P. congesta*
    3. Corolla 5–8 mm across; filaments not subtended by scales; stamens included .... 3. *P. gilioides*

1. **Phacelia bipinnatifida** Michx. Phacelia. Apr.–June. Rich, often rocky woods; occasional in the s. ½ of IL, rare elsewhere.
2. **Phacelia congesta** Hook. Crowded Phacelia. June. Native to the sw. U.S.; adventive along a railroad; Marion Co.
3. **Phacelia gilioides** A. Brand. May–June. Dry, open areas, very rare; Calhoun and St. Clair cos.
4. **Phacelia purshii** Buckl. Miami Mist. Apr.–June. Woods, thickets; occasional in the s. ½ of IL.
5. **Phacelia ranunculacea** (Nutt.) Constance. Dwarf Phacelia. Apr.–May. Rich woods; rare in the s. ⅓ of IL; also Adams Co.

85. HYPERICACEAE—ST. JOHN'S-WORT FAMILY

1. Petals yellow or orange, convolute in bud; stamens (4–) 5–many, without glands at the base.
  2. Sepals 4; petals 4; fruit between pair of enlarged bracts .... 1. *Ascyrum*
  2. Sepals 5; petals 5; fruit without bracts .... 2. *Hypericum*
1. Petals pinkish or flesh-colored, imbricate in bud; stamens 9, with 3 large glands at the base .... 3. *Triadenum*

1. **Ascyrum** L.—St. Peter's-wort

1. Ascending shrub with linear-lanceolate leaves 2–4 (–5) mm broad .... 1. *A. hypericoides*
1. Sprawling shrub with oblong-lanceolate leaves 4 mm broad or broader .... 2. *A. multicaule*

1. **Ascyrum hypericoides** L. St. Andrew's Cross. July–Aug. Dry, sandy soil, very rare; Hancock Co. *Hypericum hypericoides* (L.) Crantz.
2. **Ascyrum multicaule** Michx. St. Andrew's Cross. July–Aug. Dry woods, on slopes and ridges; occasional in the s. ¼ of IL, absent elsewhere. *Hypericum stragulum* P. Adams & Robson.

2. **Hypericum** L.—St. John's-wort

1. Either the margins or the surface of the petals streaked or dotted with black.
  2. Only the margins of the petals streaked with black; seeds rough; stems repeatedly branched .... 13. *H. perforatum*
  2. Petals black-dotted over the entire surface; seeds smooth or nearly so; stems mostly unbranched.
    3. Petals 5–7 mm long; uppermost leaves obtuse at the tip .... 16. *H. punctatum*
    3. Petals 8–12 mm long; uppermost leaves acute at the tip .... 15. *H. pseudomaculatum*
1. Petals not dotted or streaked with black.
  4. Leaves more than 2 mm wide (if less than 2 mm wide, then the leaves 3-nerved).
    5. Petals 8 mm long or longer; stamens 20 or more.
      6. Styles 5 (rarely 4).

7. Flowers 4–7 cm across; capsule 2–3 cm long; some of the leaves partly clasping .... 17. *H. pyramidatum*
7. Flowers 1.5–3.0 cm across; capsule up to 1 cm long; leaves not clasping.
8. Capsule 7–10 mm long; stem with papery whitish bark; sepals 5–15 mm long .... 9. *H. kalmianum*
8. Capsule 3.0–6.5 mm long; stem without papery whitish bark; sepals 2–5 mm long .... 10. *H. lobocarpum*
6. Styles 3, often united to appear as a single beak.
9. Stems arising from a creeping, stoloniferous base.
10. Leaves linear-lanceolate; stamens persistent .... 1. *H. adpressum*
10. Leaves elliptic-oblong; stamens deciduous .... 6. *H. ellipticum*
9. Plants without a creeping, stoloniferous base, although underground rhizomes may be present.
11. Ovary and capsule 3-locular; stems woody nearly throughout .... 14. *H. prolificum*
11. Ovary and capsule 1-locular; stems herbaceous or slightly woody only at the base.
12. Styles free; plants virgate .... 4. *H. denticulatum*
12. Styles united into a beak; plants not virgate .... 18. *H. sphaerocarpum*
5. Petals at most only 6 mm long; stamens 5–12 (rarely 20).
13. Bracts foliaceous, resembling the foliage leaves .... 2. *H. boreale*
13. Bracts linear-setaceous, much reduced from the foliage leaves.
14. Leaves ovate to orbicular.
15. Capsules 3.0–3.5 mm long; plants much branched, not virgate........ .... 12. *H. mutilum*
15. Capsules 4 mm long or longer; plants sparingly branched, virgate .... 8. *H. gymnanthum*
14. Leaves linear to lanceolate.
16. Leaves linear, 1- to 3-nerved; some or all the sepals less than 4 mm long .... 3. *H. canadense*
16. Leaves lanceolate, 5- to 7-nerved; some or all the sepals 4 mm long or longer .... 11. *H. majus*
4. None of the leaves wider than 2 mm.
17. Leaves scalelike, at most only 3 mm long; capsule at least twice as long as the calyx .... 7. *H. gentianoides*
17. Leaves linear, mostly 6–20 mm long; capsule about as long as the calyx .... 5. *H. drummondii*

1. **Hypericum adpressum** Bart. Creeping St. John's-wort. July–Aug. Wet ground, rare; scattered in IL.

2. **Hypericum boreale** (Britt.) Bickn. Northern St. John's-wort. July–Sept. Sandy marshes, very rare; Cook and Iroquois cos.

3. **Hypericum canadense** L. Canadian St. John's-wort. July–Sept. Moist sandy flats; occasional to rare in the n. ⅕ of IL; also Gallatin and Jefferson cos.

4. **Hypericum denticulatum** Walt. St. John's-wort. June–Aug. Moist woods, gravelly hills, rare; Hardin, Jackson, Massac, and Pope cos. (Including var. *recognitum* Fern. & Schub.) *Hypericum virgatum* Lam.

5. **Hypericum drummondii** (Grev. & Hook.) Torr. & Gray. Nits-and-lice. July–Sept. Bluffs, fields, dry wooded slopes; occasional to common in the s. ½ of IL; also Handcock Co.

6. **Hypericum ellipticum** Hook. St. John's-wort. May–Aug. Roadsides, rare; Fulton and St. Clair cos.

7. **Hypericum gentianoides** (L.) BSP. Pineweed; Orangeweed. June–Oct. Sandy soil, often on dry bluffs, occasional; scattered throughout the state.

8. **Hypericum gymnanthum** Engelm. & Gray. Small St. John's-wort. June–Sept. Moist soil; rare to occasional in the n. ¾ of IL; also Gallatin Co.

9. **Hypericum kalmianum** L. Kalm's St. John's-wort. June–Aug. Calcareous sand and small bogs, locally common; Cook and Lake cos.

10. **Hypericum lobocarpum** Gattinger. St. John's-wort. June–Aug. Low woods, rare; Alexander, Massac, and Pope cos. Specimens I had reported in *Guide to the Vascular Flora of Illinois* (1986) as *H. densiflorum* Pursh are actually *H. lobocarpum.*

11. **Hypericum majus** (Gray) Britt. St. John's-wort. July–Aug. Moist ground, not common; confined to the n. ½ of IL.

12. **Hypericum mutilum** L. Dwarf St. John's-wort. July–Sept. Moist soil; occasional to common in all parts of IL. (Including var. *parviflorum* [Willd.] Fern.)

13. **Hypericum perforatum** L. Common St. John's-wort. June–Sept. Native to Europe; naturalized on roadsides, dry pastures, and fields; scattered in all parts of IL.

14. **Hypericum prolificum** L. Shrubby St. John's-wort. July–Sept. Rocky stream banks, pastures, dry woods; occasional in the s. ⅗ of IL; also Cook, DuPage, Kane, and Lake cos. *Hypericum spathulatum* (Spach) Steud.

15. **Hypericum pseudomaculatum** Bush. Large Spotted St. John's-wort. May–July. Dry woods; scattered but rare in IL. *Hypericum punctatum* Lam. var. *pseudomaculatum* (Bush) Fern.

16. **Hypericum punctatum** Lam. Spotted St. John's-wort. July–Aug. Woods, roadsides, old fields; common throughout the state; in every co.

17. **Hypericum pyramidatum** Dryander. Giant St. John's-wort. July–Aug. Banks of rivers and streams, calcareous fens; occasional in the n. ½ of the state; also Fayette, Macoupin, and St. Clair cos.

18. **Hypericum sphaerocarpum** Michx. Two varieties may be recognized in Illinois:

a. Leaves narrowly oblong to narrowly elliptic, flat, with lateral nerves evident .... 18a. *H. sphaerocarpum* var. *sphaerocarpum*

a. Leaves linear, revolute, without any apparent lateral nerves .... 18b. *H. sphaerocarpum* var. *turgidum*

18a. **Hypericum sphaerocarpum** Michx. var. **sphaerocarpum.** Round-fruited St. John's-wort. June–Sept. Rocky woods, hill prairies, roadsides; scattered throughout the state.

18b. **Hypericum sphaerocarpum** Michx. var. **turgidum** (Small) Svenson. Round-fruited St. John's-wort. June–Sept. Dry fields; scattered in the s. ½ of IL.

3. **Triadenum** Raf.—Marsh St. John's-wort

1. Leaves without punctations .... 2. *T. tubulosum*

1. Leaves punctate, at least on the lower surface.

  2. Leaves petiolate .... 4. *T. walteri*

  2. Leaves sessile.

    3. Sepals obtuse, up to 5 mm long; styles up to 1.5 mm long .... 1. *T. fraseri*

    3. Sepals acute, 5–8 mm long; styles 2–3 mm long .... 3. *T. virginicum*

1. **Triadenum fraseri** (Spach) Gl. Fraser's Marsh St. John's-wort. July–Sept. Bogs, wooded swamps, rare; confined to the n. ½ of IL. *Hypericum fraseri* Spach.

2. **Triadenum tubulosum** (Walt.) Gl. Marsh St. John's-wort. Aug.–Sept. Swampy or marshy ground in woods, rare; confined to the s. ⅕ of IL. *Hypericum tubulosum* Walt.

3. **Triadenum virginicum** (L.) Raf. Marsh St. John's-wort. July–Sept. Bogs, very rare; Lake Co. *Hypericum virginicum* L.

4. **Triadenum walteri** (J. Gmel.) Gl. Walter's Marsh St. John's-wort. July–Sept. Wooded swamps, not common; confined to the s. ⅙ of IL. *Hypericum walteri* J. Gmel.

86. ITEACEAE—ITEA FAMILY

1. **Itea** L.—Virginia Sweetspire

1. **Itea virginica** L. Virginia Sweetspire; Virginia Willow. May–June. Swampy woods, rare; confined to the s. ⅙ of IL.

87. JUGLANDACEAE—WALNUT FAMILY

1. Husk of fruits splitting, at least partially, at maturity; pith not chambered .... 1. *Carya*

1. Husk of fruits not splitting; pith chambered .... 2. *Juglans*

1. **Carya** Nutt.—Hickory

1. At least some of the leaves with 9 or more leaflets (if only 7 in *C. cordiformis,* then the buds mustard-yellow and very elongated).
  2. Nut cylindric, sweet; buds often with yellow hairs but not glandular nor scurfy .... 4. *C. illinoinensis*
  2. Nut usually compressed, bitter; buds with yellow glands or yellow scurfiness.
    3. Buds with yellow glands; leaflets strongly falcate; fruit with wings extending to the base .... 1. *C. aquatica*
    3. Buds with yellow scurfiness; leaflets not usually strongly falcate; fruit with wings extending only about halfway to the base .... 2. *C. cordiformis*
1. Leaves with (3 or) 5 or 7 leaflets (occasionally 9 in the very tomentose *C. tomentosa*).
  4. Buds with yellow scales.
    5. Rachis, at least in spring, with reddish hairs .... 9. *C. texana*
    5. Rachis without reddish hairs .... 8. *C. pallida*
  4. Buds without yellow scales.
    6. Terminal bud up to 1.2 cm long; leaflets without cilia along the margin.
      7. Husk of fruit readily splitting; bark becoming platy or even shaggy at maturity; leaflets mostly 7; outermost bud scales hairy throughout on the margins .... *C. ovalis*
      7. Husk of fruit tardily splitting or not splitting at all; bark tight at maturity; leaflets mostly 5; outermost bud scales hairy only at tip, if at all .... 3. *C. glabra*
    6. Terminal bud over 1.2 cm long; leaflets often with cilia along the margins.
      8. Branchlets densely tomentose; bark tight at maturity; kernel bitter .... 10. *C. tomentosa*
      8. Branchlets glabrous or sparsely pubescent; bark shaggy at maturity; kernel sweet.
        9. Leaflets mostly 5; twigs without conspicuous raised orange lenticels; teeth of leaflets usually with a tuft of black cilia .... 7. *C. ovata*
        9. Leaflets mostly 7; twigs with conspicuous raised orange lenticels; teeth of leaflets glabrous or with white cilia .... 5. *C. laciniosa*

1. **Carya aquatica** (Michx. f.) Nutt. Water Hickory. Mar.–Apr. Swamps and wet woods, rare; confined to the s. ⅛ of IL; also Wabash Co.

2. **Carya cordiformis** (Wangenh.) K. Koch. Bitternut Hickory. May–June. Moist or dry woods; common throughout the state; in every co.

3. **Carya glabra** (Mill.) Sweet. Two varieties may be distinguished in Illinois:

a. Husk of fruit up to 2.5 mm thick .... 3a. *C. glabra* var. *glabra*
a. Husk of fruit about 3.5 mm thick .... 3b. *C. glabra* var. *megacarpa*

3a. **Carya glabra** (Mill.) Sweet var. **glabra**. Pignut Hickory. Apr.–May. Woods; occasional to common in the s. ½ of IL, becoming rare or absent elsewhere.

3b. **Carya glabra** (Mill.) Sweet var. **megacarpa** Sarg. Pignut Hickory. Apr.–May. Woods; rare in the s. ¼ of the state, apparently absent elsewhere.

4. **Carya illinoinensis** (Wangenh.) K. Koch. Pecan. Apr.–May. Bottomland woods; occasional throughout the state except for the ne. ¼.

5. **Carya laciniosa** (Michx.) Loud. Shellbark Hickory; Kingnut Hickory; Rivernut Hickory. Apr.–May. Bottomland forests; occasional to rare in the s. ⅔ of the state.

6. **Carya ovalis** (Wangenh.) Sarg. Three varieties have been found in Illinois:

a. Leaflets not glandular-viscid beneath; husk of fruit narrowly winged or wingless.
  b. Fruit ellipsoid to nearly spherical .... 6a. *C. ovalis* var. *ovalis*
  b. Fruit obovoid .... 6b. *C. ovalis* var. *obovalis*
a. Leaflets glandular-viscid beneath; husk of fruit strongly winged .... 6c. *C. ovalis* var. *odorata*

6a. **Carya ovalis** (Wangenh.) Sarg. var. **ovalis**. Sweet Pignut Hickory; Small-fruited Hickory; False Shagbark Hickory. Apr.–June. Woods; occasional to common in the s. ½ of the state. (Including var. *obcordata* [Muhl.] Sarg.)

6b. **Carya ovalis** (Wangenh.) Sarg. var. **obovalis** Sarg. Sweet Pignut Hickory. Apr.–May. Woods; occasional in the s. ½ of the state.

6c. **Carya ovalis** (Wangenh.) Sarg. var. **odorata** (Marsh.) Sarg. Sweet-scented Pignut Hickory. Apr.–May. Woods; rare in the s. tip of the state.

7. **Carya ovata** (Mill.) K. Koch. Three varieties may be recognized in Illinois:

a. Some of all of the leaflets 5 cm broad or broader.
  b. Fruit 3.5–6.0 cm long .... 7a. *C. ovata* var. *ovata*
  b. Fruit 1.5–2.0 cm long .... 7c. *C. ovata* var. *nuttallii*
a. None of the leaflets 5 cm broad or broader .... 7b. *C. ovata* var. *fraxinifolia*

7a. **Carya ovata** (Mill.) K. Koch var. **ovata.** Shagbark Hickory. Apr.–May. Rich woods, upland woods; occasional to common in IL; in every co. (Including var. *pubescens* Sarg.)

7b. **Carya ovata** (Mill.) K. Koch var. **fraxinifolia** Sarg. Ash-leaved Shagbark Hickory. Apr. Woods; apparently confined to the s. ½ of IL.

7c. **Carya ovata** (Mill.) K. Koch var. **nuttallii** Sarg. Small Shagbark Hickory. Apr. Rich woods; rare in the s. tip of IL.

8. **Carya pallida** (Ashe) Engl. & Graebn. Pale Hickory. Apr.–May. Wooded slopes, rare; Alexander, Jackson, and Union cos.

9. **Carya texana** Buckl. Black Hickory; Red Hickory. Apr.–May. Dry woods and bluffs; occasional to rare in the s. ½ of the state, rare elsewhere. (Including var. *arkansana* [Sarg.] Sarg. and var. *villosa* [Sarg.] Little.)

10. **Carya tomentosa** (Poir.) Nutt. Mockernut Hickory. May–June. Moist or dry woods; occasional in the s. ¾ of IL, rare elsewhere.

Several hybrid hickories have been reported from Illinois. Their hybrid status is due primarily to the appearance of the plants rather than to any scientific evidence that they are hybrids. The hybrids that have been found in Illinois are listed below:

Carya X lecontei Little (*C. aquatica X C. illinoinensis*) Union Co.
Carya X nussbaumeri Sarg. (*C. illinoinensis X C. laciniosa*) Union Co.
Carya X schneckii Sarg. (*C. glabra X C. illinoinensis*) Wabash Co.

2. **Juglans** L.—Walnut

1. Branchlets and husk of nuts downy; fruit ellipsoid .... 1. *J. cinerea*
1. Branchlets and husk of nuts not downy; fruit globose (very rarely ellipsoid) .... 2. *J. nigra*

1. **Juglans cinerea** L. Butternut; White Walnut. Apr.–May. Rich woods; occasional but becoming very rare because of disease; throughout the state.

2. **Juglans nigra** L. Black Walnut. Apr.–May. Woods; common throughout the state. (Including f. *oblonga* [Marsh.] Fern.)

88. LAMIACEAE—MINT FAMILY

1. Lobes of corolla almost equal, not bilabiate.
  2. All flowers borne in the axils of the leaves.
    3. Flowers white; fertile stamens 2 .... 18. *Lycopus*
    3. Flowers pink, purple, or blue; fertile stamens 4.
      4. Leaves entire; flowers blue .... 15. *Isanthus*
      4. Leaves toothed; flowers pink or purple .... 21. *Mentha*
  2. Some or all the flowers in terminal inflorescences.
    5. Leaves entire or nearly so .... 37. *Trichostema*
    5. Leaves toothed.
      6. Corolla 4-lobed .... 21. *Mentha*
      6. Corolla 5-lobed .... 26. *Perilla*
1. Lobes of corolla bilabiate, often very unequal.
  7. Ovary 4-lobed, the style not basal.
    8. Upper lip of corolla seemingly absent; leaves petiolate; stems at least 2.5 cm tall; creeping stolons absent .... 35. *Teucrium*
    8. Upper lip of corolla very short; leaves, or the upper ones, sessile; stems less than 2.5 cm tall; creeping stolons present .... 2. *Ajuga*
  7. Ovary deeply 4-lobed, the style basal.
    9. Calyx crested on the upper side.
      10. Calyx bilobed, the tips rounded .... 32. *Scutellaria*

10. Calyx 5-lobed, the tips pointed .... 24. *Ocimum*

9. Calyx not crested on the upper side.

11. Stamens completely included within the corolla tube; calyx with 5 or 10 sharp, clawlike teeth; plants white-woolly .... 19. *Marrubium*

11. Stamens exserted, or at least longer than the corolla tube, and arching up beneath the upper lip of the corolla; calyx teeth not clawlike, although they may be spiny in *Leonurus;* plants usually not white-woolly.

12. Anther bearing stamens 2.

13. Stamens exserted beyond the corolla.

14. Flowers in axillary clusters, cymes, or terminal glomerules; flowers not yellow or, if yellow, then purple-spotted.

15. Calyx very unequally lobed .... 4. *Blephilia*

15. Calyx equally 5-lobed.

16. Corolla up to 1 cm long; flowers in loose cymes .... 10. *Cunila*

16. Corolla more than 1 cm long; flowers in dense capitate clusters .... 22. *Monarda*

14. Flowers in terminal panicles; flowers light yellow with purple spots .... 9. *Collinsonia*

13. Stamens not exserted beyond the corolla, usually arching under the upper lip of the corolla.

17. Bracts longer than the cluster of flowers they subtend .... 14. *Hedeoma*

17. Bracts shorter than the cluster of flowers they subtend.

18. Flowers mostly up to 10 in a rather loose whorl ..... 30. *Salvia*

18. Flowers mostly more than 10 in dense whorls..... 4. *Blephilia*

12. Anther bearing stamens 4.

19. Leaves entire or nearly so.

20. Plants forming dense mats; leaves obtuse at apex .... 36. *Thymus*

20. Plants upright, not forming dense mats; leaves acute to acuminate at apex (sometimes more or less obtuse in *Origanum*).

21. Flowers subtended by large, purple bracts....... 25. *Origanum*

21. Flowers not subtended by large, purple bracts.

22. Leaves ovate; upper lip of corolla 4-lobed ...... 24. *Ocimum*

22. Leaves linear to lanceolate; upper lip of corolla 2-lobed, emarginate, or entire.

23. Inflorescence in dense terminal heads or cymes.......... .... 29. *Pycnanthemum*

23. Inflorescence with 1–4 flowers in the axils of the leaves.

24. Perennial with stolons; calyx glabrous, with the calyx tube much longer than the teeth; corolla 8–15 mm long .... 8. *Clinopodium*

24. Annual; calyx pubescent, with the calyx tube about the same length as the teeth; corolla 5–7 mm long .... 31. *Satureja*

19. Leaves serrate, crenate, dentate, or lobed.

25. Some or all of the cauline leaves cordate.

26. Flowers yellow, greenish yellow, white, or white dotted with purple.

27. Stems glabrous or nearly so .... 1. *Agastache*

27. Stems pubescent.

28. Stems densely canescent or puberulent; calyx 5-toothed.

29. Stems densely canescent; corolla white, dotted with purple .... 23. *Nepeta*

29. Stems puberulent; corolla bright yellow.................. .... 16. *Lamium*

28. Stems more or less hirsute; calyx 4-parted .... 34. *Synandra*
26. Flowers pink, purple, rose-purple, or blue.
30. Stems creeping or trailing .... 13. *Glechoma*
30. Stems erect.
31. All flowers borne in axillary clusters.
32. Calyx lobes as long as or longer than the calyx tube; anthers pubescent .... 16. *Lamium*
32. Calyx lobes much shorter than the calyx tube; anthers glabrous.. 3. *Ballota*
31. Some or all of the flowers in terminal inflorescences.
33. Lower leaf surface densely short-white-hairy.......... .... 1. *Agastache*
33. Lower leaf surface glabrous or pubescent but not short-white-hairy.
34. Calyx 15-nerved; inflorescence crowded, the flower clusters touching .... 1. *Agastache*
34. Calyx 5- to 10-nerved; inflorescence interrupted, most or all of the flower clusters separated .... 33. *Stachys*
25. None of the cauline leaves cordate.
35. All flowers borne in axillary clusters.
36. Shrubs; upper lip of corolla fringed ....6. *Caryopteris*
36. Herbs, or rarely with a slightly woody base; none of the petals fringed.
37. Flowers and fruits on pedicels 1–5 mm long.
38. Flowers yellow, becoming white .... 20. *Melissa*
38. Flowers pink or purple .... 8. *Clinopodium*
37. Flowers and fruits sessile.
39. Calyx regular, not 2-lipped; plants lemon-scented.. .... 3. *Ballota*
39. Calyx 2-lipped; plants not lemon-scented.
40. Upper lip of corolla woolly.
41. Leaves lobed .... 17. *Leonurus*
41. Leaves unlobed.............................7. *Chaiturus*
40. Upper lip of corolla not woolly.
42. Midnerve of calyx lobe extended as a spine .... 12. *Galeopsis*
42. Midnerve of calyx lobe not extended as a spine ....5. *Calamintha*
35. Some or all of the flowers borne in terminal inflorescences.
43. Calyx teeth spinescent.................................... 17. *Leonurus*
43. Calyx teeth not spinescent.
44. Flowers and fruits pedicellate.
45. Calyx teeth more or less equal........ 27. *Physostegia*
45. Calyx 2-lipped.
46. Stamens 4; leaves not purple-tinged; flowers in glomerules.................................. 8. *Clinopodium*
46. Stamens 2; leaves usually purple or purple-tinged; flowers in elongated racemes................ ....26. *Perilla*
44. Flowers and fruits sessile.
47. Calyx with lobes more or less equal, not bilabiate.
48. Flowers in dense heads as broad as or broader than long.............................. 29. *Pycnanthemum*
48. Flowers in dense or loose inflorescences longer than broad.
49. Flowers much longer than the subtending bracts; stems glabrous; leaves sessile.......... .... 27. *Physostegia*

49. Flowers not longer than or only a little longer than the subtending bracts; stems pubescent or, if glabrous, the leaves petiolate.. 33. *Stachys*

47. Calyx bilabiate.

50. Bracts spinescent.................. 11. *Dracocephalum*

50. Bracts not spinescent.

51. Flowers in heads as broad as or broader than long; corolla less than 1 cm long.......... .... 29. *Pycnanthemum*

51. Flowers in heads longer than broad; corolla 1 cm long or longer..................... 28. *Prunella*

1. **Agastache** Clayton—Giant Hyssop

1. Corolla yellow; calyx teeth obtuse to subacute; stems glabrous or nearly so............ .... 1. *A. nepetoides*

1. Corolla purple or blue; calyx teeth acute to acuminate; stems pubescent .... 2. *A. scrophulariaefolia*

1. **Agastache nepetoides** (L.) Ktze. Yellow Giant Hyssop. July–Sept. Open woods; occasional throughout the state.

2. **Agastache scrophulariaefolia** (Willd.) Ktze. Purple Giant Hyssop. July–Sept. Open woods; occasional in the n. ⅔ of IL; also Wabash Co. (Including var. *mollis* [Fern.] Heller.)

2. **Ajuga** L.—Bugleweed

1. Leaves densely soft-pubescent; plants tufted, without stolons .... 1. *A. genevensis*

1. Leaves glabrous or slightly pubescent; plants stoloniferous.................. 2. *A. reptans*

1. **Ajuga genevensis** L. Geneva Bugleweed. May–July. Native to Europe; occasionally cultivated, seldom escaped; Cook, DuPage, Jackson, Lake, and McHenry cos.

2. **Ajuga reptans** L. Carpet Bugleweed. May–July. Native to Europe; frequently cultivated but infrequently escaped; mostly in the ne. cos.

3. **Ballota** L.—Black Horehound

1. **Ballota nigra** L. Black Horehound. June–Sept. Native of the Mediterranean area and Asia; rarely escaped from cultivation; DuPage and Will cos.

4. **Blephilia** Raf.—Pagoda Plant

1. At least the upper leaves cuneate to the nearly sessile base..................... 1. *B. ciliata*

1. Upper leaves petiolate, rounded at the base.. 2. *B. hirsuta*

1. **Blephilia ciliata** (L.) Bernh. Pagoda Plant. May–Aug. Open woods, fields, prairies; occasional throughout the state.

2. **Blephilia hirsuta** (Pursh) Bernh. Pagoda Plant. May–Sept. Rich woods; occasional throughout the state.

5. **Calamintha** P. Mill.—Calamint

1. **Calamintha nepeta** (L.) Savi. Lesser Calamint; Basil-thyme. June–Aug. Native to Europe; rarely escaped into disturbed soil; Jackson Co. *Satureja calamintha* (L.) Scheele.

6. **Cariopteris** L.

1. **Cariopteris incana** (L.) L. Blue Spiraea. July–Aug. Native to Europe; rarely escaped from cultivation; Jackson Co.

7. **Chaiturus** Willd.—Lion's-tail

1. **Chaiturus marrubiastrum** (L.) Reichenb. Lion's-tail. June–Aug. Native to Europe and Asia; occasionally adventive in disturbed soil in the n. ⅔ of IL. *Leonurus marrubiastrum* L.

8. **Clinopodium** L.—Wild Basil

1. Leaves entire .... 1. *C. arkansanum*
1. Leaves toothed .... 2. *C. vulgare*

1. **Clinopodium arkansanum** (Nutt.) House. Low Calamint; Limestone Calamint. May–Oct. Calcareous fens, rocky soil, sand flats; occasional in the n. ¼ of the state. *Calamintha arkansana* (Nutt.) Shinners.

2. **Clinopodium vulgare** L. Wild Basil. June–Sept. Woods, rare; introduced in the ne. cos. *Satureja vulgaris* (L.) Fritsch var. *neogaea* Fern.

9. **Collinsonia** L.—Richweed

1. **Collinsonia canadensis** L. Richweed. July–Sept. Rocky woods; occasional in the s. tip of the state; also Champaign, Clark, Crawford, and Edgar cos.

10. **Cunila** L.—Dittany

1. **Cunila origanoides** (L.) Britt. Dittany. July–Nov. Dry woods, sandstone cliffs; occasional to common in the s. ½ of the state.

11. **Dracocephalum** L.—Dragonhead

1. **Dracocephalum parviflorum** Nutt. American Dragonhead. May–Aug. Native to the w. U.S. Dry soil, along railroads, not common; confined to the n. ½ of the state.

12. **Galeopsis** L.—Hemp Nettle

1. Stems appressed-pubescent at the nodes and not swollen beneath the nodes .... 1. *G. ladanum*
1. Stems bristly pubescent at the nodes and swollen beneath the nodes .... 2. *G. tetrahit*

1. **Galeopsis ladanum** L. Red Hemp Nettle. June–Sept. Native to Europe; rarely adventive in waste areas; Cook Co.

2. **Galeopsis tetrahit** L. Common Hemp Nettle. June–Sept. Native to Europe and Asia; naturalized in waste places; Boone, Cook, DeKalb, DuPage, Henderson, Kane, and Lake cos. *Galeopsis tetrahit* L. var. *bifida* (Buenn.) Les & Court.

13. **Glechoma** L.—Ground Ivy

1. **Glechoma hederacea** L. Two varieties may be distinguished in Illinois:

a. Corolla more than 1.5 cm long .... 1a. *G. hederacea* var. *hederacea*
a. Corolla up to 1.5 cm long .... 1b. *G. hederacea* var. *micrantha*

1a. **Glechoma hederacea** L. var. **hederacea.** Ground Ivy; Creeping Jenny; Gill-all-over-the-ground. Apr.–July. Native to Europe; naturalized in moist soil, not common; DuPage, Lawrence, and Peoria cos.

1b. **Glechoma hederacea** L. var. **micrantha** Moricand. Ground Ivy. Apr.–July. Native to Europe; naturalized in moist soil and lawns; occasional throughout the state, becoming more abundant northward.

14. **Hedeoma** Pers.—Pennyroyal

1. Leaves linear, entire, sessile; plants faintly aromatic .... 1. *H. hispida*
1. Leaves oblong-ovate to elliptic, more or less toothed, petiolate; plants strongly aromatic .... 2. *H. pulegioides*

1. **Hedeoma hispida** Pursh. Rough Pennyroyal. May–July. Rocky woods, prairies; occasional in the n. ½ of the state, less common in the s. ½.

2. **Hedeoma pulegioides** (L.) Pers. American Pennyroyal. July–Sept. Rocky woods, fields, roadsides; occasional to common throughout the state.

15. **Isanthus** Michx.—False Pennyroyal

1. **Isanthus brachiatus** (L.) BSP. False Pennyroyal. July–Oct. Rocky woods, prairies; occasional throughout the state. *Trichostema brachiatum* L.

16. **Lamium** L.—Dead Nettle

1. Flowers bright yellow .... 2. *L. galeobdolon*
1. Flowers pink or purple.
  2. At least the upper leaves sessile .... 1. *L. amplexicaule*
  2. All leaves petiolate.
    3. Flowers more than 15 mm long .... 3. *L. maculatum*
    3. Flowers up to 15 mm long .... 4. *L. purpureum*

1. **Lamium amplexicaule** L. Henbit. Feb.–Nov. Native to Europe, Asia, and Africa; naturalized in disturbed soil; occasional to common throughout IL. (Including f. *albiflorum* D. M. Moore.)

2. **Lamium galeobdolon** (L.) L. Golden Dead Nettle; Yellow Archangel. May–July. Native to Europe; rarely escaped from cultivation; DuPage Co. (Wilhelm, 2010.) *Lamiastrum galeobdolon* (L.) Herend. & Polatschek.

3. **Lamium maculatum** L. Spotted Dead Nettle. May–Aug. Native to Europe; adventive in disturbed soil; Cook, DuPage, Kane, and McLean cos.

4. **Lamium purpureum** L. Purple Dead Nettle. Apr.–Oct. Native to Europe and Asia; naturalized in waste ground; occasional to common in the s. ½ of IL, uncommon in the n. ½.

17. **Leonurus** L.—Motherwort

1. Calyx 5-ribbed and 5-angled .... 1. *L. cardiaca*
1. Calyx 10-ribbed, scarcely angled .... 2. *L. sibiricus*

1. **Leonurus cardiaca** L. Motherwort. May–Aug. Native to Europe and Asia; naturalized in disturbed, shaded areas; occasional in the n. ½ of IL, less common in the s. ½.

2. **Leonurus sibiricus** L. Siberian Motherwort. May–Sept. Native to Asia; rarely adventive in disturbed soil; Cook and DuPage cos.

18. **Lycopus** L.—Water Horehound; Bugleweed

1. Some or all the leaves pinnatifid to pinnate.
  2. Leaves lanceolate to narrowly oblong; nutlets up to 1 mm broad .... 1. *L. americanus*
  2. Leaves ovate to ovate-oblong; nutlets 1 mm broad or broader .... 4. *L. europaeus*
1. All leaves merely toothed, never pinnate nor pinnatifid.
  3. Calyx teeth broadly triangular, obtuse to subacute, up to 1 mm long.
    4. Stem puberulent; base of plant not tuberous; stamens more or less included within the corolla .... 7. *L. virginicus*
    4. Stem glabrous or nearly so; base of plant tuberous; stamens exserted .... 6. *L. uniflorus*
  3. Calyx teeth narrowly triangular, acute to subulate, 1–2 mm long.
    5. Middle and lower leaves petiolate; base of plant not tuberous .... 5. *L. rubellus*
    5. Middle and lower leaves sessile; base of plant tuberous.
      6. Bracts and calyx nearly equal in size; leaves scabrous .... 3. *L. asper*
      6. Bracts minute, much smaller than the calyx; leaves more or less smooth. .... 2. *L. amplectens*

1. **Lycopus americanus** Muhl. Common Water Horehound. June–Oct. Wet ground, common; in every co.

2. **Lycopus amplectens** Raf. Bugleweed. Aug.–Sept. Sandy soil, very rare; Mason Co.

3. **Lycopus asper** E. Greene. Rough Water Horehound. July–Aug. Usually in stagnant water, rare; confined to the n. ¼ of IL.

4. **Lycopus europaeus** L. Water Horehound. Aug.–Oct. Native to Europe; rarely adventive in moist, disturbed areas; DuPage, Kane, Kendall, and McHenry cos.

5. **Lycopus rubellus** Moench. Stalked Water Horehound; Reddish Water Horehound. July–Oct. Low woods, wet meadows, not common; scattered throughout the state. (Including var. *arkansanus* [Fresn.] Benner.)

6. **Lycopus uniflorus** Michx. Northern Bugleweed. Aug.–Sept. Marshes, calcareous fens, around lakes; occasional in the n. ½ of IL; also Clark Co.

7. **Lycopus virginicus** L. Bugleweed. July–Oct. Wet ground, occasional; scattered throughout the state.

19. **Marrubium** L.—Horehound

1. **Marrubium vulgare** L. Common Horehound. May–Sept. Native to Europe and Asia; naturalized in fields and pastures and along roads; occasional to common throughout the state.

20. **Melissa** L.—Balm

1. **Melissa officinalis** L. Balm. June–Aug. Native to Europe; rarely adventive in IL; DuPage, Jackson, Lake, Lawrence, Sangamon, and Wabash cos.

21. **Mentha** L.—Mint

1. Flowers in the axils of the leaves, with the bracts usually longer than the flowers.
  2. Calyx tube pubescent, at least along the veins.
    3. Leaves broadly ovate, rounded to truncate to cordate at base, rugose, densely hairy; stamens sterile; foul-smelling plant............11. *M. X verticillata*
    3. Leaves ovate to lance-ovate, cuneate at base, not rugose, sparsely hairy; stamens fertile; pleasant-smelling plant ....2. *M. arvensis*
  2. Calyx tube glabrous...5. *M. X gentilis*
1. At least some of the flowers in terminal inflorescences, the bracts often shorter than the flowers.
  4. Inflorescence a single terminal spike; calyx glabrous; plant strongly lemon-scented..3. *M. X citrata*
  4. Inflorescence of 2 or more spikes; calyx pubescent, at least on the lobes; plants without a lemon scent.
    5. Leaves petiolate.
      6. Leaves lacerate, crisped ....4. *M. crispa*
      6. Leaves sharply serrate, not crisped.
        7. Plants glabrous; leaves lance-ovate to oblong-ovate; calyx glabrous ... .7. *M. X piperita*
        7. Plants pubescent; leaves broadly oval to suborbicular; calyx pubescent...1. *M. aquatica*
    5. Leaves sessile or nearly so.
      8. Bracts and calyx green, glabrous or sparsely pubescent.
        9. Plants glabrous or sparsely hairy...9. *M. spicata*
        9. Plants downy-pubescent.. 8. *M. X rotundifolia*
      8. Bracts and calyx white-hairy, the hairs usually dense.
        10. Leaves often cordate-clasping, rugose ....10. *M. suaveolens*
        10. Leaves rounded at base but not cordate-clasping, not rugose.
          11. Leaves and stems villous... 12. *M. X villosa*
          11. Leaves and stems glabrous or somewhat pubescent, but not villous..6. *M. longifolia*

1. **Mentha aquatica** L. Mint. July–Aug. Moist ground; DuPage, Jackson, and LaSalle cos.

2. **Mentha arvensis** L. Two varieties may be distinguished in Illinois:

a. Leaves rounded at base, the petioles longer than the clusters of flowers................. .... 2a. *M. arvensis* var. *arvensis*
a. Leaves cuneate at base, the petioles equaling or shorter than the cluster of flowers .... 2b. *M. arvensis* var. *villosa*

2a. **Mentha arvensis** L. var. **arvensis.** Field Mint. July–Sept. Native to Europe and Asia; rarely adventive in disturbed areas; Crawford Co.

2b. **Mentha arvensis** L. var. **villosa** (Benth.) S. R. Stewart. Field Mint. July–Sept. Marshes, low ground; occasional to common in most of the state.

3. **Mentha X citrata** Ehrh. Lemon Mint; Bergamot Mint. July–Oct. Native to Europe; not commonly adventive in moist ground. Suggested to be a hybrid between *M. aquatica* L. and *M. piperita* L.

4. **Mentha crispa** L. Curly Mint. July–Sept. Native to Europe; adventive in disturbed soil; Cook, Jackson, Lake, and Peoria cos.

5. **Mentha X gentilis** L. Little-leaved Mint; Red Mint. July–Oct. Native to Europe; occasionally adventive in IL. Reputed to be the hybrid between *M. arvensis* L. and *M. spicata* L. This hybrid has bracts nearly or quite as large as the foliage leaves. It is commonly called red mint. Another hybrid with presumably the same parentage has bracts much smaller than the foliage leaves, and it looks very different. It has been called *M. X cardiaca* Gerarde, and it is also known from IL.

6. **Mentha longifolia** L. Mint. July–Aug. Moist, disturbed soil; DuPage Co.

7. **Mentha X piperita** L. Peppermint. June–Oct. Native to Europe; occasionally adventive in moist waste ground. Reputed to be a hybrid between *M. aquatica* L. and *M. spicata* L.

8. **Mentha X rotundifolia** (L.) Huds. Apple Mint. June–Sept. Native to Europe; rarely adventive in IL; Fayette, Hancock, and Lake cos. Reputed to be the hybrid between *M. longifolia* L. and *M. suaveolens* Ehrh.

9. **Mentha spicata** L. Spearmint. June–Oct. Native to Europe; frequently planted and occasionally escaped into disturbed areas; throughout the state.

10. **Mentha suaveolens** Ehrh. Sweet Apple Mint. July–Aug. Native to Europe; rarely adventive into waste ground; Cook, DuPage, Kankakee, and Lake cos.

11. **Mentha X verticillata** L. Whorled Mint. July–Aug. Native to Europe; rarely adventive into disturbed low ground; Cook Co. Reputed to be the hybrid between *M. aquatica* L. and *M. arvensis* L.

12. **Mentha X villosa** Huds. Foxtail Mint. July–Oct. Native to Europe; adventive and scattered in IL. Reputed to be the hybrid between *M. spicata* L. and *M. suaveolens* Ehrh. *Mentha X alopecuroides* Hull.

22. **Monarda** L.—Wild Bergamot

1. Only 1 head of flowers on each stem or stem-branch.
  2. Leaves sessile or nearly so .... 1. *M. bradburiana*
  2. Leaves on petioles 5 mm long or longer.
    3. Corolla glabrous; tip of upper lip of corolla beardless; bracts bright red or whitish; corolla bright red or white to greenish to flesh-colored.
      4. Bracts bright red; corolla bright red, 3.0–4.5 cm long............. 4. *M. didyma*
      4. Bracts whitish; corolla white to greenish to flesh-colored, 1.5–3.0 cm long .... 3. *M. clinopodia*
    3. Corolla pubescent and lavender, pink, rose, or purple; tip of upper lip of corolla bearded; bracts green and often pink-tinged or rose or purple.
      5. Corolla lavender or pink (deep purple to crimson in 1 variety); leaves firm, narrowly deltate-ovate.. 5. *M. fistulosa*
      5. Corolla rose or purple; leaves thin, broadly deltate-ovate ....6. *M. X media*
1. Heads of flowers 2 or more on each stem or stem-branch.
  6. Corolla yellow, with purple spots; calyx teeth triangular.................7. *M. punctata*
  6. Corolla white or pink; calyx teeth subulate .... 2. *M. citriodora*

1. **Monarda bradburiana** Beck. Bee Balm; Horse Balm. Apr.–June. Dry woods, bluffs, roadsides; occasional in the s. 3/5 of the state; also DuPage Co.

2. **Monarda citriodora** Cerv. Lemon Mint. May–Aug. Native to the w. U.S.; rarely escaped from cultivation; Cook Co.

3. **Monarda clinopodia** L. Bee Balm. June–July. Woods; scattered in the s. 3/4 of IL; also Grundy, Kendall, and LaSalle cos.

4. **Monarda didyma** L. Oswego Tea. June–Sept. Native to the e. U.S.; occasionally escaped from cultivation into woodlands; Cass, Cook, DuPage, Hancock, Lake, Macon, McDonough, Shelby, and Wabash cos.

5. **Monarda fistulosa** L. Three varieties have been found in Illinois:

a. Corolla deep purple to crimson; middle lobe of lower lip of corolla 4–6 mm long; bracts pink-tinged.. 5c. *M. fistulosa* var. *rubra*
a. Corolla pink or lavender; middle lobe of lower lip of corolla 2–4 mm long; bracts green except for a reddish midvein.
  b. Leaves thin, green; lower surface of leaves with hairs 1–3 mm long ....5a. *M. fistulosa* var. *fistulosa*

b. Leaves firm, pale green; lower surface of leaves canescent, the hairs less than 1 mm long ....5b. *M. fistulosa* var. *mollis*

5a. **Monarda fistulosa** L. var. **fistulosa.** Wild Bergamot. May–Aug. Dry woods, fields, prairies, roadsides; common throughout the state. An unnamed variety with a strong lemon scent has been found several times in Illinois.

5b. **Monarda fistulosa** L. var. **mollis** (L.) Benth. Canescent Wild Bergamot. May–Aug. Dry woods, prairies, fields; scattered in IL.

5c. **Monarda fistulosa** L. var. **rubra** Gray. Red Wild Bergamot. May–Aug. Fields, not common; scattered in IL.

6. **Monarda X media** Willd. Hybrid Monarda. May–Aug. Dry woods, rare; Henderson, Tazewell, and Union cos. This is the reputed hybrid between *M. fistulosa* L. and *M. clinopodia* L.

7. **Monarda punctata** L. Two varieties occur in Illinois:

a. Stems canescent, with short, curved hairs; teeth of calyx acute ....7a. *M. punctata* var. *occidentalis*

a. Stems villous or pilose; teeth of calyx acuminate.......7b. *M. punctata* var. *villicaulis*

7a. **Monarda punctata** L. var. **occidentalis** (Epling) Palmer & Steyerm. Horsemint. June–Oct. Sandy soil; Madison and St. Clair cos. *Monarda punctata* L. var. *lasiodonta* Gray.

7b. **Monarda punctata** L. var. **villicaulis** (Pennell) Shinners. Horsemint. July–Oct. Sandy fields and woods, dunes, prairies; occasional in the n. ½ of the state, rare elsewhere.

23. **Nepeta** L.—Catnip

1. **Nepeta cataria** L. Catnip. June–Sept. Native to Europe; naturalized in fields, open woods, along roads and railroads; occasional to common throughout the state.

24. **Ocimum** L.—Basil

1. **Ocimum basilicum** L. Basil. Aug.–Sept. Native to Africa and Asia; occasionally cultivated but rarely escaped into disturbed soil; Hancock, Jackson, and Sangamon cos.

25. **Origanum** L.—Oregano

1. **Origanum vulgare** L. Oregano. Aug.–Sept. Native to Europe and Asia; rarely adventive in disturbed soil; Cook, DuPage and Lake cos.

26. **Perilla** L.—Beefsteak Plant

1. **Perilla frutescens** (L.) Britt. Two varieties may be distinguished in Illinois:

a. Leaves usually purple-tinged only on lower surface, not crisped along the margin .... 1a. *P. frutescens* var. *frutescens*

a. Leaves purple throughout, crisped along the margin......1b. *P. frutescens* var. *crispa*

1a. **Perilla frutescens** (L.) Britt. var. **frutescens.** Beefsteak Plant. Aug.–Oct. Native to Asia; naturalized in disturbed habitats, particularly in damp areas; common in the s. ¼ of IL, less common northward.

1b. **Perilla frutescens** (L.) Britt. var. **crispa** (Benth.) Deane. Purple Beefsteak Plant. Aug.–Oct. Native to Asia; rarely escaped from cultivation; Jackson and Union cos.

27. **Physostegia** Benth.—False Dragonhead

1. Leaves, at least the upper, broadly rounded at base, the teeth rarely more than 1 mm long; corolla rarely longer than 1.5 cm .... 2. *P. parviflora*

1. Leaves cuneate or subcuneate at base, the teeth regularly more than 1 mm long; corolla 1.5–3.0 cm long.

2. Upper leaves abruptly reduced in size; spike appearing pedunculate.

3. Broadest leaves never exceeding a width of 1 cm; flowers remote .... 1. *P. angustifolia*

3. At least some of the leaves more than 1 cm broad; some of the flowers overlapping .... 4. *P. virginiana*
2. Upper leaves gradually reduced in size; spike appearing sessile..... 3. *P. speciosa*

1. **Physostegia angustifolia** Fern. Narrow-leaved False Dragonhead. June–Sept. Low prairies; occasional in the n. ¾ of IL.

2. **Physostegia parviflora** Nutt. Small-flowered False Dragonhead. July–Oct. Moist prairies, rare; Kane, Lake, and Lee cos.

3. **Physostegia speciosa** (Sweet) Sweet. False Dragonhead. May–Sept. Low prairies; occasional throughout the state.

4. **Physostegia virginiana** (L.) Benth. False Dragonhead; Obedience Plant. May–Sept. Moist soil, particularly prairies; occasional throughout the state. (Including f. *candida* Benke.)

28. **Prunella** L.—Self-heal

1. **Prunella vulgaris** L. Two varieties may be recognized in Illinois:

a. At least the upper leaves rounded at the base .... 1a. *P. vulgaris* var. *vulgaris*
a. At least the upper leaves cuneate at the base .... 1b. *P. vulgaris* var. *elongata*

1a. **Prunella vulgaris** L. var. **vulgaris.** Self-heal; Heal-all. May–Sept. Native to Europe; naturalized in lawns, fields, and waste ground; occasional throughout IL.

1b. **Prunella vulgaris** L. var. **elongata** Benth. Self-heal; Heal-all. May–Sept. Disturbed woods, pastures, meadows, common; in every co.

29. **Pycnanthemum** Michx.—Mountain Mint

1. Inflorescence loose, with the branchlets of the flowering clusters evident.
2. Teeth of calyx not bristle-tipped .... 1. *P. albescens*
2. Teeth of calyx bristle-tipped.
3. Teeth of calyx obtuse to acute to acuminate, less than ½ as long as the tube .... 2. *P. incanum*
3. Teeth of calyx acuminate to attenuate, more than ½ as long as the tube.
4. Nutlets smooth or with a few short hairs at tip, 0.5–1.3 mm long .... 3. *P. loomisii*
4. Nutlets rugose or pitted, densely hairy at tip, 1.2–1.5 mm long .... 5. *P. pycnanthemoides*
1. Inflorescence dense, the branchlets of the flowering clusters not evident.
5. Calyx with uniform teeth up to 1 mm long; leaves not more than 3 times longer than broad .... 4. *P. muticum*
5. Calyx bilabiate, the lower teeth 1–2 mm long; leaves usually at least 3 times longer than broad.
6. Stems glabrous; largest leaves up to 5.0 (–5.5) mm broad .... 6. *P. tenuifolium*
6. Stems pubescent, at least on the angles; largest leaves (5–) 6–40 mm broad.
7. Outermost bracts pubescent on the upper surface.......... 8. *P. verticillatum*
7. Outermost bracts glabrous or nearly so on the upper surface.
8. Calyx teeth pilose throughout; leaves short-petiolate.......... 7. *P. torrei*
8. Calyx teeth pubescent only at the tip; leaves sessile .... 9. *P. virginianum*

1. **Pycnanthemum albescens** Torr. & Gray. White Mountain Mint. July–Sept. Cherty slopes, very rare; Union Co.

2. **Pycnanthemum incanum** (L.) Michx. Gray Mountain Mint. July–Sept. Dry woods; confined to the s. ⅓ of IL.

3. **Pycnanthemum loomisii** Nutt. Loomis' Mountain Mint. July–Sept. Dry woods, very rare; Union Co.

4. **Pycnanthemum muticum** (Michx.) Pers. Mountain Mint. July–Sept. Low woods, rare; Henderson and Wabash cos.

5. **Pycnanthemum pycnanthemoides** (Leavenw.) Fern. Mountain Mint. July–Sept. Open woods; confined to the s. ⅕ of IL.

6. **Pycnanthemum tenuifolium** Schrad. Slender Mountain Mint. June–Sept. Woods, fields, prairies; occasional to common throughout the state.

7. **Pycnanthemum torrei** Benth. Torrey's Mountain Mint. June–Oct. Dry woods, very rare; Alexander, Jackson, Pope, and Pulaski cos.

8. **Pycnanthemum verticillatum** (Michx.) Pers. Hairy Mountain Mint. July–Sept. Dry woods, prairies; occasional in the central cos., less common to rare in the n. and s. cos. *Pycnanthemum pilosum* Nutt.; *P. verticillatum* (Michx.) Pers. var. *pilosum* (Nutt.) Cooperrider.

9. **Pycnanthemum virginianum** (L.) Dur. & B. D. Jacks. Common Mountain Mint. July–Sept. Marshes, calcareous fens, prairies; occasional to common in the n. ½ of IL, becoming less common southward.

30. **Salvia** L.—Sage

1. Most leaves in a basal rosette.
  2. Whorls of flowers more or less separated; leaves lyrate-pinnatifid; calyx not viscid .... 2. *S. lyrata*
  2. Whorls of flowers very remote; leaves crenate; calyx viscid .... 3. *S. pratensis*
1. Most leaves cauline.
  3. Leaves not more than 2 cm broad; upper lip of calyx without teeth.
    4. Leaves denticulate or serrate; corolla 15 mm long or longer; flowers 6 or more per whorl .... 1. *S. azurea*
    4. Leaves entire or sparsely serrate; corolla 8–12 mm long; flowers 1–3 per whorl .... 4. *S. reflexa*
  3. Leaves more than 2 cm broad; upper lip of calyx 3-toothed.
    5. Upper surface of leaves glabrous; calyx much longer than pedicel in fruit .... 5. *S. sylvestris*
    5. Upper surface of leaves pubescent; calyx about as long as the pedicel in fruit .... 6. *S. verticillata*

1. **Salvia azurea** Michx. & Lam. var. **grandiflora** Benth. Blue Sage. June–Aug. Dry woods, prairies; scattered in IL. Most records appear to be garden escapes.

2. **Salvia lyrata** L. Cancer-weed; Lyre-leaved Sage. Apr.–June. Rich woods, open woods; occasional to common in the s. ⅕ of IL.

3. **Salvia pratensis** L. Meadow Sage. June–July. Native to Europe; seldom escaped into disturbed areas; Lake, McHenry, and Piatt cos.

4. **Salvia reflexa** Hornem. Rocky Mountain Sage. May–Oct. Native to the w. U.S. Dry woods, pastures, fields; occasional throughout the state except for the southernmost cos.

5. **Salvia sylvestris** L. Wild Sage. June–Sept. Native to Europe; adventive in pastures and along roads; Cook, McHenry, and Winnebago cos. *Salvia nemorosa* L.

6. **Salvia verticillata** L. Sage. June–Sept. Native to Europe; adventive along a railroad; Scott Co.

31. **Satureja** L.—Savory

1. **Satureja hortensis** L. Summer Savory. Aug.–Oct. Native to Europe and Asia; rarely escaped from cultivation; DuPage and Peoria cos.

32. **Scutellaria** L.—Skullcap

1. Flower solitary in the axils of the leaves or of the reduced leaflike bracts.
  2. Corolla at least 1.5 cm long .... 3. *S. galericulata*
  2. Corolla up to 1.2 cm long.
    3. Most of the leaves 2 cm long or longer; stolons without moniliform tubers; stems glabrous at maturity .... 7. *S. nervosa*
    3. Most of the leaves less than 2 cm long; stolons with moniliform tubers; stems pubescent at maturity.
      4. Pubescence eglandular .... 6. *S. leonardii*
      4. Pubescence glandular.
        5. Lower leaf surface with sessile glands and long hairs .... 9. *S. parvula*
        5. Lower leaf surface with long hairs only .... 1. *S. australis*
1. Flowers 2–several in racemes.
  6. Corolla up to 1 cm long; racemes produced from the axils of the leaves and therefore lateral .... 5. *S. lateriflora*

6. Corolla 1 cm long or longer; most of the racemes terminal.
  7. Leaves cordate at the base .... 8. *S. ovata*
  7. Leaves rounded to cuneate at the base.
    8. Calyx with glandular hairs..2. *S. elliptica*
    8. Calyx with eglandular hairs...4. *S. incana*

1. **Scutellaria australis** (Fassett) Epling. Small Skullcap. May–July. Rocky woods, prairies, fields, limestone barrens; occasional in the s. ½ of IL.

2. **Scutellaria elliptica** Muhl. Hairy Skullcap. May–July. Dry, rocky woods; restricted to the s. ⅓ of the state; also Sangamon Co. (Including var. *hirsuta* [Short] Fern.)

3. **Scutellaria galericulata** L. Marsh Skullcap. June–Sept. Marshes; confined to the n. ½ of IL. *Scutellaria epilobiifolia* L.

4. **Scutellaria incana** Biehler. Downy Skullcap. June–Sept. Dry, rocky woods; occasional in the s. ⅔ of IL.

5. **Scutellaria lateriflora** L. Mad-dog Skullcap. June–Oct. Marshes, swampy woods, borders of rivers and streams; occasional to common throughout the state.

6. **Scutellaria leonardii** Epling. Small Skullcap. May–July. Prairies, rocky woods; scattered throughout the state.

7. **Scutellaria nervosa** Pursh. Two forms occur in Illinois:

a. Flowers bluish ....7a. *S. nervosa* f. *nervosa*
a. Flowers white ....7b. *S. nervosa* f. *alba*

7a. **Scutellaria nervosa** Pursh f. **nervosa.** Veiny Skullcap. Apr.–July. Low woods; occasional in the s. ¾ of IL; also Carroll, Jo Daviess, and Kankakee cos. (Including var. *calvifolia* Fern.)

7b. **Scutellaria nervosa** Pursh f. **alba** Steyerm. White Veiny Skullcap. Apr.–May. Low woods, rare; confined to the s. ¼ of IL. Almost all specimens of this species from s. IL are the white-flowered form.

8. **Scutellaria ovata** Hill. Three varieties may be recognized in Illinois:

a. Some or all of the leaves more than 4 cm long; plants generally taller than 25 cm.
  b. Uppermost bracts longer than the calyx ....8a. *S. ovata* var. *ovata*
  b. Uppermost bracts shorter than the calyx .... 8b. *S. ovata* var. *bracteata*
a. None of the leaves 4 cm long; plants less than 25 cm tall.....8c. *S. ovata* var. *rugosa*

8a. **Scutellaria ovata** Hill var. **ovata.** Heart-leaved Skullcap. May–Oct. Rocky woods; occasional throughout the state.

8b. **Scutellaria ovata** Hill var. **bracteata** (Benth.) S. F. Blake. Heart-leaved Skullcap. May–Oct. Rocky woods, rich woods; occasional throughout the state. *Scutellaria ovata* Hill var. *versicolor* (Nutt.) Fern.

8c. **Scutellaria ovata** Hill var. **rugosa** (Wood) Epling. Dwarf Heart-leaved Skullcap. May–Oct. Limestone woods, rare; Jackson, Monroe, Randolph, and Union cos.

9. **Scutellaria parvula** Michx. Small Skullcap. May–July. Rocky woods, prairies, fields, limestone barrens; occasional in the s. ½ of IL; also Tazewell Co.

33. **Stachys** L.—Hedge Nettle

1. Leaves and stems white-woolly... 2. *S. byzantina*
1. Leaves and stems glabrous or variously pubescent but not white-woolly.
  2. Teeth of the calyx ½ as long as the tube..3. *S. cordata*
  2. Teeth of the calyx ¾ as long as or equaling the tube.
    3. Stems pubescent on the sides as well as on the angles.
      4. Corolla purple; calyx with short glandular hairs; teeth of calyx long-attenuate .... 6. *S. palustris*
      4. Corolla pink or lavender; calyx with long glandular hairs; teeth of calyx acute to acuminate but not long-attenuate................................ 7. *S. pilosa*
    3. Stems glabrous on the sides although sometimes pubescent on the angles.
      5. Teeth of calyx hirsute; leaves densely hairy on the upper surface .... 4. *S. hispida*
      5. Teeth of calyx glabrous; leaves glabrous or sparsely hairy on the upper surface.
        6. Some of all of the petioles at least 8 mm long ....8. *S. tenuifolia*

6. Leaves sessile or on petioles less than 8 mm long.
7. Stems glabrous or nearly so; leaves linear to narrowly oblong, entire or nearly so; calyx glabrous or with setae on the angles............ .... 5. *S. hyssopifolia*
7. Stems retrorsely hispid on the angles; leaves oblong, with low teeth; calyx puberulent, sometimes witih setae on the angles......... ....1. *S. aspera*

1. **Stachys aspera** Michx. Hyssop Hedge Nettle. June–Aug. Moist soil; occasional in the cent. cos., rare in the n. and s. cos.

2. **Stachys byzantina** C. Koch. Lamb's-ears; Woolly Hedge Nettle. July–Aug. Native to the Mediterranean area; rarely escaped from cultivation; DuPage and Lake cos.

3. **Stachys cordata** Riddell. Heart-leaved Hedge Nettle. June–July. Rich woods, very rare; Hardin Co. *Stachys nuttallii* Shuttlw.

4. **Stachys hispida** Pursh. Hispid Hedge Nettle. June–Sept. Low woods, swamps, marshes; occasional in the n. ½ of IL, uncommon elsewhere. (Including var. *platyphylla* Fern.) *Stachys tenuifolia* Willd. var. *hispida* (Pursh) Fern.

5. **Stachys hyssopifolia** L. Hyssop-leaved Hedge Nettle. June–Aug. Wet ground, very rare; Williamson Co.

6. **Stachys palustris** L. Hedge Nettle. June–Sept. Native to Europe; occasionally naturalized in wet, disturbed areas; scattered in IL.

7. **Stachys pilosa** Nutt. Two varieties occur in Illinois:

a. Leaves obtuse to subacute, oblong to oval, sessile or on petioles less than 1 cm long.. 7a. *S. pilosa* var. *pilosa*
a. Leaves acuminate, lanceolate, on petioles more than 1 cm long............................... ....7b. *S. pilosa* var. *homotricha*

7a. **Stachys pilosa** Nutt. var. **pilosa.** Woundwort. June–Sept. Moist soil, rare; Hancock and Union cos.

7b. **Stachys pilosa** Nutt. var. **homotricha** (Fern.) Mohlenbr., comb. nov. (basionym: *Stachys palustris* L. var. *homotricha* Fern). Woundwort. June–Sept. Wet prairies, swampy or marshy soil; occasional to common in the n. ½ of IL, uncommon elsewhere. *Stachys palustris* L. var. *homotricha* Fern.

8. **Stachys tenuifolia** Willd. Smooth Hedge Nettle. June–Sept. Moist soil; occasional throughout the state.

34. **Synandra** Nutt.

1. **Synandra hispidula** (Michx.) Baill. Guyandotte Beauty. May–June. Rich woods, rare; Jackson and Williamson cos.

35. **Teucrium** L.—Germander

1. **Teucrium canadense** L. Two varieties may be distinguished in Illinois:

a. Calyx and bracts without glandular hairs; lower leaf surface appressed-pubescent ....1a. *T. canadense* var. *canadense*
a. Calyx and bracts glandular-pubescent; lower leaf surface spreading-pubescent ....1b. *T. canadense* var. *occidentale*

1a. **Teucrium canadense** L. var. **canadense.** American Germander. June–Sept. Low woods, wet prairies, wet ditches, wet fields, common; in every co.

1b. **Teucrium canadense** L. var. **occidentale** (Gray) McClintock & Epling. Western Germander. July–Sept. Low woods, wet prairies, wet fields; occasional in the n. ¾ of IL. *Teucrium canadense* L. var. *boreale* (Bickn.) Shinners.

36. **Thymus** L.—Thyme

1. **Thymus praecox** Opiz. Thyme. July–Aug. Native to Europe; occasionally cultivated but rarely escaped into disturbed soil; Hancock, Henderson, and Jackson cos.

37. **Trichostema** L.—Blue Curls

1. Leaves oblong to ovate, most of them at least 5 mm broad........... 1. *T. dichotomum*
1. Leaves linear, up to 5 mm broad, usually narrower.............................. 2. *T. setaceum*

1. **Trichostema dichotomum** L. Blue Curls. Aug.–Oct. Woods, edge of fields, not common; scattered in IL.

2. **Trichostema setaceum** Houtt. Slender-leaved Blue Curls. July–Sept. Dry soil, very rare; Johnson Co.

89. LARDIZABALACEAE—LARDIZABALA FAMILY

1. **Akebia** Decne.

1. **Akebia quinata** Decne. Akebia. June–July. Native to Asia; naturalized along fences in Kane and Pope cos.

90. LAURACEAE—LAUREL FAMILY

1. None of the leaves lobed; flowers appearing before the leaves, in axillary clusters; fruits red .... 1. *Lindera*

1. Some of the leaves lobed; flowers appearing as the leaves unfold, in modified racemes; fruits blue .... 2. *Sassafras*

1. **Lindera** Thunb.—Spicebush

1. **Lindera benzoin** (L.) Blume. Two varieties occur in Illinois:

a. Leaves glabrous .... 1a. *L. benzoin* var. *benzoin*

a. Leaves permanently short-hairy on both surfaces .... 1b. *L. benzoin* var. *pubescens*

1a. **Lindera benzoin** (L.) Blume var. **benzoin.** Spicebush. Mar.–May. Rich woods, wet woods; common in the s. ½ of IL, occasional or absent elsewhere.

1b. **Lindera benzoin** (L.) Blume var. **pubescens** (Palmer & Steyerm.) Rehder. Hairy Spicebush. Mar.–May. Swampy woods, rare; Jackson, Johnson, and Union cos.

2. **Sassafras** Nees—Sassafras

1. **Sassafras albidum** (Nutt.) Nees. Two varieties may be recognized in Illinois:

a. Leaves glabrous or nearly so on the lower surface .... 1a. *S. albidum* var. *albidum*

a. Leaves permanently pubescent on the lower surface .... 1b. *S. albidum* var. *molle*

1a. **Sassafras albidum** (Nutt.) Nees var. **albidum.** Sassafras. Apr.–May. Dry or moist woods, thickets, roadsides; common in the s. ¾ of IL, rare or absent elsewhere.

1b. **Sassafras albidum** (Nutt.) Nees var. **molle** (Raf.) Fern. Red Sassafras. Apr.–May. Woods, thickets, roadsides; occasional in the s. ¾ of IL.

91. LENTIBULARIACEAE—BLADDERWORT FAMILY

1. **Utricularia** L.—Bladderwort

1. Leaves minute, rarely seen; plants usually terrestrial, the stems creeping in mud or wet sand.

  2. Calyx enclosing the fruit; a pair of bractlets as well as bracts at base of pedicels .... 1. *U. cornuta*

  2. Calyx not enclosing the fruit; bractlets absent .... 8. *U. subulata*

1. Leaves pinnately divided, conspicuous; plants aquatic, either free floating or creeping in mud at base of plant in water.

  3. Branches in whorls or opposite; lobes of lower lip of corolla saccate; flowers purple .... 7. *U. purpurea*

  3. Branches alternate or absent; no corolla lobes saccate; flowers yellow.

    4. Divisions of stems flat.

      5. Apex of stem segments entire; bladders borne on the stem .... 5. *U. minor*

      5. Apex of stem segments serrulate; bladders borne on separate branches.

        6. Branches bearing bladders without lateral divisions; spur of flower as long as lower lip of corolla .... 3. *U. intermedia*

        6. Branches bearing bladders with small dissected divisions; spur of flower about half as long as lower lip of corolla .... 6. *U. ochroleuca*

    4. Divisions of stems terete.

      7. Stems free floating .... 4. *U. macrorhiza*

      7. Stems creeping near bottom of plant in shallow water .... 2. *U. gibba*

1. **Utricularia cornuta** Michx. Horned Bladderwort. July–Sept. Bogs and wet sands, rare; Cook, Lake, and McHenry cos.

2. **Utricularia gibba** L. Humped Bladderwort. Aug.–Oct. Swamps, ponds, lakes, and ditches, not common; in s. cos. along the Mississippi River; also Cook, Lake, and McHenry cos.

3. **Utricularia intermedia** Hayne. Flat-leaved Bladderwort. July–Aug. Shallow water, rare; Cook, Kane, Lake, McHenry, Ogle, and Tazewell cos.

4. **Utricularia macrorhiza** LeComte. Common Bladderwort. July–Aug. Shallow water, not common; scattered throughout the state. *Utricularia vulgaris* L., misapplied.

5. **Utricularia minor** L. Small Bladderwort. June–Aug. Shallow water, very rare; Clay, Cook, Lake, McHenry, and Saline cos.

6. **Utricularia ochroleuca** R. Hartm. Pale Bladderwort. July–Aug. Shallow water, very rare.

7. **Utricularia purpurea** Walt. Purple Bladderwort. June–Sept. Shallow water, very rare.

8. **Utricularia subulata** L. Subulate Bladerwort. May–Sept. Shallow water, very rare; Lake Co.

92. LIMNANTHACEAE—LIMNANTHUS FAMILY

1. **Floerkea** Willd.—False Mermaid

1. **Floerkea proserpinacoides** Willd. False Mermaid. Apr.–June. Rich woods; occasional in the n. ½ of the state, extending southward to Crawford Co.

93. LINACEAE—FLAX FAMILY

1. **Linum** L.—Flax

1. Flowers blue.
  2. Sepals acuminate; annual .... 7. *L. usitatissimum*
  2. Sepals mostly rounded at tip; perennial .... 3. *L. perenne*
1. Flowers yellow.
  3. Styles partly united; petals 8–15 mm long; annual.
    4. Petals 8–10 mm long; styles united only near base; fruit 2.5–3.5 mm long .... 6. *L. sulcatum*
    4. Petals 10–17 mm long; styles united to above the middle; fruit 4–5 mm long. .... 4. *L. rigidum*
  3. Styles free from each other; petals up to 7 mm long (to 10 mm in *L. floridanum*); perennial.
    5. Fruits ovoid to pyriform .... 1. *L. floridanum*
    5. Fruits depressed-globose.
      6. Leaves mostly opposite; outer sepals elliptic, membranaceous .... 5. *L. striatum*
      6. Leaves mostly alternate; outer sepals lanceolate, stiff.
        7. Inner sepals with conspicuous glandular cilia; leaves stiff .... 2. *L. medium*
        7. Inner sepals eglandular or nearly so; leaves thin .... 8. *L. virginianum*

1. **Linum floridanum** (Planch.) Trel. Coastal Plain Yellow Flax. June–July. Damp woods, very rare; Union Co. Specimen is deposited in the herbarium of the Misssouri Botanical Garden.

2. **Linum medium** (Planch.) Britt. var. **texanum** (Planch.) Fern. Wild Flax. May–Aug. Dry soil or calcareous pond shores; occasional in the s. ½ of IL, rare in the n. ½.

3. **Linum perenne** L. Flax. June–Sept. Native to the w. U.S.; escaped from cultivation in a few n. cos; also Coles Co. *Linum perenne* L. ssp. *lewisii* (Pursh) Hult.

4. **Linum rigidum** Pursh. Two varieties may be distinguished in Illinois:

a. Stipules glandular .... 4a. *L. rigidum* var. *rigidum*
a. Stipules eglandular .... 4b. *L. rigidum* var. *compactum*

4a. **Linum rigidum** Pursh var. **rigidum.** Stiff Flax. May–Sept. Open dry soil, very rare; Union Co.

4b. **Linum rigidum** Pursh var. **compactum** (A. Nels.) Rogers. Compact Stiff Flax. Open dry soil, very rare; Kane Co.

5. **Linum striatum** Walt. Wild Flax. June–Sept. Moist soil, sometimes in woods; occasional in the s. ⅓ of IL.

6. **Linum sulcatum** Riddell. Wild Flax. May–Sept. Dry soil, hill prairies; occasional throughout the state.

7. **Linum usitatissimum** L. Common Flax. May–Aug. Native to Europe; adventive in disturbed areas; occasional throughout the state.

8. **Linum virginianum** L. Wild Flax. June–Aug. Dry woods; occasional in the s. ½ of the state, rare in the n. ½.

94. LOASACEAE—LOASA FAMILY

1. **Mentzelia** L.

1. Petals 5, 1.0–1.5 cm long; stamens about 20; capsules up to 1 cm long, containing about 9 seeds.......... 3. *M. oligosperma*
1. Petals 10, 2–5 cm long; stamens numerous; capsules 2.5–4.0 cm long, containing more than 9 seeds.
  2. Petals 4–5 cm long; all filaments filiform; capsules about 4 cm long.......... 1. *M. decapetala*
  2. Petals 2–3 cm long; outer filaments dilated; capsules about 3 cm long.......... 2. *M. nuda*

1. **Mentzelia decapetala** (Pursh) Urban & Gilg. Blazing-star. July–Sept. Native to the w. U.S.; adventive along a railroad; Grundy Co.

2. **Mentzelia nuda** (Pursh) Torr. & Gray. Blazing-star. July–Aug. Native to the w. U.S.; adventive along a railroad; Cook Co.

3. **Mentzelia oligosperma** Nutt. Stickleaf. June–July. Limestone ledges, very rare; confined to cos. bordering the Mississippi River.

95. LOGANIACEAE—LOGANIA FAMILY

1. **Spigelia** L.—Indian Pink

1. **Spigelia marilandica** L. Indian Pink. May–June. Woods; occasional in the s. ¼ of IL; also Marion Co.

96. LYTHRACEAE—LOOSESTRIFE FAMILY

1. Plants woody.
  2. Shrubs; leaves, or some of them, whorled.......... 3. *Decodon*
  2. Trees; leaves opposite.......... 5. *Lagerstroemia*
1. Plants herbaceous.
  3. Plants viscid-pubescent; petals 6; calyx tube spurred at base.......... 2. *Cuphea*
  3. Plants glabrous or variously pubescent but not viscid; petals 4–6 or absent; calyx tube not spurred at base.
    4. Calyx with 5, 6, or 7 lobes or teeth, the tube cylindric.......... 6. *Lythrum*
    4. Calyx with 4 lobes or teeth, the tube hemispheric.
      5. Plants usually in water; petals absent; leaves up to 3 mm broad.......... 4. *Didiplis*
      5. Plants usually not in water; petals 4; leaves over 3 mm broad.
        6. Flower usually 1 per leaf axil; leaves tapering to a short petiole or subsessile base; capsule dehiscing septicidally.......... 7. *Rotala*
        6. Flowers often 2 or more per leaf axil; leaves auriculate at sessile base; capsule dehiscing irregularly.......... 1. *Ammannia*

1. **Ammannia** L.—Tooth-cup

1. Fruit 2.0–3.5 mm in diameter, longer than the calyx.......... 1. *A. auriculata*
1. Fruit 3.5–5.0 mm in diameter, shorter than the calyx or equalling the calyx.
  2. Petals rose-purple; pedicels usually 1–4 mm long.......... 2. *A. coccinea*
  2. Petals pale lavender; pedicels up to 1 mm long, or absent.......... 3. *A. robusta*

1. **Ammannia auriculata** Willd. Redstem. July–Aug. Wet ground, very rare; Alexander Co.

2. **Ammannia coccinea** Rottb. Tooth-cup. July–Aug. Wet soil; occasional to common in the s. ¾ of IL, rare elsewhere.

3. **Ammannia robusta** Heer & Regel. Tooth-cup. July–Sept. Wet soil; throughout the state, but mostly in the ne. cos.

2. **Cuphea** P. Br.

1. **Cuphea viscosissima** Jacq. Clammy Cuphea; Waxweed. July–Oct. Usually dry soil; occasional in the s. ¾ of IL, rare or absent elsewhere.

3. **Decodon** J. F. Gmel.—Water Willow

1. **Docodon verticillatus** (L.) Ell. Two varieties may be recognized in Illinois:

a. Stems, lower surface of leaves, and pedicels pubescent .... 1a. *D. verticillatus* var. *verticillatus*
a. Stems, lower surface of leaves, and pedicels glabrous .... 1b. *D. verticillatus* var. *laevigatus*

1a. **Decodon verticillatus** (L.) Ell. var. **verticillatus.** Swamp Loosestrife; Shrubby Loosestrife. July–Aug. Swamps, marshes, bogs; uncommon but scattered throughout the state.

1b. **Decodon verticillatus** (L.) Ell. var. **laevigatus** Torr. & Gray. Smooth Swamp Loosestrife. July–Aug. Swamps, marshes, bogs; uncommon but scattered throughout the state.

4. **Didiplis** Raf.—Water Purslane

1. **Didiplis diandra** (DC.) Wood. Water Purslane. June–Aug. Shallow water; uncommon but scattered in the state. *Peplis diandra* DC.

5. **Lagerstroemia** L.—Crepe Myrtle

1. **Lagerstroemia indica** L. Crepe Myrtle. Apr.–June. Native to the tropics; rarely escaped from gardens; Jackson Co.

6. **Lythrum** L.—Loosestrife

1. Flower solitary, axillary, up to 1.2 cm across; stamens 4 .... 1. *L. alatum*
1. Flowers in terminal spikes, usually at least 1.2 cm across; stamens 8 (–10).
  2. Leaves rounded or truncate at the base; flowers sessile or nearly so; appendages of calyx at least twice as long as the lobes .... 2. *L. salicaria*
  2. Leaves tapering to the base; flowers pedicellate; appendages of calyx equaling the length of the lobes .... 3. *L. virgatum*

1. **Lythrum alatum** Pursh. Winged Loosestrife. June–Aug. Moist ground; occasional to common throughout IL.

2. **Lythrum salicaria** L. Purple Loosestrife. June–Sept. Native to Europe; occasionally escaped andoften invasive into moist areas, particularly in the n. ⅔ of IL. (Including var. *tomentosum* [Mill.] DC.)

3. **Lythrum virgatum** L. Showy Purple Loosestrife. June–Sept. Native to Europe; rarely escaped from cultivation; Jackson and Williamson cos.

7. **Rotala** L.—Tooth-cup

1. **Rotala ramosior** (L.) Koehne. Tooth-cup. July–Sept. Wet ground; occasional throughout the state.

97. MAGNOLIACEAE—MAGNOLIA FAMILY

1. Leaves 4-lobed; petals with an orange blotch at the base within; seeds winged .... 1. *Liriodendron*
1. Leaves entire; petals without an orange blotch at the base within; seeds unwinged .... 2. *Magnolia*

1. **Liriodendron** L.

1. **Liriodendron tulipifera** L. Tulip Poplar; Tulip Tree. Apr.–May. Rich woods, generally common; confined to the s. ⅗ of IL.

2. **Magnolia** L.—Magnolia

1. Flowers blooming after the leaves unfold, greenish yellow, not showy; leaves green on both surfaces, short-acuminate, 15–25 cm long; sepals reflexed .... 1. *M. acuminata*
1. Flowers blooming before the leaves unfold, white, often streaked with pink, showy; leaves sometimes glaucous on the lower surface, obtuse to acute, up to 15 cm long; sepals not reflexed .... 2. *M. stellata*

1. **Magnolia acuminata** L. Cucumber Magnolia; CucumberTree. Apr.–May. Rich woods; restricted to the s. 3 tiers of cos; also Adams and Greene cos. where it may be introduced.

2. **Magnolia stellata** Maxim. Star Magnolia. Mar.–Apr. Native to Asia; rarely escaped from cultivation; DuPage Co. (Wilhelm, 2010.)

98. MALVACEAE—MALLOW FAMILY

1. Some or all the leaves lobed.
2. Calyx subtended by an involucre of bracts at the base.
3. Bracts at base of calyx 3.
4. Bracts laciniate .... 6. *Gossypium*
4. Bracts unlobed, or at least not laciniate.
5. All or most of the leaves divided nearly to the base.
6. Plants hispid; stems more or less procumbent .... 5. *Callirhoe*
6. Plants with scattered pubescence, but not hispid; stems ascending .... 9. *Malva*
5. None of the leaves divided nearly to the base.
7. Flowers less than 2 cm across .... 9. *Malva*
7. Flowers more than 2 cm across.
8. Lower leaves triangular, unlobed .... 5. *Callirhoe*
8. Lower leaves shallowly lobed or, if unlobed, not triangular.
9. Bracts at base of calyx lanceolate to ovate-lanceolate; calyx lobes acute; carpels 1-seeded .... 9. *Malva*
9. Bracts at base of calyx linear; calyx lobes acuminate; carpels 2- to 3-seeded .... 8. *Iliamna*
3. Bracts at base of calyx 6 or more.
10. Bracts triangular; fruit separating at maturity into 15–20 carpels... 3. *Alcea*
10. Bracts linear; fruit a 5-celled capsule.
11. Calyx spathelike, splitting down 1 side; seeds mucilaginous; capsule cylindrical, several times longer than broad .... 1. *Abelmoschus*
11. Calyx not spathelike, 5-lobed at the apex; seeds not mucilaginous; capsule globose to ovoid, about as broad as long .... 7. *Hibiscus*
2. Calyx without an involucre of bracts at the base.
12. Flowers unisexual, white, up to 2 cm across .... 11. *Napaea*
12. Flowers bisexual, blue or, if white, more than 2 cm across.
13. Flowers more than 4 cm across .... 5. *Callirhoe*
13. Flowers up to 4 cm across .... 4. *Anoda*
1. None of the leaves lobed.
14. Leaves as broad as long, some of them at least 3 cm broad.
15. Calyx not subtended by bracts .... 2. *Abutilon*
15. Calyx subtended by bracts.
16. Bracts subtending the calyx 6 or more .... 7. *Hibiscus*
16. Bracts subtending the calyx 3.
17. Flowers more than 2 cm across, deep purple .... 5. *Callirhoe*
17. Flowers much less than 2 cm across, pale lilac or white .... 9. *Malva*
14. Leaves longer than broad, never 3 cm broad.
18. Calyx subtended by 2–3 setaceous bracts .... 10. *Malvastrum*
18. Calyx not subtended by bracts .... 12. *Sida*

1. **Abelmoschus** Medic.—Okra

1. **Abelmoschus esculentus** (L.) Moench. Okra. July–Sept. Native to Africa; rarely escaped from cultivation.

2. **Abutilon** Mill.—Indian Mallow

1. **Abutilon theophrastii** Medic. Velvet-leaf; Butter-print. Aug.–Oct. Native to India; naturalized in disturbed areas; scattered throughout the state, often common.

3. **Alcea** L.—Hollyhock

1. **Alcea rosea** L. Hollyhock. July–Oct. Native to Europe and Asia; occasionally escaped from gardens; scattered in IL.

4. **Anoda** Cav.

1. **Anoda cristata** (L.) Schlecht. Anoda. Aug.–Oct. Native to the sw. U.S.; rarely escaped from cultivation into disturbed soil; Alexander, Douglas, Hancock, Kankakee, Massac, Monroe, Union, and Winnebago cos.

5. **Callirhoe** Nutt.—Poppy Mallow

1. Each flower subtended by 3 bracts.
  2. Some of the leaves triangular; flowers in panicles or appearing umbellate; carpels not rugose .... 4. *C. triangulata*
  2. None of the leaves triangular; flower solitary; carpels rugose .... 3. *C. involucrata*
1. Flowers without subtending bracts.
  3. Basal leaves more or less triangular; top of carpels pubescent ... 1. *C. alcaeoides*
  3. Basal leaves rounded; top of carpels glabrous .... 2. *C. digitata*

1. **Callirhoe alcaeoides** (Michx.) Gray. Poppy Mallow. May–Aug. Dry, gravelly areas, rare; Cass, Christian, DuPage, Henry, Peoria, and Winnebago cos.
2. **Callirhoe digitata** Nutt. Poppy Mallow. May–July. Native to the w. U.S.; rarely adventive into disturbed soil; DuPage, Henderson, and Kane cos.
3. **Callirhoe involucrata** (Torr. & Gray) Gray. Poppy Mallow. June–Aug. Native to the w. U.S.; naturalized in disturbed soil; confined to the cent. and n. cos.
4. **Callirhoe triangulata** (Leavenw.) Gray. Poppy Mallow. June–Sept. Sandy soil; mostly in the n. cos.

6. **Gossypium** L.—Cotton

1. **Gossypium hirsutum** L. Cotton. May–June. Native to tropical America; escaped from cultivation along roadsides and old fields; scattered in IL.

7. **Hibiscus** L.—Rose Mallow

1. Annual herbs; flowers yellow or whitish with a black or purple center .... 5. *H. trionum*
1. Perennial herbs, shrubs, or small trees; flowers not yellow.
  2. Stems and leaves glabrous.
    3. Herb 1.0–2.5 m tall; calyx glabrous or with simple hairs .... 1. *H. laevis*
    3. Shrub or small tree 3 m tall or taller; calyx stellate-pubescent .... 4. *H. syriacus*
  2. Stems and at least the lower surface of the leaves pubescent.
    4. Leaves glabrous or nearly so on the upper surface; some of the leaves often 3-lobed; capsules glabrous .... 3. *H. moscheutos*
    4. Leaves soft-hairy on both surfaces; none of the leaves lobed; capsules densely hairy .... 2. *H. lasiocarpos*

1. **Hibiscus laevis** All. Halberd-leaved Rose Mallow. July–Oct. Wet areas; occasional throughout the state except for the northernmost tier of cos. *Hibiscus militaris* L.
2. **Hibiscus lasiocarpos** Cav. Hairy Rose Mallow. July–Oct. Low, wet soil; occasional to common in the s. ½ of IL, rare or absent in most of the n. cos.
3. **Hibiscus moscheutos** L. Swamp Rose Mallow. July–Sept. Wet soil; scattered in IL.
4. **Hibiscus syriacus** L. Rose-of-Sharon. July–Sept. Native to Asia; rarely spreading from cultivation.
5. **Hibiscus trionum** L. Flower-of-an-hour. July–Oct. Native to Europe; adventive throughout the state, although uncommon in the extreme s. tip.

8. **Iliamna** Greene

1. **Iliamna remota** Greene. Kankakee Mallow. June–Aug. Dry banks, very rare; Kankakee Co.; sometimes escaped from gardens elsewhere. *Iliamna rivularis* (Dougl.) Greene, misapplied.

9. **Malva** L.—Mallow

1. Leaves deeply pinnatifid, with 5 or 7 segments.
  2. Bracts lance-ovate to oblong, more than 3 times longer than broad .... 2. *M. moschata*
  2. Bracts ovate, less than 3 times longer than broad .... 1. *M. alcea*
1. Leaves shallowly lobed or unlobed, not deeply pinnatifid.
  3. Flowers at least 2.5 cm across, usually broader; petals more than twice as long as the calyx .... 6. *M. sylvestris*
  3. Flowers up to 1.5 cm across; petals at most twice as long as the calyx.
    4. Flowers sessile or subsessile; stems erect, to 2 m tall .... 7. *M. verticillata*
    4. Flowers pedicellate; stems prostrate to ascending, to 1 m long.
      5. Petals twice as long as the sepals; carpels with rounded margins .... 3. *M. neglecta*
      5. Petals about as long as the sepals; carpels with angular margins.
        6. Margins of carpels not winged; pedicels of fruits usually at least 10 mm long .... 5. *M. rotundifolia*
        6. Margins or carpels with a narrow wing; pedicels of fruits less than 10 mm long .... 4. *M. parviflora*

1. **Malva alcea** L. Vervain Mallow. June–July. Native to Europe; rarely escaped from gardens; Champaign, Cook, Kane, and Will cos.

2. **Malva moschata** L. Musk Mallow. June–Aug. Native to Europe; rarely escaped from cultivation; Cook, DuPage, LaSalle, and Lawrence cos.

3. **Malva neglecta** Wallr. Common Mallow. June–Sept. Native to Europe; naturalized in waste areas, particularly farm lots; occasional to common throughout IL.

4. **Malva parviflora** L. Little Mallow. July–Sept. Native to Eurasia; disturbed soil; DuPage Co. (Wilhelm, 2010.)

5. **Malva rotundifolia** L. Mallow. June–Sept. Native to Europe; naturalized in waste areas; occasional and scattered in the n. ½ of IL; also St. Clair Co.

6. **Malva sylvestris** L. Two varieties may be distinguished in Illinois:

a. Lobes of leaf triangular; stems and leaves hirsute ..... 6a. *M. sylvestris* var. *sylvestris*
a. Lobes of leaf broadly rounded; stems and leaves glabrous or nearly so .... 6b. *M. sylvestris* var. *mauritiana*

6a. **Malva sylvestris** L. var. **sylvestris.** High Mallow. Aug.–Sept. Native to Europe; occasionally escaped from cultivation; scattered in IL.

6b. **Malva sylvestris** L. var. **mauritiana** (L.) Boiss. High Mallow. Aug.–Sept. Native to Europe; seldom escaped from cultivation; known from the ne. cos.

7. **Malva verticillata** L. var. **crispa** L. Curly Mallow. July–Sept. Native to Europe; rarely escaped from cultivation in a few of the n. cos.

10. **Malvastrum** Gray—Globe Mallow

1. **Malvastrum hispidum** (Pursh) Hochr. Globe Mallow. July–Aug. Dry soil, rare; Grundy, LaSalle, Randolph, Rock Island, St. Clair, and Will cos. *Sidopsis hispida* (Pursh) Rydb.; *Sphaeralcea angusta* (Gray) Fern.

11. **Napaea** L.—Glade Mallow

1. **Napaea dioica** L. Glade Mallow. June–July. Alluvial soil along streams and rivers; occasional in the n. ½ of IL, apparently absent from the s. ½.

12. **Sida** L.—Sida

1. Leaves linear, rarely more than 7 mm broad; carpels 8 or more per flower .... 1. *S. elliottii*
1. Leaves oblong to lance-ovate, all or most of them more than 8 mm broad; carpels 5 per flower .... 2. *S. spinosa*

1. **Sida elliottii** Torr. & Gray. Elliott's Sida. Aug.–Oct. Farm lots, very rare; Alexander Co.

2. **Sida spinosa** L. Prickly Sida. June–Oct. Native to tropical America; naturalized in waste areas; occasional to common in the s. 4/5 of IL, rare or absent elsewhere.

99. MELASTOMACEAE—MEADOW BEAUTY FAMILY

1. **Rhexia** L.—Meadow Beauty

1. Stems terete; plants without tubers; leaves narrowed to a short petiole; neck of hypanthium longer than the body at maturity .... 1. *R. mariana*
1. Stems 4-angled; plants with tubers; leaves rounded at the sessile base; neck of hypanthium shorter than the body at maturity .... 2. *R. virginica*

1. **Rhexia mariana** L. Meadow Beauty. June–Sept. Sandy fields, rare; in the s. 1/4 of IL; also Mason Co.

2. **Rhexia virginica** L. Meadow Beauty. July–Sept. Sandy soils, not common; scattered throughout the state.

100. MENISPERMACEAE—MOONSEED FAMILY

1. Leaves deeply 5- or 7-lobed; petals none; drupes 15–25 mm long .... 1. *Calycocarpum*
1. Leaves entire or with 3–7 angles or shallow lobes; petals 6–8; drupes 5–10 mm long.
  2. Drupes red; stamens 6; leaves not peltate...2. *Cocculus*
  2. Drupes blue-black; stamens 12–24; leaves peltate...................... 3. *Menispermum*

1. **Calycocarpum** Nutt.—Cupseed

1. **Calycocarpum lyonii** (Pursh) Gray. Cupseed. May–June. Swampy woods, rare; confined to s. IL; introduced in Champaign Co

2. **Cocculus** DC.—Snailseed

1. **Cocculus carolinus** (L.) DC. Snailseed. July–Aug. Moist woods, thickets; occasional in the s. 1/4 of IL.

3. **Menispermum** L.—Moonseed

1. **Menispermum canadense** L. Moonseed. May–July. Moist woods, thickets; common throughout the state.

101. MENYANTHACEAE—BUCKBEAN FAMILY

1. Leaves trifoliolate...1. *Menyanthes*
1. Leaves simple .... 2. *Nymphoides*

1. **Menyanthes** L.—Buckbean

1. **Menyanthes trifoliata** L. var. **minor** Raf. Buckbean. May–June. Bogs, marshes, not common; confined to the ne. cos.; also Peoria Co.

2. **Nymphoides** Hill—Floating Heart

1. **Nymphoides peltata** (S. Gmel.) Kuntze. Yellow Floating Heart. June–Sept. Native to Europe; rarely adventive in IL; Clark and Montgomery cos.

102. MIMOSACEAE—MIMOSA FAMILY

1. Flowers in clusters about 5 cm across; stamens 15 or more; tree................ 1. *Albizia*
1. Flowers in clusters up to 2.5 cm across; stamens up to 12; herbs.
  2. Flowers white or pink; stamens 5; legumes not prickly
    3. Leaves with glands; flowers white...2. *Desmanthus*
    3. Leaves without glands; flowers pink .... 3. *Mimosa*
  2. Flowers rose; stamens 10–12; legumes prickly ....4. *Schrankia*

1. **Albizia** Duraz.—Albizia

1. **Albizia julibrissin** Duraz. Mimosa. June–Aug. Native to Asia and Africa; occasionally escaped from cultivation along roads; mostly confined to the s. cos.

2. **Desmanthus** Willd.—Bundle-flower

1. **Desmanthus illinoensis** (Michx.) MacM. Illinois Mimosa; Bundle-flower. June–Aug. Prairies, along levees, moist soil, roadsides; occasional throughout the state except for the nw. cos.

3. **Mimosa** L.—Mimosa

1. **Mimosa strigillosa** Torr. & Gray. Powderpuff. June–Aug. Native to the s. U.S.; along banks of rivers, very rare; Alexander and Massac cos.

4. **Schrankia** Willd.—Sensitive Brier

1. **Schrankia nuttallii** (DC.) Standl. Sensitive Brier; Cat's-claw. June–Sept. Prairies, rare; scattered in the n. ½ of IL. *Schrankia uncinata* Willd.; *Mimosa nuttallii* (DC.) B. L. Turner.

103. MOLLUGINACEAE—CARPETWEED FAMILY

1. **Mollugo** L.—Carpetweed

1. **Mollugo verticillata** L. Carpetweed. June–Oct. Native to tropical America; naturalized in disturbed soil; common throughout the state.

104. MONOTROPACEAE—INDIAN PIPE FAMILY

1. **Monotropa** L.—Indian Pipe

1. Flowers several per stem; plants somewhat pubescent.................. 1. *M. hypopithys*
1. Flower solitary; plants glabrous...2. *M. uniflora*

1. **Monotropa hypopithys** L. Pinesap. June–Oct. Rich woods, not common; scattered throughout the state except for the nw. cos.
2. **Monotropa uniflora** L. Indian Pipe. May–Sept. Rich woods; occasional throughout the state.

105. MORACEAE—MULBERRY FAMILY

1. Plants herbaceous...2. *Fatoua*
1. Plants woody.
  2. Leaves toothed, often lobed; branches without spines; fruit 1–2 cm in diameter, white, orange, purple, or red.
    3. Leaves glabrous or short-hairy on the lower surface, not velvety; most or all the petioles less than 3 cm long, glabrous or appressed-pubescent; bark roughened... 4. *Morus*
    3. Leaves velvety on the lower surface; most or all the petioles more than 3 cm long, pilose; bark smooth .... 1. *Broussonetia*
  2. Leaves entire, unlobed; branches with short spines; fruit 10–20 cm in diameter, greenish yellow.. 3. *Maclura*

1. **Broussonetia** L'Her.—Paper Mulberry

1. **Broussonetia papyrifera** (L.) Vent. Paper Mulberry. Apr.–May. Native to Asia; occasionally naturalized in the s. ¼ of IL; also Calhoun Co.

2. **Fatoua** Gaud.

1. **Fatoua villosa** (Thunb.) Nakai. Fatoua. June–July. Native to Europe; escaped from cultivation; Cook Co. (Wilhelm, 2010.)

3. **Maclura** Nutt.—Osage Orange

1. **Maclura pomifera** (Raf.) Schneider. Osage Orange; Hedge Apple; Bois d'arc. May–June. Native to the s. cent. U.S.; apparently naturalized in IL; woods, roadsides; occasional throughout the state.

4. **Morus** L.—Mulberry

1. Lower surface of leaves glabrous or pubescent only on the veins or the hairs confined to axillary tufts; leaves shiny.
  2. Fruits white or pale pink, usually over 1 cm long .... 1. *M. alba*

2. Fruits dark red, dark purple, or black .... 4. *M. tatarica*
1. Lower surface of leaves pubescent on the blade as well as on the veins; leaves dull.
3. Some or all the leaves more than 20 cm across .... 2. *M. murrayana*
3. Most or all the leaves less than 20 cm across .... 3. *M. rubra*

1. **Morus alba** L. White Mulberry. Apr.–May. Native to Asia; naturalized throughout the state. (Including f. *skeletoniana* [Schneider] Rehder, with very deeply lobed leaves.)

2. **Morus murrayana** D. E. Saar & S. J. Galla. Large-leaved Mulberry. Apr.–May. Woods; occasional in the s. ½ of the state.

3. **Morus rubra** L. Red Mulberry. Apr.–May. Lowland or upland woods, edges of fields; common throughout the state.

4. **Morus tatarica** L. Russian Mulberry. Apr.–May. Native to Europe and Asia; naturalized in disturbed soil and along the edges of woods; occasional throughout the state. *Morus alba* L. var. *tatarica* (L.) Loudon.

106. MYRICACEAE—WAX-MYRTLE FAMILY

1. Leaves pinnately lobed, not beset with yellow glands .... 1. *Comptonia*
1. Leaves entire or with a few spinulose teeth, beset with yellow glands ..... 2. *Morella*

1. **Comptonia** L'Her.—Sweet-fern

1. **Comptonia peregrina** (L.) Coult. Sweet-fern. Apr.–May. Sand flats and barrens; confined to the n. ¼ of IL.

2. **Morella** Lour.—Wax-myrtle

1. **Morella cerifera** (L.) Small. Wax-myrtle. May–June. Native to the se. U.S.; reported from Cook Co. in 1884 by Pepoon. *Myrica cerifera* L.

107. NELUMBONACEAE—WATER LOTUS FAMILY

1. **Nelumbo** Adans.—Water Lotus

1. **Nelumbo lutea** (Willd.) Pers. Water Lotus. July–Aug. Lakes and ponds; occasional throughout IL.

108. NYCTAGINACEAE—FOUR-O'CLOCK FAMILY

1. **Mirabilis** L.—Four-o'clock

1. Anthocarp (fruit) smooth or 5-angled; involucre not papery or membranous, scarcely or not enlarged in fruit .... 3. *M. jalapa*
1. Anthocarp (fruit) prominently 5-ribbed; involucre papery or membranous, greatly enlarged in fruit.
2. Leaves petiolate, ovate, rounded or cordate at the base .... 5. *M. nyctaginea*
2. Leaves sessile or nearly so, linear to lanceolate to oblong, not rounded or cordate at the base.
3. Leaves linear to linear-lanceolate, 1–5 mm wide .... 4. *M. linearis*
3. Leaves lanceolate to oblong, some or all of them more than 5 mm wide.
4. Stems densely hirsute, at least at the base and at the nodes .... 2. *M. hirsuta*
4. Stems glabrous or puberulent in lines .... 1. *M. albida*

1. **Mirabilis albida** (Walt.) Heimerl. White Wild Four-o'clock. July–Aug. Native to the s. U.S.; adventive in disturbed soil; Cook, Grundy, Logan, and Sangamon cos.

2. **Mirabilis hirsuta** (Pursh) MacM. Hairy Wild Four-o'clock. July–Aug. Apparently adventive in disturbed soil, including railroad ballast, in Cook, DuPage, St. Clair, Tazewell, and Will cos.; native in a hill prairie in Jo Daviess Co.

3. **Mirabilis jalapa** L. Garden Four-o'clock. June–Oct. Native to tropical America; occasionally planted as an ornamental but rarely escaped; Grundy Co.

4. **Mirabilis linearis** (Pursh) Heimerl. Linear-leaved Four-o'clock. June–Aug. Native to the w. U.S.; adventive in disturbed sandy soil and in railroad ballast; Cook, Madison, St. Clair, and Will cos.

5. **Mirabilis nyctaginea** (Michx.) MacM. Wild Four-o'clock; Umbrella-wort. June–Aug. Native to the w. U.S.; naturalized in IL, particularly along railroads; common throughout the state.

109. NYMPHAEACEAE—WATER LILY FAMILY

1. Flowers yellow; sepals 5–6; leaves oval, not peltate.................................... 1. *Nuphar*
1. Flowers white or pinkish; sepals 4; leaves orbicular, peltate .... 2. *Nymphaea*

1. **Nuphar** Smith—Pond Lily

1. Petioles terete or more or less flattened; sepals green .... 1. *N. advena*
1. Petioles conspicuously flattened; sepals red-tinged .... 2. *N. variegatum*

1. **Nuphar advena** (Ait.) Ait. f. Yellow Pond Lily; Cow Lily; Spatterdock. May–Aug. Ponds, swamps; occasional throughout the state. *Nuphar luteum* (L.) Sibth. & Smith ssp. *macrophyllum* (Dur.) Beal.

2. **Nuphar variegatum** Engelm. Bullhead Lily. May–Aug. Ponds, rare; Cook, Kendall, Lake, and McHenry cos. *Nuphar luteum* (L.) Sibth. & Smith ssp. *variegatum* (Engelm.) Beal.

2. **Nymphaea** L.—Water Lily

1. Petals subacute at tip; flowers fragrant; seeds 2 mm long...................... 1. *N. odorata*
1. Petals rounded at tip; flowers not fragrant; seeds 3–4 mm long .... 2. *N. tuberosa*

1. **Nymphaea odorata** Sol. Fragrant Water Lily. June–Sept. Lakes and ponds, rare; scattered in IL. (Including var. *gigantea* Tricker.)

2. **Nymphaea tuberosa** Paine. White Water Lily. June–Aug. Ponds, lakes, and streams; occasional in IL.

110. NYSSACEAE—NYSSA FAMILY

1. **Nyssa** L.—Tupelo

1. Most or all of the petioles 3 cm long or longer; pistillate flowers borne singly; fruit 2–3 cm long.. 1. *N. aquatica*
1. None of the petioles over 2.5 cm long; pistillate flowers (1–) 2 or more in a cluster; fruit up to 1.5 cm long.
  2. Flowers and fruits (2–) 3–5 in a cluster; largest leaves usually more than 6 cm long, some of them often with irregular teeth .... 3. *N. sylvatica*
  2. Flowers (1–) 2 (–3) in a cluster; largest leaves usually less than 6 cm long, none of them with irregular teeth .... 2. *N. biflora*

1. **Nyssa aquatica** L. Tupelo Gum. Apr.–May. Swamps, not common; in the s. 1/6 of the state.

2. **Nyssa biflora** Walt. Swamp Gum. Apr.–May. Wet soil, very rare. Massac Co. Specimen in the herbarium of the Missouri Botanical Garden.

3. **Nyssa sylvatica** Marsh. Two varieties may be distinguished in Illinois:

a. Leaves obtuse or acute; lower surface of leaves not papillose.................................. ....3a. *N. sylvatica* var. *sylvatica*
a. Leaves acuminate; lower surface of leaves papillose... .... 3b. *N. sylvatica* var. *caroliniana*

3a. **Nyssa sylvatica** Marsh. var. **sylvatica**. Sour Gum; Black Gum. Apr.–June. Bogs, swamps, not common; in the extreme s. tip of the state and extending up the far e. cos. to Cook Co.

3b. **Nyssa sylvatica** Marsh. var. **caroliniana** (Poir.) Fern. Black Gum. Apr.–June. Upland woods; occasional in the s. 1/3 of IL.

111. OLEACEAE—ASH FAMILY

1. Leaves pinnately compound; fruit a samara; trees .... 4. *Fraxinus*
1. Leaves simple; fruit a capsule, drupe, or berry; shrubs or small trees.
  2. Leaves serrate or somewhat serrulate.
    3. Petals absent; fruit an ellipsoid drupe .... 2. *Forestiera*
    3. Petals 4, yellow or greenish yellow; fruit a capsule........................ 3. *Forsythia*
  2. Leaves entire.
    4. Leaves cuneate at base; flowers white; fruit a drupe.
      5. Corolla tubular, the lobes lanceolate to lance-ovate; fruit green or black... ....5. *Ligustrum*

5. Corolla free except at the very base, the segmants linear; fruit purple or pink-purple .... 1. *Chionanthus*
4. Leaves truncate to cordate at base; flowers lilac (rarely white in *Syringa vulgaris*); fruit a capsule .... 6. *Syringa*

1. **Chionanthus** L.—Fringe-tree

1. **Chionanthus virginicus** L. Fringe-tree. May. Native to the se. U.S.; rarely escaped from cultivation; DuPage Co.

2. **Forestiera** Poir.—Swamp Privet

1. **Forestiera acuminata** (Michx.) Poir. Swamp Privet. Mar.–Apr. Swamps, low woods, along rivers; occasional in the s. ⅗ of IL.

3. **Forsythia** Vahl—Forsythia

1. Branches hollow, except at the nodes .... 2. *F. suspensa*
1. Branches solid.
2. Branches more or less erect; flowers greenish yellow; leaves never 3-parted .... 3. *F. viridissima*
2. Branches distinctly arching; flowers bright yellow; leaves sometimes 3-parted .... 1. *F. X intermedia*

1. **Forsythia X intermedia** Zabel. Hybrid Forsythia. Apr.–May. Cultivar; rarely escaped from cultivation; DuPage Co. This is reputed to be the hybrid between *F. suspensa* (Thunb.) Vahl and *F. viridissima* Lindl.

2. **Forsythia suspensa** (Thunb.) Vahl. Forsythia. Apr.–May. Native to Europe; rarely escaped from cultivation; Jackson, Jersey, and LaSalle cos.

3. **Forsythia viridissima** Lindl. Yellow-green Forsythia. Apr.–May. Native to China; rarely escaped from cultivation; Coles Co.

4. **Fraxinus** L.—Ash

1. Leaflets sessile (except for the terminal one) .... 4. *F. nigra*
1. Leaflets petiolulate.
2. Twigs 4-sided; body of samara flat .... 7. *F. quadrangulata*
2. Twigs terete or somewhat angular, not 4-sided; body of samara terete.
3. Leaflets more or less green on both sides; wing of samara covering at least ½ of the body.
4. Twigs, leaflets, and petioles densely pubescent .... 5. *F. pennsylvanica*
4. Twigs, leaflets, and petioles glabrous or nearly so .... 3. *F. lanceolata*
3. Leaflets whitened or more or less brownish beneath; wing of samara covering at most ⅓ of the body.
5. Wing of samara 7–15 mm broad; petiolules 8–15 mm long .... 6. *F. profunda*
5. Wing of samarra 4–6 mm broad; petioles up to 5 mm long.
6. Twigs, leaflets, and petioles densely velvety-tomentose .... 2. *F. biltmoreana*
6. Twigs, leaflets, and petioles glabrous or pubescent but not velvety-tomentose .... 1. *F. americana*

1. **Fraxinus americana** L. White Ash. Apr.–May. Upland woods; common throughout the state.

2. **Fraxinus biltmoreana** Beadle. Biltmore Ash. Apr.–May. Upland woods, not common; in the s. ¼ of IL; also DuPage and Kane cos.

3. **Fraxinius lanceolata** Borkh. Green Ash. Apr.–May. Moist woods, bottomland forests; common throughout the state.

4. **Fraxinus nigra** Marsh. Black Ash. May–June. Moist woods, springy slopes; occasional in the n. ½ of IL; also Fayette, Jersey, Lawrence, and Wabash cos.

5. **Fraxinus pennsylvanica** Marsh. Red Ash. Apr.–May. Wet woods; occasional throughout the state.

6. **Fraxinus profunda** (Bush) Bush. Pumpkin Ash. Apr.–May. Swamps, low woods, bottomland forests; occasional or rare in the s. ⅖ of IL; also Iroquois Co. *Fraxinus tomentosa* Michx. f.

7. **Fraxinus quadrangulata** Michx. Blue Ash. Square-stemmed Ash. Mar.–Apr. Moist woods, limestone cliffs; occasional in the n. ⅖ of IL, rare elsewhere.

5. **Ligustrum** L.—Privet

1. Stems and leaves densely pubescent .... 1. *L. obtusifolium*
1. Stems and leaves glabrous or nearly so .... 2. *L. vulgare*

1. **Ligustrum obtusifolium** Sieb. & Zucc. Blunt-leaved Privet. June–July. Native to Japan; occasionally escaped from cultivation in IL.
2. **Ligustrum vulgare** L. Common Privet. Native to Europe; seldom occasionally scaped from cultivation in IL.

6. **Syringa** L.—Lilac

1. Leaves pubescent beneath, at least on the veins .... 2. *S. pubescens*
1. Leaves glabrous.
  2. Leaves truncate or cordate at base .... 3. *S. vulgaris*
  2. Leaves tapering to base .... 1. *S. X chinensis*

1. **Syringa X chinensis** Willd. Chinese Lilac. May–June. Cultivar; rarely escaped from cultivation; Cook Co. This is reputed to be the hybrid between *S. persica* L. and *S. vulgaris* L.
2. **Syringa pubescens** Turcz. ssp. **patens** (Palib.) M. C. Chang & X.Y. Chen. Lilac. May–June. Rarely escaped from cultivation; DuPage Co.
3. **Syringa vulgaris** L. Common Lilac. May–June. Native to Europe; occasionally persisting around old home sites in IL.

112. ONAGRACEAE—EVENING PRIMROSE FAMILY

1. Calyx 2-lobed; petals 2; stamens 2 .... 3. *Circaea*
1. Calyx 4- to 5-lobed; petals 4–5 or absent; stamens 4–12.
  2. Leaves opposite.
    3. Leaves entire .... 6. *Ludwigia*
    3. Leaves serrate .... 4. *Epilobium*
  2. Leaves alternate and/or basal.
    4. Petals yellow or greenish.
      5. Calyx divided to the tip of the ovary, persistent on the fruit ..... 6. *Ludwigia*
      5. Calyx not divided to the tip of the ovary, deciduous from the fruit.
        6. Stigma discoid, not distinctly 4-lobed .... 1. *Calylophus*
        6. Stigma distinctly 4-lobed .... 7. *Oenothera*
    4. Petals white or pink.
      7. Petals at least 1 cm long.
        8. Calyx tube prolonged beyond the ovary; seeds without a coma .... 7. *Oenothera*
        8. Calyx tube not prolonged beyond the ovary; seeds with a coma .... 2. *Chamerion*
      7. Petals up to 1 cm long.
        9. Fruit indehiscent, with 1–4 seeds; seeds without a coma .... 5. *Gaura*
        9. Fruit dehiscent, with several seeds; seeds with a coma .... 4. *Epilobium*

1. **Calylophus** Spach

1. **Calylophus serrulatus** (Nutt.) Raven. Toothed Evening Primrose. May–Sept. Native to the w. U.S.; rarely adventive in disturbed soil; Fulton, Ogle, Randolph and Winnebago cos. *Oenothera serrulata* Nutt.

2. **Chamerion** Raf.

1. **Chamerion angustifolium** (L.) Holub. Fireweed. June–Aug. Dunes, bogs, often after a fire; confined to the n. ¼ of IL. *Epilobium angustifolium* L.

3. **Circaea** L.—Enchanter's Nightshade

1. Stems weak, up to 30 cm tall; calyx lobes up to 1.2 mm long; bristles of fruit weak; fruit 1-locular .... 1. *C. alpina*
1. Stems firm, often 40 cm tall or taller; calyx lobes (1.8–) 2.0–2.5 mm long; bristles of fruit stiff; fruit 2-locular .... 2. *C. lutetiana*

1. **Circaea alpina** L. Small Enchanter's Nightshade. June–July. Cool ravines, rare; Cook, Jo Daviess, Kane, and Lake cos.

2. **Circaea lutetiana** Aschers. & Magnus ssp. **canadensis** (L.) Aschers. & Magnus. Enchanter's Nightshade. June–Aug. Moist woods; common throughout the state. *Circaea quadrisulcata* (Maxim.) Franch. & Sav. var. *canadensis* (L.) Hara.

4. **Epilobium** L.—Willow Herb

1. Stigmas deeply 4-parted.
  2. Petals deeply notched, at least 1 cm long; some of the leaves usually clasping .... 3. *E. hirsutum*
  2. Petals shallowly notched, up to 9 mm long; none of the leaves clasping .... 6. *E. parviflorum*
1. Stigmas not 4-parted.
  3. Leaves entire or nearly so, often revolute, up to 1 cm wide (to 1.5 cm wide in *E. palustre*).
    4. Stems and leaves velvety-pubescent .... 7. *E. strictum*
    4. Stems and leaves canescent or strigose, not velvety-pubescent.
      5. Leaves up to 7 mm wide, finely pubescent on the upper surface .... 4. *E. leptophyllum*
      5. Leaves (5–) 7–15 mm wide, glabrous on the upper surface .... 5. *E. palustre*
  3. Leaves serrulate or denticulate, not revolute, usually more than 1 cm wide.
    6. Inflorescence and capsules glandular-pubescent; seeds short-beaked; coma white.
      7. Petioles up to 5 mm long; coma persistent .... 8. *E. X wisconsinensis*
      7. Petioles up to 10 mm long; coma deciduous .... 1. *E. ciliatum*
    6. Inflorescence and capsules not glandular-pubescent; seeds beakless; coma reddish brown .... 2. *E. coloratum*

1. **Epilobium ciliatum** Raf. Northern Willow Herb. July–Sept. Bogs, calcareous springy places; occasional in the n. ¼ of IL. *Epilobium adenocaulon* Haussk.; *E. glandulosum* Lehm. var. *adenocaulon* (Haussk.) Fern.

2. **Epilobium coloratum** Spreng. Cinnamon Willow Herb. Aug.–Sept. Bogs, marshes, wet ground; occasional to common throughout IL.

3. **Epilobium hirsutum** L. Hairy Willow Herb. Aug. Native to Europe; adventive in springy habitats and borders of ditches; confined to the ne. cos. of IL.

4. **Epilobium leptophyllum** Raf. Bog Willow Herb. Aug.–Sept. Bogs, springy habitats, not common; confined to the n. ½ of IL.

5. **Epilobium palustre** L. Marsh Willow Herb. July–Aug. Wet soil, very rare; Cook Co.

6. **Epilobium parviflorum** Schreb. Small-flowered Hairy Willow Herb. July–Aug. Native to Europe; adventive in wet areas; DuPage Co.

7. **Epilobium strictum** Muhl. Downy Willow Herb. July–Sept. Bogs, springy habitats, not common; confined to the n. ⅓ of IL.

8. **Epilobium X wisconsinensis** Ugent. Wisconsin Willow Herb. Aug.–Sept. Wet ground, rare; Cook, Kankakee, and Kendall cos. Reputed to be the hybrid between *E. ciliatum* Raf. and *E. coloratum* Spreng.

5. **Gaura** L.

1. Ovary and capsule stipitate .... 2. *G. filipes*
1. Ovary and capsule sessile.
  2. Petals 1.5–3.0 mm long; calyx lobes 2.0–3.5 mm long .... 4. *G. parviflora*
  2. Petals 5.5–15.0 mm long; calyx lobes 6–13 mm long.
    3. Stems densely villous .... 1. *G. biennis*
    3. Stems strigulose to short-villous .... 3. *G. longiflora*

1. **Gaura biennis** L. Butterfly-weed. July–Sept. Clearings in woods, roadsides, other disturbed areas; occasional throughout the state.

2. **Gaura filipes** Spach. Slender Gaura. July–Sept. Dry woods, very rare; Hardin Co.; adventive in LaSalle and Will cos. (Including var. *major* Torr. & Gray.)

3. **Gaura longiflora** Spach. Gaura. July–Oct. Disturbed areas, often along roadsides

and railway embankments; occasional throughout the state. *Gaura biennis* L. var. *pitcheri* Torr. & Gray.

4. **Gaura parviflora** Dougl. Small-flowered Gaura. June–July. Native to the w. U.S.; adventive in fields, roadsides, edge of woods; occasional in the n. ½ of the state, less common elsewhere. *Oenothera curtifolia* W. L. Wagner & Hoch.

6. **Ludwigia** L.

1. Leaves opposite .... 5. *L. palustris*
1. Leaves alternate.
  2. Stamens 4; fruit up to 1 cm long.
    3. Flowers at least 1 cm across, yellow, short-pedicellate .... 1. *L. alternifolia*
    3. Flowers up to 3 mm across, greenish, sessile.
      4. Calyx tube 6–10 mm long, about 4 times longer than the calyx lobes .... 3. *L. glandulosa*
      4. Calyx tube up to 5 mm long, at most about 2 times longer than the calyx lobes.
        5. Bracts 2–5 mm long .... 7. *L. polycarpa*
        5. Bracts up to 1 mm long .... 8. *L. sphaerocarpa*
  2. Stamens 8–12; fruit 1.5 cm long or longer.
    6. Stems and leaves pubescent .... 4. *L. leptocarpa*
    6. Stems and leaves glabrous.
      7. Plants more or less erect; petals 4; leaves decurrent; capsule 4-sided .... 2. *L. decurrens*
      7. Plants creeping; petals 5; leaves not decurrent; capsule cylindrical .... 6. *L. peploides*

1. **Ludwigia alternifolia** L. Two varieties may be recognized in Illinois:

a. Stems, leaves, and calyx glabrous .... 1a. *L. alternifolia* var. *alternifolia*
a. Stems, leaves, and calyx pubescent .... 1b. *L. alternifolia* var. *pubescens*

1a. **Ludwigia alternifolia** L. var. **alternifolia**. Seedbox. June–Aug. Wet ground; occasional throughout the state.

1b. **Ludwigia alternifolia** L. var. **pubescens** Palmer & Steyerm. Hairy Seedbox. June–Aug. Wet ground; occasional in the southernmost cos.

2. **Ludwigia decurrens** Walt. Erect Primrose Willow. July–Sept. Wet ground, not common; mostly in the s. ⅙ of the state; also Cass, Jasper, and McDonough cos. *Jussiaea decurrens* (Walt.) DC.

3. **Ludwigia glandulosa** Walt. False Loosestrife. June–Sept. Swamps, low woods, not common; confined to the s. ⅕ of the state.

4. **Ludwigia leptocarpa** (Nutt.) Hara. Hairy Primrose Willow. July–Sept. Wet ground, particularly along the Mississippi River; confined to the s. ⅕ of the state. *Jussiaea leptocarpa* Nutt.

5. **Ludwigia palustris** (L.) Ell. var. **americana** (DC.) Fern. & Grisc. Marsh Purslane. July–Aug. Wet soil, particularly around ponds, along streams, and in ditches; common throughout the state.

6. **Ludwigia peploides** (HBK.) Raven var. **glabrescens** (Ktze.) Raven. Creeping Primrose Willow. May–Oct. In shallow water or on muddy banks and shores; occasional to common in the s. ⅔ of the state; also Cook, DuPage, and Marshall cos. *Jussiaea repens* L. var. *glabrescens* Ktze.

7. **Ludwigia polycarpa** Short & Peter. False Loosestrife. June–Sept. Wet soil, particularly in ditches; occasional throughout the state.

8. **Ludwigia sphaerocarpa** Ell. var. **deamii** Fern. & Grisc. Round-fruited Loosestrife. July–Sept. Swamp borders and in shallow water, very rare; Cook and Will cos.

7. **Oenothera** L.—Evening Primrose

1. Plants without leafy stems; flowers basal .... 17. *O. triloba*
1. Plants with leafy stems; flowers not basal.
  2. Petals white, pink, or rosy.
    3. Stems white; capsule 2–4 cm long.

4. Stem with peeling epidermis; leaves not pinnatifid; petals not deeply notched; perennial .... 10. *O. nuttallii*
4. Stem without peeling epidermis; leaves pinnatifid; petals deeply notched; annual .... 1. *O. albicaulis*
3. Stems green; capsule up to 2 cm long .... 16. *O. speciosa*
2. Petals yellow, sometimes fading reddish.
5. Leaves filiform, up to 1 mm broad .... 8. *O. linifolia*
5. Leaves linear or broader, never filiform, more than 1 mm broad.
6. Some or all the petals 4 cm long or longer; capsule at least 4 cm long.
7. Fruit and ovary terete or nearly so, not winged; plants up to 1.5 m tall .... 5. *O. glazioviana*
7. Fruit and ovary 4-angled and winged; plants up to 50 cm tall .... 9. *O. macrocarpa*
6. None of the petals 4 cm long; capsule up to 4 cm long.
8. Some of the leaves sinuate-pinnatifid.
9. Petals up to 1.5 (–2.5) cm long; style exserted 0.3–2.0 (–2.5) cm beyond the floral tube .... 7. *O. laciniata*
9. Petals 2.5–4.0 cm long; style exserted 1.5–2.0 cm beyond the floral tube .... 6. *O. grandis*
8. None of the leaves sinuate-pinnatifid.
10. Ovary and capsule sharply 4-angled.
11. Petals up to 1 cm long; flower buds nodding; anthers to 2.5 mm long .... 13. *O. perennis*
11. Petals at least 1 cm long; flower buds erect; anthers 4 mm long or longer.
12. Capsule glandular-pubescent .... 4. *O. fruticosa*
12. Capsule eglandular.
13. Stems with spreading hairs; calyx tube 1.5–2.5 cm long .... 14. *O. pilosella*
13. Stems with appressed hairs; calyx tube 0.5–1.5 cm long .... 4. *O. fruticosa*
10. Ovary and capsule more or less terete.
14. Capsule at least 4 mm thick at base; most of the main cauline leaves 10 mm broad or broader.
15. Petals 3.5 mm wide or wider.
16. Plants white-hairy .... 18. *O. villosa*
16. Plants variously pubescent but not white-hairy .... 2. *O. biennis*
15. Petals up to 3 mm wide.
17. Pubescence of both short and long hairs, some of them gland-tipped; leaves bright green .... 12. *O. parviflora*
17. Plants canescent, with all short hairs, none of which are gland-tipped; leaves gray-green to dull green .... 11. *O. oakesiana*
14. Capsule up to 3 mm thick at base; most of the main cauline leaves up to 10 mm broad.
18. Sepals 1.5–2.5 cm long; petals 1.5–3.5 cm long .... 15. *O. rhombipetala*
18. Sepals 0.5–1.5 cm long; petals 0.5–1.5 (–1.7) long .... 3. *O. clelandii*

1. **Oenothera albicaulis** Pursh. Prairie Evening Primrose. June–July. Native to the w. U.S.; rarely adventive in disturbed soil; McDonough Co.

2. **Oenothera biennis** L. Evening Primrose. June–Oct. Fields, prairies, disturbed soil, common; in every co. (Including var. *pycnocarpa* [Atkinson & Bartlett] Wieg.)

3. **Oenothera clelandii** W. Dietr., Raven, & W. L. Wagner. Sand Evening Primrose. June–Oct. Disturbed sandy soils, black oak savannas; scattered throughout the state but more common in the n. cos.

4. **Oenothera fruticosa** L. Two subspecies may be distinguished in Illinois:

a. Capsule eglandular .... 4a. *O. fruticosa* ssp. *fruticosa*
a. Capsule glandular .... 4b. *O. fruticosa* ssp. *glauca*

4a. **Oenothera fruticosa** L. ssp. **fruticosa**. Shrubby Sundrops. May–Aug. Rocky woods, rare; confined to the s. ⅙ of IL. (Including var. *linearis* [Michx.] S. Wats.)

4b. **Oenothera fruticosa** L. ssp. **glauca** (Michx.) Straley. Four-angled Sundrops. June–Aug. Dry woods, rare; confined to the s. ⅙ of IL. *Oenothera tetragona* Roth.

5. **Oenothera glazioviana** Micheli. Showy Evening Primrose. June–July. Native to Europe; rarely escaped from cultivation; Cook Co.

6. **Oenothera grandis** (Britt.) Smyth. Showy Ragged Evening Primrose. May–Aug. Native to the sw. U.S.; probably adventive in IL; Champaign and Cook cos.

7. **Oenothera laciniata** Hill. Ragged Evening Primrose. May–Aug. Fields, prairies, roadsides; occasional to common throughout IL.

8. **Oenothera linifolia** Nutt. Thread-leaved Sundrops. May–July. Sandstone cliffs, prairies; occasional in the s. ⅕ of IL.

9. **Oenothera macrocarpa** Nutt. Missouri Evening Primrose. May–Aug. Rocky glades, very rare; St. Clair Co; reported as an adventive in DuPage Co. *Oenothera missouriensis* Sims.

10. **Oenothera nuttallii** Sweet. White Evening Primrose. June–July. Native to the w. U.S.; rarely adventive in IL; Boone, Christian, Lake, and McHenry cos.

11. **Oenothera oakesiana** L. Oakes' Evening Primrose. July–Aug. Sandy soil, very rare; a few ne. cos.

12. **Oenothera parviflora** L. Small-flowered Evening Primrose. July–Aug. Native to the ne. U.S.; adventive in dry soil; adventive in the n. ½ of IL.

13. **Oenothera perennis** L. Small Sundrops. June–Aug. Moist, open ground, very rare; Cook, DuPage, Lake, McHenry, Will, and Winnebago cos.

14. **Oenothera pilosella** Raf. Prairie Sundrops. May–July. Prairies, fields; occasional throughout the state.

15. **Oenothera rhombipetala** Nutt. Sand Evening Primrose. June–Sept. Disturbed sandy soil; occasional in the s. ¾ of IL, rare or absent elsewhere.

16. **Oenothera speciosa** Nutt. Showy Evening Primrose. June–July. Native to the w. U.S.; occasionally or commonly adventive along roadsides throughout IL.

17. **Oenothera triloba** Nutt. Stemless Evening Primrose. Apr.–May. Native to thes. U.S.; adventive in dry soil; Bureau and Jackson cos.

18. **Oenothera villosa** Thunb. Gray Evening Primrose. June–Oct. Fields, prairies, waste ground; occasional throughout IL. *Oenothera biennis* L. var. *canescens* Torr. & Gray. (Including *O. biennis* L. var. *hirsutissima* Gray.)

113. OROBANCHACEAE—BROOMRAPE FAMILY

1. Calyx deeply split on the lower side, 3- to 4-toothed on the upper side; stamens exserted .... 1. *Conopholis*
1. Calyx more or less equally 4- or 5-toothed; stamens included.
   2. Corolla 5-lobed; flower solitary .... 3. *Orobanche*
   2. Corolla 4-lobed; flowers spicate.
      3. Calyx 4-toothed .... 3. *Orobanche*
      3. Calyx 5-toothed.
         4. Corolla up to 1 cm long .... 2. *Epifagus*
         4. Corolla 1.5–2.0 cm long .... 3. *Orobanche*

1. **Conopholis** Wallr.—Squaw-root

1. **Conopholis americana** (L.) Wallr. Cancer-root; Squaw-root. May–July. Wooded ravines; parasitic on oak roots, rare; scattered in the n. ½ of IL.

2. **Epifagus** Nutt.—Beech-drops

1. **Epifagus virginiana** (L.) Bart. Beech-drops. Aug.–Oct. Parasitic on beech roots; restricted to the s. ¼ of IL; also Clark, Crawford, Edgar, Lawrence, and Vermilion cos.

3. **Orobanche** L.—Broomrape

1. Flower solitary, not subtended by bracts; corolla 5-lobed.
   2. Corolla purple; calyx lobes deltoid, equaling or shorter than the tube .... 1. *O. fasciculata*

2. Corolla creamy-white to lilac; calyx lobes lance-subulate, longer than the tube.. .... 5. *O. uniflora*

1. Flowers in spikes; each flower subtended by 1–3 bracts; corolla 4-lobed.
    3. Calyx 5-lobed; each flower subtended by 1–2 bracts.
        4. Lobes of corolla obtuse, 4–8 mm long; tube of corolla slightly curved; flowers dense in a compact raceme; perennial of dry soil, flowering Apr.–Aug. .... 2. *O. ludoviciana*
        4. Lobes of corolla acute, 4–5 mm long; tube of corolla strongly curved; flowers loose in an open raceme; annual of wet soil, flowering Aug.–Oct. .... 4. *O. riparia*
    3. Calyx 4-lobed; each flower subtended by 3 bracts .... 3. *O. ramosa*

1. **Orobanche fasciculata** Nutt. Clustered Broomrape. Apr.–Aug. Parasitic on roots of various members of the Asteraceae, not common; confined to the n. ½ of IL.

2. **Orobanche ludoviciana** Nutt. Broomrape. Apr.–Aug. Parasitic on roots of various members of the Asteraceae in dry soil; rare in the n. ½ of IL; also Fayette, Wabash, and White cos.

3. **Orobanche ramosa** L. Broomrape. June–Aug. Native to Europe; parasitic on hemp roots, very rare; Champaign Co.

4. **Orobanche riparia** L.T. Collins. Riverbank Broomrape. Aug.–Oct. Parasitic on roots of various members of the Asteraceae in wet soil; rare in c. IL; Cumberland, Mason, Menard, and Wabash cos.

5. **Orobanche uniflora** L. One-flowered Broomrape. May–June. Parasitic on roots of various plants; scattered throughout the state.

114. OXALIDACEAE—SORREL FAMILY

1. **Oxalis** L.—Wood Sorrel

1. Leaves all from the base of the plant; flowers purple to lavender (rarely white); plants with bulbs .... 5. *O. violacea*
1. Leaves cauline; flowers yellow; plants without bulbs (except *O. illinoensis*).
    2. Plants creeping, rooting at the nodes .... 1. *O. corniculata*
    2. Plants erect or ascending, not rooting at the nodes.
        3. Flowers 12–18 mm long; seeds 1.8–2.2 mm long .... 3. *O. illinoensis*
        3. Flowers up to 12 mm long; seeds 1.0–1.6 mm long.
            4. Septate hairs absent on all parts; capsules appressed-pubescent; flowers in umbels; stipules oblong .... 4. *O. stricta*
            4. Septate hairs present on stems, petioles, and pedicels; capsules glabrous or with a few scattered hairs; flowers in cymes; stipules linear .... 2. *O. fontana*

1. **Oxalis corniculata** L. Creeping Wood Sorrel. Apr.–Oct. Native to the tropics; adventive in waste areas, often in greenhouses; occasional in IL.

2. **Oxalis fontana** Bunge. Yellow Wood Sorrel. May–Oct. Woods, fields, roadsides, common; in every co. *Oxalis stricta* L., misapplied.

3. **Oxalis illinoensis** Schwegm. Illinois Wood Sorrel. June–Sept. Rich woods, very rare; Hardin and Pope cos.

4. **Oxalis stricta** L. Yellow Wood Sorrel. May–Nov. Fields, prairies, roadsides, common; in every co. *Oxalis dillenii* Jacq.

5. **Oxalis violacea** L. Purple Oxalis; Purple Wood Sorrel. Apr.–June. Woods, bluffs, prairies, common; in every co.

115. PAPAVERACEAE—POPPY FAMILY

1. Flowers basically white (pink in a rare form of *Sanguinaria*).
    2. Perianth parts 2 .... 4. *Macleaya*
    2. Perianth parts 6 or more.
        3. Plants without aerial stems; petals 6 or more .... 6. *Sanguinaria*
        3. Plants with aerial stems; petals 4–6.
            4. Leaves prickly .... 1. *Argemone*
            4. Leaves not prickly .... 5. *Papaver*
1. Flowers not white.
    5. Leaves pinnatifid.

6. Leaves prickly .... 1. *Argemone*
6. Leaves not prickly.
  7. Flowers red or orange; sap milky; capsules opening by pores along the edge .... 5. *Papaver*
  7. Flowers yellow; sap yellow; capsules opening from the bottom upward.
    8. Capsules bristly; styles distinct; petals 2–3 cm long .... 7. *Stylophorum*
    8. Capsules smooth; style inconspicuous or absent; petals about 1 cm long .... 2. *Chelidonium*
5. Leaves ternately divided .... 3. *Eschscholtzia*

1. **Argemone** L.—Prickly Poppy

1. Flowers yellow or orange or cream; leaves with patches of pale green .... 2. *A. mexicana*
1. Flowers white or pink; leaves of a uniform color.
  2. Peduncles leafy .... 3. *A. polyanthemos*
  2. Peduncles leafless .... 1. *A. albiflora*

1. **Argemone albiflora** Hornem. June–Sept. White Prickly Poppy. Native to the w. U.S.; cultivated and occasionally escaped in IL. *Argemone intermedia* Sweet.

2. **Argemone mexicana** L. Mexican Poppy. May–Sept. Introduced from Mexico; rarely escaped in IL; DuPage, Henderson, Mason, Menard, Stephenson, and Tazewell cos.

3. **Argemone polyanthemos** (Fedde) G. Ownbey. June–Sept. White Prickly Poppy. Native to the w. U.S.; occasional as an adventive; the Morgan Co. collection may represent a native population.

2. **Chelidonium** L.—Celandine

1. **Chelidonium majus** L. Celandine. May–Aug. Native to Europe; occasionally escaped from cultivation; scattered in IL.

3. **Eschscholtzia** Cham.—California Poppy

1. **Eschscholtzia californica** Cham. California Poppy. June–Sept. Native to the w. U.S.; rarely escaped from cultivation; Kane and Wabash cos.

4. **Macleaya** R. Br.—Plume Poppy

1. **Macleaya cordata** (Willd.) R. Br. Plume Poppy. June–July. Native to Asia; seldom planted as an ornamental and rarely escaped into disturbed soil; Cook, Henry, Madison, Schuyler, and Will cos.

5. **Papaver** L.—Poppy

1. Stems glabrous; cauline leaves cordate-clasping; plants glaucous; capsules nearly spherical .... 3. *P. somniferum*
1. Stems hirsute; cauline leaves not cordate-clasping; plants not glaucous; capsules longer than broad.
  2. Capsules narrowly ovoid .... 1. *P. dubium*
  2. Capsules broadly obovoid .... 2. *P. rhoeas*

1. **Papaver dubium** L. Poppy. May–Aug. Native to Europe; rarely escaped from cultivation; scattered in IL.

2. **Papaver rhoeas** L. Corn Poppy. May–Sept. Native to Europe; occasionally escaped from cultivation; scattered in IL.

3. **Papaver somniferum** L. Opium Poppy. June–Aug. Native to Europe; rarely escaped from cultivation; scattered in IL.

6. **Sanguinaria** L.—Bloodroot

1. **Sanguinaria canadensis** L. Bloodroot. Mar.–Apr. Rich woods; common throughout the state. (Including f. *colbyorum* Benke, with pink petals, and var. *rotundifolium* [Greene] Fedde, with flowers elevated above the leaves.)

7. **Stylophorum** Nutt.—Celandine Poppy

1. **Stylophorum diphyllum** (Michx.) Nutt. Celandine Poppy. Mar.–May. Rich woods,

not common; restricted to the s. cos.; also Cook, DuPage, Lake, Macon, and Vermilion cos.

116. PARNASSIACEAE—GRASS-OF-PARNASSUS FAMILY

1. **Parnassia** L.—Grass-of-Parnassus

1. **Parnassia glauca** Raf. Grass-of-Parnassus. July–Sept. Calcareous springs, dune flats; restricted to the n. ½ of IL.

117. PASSIFLORACEAE—PASSION-FLOWER FAMILY

1. **Passiflora** L.—Passion-flower

1. Leaves deeply 3- or 5-lobed, the lobes acute to acuminate, serrulate; flowers usually purple and white, 4 cm or more across; fruit yellow .... 1. *P. incarnata*
1. Leaves shallowly 3-lobed, the lobes obtuse, entire; flowers greenish yellow, up to 2.5 cm across; fruit purple .... 2. *P. lutea*

1. **Passiflora incarnata** L. Large Purple Passion-flower; Maypops. May–Sept. Dry soil, often in fields, roadsides; occasional in the s. ⅓ of the state.

2. **Passiflora lutea** L. var. **glabriflora** Fern. Small Passion-flower. May–Sept. Woods, thickets; occasional to common in the s. ½ of the state, rare or absent elsewhere.

118. PAULOWNIACEAE—PRINCESS TREE FAMILY

1. **Paulowina** Sieb. & Zucc.—Princess Tree

1. **Paulownia tomentosa** (Thunb.) Steud. Princess Tree. Apr.–May. Native to Asia; adventive in the s. ⅓ of IL. This genus is sometimes placed in the Bignoniaceae or Scrophulariaceae.

119. PEDALIACEAE—SESAME FAMILY

1. **Proboscidea** Schmidel—Unicorn-plant

1. **Proboscidea louisianica** (Mill.) Thell. Proboscis-flower; Unicorn-plant. June–Sept. Low ground, particularly along rivers; occasional throughout the state, where it is probably naturalized.

120. PENTHORACEAE—DITCH STONECROP FAMILY

1. **Penthorum** L.—Ditch Stonecrop

1. **Penthorum sedoides** L. Ditch Stonecrop. July–Oct. Wet ground, common; in every co. This genus is sometimes placed in the Crassulaceae or in the Saxifragaceae.

121. PHILADELPHACEAE—MOCK ORANGE FAMILY

1. Stamens 10; pubescence stellate .... 1. *Deutzia*
1. Stamens 15 or more; pubescence not stellate .... 2. *Philadelphus*

1. **Deutzia** Thunb.—Deutzia

1. **Deutzia scabra** Thunb. Pride-of-Rochester. May–June. Native to China and Japan; rarely escaped from cultivation; Jackson Co.

2. **Philadelphus** L.—Mock Orange

1. Flowers 1–4 in a terminal inflorescence.
  2. Flowers with a sweet fragrance .... 2. *P. floridus*
  2. Flowers without a fragrance .... 3. *P. inodorus*
1. Flowers 5–7 in both terminal and lateral racemes.
  3. Calyx glabrous; lower surface of leaves pubescent only on the nerves; flowers very fragrant .... 1. *P. coronarius*
  3. Calyx pubescent; lower surface of leaves pubescent throughout; flowers faintly fragrant.
    4. Twigs with close gray bark .... 4. *P. pubescens*
    4. Twigs with flaky reddish brown bark .... 5. *P. verrucosus*

1. **Philadelphus coronarius** L. Sweet Mock Orange. May–June. Native to Europe; sometimes cultivated but rarely escaped.

2. **Philadelphus floridus** Beadle. Few-flowered Mock Orange. July. Native to the s. U.S.; seldom cultivated and rarely escaped; Cook Co.

3. **Philadelphus inodorus** L. Scentless Mock Orange. May–June. Native to the se. U.S.; occasionally cultivated but rarely escaped; Cook, Jackson, Pulaski, Tazewell, Union, and Will cos.

4. **Philadelphus pubescens** Loisel. Downy Mock Orange. June. Native to the s. U.S.; sometimes cultivated but rarely escaped; Cook, DuPage, Grundy, and Madison cos.

5. **Philadelphus verrucosus** Schrad. Native Mock Orange. May–June. Wooded bluffs along the Ohio River, very rare; Pope Co.

122. PHRYMACEAE—LOPSEED FAMILY

1. **Phryma** L.—Lopseed

1. **Phryma leptostachya** L. Lopseed. June–Sept. Rich woods, occasional; scattered throughout IL.

123. PHYTOLACCACEAE—POKEWEED FAMILY

1. **Phytolacca** L.—Pokeweed

1. **Phytolacca americana** L. Pokeweed; Pokeberry. July–Oct. Fields, woods, waste areas; occasional to common throughout the state.

124. PLANTAGINACEAE—PLANTAIN FAMILY

Plantago L.—Plantain

1. Leaves opposite, cauline .... 8. *P. psyllium*
1. Leaves all basal.
  2. Leaves linear, never broader than 1 cm.
    3. Bracts much exceeding the flowers, very conspicuous .... 1. *P. aristata*
    3. Bracts shorter than to barely exceeding the flowers, not conspicuous.
      4. Bracts, sepals, and inflorescence tomentose .... 7. *P. patagonica*
      4. Bracts, sepals, and inflorescence glabrous or nearly so.
        5. Capsules with 4 seeds; sepals obovate .... 9. *P. pusilla*
        5. Capsules with 10 or more seeds; sepals ovate .... 3. *P. heterophylla*
  2. Leaves lanceolate, elliptic, oval, or ovate, some or all of them usually over 1 cm broad.
    6. Leaves cordate at the base, with lateral veins arising from the midvein; flowering stalks hollow .... 2. *P. cordata*
    6. Leaves rounded or cuneate at the base, not cordate, with few or no veins appearing to arise from the midvein; flowering stalks solid.
      7. Leaves lanceolate, oblanceolate, or elliptic; seeds 1–3 (–4) per capsule.
        8. Spikes ellipsoid; leaves lanceolate .... 4. *P. lanceolata*
        8. Spikes cylindrical; leaves obovate to elliptic.
          9. Flowers fragrant; corolla spreading, leaving the capsule exposed .... 6. *P. media*
          9. Flowers not fragrant; corolla ascending, closed over the concealed capsule.
            10. Leaves mostly entire; sepals obtuse; seeds pale brown, up to 2 mm long .... 12. *P. virginica*
            10. Leaves with a few coarse teeth; sepals attenuate at the tip; seeds red, 2.5–3.0 mm long .... 10. *P. rhodosperma*
      7. Leaves ovate or oval; seeds usually 4 or more per capsule.
        11. Seeds flat; leaves densely canescent; seeds never more than 4 per capsule .... 6. *P. media*
        11. Seeds turgid; leaves glabrous or nearly so on at least 1 surface, not densely canescent.
          12. Base of petioles purplish; sepals acute or subacute, 2.5–3.0 mm long .... 11. *P. rugelii*
          12. Base of petioles green; sepals obtuse, up to 2 mm long .... 5. *P. major*

1. **Plantago aristata** Michx. Bracted Plantain. May–Nov. Pastures, fields, upland woods, cliffs, waste ground; common throughout the state.

2. **Plantago cordata** Lam. Heart-leaved Plantain. Apr.–July. Along streams in woods, rare; scattered in IL.

3. **Plantago heterophylla** Nutt. Small Plantain. Apr.–May. Sandy soil, very rare; Pulaski and Union cos.

4. **Plantago lanceolata** L. Buckhorn Plantain. Apr.–Oct. Native to Europe; naturalized in disturbed areas, common; in every co.

5. **Plantago major** L. Common Plantain. May–Oct. Native to Europe; naturalized in disturbed areas; common in the n. ⅓ of IL, less common elsewhere.

6. **Plantago media** L. Hoary Plantain. June–Sept. Native to Europe; rarely adventive in disturbed soil; Cook Co.

7. **Plantago patagonica** Jacq. Salt-and-pepper Plant. May–Aug. Sandy soil; not common in the n. ½ of IL; also Christian Co. *Plantago purshii* Roem. & Schultes.

8. **Plantago psyllium** L. Whorled Plantain. July–Oct. Native to Europe; adventive in disturbed areas; scattered in the n. ⅖ of IL. *Plantago arenaria* Waldst. & Kit.; *P. indica* L.

9. **Plantago pusilla** Nutt. Two varieties may be distinguished in Illinois:

a. Leaves more or less entire; spikes up to 6 cm long; seeds 1.2 mm long .... 9a. *P. pusilla* var. *pusilla*

a. Leaves usually toothed; spikes more than 6 cm long; seeds 1.7–1.8 mm long........... 9b. *P. pusilla* var. *major*

9a. **Plantago pusilla** Nutt. var. **pusilla.** Small Plantain. Apr.–June. Fields, pastures, sandstone cliffs; scattered in IL.

9b. **Plantago pusilla** Nutt. var. **major** Engelm. Small Plantain. Apr.–June. Sandstone cliffs, rare; confined to the s. ¼ of IL.

10. **Plantago rhodosperma** Decne. Red-seeded Plantain. May–June. Native to the sw. U.S.; rarely adventive in IL; old field in Williamson Co.

11. **Plantago rugelii** Decne. Rugel's Plantain. May–Oct. Fields, woods, disturbed soil, common; in every co.

12. **Plantago virginica** L. Virginia Plantain. Apr.–June. Fields, roadsides, sandstone cliffs; occasional to common in the s. ⅔ of IL, rare elsewhere.

125. PLATANACEAE—SYCAMORE FAMILY

1. **Platanus** L.—Sycamore

1. **Platanus occidentalis** L. Sycamore. Apr.–May. Moist woods and along streams and rivers; common in the s. ⅔ of IL, occasional elsewhere; in every co.

126. POLEMONIACEAE—PHLOX FAMILY

1. Leaves pinnately compound or deeply pinnatifid.

  2. Leaves pinnately compound and with lanceolate to oval leaflets; flowers blue, campanulate, to 1.5 cm long .... 6. *Polemonium*

  2. Leaves deeply pinnatifid, the segments filiform; flowers red or pink, narrowly funnelform or salverform, at least 2.5 cm long.

    3. Flowers funnelform, only the clusters subtended by a bract; seeds spherical .... 2. *Gilia*

    3. Flowers salverform, each subtended by a bract; seeds long, slender, curved .... 3. *Ipomopsis*

1. Leaves simple, entire.

  4. All leaves alternate .... 1. *Collomia*

  4. At least the lower cauline leaves opposite.

    5. Usually all leaves opposite; calyx actinomorphic; corolla 1.5 cm long or longer .... 5. *Phlox*

    5. Uppermost leaves alternate; calyx slightly zygomorphic; corolla 8–12 mm long .... 4. *Microsteris*

1. **Collomia** Nutt.—Collomia

1. **Collomia linearis** Nutt. Colomia. June–Aug. Native to the w. U.S.; adventive along railroads and in dry, disturbed areas; occasional in the n. ⅓ of IL.

2. **Gilia** R. & P.—Gilia

1. **Gilia capitata** Sims. Gilia. June–Sept. Native to the w. U.S.; adventive as a garden escape; Putnam Co.

3. **Ipomopsis** Michx.—Standing Cypress

1. **Ipomopsis rubra** (L.) Wherry. Standing Cypress. June–Aug. Native to the s. U.S.; adventive as a garden escape; scattered in IL. *Gilia rubra* (L.) Heller.

4. **Microsteris** Greene—Microsteris

1. **Microsteris gracilis** (Dougl.) Greene. Microsteris. June–Aug. Native to the w. U.S.; adventive along a road; Macon Co. *Phlox gracilis* Dougl.

5. **Phlox** L.—Phlox

1. Annuals; upper leaves alternate; stems glandular-viscid; some or all the leaves clasping .... 4. *P. drummondii*
1. Perennials; upper leaves opposite; stems not glandular-viscid; none of the leaves clasping.
  2. Petals deeply notched or emarginate at the tip.
    3. Leaves lanceolate to narrowly ovate; plants erect or ascending, not diffusely branched; petals emarginate .... 3. *P. divaricata*
    3. Leaves linear to subulate; plants diffuse and much branched; petals deeply notched.
      4. Leaves linear, usually without fascicles of smaller leaves in the axils .... 1. *P. bifida*
      4. Leaves subulate, with fascicles of smaller leaves in the axils .... 9. *P. subulata*
  2. Petals not notched or emarginate at the tip.
    5. Calyx lobes longer than the tube; stamens and style about ½ as long as the corolla tube.
      6. Leaves lanceolate to elliptic to narrowly ovate; plants often with sterile leafy shoots .... 3. *P. divaricata*
      6. Leaves linear to narrowly lanceolate; plants without sterile leafy shoots .... 8. *P. pilosa*
    5. Calyx lobes equaling or shorter than the tube; stamens and style about as long as the corolla tube.
      7. Leaves with conspicuous lateral veins and reticulations; calyx teeth subulate .... 7. *P. paniculata*
      7. Leaves without conspicuous lateral veins and reticulations; calyx teeth lanceolate.
        8. Flowers in panicles usually longer than broad; stems usually purple-spotted .... 6. *P. maculata*
        8. Flowers in corymbs nearly as broad as long; stems green or purplish, not spotted.
          9. Calyx to 7.5 mm long .... 5. *P. glaberrima*
          9. Calyx 9–12 mm long .... 2. *P. carolina*

1. **Phlox bifida** Beck. Cleft Phlox. Mar.–June. Dry, rocky woods, cliffs; throughout the state except for some of the cent. cos.

2. **Phlox carolina** L. ssp. **angusta** Wherry. Carolina Phlox. May–June. Open habitat (in Illinois), very rare; Jefferson Co.

3. **Phlox divaricata** L. ssp. **laphamii** (Wood) Wherry. Common Phlox. Apr.–June. Woods; common throughout the state. *Phlox divaricata* L. var. *laphamii* Wood.

4. **Phlox drummondii** Hook. Annual Phlox. Apr.–June. Native to the sw. U.S.; disturbed soil; Cook Co.

5. **Phlox glaberrima** L. ssp. **interior** (Wherry) Wherry. Smooth Phlox. May–Aug. Woods, prairies; occasional throughout the state. *Phlox glaberrima* L. var. *interior* Wherry.

6. **Phlox maculata** L. Two subspecies occur in Illinois:

a. Inflorescence narrow-conical .... 6a. *P. maculata* ssp. *maculata*
a. Inflorescence broadly pyramidal .... 6b. *P. maculata* ssp. *pyramidalis*

6a. **Phlox maculata** L. ssp. **maculata.** Wild Sweet William. June–Aug. Wet prairies, near bogs and swamps, in moist woodlands and meadows, along wooded streams; occasional in the n. ½ of the state.

6b. **Phlox maculata** L. ssp. **pyramidalis** (J. E. Smith) Wherry. Wild Sweet William. June–Aug. Drier habitats than ssp. *maculata;* occasional in the n. ½ of IL. *Phlox maculata* L. var. *purpurea* Michx.

7. **Phlox paniculata** L. Garden Phlox. July–Sept. Rich woods; scattered in IL, except for the nw. cos.

8. **Phlox pilosa** L. Three subspecies occur in Illinois:

a. Stems, leaves, and calyx pubescent.
 b. Pubescence glandular .... 8a. *P. pilosa* ssp. *pilosa*
 b. Pubescence eglandular .... 8b. *P. pilosa* ssp. *fulgida*
a. Stems, leaves, and calyx glabrous .... 8c. *P. pilosa* ssp. *sangamonensis*

8a. **Phlox pilosa** L. ssp. **pilosa.** Downy Phlox. May–Aug. Dry, rocky woods, prairies; occasional to common throughout the state.

8b. **Phlox pilosa** L. ssp. **fulgida** (Wherry) Wherry. Downy Phlox. May–Aug. Prairies; occasional throughout the state. *Phlox pilosa* L. var. *fulgida* Wherry.

8c. **Phlox pilosa** L. ssp. **sangamonensis** Levin & Smith. Sangamon Phlox. May–Aug. Roadsides, fields, woods; Champaign and Piatt cos.

9. **Phlox subulata** L. Moss Pink. Mar.–May. Native e. and s. of IL; adventive in sandy soil along a road; Coles, Cook, DuPage, and Mason cos.

6. **Polemonium** L.—Jacob's-ladder

1. **Polemonium reptans** L. Jacob's-ladder. Apr.–June. Woods, prairies, fens; common throughout the state.

127. POLYGALACEAE—MILKWORT FAMILY

1. **Polygala** L—Milkwort

1. Flowers 1.5 cm long or longer .... 4. *P. paucifolia*
1. Flowers up to 1 cm long.
 2. At least the lowermost leaves in whorls.
  3. Leaves mostly obtuse; spikes at least 1 cm thick .... 2. *P. cruciata*
  3. Leaves mostly acute; spikes up to 7 mm thick.
   4. All leaves whorled except near the inflorescence .... 8. *P. verticillata*
   4. Only the lowest 1–3 nodes with whorled leaved .... 1. *P. ambigua*
 2. All leaves alternate.
  5. Petals conspicuously united into a cleft tube; stems glaucous .... 3. *P. incarnata*
  5. Petals not united into a cleft tube; stems not glaucous.
   6. Flowers in racemes, the pedicels at least 2 mm long .... 5. *P. polygama*
   6. Flowers in spikes, the pedicels absent or up to 1 mm long.
    7. Leaves serrulate; wings 3.0–3.7 mm long; plants with several stems from a thick perennial crown .... 7. *P. senega*
    7. Leaves entire or nearly so; wings 4.5 mm long or longer; annuals with a solitary stem .... 6. *P. sanguinea*

1. **Polygala ambigua** Nutt. Whorled Milkwort. July–Sept. Fields, dry woods, occasional; apparentlhy confined to the s. ½ of IL. *Polygala verticillata* L. var. *ambigua* (Nutt.) Wood.

2. **Polygala cruciata** L. var. **aquilonia** Fern. & Schub. Cross Milkwort. July–Sept. Acid, sandy soils, not common; mostly in the n. ¼ of IL; also Clark, Mason, and Menard cos.

3. **Polygala incarnata** L. Pink Milkwort. July–Sept. Prairies and gravel hills, not common; in the n. ½ of the state; also Massac and Pope cos.

4. **Polygala paucifolia** Willd. Flowering Wintergreen. May–June. Moist woods, very rare; Cook Co.

5. **Polygala polygama** Walt. var. **obtusata** Chod. Purple Milkwort. June–Aug. Sandy waste ground and open woods, occasional; confined to the n. ½ of the state.

6. **Polygala sanguinea** L. Field Milkwort. July–Sept. Fields, woods, prairies; occasional throughout the state. (Including f. *viridescens* [L.] Farw. and f. *albiflora* [Wheelock] Millsp.)

7. **Polygala senega** L. Two varieties may be recognized in Illinois:

a. Upper leaves linear-lanceolate to lance-ovate, finely serrulate; capsules 2.0–3.5 mm long, 3–4 mm wide; seeds 2.5 mm long .... 7a. *P. senega* var. *senega*
a. Upper leaves oval, strongly serrulate; capsules 3.5–4.2 mm long, 4.0–4.3 mm wide; seeds 3.0–3.5 mm long .... 7b. *P. senega* var. *latifolia*

7a. **Polygala senega** L. var. **senega.** Seneca Snakeroot. May–Sept. Prairies and dry woods; occasional in the n. ¾ of IL, apparently absent in the s. ¼.

7b. **Polygala senega** L. var. **latifolia** Torr. & Gray. Seneca Snakeroot. May–Sept. Rich woods; rare in the n. ½ of IL, absent elsewhere.

8. **Polygala verticillata** L. Two varieties may be recognized in Illinois:

a. Pedicels at least 0.5 mm long .... 8a. *P. verticillata* var. *verticillata*
a. Pedicels up to 0.3 mm long .... 8b. *P. verticillata* var. *isocycla*

8a. **Polygala verticillata** L. var. **verticillata.** Whorled Milkwort. July–Sept. Dry, often sandy soil, not common; scattered in IL.

8b. **Polygala verticillata** L. var. **isocycla** Fern. Whorled Milkwort. July–Sept. Dry, acid soil, occasional; throughout most of IL. (Including var. *sphaenostachya* Pennell.)

128. POLYGONACEAE—SMARTWEED FAMILY

1. Climbing shrubs with tendrils; pedicels winged on 1 side by the calyx .... 2. *Brunnichia*
1. Prostrate or ascending or erect herbs or, if climbing, then without tendrils; pedicels not winged by the calyx.
  2. Leaves very narrow, needlelike .... 6. *Polygonella*
  2. Leaves linear or broader, not needlelike.
    3. Sepals 6; stamens 6.
      4. All sepals similar; fruit 3-winged .... 9. *Rheum*
      4. Outer sepals narrower than inner sepals; fruit 3-angled .... 10. *Rumex*
    3. Sepals usually 5; stamens 3–8.
      5. Stems retrorse prickly.
        6. Leaves tapering to base .... 5. *Persicaria*
        6. Leaves hastate or sagittate .... 11. *Tracaulon*
      5. Stems not retrorse prickly.
        7. Leaves hastate-deltoid .... 3. *Fagopyrum*
        7. Leaves not hastate-deltoid.
          8. Stems twining or trailing, neither erect nor prostrate ...... 4. *Fallopia*
          8. Stems erect or prostrate, neither twining nor trailing.
            9. Flowers borne in small axillary clusters .... 7. *Polygonum*
            9. Flowers borne in terminal and/or axillary spikes or racemes.
              10. Outer sepals broadly winged; plants bushy, more than 1 m tall .... 8. *Reynoutria*
              10. Outer sepals unwinged; plants not bushy, often less than 1 m tall.
                11. Style persistent as a beak on the achene; calyx usually 4-parted .... 1. *Antenoron*
                11. Style deciduous; calyx usually 5-parted .... 5. *Persicaria*

1. **Antenoron** Raf.

1. **Antenoron virginianum** (L.) Roberty & Vautier. Virginia Knotweed; Jumpseed. July–Sept. Woods; common throughout the state. *Polygonum virginianum* L.; *Tovara virginiana* (L.) Raf.

2. **Brunnichia Banks**—Ladies' Ear-drops

1. **Brunnichia ovata** (Walt.) Shinners. Ladies' Ear-drops. June–Aug. Swampy woods; confined to the s. ⅙ of IL; also Franklin and Richland cos. *Brunnichia cirrhosa* Banks.

3. **Fagopyrum** Mill.—Buckwheat

1. **Fagopyrum esculentum** Moench. Buckwheat. June–Sept. Native to Asia; adventive in disturbed soil; occasional throughout the state.

4. **Fallopia** Adans.—Climbing Buckwheat

1. Base of ocreal sheaths with bristles and reflexed hairs.......................... 1. *F. cilinodis*
1. Base of ocreal sheaths glabrous or scabrous.
  2. Achenes dull, granular; outer sepals keeled................................2. *F. convolvulus*
  2. Achenes shiny, smooth; outer sepals winged.
    3. Wings of fruit decurrent on a stipelike base, usually not flat, the margins of the wings crenulate to toothed.
      4. Calyx in fruit up to 10 mm long; achenes up to 3 mm long ......3. *F. cristata*
      4. Calyx in fruit more than 10 mm long; achenes more than 3 mm long........ ....5. *F. scandens*
    3. Wings of fruit truncate at base, usually flat, the margins of the wings usually entire..4. *F. dumetorum*

1. **Fallopia cilinodis** (Michx.) Holub. Bristly Climbing Buckwheat. June–Aug. Edge of woods, very rare; Cook and Jackson cos. *Polygonum cilinode* Michx.

2. **Fallopia convolvulus** (L.) A. Love. Black Bindweed. May–Oct. Native to Europe; fields and edges of woods; occasional to common throughout the state. *Polygonum convolvulus* L.

3. **Fallopia cristata** (Engelm. & Gray) Holub. Crested Bindweed. Aug.–Sept. Woods; occasional in the s. ¾ of IL, rare elsewhere. *Polygonum cristatum* Engelm. & Gray; *P. scandens* L. var. *cristatum* (Engelm. & Gray) Gl.

4. **Fallopia dumetorum** (L.) Holub. Climbing Buckwheat. June–Oct. Thickets, edge of woods; scattered in a few cos., mostly on the w. side of IL. *Polygonum dumetorum* L.

5. **Fallopia scandens** (L.) Holub. Climbing Buckwheat. Aug.–Oct. Woods, thickets; occasional to common throughout the state. *Polygonum scandens* L.

5. **Persicaria** (Turcz.) Nakai—Smartweed

1. Stems with retrorse bristles ....3. *P. bungeana*
1. Stems without retrorse bristles.
  2. Sheaths with bristles.
    3. Perennials with thick rhizomes or stolons; spikes 1–2 per stem.
      4. Flowers white ....6. *P. glabra*
      4. Flowers pink or red.
        5. Spikes cylindrical, usually more than 4 cm long; peduncles densely pubescent; stipules without a flange ....5. *P. coccinea*
        5. Spikes conical to ovoid, usually less than 4 cm long; peduncles sparsely pubescent; stipules often with a flange.............. 1. *P. amphibia*
    3. Annuals with fibrous roots or perennials with slender rhizomes; spikes more than 2 per stem.
      6. Peduncles with stipitate glands ....4. *P. careyi*
      6. Peduncles without stipitate glands.
        7. Apex of sheaths with an expanded rim .... 14. *P. orientalis*
        7. Apex of sheaths rimless.
          8. Sepals punctate or glandular-dotted.
            9. Sepals white; achenes shiny.
              10. Racemes continuous; stems stout, to 2 m tall; leaves to 20 cm long, to 4.5 cm wide................................17. *P. robustior*
              10. Racemes usually interrupted; stems slender, to 1 m tall; leaves to 20 cm long, to 2 cm wide............... 16. *P. punctatum*
            9. Sepals often with a pinkish or greenish tinge; achenes dull........ ....7. *P. hydropiper*
          8. Sepals neither punctate nor glandular-dotted.
            11. Racemes uninterrupted.
              12. Flowers bright rose-pink; racemes up to 7 mm thick; leaves without a dark blotch... 10. *P. longiseta*

12. Flowers pink or pale pink; racemes more than 7 mm thick; leaves often with a dark blotch .... 11. *P. maculosa*

11. Racemes interrupted, at least near the tip.

13. Annuals without slender rhizomes; flowers rose to pink .... 12. *P. minor*

13. Perennials with slender rhizomes; flowers white, pink, or greenish purple.

14. Leaves strigose on the upper surface; sepals white .... 19. *P. setacea*

14. Leaves glabrous or scabrous on the upper surface; sepals mostly pink or greenish.

15. Achenes partly exserted; sepals greenish or greenish purple .... 13. *P. opelousana*

15. Achenes enclosed by the perianth; sepals rosy or pinkish .... 8. *P. hydropiperoides*

2. Sheaths without bristles

16. Perennials with rhizomes or stolons.

17. Sepals white .... 6. *P. glabra*

17. Sepals red or pink.

18. Racemes cylindrical, usually more than 4 cm long; peduncles densely pubescent; sheaths without a flange .... 5. *P. coccinea*

18. Racemes conical or ovoid, usually less than 4 cm long; peduncles sparsely pubescent; sheaths often with a flange .... 1. *P. amphibia*

16. Annuals with fibrous roots.

19. Some or all the flowers with long exserted stamens and styles .... 2. *P. bicornis*

19. None of the flowers with exserted stamens or styles.

20. Sepals greenish .... 18. *P. scabra*

20. Sepals pink or rose or white.

21. Spikes arching or pendulous; achenes up to 2 mm broad; sepals usually white .... 9. *P. lapathifolia*

21. Spikes not arching; achenes broader than 2 mm; sepals pink .... 15. *P. pensylvanica*

1. **Persicaria amphibia** (L.) S. F. Gray. Water Smartweed. June–Sept. Wet ground; scattered throughout the state, although not common in the s. ¼ of IL. (Including var. *stipulaceum* Coleman.) *Polygonum amphibium* L.

2. **Persicaria bicornis** (Raf.) Nieuwl. Smartweed. July–Oct. Wet ground, rare; s. ⅓ of IL; also Kendall Co. *Polygonum bicorne* Raf.; *Persicaria longistyla* (Small) Small.

3. **Persicaria bungeana** (Turcz.) Moldenke. Prickly Smartweed. June–July. Native to Europe; adventive in old fields; DuPage, Kane, and Will cos. *Polygonum bungeanum* Turcz.

4. **Persicaria careyi** (Olney) Greene. Carey's Smartweed. July–Sept. Sandy soil, very rare; Cook, Grundy, Iroquois, and Kankakee cos. *Polygonum careyi* Olney.

5. **Persicaria coccinea** (Muhl.) Greene. Scarlet Smartweed. June–Sept. Wet ground; scattered throughout the state. *Polygonum coccineum* Muhl.

6. **Persicaria glabra** (Willd.) M. Gomez. Smartweed. June–Aug. Wet roadside ditch, very rare; Alexander Co. *Polygonum densiflorum* Meisn.; *Persicaria densiflora* (Meisn.) Moldenke; *Polygonum glabrum* Willd.

7. **Persicaria hydropiper** (L.) Opiz. Water Pepper. June–Oct. Native to Europe; adventive in wet ground; occasional to common throughout IL. *Polygonum hydropiper* L.

8. **Persicaria hydropiperoides** (Michx.) Small. Mild Water Pepper. June–Oct. Wet ground, sometimes in shallow water; occasional to common throughout IL. *Polygonum hydropiperoides* Michx.

9. **Persicaria lapathifolia** (L.) S. F. Gray. Pale Smartweed. July–Oct. Wet ground, common; throughout the state. *Polygonum lapathifolium* L.

10. **Persicaria longiseta** (DeBruyn) Kitagawa. Creeping Smartweed. June–Oct. Native to Asia; adventive in shaded ground and in lawns; scattered throughout the state. *Persicaria cespitosa* (Blume) Nakai; *Polygonum cespitosum* Blume var. *longisetum* (DeBruyn) Steward.

11. **Persicaria maculosa** S. F. Gray. Lady'sThumb-print. May–Sept. Native to Europe; naturalized in disturbed ground; common throughout IL. *Polygonum persicaria* L.; *Persicaria vulgaris* Webb. & Moq.

12. **Persicria minor** (Hudson) Opiz. Small Water Pepper. July–Sept. Native to Europe; scattered in s. IL. *Polygonum minor* Hudson.

13. **Persicaria opelousana** (Riddell) Small. Water Pepper. July–Oct. Wet ground, not common; confined to the s. ⅓ of IL; also Kankakee and Will cos. *Polygonum opelousanum* Riddell; *P. hydropiperoides* Michx. var. *opelousanum* (Riddell) Stone.

14. **Persicaria orientalis** (L.) Spach. Prince's Feather. June–Oct. Native to Europe and Asia; escaped into waste ground; scattered in IL. *Polygonum orientale* L.

15. **Persicaria pensylvanica** (L.) Small. Three varieties may be distinguished in Illinois:

a. Leaves strigose.
  b. Peduncles with spreading hairs .... 15a. *P. pensylvanica* var. *pensylvanica*
  b. Peduncles with appressed hairs .... 15b. *P. pensylvanica* var. *dura*
a. Leaves glabrous .... 15c. *P. pensylvanica* var. *laevigata*

15a. **Persicaria pensylvanica** (L.) Small var. **pensylvanica.** Pinkweed; Pink Smartweed. July–Oct. Wet ground; occasional in IL. *Polygonum pensylvanicum* L.

15b. **Persicaria pensylvanica** (L.) Small var. **dura** (Stanford) C. F. Reed. Pinkweed. July–Sept. Wet ground; scattered throughout IL. *Polygonum pensylvanicum* L. var. *durum* Stanford.

15c. **Persicaria pensylvanica** (L.) Small var. **laevigata** (Fern.) Mohlenbr., comb. nov. (basionym: *Polygonum pensylvanicum* L. var. *laevigatum* Fern). Pinkweed. July–Oct. Wet ground, common; in every co. *Polygonum pensylvanicum* L. var. *laevigatum* Fern.

16. **Persicaria punctata** (Ell.) Small. Dotted Smartweed. July–Oct. Wet ground, common; throughout the state. (Including var. *leptostachyum* [Meisn.] Small.) *Polygonum punctatum* Ell.

17. **Persicaria robustior** (Small) E. P. Bickn. Stout Smartweed. July–Oct. Edge of lake in shallow water, fen; very rare; Henry and Jackson cos. *Polygonum robustior* Small.

18. **Persicaria scabra** (Moench) Moldenke. Rough Smartweed. June–Aug. Native to Europe; rarely adventive along railroads; Champaign, Lake, and Lee cos. *Polygonum scabrum* Moench; *P. tomentosum* Schrank.

19. **Persicaria setacea** (Baldw.) Small. Slender Smartweed. July–Oct. Wet ground, occasional; mostly in the s. ⅓ of IL. (Including var. *interjectum* Fern.) *Polygonum setaceum* Baldw.; *P. hydropiperoides* Michx. var. *setaceum* (Baldw.) Gl.

6. **Polygonella** Michx.—Jointweed

1. **Polygonella articulata** (L.) Meisn. Jointweed. July–Nov. Sandy soil; restricted to the n. ½ of IL.

7. **Polygonum** L.—Knotweed

1. Leaves plicate; stems angular .... 12. *P. tenue*
1. Leaves flattened; stems terete or nearly so.
  2. Achenes usually included and enclosed by the perianth.
    3. Plants erect, not mat-forming.
      4. Leaves linear to lanceolate, to 10 mm wide, at least 4 times longer than wide; achenes shiny, more or less smooth .... 10. *P. ramosissimum*
      4. Leaves elliptic to oval, 10–30 mm wide, less than 4 times longer than wide; achenes dull, striate or punctate .... 5. *P. erectum*
    3. Plants mat-forming, with some of the stems ultimately ascending.
      5. Leaves all similar in shape and size; achenes shiny, 2.0–2.5 (–2.8) mm long.
        6. Leaves obtuse at apex, gray-green; ocreae red-brown; achenes light brown to brown .... 4. *P. buxiforme*
        6. Leaves acute at apex, green to blue-green; ocreae silvery; achenes dark brown .... 2. *P. arenastrum*

5. Upper leaves conspicuously smaller and narrower than lower leaves; achenes dull, 2.5–3.0 mm long .... 3. *P. aviculare*

2. Achenes usually exserted beyond the tips of the perianth.

7. Plants mat-forming, with some of the stems often ascending at maturity .... 2. *P. arenastrum*

7. Plants ascending to erect, not mat-forming.

8. Leaves obtuse at apex, the largest ones 8–15 mm wide ...... 1. *P. achoreum*

8. Leaves acute at apex, the largest ones up to 8 mm wide.

9. Leaves cuspidate at tip; achenes long-exserted beyond the perianrh, the achenes 4–6 mm long .... 6. *P. exsertum*

9. Leaves not cuspidate at tip; achenes exserted, but not greatly so, the achenes up to 4 mm long.

10. All leaves similar in size and shape; achenes shiny, black; sepals up to 2 mm long .... 9. *P. prolificum*

10. Upper leaves smaller and narrower than lower leaves; achenes dull, brown to dark brown (sometimes nearly black in *P. rurivagum*); sepals 2–3 mm long.

11. Achenes striate and tuberculate; leaves blue-green .... 8. *P. patulum*

11. Achenes striate or minutely punctate, not tuberculate; leaves green.

12. Achenes striate, 2.5–4.0 mm long, dark brown to black; ocreae veiny .... 11. *P. rurivagum*

12. Achenes minutely punctate, 2–3 mm long, brown; ocreae not veiny .... 7. *P. neglectum*

1. **Polygonum achoreum** S. F. Blake. Knotweed. Aug.–Oct. Waste ground, rare; scattered in IL.

2. **Polygonum arenastrum** Boreau. Knotweed. June–Oct. Native to Europe; particularly common in sidewalk cracks; common throughout the state. *Polygonum aviculare* L., misapplied.

3. **Polygonum aviculare** L. Knotweed. June–Oct. Native to Europe; adventive in disturbed soil; scattered in IL.

4. **Polygonum buxiforme** Small. Knotweed. June–Oct. Waste ground, not common; scattered in IL. *Polygonum aviculare* L. var. *littorale* (Link) W. D. J. Koch.

5. **Polygonum erectum** L. Knotweed. Aug.–Oct. Disturbed soil; occasional throughout the state.

6. **Polygonum exsertum** Small. Knotweed. Aug.–Oct. Banks and shores, usually in sand; scattered throughout the state.

7. **Polygonum neglectum** Besser. Knotweed. June–Oct. Native to Europe and Asia; scattered in IL. *Polygonum aviculare* L., misapplied.

8. **Polygonum patulum** M. Bieberstein. Tubercled Knotweed. July–Oct. Native to Eurasia and Africa; disturbed soil; Jackson, St. Clair, and Union cos.

9. **Polygonum prolificum** (Small) Robins. Knotweed. July–Oct. Waste ground, rare; scattered in IL. *Polygonum ramosissimum* Michx. var. *prolificum* Small.

10. **Polygonum ramosissimum** Michx. Knotweed. July–Oct. Sandy soil; occasional throughout IL.

11. **Polygonum rurivagum** Jordan ex Bor. Narrow-leaved Knotweed. July–Oct. Native to Europe; disturbed areas; Jackson Co.

12. **Polygonum tenue** Michx. Slender Knotweed. July–Sept. Dry, sandy soil; crevices of sandstone bluffs; occasional throughout the state.

8. **Reynoutria** Houtt.

1. Hairs on the veins of the leaves multicellular; base of leaves cordate .... 3. *R. sachalinensis*

1. Hairs on the veins of the leaves unicellular, or veins glabrous or scabrous; base of leaves cordate or truncate.

2. Veins of the leaves glabrous or scabrous; base of leaves truncate .... 2. *R. japonica*

2. Hairs on the veins of the leaves unicellular; base of leaves cordate .... 1. *R. X bohemica*

1. **Reynoutria X bohemica** Chrtek & Chrtkova. Bohemian Knotweed. July–Sept. Native to Europe; disturbed areas; scattered in IL. *Polygonum X bohemicum* (Chrtek & Chrtkova) Zika & Jacobson.

2. **Reynoutria japonica** Houtt. Japanese Knotweed; Japanese Bamboo. Aug.–Sept. Native to e. Asia; occasionally spreading from cultivation in IL. *Polygonum cuspidatum* Sieb. & Zucc.

3. **Reynoutria sachalinensis** (F. Schmidt) Nakai. Giant Knotweed; Japanese Bamboo. Aug.–Oct. Native to e. Asia; rarely adventive in IL; Jackson, Lake, and Logan cos. *Polygonum sachalinenses* F. Schmidt.

9. **Rheum** L.—Rhubarb

1. **Rheum rhabarbarum** L. Rhubarb. July–Sept. Native to Asia; rarely escaped and persistent; Champaign, Cook, and Sangamon cos. *Rheum raponticum* L.

10. **Rumex** L.—Dock

1. Some of the leaves hastate (rarely entire in 1 form of *R. acetosella*); plants dioecious.
  2. Plants with stolons or slender rhizomes; achenes exserted from the calyx .... 1. *R. acetosella*
  2. Plants with a taproot; achenes enclosed by the calyx.
    3. Leaves not sagittate at base; valves of fruit 3–4 mm wide ...... 9. *R. hastatulus*
    3. Leaves sagittate at base; valves of fruit 4–5 mm wide .... 14. *R. thyrsiflorus*
1. None of the leaves hastate; plants monoecious.
  4. All fruiting valves lacking tubercles .... 10. *R. longifolius*
  4. At least 1 of the fruiting valves with a tubercle.
    5. Fruiting valves with spinulose bristles or conspicuously dentate.
      6. Fruiting valves with spinulose bristles.
        7. Stems hollow; tubercle of fruit long and slender; bristles of fruiting sepals longer than the width of the sepals .... 8. *R. fueginus*
        7. Stems firm; tubercle of fruit only slightly longer than broad; bristles of fruiting sepals not longer than width of the sepals .... 11. *R. obtusifolius*
      6. Fruiting valves conspicuously dentate.
        8. Teeth of fruiting valves broad, 4–5 mm long; leaves flat or slightly undulate .... 7. *R. dentatus*
        8. Teeth of fruiting valves narrow, less than 4 mm long; leaves conspicuously crispate .... 13. *R. stenophyllus*
    5. Fruiting valves either entire or minutely toothed or erose.
      9. Valves about as wide as the face of the achene; fruit not appearing 3-winged .... 4. *R. conglomeratus*
      9. Valves much wider than the face of the achene; fruit appearing 3-winged.
        10. Only 1 of the 3 fruiting valves with a tubercle.
          11. Some of the leaves over 10 cm broad; each fruiting sepal 8–10 mm broad .... 12. *R. patientia*
          11. None of the leaves 10 cm broad; each fruiting sepal up to 6 mm broad.
            12. Leaves crispate along the margins .... 5. *R. crispus*
            12. Leaves flat .... 2. *R. altissimus*
        10. Each of the fruiting valves with a tubercle.
          13. Leaves with conspicuous crispate and undulate margins .... 5. *R. crispus*
          13. Leaves flat, entire to crenulate.
            14. Leaves crenulate; lateral veins of leaves forming right angles with the vertical veins .... 3. *R. brittanica*
            14. Leaves entire; lateral veins of leaves ascending.
              15. Fruiting pedicels 2–5 times longer than the calyx .... 16. *R. verticillatus*

15. Fruiting pedicels shorter than to about twice as long as the calyx.
16. Pedicels 1 ½–2 times longer than the calyx; achenes 3.0–3.2 mm long .... 6. *R. cristatus*
16. Pedicels shorter than or equaling the calyx; achenes 1.5–2.8 mm long.
17. Leaves narrowly lanceolate, never more than 3 cm broad .... 15. *R. triangulivalvis*
17. Leaves broadly lanceolate, at least some of them more than 3 cm broad .... 2. *R. altissimus*

1. **Rumex acetosella** L. Two forms occur in Illinois:

a. Leaves hastate .... 1a. *R. acetosella* f. *acetosella*
a. Leaves unlobed .... 1b. *R. acetosella* f. *integrifolius*

1a. **Rumex acetosella** L. f. **acetosella.** Sour Dock. Apr.–Aug. Native to Europe and Asia; adventive in fields, roadsides, disturbed soil, common; in every co.

1b. **Rumex acetosella** L. f. **integrifolius** (Wallr.) G. Beck. Sour Dock. Apr.–Aug. Native to Europe and Asia; adventive in IL; Kane Co.

2. **Rumex altissimus** Wood. Pale Dock; Smooth Dock. Apr.–May. Moist soil; occasional to common throughout the state.

3. **Rumex brittanica** L. Great Water Dock. June–Sept. Moist soil, bogs, deep marshes; mostly restricted to the n. ⅗ of IL; also Hamilton Co. *Rumex orbiculatus* Gray.

4. **Rumex conglomeratus** L. Dock. June–July. Native to Europe; rarely adventive in disturbed soil in IL; Cook and DuPage cos.

5. **Rumex crispus** L. Curly Dock. Apr.–May. Native to Europe; adventive in waste ground, common; probably in every co. (Including f. *unicallosus* Peterm., with only 1 of the valves bearing a tubercle.)

6. **Rumex cristatus** DC. Crested Dock. July–Sept. Native to Europe; rarely adventive in disturbed soil; Macon, Madison, St. Clair, and Stark cos.

7. **Rumex dentatus** L. Toothed Dock. July–Aug. Native to Europe; rarely adventive in disturbed soil in IL; Madison and St. Clair cos.

8. **Rumex fueginus** Phil. Dock. July–Oct. Sandy shores; not common; scattered throughout the state. *Rumex maritimus* L. var. *fueginus* (Phil.) Dusen.

9. **Rumex hastatulus** Baldw. Sour Dock. Apr.–Aug. Sandy soil, very rare; Madison and St. Clair cos.

10. **Rumex longifolius** DC. Dock. July–Sept. Native to Europe; rarely adventive in disturbed soil; Peoria and Richland cos.

11. **Rumex obtusifolius** L. Bitter Dock; Broad-leaved Dock. Apr.–May. Native to Europe; adventive in disturbed soil; occasional to common throughout the state.

12. **Rumex patientia** L. Patience Dock. Apr.–June. Native to Europe and Asia; adventive in waste ground; scattered throughout the state.

13. **Rumex stenophyllus** Ledeb. Narrow-leaved Curly Dock. May–Sept. Native to Europe and Asia; wet, disturbed areas; St. Clair Co.

14. **Rumex thyrsiflorus** Fingerh. Sagittate-leaved Dock. Native to Europe; wet, disturbed soil; Cook Co.

15. **Rumex triangulivalvis** (Danser) Rech. f. Dock. July–Sept. Moist soil; scattered throughout the state. *Rumex mexicanus* Meisn.

16. **Rumex verticillatus** L. Swamp Dock. Apr.–Sept. Wet ground, sometimes in standing water; common throughout the state.

11. **Tracaulon** Raf.—Tear Thumb

1. Leaves hastate; achenes lenticular .... 1. *T. arifolium*
1. Leaves sagittate; achenes trigonous .... 2. *T. sagittatum*

1. **Tracaulon arifolium** (L.) Raf. var. **pubescens** (Keller) Mohlenbr., comb. nov. (basionym: *Polygonum sagittatum* L. var. *pubescens* Keller). Hairy Hastate-leaved Tear Thumb. July–Oct. Wet ground, very rare; Jasper, Lawrence, Macon, and McHenry cos. *Polygonum arifolium* L. var. *pubescens* (Keller) Fern.

2. **Tracaulon sagittatum** (L.) Small. TearThumb. July–Oct. Swampy woods, marshes, wet ground; occasional throughout the state. *Polygonum sagittatum* L.

129. PORTULACACEAE—PURSLANE FAMILY

1. Ovary inferior or partly so; capsule circumscissile; cauline leaves numerous .... 3. *Portulaca*
1. Ovary superior; capsule dehiscing vertically; cauline leaves 2 or absent.
   2. Cauline leaves absent; all leaves basal or nearly so, terete, very fleshy; inflorescence cymose; flowers pink or rose .... 2. *Phemeranthus*
   2. Cauline leaves 2; all leaves flat, not particularly fleshy; inflorescence racemose; flowers white, rarely faint pink .... 1. *Claytonia*

1. **Claytonia** L.—Spring Beauty

1. **Claytonia virginica** L. Spring Beauty. Mar.–June. Woods, both rich and degraded, lawns; common throughout the state; in every co. (Includes var. *robusta* Somes and var. *simsii* [Sweet] R. J. Davis.)

2. **Phemeranthus** Raf. Flower-of-an-hour; Fame-flower

1. Stamens 4–8; flowers pale pink .... 2. *P. parviflorus*
1. Stamens 10–45; flowers bright pink to rose.
   2. Petals 6–8 mm long; stamens 10–25; capsules 4–5 mm long .... 3. *P. rugospermus*
   2. Petals 12–16 mm long; stamens 30–45; capsules 6–8 mm long .... 1. *P. calycinus*

1. **Phemeranthus calycinus** (Engelm.) Kiger. Large-flowered Flower-of-an-hour. June–Sept. Exposed edge of sandstone cliffs, very rare; Randolph Co. *Talinum calycinum* Engelm.

2. **Phemeranthus parviflorus** (Nutt.) Kiger. Small-flowered Flower-of-an-hour. June–July. Exposed sandstone cliffs, rare; Calhoun, Johnson, Pope, and Union cos. *Talinum parviflorum* Nutt.

3. **Phemeranthus rugospermum** (Holz.) Kiger. Wrinkle-seeded Flower-of-an-hour. June–Aug. Sandy savannas, sandstone cliffs; scattered and generally not common in the n. ½ of IL. *Talinum rugospermum* Holz.

3. **Portulaca** L.—Purslane

1. Leaves terete, linear; stems pubescent at the nodes; flowers 2–4 cm across .... 1. P. *grandiflora*
1. Leaves flat, spatulate to ovate; stems glabrous; flowers up to 1 cm across .... 2. P. *oleracea*

1. **Portulaca grandiflora** Hook. Rose Moss; Garden Purslane. June–Oct. Native to S. America; sparingly escaped from cultivation.

2. **Portulaca oleracea** L. Common Purslane. June–Oct. Native to Europe; disturbed soil, often in cracks of sidewalks; common throughout IL; in every co.

130. PRIMULACEAE—PRIMROSE FAMILY

1. Leaves deeply dissected into threadlike divisions .... 4. *Hottonia*
1. Leaves entire or toothed.
   2. Leaves all basal.
      3. Lobes of corolla reflexed; anthers exserted as a cone; calyx deeply divided; leaves entire, more than 2.5 cm long .... 3. *Dodecatheon*
      3. Lobes of corolla spreading; anthers included; calyx tubular; leaves dentate or, if entire, less than 2.5 cm long.
         4. Leaves usually more than 2 cm long, dentate; corolla tube at least as long as the calyx, open at the throat; perennial .... 7. *Primula*
         4. Leaves up to 2 cm long, entire; corolla tube shorter than the calyx, closed at the throat; annual .... 2. *Androsace*
   2. Leaves cauline.
      5. Leaves, or most of them, alternate.
         6. Ovary superior; flowers generally 4-merous; pedicels absent or less than 5 mm long; flower solitary .... 1. *Anagallis*

6. Ovary inferior; flowers 5-merous; pedicels 1–2 cm long; flowers in racemes .... 8. *Samolus*

5. Leaves, or most of them, opposite or whorled (occasional alternate leaves may be found in *Trientalis,* which also always has some whorled leaves, *Lysimachia quadrifolia,* and *L. terrestris*).

7. Flowers scarlet (rarely blue; if white, the petals rounded at the tip); capsules circumscissile .... 1. *Anagallis*

7. Flowers yellow or, if white, the petals usually pointed at the tip; capsules splitting lengthwise.

8. Flowers yellow (white in *Lysimachia clethroides*); petals usually 5 or 6.

9. Flowers white .... 5. *Lysimachia*

9. Flowers yellow.

10. Leaves epunctate; corolla lobes erose and apiculate; staminodia present .... 9. *Steironema*

10. Leaves punctate; corolla lobes entire; staminodia absent.

11. Flowers in dense axillary racemes; corolla lobes linear .... 6. *Naumbergia*

11. Flowers in terminal panicles or racemes or. of axillary, solitary or in panicles; corolla lobes lanceolate to orbicular (rarely linear in *Lysimachia X commixta*) .... 5. *Lysimachia*

8. Flowers white; petals usually 7 .... 10. *Trientalis*

1. **Anagallis** L.—Pimpernel

1. Leaves opposite; flowers in racemes, 4 mm or more across; corolla lobes 5; leaves 1 cm long or longer .... 1. *A. arvensis*

1. Leaves, or most of them, alternate; flower solitary, about 1 mm across; corolla lobes usually 4; leaves less than 1 cm long .... 2. *A. minima*

1. **Anagallis arvensis** L. Scarlet Pimpernel. June–Aug. Native to Europe; naturalized in waste ground, not common; scattered in IL. (Including f. *caerulea* [Schreb.] Gren. & Godr., with blue flowers.)

2. **Anagallis minima** (L.) Krause. Chaffweed. May–Aug. Moist soil, rare; mostly in the ne. cos. *Centunculus minimus* L.

2. **Androsace** L.

1. **Androsace occidentalis** Pursh. Mar.–Apr. Rock Jasmine. Cliffs, sandy soil, rocky woods; scattered in most parts of IL, except the se. cos.

3. **Dodecatheon** L.—Shooting-star

1. Leaves abruptly contracted to a distinct petiole, the blades broadly ovate .... 2. *D. frenchii*

1. Leaves tapering to the petiole, oblanceolate.

2. Capsule pale brown to yellow, with thin, papery walls .... 1. *D. amethystinum*

2. Capsule dark reddish brown, with rather thick, woody walls .... 3. *D. meadia*

1. **Dodecatheon amethystinum** Fassett. Jeweled Shooting-star. May–June. Bluffs and wooded slopes, rare; Calhoun, Carroll, Fulton, Greene, Jersey, Jo Daviess, Pike, St. Clair, and Whiteside cos. *Dodecatheon radicatum* Greene.

2. **Dodecatheon frenchii** (Vasey) Rydb. French's Shooting-star. Apr.–May. Under overhanging sandstone cliffs, locally abundant; confined to the s. ⅙ of IL. *Dodecatheon meadia* L. var. *frenchii* Vasey.

3. **Dodecatheon meadia** L. Two varieties may be distinguished in Illinois:

a. Capsules 10.5–18.0 mm long .... 3a. *D. meadia* var. *meadia*

a. Capsules 7.5–10.0 mm long .... 3b. *D. meadia* var. *brachycarpum*

3a. **Dodecatheon meadia** L. var. **meadia.** Shooting-star. Apr.–June. Woods, bluffs, meadows, prairies; occasional to common throughout IL.

3b. **Dodecatheon meadia** L. var. **brachycarpum** (Small) Fassett. Shooting-star. Apr.–June. Woods, prairies; scattered in IL.

4. **Hottonia** L.—Featherfoil

1. **Hottonia inflata** Ell. Featherfoil. June–Aug. Swamps, rare; Jackson, Johnson, Massac, Pope, Saline, St. Clair, and Union cos.

5. **Lysimachia** L.—Loosestrife

1. Flowers white .... 1. *L. clethroides*
1. Flowers yellow.
  2. Plants creeping, rooting at the nodes; leaves evergreen .... 4. *L. nummularia*
  2. Plants erect, not rooting at the nodes; leaves deciduous.
    3. Corolla without dark markings, crateriform.
      4. Calyx dark glandular along the margin; corolla lobes not glandular-ciliate; flowers in terminal or axillary panicles.........................9. *L. vulgaris*
      4. Calyx not dark glandular along the margin; corolla lobes glandular-ciliate; flowers in the axils of the upper leaves .... 6. *L. punctata*
    3. Corolla usually streaked with dark markings, rotate or saucer-shaped.
      5. Leaves, or some of them, in whorls of 3–7.
        6. Flowers in terminal panicles.. 3. *L. fraseri*
        6. At least some flowers in the axils of the leaves.
          7. All flowers in the axils of the leaves..........................7. *L. quadrifolia*
          7. Some flowers in terminal racemes, others in the axils of the leaves .... 5. *L. X producta*
      5. Leaves opposite or alternate.
        8. Most or all the leaves (except the lowermost scalelike ones) opposite; style 3–4 mm long; capsule 2.8–3.5 mm in diameter .... 8. *L. terrestris*
        8. Most or all the leaves (except the lowermost scalelike ones) alternate; style 5–6 mm long; capsule about 2 mm in diameter ....2. *L. X commixta*

1. **Lysimachia clethroides** Duby. White Loosestrife. July–Aug. Native to Japan; often planted as an ornamental, occasionally escaped.

2. **Lysimachia X commixta** Fern. Hybrid Loosestrife. June–Aug. Wet areas, rare; scattered in IL. Reputed to be the hybrid between *L. terrestris* (L.) BSP. and *Waumbergia thyrsiflora* (L.) Reich

3. **Lysimachia fraseri** Duby. Fraser's Loosestrife. July–Aug. Mesic woods, very rare; Hardin and Pope cos.

4. **Lysimachia nummularia** L. Moneywort. June–Aug. Native to Europe; naturalized in moist, shady areas; common throughout the state.

5. **Lysimachia X producta** (Gray) Fern. Hybrid Loosestrife. May–Aug. High bank of the Ohio River, Hardin Co. This is the hybrid beween *L. quadrifolia* L. and *L. terrestris* (L.) BSP.

6. **Lysimachia punctata** L. Dotted Loosestrife. May–June. Native to Europe; rarely escaped along edges of woods; Champaign, Cook, and Sangamon cos.

7. **Lysimachia quadrifolia** L. Whorled Loosestrife. May–Aug. Woods, roadsides, fields; confined to the n. ¼ of IL.

8. **Lysimachia terrestris** (L.) BSP. Swamp Candles. June–Aug. Swamps, bogs; nearly confined to the n. ½ of IL; also Union Co.

9. **Lysimachia vulgaris** L. Loosestrife. June–Sept. Native to Europe; naturalized in moist shady areas; occasional to common throughout the state.

6. **Naumbergia** Moench—Tufted Loosestrife

1. **Naumbergia thyrsiflora** (L.) Reich. Tufted Loosestrife. May–July. Bottomland woods, swamps, marshes, bogs; occasional in the n. ½ of IL; also Wabash Co.

7. **Primula** L.—Primrose

1. **Primula mistassinica** Michx. Bird's-eye Primrose. May–June. Limestone cliffs, very rare; Jo Daviess Co.

8. **Samolus** L.—Brookweed

1. **Samolus parviflorus** Raf. Brookweed. May–June. Moist soil; occasional throughout the state except for the nw. cos. *Samolus floribundus* Kunth; *S. valerandi* L.

9. **Steironema** Raf.—Loosestrife

1. Lateral nerves of blades obscure; blades firm .... 4. *S. quadriflorum*
1. Lateral nerves of blades conspicuous; blades thin.
  2. Plants decumbent, rooting at the nodes .... 5. *S. radicans*
  2. Plants erect, not rooting at the nodes.
    3. Leaves ovate to ovate-lanceolate, rounded or subcordate at the base; petioles strongly ciliate .... 1. *S. ciliatum*
    3. Leaves narrowly lanceolate, elliptic, or linear, tapering to the base (rarely sometimes rounded); petioles not ciliate
      4. Basal rosettes developing from slender rhizomes; leaves pale beneath .... 3. *S. lanceolatum*
      4. Basal rosettes not developing from slender rhizomes; leaves green beneath .... 2. *S. hybridum*

1. **Steironema ciliatum** (L.) Baudo. Fringed Loosestrife. June–Aug. Moist woods, bottomlands, thickets; common throughout the state. *Lysimachia ciliata* L.

2. **Steironema hybridum** (Michx.) Raf. ex Small. Loosestrife. July–Aug. Swampy woods, thickets, meadows; occasional throughout the state. *Lysimachia hybrida* Michx.; *Lysimachia lanceolata* Walt. var. *hybrida* (Michx.) Gray.

3. **Steironema lanceolatum** (Walt.) Gray. Loosestrife. June–Aug. Moist woods, stream banks, thickets; common throughout the state. *Lysimachia lanceolata* Walt.

4. **Steironema quadriflorum** (Sims) Hitchc. Loosestrife. June–Aug. Marshes, moist prairies, bogs; occasional in the n. cos., becoming rare in the s. cos. *Lysimachia quadriflora* Sims.

5. **Steironema radicans** (Hook.) Gray. Creeping Loosestrife. June–Aug. Swampy woods, rare; Johnson, Pulaski, and St. Clair cos. *Lysimachia radicans* Hook.

10. **Trientalis** L.—Star-flower

1. **Trientalis borealis** Raf. Star-flower. June–July. Moist woods, rare; Cook, Lake, LaSalle, Lee, McHenry, Ogle, and Winnebago cos.

131. PYROLACEAE—PYROLA FAMILY

1. Leaves cauline; inflorescence corymbose; stigmas broad and 5-toothed, peltate, with a very short style; filaments hairy at about the middle .... 1. *Chimaphila*
1. Leaves basal or nearly so; inflorescence racemose; stigmas 5-lobed, with an elongated style; filaments glabrous.
  2. Racemes secund; hypogynous disk 10-lobed; corolla longer than broad............. .... 2. *Orthilia*
  2. Racemes spiral; hypogynous disk not 10-lobed; corolla broader than long.......... .... 3. *Pyrola*

1. **Chimaphila** Pursh—Pipsissewa

1. Leaves mottled with white, lanceolate, more or less rounded at the base .... 1. *C. maculata*
1. Leaves green throughout, oblanceolate, tapering to the base............ 2. *C. umbellata*

1. **Chimaphila maculata** (L.) Pursh. Spotted Wintergreen. June–July. Woods, very rare; Cook, Hardin, and Pope cos.

2. **Chimaphila umbellata** (L.) Bart. ssp. **cisatlantica** (Blake) Hulten. Pipsissewa. June–Aug. Dry woods, very rare; Lake, McHenry, and Winnebago cos. *Chimaphila umbellata* (L.) Bart. var. *cisatlantica* Blake.

2. **Orthilia** Raf.

1. **Orthilia secunda** (L.) House. One-sided Pyrola. June–Aug. Boggy areas, very rare; Cook Co. *Pyrola secunda* L.

3. **Pyrola** L.—Wintergreen

1. Leaves coriaceous, shiny, the blade about the same length as the petiole; lobes of calyx more than 2 mm long .... 1. *P. americana*

1. Leaves thin, dull, the blade longer than the petiole; lobes of calyx up to 2 mm long .... 2. *P. elliptica*

1. **Pyrola americana** Sweet. Wild Lily-of-the-valley. June–Aug. Shaded, mossy, wooded slope, very rare; Ogle Co. *Pyrola rotundifolia* L. var. *americana* (Sweet) Fern.

2. **Pyrola elliptica** Sweet. Shinleaf. June–Aug. Moist woods, rare; confined to extreme n. IL.

132. RANUNCULACEAE—BUTTERCUP FAMILY

1. Flowers yellow, white, or occasionally pinkish.
2. Stems climbing; leaves opposite........ 8. *Clematis*
2. Stems erect, creeping, or floating, not climbing; leaves basal or alternate.
3. Flowers yellow.
4. Sepals and petals present, flat, and differentiated; fruit an achene........ 18. *Ranunculus*
4. Sepals yellow and petal-like; petals absent or reduced to small 2-lobed nectaries; fruit a follicle.
5. Petals absent; leaves crenate........ 6. *Caltha*
5. Petals reduced to small 2-lobed nectaries; leaves palmately cleft........ 12. *Eranthis*
3. Flowers white or occasionally pinkish.
6. Plants aquatic; sepals and petals each 5, differentiated ..... 18. *Ranunculus*
6. Plants not true aquatics; petals absent or, if present, either stamenlike or with a spur.
7. Flowers spurred.
8. Pistil 1; follicle 1; annual .... 9. *Consolida*
8. Pistils 3–5; follicles 3–5; perennial........ 10. *Delphinium*
7. Flowers without a spur.
9. Flowers numerous in racemes, panicles, or corymbs; sepals inconspicuous, falling away as the flower opens.
10. Leaves simple, although deeply lobed .... 20. *Trautvetteria*
10. Leaves variously compound.
11. Flowers unisexual, arranged in much branched panicles........ 19. *Thalictrum*
11. Flowers perfect, arranged in racemes or corymbs.
12. Fruit fleshy, berrylike; raceme simple........ 2. *Actaea*
12. Fruit dry, follicular; raceme sparingly branched........ 7. *Cimicifuga*
9. Flowers 1–4, never in a raceme; sepals showy, petal-like, persistent.
13. Leaves 3-lobed, all basal .... 14. *Hepatica*
13. Leaves either not 3-lobed or not all basal.
14. Cauline leaves alternate.
15. Leaves ternately compound; sepals 5; pedicels glabrous .... 11. *Enemion*
15. Leaves simple, palmately lobed; sepals 3; pedicels pubescent .... 15. *Hydrastis*
14. Cauline leaves opposite or whorled.
16. Leaves ternately compound; roots tuberous-thickened .... 4. *Anemonella*
16. Leaves deeply to shallowly palmately lobed but not ternately divided; plants rhizomatous or with a woody caudex........ 3. *Anemone*
1. Flowers red, blue, purple, or green.
17. Leaves all basal.
18. Leaves linear, entire; sepals spurred; flowers greenish; receptacle elongated .... 16. *Myosurus*
18. Leaves palmately lobed; sepals not spurred; flowers light purplish or sometimes pinkish; receptacle not elongated .... 14. *Hepatica*
17. At least some of the leaves cauline.

19. 1 or more petals or sepals prolonged backward into 1 or more spurs.
  20. Flowers red and yellow, rarely blue, purple, or pink, with each of the 5 petals prolonged backward into a long spur .... 5. *Aquilegia*
  20. Flowers purple, blue, or greenish; 1 of the petal-like sepals prolonged backward into a spur.
    21. Pistil 1; follicle 1; annual .... 9. *Consolida*
    21. Pistils 3–5; follicles 3–5; perennial.............................. 10. *Delphinium*
19. None of the perianth parts spurred.
  22. Flowers strongly asymmetrical; 1 of the sepals hooded..... 1. *Aconitum*
  22. Flowers radially symmetrical; none of the sepals hooded.
    23. Cauline leaves opposite or whorled.
      24. Sepals 4, thick and fleshy; leaves opposite................. 8. *Clematis*
      24. Sepals 5–20, thin; leaves usually whorled ....3. *Anemone*
    23. Cauline leaves alternate.
      25. Flowers green... 13. *Helleborus*
      25. Flowers purplish or blue.
        26. Flowers blue; inflorescence subtended by deeply divided bracts; flowers perfect... 17. *Nigella*
        26. Flowers purplish; inflorescence without bracts; flowers unisexual .... 19. *Thalictrum*

1. **Aconitum** L.—Monkshood

1. **Aconitum uncinatum** L. Wild Monkshood. Aug.–Oct. Woods, very rare; Lake and DuPage cos.

2. **Actaea** L.—Baneberry

1. Pedicels about as thick as the peduncles in fruit, bright red; seeds 3–9 (–10) per berry, more than 4 mm long .... 1. *A. pachypoda*
1. Pedicels much narrower than the peduncles in fruit, dull green or brown; seeds 10 or more per berry, up to 4 mm long .... 2. *A. rubra*

1. **Actaea pachypoda** Ell. Doll's-eyes. Apr.–June (fruiting July–Oct.). Rich woods; occasional throughout most of IL. *Actaea alba* (L.) Mill.

2. **Actaea rubra** (Ait.) Willd. Two forms occur in Illinois:

a. Berries red... 2a. *A. rubra* f. *rubra*
a. Berries white...2b. *A. rubra* f. *neglecta*

2a. **Actaea rubra** (Ait.) Willd. f. **rubra**. Red Baneberry. Apr.–July (fruiting Aug.–Oct.). Rich woods, not common; confined to the n. ¼ of IL.

2b. **Actaea rubra** (Ait.) Willd. f. **neglecta** (Gillman) Robins. White Baneberry. Apr.–July (fruiting Aug.–Oct.). Rich woods; apparently not as common in IL as the preceding.

3. **Anemone** L.—Anemone

1. Styles 2–4 cm long, plumose; staminodia present.................................. 4. *A. patens*
1. Styles up to 4 mm long, not plumose; staminodia absent.
  2. Plants arising from a tuber; sepals 10–20 ....2. *A. caroliniana*
  2. Plants arising from rhizomes; sepals 5 (–6).
    3. Leaves of the involucre sessile; beak of achene 2–5 mm long .... 1. *A. canadensis*
    3. Leaves of the involucre petiolate; beak of achene less than 2 mm long.
      4. Basal leaf solitary; plants at maturity less than 30 cm tall; achenes hirsutulous but not woolly .... 5. *A. quinquefolia*
      4. Basal leaves 2–several; plants at maturity more than 30 cm tall; achenes woolly.
        5. Leaves of the involucre 5–9; fruiting heads more than twice as long as wide; styles less than 1 mm long .... 3. *A. cylindrica*
        5. Leaves of the involucre 3; fruiting heads less than twice as long as wide; styles 1 mm long or longer.................................... 6. *A. virginiana*

1. **Anemone canadensis** L. Meadow Anemone. May–July. Open woods, moist prairies; occasional in the n. ⅔ of IL; also Jackson and Gallatin cos.

2. **Anemone caroliniana** Walt. Carolina Anemone. Apr.–May. Prairie soil, bluffs, not common; confined to the n. ½ of IL; also St. Clair Co.

3. **Anemone cylindrica** Gray. Thimbleweed. May–Aug. Open woods, prairies, occasional; limited to the n. ¾ of IL; also Jackson Co.

4. **Anemone patens** L. var. **multifida** Pritz. Pasque-flower. Mar.–Apr. Prairies; confined to the extreme n. cos. *Pulsatilla patens* (L.) P. Mill. ssp. *multifida* (Pritz.) Zamels; *Anemone patens* L. var. *wolfgangiana* (Bess.) Koch; *Anemone ludoviciana* Nutt.

5. **Anemone quinquefolia** L. Two varieties may be distinguished in Illinois:

a. Stems glabrous or nearly so .... 5a. *A. quinquefolia* var. *quinquefolia*
a. Stems spreading villous .... 5b. *A. quinquefolia* var. *interior*

5a. **Anemone quinquefolia** L. var. **quinquefolia.** Wood Anemone. Apr.–May. Rich woods, not common; confined to the n. ¼ of IL; also Hardin and Menard cos.

5b. **Anemone quinquefolia** L. var. **interior** Fern. Wood Anemone. Apr.–May. Rich woods, not common; confined to the n. ¼ of IL.

6. **Anemone virginiana** L. Tall Anemone. June–Aug. Open, usually dry woods; common throughout IL.

4. **Anemonella** Spach—Rue Anemone

1. **Anemonella thalictroides** (L.) Spach. Rue Anemone. Apr.–May. Dry or moist open woods; occasional throughout the state. *Thalictrum thalictroides* (L.) Eaves & Boivin.

5. **Aquilegia** L.—Columbine

1. Spurs of flower straight; flowers red or yellow .... 1. *A. canadensis*
1. Spurs of flower hooked; flowers blue, purple, pink, or white .... 2. *A. vulgaris*

1. **Aquilegia canadensis** L. Columbine. Apr.–July. Rocky woods; occasional to common throughout the state. (Including var. *coccinea* [Small] Munz.)

2. **Aquilegia vulgaris** L. Garden Columbine. May–July. Escaped from cultivation; rarely observed outside of gardens.

6. **Caltha** L.—Marsh Marigold

1. **Caltha palustris** L. Marsh Marigold. Apr.–June. Wet meadows; confined to the n. ⅔ of IL.

7. **Cimicifuga** L.—Bugbane

1. Pistils 3 or more, stipitate, glabrous .... 1. *C. americana*
1. Pistil 1, not stipitate, pubescent or sparsely glandular.
  2. All leaflets truncate to subcordate; seeds not scaly; pistils pubescent .... 2. *C. racemosa*
  2. At least the terminal leaflet deeply cordate; seeds scaly; pistils sparsely glandular.. 3. *C. rubifolia*

1. **Cimicifuga americana** Michx. American Bugbane. Aug.–Sept. Woods, very rare; Carroll Co.

2. **Cimicifuga racemosa** (L.) Nutt. Black Cohosh. June–July. Woods, rare; confined to the n. ⅙ and the s. ⅓ of IL; absent elsewhere.

3. **Cimicifuga rubifolia** Kearney. Appalachian Bugbane. July–Sept. Rich woods, very rare; confined to the s. ⅙ of IL.

8. **Clematis** L.—Clematis

1. Inflorescence paniculate; sepals white.
  2. Leaves primarily 5-foliolate; anthers at least 2 mm long; achenes with appressed silky hairs; flowers perfect... 5. *C. terniflora*
  2. Leaves primarily 3-foliolate; anthers up to 1.5 mm long; achenes with spreading hairs; flowers unisexual...7. *C. virginiana*
1. Flower solitary; sepals blue to purple.

3. Sepals thin, membranous; outer stamens sterile, petaloid ....3. *C. occidentalis*
3. Sepals thick, leathery; all stamens fertile, not petaloid.
4. Sepals recurved.
5. Tails of fruits densely plumose; only the tips of the sepals recurved.......... ....6. *C. viorna*
5. Tails of fruits glabrous or pubescent but not plumose; upper ½ of the sepals recurved.
6. Leaves thick, conspicuously reticulate beneath; sepals less than 25 mm long, the margins not crisped.................................4. *C. pitcheri*
6. Leaves thin, not conspicuously reticulate beneath; sepals more than 25 mm long, the margins crisped ....1. *C. crispa*
4. Sepals spreading.
7. Margins of sepals crisped; flowers blue-purple..........................1. *C. crispa*
7. Margins of sepals not crisped; flowers blue....................... 2. *C. integrifolia*

1. **Clematis crispa** L. Blue Jasmine. Apr.–July. Swampy woods, wet ditches, rare; Alexander, Pulaski, and St. Clair cos.

2. **Clematis integrifolia** L. Blue Virgin's-bower. July–Aug. Escaped form cultivation; DuPage Co. (Wilhelm, 2010.)

3. **Clematis occidentalis** (Hornem.) DC. Mountain Clematis. May–June. Algific slopes, very rare; Jo Daviess Co.

4. **Clematis pitcheri** Torr. & Gray. Leatherflower. May–Sept. Woods and thickets, occasional; throughout IL.

5. **Clematis terniflora** DC. Virgin's Bower. July–Oct. Native to Asia; naturalized along roads; scattered in IL. *Clematis dioscoreifolia* Levl. & Vaniot.

6. **Clematis viorna** L. Leatherflower. May–July. Along streams, rare; Jasper, Johnson, Pike, Richland, and Wayne cos.

7. **Clematis virginiana** L. Virgin's Bower. July–Sept. Moist soil, particularly at the edges of woods; occasional to common throughout the state.

9. **Consolida** (DC.) S. F. Gray—Rocket Larkspur

1. All of the lowermost bracts dissected into 3–several lobes.
2. Stems pubescent; lowest follicles at least half as long or longer than the pedicels..2. *C. pubescens*
2. Stems usually glabrous; lowest follicles less than half as long as the pedicels ... .1. *C. ajacis*
1. Some or all of the lowermost bracts not dissected, or at most 3-lobed..................... ....3. *C. regalis*

1. **Consolida ajacis** (L.) Schur. Rocket Larkspur. June–Aug. Native to Europe; planted in gardens and occasionally escaped in IL. *Consolida ambigua* (L.) Ball & Heywood; *Delphinium ajacis* L.

2. **Consolida pubescens** (DC.) Soo. Hairy Larkspur. Native to Europe and Africa; rarely escaped from cultivation; Champaign and Cook cos. *Delphinium pubescens* DC.

3. **Consolida regalis** S. F. Gray. Larkspur. June–July. Native to Europe; rarely escaped from cultivation in IL. *Delphinium consolida* L.

10. **Delphinium** L.—Larkspur

1. Racemes virgate; follicles erect; seeds rugose...............................1. *D. carolinianum*
1. Racemes open; follicles spreading at maturity; seeds smooth............. 2. *D. tricorne*

1. **Delphinium carolinianum** Walt. Two subspecies occur in Illinois:

a. Basal leaves absent at anthesis; segments of cauline leaves not distinctly 3-parted ....1a. *D. carolinianum* ssp. *carolinianum*
a. Basal leaves present at anthesis; segments of cauline leaves distinctly 3-parted...... .... 1b. *D. carolinianum* ssp. *virescens*

1a. **Delphinium carolinianum** Walt. ssp. **carolinianum.** Wild Blue Larkspur. May–June. Dry, often sandy soil, rare; mostly in the w. cent. cos. *Delphinium carolinianum* Walt. var. *crispum* Perry.

1b. **Delphinium carolinianum** Walt. ssp. **virescens** (Nutt.) R. E. Brooks. Wild Blue Larkspur. May–June. Dry soil, very rare; Henderson Co. *Delphinium carolinianum* Walt. var. *penardii* (Huth) Warnock; *D. virescens* Nutt. var. *penardii* (Huth) Perry.

2. **Delphinium tricorne** Michx. Dwarf Larkspur. Apr.–May. Rich woods; occasional to common in the s. 4/5 of IL, absent or rare elsewhere. White-flowered forms may be known as f. *albiflora* Millsp.

11. **Enemion** Raf.—False Rue Anemone

1. **Enemion biternatum** Raf. False Rue Anemone. Mar.–May. Rich woods; common throughout the state. *Isopyrum biternatum* (Raf.) Torr. & Gray.

12. **Eranthis** Salisb.—Winter Aconite

1. **Eranthis hyemalis** (L.) Salisb. Winter Aconite. Mar.–Apr. Native to Europe; occasionally planted in IL, but rarely escaped; Piatt and St. Clair cos.

13. **Helleborus** L.—Hellebore

1. **Helleborus viridis** L. Green Hellebore. Mar.–Apr. Native to Europe; rarely escaped from gardens in IL.

14. **Hepatica** Mill.—Hepatica

1. Lobes of the leaves, as well as the bracts, acute at the tip .... 1. *H. acutiloba*
1. Lobes of the leaves, as well as the bracts, rounded at the tip .... 2. *H. americana*

1. **Hepatica acutiloba** DC. Sharp-lobed Hepatica. Mar.–May. Rich woods; occasional throughout the state. *Hepatica nobilis* Schreb. var. *acuta* (Pursh) Steyerm.

2. **Hepatica americana** (DC.) Ker. Round-lobed Hepatica. Mar.–Apr. Rich woods, not common; confined to the ne. cos. of IL. *Hepatica nobilis* Schreb. var. *obtusa* (Pursh) Steyerm.

15. **Hydrastis** Ellis—Goldenseal

1. **Hydrastis canadensis** L. Goldenseal. Apr.–May. Rich woods, occasional; scattered throughout IL.

16. **Myosurus** L.—Mousetail

1. **Myosurus minimus** L. Mousetail. Apr.–June. Moist ground in woods and fields; occasional to common in the s. cos., rare northward.

17. **Nigella** L.—Love-in-a-mist

1. **Nigella damascena** L. Love-in-a-mist. June–Aug. Native to Europe; occasionally grown as an ornamental but rarely escaped; Jackson Co.

18. **Ranunculus** L.—Buttercup

1. Petals white; achenes covered by horizontal wrinkles.
  2. Leaves remaining firm after removal from water; beak of achene about 1 mm long .... 15. *R. longirostris*
  2. Leaves becoming limp after removal from water; beak of achene less than 1 mm long or absent .... 27. *R. trichophyllus*
1. Petals yellow; achenes smooth or variously marked but not with horizontal wrinkles.
  3. At least some of the leaves simple and unlobed.
    4. All leaves simple and unlobed.
      5. Leaves reniform, ovate, or cordate.
        6. Sepals 3; petals usually more than 5, 10 mm long or longer; plants without stolons .... 9. *R. ficaria*
        6. Sepals and petals usually 5, 3–5 mm long; plants with stolons .... 7. *R. cymbalaria*
      5. Leaves linear to lanceolate, tapering to the base.
        7. Petals 5–7 in number, 3–9 mm long; stamens 12–50.
          8. Perennial; stamens 25–50; achenes flattened, about 2 mm long .... 3. *R. ambigens*

8. Annual; stamens 12–25; achenes plump, about 1 mm long .... 14. *R. laxicaulis*
7. Petals 1–3 in number, 1.0–2.5 mm long; stamens 3–10 .... 19. *R. pusillus*
4. At least some of the leaves lobed or divided.
9. Petals longer than the sepals.
10. Stamens in 1 or 2 series; fruiting head less than 6 mm thick; sepals without long white hairs .... 12. *R. harveyi*
10. Stamens in 3–5 series; fruiting head 6–10 mm thick; sepals with long white hairs .... 22. *R. rhomboideus*
9. Petals equaling or shorter than the sepals.
11. Plants more or less fleshy; achenes with corky thickenings at base; stems often hollow .... 24. *R. sceleratus*
11. Plants not fleshy; achenes without corky thickenings at base; stems not hollow.
12. Achenes shiny; receptacle pubescent; roots slender .... 1. *R. abortivus*
12. Achenes dull; receptacle glabrous except sometimes near the tip; roots thickened .... 16. *R. micranthus*
3. None of the leaves simple and unlobed.
13. Plants aquatic or, if creeping in mud, some of the leaves finely dissected.
14. Achenes rugose on the sides, corky-thickened at base; beak of achenes about 1.5 mm long .... 10. *R. flabellaris*
14. Achenes smooth on the sides, not corky-thickened at base; beak of achenes up to 0.8 mm long .... 11. *R. gmelinii*
13. Plants not truly aquatics; leaves not finely dissected.
15. Achenes with spines, although some of them may be very short and sparse.
16. Petals 1–2 mm long; achenes up to 3.5 mm long .... 17. *R. parviflorus*
16. Petals 5–7 mm long; achenes 4.5 mm long or longer........ 4. *R. arvensis*
15. Achenes smooth or papillate on the sides.
17. Petals up to 6 mm long.
18. Fruiting receptacle cylindrical, 9–27 mm long; achenes tomentose .... 26. *R. testiculatus*
18. Fruiting receptacle ovoid to globose, not cylindrical, up to 6 mm long; achenes not tomentose.
19. Petals about equaling the sepals; achenes flat, with strongly recurved beaks; terminal lobe of leaves not stalked.. 20. *R. recurvatus*
19. Petals distinctly shorter than the sepals; achenes not flat, with nearly straight beaks; terminal lobe of leaves stalked.. 18. *R. pensylvanicus*
17. Petals 6 mm long or longer.
20. Achenes papillate on the sides........................ 23. *R. sardous*
20. Achenes smooth on the sides.
21. Petals at least ½ as broad as long.
22. Terminal segment of leaves not stalked........................ 2. *R. acris*
22. Terminal segment of leaves stalked.
23. Some of the roots tuberous-thickened, or plants with a bulbous base.
24. Some of the roots tuberous-thickened .... 8. *R. fascicularis*
24. Plants with a bulbous base........................ 5. *R. bulbosus*
23. All roots fibrous, not thickened............... 6. *R. carolinianus*
21. Petals less than ½ as broad as long.
25. Stems erect or ascending, not rooting at the nodes .... 13. *R. hispidus*
25. Stems creeping and rooting at the nodes.
26. Achenes plump, the beak up to 1.5 mm long .... 21. *R. repens*

26. Achenes flattened, the beak 1.5–3.0 mm long.
  27. Achenes up to 3.5 (–4.5) mm long, with a low narrow keel near the margin .... 25. *R. septentrionalis*
  27. Achenes 3.5–5.0 mm long, with a high broad keel near the margin .... 6. *R. carolinianus*

1. **Ranunculus abortivus** L. Two varieties may be distinguished in Illinois:

a. Leaves, stems, and peduncles glabrous .... 1a. *R. abortivus* var. *abortivus*
a. Leaves, stems, and peduncles pilose .... 1b. *R. abortivus* var. *acrolasius*

1a. **Ranunculus abortivus** L. var. **abortivus.** Small-flowered Crowfoot. Apr.–June. Fields, moist woods, common; throughout the state.

1b. **Ranunculus abortivus** L. var. **acrolasius** Fern. Small-flowered Crowfoot. Apr.–June. Fields, moist woods, much less common than the preceding; scattered throughout IL.

2. **Ranunculus acris** L. Tall Buttercup. May–Aug. Native to Europe; naturalized along roads and in fields; occasional in the n. ½ of IL, rare elsewhere.

3. **Ranunculus ambigens** Wats. Spearwort. June–Sept. Swampy woods and ditches, rare; Fulton, Hancock, St. Clair, and Wabash cos.

4. **Ranunculus arvensis** L. Spiny-fruited Buttercup. May–June. Native to Europe; rarely adventive in IL; Jackson Co.

5. **Ranunculus bulbosus** L. Bulbous Buttercup. Apr.–July. Native to Europe; rarely adventive in disturbed soil; scattered in IL.

6. **Ranunculus carolinianus** DC. Carolina Buttercup. Apr.–May. Lowland woods, rare; scattered in the s. ½ of IL. *Ranunculus septentrionalis* Poir. var. *pterocarpus* Benson.

7. **Ranunculus cymbalaria** Pursh. Seaside Crowfoot. May–Aug. Wet soil, rare; restricted to the n. ¼ of IL.

8. **Ranunculus fascicularis** Bigel. Early Buttercup. Apr.–May. Open woods and meadows; occasional to common throughout the state. (Including var. *apricus* [Greene] Fern.)

9. **Ranunculus ficaria** L. Three subspecies occur in Illinois:

a. Leaves not crowded at base of stem, with numerous leaves on an elongated stem.
  b. Bulbils not present in axils of leaves; achenes well developed .... 9a. *R. ficaria* ssp. *ficaria*
  b. Bulbils present in axils of leaves; achenes poorly developd .... 9b. *R. ficaria* ssp. *bulbilifer*
a. Leaves crowded at base of stem, with few leaves on a short stem .... 9c. *R. ficaria* ssp. *calthifolius*

9a. **Ranunculus ficaria** L. ssp. **ficaria.** Lesser Celandine. Apr.–May. Native to Europe; sometimes planted as an ornamental but seldom escaped; Jackson and Lake cos. *Ficaria verna* Huds.

9b. **Ranunculus ficaria** L. ssp. **bulbilifer** Lambinon. Lesser Celandine. Apr.–Nov. Native to Europe; sometimes planted as an ornamental but seldom escaped; Cook, DuPage, and Lake cos.

9c. **Ranunculus ficaria** L. ssp. **calthifolius** (Reichenb.) Arcangelis. Lesser Celandine. Apr.–Nov. Native to Europe; sometimes planted as an ornamental but seldom escaped; Cook and Lake cos.

10. **Ranunculus flabellaris** Raf. Yellow Water Crowfoot. Apr.–June. Swamps and ponds; occasional throughout the state.

11. **Ranunculus gmelinii** DC. var. **hookeri** (D. Don) Benson. Small Yellow Water Crowfoot. July–Aug. Ponds, very rare; Cook and Menard cos.

12. **Ranunculus harveyi** (Gray) Britt. Two forms occur in Illinois:

1. Stems and leaves glabrous .... 12a. *R. harveyi* f. *harveyi*
1. Stems and leaves pilose .... 12b. *R. harveyi* f. *pilosus*

12a. **Ranunculus harveyi** (Gray) Britt. f. **harveyi.** Harvey's Buttercup. Apr.–May. Sandstone ravines, rare; Effingham, Fayette, Jackson, Macoupin, and Randolph cos.

12b. **Ranunculus harveyi** (Gray) Britt. f. **pilosus** (Benke) Palmer & Steyerm. Harvey's Buttercup. Apr.–May. Sandstone ravine, rare; Randolph Co.

13. **Ranunculus hispidus** Michx. Two varieties may be recognized in Illinois:

a. Pubescence spreading; achenes 3.0–3.5 mm long (excluding the beak) .... 13a. *R. hispidus* var. *hispidus*
a. Pubescence generally appressed; achenes 2.0–2.5 mm long (excluding the beak) ... . 13b. *R. hispidus* var. *marilandicus*

13a. **Ranunculus hispidus** Michx. var. **hispidus.** Bristly Buttercup. Apr.–May. Dry woods; occasional throughout IL.

13b. **Ranunculus hispidus** Michx. var. **marilandicus** (Poir.) L. Benson. Bristly Buttercup. Apr.–May. Dry woods; rare in s. IL. (Including var. *falsus* Fern.)

14. **Ranunculus laxicaulis** (Torr. & Gray) Darby. Spearwort. May–July. Wet woods and wet ditches; generally in s. IL; also Fulton Co.

15. **Ranunculus longirostris** Godr. White Water Crowfoot. May–Aug. Ponds and slow streams; mostly in the n. cos. *Ranunculus circinatus* Sibth.

16. **Ranunculus micranthus** Torr. & Gray. Small-flowered Buttercup. Mar.–May. Moist or dry woods; occasional to common throughout the state. (Including var. *delitescens* [Greene] Fern.)

17. **Ranunculus parviflorus** L. Small-flowered Buttercup. Apr.–June. Native to Europe; adventive in disturbed soil; Jackson Co.

18. **Ranunculus pensylvanicus** L. f. Bristly Crowfoot. July–Sept. Wet ground, not common; scattered throughout the state.

19. **Ranunculus pusillus** Poir. Small Spearwort. May–June. Swamps, wet woods, wet ditches; confined to the s. ⅓ of IL; also Iroquois and Macoupin cos.

20. **Ranunculus recurvatus** Poir. Recurved Buttercup. Apr.–June. Wet woods; occasional throughout IL.

21. **Ranunculus repens** L. Two varieties occur in Illinois:

a. Petals 10 or fewer .... 21a. *R. repens* var. *repens*
a. Petals more than 10 .... 21b. *R. repens* var. *degeneratus*

21a. **Ranunculus repens** L. var. **repens.** Creeping Buttercup. Apr.–Aug. Native to Europe; adventive along roads and in fields; rare in the n. ⅔ of IL.

21b. **Ranunculus repens** L. var. **degeneratus** Schur. Double-flowered Creeping Buttercup. Planted in gardens but rarely escaped; Grundy, Jackson, and Kendall cos. *Ranunculus repens* L. var. *pleniflorus* Fern.

22. **Ranunculus rhomboideus** Goldie. Prairie Buttercup. May. Prairies, rare; restricted to the extreme n. cos.; also Macoupin Co.

23. **Ranunculus sardous** Crantz. Buttercup. May–July. Native to Europe; naturalized in low fields and disturbed areas; confined to a few cos. in the southernmost part of IL; also Clay, Cook, and DeKalb cos.

24. **Ranunculus sceleratus** L. Cursed Crowfoot. May–Aug. Wet meadows, ditches, river banks; occasional throughout IL.

25. **Ranunculus septentrionalis** Poir. Two varieties may be distinguished in Illinois:

a. Petioles and lower part of stems glabrous or appressed-pubescent .... 25a. *R. septentrionalis* var. *septentrionalis*
a. Petioles and lower part of stems with retrorse pubescence .... 25b. *R. septentrionalis* var. *caricetorum*

25a. **Ranunculus septentrionalis** Poir. var. **septentrionalis.** Marsh Buttercup. Apr.–July; Sept.–Oct. Low woods, wet ditches, swampy areas; common throughout IL. *Ranunculus hispidus* Michx. var. *nitidus* (Ell.) Duncan.

25b. **Ranunculus septentrionalis** Poir. var. **caricetorum** (Greene) Fern. Marsh Buttercup. Apr.–July. Low woods; much less common than the preceding variety; scattered in IL. *Ranunculus hispidus* Michx. var. *caricetorum* (Greene) Duncan.

26. **Ranunculus testiculatus** L. Cylindric-fruited Buttercup. May–June. Native to the w. U.S.; naturalized in waste areas, particularly in campgrounds; ne. cos. *Ceratocephalus testiculatus* (L.) Roth.

27. **Ranunculus trichophyllus** Chaix. White Water Crowfoot. May–Aug. Ponds and slow streams; mostly in the n. cos. *Ranunculus aquatilis* L. var. *capillaceus* (Thuill.) DC.

19. **Thalictrum** L.—Meadow Rue

1. Middle and upper leaves petiolate; leaflets 3- to 12-lobed at the apex, their margins crenate .... 1. *T. dioicum*
1. Middle and upper leaves sessile; leaflets 3-lobed at the apex, their margins entire.
  2. Leaves with glandular hairs on the lower surface .... 4. *T. revolutum*
  2. Leaves glabrous or pubescent on the lower surface but without glandular hairs.
    3. Achenes stipitate; anthers constricted at the tip; stigmas curved .... 3. *T. pubescens*
    3. Achenes not stipitate; anthers not constricted at the tip; stigmas more or less straight .... 2. *T. dasycarpum*

1. **Thalictrum dioicum** L. Early Meadow Rue. Apr.–May. Rich woods, prairies; occasional throughout IL.

2. **Thalictrum dasycarpum** Fisch. & Lall. Two varieties may be distinguished in Illinois:

a. Leaflets firm, pubescent on the lower surface... 2a. *T. dasycarpum* var. *dasycarpum*
a. Leaflets thin, glabrous beneath .... 2b. *T. dasycarpum* var. *hypoglaucum*

2a. **Thalictrum dasycarpum** Fisch. & Lall. var. **dasycarpum.** Purple Meadow Rue. May–June. Moist, wooded ravines, occasional; scattered throughout IL.

2b. **Thalictrum dasycarpum** Fisch. & Lall. var. **hypoglaucum** (Rydb.) Boivin. Meadow Rue. June–July. Moist, wooded ravines, occasional; scattered throughout IL.

3. **Thalictrum pubescens** Pursh. Appalachian Meadow Rue. May–June. Wet meadows, very rare; Hardin Co.

4. **Thalictrum revolutum** DC. Waxy Meadow Rue. May–June. Prairies, open woods; occasional throughout IL.

20. **Trautvetteria** Fisch. & Mey.—False Bugbane

1. **Trautvetteria caroliniensis** (Walt.) Vail. False Bugbane. June–July. Moist ground along stream, very rare; Cass Co.

133. RESEDACEAE—MIGNONETTE FAMILY

1. **Reseda** L.—Mignonette

1. Leaves pinnately divided or pinnately lobed; petals 5 or 6.
  2. Flowers greenish white; carpels 4; stamens 12–15 .... 1. *R. alba*
  2. Flowers yellow; carpels 3; stamens 15–20 .... 2. *R. lutea*
1. Leaves entire; petals 4 .... 3. *R. luteola*

1. **Reseda alba** L. Mignonette. June–Oct. Sometimes grown as an ornamental but rarely escaped into disturbed soil; Cook and DuPage cos.

2. **Reseda lutea** L. Yellow Mignonette. June–Sept. Native to Europe; rarely escaped from cultivation; Cook Co.

3. **Reseda luteola** L. Dyer's Rocket. June–Oct. Native to Europe; occasionally planted as an ornamental but rarely escaped; Cook Co.

134. RHAMNACEAE—BUCKTHORN FAMILY

1. Woody vines .... 1. *Berchemia*
1. Trees or shrubs.
  2. Leaves 3-veined from the base; fruit a capsule .... 2. *Ceanothus*
  2. Leaves pinnately veined; fruit a drupe.
    3. Winter buds without scales; flowers perfect; nutlets without a groove .... 3. *Frangula*
    3. Winter buds scaly; flowers unisexual; nutlets usually with a groove .... 4. *Rhamnus*

1. **Berchemia** Neck.—Supple-jack

1. **Berchemia scandens** (Hill) K. Koch. Supple-jack; Rattan Vine. Apr.–June. Edge of pine plantation; Pope Co.

2. **Ceanothus** L.

1. Leaves ovate to ovate-oblong, acute to acuminate (obtuse in var. *pitcheri*) .... 1. *C. americanus*
1. Leaves elliptic to elliptic-lanceolate, obtuse to subacute .... 2. *C. herbaceus*

1. **Ceanothus americanus** L. Two varieties may be recognized in Illinois:

a. Leaves acute to acuminate, glabrous or nearly so on the upper surface .... 1a. *C. americanus* var. *americanus*
a. Leaves obtuse to subacute, pilose on the upper surface .... 1b. *C. americanus* var. *pitcheri*

1a. **Ceanothus americanus** L. var. **americanus.** New Jersey Tea. June–Aug. Woods; occasional throughout the state. (Including var. *intermedius* [Pursh] K. Koch.)

1b. **Ceanothus americanus** L. var. **pitcheri** Torr. & Gray. New Jersey Tea. June–Aug. Woods, not common; scattered in IL.

2. **Ceanothus herbaceus** Raf. May–June. Ceanothus. Low dunes, sandy soil, rare; Carroll, Cook, Jo Daviess, Lake, Ogle, Whiteside, and Winnebago cos. *Ceanothus ovatus* Desf.

3. **Frangula** L.—Buckthorn

1. Umbels peduncled; leaves more or less acute; most or all the petioles 5 mm long or longer; pedicels pubescent .... 2. *F. caroliniana*
1. Umbels sessile; leaves more or less obtuse; most or all the petioles less than 5 mm long; pedicels glabrous or nearly so .... 1. *F. alnus*

1. **Frangula alnus** Mill. Two varieties occur in Illinois:

a. Leaves obovate .... 1a. *F. alnus* var. *alnus*
a. Leaves narrowly lanceolate .... 1b. *F. alnus* var. *angustifolia*

1a. **Frangula alnus** Mill. var. **alnus.** Glossy Buckthorn. May–July. Native to Europe; naturalized in woods and bogs; occasional in the n. ½ of the state; also Crawford and St. Clair cos. *Rhamnus frangula* L. var. *frangula.*

1b. **Frangula alnus** Mill. var. **angustifolia** (Loud.) Mohlenbr., comb. nov. (basionym: *Rhamnus frangula* L. var. *angustifolia* Loud.). Narrow-leaved Glossy Buckthorn. May–July. Native to Europe; rarely naturalized in IL; Cook Co. *Rhamnus frangula* L. var. *angustifolia* Loud.

2. **Frangula caroliniana** (Walt.) Gray. Two varieties may be recognized in Illinois:

a. Leaves glabrous or nearly so .... 2a. *F. caroliniana* var. *caroliniana*
a. Leaves densely pubescent .... 2b. *F. caroliniana* var. *mollis*

2a. **Frangula caroliniana** (Walt.) Gray var. **caroliniana.** Carolina Buckthorn. May–June. Woods, not common; in the s. ⅓ of IL. *Rhamnus caroliniana* Walt. var. *caroliniana.*

2b. **Frangula caroliniana** (Walt.) Gray var. **mollis** (Fern.) Mohlenbr., comb. nov. (basionym: *Rhamnus caroliniana* Walt. var. *mollis* Fern.). Hairy Carolina Buckthorn. May–June. Woods, very rare; Jackson Co. *Rhamnus caroliniana* Walt. var. *mollis* Fern.

4. **Rhamnus** L.—Buckthorn

1. Leaves with 2–4 pairs of distinct lateral veins, obtuse to abruptly acuminate, not as much as 6 times as long as the petioles.
    2. Leaves ovate to elliptic, dull on the upper surface, more or less rounded at the base .... 3. *R. cathartica*
    2. Leaves obovate to oblong, shiny on the upper surface, tapering to the base .... 4. *R. davurica*
1. Leaves with (4–) 5 or more pairs of distinct lateral veins, acuminate, more than 6 times longer than the petioles.
    3. Leaves less than 8 cm long and less than 3.3 cm broad .... 6. *R. lanceolata*
    3. Leaves more than 8 cm long or more than 3.3 cm broad or both.

4. All leaves alternate.
5. Leaves round-toothed, glabrous or nearly so beneah; shrubs to 1 m tall .... 1. *R. alnifolia*
5. Leaves sharply toothed, velutinous beneath; trees to 7 m tall .... 2. *R. arguta*
4. Some of the leaves appearing to be subopposite.
6. Leaves broadest at or below the middle .... 5. *R. japonica*
6. Leaves broadest above the middle .... 7. *R. utilis*

1. **Rhamnus alnifolia** L'Her. Alder Buckthorn. May–July. Bogs, wooded swamps, rare; confined to the n. ½ of IL; also Richland Co.

2. **Rhamnus arguta** Maim. var. **velutina** Hand.-Mazz. Saw-toothed Buckthorn. May–July. Native to Europe; disturbed soil; Will Co.

3. **Rhamnus cathartica** L. Common Buckthorn. May–June. Native to Europe; naturalized in disturbed areas and woodlands; common in the n. ⅗ of IL.

4. **Rhamnus davurica** Pall. Dahurian Buckthorn. May–June. Native to Asia; naturalized in disturbed areas and woodlands; occasional in the ne. cos.

5. **Rhamnus japonica** Maxim. Japanese Buckthorn. May–June. Native to Japan; rarely escaped from cultivation in degraded open ground; DuPage Co.

6. **Rhamnus lanceolata** Pursh. Two varieties occur in Illinois:

a. Branches and lower leaf surfaces pubescent .... 6a. *R. lanceolata* var. *lanceolata*
a. Branches and lower leaf surfaces glabrous or nearly so .... 6b. *R. lanceolata* var. *glabrata*

6a. **Rhamnus lanceolata** Pursh var. **lanceolata.** Lance-leaved Buckthorn. Apr.–June. River banks, bluffs, calcareous fens; rare to occasional in the n. ⅔ of IL, rare or absent elsewhere.

6b. **Rhamnus lanceolata** Pursh var. **glabrata** Gl. Lance-leaved Buckthorn. Apr.–June. Calcareous fens; apparently not as common as var. *lanceolata;* ne. cos.

7. **Rhamnus utilis** Decne. Chinese Buckthorn. May–June. Native to China; rarely escaped from cultivation in degraded woods and meadows; ne. cos.

135. ROSACEAE—ROSE FAMILY

1. Plants woody, either trees, shrubs, or woody arching brambles.
2. Leaves simple.
3. Ovary or ovaries superior.
4. Flowers purple .... 27. *Rubus*
4. Flowers white, pink, rose, or yellow.
5. Ovary 1; fruit fleshy .... 22. *Prunus*
5. Ovaries 3–8; fruit dry.
6. Leaves opposite; sepals 4; petals 4 .... 24. *Rhodotypos*
6. Leaves alternate; sepals 5; petals 5.
7. Leaves palmately lobed; ovaries 3–5 .... 19. *Physocarpus*
7. Leaves unlobed; ovaries 5–8.
8. Flowers golden yellow; stipules present .... 16. *Kerria*
8. Flowers white or rose; stipules absent .... 31. *Spiraea*
3. Ovary inferior, enclosed, at least in part, by the hypanthium.
9. Leaves entire.
10. Fruit yellow, fuzzy, pyriform, 6–10 cm in diameter; flowers 4–7 cm across .... 9. *Cydonia*
10. Fruit red or black, glabrous, subglobose, 7–10 mm in diameter; flowers up to 3 cm across .... 7. *Cotoneaster*
9. Leaves toothed or lobed.
11. Plants without thorns.
12. Petals at least twice as long as broad; ovary 6- to 10-locular; leaves often subcordate at base .... 2. *Amelanchier*
12. Petals less than twice as long as broad; ovary 2- to 5-locular; leaves seldom or never subcordate at base.
13. Petals red .... 5. *Chaenomeles*
13. Petals white, pink, or pinkish white.
14. Petals 1 cm long or longer; small trees; midrib of leaves eglandular .... 17. *Malus*

14. Petals less than 1 cm long; shrubs; midrib of leaves glandular .... 18. *Photinia*
11. Plants with thorns.
15. Petals red .... 5. *Chaenomeles*
15. Petals white, pink, or pinkish white.
16. Styles united at base .... 17. *Malus*
16. Styles free to base.
17. Some or all the flowers 2.5 cm broad or broader; fruit with grit cells .... 23. *Pyrus*
17. Flowers up to 2.5 cm broad; fruit without grit cells .... 8. *Crataegus*
2. Leaves compound.
18. Plants spiny or bristly, usually arching or trailing.
19. Leaves palmately compound; flowers white; fruit a blackberry or raspberry .... 26. *Rubus*
19. Leaves pinnately compound; flowers usually pink or rose (sometimes white in some species and forms); fruit enclosed in the hypanthium .... 25. *Rosa*
18. Plants without spines or bristles, usually erect.
20. Ovary inferior, enclosed in the hypanthium; flowers white .... 30. *Sorbus*
20. Ovary superior; flowers yellow or white.
21. Flowers yellow, solitary or few in clusters; plants to 1.5 m tall .... 10. *Dasiphora*
21. Flowers white, many in a panicle; plants usually more than 1.5 m tall .... 29. *Sorbaria*
1. Plants herbaceous.
22. Leaves confined to or near the base of the plant.
23. Flowers yellow .... 32. *Waldsteinia*
23. Flowers white.
24. Leaflets evergreen, 3-toothed at the apex, entire; fruit not fleshy .... 28. *Sibbaldiopsis*
24. Leaflets deciduous, serrate throughout; fruit fleshy .... 14. *Fragaria*
22. Leaves borne all along the stem.
25. Leaves 2- to 3-pinnate; plants dioecious .... 4. *Aruncus*
25. Leaves once-pinnate or palmate; plants with perfect flowers.
26. Flowers pink, rose, or purple.
27. Flowers pink, many in a panicle, each flower less than 1 cm across; stems glabrous; terminal leaflet several-lobed; plants 1–2 m tall .... 13. *Filipendula*
27. Flowers purple, solitary or few in a cyme or corymb, each flower more than 1 cm across; stems pubescent; terminal leaflet usually not lobed; plants up to 1 m tall.
28. Flowers erect; leaflets 5–7; style not persistent on the achene, not plumose .... 6. *Comarum*
28. Flowers often nodding; leaflets 7–17; style persistent on the achene, plumose .... 15. *Geum*
26. Flowers white, cream, or yellow.
29. Petals absent; calyx 4-parted .... 27. *Sanguisorba*
29. Petals present; calyx 5-parted.
30. Petals white.
31. Petals strap-shaped, usually 2–3 times longer than broad; each leaf often subtended by a foliaceous pair of stipules .... 20. *Porteranthus*
31. Petals not strap-shaped, never 2–3 times longer than broad; foliaceous stipules may or may not be present.
32. Plants trailing or ascending, nearly glabrous; leaves pedately 3- or 5-foliolate .... 26. *Rubus*
32. Plants erect, pubescent; leaves pinnately 3- to 11-foliolate.
33. Flowers in panicles; pistils 5–15 in a ring; terminal leaflet several-lobed; fruit a twisted follicle; plants 1–2 m tall .... 13. *Filipendula*

33. Flowers solitary or in cymes or corymbs; pistils on an elongated receptacle; terminal leaflet usually not lobed; fruit an achene; plants up to 1 m tall.
34. Style persistent on the achene; petals 3–5 mm long; stems not glandular-villous; uppermost leaf often unlobed...15. *Geum*
34. Style not persistent on the achene; petals 5–8 mm long; stems glandular-villous; uppermost leaf usually lobed...11. *Drmocallis*
30. Petals yellow.
35. All leaves trifoliolate.
36. Receptacle in fruit spongy, greatly enlarged; bractlets between the calyx lobes longer than the lobes ....12. *Duchesnea*
36. Receptacle in fruit dry, not enlarged; bractlets between the calyx lobes about the same size as the lobes .....21. *Potentilla*
35. At least some of the leaves 5-foliolate or more.
37. Leaves pinnately compound.
38. Bractlets present between calyx lobes; ovaries at least 30 per flower.
39. Style deciduous from the achene.
40. Leaflets silvery-silky beneath; flower solitary... 3. *Argentina*
40. Leaflets not silvery-silky beneath; flowers in cymes ....21. *Potentilla*
39. Style persistent on the achene ....15. *Geum*
38. Bractlets absent between calyx lobes; ovaries 1–4 per flower ....1. *Agrimonia*
37. Leaves palmately compound.
41. Style deciduous from the achene ....21. *Potentilla*
41. Style persistent on the achene.............................15. *Geum*

1. **Agrimonia** L.—Agrimony

1. Leaflets (excluding smaller interposed ones) 11–17............................2. *A. parviflora*
1. Most of the leaflets (excluding smaller interposed ones) 5–9.
2. Branches of the inflorescence eglandular.
3. Roots fibrous; leaflets glandular and sparsely pubescent on the lower surface; fruit 4–5 mm long ....5. *A. striata*
3. Roots tuberous-thickened; leaflets mostly glandular and velvety-pubescent on the lower surface; fruit 2.5–3.0 mm long ....3. *A. pubescens*
2. Branches of the inflorescence glandular.
4. Roots tuberous; fruit 3–4 mm long ....4. *A. rostellata*
4. Roots not tuberous; fruit 5–8 mm long.................................1. *A. gryposepala*

1. **Agrimonia grypsosepala** Wallr. Tall Agrimony. June–Aug. Woods and thickets; occasional in the n. ¾ of IL, absent elsewhere.

2. **Agrimonia parviflora** Sol. Swamp Agrimony. July–Sept. Low ground; occasional throughout the state.

3. **Agrimonia pubescens** Wallr. Soft Agrimony. July–Sept. Dry or moist woods; occasional throughout the state. (Including *A. mollis* [Torr. & Gray] Britt.) *A. microcarpa* Wallr., misapplied.

4. **Agrimonia rostellata** Wallr. Woodland Agrimony. July–Sept. Woods; occasional in the s. ⅔ of IL; also Cook, DuPage, Kankakee, and Will cos.

5. **Agrimonia striata** Michx. Roadside Agrimony. June–Aug. Woods, rare; Boone, Cook, and DeKalb cos.

2. **Amelanchier** Medic.—Shadbush

1. Stoloniferous shrubs; leaves toothed only in the upper ⅔; petals up to 1 cm long ....3. *A. humilis*
1. Non-stoloniferous shrubs or small trees; leaves toothed throughout; petals 1 cm long or longer.

2. Leaves white-tomentose beneath (at least at first); pedicles of lowermost flowers in each raceme up to 2.5 cm long.
3. Shrubs to 3 m tall; leaves unfolded at flowering time; fruit dark purple.......... ....6. *A. sanguinea*
3. Small trees usually more than 3 m tall; leaves folded at flowering time; fruit red-purple to dark purple to black.
4. Racemes pendulous; petals 10 mm long or longer; sepals strongly reflexed on the fruit ....1. *A. arborea*
4. Racemes erect to ascending; petals up to 10 mm long; sepals spreading or recurved, but not reflexed on the fruit........................2. *A. canadensis*
2. Leaves glabrous or sparsely pubescent beneath; pedicels of lowermost flowers in each raceme 2.5 cm long or longer.
5. Leaves with up to 10 pairs of veins; ovary and usually the young fruit pubescent........................4. *A. interior*
5. Leaves with more than 10 pairs of veins; ovary and young fruit glabrous ....5. *A. laevis*

1. **Amelanchier arborea** (Michx. f.) Fern. Shadbush; Serviceberry; Juneberry. Mar.–May. Wooded slopes and bluffs; occasional throughout the state.
2. **Amelanchier canadensis** (L.) Medic. Eastern Shadbush. Apr.–May. Woods, very rare; DuPage Co.
3. **Amelanchier humilis** Wieg. Low Shadbush. May–June. Rocky or sandy soil; restricted to the n. ¼ of IL; also Adams and Lawrence cos.; *Amelanchier stolonifera* Wieg.
4. **Amelanchier interior** Nielsen. Shadbush. June. Bogs and wet woods, rare; Cook, DuPage, Jo Daviess, Lake, Will, and Winnebago cos.
5. **Amelanchier laevis** Wieg. Shadbush. Mar.–June. Wooded slopes; confined to the n. ¼ of IL; also Champaign, Clark, and Vermilion cos.
6. **Amelanchier sanguinea** (Pursh) DC. Shadbush. May–June. Woods, rare; n. ½ of IL; also Adams Co.

3. **Argentina** Hill

1. **Argentina anserina** (L.) Rydb. Silverweed. May–Aug. Sandy beaches, interdunal ponds, meadows, gravel bars; restricted to the ne. cos. *Potentilla anserina* L.

4. **Aruncus** Adans.—Goat's-beard

1. **Aruncus dioicus** (Walt.) Fern. Two varieties occur in Illinois:

a. Leaflets green, glabrous; follicles 1.5–2.0 mm long .... 1a. *A. dioicus* var. *dioicus*
a. Leaflets grayish, usually pubescent; follicles 1.7–2.5 mm long........................ ....1b. *A. dioicus* var. *pubescens*

1a. **Aruncus dioicus** (Walt.) Fern. var. **dioicus.** Goat's-beard. May–June. Rich woods; occasional throughout the state; absent in the ne. cos., except Grundy and LaSalle cos.

1b. **Aruncus dioicus** (Walt.) Fern. var. **pubescens** (Rydb.) Fern. Gray Goat's-beard. May–June. Rich woods, rare; scattered in the s. ½ of IL.

5. **Chaenomeles** L.—Japanese Quince

1. Plants without thorns, rarely more than 1.5 m tall ....1. *C. japonica*
1. Plants thorny, always more than 1.5 m tall when mature........................2. *C. speciosa*

1. **Chaenomeles japonica** (Thunb.) Lindl. Spineless Japanese Quince. Apr.–May. Native to Asia; adventive at the edge of a thicket; Jackson Co.
2. **Chaenomeles speciosa** (Sweet) Nakai. Japanese Quince; Japonica. Mar.–Apr. Frequently grown as an ornamental but rarely escaped; Champaign, DuPage, and Jackson cos. *Chaenomeles lagenaria* (Loisel.) Koidz., misapplied.

6. **Comarum** L.

1. **Comarum palustre** L. Marsh Cinquefoil. June–July. Bogs, very rare; Cook, Lake, and McHenry cos. *Potentilla palustris* (L.) Scop.

7. **Cotoneaster** Medic.

1. Leaves up to 15 mm long .... 2. *C. divaricata*
1. All or most of the leaves more than 15 mm long.
  2. Fruit black; leaves more or less tapering to the base; flowers pinkish, 2–5 in a cluster; calyx pubescent .... 1. *C. acutifolia*
  2. Fruit red; leaves rounded at the base; flowers white, several in a cluster; calyx glabrous .... 3. *C. multiflora*

1. **Cotoneaster acutifolia** Turcz. Peking Cotoneaster. June–July. Native to China; rarely escaped from gardens; Cook and DuPage cos. *Cotoneaster acutifolia* Turcz. var. *lucida* (Schltdl.) L. T. Lu; *Cotoneaster lucida* Schltdl.

2. **Cotoneaster divaricata** Rehd. & Wilson. Spreading Cotoneaster. June–July. Native to China; rarely escaped from gardens; Will Co.

3. **Cotoneaster multiflora** Bunge. Cotoneaster. May–June. Native to China; occasionally planted but seldom escaped into old fields and along roadsides; DuPage Co.

8. **Crataegus** L.—Hawthorn

1. Veins of larger leaves running to the sinuses as well as to the points of the teeth and lobes.
  2. Leaves thick, persisting late into the autumn; thorns slender, up to 2 cm long .... 35. *C. monogyna*
  2. Leaves thin, early deciduous; thorns usually stout, some or all over 2 cm long.
    3. Most of the leaves 3.0–4.5 cm broad, acute to acuminate at the apex, cordate at the base .... 43. *C. phaenopyrum*
    3. Most of the leaves 1.5–3.5 cm broad, rounded at the apex, cuneate or rounded at the base.
      4. Leaves obovate or spatulate, longer than broad, unlobed or 3-lobed; fruit subglobose .... 52. *C. spathulata*
      4. Leaves broadly obovate or deltoid, broader than long, deeply several-lobed; fruit oblongoid .... 33. *C. marshallii*
1. Veins of larger leaves running only to the points of the teeth and lobes.
  5. Leaves broadest at the middle or near the apex, the base cuneate.
    6. Leaves broadest above the middle, often nearest the apex.
      7. Leaves impressed-veined above.
        8. Leaves pubescent beneath at maturity, at least on the veins, dull gray-green, deeply lobed on the vegetative shoots; fruits 12–20 mm thick .... 47. *C. punctata*
        8. Leaves glabrous beneath at maturity, bright green or yellow-green, unlobed or shallowly lobed on the vegetative shoots; fruits 7–12 mm thick.
          9. Most of the leaves of the flowering branches lobed.
            10. Leaves of flowering branches with conspicuous, spreading lobes .... 5. *C. chrysocarpa*
            10. Leaves of flowering branches obscurely shallowly lobed or, if somewhat conspicuously lobed, the lobes ascending.
              11. Leaves yellow-green .... 10. *C. cuneiformis*
              11. Leaves bright green.
                12. Leaves with obtuse lobes and crenate margins; stamens 20 .... 32. *C. margarettiae*
                12. Leaves with acute lobes and serrate margins; stamens 10 .... 17. *C. dodgei*
          9. Most of the leaves of the flowering branches unlobed or sparsely shallowly lobed.
            13. Leaves yellow-green; fruits 7–8 mm in diameter .... 23. *C. hannibalensis*
            13. Leaves green; fruits 8–12 mm in diameter.
              14. Leaves pubescent when young, becoming glabrous or nearly so at maturity; leaves dull green .... 7. *C. collina*
              14. Leaves glabrous, even when young; leaves dark green, shiny.

15. Leaves obtuse at the apex, the lower surface pale; fruit 9–12 mm in diameter .... 44. *C. pratensis*
15. Leaves acute to acuminate at the apex, the lower surface not pale; fruit 8–10 mm in diameter .... 13. *C. disperma*
7. Leaves not impressed-veined above.
16. None of the leaves lobed.
17. Leaves glabrous, not reticulate-veined; petioles glabrous; small tree .... 9. *C. crus-galli*
17. Leaves pubescent, reticulate-veined; petioles villous; shrub ....18. *C. engelmannii*
16. At least the terminal shoot leaves with a few shallow, obscure lobes.
18. Leaves pubescent when young; pedicels of flowers villous .... 20. *C. fecunda*
18. Leaves glabrous, even when young; pedicels of flowers glabrous.
19. Terminal shoot leaves at least twice as broad as those of the flowering branches; stamens 5, 10, or 15 ....1. *C. acutifolia*
19. Terminal shoot leaves usually less than twice as broad as those of the flowering branches; stamens 20.
20. Fruit obovoid, 8–12 mm in diameter, the flesh firm; pedicels of flowers glabrous; anthers pink ....42. *C. permixta*
20. Fruit subglobose, 7–8 mm in diameter, the flesh dry; pedicels of flowers villous; anthers yellow ....26. *C. laxiflora*
6. Leaves broadest at the middle.
21. Leaves impressed-veined above.
22. Leaves bright green, dark green, or blue-green, shiny or dull, thick; young branchlets glabrous or sparsely villous.
23. Leaves bright green, shiny ....54. *C. succulenta*
23. Leaves dark green or blue-green, dull or shiny, thick.
24. Leaves dark green.
25. Leaves dull; flowers 15–16 mm across; fruits globose, 5–8 mm in diameter..29. *C. macracantha*
25. Leaves shiny; flowers 10–12 mm across; fruits oblongoid, 8–12 mm in diameter ....41. *C. peoriensis*
24. Leaves blue-green...2. *C. apiomorpha*
22. Leaves yellow-green, dull, thin; young branchlets tomentose.
26. Flowers 12–15 mm across, in large, open corymbs; fruit oblongoid to obovoid, 7–9 mm thick .... 3. *C. calpodendron*
26. Flowers 15–20 mm across, in small, compact corymbs; fruit subglobose, 10–12 mm in diameter............................ 57. *C. X whittakeri*
21. Leaves not impressed-veined above.
27. Petioles and usually the leaves glandular.
28. Anthers pink or red; fruit subglobose.......................36. *C. neobushii*
28. Anthers white; fruit obovoid ....25. *C. intricata*
27. Petioles and leaves eglandular.
29. Leaves at maturity with tufts of tomentum in the axils beneath; pedicels of flowers glabrous.
30. Leaves thin, scarcely lustrous above; fruit 5–8 mm thick .... 46. *C. viridis*
30. Leaves thick, lustrous above; fruit 8–10 mm thick......31. *C. nitida*
29. Leaves at maturity glabrous beneath or at least without tufts of tomentum in the axils; pedicels of flowers pubescent.
31. Leaves uniformly pubescent on the upper surface, but not scabrous; anthers yellow ....19. *C. faxonii*
31. Leaves glabrous or scabrous above, but not uniformly pubescent, even when young; anthers pink or red.
32. Leaves dark green, thin.
33. Leaves ovate to obovate, dull, scabrous on the upper surface; fruit obovoid...................................28. *C. lucorum*
33. Leaves elliptic, shiny, glabrous on the upper surface; fruit subglobose .... 57. *C. vegeta*

32. Leaves dark yellow-green, thick .... 27. *C. longipes*

5. Leaves broadest below the middle, the base usually rounded or truncate or cordate.

34. Leaves glandular-pubescent .... 22. *C. fulleriana*

34. Leaves not glandular-pubescent.

35. Leaves blue-green.

36. Leaves scabrous on the upper surface.

37. Fruits pyriform, pruinose .... 39. *C. paucispinus*

37. Fruits subglobose, not pruinose.

38. Leaves broadly ovate.

39. Flowers 10–15 mm in diameter; stamens 10 ...... 51. *C. sextilis*

39. Flowers 18–24 mm in diameter; stamens 20 .... 54. *C. tarda*

38. Leaves oblong to oval.

40. Fruit subglobose; stamens usually 10 .... 16. *C. divida*

40. Fruit obovoid; stamens 5 .... 56. *C. trachyphylla*

36. Leaves glabrous on the upper surface, not scabrous.

41. Leaves thick; fruits pruinose.

42. Calyx pedicellate; stamens 20 .... 46. *C. pruinosa*

42. Calyx sessile; stamens 10 .... 15. *C. dissona*

41. Leaves thin; fruits succulent or dry, not pruinose.

43. Fruit succulent .... 55. *C. tortilis*

43. Fruit dry .... 11. *C. cyanophylla*

35. Leaves yellow-green, green, or dark green.

44. Leaves yellow-green.

45. Leaves scabrous on the upper surface.

46. Leaves broadly ovate; fruits 15–17 mm in diameter .... 48. *C. putnamiana*

46. Leaves oblong-ovate; fruits 5–15 mm in diameter.

47. Pedicels of flowers glabrous; flowers 25–30 mm across .... 31. *C. magniflora*

47. Pedicels of flowers villous; flowers up to 22 mm across.

48. Fruits 7–10 mm in diameter; stamens 20 .... 40. *C. pedicellata*

48. Fruits 10–14 mm in diameter; stamens 5 or 10 .... 50. *C. sertata*

45. Leaves glabrous on the upper surface, not scabrous.

49. Pedicels of flowers villous; stamens 5 or 10; fruit oblongoid, 8–12 mm in diameter .... 45. *C. pringlei*

49. Pedicels of flowers glabrous; stamens 20; fruit subglobose, 12–15 mm in diameter .... 8. *C. corusca*

44. Leaves green or dark green.

50. Margins of leaves often crispate at maturity; flowers 24–26 mm across; stamens 20; fruit subglobose .... 6. *C. coccinioides*

50. Margins of leaves not crispate; flowers 12–18 mm across; stamens 20 or fewer; fruit obovoid or oblongoid or subglobose.

51. Leaves on vegetative shoots deeply laciniate; stamens 10; fruit 9–10 mm in diameter .... 21. *C. flabellata*

51. Leaves on vegetative shoots shallowly lobed or unlobed; stamens 5, 10, or 20; fruit 5–15 mm in diameter.

52. Leaves densely pubescent above.

53. Leaves lobed on vegetative shoots; fruits subglobose, 13–18 mm thick .... 34. *C. mollis*

53. Leaves unlobed on vegetative shoots; fruits globose, 10–12 mm thick .... 14. *C. dispessa*

52. Leaves more or less glabrous on the upper surface.

54. Leaves scabrous on the upper surface.

55. Leaves thick; pedicels of the flowers glabrous; stamens 10 (–20); fruit shiny, 6–7 mm in diameter .... 24. *C. holmesiana*

55. Leaves thin; pedicels of the flowers villous; stamens 5; fruit dull, 9–10 mm in diameter .... 3. *C. assurgens*

54. Leaves glabrous on the upper surface, not scabrous.

56. Leaves with tufts of tomentum in the axils of the veins on the lower surface; fruit 5–8 mm in diameter .... 38. *C. ovata*

56. Leaves without tufts of tomentum in the axils of the veins on the lower surface; fruit (6–) 8–10 mm in diameter.

57. Fruit oblongoid to obovoid...... 30. *C. macrosperma*

57. Fruit globose to subglobose.

58. Shrub; stamens 10; fruit globose, 6–9 mm thick ....12. *C. demissa*

58. Tree; stamens 20; fruit subglobose, about 10 mm thick ....49. *C. schuettei*

1. **Crataegus acutifolia** Sarg. Hawthorn. May. Low woods, rare; DuPage, Richland, St. Clair, and Will cos. (Including *C. erecta* Sarg. and var. *insignis* [Sarg.] Palmer; *C. arduennae* Sarg.; *C. subrotundifolia* Sarg.)

2. **Crataegus apiomorpha** Sarg. Hawthorn. May. Woods and along streams; Cook, Lake, and Will cos.

3. **Crataegus assurgens** Sarg. Hawthorn. May. Thickets; Cook Co.

4. **Crataegus calpodendron** (Ehrh.) Medic. Hawthorn. May. Woods and thickets; occasional throughout the state. (Including *C. chapmanii* [Beadle] Ashe; *C. hispidula* Sarg.; *C. mollicula* Sarg.; *C. pertomentosa* Ashe; *C. structilis* Ashe; *C. tomentosa* DuRoi.)

5. **Crataegus chrysocarpa** Ashe. Fireberry Hawthorn. May. Woods, thickets, pastures; occasional in the ne. cos.

6. **Crataegus coccinioides** Ashe. Scarlet Hawthorn. May. Thickets, rocky woods; confined to the s. ⅓ of IL; also Vermilion Co.

7. **Crataegus collina** Chapm. Hawthorn. Apr.–May. Low woods, very rare; St. Clair and Union cos. (Including *C. lettermanii* Sarg.)

8. **Crataegus corusca** Sarg. Hawthorn. May. Woods and thickets; occasional in the ne. cos.

9. **Crataegus crus-galli** L. Two varieties may be distinguished in Illinois:

a. Leaves thick...9a. *C. crus-galli* var. *crus-galli*

a. Leaves thin..9b. *C. crus-galli* var. *barrettiana*

9a. **Crataegus crus-galli** L. var. **crus-galli**. Cock-spur Thorn. May–June. Open woods, thickets; occasional to common throughout the state, except the nw. cos. (Including *C. acanthocolonensis* Laughlin; *C. attenuata* Ashe; *C. farwellii* Sarg.; *C. pachyphylla* Sarg.)

9b. **Crataegus crus-galli** L. var. **barrettiana** (Sarg.) Palmer. Cock-spur Thorn. May. Thickets, rare; St. Clair Co.

10. **Crataegus cuneiformis** (Marsh.) Egglest. Hawthorn. May. Open woods; occasional in the n. ½ of the state.

11. **Crataegus cyanophylla** Sarg. Blue-leaf Hawthorn. May. Woods; Will Co.

12. **Crataegus demissa** Sarg. Hawthorn. May. Thickets, rare; apparently only in the ne. cos.

13. **Crataegus disperma** Ashe. Hawthorn. May. Open areas; scattered in IL.

14. **Crataegus dispessa** Ashe. May. Hawthorn. Thickets, rare; confined to the s. ¼ of IL.

15. **Crataegus dissona** Sarg. Hawthorn. May. Thickets, rare; scattered in IL.

16. **Crataegus divida** Sarg. Hawthorn. May. Woods; Lake Co.

17. **Crataegus dodgei** Ashe. Hawthorn. May. Thickets, rare; McDonough Co.

18. **Crataegus engelmannii** Sarg. Barberry-leaved Hawthorn. May. Open woods, bluffs, pastures; occasional in the s. ⅙ of IL.

19. **Crataegus faxonii** Sarg. Hawthorn. May. Rocky woods and thickets; St. Clair Co.

20. **Crataegus fecunda** Sarg. Fruitful Thorn. May. Moist woods, rare; Gallatin and St. Clair cos. (Including *C. pilifera* Sarg.)

21. **Crataegus flabellata** (Spach) Kirchn. Hawthorn. May. Thickets, very rare.

22. **Crataegus fulleriana** Sarg. Fuller's Hawthorn. May. Thickets, rare; scattered in IL.

23. **Crataegus hannibalensis** Palmer. Hawthorn. May. Thickets, rare; Adams Co.

24. **Crataegus holmesiana** Ashe. Holmes' Hawthorn. May. Woods and thickets; confined to the n. ½ of IL. (Including *C. amicta* Ashe.)

25. **Crataegus intricata** Lange. Hawthorn. May. Woods; Champaign, Madison, and St. Clair cos.

26. **Crataegus laxiflora** Sarg. Lax-flowered Hawthorn. June. Thickets, rare; Will Co.

27. **Crataegus longispina** Sarg. Long-spined Hawthorn. May. Woods; Lake Co.

28. **Crataegus lucorum** Sarg. Hawthorn. May. Woods and stream banks, rare; apparently confined to the ne. cos.

29. **Crataegus macracantha** Sarg. Hawthorn. Woods; ne. cos. (*Crataegus gaultii* Sarg.).

30. **Crataegus macrosperma** Ashe. Hawthorn. May. Thickets and woods; occasional in the ne. ¼ of IL. (Including *C. blothra* Laughlin; *C. depilis* Sarg.; *C. eganii* Ashe; *C. ferrissii* Ashe; *C. hillii* Sarg.; *C. taetrica* Sarg.)

31. **Crataegus magniflora** Sarg. Large-flowered Hawthorn. May–June. Woods; Cook Co.

32. **Crataegus margarettiae** Ashe. Hawthorn. Apr.–May. Open woods, thickets; occasional in the n. ⅗ of IL.

33. **Crataegus marshallii** Egglest. Parsley Haw. Apr. Wet, pin oak woods, very rare; Jackson Co.

34. **Crataegus mollis** (Torr. & Gray) Scheele. Red Haw. May. Moist woods; occasional to common throughout the state. (Including *C. altrix* Ashe; *C. declivitatis* Sarg.; *C. lanigera* Sarg.; *C. nupera* Ashe; *C. pachyphylla* Sarg.; *C. ridgwayi* Sarg.; *C. sera* Sarg.; *C. valens* Ashe; *C. venosa* Ashe; *C. verna* Ashe.)

35. **Crataegus monogyna** Jacq. English Hawthorn. May. Native to Europe and Asia; rarely escaped in IL; Cook, DuPage, and LaSalle cos.

36. **Crataegus neobushii** Sarg. Hawthorn. May. Rocky woods, wet woods; Jackson, Madison, St. Clair, and Union cos.

37. **Crataegus nitida** (Engelm.) Sarg. Shiny Hawthorn. May. Low woods; Gallatin, Henderson, Johnson, Pulaski, Saline, and St. Clair cos. Sometimes reputed to be the hybrid between *C. crus-galli* L. and *C. viridis* L.

38. **Crataegus ovata** Sarg. Hawthorn. May. Wet woods, rare; Jackson Co.

39. **Crataegus paucispina** Sarg. Few-spined Hawthorn. May. Woods; Cook Co.

40. **Crataegus pedicellata** Sarg. Hawthorn. May. Woods and thickets; n. ½ of IL. (Including *C. albicans* Ashe; *C. coccinea* L.; *C. elongata* Sarg.; *C. robesoniana* Sarg.; *C. subrotundifolia* Sarg.) *Crataegus coccinea* L., misapplied.

41. **Crataegus peoriensis** Sarg. Peoria Hawthorn. June. Thickets, rare; Peoria and Will cos.

42. **Crataegus permixta** Palmer. Hawthorn. May. Open woods, rare; Peoria Co.

43. **Crataegus phaenopyrum** (L. f.) Medic. Washington Thorn. May–June. Open woods; occasional in the s. ⅓ of the state; also Clay and Cook cos.

44. **Crataegus pratensis** Sarg. Hawthorn. May. Woods, not common; Marion, Peoria and Stark cos.

45. **Crataegus pringlei** Sarg. Pringle's Hawthorn. May. Thickets and woods; Henry and Lake cos.

46. **Crataegus pruinosa** (Wendl.) K. Koch. Hawthorn. May. Thickets and rocky woods; occasional throughout the state. (Including *C. conjuncta* Sarg.; *C. gattingeri* Ashe; *C. platycarpa* Sarg.)

47. **Crataegus punctata** Jacq. Two varieties occur in Illinois:

a. Fruits red .... 47a. *C. punctata* var. *punctata*
a. Fruits yellow .... 47b. *C. punctata* var. *aurea*

47a. **Crataegus punctata** Jacq. var. **punctata.** Dotted Thorn. May–June. Open woods, pastures; occasional throughout the state. (Including *C. mortonis* Laughlin; *C. sucida* Sarg.)

47b. **Crataegus punctata** Jacq. var. **aurea** Ait. Dotted Thorn. May–June. Open woods, pastures; occasional throughout the state.

48. **Crataegus putnamiana** Sarg. Putman's Hawthorn. June. Thickets, rare.

49. **Crataegus schuettei** Ashe. Hawthorn. May. Woods; rare in the extreme n. cos.

50. **Crataegus sertata** Sarg. Hawthorn. May. Woods; Cook, Lake, and Will cos.

51. **Crataegus sextilis** Sarg. Hawthorn. May. Woods; Cook and Lake cos.

52. **Crataegus spathulata** Michx. Three-lobed Hawthorn. Apr.–May. Wet woods, rare; Alexander and Pulaski cos.

53. **Crataegus succulenta** Schrad. Hawthorn. May. Dry or moist woods and thickets; occasional in the n. ½ of IL. (Including *C. corporea* Sarg.; *C. gemmosa* Sarg.; *C. illinoensis* Ashe; *C. leucantha* Laughlin.)

54. **Crataegus tarda** Sarg. Hawthorn. May. Woods; Cook, Lake, and Will cos.

55. **Crataegus tortilis** Ashe. Hawthorn. May. Woods and thickets; Cook Co.

56. **Crataegus trachyphylla** Sarg. Hawthorn. May. Thickets; Will Co.

57. **Crataegus vegetus** Sarg. Hawthorn. May. Woods; Cook Co.

58. **Crataegus viridis** L. Green Haw. May. Low woods; occasional in the s. ½ of the state. (Including *C. dawsoniana* Sarg.; *C. durifolia* Ashe; *C. lanceolata* Sarg.; *C. mitis* Sarg.; *C. pechiana* Sarg.; *C. schneckii* Ashe.)

59. **Crataegus X whittakeri** Sarg. Whittaker's Hawthorn. May. Thickets, very rare; Richland Co. Reputed to be the hybrid between *C. calpodendron* (Ehrh.) Medic. and *C. viridis* L.

9. **Cydonia** Mill.—Quince

1. **Cydonia oblonga** Mill. Common Quince. May–June. Native to Asia; rarely escaped from cultivation, but established in Union Co.

10. **Dasiphora** Rydb.—Shrubby Cinquefoil

1. **Dasiphora fruticosa** (L.) Rydb. ssp. **floribunda** (Pursh) Kartesz. Shrubby Cinquefoil. June–Aug. Interdunal ponds, hill prairies, calcareous fens, not common; scattered in the n. ½ of IL. *Potentilla fruticosa* L.; *Pentaphylloides floribunda* (Pursh) A. Love.

11. **Drymocallis** Fourr. ex Rydb.

1. **Drymocallis arguta** (Pursh) Rydb. Prairie Cinquefoil. June–July. Prairies; occasional in the n. ½ of IL; also Shelby Co. *Potentilla arguta* Pursh.

12. **Duchesnea** Sm.—Indian Strawberry

1. **Duchesnea indica** (Andrews) Focke. Indian Strawberry. Apr.–June. Native to Asia; adventive in disturbed areas; scattered throughout IL.

13. **Filipendula** Mill.

1. Flowers pink; terminal leaflet divided into 7–9 segments .... 1. *F. rubra*
1. Flowers white; terminal leaflet divided into 3–5 segments .... 2. *F. ulmaria*

1. **Filipendula rubra** (Hill) Robins. Queen-of-the-prairie. June–July. Springy fens, occasional; confined to the n. ½ of IL.

2. **Filipendula ulmaria** (L.) Maxim. Queen-of-the-meadow. June–July. Native to Europe; adventive in disturbed areas; DuPage and Kane cos.

14. **Fragaria** L.—Strawberry

1. Calyx lobes appressed to the fruit; inflorescence umbellate or cymose.
  2. Fruits more than 1.5 cm across .... 2. *F. X ananassa*
  2. Fruits less than 1.5 cm across .... 4. *F. virginiana*
1. Calyx lobes spreading or reflexed; inflorescence racemose.
  3. Pubescence of stems and leaves spreading .... 3. *F. vesca*
  3. Pubescence of stems and leaves appressed ....1. *F. americana*

1. **Fragaria americana** (Porter) Britt. Hillside Strawberry. May–Aug. Wooded slopes; occasional to rare; confined to the n. ¾ of IL.

2. **Fragaria X ananassa** Duchesne. Cultivated Strawberry. Apr.–May. Rarely escaped from cultivation. Reputed to be a hybrid between *F. chiloensis* (L.) Duchesne and *F. virginiana* Duchesne.

3. **Fragaria vesca** L. Strawberry. June. Native to Europe; rarely escaped into dis-

turbed soil; n. ½ of IL. The Illinois specimens with whitish fruits may be referred to as f. *alba* (Ehrh.) Rydb.

4. **Fragaria virginiana** Duchesne. Wild Strawberry. Apr.–July. Woods, prairies, fields; occasional to common throughout the state. (Including var. *illinoensis* [Prince] Gray.)

15. **Geum** L.—Avens

1. Flowers purple.
   2. Styles not jointed, 2 cm long or longer; flowers mostly less than 2 cm across .... 5. *G. triflorum*
   2. Styles jointed, up to 1 cm long; flowers mostly at least 2 cm across .... 4. *G. rivale*
1. Flowers white, cream, or yellow.
   3. Petals white.
      4. Most of the petals 5 mm long or longer; peduncles puberulent; fruiting receptacle densely white-villous .... 2. *G. canadense*
      4. Petals up to 5 mm long; peduncles hirsute; fruiting receptacle glabrous or nearly so .... 3. *G. laciniatum*
   3. Petals yellow or cream.
      5. Receptacle stalked in the calyx; calyx not subtended by bractlets; petals about 2 mm long, yellow .... 7. *G. vernum*
      5. Receptacle sessile; calyx subtended by bractlets; petals (or some of them) more than 2 mm long, cream, orange, or deep yellow.
         6. Petals 2–4 mm long, shorter than the calyx.
            7. Petals bright yellow; style glabrous or with short bristles .... 6. *G. urbanum*
            7. Petals pale cream; style with long bristles .... 8. *G. virginianum*
         6. Petals 5–10 mm long, as long as the calyx .... 1. *G. aleppicum*

1. **Geum aleppicum** Jacq. Yellow Avens. June–July. Bogs, moist thickets; occasional in the n. ½ of IL; also St. Clair Co.

2. **Geum canadense** Jacq. Two varieties occur in Illinois:

a. Fruiting carpels pubescent throughout .... 2a. *G. canadense* var. *canadense*
a. Fruiting carpels hispid above, glabrous below .... 2b. *G. canadense* var. *grimesii*

2a. **Geum canadense** Jacq. var. **canadense.** White Avens. June–Aug. Woods, common; in every co.

2b. **Geum canadense** Jacq. var. **grimesii** Fern. & Weath. White Avens. June–Aug. Woods, rare; confined to the s. ⅕ of the state.

3. **Geum laciniatum** Murr. Two varieties occur in Illinois:

a. Achenes glabrous .... 3a. *G. laciniatum* var. *laciniatum*
a. Achenes bristly at the summit .... 3b. *G. laciniatum* var. *trichocarpum*

3a. **Geum laciniatum** Murr. var. **laciniatum.** Rough Avens. June–July. Meadows and thickets, not common; restricted to the n. ¾ of IL.

3b. **Geum laciniatum** Murr. var. **trichocarpum** Fern. Rough Avens. June–July. Meadows and thickets; occasional in the n. ¾ of the state.

4. **Geum rivale** L. Purple Avens. May–Aug. Moist soil, very rare; Kane, McHenry, and Winnebago cos.

5. **Geum triflorum** Pursh. Prairie Avens. May–June. Dry prairies; occasional in the n. ⅙ of the state, absent elsewhere.

6. **Geum urbanum** L. City Avens. May–June. Native to Europe and Asia; naturalized along a path; DuPage and Lake cos.

7. **Geum vernum** (Raf.) Torr. & Gray. Spring Avens. Apr.–May. Moist woods; common in the s. ½ of IL, occasional or rare in the n. ½.

8. **Geum virginianum** L. Pale Avens. June–July. Dry woods, rare; restricted to the s. ½ of IL.

16. **Kerria** DC.

1. **Kerria japonica** DC. Yellow Rose. June–July. Native to Japan; rarely escaped from cultivation; Jackson Co.

17. **Malus** Mill.—Apple

1. Leaves irregularly doubly serrate to dentate, those on the shoots often lobed.
  2. Petioles, pedicels, calyx lobes, and hypanthia densely and permanently tomentose; leaves thick, impressed-veined..4. *M. ioensis*
  2. Petioles, pedicels, calyx lobes, and hypanthia glabrous or sparsely pilose; leaves thin, not impressed-veined.
    3. Leaves oblong to narrowly elliptic.. 1. *M. angustifolia*
    3. Leaves broadly lanceolate to ovate ....3. *M. coronaria*
1. Leaves singly serrate or entire, never dentate, those on the shoots occasionally lobed in some species.
  4. Pedicels and hypanthia densely tomentose; fruits at least 3 cm across.
    5. Most or all of the leaves unlobed; anthers yellow.........................6. *M. pumila*
    5. Some of the leaves usually shallowly lobed; anthers red ...... 7. *M. X soulardii*
  4. Pedicels and hypanthia glabrous or sparsely tomentose; fruits up to 3 cm across.
    6. Petals pink or rose-tinged; some of the leaves often divided; fruits 4–7 mm across ....8. *M. toringa*
    6. Petals white; none of the leaves divided; fruits 7 mm across or wider.
      7. Calyx glabrous, the lobes long-acuminate; fruit about 1 cm across ....2. *M. baccata*
      7. Calyx tomentose, the lobes lanceolate; fruits 1.5–2.5 cm across................. .... 5. *M. prunifolia*

1. **Malus angustifolia** (Ait.) Michx. Narrow-leaved Crab Apple. May. Low woods, rare; Hardin, Jackson, Massac, and Pope cos. *Pyrus angustifolia* Ait.

2. **Malus baccata** (L.) Borkh. Siberian Crab Apple. May. Native to Asia; often planted as an ornamental, but seldom escaped; DuPage, Kane, McHenry, and St. Clair cos. *Pyrus baccata* L.

3. **Malus coronaria** (L.) Mill. Two varieties occur in Illinois:

a. Leaves at least ½ as broad as long, sometimes lobulate .... 3a. *M. coronaria* var. *coronaria*
a. Leaves less than ½ as broad as long, not lobulate.. .... 3b. *M. coronaria* var. *dasycalyx*

3a. **Malus coronaria** (L.) Mill. var. **coronaria.** Wild Sweet Crab Apple. Apr.–May. Woods; occasional in the s. ½ of the state, less common in the n. ½. *Pyrus coronaria* L.

3b. **Malus coronaria** (L.) Mill. var. **dasycalyx** Rehd. Wild Sweet Crab Apple. Apr.–May. Woods, not common; Gallatin, Jackson, Pope, and St. Clair cos. *Pyrus coronaria* L. var. *dasycalyx* (Rehd.) Fern.

4. **Malus ioensis** (Wood) Britt. Iowa Crab Apple. May. Woods and thickets; occasional to common throughout the state; in every co. *Pyrus ioensis* (Wood) Bailey.

5. **Malus prunifolia** (Willd.) Borkh. Plum-leaved Crab Apple. May. Native to Asia; occasionally planted as an ornamental but seldom escaped; disturbed mesic woods; Cook, DuPage, Kane, and Will cos. *Pyrus prunifolia* Willd.

6. **Malus pumila** Mill. Apple. May. Introduced from Asia; occasionally escaped from cultivation; scattered in IL. *Pyrus malus* L.

7. **Malus X soulardii** (Bailey) Britt. Soulard Crab Apple. May. Originally cultivated at Galena, IL; rarely escaped in IL; Cook Co. Reputed to be the hybrid between *M. ioensis* (Wood) Britt. and *M. pumila* Mill. *Pyrus X soulardii* Bailey.

8. **Malus toringa** (Sieb.) Sieb. ex deVereis. Japanese Crab Apple. Apr.–May. Native to Asia; occasionally escaped from cultivation. (Including *M. floribunda* Siebold; *M. sargentii* Rehd.; *M. zumi* [Matsum.] Rehd.) *Malus sieboldii* (Regel) Rehd. *Pyrus sieboldii* Regel.

18. **Photinia** Raf.—Chokeberry

1. Axis of inflorescence and lower surface of leaves glabrous or nearly so .... 1. *P. melanocarpa*
1. Axis of inflorescence and lower surface of leaves pubescent.............. 2. *P. prunifolia*

1. **Photinia melanocarpa** (Michx.) Robertson & Phipps. Black Chokeberry. May–

June. Bogs, moist woods, sandstone ledge (in Saline Co.); occasional in the ne. ¼ of IL; also Johnson, Saline, and St. Clair. cos. *Pyrus melanocarpa* (Michx.) Willd.; *Aronia melanocarpa* (Michx.) Ell.

2. **Photinia prunifolia** (Marsh.) Robertson & Phipps. Purple Chokeberry. May–June. Occasional in bogs, otherwise rare; restricted to a few cos. in the ne. corner of IL; also Winnebago Co. *Pyrus floribunda* Lindl.; *Aronia prunifolia* (Marsh.) Rehd.

19. **Physocarpus** Maxim.—Ninebark

1. **Physocarpus opulifolius** (L.) Maxim. Two varieties may be distinguished in Illinois:

a. Capsules glabrous or nearly so .... 1a. *P. opulifolius* var. *opulifolius*
a. Capsules pubescent .... 1b. *P. opulifolius* var. *intermedius*

1a. **Physocarpus opulifolius** (L.) Maxim. var. **opulifolius.** Ninebark. May–June. Rocky slopes, rocky banks, moist swales; occasional in the n. ½ of IL; also Jackson, Madison, Monroe, Pope, and Wabash cos.

1b. **Physocarpus opulifolius** (L.) Maxim. var. **intermedius** (Rydb.) Robins. Ninebark. May–June. Rocky slopes, rocky banks, moist swales; scattered in the n. ½ of IL.

20. **Porteranthus** Britt.—Indian Physic

1. Stipules large, leaflike .... 1. *P. stipulatus*
1. Stipules small, subulate .... 2. *P. trifoliatus*

1. **Porteranthus stipulatus** (Muhl.) Britt. Indian Physic. May–July. Woods; occasional in the s. ½ of the state; also LaSalle Co. *Gillenia stipulata* (Muhl.) Baill.

2. **Porteranthus trifoliatus** (L.) Britt. Three-leaved Indian Physic. May–July. Woods, very rare; Wabash Co. *Gillenia trifoliata* (L.) Moench.

21. **Potentilla** L—Cinquefoil

(See also Argentina, Comarum, Dasiphora, Drymocallis, Sibbaldiopsis.)

1. Flower solitary.
  2. Flower 1.5–2.0 cm across; plants without tuberous rhizomes; calyx lobes ovate .... 9. *P. reptans*
  2. Flower up to 1.5 cm across; plants with tuberous rhizomes; calyx lobes lanceolate .... 11. *P. simplex*
1. Flowers in cymes.
  3. Lower leaves ternate or digitate.
    4. Lower leaves with 5–7 leaflets.
      5. Achenes rugose; flowers 1.5–2.5 cm across, the petals emarginate; stems and leaves hirsute .... 8. *P. recta*
      5. Achenes smooth; flowers up to 1.5 cm across, the petals entire; stems and leaves white-woolly or gray-tomentose.
        6. Stems and leaves white-woolly .... 1. *P. argentea*
        6. Stems and leaves gray-tomentose.
          7. Lower surface of leaflets with short tomentum and long hairs .... 2. *P. inclinata*
          7. Lower surface of leaflets only with short tomentum .... 3. *P. intermedia*
    4. Lower leaves with 3 leaflets.
      8. Plants softly villous; flowers up to 4 mm across; stamens about 10 .... 4. *P. millegrana*
      8. Plants rough-hirsute; flowers usually at least 6 mm across; stamens 15–20 .... 5. *P. norvegica*
  3. Lower leaves pinnate.
    9. Cauline leaves palmate; achenes smooth .... 10. *P. rivalis*
    9. Cauline leaves pinnate; achenes ribbed.
      10. Annual or biennial; hypanthium hirsute; sepals 3–4 mm long .... 6. *P. paradoxa*
      10. Perennial; hypanthium strigose or tomentose; sepals 4–6 mm long .... 7. *P. pensylvanica*

1. **Potentilla argentea** L. Silvery Cinquefoil. May–Sept. Native to Europe; adventive in disturbed sandy soil; occasional in the n. ½ of IL.

2. **Potentilla inclinata** Vill. Gray Cinquefoil. May–Aug. Native to Europe; rarely adventive in disturbed soil; ne. cos. *Potentilla canescens* Bess.

3. **Potentilla intermedia** L. Intermediate Cinquefoil. June–Aug. Native to Europe; adventive along railroads; Champaign, DuPage, Hamilton, and McDonough cos.

4. **Potentilla millegrana** Engelm. Cinquefoil. June–July. Moist soil, rare; Adams, Johnson, St. Clair, and Union cos.

5. **Potentilla norvegica** L. Rough Cinquefoil. June–July. Native to Europe; adventive in disturbed areas; occasional throughout the state.

6. **Potentilla paradoxa** Nutt. Cinquefoil. May–Sept. Wet soil, usually along rivers; Alexander, Jackson, Mason, Monroe, Randolph, and St. Clair cos.

7. **Potentilla pensylvanica** L. var. **bipinnatifida** (Dougl.) Torr & Gray. Prairie Cinquefoil. June–July. Prairie, very rare; McHenry Co.

8. **Potentilla recta** L. Sulfur Cinquefoil. May–July. Native to Europe; naturalized in fields and along roads; occasional to common throughout IL.

9. **Potentilla reptans** L. Creeping Cinquefoil. May–July. Native to Europe; rarely adventive in disturbed soil; DeKalb Co.

10. **Potentilla rivalis** Nutt. Brook Cinquefoil. June–Aug. Native to the w. U.S.; adventive in Cook, Johnson, St. Clair, and Union cos.

11. **Potentilla simplex** Michx. Three varieties may be distinguished in Illinois:

a. Stems villous or hirsute.
  b. Lower surface of leaflets strigose .... 11a. *P. simplex* var. *simplex*
  b. Lower surface of leaflets densely silvery-silky .... 11b. *P. simplex* var. *argyrisma*
a. Stems glabrous or strigillose .... 11c. *P. simplex* var. *calvescens*

11a. **Potentilla simplex** Michx. var. **simplex.** Common Cinquefoil. May–June. Dry woods, prairies, fields, common; in every co.

11b. **Potentilla simplex** Michx. var. **argyrisma** Fern. Common Cinquefoil. May–June. Dry woods, rare; apparently confined to the s. cos.

11c. **Potentilla simplex** Michx. var. **calvescens** Fern. Common Cinquefoil. May–June. Dry woods, fields, prairies; occasional throughout the state.

22. **Prunus** L.—Plum; Cherry

1. Flowers solitary or in corymbs or umbels.
  2. Shrubs branched from the base, generally less than 3 m tall; flowers "double" .... 18. *P. triloba*
  2. Trees or shrubs more than 3 m tall; flowers not "double."
    3. Ovary and fruit pubescent or glaucous.
      4. Ovary and fruit pubescent, not glaucous.
        5. Flowers sessile.
          6. Ovary densely tomentose; fruits more than 3 cm in diameter; petals bright pink to rose .... 13. *P. persica*
          6. Ovary sparsely hairy; fruits about 1 cm in diameter; petals white or pale pink .... 17. *P. tomentosa*
        5. Flowers pedicellate .... 3. *P. armeniaca*
      4. Ovary and fruit glabrous, glaucous.
        7. Lobes of calyx glandular along the margins.
          8. Flowers at least 2 cm across; lobes of calyx 3–5 mm long, reflexed from the beginning of flowering .... 10. *P. nigra*
          8. Flowers 1.0–1.5 cm across; lobes of calyx less than 3 mm long, reflexed only toward the end of flowering.
            9. Flowers appearing after the expansion of the leaves .... 6. *P. hortulana*
            9. Flowers appearing before the expansion of the leaves .... 9. *P. munsoniana*
        7. Lobes of calyx eglandular along the margins.
          10. Flowers up to 1 cm across; fruits 1–2 cm in diameter, the stone not compressed .... 2. *P. angustifolia*
          10. Flowers 1.5 cm or more across; fruits at least 2 cm in diameter, the stone compressed.

11. Petioles with 1–2 glands near summit; leaves usually broadly rounded at the base .... 8. *P. mexicana*
11. Petioles without glands near summit; leaves mostly narrowed at the base .... 1. *P. americana*

3. Ovary and fruit neither pubescent nor glaucous.
12. Dwarf shrubs to 2 m tall; leaves entire in the lower ½; fruits 1.0–1.5 cm in diameter.. 14. *P. pumila*
12. Trees to 12 m tall; leaves toothed to the base; fruits either 5–7 mm in diameter or more than 1.5 cm in diameter.
13. Flowers up to 1.5 cm across; fruits 5–7 mm in diameter......................... ....12. *P. pensylvanica*
13. Flowers at least 2 cm across; fruits 8–25 mm in diameter.
14. Flowers white; sepals reflexed; veins on lower surface of leaves pubescent or glabrous; petioles glandular.
15. Leaves glabrous beneath; fruits sour. ....5. *P. cerasus*
15. Leaves pubescent on the nerves beneath; fruits sweet................ .... 4. *P. avium*
14. Flowers pink; sepals erect; veins on lower surface of leaves pubescent; petioles eglandular .... 16. *P. subhirtella*

1. Flowers in racemes.
16. Inflorescence 4- to 10-flowered; flowers about 1.5 cm across ....7. *P. mahaleb*
16. Inflorescence more than 10-flowered; flowers about 1 cm across.
17. Leaves firm, crenulate-serrate; lobes of calyx acute; small to large tree, often at least 10 m tall ....15. *P. serotina*
17. Leaves thin, sharply serrulate; lobes of calyx obtuse; shrub or small tree rarely exceeding 10 m tall.
18. Racemes drooping; petals twice as long as the stamens; fruits black .... 11. *P. padus*
18. Racemes ascending (at least at first); petals about as long as the stamens; fruits red-purple... 19. *P. virginiana*

1. **Prunus americana** Marsh. Two varieties may be recognized in Illinois:

a. Leaves glabrous or sparsely pubescent beneath ..... 1a. *P. americana* var. *americana*
a. Leaves soft-pubescent beneath... 1b. *P. americana* var. *lanata*

1a. **Prunus americana** Marsh. var. **americana.** Wild Plum. Apr.–May. Thickets, woodlands; occasional throughout the state.

1b. **Prunus americana** Marsh. var. **lanata** Sudw. Hairy Wild Plum. Apr.–May. Thickets, woodlands; occasional throughout the state.

2. **Prunus angustifolia** Marsh. Chickasaw Plum. Apr.–May. Thickets; occasional in the s. ½ of the state, uncommon and probably adventive in the n. ½.

3. **Prunus armeniaca** L. Apricot. May. Native to Europe; rarely escaped from cultivation; Champaign, Cook, Ford, and Mason cos.

4. **Prunus avium** (L.) L. Sweet Cherry. Apr.–May. Native to Europe and Asia; rarely escaped from cultivation; DuPage, Jackson, and Lake cos.

5. **Prunus cerasus** L. Pie Cherry. Apr.–May. Native to Europe; rarely escaped from cultivation; Champaign, Cook, Jackson, Lawrence, Pope, Union, and Wayne cos.

6. **Prunus hortulana** Bailey. Wild Goose Plum. Mar.–Apr. Thickets; occasional in the s. ⅔ of the state, introduced elsewhere.

7. **Prunus mahaleb** L. Mahaleb Cherry. May. Native to Europe; rarely escaped from cultivation; scattered in IL.

8. **Prunus mexicana** S. Wats. Big Tree Plum. Apr. Base of cherty cliffs, woods; scattered in IL.

9. **Prunus munsoniana** Wright & Hedrick. Munson's Plum. Mar.–Apr. Thickets; occasional in the s. ⅔ of the state.

10. **Prunus nigra** Ait. Canada Plum. May. Mostly along rivers and streams; occasional in the n. ½ of the state; also Union Co.

11. **Prunus padus** L. European Bird Cherry. May. Native to Europe and Asia; rarely escaped from cultivation; Cook, DuPage, Kane, and Lake cos.

12. **Prunus pensylvanica** L. f. Pin Cherry. Apr. Sandy soil, particularly along Lake Michigan; confined to the n. ½ of the state; also Jersey Co.

13. **Prunus persica** (L.) Batsch. Peach. Apr. Native to Asia; occasionally escaped along roadsides.

14. **Prunus pumila** L. Sand Cherry. May–June. Sandy areas, particularly along Lake Michigan; confined to the n. ¼ of the state. *Prunus susquehanae* Willd.

15. **Prunus serotina** Ehrh. Wild Black Cherry. May. Woods; common throughout the state; in every co.

16. **Prunus subhirtella** Miq. Rosebud Cherry. May–June. Not known from the wild; rarely escaped from cultivation; DuPage Co.

17. **Prunus tomentosa** Thunb. Nanking Cherry. May. Native to Asia; rarely escaped from cultivation; Cook, DuPage, and Lake cos.

18. **Prunus triloba** Lindl. Flowering Almond. May. Native to Asia; rarely escaped from cultivation around old homesteads; Jackson Co.

19. **Prunus virginiana** L. Common Chokecherry. May. Woods and thickets; occasional in the n. ½ of the state, rare in the s. ½.

23. **Pyrus** L.—Pear

1. Leaves serrate-setose .... 4. *P. pyrifolia*
1. Leaves serrate or crenate.
  2. Flowers up to 2 cm across; styles 2 or 3; leaves crenate; fruits globose .... 2. *P. calleryana*
  2. Flowers 2 cm or more across; styles usually 5; leaves serrate; fruits subglobose or pyriform.
    3. Leaves acute, dull; fruits more or less pyriform .... 3. *P. communis*
    3. Leaves acuminate, glossy; fruits subglobose .... 1. *P. betulaefolia*

1. **Pyrus betulaefolia** Bunge. Birch-leaved Pear. May–June. Escaped from cultivation; DuPage Co.

2. **Pyrus calleryana** Decne. Ornamental Pear. Native to China; rarely escaped from cultivation; scattered in IL.

3. **Pyrus communis** L. Pear. Apr.–May. Native to Europe and Asia; escaped from cultivation throughout IL.

4. **Pyrus pyrifolia** (Burm. f.) Nakai. Chinese Pear. Apr. Native to Asia; escaped from cultivation along a stream; Cook and Union cos.

24. **Rhodotypos** Sieb. & Zucc.—Jetbead

1. **Rhodotypos scandens** (Thunb.) Makino. Jetbead. June–July. Native to Japan; rarely escaped from cultivation; ne. cos.

25. **Rosa** L.—Rose

1. Styles much exserted beyond the hypanthium, united into a column.
  2. Leaflets 3 (–5); stipules entire or merely ciliate; flowers deep rose .... 18. *R. setigera*
  2. Leaflets (5–) 7–9; stipules jagged-toothed or pectinate; flowers white (rarely pink).
    3. Plants trailing; flowers 4–5 cm across .... 22. *R. wichuriana*
    3. Plants erect or climbing; flowers up to 4 cm across.
      4. Flowers up to 2.5 cm across; stipules pectinate .... 14. *R. multiflora*
      4. Flowers 2.5–4.0 cm across; stipules toothed .... 13. *R. moschata*
1. Styles not exserted beyond the hypanthium, free.
  5. Flowers double.
    6. Flowers pink or rose, nodding .... 6. *R. centifolia*
    6. Flowers yellow, not nodding .... 11. *R. X harisonii*
  5. Flowers single, usually not nodding.
    7. Flower solitary, not subtended by a bract.
      8. Leaflets 3–7, at least 2 cm long and 2 cm broad; petals (2.5–) 3.0–4.5 mm long, deep pink .... 10. *R. gallica*
      8. Leaflets 7–11, up to 2 cm long, up to 1 cm broad; petals 1.0–2.0 (–2.5) mm long, white (rarely pink) .... 19. *R. spinosissima*

7. Flowers in corymbs or, if solitary, subtended by a bract.
    9. Some of the sepals pinnatifid with 5 or more segments; orifice of hypanthium about 1 mm across.
        10. Lower surface of leaflets stipitate-glandular.
            11. Style pubescent; sepals persistent on the fruit..........8. *R. eglanteria*
            11. Style glabrous; sepals deciduous ....12. *R. micrantha*
        10. Lower surface of leaflets glabrous or nearly so, not stipitate-glandular.
            12. Leaflets with a reddish tint .... 9. *R. ferruginea*
            12. Leaflets green, without a reddish tint.
                13. Leaflets green beneath, glandular; flower not cinnamon-scented...4. *R. canina*
                13. Leaflets glaucous beneath, eglandular; flower cinnamon-scented... 7. *R. cinnamomea*
    9. All sepals entire or with only 1–4 linear segments; orifice of hypanthium more than 1 mm across.
        14. Twigs tomentose; petals 3–5 cm long.................................17. *R. rugosa*
        14. Twigs glabrous or nearly so; petals usually less than 3 cm long.
            15. Hypanthium and petioles glandular-stipitate; sepals spreading or reflexed.
                16. Leaflets finely toothed, each tooth about 0.5 mm high .... 15. *R. palustris*
                16. Leaflets coarsely toothed, each tooth about 1 mm high.
                    17. Leaflets mostly 5.. 5. *R. carolina*
                    17. Leaflets mostly 9.. 21. *R. virginiana*
            15. Hypanthium and petioles eglandular; sepals erect.
                18. Leaflets mostly 9 or 11.
                    19. Stipules glandular .... 16. *R. X rudiuscula*
                    19. Stipules eglandular.
                        20. Leaflets glabrous on both surfaces .... 2. *R. arkansana*
                        20. Leaflets softly pilose .... 20. *R. suffulta*
                18. Leaflets mostly 5 or 7.
                    21. Stems and branchlets without prickles, although sometimes slightly bristly when young .... 3. *R. blanda*
                    21. Stems and branchlets densely bristly.............. 1. *R. acicularis*

1. **Rosa acicularis** Lindl. Needle Rose. June–July. Algific slopes, very rare; Jo Daviess Co.

2. **Rosa arkansana** Porter. Wild Rose. June–July. Native to the w. U.S.; adventive along a railroad, disturbed areas; scattered in IL.

3. **Rosa blanda** Ait. Meadow Rose. May–June. Thickets, woods, open areas; occasional in the n. ½ of the state, rare elsewhere.

4. **Rosa canina** L. Dog Rose. May–July. Native to Europe; occasionally escaped from cultivation; scattered in IL.

5. **Rosa carolina** L. Two varieties may be distinguished in Illinois:

a. Lower surface of leaflets glabrous or sparsely pubescent .... 5a. *R. carolina* var. *carolina*
a. Lower surface of leaflets softly pubescent..........................5b. *R. carolina* var. *villosa*

5a. **Rosa carolina** L. var. **carolina**. Carolina Rose; Pasture Rose. May–July. Prairies, fields, dry woods; common throughout the state. (Including var. *grandiflora* [Baker] Rehd.)

5b. **Rosa carolina** L. var. **villosa** (Best) Rehd. Carolina Rose; Pasture Rose. May–July. Prairies, fields; occasional throughout the state.

6. **Rosa centifolia** L. Cabbage Rose. June–July. Native to the Caucasus; rarely escaped from gardens; DuPage Co.

7. **Rosa cinnamomea** L. Cinnamon Rose. June–July. Native to Europe; rarely escaped from cultivation; Cook and DuPage cos. *Rosa pendulina* L.

8. **Rosa eglanteria** L. Sweet-brier. May–July. Native to Europe; occasionally escaped from cultivation.

9. **Rosa ferruginea** L. Red-leaved Rose. May–June. Native to Europe; rarely escaped from cultivation; Jackson Co. *Rosa rubrifolia* Vill.

10. **Rosa gallica** L. French Rose. June–July. Native to Europe; rarely escaped from cultivation; DuPage, Jackson, and Piatt cos.

11. **Rosa X harisonii** Rivers. Harison's Yellow Rose. May–June. Not known from the wild; rarely escaped from gardens; DuPage Co. This is reputed to be the hybrid between *R. foetida* Herrm. and *R. spiniosissima* L.

12. **Rosa micrantha** Sm. Small Sweet-brier. May–July. Native to Europe; rarely escaped from cultivation; Kane and Winnebago cos.

13. **Rosa moschata** Herrm. Musk Rose. May–June. Native to Europe and Africa; rarely escaped from cultivation; Perry Co.

14. **Rosa multiflora** Thunb. Multiflora Rose. May–June. Native to Asia; frequently planted and often escaped throughout the state; probably in every co.

15. **Rosa palustris** Marsh. Swamp Rose. June–Aug. Swamps, bogs, moist thickets; occasional throughout the state.

16. **Rosa X rudiuscula** Greene. Rough Rose. June. Dry prairie remnants and along a railroad; ne. cos and Jackson Co. Reputed to be the hybrid between *R. carolina* L. and *R. suffulta* Greene.

17. **Rosa rugosa** Thunb. Rugose Rose. June–Sept. Native to Asia; rarely escaped from cultivation; Kane Co.

18. **Rosa setigera** Michx. Two varieties may be recognized in Illinois:

a. Leaflets glabrous beneath or pilose on the nerves .... 18a. *R. setigera* var. *setigera*
a. Leaflets tomentose beneath .... 18b. *R. setigera* var. *tomentosa*

18a. **Rosa setigera** Michx. var. **setigera.** Prairie Rose. June–July. Woods, thickets, clearings; occasional to common throughout the state.

18b. **Rosa setigera** Michx. var. **tomentosa** Torr. & Gray. Prairie Rose. June–July. Woods, thickets, clearings; occasional throughout the state.

19. **Rosa spinosissima** L. Scotch Rose. May–July. Native to Europe; occasionally escaped from cultivation. *Rosa pimpinellifolia* L.

20. **Rosa suffulta** Greene. Sunshine Rose. Thickets, woods; occasional in the n. ⅔ of IL, apparently absent elsewhere. (Including *R. relicta* Erlanson.) *Rosa arkansana* Porter var. *suffulta* (Greene) Cockerell.

21. **Rosa virginiana** Mill. Virginia Rose. June–July. Native to the e. U.S.; adventive in IL; Cook, DuPage, and Lake cos.

22. **Rosa wichuriana** Crep. Memorial Rose. May–June. Native to Asia; rarely escaped from cultivation; Jackson Co.

26. **Rubus** L.—Blackberry; Raspberry

1. Leaves simple.
  2. Petals purplish; fruit dry, inedible; calyx lobes densely covered with glandular hairs .... 30. *R. odoratus*
  2. Petals white; fruit juicy, edible; calyx lobes without glandular hairs or, at most, with only a few glandular hairs .... 31. *R. parviflorus*
1. Leaves compound.
  3. Stems without prickles or bristles, herbaceous .... 37. *R. pubescens*
  3. Stems prickly, bristly, or both, shrubby.
    4. Leaves white-tomentose beneath.
      5. Petals red or pink .... 32. *R. parvifolius*
      5. Petals white.
        6. Stems glaucous; fruit purple-black at maturity.
          7. Stems strongly glaucous; when the glaucous wax is removed, the stem is green; fruit purple-black .... 28. *R. occidentalis*
          7. Stems sparsely glaucous; when the glaucous wax is removed, the stem is pinkish; fruit pale purple........................ 27. *R. X neglectus*
        6. Stems not glaucous; fruit red at maturity (except *R. armeniacus*).
          8. Stems, pedicels, and sepals with dense, red or purple, gland-tipped hairs .... 35. *R. phoenicolasius*
          8. Stems, pedicels, and sepals prickly, bristly, or smooth, but without dense, red or purple, gland-tipped hairs, or sparsely so.

9. Plants with prickles on stem and branches of inflorescence; fruit a blackberry .... 6. *R. armeniacus*
9. Plants with bristles or broad-based prickles only on the stem; fruit a raspberry.
10. Sepals and pedicels without gland-tipped hairs; prickles broad based or not.
11. Some or all of the prickles broad-based ....18. *R. idaeus*
11. None of the prickles broad-based...........27. *R. X neglectus*
10. Sepals and pedicels with gland-tipped hairs; prickles not broad-based .... 46. *R. strigosus*
4. Leaves not white-tomentose beneath nor glaucous beneath.
12. Leaflets laciniate...21. *R. laciniatus*
12. Leaflets not laciniate.
13. Young twigs strongly glaucous .... 10. *R. caesius*
13. Young twigs not glaucous.
14. Stems trailing, except for the floral branches. (Dewberries, except *R. fulleri.*)
15. Stems beset with bristles but without curved prickles.
16. Stems rooting at the tips .... 16. *R. fulleri*
16. Stems not rooting at the tips............................ 17. *R. hispidus*
15. Stems with curved prickles (bristles may also be intermixed in *R. trivialis*).
17. Stems with both bristles and prickles................47. *R. trivialis*
17. Stems with only prickles.
18. Leaflets velvety on the lower surface.
19. Prickles more or less straight, not decidedly curved; inflorescence paniculate or cymiform.
20. Inflorescence paniculate, longer than broad .... 19. *R. impar*
20. Inflorescence cymiform, as broad as long .... 20. *R. ithacanus*
19. Prickles strongly curved; inflorescence paniculate, or reduced to a single flower.
21. Primocane leaves with 3 leaflets; flower 1 per inflorescence......................................26. *R. mundus*
21. Primocane leaves mostly with 5 leaflets; flowers more than 1 per inflorescence (rarely 1 in *R. aboriginus.*)
22. Leaflets irregularly serrate or lobulate; flowers 1–3 per inflorescence..................2. *R. aboriginus*
22. Leaflets regularly serrate, never lobulate; flowers 3 or more per inflorescence.
23. Central leaflet of primocane leaves ovate, rounded or somewhat cordate at base.
24. Central leaflet of primocane leaves acute .... 40. *R. roribaccus*
24. Central leaflet of primocane leaves long-acuminate .... 42. *R. satis*
23. Central leaflet of primocane leaves elliptic, tapering to the base.
25. Petioles with stout prickles....................... ....29. *R. occidualis*
25. Petioles with slender prickles.
26. Pedicels with several short prickles..... .... 23. *R. meracus*
26. Pedicels with 1–2 short prickles .... 12. *R. curtipes*
18. Leaflets glabrous or variously pubescent on the lower surface but not velvety.

27. Flower solitary.
28. Stems slender, up to 3 mm thick; prickles often absent or, if present, up to 2 mm long ....13. *R. enslenii*
28. Stems more stout, 2–5 mm thick; prickles present, 2–3 mm long .... 8. *R. baileyanus*
27. Flowers 2 or more in corymblike racemes.
29. Pedicels and petioles of leaflets of the flowering branches glabrous or appressed-pubescent.. 14. *R. flagellaris*
29. Pedicels and petioles of leaflets of the flowering branches spreading-villous, or with prickles.
30. Primocane leaves with 3 leaflets ....45. *R. steelei*
30. Primocane leaves witih 5 leaflets.
31. Pedicels and petioles spreading-villous, without prickles.............................. 11. *R. celer*
31. Pedicels and petioles not spreading-villous, with prickles.
32. Leaflets abruptly acuminate; veins of leaflets not pleated............. 25. *R. multifer*
32. Leaflets gradually acuminate; veins of leaflets pleated .... 36. *R. plicatifolius*
14. All stems arched-ascending to erect.
33. Stems with bristles or slender, usually straight prickles. (Bristleberries.)
34. Leaflets distinctly pubescent to the touch on the lower surface.
35. Petioles and pedicels glandular-hairy.
36. Central leaflet of primocane leaves acute to short-acuminate, rounded at the base......44. *R. semisetosus*
36. Central leaflet of primocne leaves long-acuminate, tapering to the base............................ 43. *R. schneideri*
35. Petioles and pedicels eglandular .... 24. *R. missouricus*
34. Leaflets glabrous or sparsely pubescent on the lower surface but not noticeably pubescent when touched.
37. Pedicels with glands ....49. *R. wisconsinensis*
37. Pedicels without glands..............................48. *R. wheeleri*
33. Stems with broad-based prickles, these either straight or curved. (Blackberies.)
38. Flowers borne in elongated racemes; petioles, pedicels, and calyx lobes usually stipitate-glandular.
39. Flowers in corymbiform racemes, broadest at summit..... ....4. *R. alumnus*
39. Flowers in racemes much longer than broad and not broadest at summit.
40. Racemes at least 3 times longer than broad .... 3. *R. allegheniensis*
40. Racemes at most about twice as long as broad.
41. Leaflets of sterile stems overlapping, at least ⅔ as wide as long..41. *R. rosa*
41. Leaflets of sterile stems not overlapping, less than ⅔ as wide as long...................... 3. *R. allegheniensis*
38. Flowers borne in corymbs or very short racemes; petioles, pedicels, and calyx lobes rarely stipitate-glandular.
42. Leaflets of the sterile stems narrow, the middle one less than ½ as wide as long ....5. *R. argutus*
42. Leaflets of the sterile stems wider, the middle one at least ½ as wide as long.
43. Inflorescence a raceme surpassing the foliage.

44. Central leaflet of the primocane leaves gradually tapering to the acuminate tip.
45. Central leaflet of the primocane leaves long-acuminate .... 1. *R. ablatus*
45. Central leaflet of the primocane leaves short-acuminate .... 22. *R. laudatus*
44. Central leaflet of the primocane leaves abruptly tapering to the acuminate tip.
46. Racemes more than twice as long as wide......... .... 34. *R. pergratus*
46. Racemes less than twice as long as wide.
47. Central leaflet of the primocane leaves rounded or truncate at the base..................... ....38. *R. pubifolius*
47. Central leaflet of the primocane leaves cordate.
48. Leaflets velvety on the lower surface....... ....9. *R. bellobatus*
48. Leaflets not velvety on the lower surface ....7. *R. avipes*
43. Inflorescence a corymb not surpassing the foliage.
49. Prickles on the stems numerous, about 2–5 mm apart ....39. *R. recurvans*
49. Prickles on the stems fewer, at least 1 cm apart.
50. Central leaflet of the primocane leaves ovate, rounded or cordate at the base ....15. *R. frondosus*
50. Central leaflet of the primocane leaves elliptic or obovate, tapering to the base ....33. *R. pensylvanicus*

1. **Rubus ablatus** Bailey. Broad-petalled Blackberry. May–June. Thickets; n. ½ of IL, s. to St. Clair Co.

2. **Rubus aboriginus** Rydb. Garden Dewberry. May–June. Thickets; scattered in IL.

3. **Rubus allegheniensis** Porter. Common Blackberry. May–June. Roadsides, thickets, woods; common throughout the state. (Including *R. impos* Bailey.)

4. **Rubus alumnus** Bailey. Blackberry. May–June. Along railroads, dry soil; scattered in IL.

5. **Rubus argutus** Link. Highbush Blackberry. May–June. Thickets, woods; occasional throughout the state. (Including *R. abactus* Bailey; *R. blakei* Bailey; *R. ostryifolius* Rydb.; *R. schneckii* Bailey; *R. virilis* Bailey.)

6. **Rubus armeniacus** Focke. Himalayan Blackberry. Native to Asia; rarely escaped from gardens; Madison, St. Clair, and Will cos. *Rubus discolor* Weihe & Nees.

7. **Rubus avipes** Bailey. Blackberry. May–June. Thickets, rare; confined to n. IL.

8. **Rubus baileyanus** Britt. Bailey's Dewberry. May–June. Thickets; Hamilton and Winnebago cos.

9. **Rubus bellobatus** Bailey. Blackberry. May–June. Thickets; Boone, Champaign, Kankakee, Piatt, and Rock Island cos.

10. **Rubus caesius** L. European Dewberry. May–June. Native to Europe; rarely escaped from gardens; DuPage Co.

11. **Rubus celer** Bailey. Dewberry. May–June. Thickets; Perry Co.

12. **Rubus curtipes** Bailey. Dewberry. May–June. Thickets; s. IL.

13 **Rubus enslenii** Tratt. Arching Dewberry. May–June. Rocky woods, thickets; occasional in s. IL; also Adams Co.

14. **Rubus flagellaris** Willd. Common Dewberry. Apr.–June. Fields, edge of woods, roadsides; common to occasional throughout most of the state.

15. **Rubus frondosus** Bigel. Blackberry. May–June. Woods and thickets; occasional throughout the state.

16. **Rubus fulleri** Bailey. Fuller's Bristleberry. May–June. Thickets; n. ¼ of IL.

17. **Rubus hispidus** L. Swamp Dewberry. June–July. Bogs, black oak woods,

marshes, thickets; restricted to the n. ½ of the state; also St. Clair Co. (Including *R. insignis* Bailey.)

18. **Rubus idaeus** L. Cultivated Raspberry. May–June. Native to Europe; occasionally escaped and persistent from cultivation.

19. **Rubus impar** Bailey. Dewberry. May–June. Thickets; s. IL.

20. **Rubus ithacanus** Bailey. Dewberry. May–June. Thickets; Kankakee, Rock Island, and Vermilion cos.

21. **Rubus laciniatus** Willd. Evergreen Blackberry. June–July. Native to Europe; scattered in IL, but not common.

22. **Rubus laudatus** Berger. Blackberry. May–June. Thickets; Champaign, Franklin, Schuyler, Union, and Wabash cos.

23. **Rubus meracus** Bailey. Dewberry. May–June. Thickets; Champaign, Hamilton, Jackson, and Union cos.

24. **Rubus missouricus** Bailey. Missouri Bristleberry. May–June. Thickets; scattered throughout the state, except for the se. cos.

25. **Rubus multifer** Bailey. Fruitful Dewberry. May–June. Thickets; confined to the ne. cos.

26. **Rubus mundus** Bailey. Mounded Dewberry. May–June. Thickets, rare; known only from the type locality in Hamilton Co.

27. **Rubus X neglectus** Peck. Hybrid Raspberry. May–June. Lake Co. This is reputed to be the hybrid between *R. occidentalis* L. and *R. strigosus* Michx.

28. **Rubus occidentalis** L. Black Raspberry. May–June. Edge of woods, thickets, roadsides, bluffs; common throughout the state.

29. **Rubus occidualis** Bailey. Velvet-leaved Dewberry. May–June. Thickets; Jackson Co.

30. **Rubus odoratus** L. Purple-flowering Raspberry. May–June. Woods; confined to the n. ⅕ of IL.

31. **Rubus parviflorus** Nutt. Small-flowered Raspberry. May–June. Thickets; Mason Co.

32. **Rubus parvifolius** L. Japanese Raspberry. May–June. Native to Asia; rarely escaped form gardens; DuPage Co.

33. **Rubus pensilvanicus** Poir. Blackberry. May–June. Fields, thickets, woods; occasional to common throughout the state.

34. **Rubus pergratus** Blanch. Blackberry. May–June. Fields; Rock Island Co.

35. **Rubus phoenicolasius** Maxim. Wineberry. May–July. Native to Asia; escaped in Jersey, Lake, Massac, and Morgan cos.

36. **Rubus plicatifolius** Blanch. Plicate-leaved Dewberry. May–June. Thickets; n. ¼ and s. ¼ of IL, apparently absent elsewhere.

37. **Rubus pubescens** Raf. Dwarf Raspberry. May–June. Bogs and springy places; Cook, DeKalb, DuPage, Kane, Lake, Vermilion, and Winnebago cos.

38. **Rubus pubifolius** Bailey. Blackberry. May–June. Thickets; Cass and Logan cos.

39. **Rubus recurvans** Blanch. Blackberry. May–June. Thickets; Kankakee and Winnebago cos.

40. **Rubus roribaccus** (Bailey) Rydb. Velvet-leaved Dewberry. May–June. Edge of woods, thickets; scattered in s. IL.

41. **Rubus rosa** Bailey. Blackberry. May–June. Thickets, fields; DeWitt, Jo Daviess, Kankakee, Morgan, and Ogle cos.

42. **Rubus satis** Bailey. Dewberry. May–June. Thickets; ne. cos., south to Iroquois Co.

43. **Rubus schneideri** Bailey. Schneider's Bristleberry. June. Sandy swales; Iroquois, Kankakee, and Will cos. (Including *R. offectus* Bailey.)

44. **Rubus semisetosus** Bailey. Bristleberry. May–June. Thickets; apparently confined to the nw. cos.

45. **Rubus steelei** Bailey. Steele's Dewberry. May–June. Thickets; scattered throughout the state.

46. **Rubus strigosus** Michx. Wild Red Raspberry. May–June. Bogs and swampy woods; confined to the n. ½ of the state; also Union Co.

47. **Rubus trivialis** Michx. Southern Dewberry; Scratch-ankle. Apr.–May. Fields, disturbed areas; confined to the s. ⅓ of the state.

48. **Rubus wheeleri** (Bailey) Bailey. Wheeler's Bristleberry. May–June. Thickets; Henry and Kankakee cos.

49. **Rubus wisconsinensis** Bailey. Wisconsin Bristleberry. May–June. Thickets, fields; Jo Daviess and Winnebago cos.

27. **Sanguisorba** L.—Burnet

1. Most or all the leaflets 2.5 cm long or longer; stamens 4, erect .... 1. *S. canadensis*
1. Leaflets up to 2.5 cm long; stamens 12 or more, drooping .... 2. *S. minor*

1. **Sanguisorba canadensis** L. American Burnet. July–Oct. Moist prairies, very rare; Cass, LaSalle, and Will cos.

2. **Sanguisorba minor** Scop. Garden Burnet. May–July. Native to Europe and Asia; adventive in Cook, Lake, Will, and Woodford cos.

28. **Sibbaldiopsis** Rydb. in Britt.

1. **Sibbaldiopsis tridentata** (Ait.) Rydb. Three-toothed Cinquefoil. June–Sept. Gravel ridge, very rare; Cook Co. *Potentilla tridentata* Ait.

29. **Sorbaria** (Seringe) A. Braun

1. **Sorbaria sorbifolia** (L.) A. Braun. False Spiraea. June–July. Native to Asia; rarely escaped from cultivation; Carroll, DeKalb, Jo Daviess, and Ogle cos.

30. **Sorbus** L.—Mountain Ash

1. Winter buds glutinous, the outer bud scales glabrous or ciliate.
   2. Inner bud scales glabrous or ciliate; leaves yellow-green; flowers 5–6 mm across; fruits 4–6 mm in diameter, not glaucous .... 1. *S. americana*
   2. Inner bud scales villous; leaves blue-green; flowers about 10 mm across; fruits 8–12 mm in diameter, glaucous .... 3. *S. decora*
1. Winter buds not glutinous, the outer bud scales pubescent.
   3. Leaves odd-pinnate .... 2. *S. aucuparia*
   3. Leaves simple, or some of them even-pinnate .... 4. *S. X thuringiaca*

1. **Sorbus americana** Marsh. Mountain Ash. May–July. Rocky woods; Cook and Ogle cos. *Pyrus americana* (Marsh.) DC.

2. **Sorbus aucuparia** L. European Mountain Ash. May–July. Native to Europe and Asia; escaped to bogs and swamps in the n. ⅕ of IL. *Pyrus aucuparia* (L.) Gaertn.

3. **Sorbus decora** (Sarg.) C. K. Schneid. Showy Mountain Ash. May–July. Cool woods, very rare; Cook and Lake cos; introduced in Macoupin Co. *Pyrus decora* (Sarg.) Hyland.

4. **Sorbus X thuringiaca** (Nymar) Schorach. Oak-leaved Mountain Ash. May–July. Hybrid cultivar; rarelly escaped from gardens; Coles Co.

31. **Spiraea** L.—Spiraea

1. Flowers rather few in umbels or racemes.
   2. Flowers sessile or on very short pedicels in leafy umbels.
      3. Leaves ovate to ovate-oblong, finely serrulate; pedicels, if present, pubescent; flowers often "double" .... 6. *S. prunifolia*
      3. Leaves lance-linear, sparsely serrate; pedicels, if present, glabrous; flowers usually not "double" .... 7. *S. thunbergii*
   2. Flowers in pedicellate leafy racemes .... 9. *S. X vanhouttei*
1. Flowers several in corymbose panicles.
   4. Flowers in corymbs .... 2. *S. japonica*
   4. Flowers in panicles.
      5. Petals pink to purple; inflorescence narrow.
         6. Leaves tomentose beneath at maturity.
            7. Leaves ovate, tawny-tomentose beneath; follicles pubescent .... 8. *S. tomentosa*
            7. Leaves oblong, white-tomentose beneath; follicles glabrous .... 3. *S. douglasii*
         6. Leaves glabrous or nearly so at maturity .... 2. *S. X billardii*
      5. Petals white; inflorescence broad.

8. Inflorescence tomentulose; stamens white .... 1. *S. alba*
8. Inflorescence glabrous; stamens pink-purple .... 5. *S. latifolia*

1. **Spiraea alba** DuRoi. Meadow-sweet. July–Aug. Wet soil; occasional in the n. 4/5 of the state.
2. **Spiraea X billardii** Hort. ex K. Koch. Bright Pink Spiraea. May. Native to Asia; rarely escaped from cultivation; Montgomery Co.
3. **Spiraea douglasii** Hook. Douglas' Spiraea. June–July. Native to the w. U.S.; rarely escaped from cultivation; DuPage and Lake cos.
4. **Spiraea japonica** L. f. Spiraea. May. Native to Asia; rarely escaped from cultivation. *Spiraea japonica* L. f. var. *fortunei* (Planch.) Rehd.
5. **Spiraea latifolia** (Ait.) Borkh. Meadow-sweet. June–July. Native to ne. N. America; rarely escaped from cultivation; Champaign and Hancock cos.
6. **Spiraea prunifolia** Sieb. & Zucc. Bridal-wreath. May–June. Native to Europe and Asia; rarely escaped from cultivation; DuPage, Jackson, and Williamson cos.
7. **Spiraea thunbergii** Sieb. ex Blumé. Common Spiraea. May–June. Native to Asia; rarely escaped from cultivation; DuPage Co.
8. **Spiraea tomentosa** L. Hardhack. July–Aug. Bogs, moist thickets, rare; n. 1/3 of IL. (Including var. *rosea* [Raf.] Fern.)
9. **Spiraea X vanhouttei** (Briot) Zabel. Bridal-wreath. May–June. Rarely escaped from cultivation; Adams, DuPage, Jackson, and Mason cos. Reputed to be the hybrid between *S. cantoniensis* L. and *S. trilobata* L.

32. **Waldsteinia** Willd.—Barren Strawberry

1. **Waldsteinia fragarioides** (Michx.) Tratt. Barren Strawberry. Apr.–May. Sandstone ledge, very rare; Pope Co.

136. RUBIACEAE—MADDER FAMILY

1. Shrubs; flowers in dense, globose, pedunculate heads .... 1. *Cephalanthus*
1. Herbs; flowers solitary, 1–3 together, in cymes, or, if in heads, the heads not pedunculate.
  2. Leaves in whorls of 4–8 (–9).
    3. Flowers pink or blue, subtended by an involucre .... 8. *Sherardia*
    3. Flowers white, yellow, greenish, or maroon, not subtended by an involucre .... 3. *Galium*
  2. Leaves opposite.
    4. Flowers sessile.
      5. Flowers 1–3 per leaf axil .... 2. *Diodia*
      5. Flowers several in an axillary glomerule .... 9. *Spermacoce*
    4. Flowers pedicellate.
      6. Leaves evergreen, about as broad as long; fruit a red berry; flowers in pairs, the ovaries united .... 6. *Mitchella*
      6. Leaves deciduous, longer than broad; fruit a capsule; flowers solitary or in cymes, the ovaries not united.
        7. Calyx lobes shorter than lower part of the ovary; capsule longer than broad; stipules of cauline leaves ciliate or composed of bristles .... 4. *Hedyotis*
        7. Calyx lobes equaling or longer than the lower part of the ovary; capsule not longer than broad; stipules of cauline leaves neither ciliate nor composed of bristles.
          8. Corolla salverform or funnelform; flowers solitary or in cymes, panicles, or corymbs; seeds up to 20 per locule .... 5. *Houstonia*
          8. Corolla rotate; flowers in glomerules; seeds numerous in each locule .... 7. *Oldenlandia*

1. **Cephalanthus** L.—Buttonbush

1. **Cephalanthus occidentalis** L. Two varieties may be distinguished in Illinois:

a. Branchlets and leaves glabrous or nearly so ...... 1a. *C. occidentalis* var. *occidentalis*
a. Branchlets and leaves soft-pubescent .... 1b. *C. occidentalis* var. *pubescens*

1a. **Cephalanthus occidentalis** L. var. **occidentalis.** Buttonbush. June–Aug. Wet ground; common throughout the state.

1b. **Cephalanthus occidentalis** L. var. **pubescens** Raf. Hairy Buttonbush. June–Aug. Wet ground; less common than the preceding variety; scattered in IL.

2. **Diodia** L.—Buttonweed

1. Corolla 4–6 mm long, white or pink; fruits 2.5–4.0 mm long........................1. *D. teres*
1. Corolla 7–10 mm long, white; fruits 7–10 mm long..............................2. *D. virginiana*

1. **Diodia teres** Walt. Two varieties may be distinguished in Illinois:

a. Young leaves without bristles at the tip; stems glabrous or puberulent...1a. *D. teres* var. *teres*
a. Young leaves with bristles at the tip; stems hirsute.............. 1b. *D. teres* var. *setifera*

1a. **Diodia teres** Walt. var. **teres.** Rough Buttonweed. July–Aug. Fields, roadsides, woods; occasional to common throughout IL.

1b. **Diodia teres** Walt. var. **setifera** Fern. & Grisc. Rough Buttonweed. July–Aug. Fields, roadsides, woods; not common in the s. ¼ of IL.

2. **Diodia virginiana** L. Large Buttonweed. June–Aug. Swamps, wet woods, wet ground, occasional; confined to the s. ¼ of IL.

3. **Galium** L.—Bedstraw

1. Fruits and ovaries bristly or pubescent.
  2. Leaves in whorls of 4.
    3. Leaves up to 1 cm long; flower solitary in the leaf axils.......... 19. *G. virgatum*
    3. Leaves, or some of them, more than 1 cm long; flowers in cymes or panicles.
      4. Stems and leaves glabrous or the leaf margins slightly scabrous.
        5. Leaves linear-lanceolate; flowers bright white, in panicles...3. *G. boreale*
        5. Leaves ovate-lanceolate to lanceolate; flowers greenish white, yellowish, or becoming purplish, in cymes.
          6. Leaves ovate-lanceolate; flowers greenish white, the lobes of the corolla usually more or less hairy..............................5. *G. circaezans*
          6. Leaves lanceolate; flowers yellow, becoming purple, the lobes of the corolla glabrous .... 8. *G. lanceolatum*
      4. Stems and leaves pilose.
        7. All flowers sessile; leaves 3-nerved ....5. *G. circaezans*
        7. Some or all flowers pedicellate; leaves 1-nerved ....13. *G. pilosum*
  2. Leaves in whorls of 6 or 8 (–10).
    8. Stems retrorsely bristly throughout; margins of leaves retrorsely scabrous ....1. *G. aparine*
    8. Stems glabrous or minutely retrorsely scabrous, or with retrorse bristles only at the nodes; margins of leaves not retrorsely scabrous.
      9. Leaves in whorls of 6, mucronate; stems smooth or minutely retrorsely scabrous; flowers rotate, 2–3 mm wide .... 17. *G. triflorum*
      9. Leaves in whorls of 6–10, not mucronate; stems smooth except for retrorse bristles at the nodes; flowers funnelform, 4–7 mm wide.. 11. *G. odoratum*
1. Fruits and ovaries smooth.
  10. Plants erect.
    11. Leaves in whorls of 4 ....3. *G. boreale*
    11. Leaves in whorls of 6 or 8.
      12. Corolla yellow; leaves scabrous .... 18. *G. verum*
      12. Corolla white; leaves not scabrous.
        13. Leaves with strongly ciliate margins .... 14. *G. sylvaticum*
        13. Leaves not ciliate .... 9. *G. mollugo*
  10. Plants matted or loosely ascending.
    14. Some of the corollas 3-lobed; upper part of stems scabrous but not retrorsely bristly.

15. Pedicels smooth; leaves in whorls of 4–6 on the same plant .... 15. *G. tinctorium*
15. Pedicels scabrous; leaves almost always in whorls of four.
16. Pedicels 5–30 mm long; fruit 1.25–1.75 mm in diameter .... 16. *G. trifida*
16. Pedicels 0.5–4.0 mm long; fruit 0.8–1.0 mm in diameter .... 4. *G. brevipes*
14. Corolla always 4-lobed; stems retrorsely bristly or smooth.
17. Stems retrorsely bristly.
18. Leaves mostly in whorls of 6; flowers white .... 2. *G. asprellum*
18. Leaves mostly in whorls of 4; flowers yellow .... 12. *G. pedemontanum*
17. Stems smooth.
19. Leaves in whorls of 6, mostly linear .... 6. *G. concinnum*
19. Leaves in whorls of 4, mostly elliptic-oblong to oblanceolate.
20. Leaves conspicuously reflexed; inflorescence overtopped by lateral branches; fruits up to 1.5 mm in diameter .... 7. *G. labradoricum*
20. Leaves ascending; inflorescence not overtopped by lateral branches; fruits more than 2 mm in diameter .... 10. *G. obtusum*

1. **Galium aparine** L. Two varieties occur in Illinois:

a. All leaves more than 2 cm long .... 1a. *G. aparine* var. *aparine*
a. Leaves up to 2 cm long .... 1b. *G. aparine* var. *echinosperma*

1a. **Galium aparine** L. var. **aparine.** Goosegrass; Cleavers. May–July. Woods; common throughout the state; in every co.

1b. **Galium aparine** L. var. **echinosperma** (Wallr.) Farw. Spiny Cleavers. May–July. Woods, rare; DeKalb Co. (Including var. *vaillantii* [DC.] W. J. D. Koch.)

2. **Galium asprellum** Michx. Rough Bedstraw. July–Aug. Moist thickets; occasional in the n. ¼ of IL; also Clark and Mason cos.

3. **Galium boreale** L. Northern Bedstraw. May–July. Moist meadows, prairies, and fields; occasional in the n. ¼ of IL; also Mason and McDonough cos. (Including var. *hyssopifolium* [Hoffm.] DC. and var. *intermedium* DC.)

4. **Galium brevipes** Fern. & Wieg. Short-stalked Bedstraw. July–Aug. Moist soil, rare; Cook Co.

5. **Galium circaezans** Michx. Two varieties occur in Illinois:

a. Stems and leaves glabrous .... 5a. *G. circaezans* var. *circaezans*
a. Stems and leaves pilose .... 5b. *G. circaezans* var. *hypomalacum*

5a. **Galium circaezans** Michx. var. **circaezans.** Wild Licorice. May–July. Rich, often rocky woods; occasional to common in the s. ½ of IL, rare elsewhere.

5b. **Galium circaezans** Michx. var. **hypomalacum** Fern. Wild Licorice. May–July. Rich woods; apparently confined to the n. ½ of the state.

6. **Galium concinnum** Torr. & Gray. Shining Bedstraw. June–July. Dry woods; common throughout the state.

7. **Galium labradoricum** (Wieg.) Wits. Bedstraw. Bogs, marshes, rare; Boone, Kane, Lake, and McHenry cos.

8. **Galium lanceolatum** Torr. Wild Licorice. June–July. Rich woods, rare; Cook, Kane, and Lake cos.

9. **Galium mollugo** L. White Bedstraw. June–Sept. Native to Europe; rarely adventive in waste ground; mostly in the n. ⅓ of IL; also Champaign and Crawford cos.

10. **Galium obtusum** Bigel. Wild Madder. May–July. Moist ground; common throughout the state.

11. **Galium odoratum** (L.) Scop. Sweet Woodruff. May–June. Native to Europe; rarely escaped from gardens; DuPage Co. *Asperula odoratum* L.

12. **Galium pedemontanum** (Bellardi) All. Yellow-flowered Bedstraw. June–July. Native to Europe; adventive in disturbed soil; scattered in IL. *Cruciena pedemontana* (Bellardi) Ehrend.

13. **Galium pilosum** Ait. Hairy Bedstraw. June–Aug. Sandy woods; occasional to common throughout the state.

14. **Galium sylvaticum** L. Scotch Mist. July–Aug. Native to Europe; rarely escaped from gardens; Kane Co.

15. **Galium tinctorium** L. Stiff Bedstraw. May–Sept. Swamps, marshes; occasional to common throughout the state.

16. **Galium trifidum** L. Small Bedstraw. July–Aug. Moist soil, not common; apparently confined to the ne. cos. and a few cos. in cent. IL along the Illinois River.

17. **Galium triflorum** Michx. Sweet-scented Bedstraw. May–Sept. Moist woods; occasional to common throughout the state.

18. **Galium verum** L. Yellow Bedstraw. June–Aug. Native to Europe; rarely adventive in disturbed soil; DuPage, Kendall, Lake, and LaSalle cos.

19. **Galium virgatum** Nutt. Dwarf Bedstraw. Apr.–June. Limestone ledges, rare; Monroe Co.

4. **Hedyotis** L.—Bluets

1. **Hedyotis nigricans** (Lam.) Fosb. Slender-leaved Bluets. May–Sept. Rocky ledges, hill prairies; confined to the s. ½ of IL along the Mississippi River. *Stenaria nigricans* (Lam.) Turrell.

5. **Houstonia** L.—Bluets

1. Corolla rotate or salverform; plants annual.
  2. Corolla lobes 2.5–5.0 mm broad, very pale blue .... 1. *H. caerulea*
  2. Corolla lobes up to 2.5 mm broad, usually dark purple.
    3. Calyx lobes 2–4 mm long, about as long as the corolla tube .... 3. *H. crassifolia*
    3. Calyx lobes 1–2 mm long, up to ⅔ as long as the corolla tube .... 7 *H. pusilla*
1. Corolla funnelform; plants perennial.
  4. Lowermost leaves 3- or 5-nerved.
    5. Leaves ovate-lanceolate to broadly ovate, mostly 10 mm broad or broader .... 6. *H. purpurea*
    5. Leaves broadly lanceolate, mostly less than 10 (–12) mm broad .... 4. *H. lanceolata*
  4. Lowermost leaves 1-nerved.
    6. Capsules as high as or higher than broad; cauline leaves oblong to oblanceolate, mostly 2–5 mm broad.
      7. Basal leaves eciliate .... 5. *H. longifolia*
      7. Basal leaves ciliate .... 2. *H. canadensis*
    6. Capsules broader than high; cauline leaves linear, 1.0–2.5 mm broad .... 8. *H. tenuifolia*

1. **Houstonia caerulea** (L.) Hook. Bluets. Apr.–June. Sandy soil of fields, prairies, and open woods; not common but scattered in all parts of the state except the nw. cos.; rare in the s. cos. *Hedyotis caerulea* (L.) Hook.

2. **Houstonia canadensis** Willd. Bluets. Apr.–July. Rocky woods, prairies, fields; occasional in the southernmost and northernmost cos., absent elsewhere. *Hedyotis longifolia* (Gaertn.) Hook. var. *ciliolata* (Torr.) Mohlenbr.

3. **Houstonia crassifolia** Raf. Tiny Bluets. Apr.–May. Rocky woods, prairies, fields; scattered throughout the state. *Hedyotis crassifolia* Raf.

4. **Houstonia lanceolata** (Poir.) Britt. Lance-leaved Bluets. May–June. Rocky woods, occasional; confined to the s. ½ of the state. *Houstonia purpurea* (L.) Torr. & Gray var. *calycosa* (Gray) Fosberg.

5. **Houstonia longifolia** Gaertn. Long-leaved Bluets. Apr.–July. Rocky woods, prairies, fields; occasional in the southernmost and northernmost cos., absent elsewhere. *Hedyotis longifolia* (Gaertn.) Hook.

6. **Houstonia purpurea** L. Broad-leaved Bluets. May–June. Rocky woods, rare; Jackson and Pope cos. *Hedyotis purpurea* (L.) Torr. & Gray.

7. **Houstonia pusilla** Schoepf. Star Violet; Small Bluets. Mar.–Apr. Rocky woods, fields; scattered in IL. *Hedyotis pusilla* (Schoepf) Mohlenbr.

8. **Houstonia tenuifolia** Nutt. Slender-leaved Bluets. Apr.–Oct. Rocky woods; occasional in the s. ½ of IL. *Hedyotis nuttalliana* Fosb.

6. **Mitchella** L.—Partridge-berry

1. **Mitchella repens** L. Partridge-berry. May–June. Rocky woods, swampy woods; occasional in the s. and n. cos., rare in the cent. cos.

7. **Oldenlandia** L.

1. **Oldenlandia uniflora** L. One-flowered Oldenlandia. Apr.–May. Muddy shore, very rare; Pope Co. *Hedyotis uniflora* (L.) Lam.

8. **Sherardia** L.—Sherardia

1. **Sherardia arvensis** L. Field Madder. May–Sept. Native to Europe; rarely adventive in disturbed ground; Jackson Co.

9. **Spermacoce** L.—Smooth Buttonweed

1. **Spermacoce glabra** Michx. Smooth Buttonweed. June–Oct. Wet ground; occasional in the s. ⅔ of the state, rare or absent elsewhere.

137. RUTACEAE—CITRUS FAMILY

1. Plants herbaceous; leaves 2- to 3-pinnate; all flowers perfect; fruit a capsule .... 4. *Ruta*
1. Plants woody; leaves once-pinnate or trifoliolate; at least some of the flowers unisexual; fruit a follicle, samara, berrylike drupe, or sour orange.
  2. Leaves opposite; fruit a berrylike drupe .... 1. *Phellodendron*
  2. Leaves alternate; fruit a follicle, samara, or sour orange.
    3. Plants not prickly or thorny; fruit an orbicular samara .... 3. *Ptelea*
    3. Plants prickly or thorny; fruit a follicle or sour orange.
      4. Spines on the branches; leaflets 5–9; petioles unwinged; fruit an ellipsoid follicle .... 5. *Zanthoxylum*
      4. Thorns axillary; leaflets 3; petioles winged; fruit a sour orange .... 2. *Poncirus*

1. **Phellodendron** Rupr.—Cork Tree

1. **Phellodendron amurense** Rupr. Amur Cork Tree. May. Native to Asia; rarely escaped from cultivation; Cook, DuPage, and Will cos.

2. **Poncirus** Raf.—Sour Orange

1. **Poncirus trifoliata** (L.) Raf. Sour Orange. May–June. Native to Asia; sometimes planted as an ornamental; rarely escaped into thickets; very rare; Jackson and Williamson cos.

3. **Ptelea** L.—Hop-tree

1. **Ptelea trifoliata** L. Two varieties occur in Illinois:

a. Branchlets glabrous .... 1a. *P. trifoliata* var. *trifoliata*
a. Branchlets densely pubescent .... 1b. *P. trifoliata* var. *mollis*

1a. **Ptelea trifoliata** L. var. **trifoliata.** Hop-tree; Wafer Ash. May–July. Open woods; occasional throughout the state although less common in the s. cos.

1b. **Ptelea trifoliata** L. var. **mollis** Torr. & Gray. Hop-tree; Wafer Ash. May–July. Sand dunes, rare; Cook and Lake cos.

4. **Ruta** L.—Rue

1. **Ruta graveolens** L. Common Rue. May–Aug. Native to Europe; rarely escaped from cultivation; Champaign, Jackson, and Lake cos.

5. **Zanthoxylum** Gmel.—Prickly Ash

1. **Zanthoxylum americanum** Mill. Prickly Ash; Tooth-ache Tree. May–June. Woods; occasional in the n. ½ of the state, rare in the s. ½.

138. SALICACEAE—WILLOW FAMILY

1. Leaves never twice as long as broad; bud scales several; catkins pendulous .... 1. *Populus*

1. Leaves twice as long as broad or longer; bud scale 1; catkins not pendulous .... 2. *Salix*

1. **Populus** L.—Poplar

1. Part or all of the petiole flattened.
  2. Leaves triangular-ovate to rhombic.
    3. Leaves triangular-ovate .... 4. *P. deltoides*
    3. Leaves rhombic .... 8. *P. nigra*
  2. Leaves ovate to suborbicular.
    4. Margin of leaf dentate, with 5–25 teeth (averaging 10–20); buds pubescent.
      5. Margin of leaf with 5–15 teeth (averaging 10); petiole 5–10 cm long (averaging 7 cm) .... 6. *P. grandidentata*
      5. Margin of leaf with 12–25 teeth (averaging 20); petiole 3–6 cm long (averaging 5.5 cm) .... 9. *P. X smithii*
    4. Margin of leaf finely crenate, with 20 or more teeth (averaging 31); buds glabrous or nearly so .... 10. *P. tremuloides*
1. Petiole round in cross-section throughout.
  6. Leaves covered with a white felt on the lower surface .... 1. *P. alba*
  6. Leaves glabrous or variously pubescent beneath, but not covered with a white felt.
    7. Leaves rounded or truncate at base or, if cordate, the buds heavily resinous.
      8. Leaves sinuate-dentate; buds to 6 mm long, not resinous .... 3. *P. X canescens*
      8. Leaves serrulate or crenulate; buds to 25 mm long, resinous.
        9. Twigs usually glabrous; leaves glabrous or with a few hairs on the midvein .... 2. *P. balsamifera*
        9. Twigs usually pubescent; leaves pubescent beneath .... 5. *P. X giliadensis*
    7. Leaves cordate at base; buds not heavily resinous .... 7. *P. heterophylla*

1. **Populus alba** L. White Poplar. Apr.–May. Native to Europe; frequently spreading from cultivation. A hybrid between *P. alba* and *P. tremuloides,* called *Populus X heimburgeri* Boivin, is known from Lake Co.

2. **Populus balsamifera** L. Balsam Poplar. Apr. Dunes, sandy soil; Cook, DuPage, Hancock, Lake, and McHenry cos.

3. **Populus X canescens** (Ait.) Sm. Gray Poplar. Apr.–May. Introduced from Europe; infrequently escaped from cultivation; scattered in IL. Reputedly the hybrid between *P. alba* L. and *P. tremuloides* Michx.

4. **Populus deltoides** Marsh. Eastern Cottonwood. Mar.–May. Along rivers and streams, common; probably in every co. (Including var. *missouriense* [Henry] Rehd.)

5. **Populus X gileadensis** Rouleau. Balm-of-Gilead. Apr. Rarely escaped from cultivation; Cook and Lake cos. Reputed to be the hybrid between *P. deltoides* Marsh. and *P. balsamifera* L. *Populus X jackii* Sarg.

6. **Populus grandidentata** Michx. Large-toothed Aspen. Apr. Disturbed areas in and around woodlands; occasional to common in the n. ½ of the state, extending s. to Marion and Wabash cos.

7. **Populus heterophylla** L. Swamp Cottonwood. Apr.–May. Swamps and low woods, often in standing water; occasional in the s. ⅖ of IL; also Iroquois and Kankakee cos.

8. **Populus nigra** L. Two varieties occur in Illinois:

a. Crown rounded; branchlets spreading .... 8a. *P. nigra* var. *nigra*
a. Crown narrow; branchlets strongly ascending .... 8b. *P. nigra* var. *italica*

8a. **Populus nigra** L. var. **nigra.** Black Poplar. Mar.–May. Native to Europe; rarely escaped from cultivation; Pope Co.

8b. **Populus nigra** L. var. **italica** Muenchh. Lombardy Poplar. Mar.–May. Introduced from Europe; rarely persistent around old homesites; scattered in IL.

9. **Populus X smithii** Boivin. Barnes' Aspen. Apr.–May. Low woods; LaSalle and Peoria cos. Reputed to be the hybrid between *P. grandidentata* Michx. and *P. tremuloides* Michx.

10. **Populus tremuloides** Michx. Quaking Aspen. Apr.–May. Low ground of woods, marshes, and bogs; common in the n. cos., absent in the extreme s. cos.

2. **Salix** L.—Willow

1. Twigs not contorted.
 2. 1 or more glands present at upper end of petiole.
  3. Leaves whitish beneath.
   4. Leaves with some pubescence usually on the nerves below; petioles and branchlets often pubescent.
    5. Teeth of leaves pointed.
     6. Leaves lanceolate to lance-ovate, sometimes falcate; vegetative sprouts often with foliaceous stipules...................... 5. *S. caroliniana*
     6. Leaves elliptic to narrowly lanceolate, not falcate; vegetative sprouts with very small stipules, or stipules absent.
      7. Young leaves with ferruginous hairs..................... 18. *S. X jesupii*
      7. Young leaves with long, silky white hairs.
       8. Branchlets brittle at base, easily broken; twigs glabrous or nearly so.. 14. *S. X fragilis*
       8. Branchlets not brittle at base; twigs sericeous .... 1. *S. alba*
    5. Teeth of leaves blunt .... 30. *S. X rubens*
   4. Leaves glabrous; petioles and branchlets glabrous............ 33. *S. serissima*
  3. Leaves green beneath.
   9. Leaf blades linear-lanceolate to ovate-lanceolate, often falcate, at least 5 times longer than broad; buds pointed.
    10. All or most of the leaves at least 7 times longer than broad..... 22. *S. nigra*
    10. Leaves 5–7 times longer than broad .... 15. *S. X glatfelteri*
   9. Leaf blades lanceolate to ovate, usually not falcate, up to 5 times longer than broad; buds rounded.
    11. Leaves with a caudate tip; stipules large and persistent..... 19. *S. lucida*
    11. Leaves acute or acuminate, not caudate; stipules minute or absent.
     12. Leaves coriaceous, some of them usually at least 3 times longer than broad.. 33. *S. serissima*
     12. Leaves membranous, less than 3 times longer than broad .... 25. *S. pentandra*
 2. Petioles not glandular.
  13. Leaves purplish, at least some of them opposite...................... 27. *S. purpurea*
  13. Leaves not purplish, all of them alternate.
   14. Leaves glabrous or sparsely pubescent only on the veins beneath.
    15. Leaves remotely denticulate... 17. *S. interior*
    15. Leaves closely serrate, crenulate, undulate, or entire.
     16. Leaves entire, revolute; some of the stems creeping........................ .... 23. *S. pedicellaris*
     16. Leaves serrate, crenate, or undulate, flat; all of the stems upright.
      17. Stipules absent or minute and falling away early on vegetative sprouts and young branchlets.
       18. Leaves ovate-lanceolate, long-attenuate to caudate at the tip; buds pointed....................................... 2. *S. amygdaloides*
       18. Leaves linear to lanceolate, acute to acuminate but not caudate; buds rounded.
        19. Leaves glabrous.
         20. Leaves linear to lanceolate, most or all of them distinctly toothed; petioles at least 7 mm long; branches "weeping."
          21. Stems strongly pendulous; leaves thin, green on the upper surface.
           22. Leaves glaucous below; all leaves serrulate; capsules 1–2 mm long.......... 31. *S. X sepulcralis*
           22. Leaves green below; uppermost leaves usually entire; capsules 2.0–2.7 mm long........................ .... 3. *S. babylonica*

21. Stems moderately pendulous; leaves thick, dark green on the upper surface .... 24. *S. X pendulina*
20. Leaves ovate, obscurely toothed; petioles up to 5 mm long; branches not "weeping" .... 28. *S. pyrifolia*
19. Leaves usually with a few hairs on the veins beneath.
23. Teeth of the leaves pointed; hairs reddish .... 26. *S. petiolaris*
23. Teeth of the leaves blunt; hairs white .... 30. *S. X rubens*
17. Stipules present on vegetative sprouts and young branchlets.
24. Leaves green on both sides .... 29. *S. rigida*
24. Leaves pale or white on the lower surface.
25. Leaves irregularly crenate; capsules pubescent .... 12. *S. discolor*
25. Leaves finely serrulate; capsules glabrous or granular .... 21. *S. myricoides*
14. Leaves pubescent on the lower surface (if only on veins, go back to couplet 21).
26. Young branchlets and new leaves covered with white wool .... 6. *S. candida*
26. Young branchlets and new leaves not covered with white wool.
27. Leaves entire or undulate.
28. Young branchlets densely tomentose or pilose ..... 16. *S. humilis*
28. Young branchlets sericeous, sparsely pilose, or glabrate.
29. Leaves revolute .... 9. *S. cinerea*
29. Leaves flat.
30. Capsules up to 6 mm long; bracts brown, with a black tip .... 34. *S. X subsericea*
30. Capsules 6–10 mm long; bracts yellow .... 4. *S. bebbiana*
27. Leaves serrulate or crenate or denticulate.
31. Leaves silvery-silky on lower surface.
32. Petiole up to 3 mm long; margin of blade remotely denticulate .... 17. *S. interior*
32. Petiole 3 mm long or longer; margin of blade finely serrulate.
33. Teeth along margin of blade not extending all the way to the base .... 26. *S. petiolaris*
33. Teeth along margin of blade extending all the way to the base .... 32. *S. sericea*
31. Leaves pubescent beneath, but not with silvery-silky hairs.
34. Leaves narrowed or rounded at base, not subcordate; stipules on sprouts and young branchlets inconspicuous and falling away early.
35. Young branches densely tomentose or pilose.................. .... 16. *S. humilis*
35. Young branches sparsely pilose or glabrous.
36. Upper surface of leaves shiny; flowers appearing before the leaves.
37. Capsule 7–12 mm long, minutely pubescent.
38. Branches glabrous or nearly so ... 12. *S. discolor*
38. Branches permanently pubescent .... 10. *S. X conifera*
37. Capsule 6–7 mm long, densely gray-hairy .... 7. *S. caprea*
36. Upper surface of leaves dull; flowers appearing with the leaves .... 4. *S. bebbiana*
34. Leaves subcordate at base (tapering in *S. eriocephala*); stipules on sprouts and young branchlets large and persistent.
39. Branchlets permanently tomentulose.

40. Leaves shiny above, acute to short-acuminate .... 11. *S. cordata*
40. Leaves dull above, tapering to a caudate tip .... 13. *S. eriocephala*
39. Branchlets glabrous or glabrate.
41. Leaves glaucous beneath, caudate at tip .... 5. *S. X bebbii*
41. Leaves green beneath, acute to short-acuminate .... 29. *S. rigida*
1. Twigs contorted .... 20. *S. matsudana*

1. **Salix alba** L. Three varieties occur in Illinois:

a. Leaves silky beneath .... 1a. *S. alba* var. *alba*
a. Leaves glabrate.
b. Branchlets brown .... 1b. *S. alba* var. *caerulea*
b. Branchlets yellow .... 1c. *S. alba* var. *vitellina*

1a. **Salix alba** L. var. **alba.** White Willow. Apr.–May. Native to Europe; occasionally escaped from cultivation in most parts of IL.

1b. **Salix alba** L. var. **caerulea** (Smith) Smith. White Willow. Apr.–May. Native to Europe; occasionally escaped in IL.

1c. **Salix alba** L. var. **vitellina** (L.) Stokes. White Willow. Apr.–May. Native to Europe; rarely escaped from cultivation; Champaign, Kankakee, and Vermilion cos.

2. **Salix amygdaloides** Anderss. Peach-leaved Willow. Apr.–May. Along streams, wet ground; occasional throughout the state.

3. **Salix babylonica** L. Weeping Willow. Apr.–May. Native to Europe and Asia; rarely escaped from cultivation throughout IL.

4. **Salix bebbiana** Sarg. Bebb's Willow. May. Bogs, occasional; confined to the n. ¼ of IL.

5. **Salix X bebbii** Gand. Bebb's Hybrid Willow. May. Moist soil, rare; Jackson Co. Reputed to be the hybrid between *S. rigida* Muhl. and *S. sericea* Marsh.

6. **Salix candida** Fluegge. Hoary Willow. Apr.–May. Bogs; confined to the n. ⅓ of IL.

7. **Salix caprea** L. Goat Willow; Pussy Willow. Feb.–Apr. Native to Europe; frequently planted, but seldom escaped from cultivation; Jackson and Lake cos.

8. **Salix caroliniana** Michx. Ward's Willow; Carolina Willow. May. Wet soil, not common; confined to the s. ⅗ of IL.

9. **Salix cinerea** L. Gray Willow. May. Native to Europe; adventive in low ground; DuPage, Kane, and McHenry cos.

10. **Salix X conifera** Wangenh. Hybrid Pussy Willow. Apr.–May. Moist soil; occasional in the n. cos. *Salix discolor* Muhl. var. *latifolia* Anders. This is the hybrid between *S. discolor Muhl.* and *S. humilis Marsh.*

11. **Salix cordata** Michx. Sand-dune Willow. June. Sand dunes, rare; Cook and Lake cos. *Salix syrticola* Fern.

12. **Salix discolor** Muhl. Pussy Willow. Mar.–May. Marshes and swamps; occasional throughout the state but less common in the s. cos.

13. **Salix eriocephala** Michx. Broad-leaved Willow. Apr. Low woods, rare; confined to the s. ¼ of IL; also Mason Co.

14. **Salix X fragilis** L. Crack Willow. Apr.–May. Native to Europe; occasionally scattered in IL. This is reputed to be the hybrid between *S. alba* L. and *S. euxina* I. V. Belgaeva.

15. **Salix X glatfelteri** Schneid. Hybrid Black Willow. Apr.–May. Low ground; scattered throughout the state. Reputed to be the hybrid between *S. nigra* Marsh. and *S. amygdaloides* Anderss.

16. **Salix humilis** Marsh. Three varieties occur in Illinois:

a. Branchlets 2 mm or more thick; petioles 3 mm long or longer.
b. Leaves pilose beneath .... 16a. *S. humilis* var. *humilis*
b. Leaves glabrate beneath .... 16b. *S. humilis* var. *hyporhysa*
a. Branchlets less than 2 mm thick; petioles less than 3 mm long .... 16c. *S. humilis* var. *microphylla*

16a. **Salix humilis** Marsh. var. **humilis.** Prairie Willow. Mar.–May. Prairies; occasional throughout the state.

16b. **Salix humilis** Marsh. var. **hyporhysa** Fern. Prairie Willow. Mar.–May. Prairies; occasional throughout the state.

16c. **Salix humilis** Marsh. var. **microphylla** (Anderss.) Fern. Sage Willow. Mar.–Apr. Prairies; confined to the n. ½ of IL.

17. **Salix interior** Rowlee. Two forms occur in Illinois:

a. Leaves glabrous or pubescent but not sericeous .... 17a. *S. interior* f. *interior*
a. Leaves sericeous .... 17b. *S. interior* f. *wheeleri*

17a. **Salix interior** Rowlee f. **interior.** Sandbar Willow. Apr.–May. Sandy shores of rivers and streams; common throughout the state. *Salix exigua* Nutt., misapplied; *S. exigua* Nutt. ssp. *interior* (Rowlee) Cronq.

17b. **Salix interior** Rowlee f. **wheeleri** (Rowlee) Rouleau. Sandbar Willow. Apr.–May. Sandy shores of rivers and streams; common throughout the state.

18. **Salix X jesupii** Fern. Jesup's Willow. Apr.–May. Native to Europe; scattered in IL. This is reputed to be the hybrid between *S. lucida* Muhl. and *S. alba* L.

19. **Salix lucida** Muhl. Shining Willow. May. Bogs and wet sandy areas, not common; confined to the n. ¼ of IL; also Calhoun Co.

20. **Salix matsudana** Koidz. Twisted Willow. May–June. Native to Asia; rarely escaped from cultivation; Cole and DuPage cos.

21. **Salix myricoides** Muhl. Blue-leaf Willow. May. Open sand, calcareous pond borders, not common; confined to the n. ½ of IL; also St. Clair Co. *Salix glaucophylloides* Fern. var. *glaucophylla* (Bebb) Schneid.

22. **Salix nigra** Marsh. Black Willow. Apr.–May. Along streams, low woods, common; in every co.

23. **Salix pedicellaris** Pursh var. **hypoglauca** Fern. Bog Willow. Apr.–May. Bogs, rare; confined to the n. ⅓ of the state.

24. **Salix X pendulina** Wender. Hybrid Weeping Willow. Apr.–May. Occasional throughout IL. This is the hybrid between *S. X fragilis* L. and *S. X sepulcralis* Simoski.

25. **Salix pentandra** L. Bay-leaved Willow. May. Native to Europe; escaped into wet soil; Champaign, DuPage, Lake, and Winnebago cos.

26. **Salix petiolaris** Sm. Meadow Willow. Apr.–June. Low prairies, rich woods, marshes, bogs; confined to the n. ⅓ of IL.

27. **Salix purpurea** L. Purple Osier. Apr.–May. Native to Europe; occasionally cultivated but rarely escaped in IL: Cook, DuPage, Lake, and Will cos.

28. **Salix pyrifolia** Anderss. Balsam Willow. May–July. Calcareous fen, very rare; Kane Co.

29. **Salix rigida** Muhl. Heart-leaved Willow. Apr.–May. Moist ground; occasional throughout the state. (Including var. *angustata* [Pursh] Fern.) *Salix eriocephala* Michx., misapplied.

30. **Salix X rubens** Schrank. Hybrid Crack Willow. Apr.–May. Low fields, along streams, disturbed soil; occasional throughout the state. Reputed to be the hybrid between *S. alba* L. and *S. X fragilis* L.

31. **Salix X sepulcralis** Simoski. Glaucous Weeping Willow. Apr.–May. Scattered in IL. This is the hybrid between *S. alba* L. and *S. X pendulina* Wender.

32. **Salix sericea** Marsh. Silky Willow. Mar.–May. Low ground, along streams, in bogs; occasional and scattered throughout the state.

33. **Salix serissima** (Bailey) Fern. Autumn Willow. June. Marshes, bogs, very rare; Cook, Lake, and McHenry cos.

34. **Salix X subsericea** (Anderss.) Schneid. Hybrid Silky Willow. Apr.–June. Low ground, rare; Cook and Kankakee cos. Reputed to be the hybrid between *S. petiolaris* Sm. and *S. sericea* Marsh.

139. SANTALACEAE—SANDALWOOD FAMILY

1. **Comandra** Nutt.—False Toadflax

1. **Comandra umbellata** (L.) Nutt. False Toadflax; Bastard Toadflax. May–Aug. Woods, prairies; occasional throughout the state. (Including *C. richardsiana* Fern.)

140. SAPINDACEAE—SOAPBERRY FAMILY

1. Annual climber with tendrils; petals white; leaves biternate ....1. *Cardiospermum*
1. Tree without tendrils; petals yellow; leaves bipinnate..........................2. *Koelreuteria*

1. **Cardiospermum** L.—Balloon-vine

1. **Cardiospermum halicacabum** L. Balloon-vine. July–Sept. Native to tropical America; adventive in thickets near streams and rivers; Alexander, Cass, Jackson, Monroe, Randolph, and St. Clair cos.

2. **Koelreuteria** Laxm.—China-tree

1. **Koelreuteria paniculata** Laxm. Golden-rain Tree. July–Sept. Native to Asia; sometimes planted and rarely escaped into disturbed areas.

141. SAPOTACEAE—SAPOTE FAMILY

1. **Bumelia** Sw.—Buckthorn

1. Branchlets and lower surface of leaves rusty-woolly; pedicels pubescent; fruit oblongoid to globose.. 1. *B. lanuginosa*
1. Branchlets and lower surface of leaves silky at first, becoming glabrous at maturity; pedicels glabrous; fruit ovoid ....2. *B. lycioides*

1. **Bumelia lanuginosa** (Michx.) Pers. var. **oblongifolia** (Nutt.) R. B. Clark. Woolly Buckthorn. June–July. Dry woods, rare; Monroe and St. Clair cos. *Sideroyxlon lanuginosa* Michx.

2. **Bumelia lycioides** (L.) Pers. Southern Buckthorn. June–Aug. Moist woods or rocky cliffs, rare; apparently confined to the s. ¼ of IL; also St. Clair Co. *Sideroxylon lycioides* L.

142. SARRACENIACEAE—PITCHER-PLANT FAMILY

1. **Sarracenia** L.—Pitcher-plant

1. **Sarracenia purpurea** L. Pitcher-plant. May–June. Bogs, rare; Cook, Lake, and McHenry cos.

143. SAURURACEAE—LIZARD'S-TAIL FAMILY

1. **Saururus** L.—Lizard's-tail

1. **Saururus cernuus** L. Lizard's-tail. May–Sept. Swampy woods, sometimes in standing water; occasional to common in the s. ⅔ of IL, rare elsewhere.

144. SAXIFRAGACEAE—SAXIFRAGE FAMILY

1. Stem with a single pair of leaves; petals deeply divided..............................3. *Mitella*
1. Stem with a single leaf or leafless; petals not deeply divided.
  2. Leaves longer than broad; stamens 10 .... 2. *Micranthes*
  2. Leaves as broad as long; stamens 5.
    3. Leaves pubescent; ovary 1-locular; seeds unwinged ....1. *Heuchera*
    3. Leaves glabrous or nearly so; ovary 2-locular; seeds winged..... 4. *Sullivantia*

1. **Heuchera** L.—Alumroot

1. Calyx to 2 mm long; leaves and stems villous .... 2. *H. parviflora*
1. Calyx 2.5 mm long or longer; leaves and stems hispid or strigose.
  2. Leaves strigose or nearly glabrous; calyx symmetrical or nearly so .... 1. *H. americana*
  2. Leaves hispid beneath; calyx asymmetrical ....3. *H. richardsonii*

1. **Heuchera americana** L. Two varieties occur in Illinois:

a. Perianth 4.0–4.5 mm long; petals 2–3 mm long.... 1a. *H. americana* var. *hirsuticaulis*
a. Perianth 3.0–3.5 mm long; petals 1.0–1.5 mm long ...... 1b. *H. americana* var. *interior*

1a. **Heuchera americana** L. var. **hirsuticaulis** (Wheelock) Rosend., Butt., & Lak. Tall Alumroot. May–June. Dry woods; occasional in the s. ⅔ of the state; also Cook and Jo Daviess cos.

1b. **Heuchera americana** L. var. **interior** Rosend., Butt., & Lak. Tall Alumroot. May–June. Dry woods; scattered in the s. ½ of IL.

2. **Heuchera parviflora** Bartl. var. **rugelii** (Shuttlw.) Rosend., Butt., & Lak. Late Alumroot. July–Nov. Moist sandstone cliffs; common in the s. 1/6 of the state, absent elsewhere.

3. **Heuchera richardsonii** R. Br. Two varieties occur in Illinois:

a. Perianth 5–7 mm long; calyx tube slightly asymmetrical .... 3a. *H. richardsonii* var. *affinis*

a. Perianth 6–10 mm long; calyx tube strongly asymmetrical.. 3b. *H. richardsonii* var. *grayana*

3a. **Heuchera richardsonii** R. Br. var. **affinis** Rosend., Butt., & Lak. Prairie Alumroot. May–July. Prairies; occasional in the n. 3/4 of IL; also Jackson Co.

3b. **Heuchera richardsonii** R. Br. var. **grayana** Rosend., Butt., & Lak. Prairie Alumroot. May–July. Prairies; occasional in the n. 3/4 of IL.

2. **Micranthes** Small—American Saxifrage

1. Some or all the leaves more than 8 cm long; calyx lobes at least twice as long as the calyx tube; petals greenish white or greenish yellow, 2–3 mm long.
- **2.** Leaves pilose on the lower surface; petals longer than the sepals .... 1. *M. forbesii*
- **2.** Leaves glabrous or nearly so on the lower surface; petals about as long as the sepals .... 2. *M. pensylvanicus*

1. None of the leaves more than 8 cm long; calyx lobes about as long as the calyx tube; petals white, at least 4 mm long .... 3. *M. virginiensis*

1. **Micranthes forbesii** (Vasey) Mohlenbr., comb. nov. (basionym: *Saxifraga forbesii* Vasey). Forbes' Saxifrage. May. Moist sandstone cliffs, not common; known from extreme s. IL and extreme n. IL; also Monroe and St. Clair cos. *Saxifraga forbesii* Vasey.

2. **Micranthes pensylvanica** (L.) Haworth. Swamp Saxifrage. May–June. Springs, bogs, moist meadows; occasional in the n. 1/2 of the state; also Fayette and Wabash cos. (Including ssp. *interior* Burns.) *Saxifraga pensylvanica* L.

3. **Micranthes virginiensis** (Michx.) Small. Early Saxifrage. Apr.–June. Rocky wooded ravine, very rare; Hardin Co. *Saxifraga virginiensis* Michx.

3. **Mitella** L.—Miterwort

1. **Mitella diphylla** L. Bishop's-cap. Apr.–May. Rich, often rocky woods; occasional in the n. 1/2 of the state, rare in the s. 1/2.

4. **Sullivantia** Torr. & Gray—Sullivantia

1. **Sullivantia sullivantii** (Torr. & Gray) Britt. Sullivantia. June–July. Moist shaded cliffs, rare; confined to the extreme nw. cos. of IL. *Sullivantia renifolia* Rosend.

145. SCROPHULARIACEAE—FIGWORT FAMILY

1. Leaves whorled .... 31. *Veronicastrum*

1. Leaves opposite or alternate.
- **2.** All leaves opposite (excluding bracteal leaves.)
  - **3.** Anther-bearing stamens 2.
    - **4.** Calyx and corolla lobes each 4; fruit heart-shaped .... 29. *Veronica*
    - **4.** Calyx and corolla lobes each 5; fruit subacute or rounded at tip, not heart-shaped.
      - **5.** Sterile stamens 2; bractlets absent .... 18. *Lindernia*
      - **5.** Sterile stamens absent or minute; each flower subtended by 2 bractlets.......................... 14. *Gratiola*
  - **3.** Anther-bearing stamens 4.
    - **6.** One elongated sterile stamen present.
      - **7.** Flowers sessile.......................... 9. *Chelone*
      - **7.** Flowers pedicellate.......................... 26. *Penstemon*
    - **6.** Sterile stamen absent or represented only by a gland as broad as long.
      - **8.** Some or all of the leaves pinnatifid or pinnately lobed (excluding *Tomanthera auriculata,* which has 2 small basal lobes).

9. Corolla lavender, to 5 mm long; plants never exceeding 20 cm in height .... 16. *Leucospora*
9. Corolla yellow, at least 15 mm long; plants more than 20 cm tall.
 10. Stamens included within the corolla.
  11. Corolla bilabiate; calyx asymmetrical; capsule flattened .... 25. *Pedicularis*
  11. Corolla not bilabiate; calyx usually symmetrical; capsule globose to ovoid .... 3. *Aureolaria*
 10. Stamens exserted from the corolla .... 12. *Dasistoma*
8. Leaves toothed or entire, not pinnatifid.
 12. Flowers 1–2 in the axils of the cauline leaves.
  13. Leaves entire.
   14. Leaves orbicular .... 4. *Bacopa*
   14. Leaves filiform to linear to elliptic.
    15. Calyx lobes 5; corolla slightly 2-lipped; seeds usually more than 4 per capsule .... 28. *Tomanthera*
    15. Calyx lobes 4; corolla strongly 2-lipped; seeds 2–4 per capsule .... 21. *Melampyrum*
  13. Leaves toothed.
   16. Calyx 5-parted nearly to the base; flowers up to 1 cm long, white with purple lines.................................... 20. *Mecardonia*
   16. Calyx tubular, with 5 lobes; flowers more than 1 cm long, violet or, if less than 1 cm long, yellow .... 22. *Mimulus*
 12. Flowers 3 or more in spikes, racemes, panicles, or corymbs.
  17. Flowers sessile.
   18. Corolla campanulate; leaves usually entire .... 1. *Agalinis*
   18. Corolla salverform; leaves toothed .... 6. *Buchnera*
  17. Flowers pedicellate.
   19. Stems creeping; sterile stamens absent................. 19. *Mazus*
   19. Stems erect; sterile stamen reduced to a gland or scale.
    20. All leaves petiolate; corolla purple or greenish purple; coarse perennials .... 27. *Scrophularia*
    20. Middle and upper leaves sessile; corolla blue and white or violet-purple; annuals .... 10. *Collinsia*
2. At least some of the leaves alternate (excluding bracteal leaves.)
 21. Corolla strongly spurred at base.
  22. Stems glabrous.
   23. Plants trailing or twining; leaves suborbicular, lobed or crenulate........ .... 11. *Cymbalaria*
   23. Plants erect; leaves linear to lanceolate, entire.
    24. Corolla yellow or yellow and orange, 2–4 cm long .... 17. *Linaria*
    24. Corolla blue or violet, up to 1.7 cm long .... 24. *Nuttallanthus*
  22. Stems pubescent.
   25. Stems spreading or prostrate; fruit glabrous; leaves ovate to oblong .. .. 15. *Kickxia*
   25. Stems erect; fruit puberulent; leaves linear to linear-lanceolate .... 8. *Chaenorrhinum*
 21. Corolla not spurred at base (occasionally saccate at base in *Antirrhinum*).
  26. Stamens 5; corolla more or less actinomorphic.................. 29. *Verbascum*
  26. Stamens 2 or 4; corolla zygomorphic.
   27. Leaves and lower part of stems glabrous; leaves entire.
    28. Perennial; leaves oblong to ovate; corolla at least 2 cm long .... 2. *Antirrhinum*
    28. Annual; leaves lanceolate; corolla up to 1.2 cm long .... 23. *Misopates*
   27. Leaves and stems pubescent; leaves toothed or lobed.
    29. Leaves serrate or dentate, never pinnatifid; calyx as long as or shorter than the greenish yellow or bright yellow corolla.
     30. Calyx about as long as the corolla; corolla greenish yellow........ .... 5. *Besseya*

30. Calyx shorter than the corolla; corolla bright yellow .... 13. *Digitalis*

29. Some or all of the leaves pinnatifid or deeply lobed; calyx shorter than the scarlet, red, or yellow corolla.

31. Margins of the lobed leaves toothed; leaves divided into 12 or more lobes .... 25. *Pedicularis*

31. Margins of the lobed leaves entire; leaves divided into 3–7 lobes .... 7. *Castilleja*

1. **Agalinis** Raf.—False Foxglove

1. Plants yellow-green, usually not turning black upon drying; calyx strongly reticulate-nerved; seeds yellow.

2. Branches with solitary flowers; stems terete .... 4. *A. gattingeri*

2. Branches with 2 or more flowers; stems more or less angled ..... 7. *A. skinneriana*

1. Plants deep green or purplish, usually turning black upon drying; calyx weakly reticulate-nerved; seeds brown or black.

3. Pedicels more than twice as long as the calyx.

4. Calyx up to 1 mm long; axillary fascicles of leaves rarely present .... 8. *A. tenuifolia*

4. Calyx 1–2 mm long; axillary fascicles of leaves commonly present .... 2. *A. besseyana*

3. Pedicels less than twice as long as the calyx.

5. Capsules ellipsoid, longer than broad; pedicels 5–8 mm long .... 1. *A. aspera*

5. Capsules globose or subglobose, about as broad as long; pedicels up to 5 mm long.

6. Axillary fascicles of leaves as long as the main leaves; stems scabrous .... 3. *A. fasciculata*

6. Axillary fascicles of leaves much shorter than the main leaves or lacking; stems smooth or nearly so.

7. Calyx up to ½ as long as the capsule; style to 1 cm long; flowers up to 2 cm long .... 5. *A. paupercula*

7. Calyx ½ as long as the capsule or longer; style 1.5–2.0 cm long; flowers usually more than 2 cm long .... 6. *A. purpurea*

1. **Agalinis aspera** (Doug.) Britt. Rough False Foxglove. Aug.–Sept. Prairies, not common; scattered throughout the state.

2. **Agalinis besseyana** Britt. Bessey's False Foxglove. Aug.–Oct. Moist soil; occasional throughout the state. (Including var. *parviflora* Nutt.)

3. **Agalinis fasciculata** (Ell.) Raf. Tufted False Foxglove. Aug.–Sept. Edge of woods, sandy open ground, not common; nearly confined to the s. ⅙ of the state; also Mason Co.

4. **Agalinis gattingeri** (Small) Small. Gattinger's False Foxglove; Rough-stemmed False Foxglove. Aug.–Oct. Dry, often rocky woods; uncommon throughout the state.

5. **Agalinis paupercula** (Gray) Britt. Two varieties occur in Illinois:

a. Summit of flower 1.0–1.5 cm across; styles 8–10 mm long .... 5a. *A. paupercula* var. *paupercula*

a. Summit of flower 0.5–1.0 cm across; styles 6–8 mm long .... 5b. *A. paupercula* var. *borealis*

5a. **Agalinis paupercula** (Gray) Britt. var. **paupercula.** False Foxglove. Aug.–Sept. Moist soil, rare; scattered in IL.

5b. **Agalinis paupercula** (Gray) Britt. var. **borealis** (Pennell) Deam. Northern False Foxglove. Aug.–Sept. Moist soil, rare; confined to the n. ¼ of IL.

6. **Agalinis purpurea** (L.) Pennell. Purple False Foxglove. July–Sept. Moist soil; occasional to common throughout the state.

7. **Agalinis skinneriana** (Wood) Britt. Pale False Foxglove. July–Sept. Dry prairies, rocky woods, calcareous fens; not common but scattered throughout the state.

8. **Agalinis tenuifolia** (Vahl) Raf. Narrow-leaved False Foxglove. Aug.–Oct. Moist soil, calcareous fens, wet meadows; occasional to common throughout the state.

2. **Antirrhinum** L.—Snapdragon

1. **Antirrhinum majus** L. Common Snapdragon. July–Sept. Native to Europe; rarely escaped from cultivation.

3. **Aureolaria** Raf.—False Foxglove

1. Stems glabrous and glaucous..1. *A. flava*
1. Stems pubescent and usually not glaucous.
  2. Corolla 4.5 cm long or longer; pubescence not glandular; capsules rusty-pubescent, 1.2–2.5 cm long...2. *A. grandiflora*
  2. Corolla 3–4 cm long; pubescence glandular; capsules not rusty-pubescent, up to 1.2 cm long...3. *A. pedicularia*

1. **Aureolaria flava** (L.) Farw. var. **macrantha** (Pennell) Fern. Smooth False Foxglove. June–Sept. Rocky woods; occasional in the s. ½ of IL, less common in the n. ½.

2. **Aureolaria grandiflora** (Benth.) Pennell var. **pulchra** Pennell. Yellow False Foxglove. July–Sept. Woods; occasional in the n. ⅔ of IL, rare elsewhere.

3. **Aureolaria pedicularia** L. var. **ambigens** (Fern.) Farw. Clammy False Foxglove. Aug.–Sept. Sandy woods; occasional in the ne. cos.

4. **Bacopa** Aubl.—Water Hyssop

1. **Bacopa rotundifolia** (Michx.) Wettst. Water Hyssop. May–Sept. Wet ground or in shallow water; occasional in the s. ½ of the state, rare or absent elsewhere.

5. **Besseya** Rydb.

1. **Besseya bullii** (Eat.) Rydb. Kitten Tails. May–June. Sandy soil, prairies, not common; primarily along the Illinois and Mississippi rivers in the nw. ¼ of IL.

6. **Buchnera** L.

1. **Buchnera americana** L. Blue Hearts. July–Sept. Moist calcareous prairies, fields, not common; w. cent. IL; also Cook and Pope cos.

7. **Castilleja** Mutis—Paintbrush

1. Bracts scarlet, orange, or yellow; corolla up to 3.5 cm long .... 1. *C. coccinea*
1. Bracts green; corolla 4 cm long or longer...2. *C. sessiliflora*

1. **Castilleja coccinea** (L.) Spreng. Two forms occur in Illinois:

a. Bracts scarlet or orange... 1a. *C. coccinea* f. *coccinea*
a. Bracts yellow ....1b. *C. coccinea* f. *lutescens*

1a. **Castilleja coccinea** (L.) Spreng. f. coccinea. Indian Paintbrush. Apr.–July. Moist calcareous prairies, wet meadows, sandy woods; occasional in the n. ¾ of IL, absent elsewhere.

1b. **Castilleja coccinea** (L.) Spreng. f. lutescens Farw. Indian Paintbrush. Apr.–July. Prairies, sandy woods; scattered in the n. ¾ of IL.

2. **Castilleja sessiliflora** Pursh. Downy Yellow Painted Cup. May–July. Prairies, rare; DuPage, Jo Daviess, Lake, Lee, McHenry, Rock Island, Stephenson, and Winnebago cos.

8. **Chaenorrhinum** Reichenb.

1. **Chaenorrhinum minus** (L.) Lange. Dwarf Snapdragon. June–Sept. Native to Europe; adventive in IL, particularly along railroads.

9. **Chelone** L.—Turtlehead

1. Corolla creamy-white; leaves sessile or on petioles up to 5 mm long ..... 1. *C. glabra*
1. Corolla rose-purple; leaves on petioles 5 mm long or longer................. 2. *C. obliqua*

1. **Chelone glabra** L. Two varieties occur in Illinois:

a. Leaves lanceolate to ovate, most or all of them 2 cm broad or broader.. 1a. *C. glabra* var. *glabra*

a. Leaves linear-lanceolate, most or all of them less than 2 cm broad .... 1b. *C. glabra* var. *linifolia*

1a. **Chelone glabra** L. var. **glabra.** White Turtlehead. July–Oct. Wet soil, bogs, fens; occasional in the n. ½ of IL, rare in the s. ½. (Including f. *tomentosa* [Raf.] Pennell.)

1b. **Chelone glabra** L. var. **linifolia** Coleman. White Turtlehead. July–Oct. Bogs, fens, marshes; scattered throughout IL, except for the southernmost cos. (Including f. *velutina* Pennell & Wherry.)

2. **Chelone obliqua** L. var. **speciosa** Pennell & Wherry. Pink Turtlehead. Aug.–Oct. Low woods, swampy meadows, fens, not common; scattered in IL.

10. **Collinsia** Nutt.

1. Lower lip of corolla blue or violet or purple; corolla 9–16 mm long; leaves broadest at or above the middle; pedicels as long as or longer than the lobes of the calyx.
  2. Lower lip of corolla blue; at least some of the upper leaves clasping; capsules with 2–4 seeds .... 2. *C. verna*
  2. Lower lip of corolla violet or purple; none of the leaves clasping; capsules with 6 or more seeds .... 3. *C. violacea*
1. Lower lip of corolla rose-purple; corolla 10–20 mm long; leaves broadest near the base; pedicels shorter than the lobes of the corolla .... 1. *C. heterophylla*

1. **Collinsia heterophylla** Graham. Chinese Houses. May. Native to the w. U.S.; escaped along a railroad; DeKalb Co.

2. **Collinsia verna** Nutt. Blue-eyed Mary. Apr.–June. Rich woods; occasional throughout the state, but often abundant where found.

3. **Collinsia violacea** Nutt. Violet Collinsia. May–June. Sandy hillside, very rare; Shelby Co.

11. **Cymbalaria** Hill—Kenilworth Ivy

1. **Cymbalaria muralis** Gaertn., Mey., & Scherb. Kenilworth Ivy. June–Oct. Native to Europe and Asia; occasionally escaped from cultivation.

12. **Dasistoma** Raf.—Mullein Foxglove

1. **Dasistoma macrophylla** (Nutt.) Raf. Mullein Foxglove. June–Sept. Rich woods, rocky slopes, dry woods, thickets; occasional throughout the state.

13. **Digitalis** L.- Foxglove

1. **Digitalis grandiflora** Mill. Yellow Foxglove. June–July. Native to Europe and Asia; rarely escaped from gardens; Lake Co.

14. **Gratiola** L.—Hedge Hyssop

1. Corolla yellow; sterile stamens 2 .... 1. *G. aurea*
1. Corolla pale yellow or white; sterile stamens 4.
  2. Pedicels over 10 mm long (in fruit); stems more or less pubescent, not fleshy.
    3. Leaves regularly toothed; stems glandular-pubescent .... 2. *G. neglecta*
    3. Leaves sparingtly toothed or entire; stems glabrous........ 3. *G. quartermaniae*
  2. Pedicels up to 10 mm long (in fruit); stems glabrous, more or less fleshy .... 4. *G. virginiana*

1. **Gratiola aurea** Pursh. Goldenpert. June–Sept. Wet ground, very rare; Cook Co.

2. **Gratiola neglecta** Torr. Clammy Hedge Hyssop. May–Oct. Moist soil, sometimes in standing water; occasional to common throughout the state.

3. **Gratiola quartermaniae** D. Estes. Limestone Hedge Hyssop. June–Sept. Dolomite prairie, very rare; Will Co.

4. **Gratiola virginiana** L. Round-fruited Hedge Hyssop. Apr.–Oct. Moist soil, sometimes in standing water; occasional throughout the state.

15. **Kickxia** Dumort.—Canker-root

1. **Kickxia elatine** (L.) Dumort. Canker-root. May–Oct. Native to Europe; adventive in waste ground; not common but scattered in IL.

16. **Leucospora** Nutt.

1. **Leucospora multifida** (Michx.) Nutt. May–Oct. Moist soil, fields, prairies; occasional in the s. 4/5 of IL, rare or absent elsewhere. *Conobea multifida* Michx.

17. **Linaria** Mill.—Toadflax

1. Leaves ovate to oblong; corolla 3.5–4.0 cm long, yellow .... 1. *L. genistifolia*
1. Leaves linear to linear-lanceolate; corolla 2–3 cm long, yellow and orange .... 2. *L. vulgaris*

1. **Linaria genistifolia** (L.) Mill. ssp. **dalmatica** (L.) Maire & Petitmengin. Toadflax. June–Sept. Native to Europe; rarely escaped from cultivation in the n. 1/4 of IL. *Linaria dalmatica* L.

2. **Linaria vulgaris** Mill. Butter-and-eggs. May–Nov. Native to Europe; commonly escaped from cultivation throughout the state.

18. **Lindernia** All.—False Pimpernel

1. Lower or all pedicels shorter than the subtending leaves .... 2. *L. dubia*
1. All pedicels longer than the subtending leaves .... 1. *L. anagallidea*

1. **Lindernia anagallidea** (Michx.) Pennell. Slender False Pimpernel. June–Oct. Moist soil, sometimes in standing water; occasional throughout the state, except for the nw. cos., where it is rare or absent.

2. **Lindernia dubia** (L.) Pennell. Two varieties occur in Illinois:

a. All pedicels shorter than the subtending leaves .... 2a. *L. dubia* var. *dubia*
a. Only the lowermost pedicels shorter than the subtending leaves.............................. .... 2b. *L. dubia* var. *riparia*

2a. **Lindernia dubia** (L.) Pennell var. **dubia**. False Pimpernel. June–Oct. Moist ground, sometimes in standing water; common throughout the state.

2b. **Lindernia dubia** (L.) Pennell var. **riparia** (Raf.) Fern. False Pimpernel. June–Oct. Moist ground; common throughout the state.

19. **Mazus** Lour.—Mazus

1. **Mazus pumilus** (Burm. f.) Steenis. Mazus. June–Oct. Native to Asia; adventive in lawns; Alexander, Jackson, Lake, and Cook cos. *Mazus japonicus* (Thunb.) Kuntze.

20. **Mecardonia** Ruiz & Pavon—Mecardonia

1. **Mecardonia acuminata** (Walt.) Small. Mecardonia. July–Aug. Native to the s. U.S.; wet roadside ditch; rarely adventive in IL; Wabash Co.

21. **Melampyrum** L.—Cow-wheat

1. **Melampyrum lineare** Desr. var. **latifolium** Bart. Cow-wheat. June–Aug. Bogs, marshes, dunes, very rare; Cook Co.

22. **Mimulus** L.—Monkey-flower

1. Corolla purple, at least 2 cm long; leaves lanceolate to elliptic to ovate, serrate.
  2. Leaves petiolate; stems slightly winged .... 1. *M. alatus*
  2. Leaves sessile or even clasping; stems wingless .... 3. *M. ringens*
1. Corolla yellow, up to 1.2 cm long; leaves oval to suborbicular, entire or denticulate .... 2. *M. glabratus*

1. **Mimulus alatus** Sol. Winged Monkey-flower. June–Sept. Wet ground; common in the s. 1/2 of IL, becoming less common in the n. 1/2. An unnamed hybrid between this species and *M. ringens* has been collected in Cass Co.

2. **Mimulus glabratus** HBK. var. **fremontii** (Benth.) A. L. Grant. Yellow Monkey-flower. June–Oct. Around streams, springs, fens, rare; confined to the n. 1/2 of IL.

3. **Mimulus ringens** L. Two varieties occur in Illinois:

a. Most or all the cauline leaves rounded or clasping at base .... 3a. *M. ringens* var. *ringens*
a. Most or all the cauline leaves tapering to base .... 3b. *M. ringens* var. *minthodes*

3a. **Mimulus ringens** L. var. **ringens.** Sessile Monkey-flower. June–Sept. Wet ground; occasional in the n. ¾ of IL, rare elsewhere.

3b. **Mimulus ringens** L. var. **minthodes** (Greene) A. L. Grant. Sessile Monkey-flower. June–Sept. Wet ground; occasional in the westernmost cos. of the state.

23. **Misopates** Raf.

1. **Misopates orontium** (L.) Raf. Lesser Snapdragon. Sept. Native to Europe; escaped into a cultivated oat field; DuPage Co.

24. **Nuttallanthus** D. Sutton

1. Flowers 5–10 mm long; seeds smooth .... 1. *N. canadensis*
1. Flowers 10–17 mm long; seeds rugulose .... 2. *N. texanus*

1. **Nuttallanthus canadensis** (L.) D. Sutton. Blue Toadflax. Apr.–Sept. Sandy soil; occasional in the n. ½ of IL, uncommon in the s. ½. *Linaria canadensis* (L.) Dum.-Cours.

2. **Nuttallanthus texanus** (Scheele) D. Sutton. Southern Blue Toadflax. Apr.–Sept. Sandy soil, very rare; Alexander Co. *Linaria texana* Scheele.

25. **Pedicularis** L.—Lousewort

1. Most of the cauline leaves alternate, petiolate; corolla yellow; flowers appearing from Apr. to June .... 1. *P. canadensis*
1. Most of the cauline leaves opposite, sessile; corolla creamy or yellowish white; flowers appearing from Aug. to Oct. .... 2. *P. lanceolata*

1. **Pedicularis canadensis** L. Two varieties occur in Illinois:

a. Flowering stems clustered; stolons absent .... 1a. *P. canadensis* var. *canadensis*
a. Flowering stem solitary; basal offshoots stoloniferous .... 1b. *P. canadensis* var. *dobbsii*

1a. **Pedicularis canadensis** L. var. **canadensis.** Lousewort. Apr.–June. Woods, prairies; occasional throughout the state.

1b. **Pedicularis canadensis** L. var. **dobbsii** Fern. Lousewort. Apr.–June. Woods, prairies, not common; scattered in IL.

2. **Pedicularis lanceolata** Michx. Swamp Wood Betony. Aug.–Oct. Calcareous fens, wet meadows, bogs, swamps; occasional in the n. ½ of the state, uncommon in the s. ½.

26. **Penstemon** Mitchell—Beardstongue

1. Upper and bracteal leaves clasping; corolla more than 3.2 cm long .... 10. *P. grandiflorus*
1. All leaves sessile to petiolate; corolla less than 3.2 cm long (except in *P. cobaea*).
  2. Corolla more than 3.2 cm long .... 6. *P. cobaea*
  2. Corolla less than 3.2 cm long.
    3. Corolla glandular-pubescent within .... 13. *P. tubaeflorus*
    3. Corolla eglandular within.
      4. Anterior lobes of the corolla equaling or barely exceeding the posterior lobe; sterile filament slightly to moderately bearded.
        5. Anthers glabrous; corolla more or less purple .... 4. *P. calycosus*
        5. Anthers with short, stiff hairs; corolla white or slightly tinged or lined with purple.
          6. Corolla 23–30 mm long; stem glabrous, shiny and slightly glaucous .... 8. *P. digitalis*
          6. Corolla 16–23 mm long; stem dull, finely pubescent to glabrous, not shiny.
            7. Sepals with a scarious margin, 3–4 mm long in fruit; lower cauline leaves rounded at the tip, entire or denticulate .... 7. *P. deamii*
            7. Sepals with a very narrow scarious margin, 5–9 mm long in fruit; lower cauline leaves acuminate, sharply serrate .... 1. *P. alluviorum*

4. Anterior lobes of the corolla much exceeding the posterior lobe; sterile filament densely bearded (moderately bearded in *P. pallidus*).
  8. Stems pubescent with glandular hairs; anterior lobes of the corolla arching upward to the posterior lobe ....11. *P. hirsutus*
  8. Stems pubescent with eglandular hairs; anterior lobes of the corolla not arching upward.
    9. Cauline leaves ovate-lanceolate to ovate...................5. *P. canescens*
    9. Cauline leaves linear-lanceolate to oblanceolate.
      10. Sepals from ½ to nearly equaling the length of the capsule........ ....9. *P. gracilis*
      10. Sepals less than ½ the length of the capsule.
        11. Corolla unlined, its anterior lobes upcurved against the posterior lobe so that the orifice to the throat is closed ....11. *P. hirsutus*
        11. Corolla lined with deeper color, the anterior lobes of the corolla projecting beyond the posterior lobe.
          12. Leaves glabrous or nearly so; throat of the corolla equaling the length of the tube or scarcely longer..................... ....2. *P. arkansanus*
          12. Leaves pubescent; throat of the corolla much longer than the tube.
            13. Corolla purplish; sterile filament strongly bearded...... ....3. *P. brevisepalus*
            13. Corolla white, lined with purple; sterile filament moderately bearded .... 12. *P. pallidus*

1. **Penstemon alluviorum** Pennell. Lowland Beardstongue. May–June. Swampy woods, not common; confined to the s. tip of the state; also St. Clair Co.

2. **Penstemon arkansanus** Pennell. Ozark Beardstongue. Apr.–June. Wooded slopes, rare; Jackson, Perry, and Randolph cos.

3. **Penstemon brevisepalus** Pennell. Short-sepaled Beardstongue. May–June. Dry, usually rocky woods, rare; Pope and Union cos.

4. **Penstemon calycosus** Small. Smooth Beardstongue. May–July. Wooded slopes, edge of woods; occasional throughout the state.

5. **Penstemon canescens** Britt. f. **brittonorum** (Pennell) Fern. Ashy Beardstongue. Woods, very rare; Franklin Co.

6. **Penstemon cobaea** Nutt. Large Beardstongue. June. Native s. and w. of IL; adventive in a flat, dry meadow; Kane Co.

7. **Penstemon deamii** Pennell. Deam's Beardstongue. May–June. Woods, not common; confined to the s. ¼ of IL.

8. **Penstemon digitalis** Nutt. Foxglove Beardstongue. May–July. Woods, thickets; occasional to common throughout the state.

9. **Penstemon gracilis** Nutt. var. **wisconsinensis** (Pennell) Bennett. Slender Beardstongue. June–July. Native to the n. of IL; adventive in Kane Co.

10. **Penstemon grandiflorus** Nutt. Large-flowered Beardstongue. May–June. Sandy soil, rare; apparently native in Henderson, Whiteside, and Winnebago cos.; adventive in McHenry Co.

11. **Penstemon hirsutus** (L.) Willd. Hairy Beardstongue. June–July. Gravelly prairies, wooded slopes; occasional throughout the state.

12. **Penstemon pallidus** Small. Pale Beardstongue. Apr.–July. Dry woods, prairies; common in the s. ½ of IL, occasional in the n. ½.

13. **Penstemon tubaeflorus** Nutt. Beardstongue. May–June. Prairies, dry woods; not common in the s. ½ of IL, rare in the n. ½.

27. **Scrophularia** L.—Figwort

1. Petioles winged; sterile stamen greenish yellow; flowering from May to June......... ....1. *S. lanceolata*
1. Petioles unwinged; sterile stamen purple or brown; flowering from July to Oct. ....2. *S. marilandica*

1. **Scrophularia lanceolata** Pursh. Early Figwort. May–June. Woods, fence rows; occasional in the n. ½ of the state, otherwise very rare.

2. **Scrophularia marilandica** L. Late Figwort. July–Oct. Usually rich woods; occasional throughout the state. (Including f. *neglecta* [Rydb.] Pennell.)

28. **Tomanthera** Raf.

1. **Tomanthera auriculata** (Michx.) Raf. Auriculate False Foxglove. Aug.–Sept. Dry prairies; occasional in the n. ¾ of IL, absent elsewhere.

29. **Verbascum** L.—Mullein

1. Plants woolly-tomentose, not glandular.
  2. Flowers in panicles or racemes.
    3. Flowers in panicles; leaves not cordate at base .... 4. *V. speciosum*
    3. Flowers in racemes; leaves cordate at base.................................... 2. *V. nigrum*
  2. Flowers in spikes.
    4. Flowers up to 2.0 (–2.5) cm across; spike crowded; leaves usually not clasping .... 5. *V. thapsus*
    4. Flowers over 2.5 cm across; spike interrupted; some of the leaves usually clasping .... 3. *V. phlomoides*
1. Plants glabrous or glandular-pubescent.
  5. Pedicels 1 cm long or longer, longer than the capsule....................... 1. *V. blattaria*
  5. Pedicels up to 5 mm long, shorter than the capsule......................... 6. *V. virgatum*

1. **Verbascum blattaria** L. Moth Mullein. May–Sept. Native to Europe; often adventive in disturbed soil; scattered throughout the state. Reddish-flowered specimens may be called f. *erubescens* Brugger; white-flowered specimens may be called f. *albiflora* (Don) House.

2. **Verbascum nigrum** L. Black Mullein. June–July. Native to Europe; rarely escaped from cultivation; Champaign Co. Not collected since 1893.

3. **Verbascum phlomoides** L. Clasping Mullein. June–July. Native to Europe; rarely adventive in IL; Cook, DuPage, Franklin, and Macon cos.

4. **Verbascum speciosum** Schrad. Mullein. July–Sept. Native to Europe; rarely adventive in disturbed areas; Lake Co.

5. **Verbascum thapsus** L. Woolly Mullein. May–Sept. Native to Europe; commonly adventive in disturbed areas; in every co.

6. **Verbascum virgatum** Stokes. Mullein. June–July. Native to Europe; rarely adventive in IL; Clark, Coles, and Pulaski cos.

30. **Veronica** L.—Speedwell

1. Flower solitary in the axils of the cauline leaves.
  2. Corolla white; leaves glabrous .... 11. *V. peregrina*
  2. Corolla blue, lilac, or, if white, then lined with blue; leaves pubescent.
    3. Corolla 6–12 mm across; capsule (5–) 6–10 mm long.
      4. Leaves sharply toothed; corolla 8–12 mm across; capsule 6 mm wide or wider .... 12. *V. persica*
      4. Leaves crenately toothed; corolla 6–8 mm across; capsule about 5 mm wide... 4. *V. biloba*
    3. Corolla up to 5 mm across; capsule less than 5 mm long.
      5. Leaves reniform-suborbicular, 3- to 7-lobed, broader than long; capsule not compressed.. 8. *V. hederaefolia*
      5. Leaves ovate to oval or the upper ones linear to lanceolate, toothed or incised-pinnatifid, longer than broad; capsule compressed.
        6. Pedicels 4–12 mm long; seeds rugulose................................ 13. *V. polita*
        6. Pedicels 1–2 mm long; seeds smooth.
          7. Leaves toothed .... 3. *V. arvensis*
          7. Leaves incised-pinnatifid .... 7. *V. dillenii*
1. Flowers in racemes or spikelike inflorescences (rarely solitary in *V. serpyllifolia*).
  8. Stems glabrous or with upcurved hairs less than 0.5 mm long.
    9. Racemes terminal; leaves up to 1.5 cm long; stems with upcurved hairs less than 0.5 mm long .... 15. *V. serpyllifolia*

9. Racemes axillary; leaves, or most of them, more than 1.5 cm long; stems glabrous.
  10. Leaves linear to linear-lanceolate; axis of raceme zigzag .... 14. *V. scutellata*
  10. Leaves lanceolate to oblong to oval; axis of raceme straight.
    11. Leaves petiolate; pedicels eglandular .... 1. *V. americana*
    11. Leaves sessile, sometimes clasping; pedicels glandular-pubescent.
      12. Fruiting pedicels ascending; corolla bluish, 4–5 mm across; capsule barely notched .... 2. *V. anagallis-aquatica*
      12. Some of the fruiting pedicels horizontally spreading; corolla white to pale rose, 3–4 mm across; capsule deeply notched .... 6. *V. comosa*

8. Stems pubescent with hairs longer than 0.5 mm.
  13. Racemes terminal; leaves acuminate .... 9. *V. longifolia*
  13. Racemes axillary; leaves obtuse to subacute.
    14. Leaves petiolate; racemes densely flowered; pedicels shorter than the calyx .... 10. *V. officinalis*
    14. Leaves sessile or very short-petiolate; racemes loosely flowered; pedicels about as long as the calyx or longer........................... 5. *V. chamaedrys*

1. **Veronica americana** (Raf.) Schwein. American Brooklime. June–Aug. Swampy ground, rare; confined to the n. ½ of the state.

2. **Veronica anagallis-aquatica** L. Water Speedwell. June–Aug. Shallow springwater, very rare; Bureau and Kane cos.

3. **Veronica arvensis** L. Corn Speedwell. Mar.–Aug. Native to Europe; fields, woods, disturbed areas, common; in every co.

4. **Veronica biloba** L. Speedwell. May–June. Native to Europe and Asia; rarely adventive in disturbed areas. *Veronica agrestis* L., misapplied

5. **Veronica chamaedrys** L. Germander Speedwell. May–July. Native to Europe; rarely adventive in IL; Cook, DuPage, Kane, and Lake cos.

6. **Veronica comosa** Richter. Water Speedwell. June–Sept. Wet ditches, shores, springs; occasional in the n. ½ of the state, rare or absent elsewhere. *Veronica catenata* Pennell.

7. **Veronica dillenii** Crantz. Speedwell. May–June. Native to Europe; rarely adventive in disturbed soil; Jo Daviess Co.

8. **Veronica hederaefolia** L. Ivy-leaved Speedwell. Mar.–Apr. Native to Europe; adventive in grassy meadows; Jackson and Union cos.

9. **Veronica longifolia** L. Garden Speedwell. May–Sept. Native to Europe; rarely escaped from cultivation; Cook, DuPage, Lake, and Winnebago cos. *Pseudolysimachion longifolium* (L.) Opiz.

10. **Veronica officinalis** L. Common Speedwell. May–Sept. Native to Europe; adventive in grassy areas and open woods; Cook, DuPage, Peoria, Richland, Tazewell, and Will cos.

11. **Veronica peregrina** L. Two varieties occur in Illinois:

a. Stems and capsules glabrous .... 11a. *V. peregrina* var. *peregrina*
a. Stems and capsules glandular-pubescent................ 11b. *V. peregrina* var. *xalapensis*

11a. **Veronica peregrina** L. var. **peregrina.** White Speedwell. Apr.–Aug. Fields, disturbed and usually moist soil, common; in every co.

11b. **Veronica peregrina** L. var. **xalapensis** (HBK.) St. John. White Speedwell. Apr.–Aug. Fields, disturbed soil; occasional but not as common as the preceding. *Veronica peregrina* L. ssp. *xalapensis* (HBK.) Pennell.

12. **Veronica persica** Poir. Bird's-eye Speedwell. Apr.–Aug. Native to Europe and Asia; occasionally naturalized and scattered in IL.

13. **Veronica polita** Fries. Speedwell. Mar.–May. Native to Europe; adventive in lawns; scattered in IL. *Veronica didyma* Ten.

14. **Veronica scutellata** L. Marsh Speedwell. Apr.–July. Moist soil, often around ponds, bogs, rare; confined to the n. ½ of the state.

15. **Veronica serpyllifolia** L. Thyme-leaved Speedwell. Apr.–July. Native to Europe; adventive in lawns and meadows; scattered throughout the state.

31. **Veronicastrum** Fab.—Culver's-root

1. **Veronicastrum virginicum** (L.) Farw. Two forms occur in Illinois:

a. Stems glabrous or short-pubescent .... 1a. *V. virginicum* f. *virginicum*
a. Stems densely villous .... 1b. *V. virginicum* f. *villosum*

1a. **Veronicastrum virginicum** (L.) Farw. f. **virginicum**. Culver's-root; Candelabra Plant. June–Sept. Woods, prairies, wet meadows; occasional to common throughout the state.

1b. **Veronicastrum virginicum** (L.) Farw. f. **villosum** (Raf.) Pennell. Culver's-root; Candelabra Plant. June–Sept. Woods, prairies; occasional throughout IL.

146. SIMAROUBACEAE—SIMAROUBA FAMILY

1. **Ailanthus** Desf.—Tree-of-heaven

1. **Ailanthus altissima** (Mill.) Swingle. Tree-of-heaven; Ailanthus. June–July. Native to Asia; frequently escaped in woods, fields, and cities; occasional to common throughout most of IL.

147. SOLANACEAE—NIGHTSHADE FAMILY

1. Plants woody, at least at the base, often climbing or trailing.
2. At least some of the leaves with lobes near the base; flowers in terminal or axillary cymes .... 9. *Solanum*
2. None of the leaves lobed near the base; flower solitary in the axils of the leaves .... 3. *Lycium*
1. Plants herbaceous throughout, not climbing nor trailing.
3. Leaves compound (some deeply lobed species of *Solanum* may be sought here).
4. Stems prickly .... 9. *Solanum*
4. Stems glabrous, appressed-pubescent, or pilose.
5. Flowers yellow; stems pilose .... 4. *Lycopersicon*
5. Flowers blue, purple, or white; stems glabrous or appressed-pubescent .... 9. *Solanum*
3. Leaves simple although sometimes deeply lobed.
6. Stems and sometimes the leaves prickly .... 9. *Solanum*
6. Stems and leaves without prickles.
7. All leaves sessile with the uppermost clasping .... 2. *Hyoscyamus*
7. Some or all the leaves petiolate, the uppermost not clasping (although the uppermost broadly rounded and sessile in some species of *Petunia* and *Nicotiana*).
8. Leaves lobed at least halfway to midvein .... 9. *Solanum*
8. Leaves entire or toothed, never lobed halfway to midvein.
9. Corolla pale blue, campanulate .... 5. *Nicandra*
9. Corolla white, yellow, purple, or pink (rarely pale blue in some forms of *Petunia*), rotate or funnelform.
10. Stamens exserted .... 9. *Solanum*
10. Stamens included.
11. Lobes of corolla with a slender, tail-like projection at the tip; fruits prickly .... 1. *Datura*
11. Lobes of corolla rounded or acute, never with a tail-like projection at the tip; fruits not prickly.
12. Corolla rotate; fruit a berry .... 8. *Physalis*
12. Corolla funnelform; fruit a capsule.
13. Flowers some shade of yellow; stamens all the same size .... 6. *Nicotiana*
13. Flowers usually some color other than a shade of yellow; one of the stamens considerably smaller than the others .... 7. *Petunia*

1. **Datura** L.—Jimsonweed

1. Corolla up to 10 cm long; calyx 3–5 cm long; leaves coarsely toothed with several teeth .... 2. *D. stramonium*

1. Corolla 12–20 cm long; calyx 6–15 cm long; leaves entire or undulate or with a few coarse teeth.
  2. Stems and leaves gray-pubescent; leaves entire or undulate; rim of corolla 5-toothed .... 3. *D. wrightii*
  2. Stems and leaves glabrous or nearly so; leaves witih a few coarse teeth; rim of corolla 10-toothed .... 1. *D. inoxia*

1. **Datura inoxia** Mill. Mexican Jimsonweed; Prickly Bur. Aug.–Oct. Native to Mexico; disturbed areas; Cook Co.

2. **Datura stramonium** L. Two varieties occur in Illinois:

a. Stems green; corolla white .... 2a. *D. stramonium* var. *stramonium*
a. Stems purple; corolla purplish .... 2b. *D. stramonium* var. *tatula*

2a. **Datura stramonium** L. var. **stramonium.** Jimsonweed. July–Oct. Native to Asia; naturalized in barnyards, fields, disturbed areas; in every co.

2b. **Datura stramonium** L. var. **tatula** (L.) Torr. Purple Jimsonweed. July–Oct. Native to Asia; naturalized in disturbed areas; occasional throughout the state.

3. **Datura wrightii** Regel. Hairy Jimsonweed; Sacred Datura. Aug.–Oct. Native to tropical America; naturalized in disturbed areas; scattered throughout the state.

2. **Hyoscyamus** L.—Henbane

1. **Hyoscyamus niger** L. Black Henbane. June–Aug. Native to Europe; rarely escaped in IL; McHenry and Peoria cos.

3. **Lycium** L.—Matrimony Vine

1. Leaves gray-green; lobes of calyx obtuse .... 1. *L. barbarum*
1. Leaves dark green; lobes of calyx acute .... 2. *L. chinense*

1. **Lycium barbarum** L. Common Matrimony Vine. June–Sept. Native to Europe; occasionally escaped from cultivation into disturbed areas; scattered throughout the state.

2. **Lycium chinense** Mill. Chinese Matrimony Vine. June–Sept. Native to Asia; rarely adventive in e. N. America; Champaign, Cook, Macon, and Mason cos.

4. **Lycopersicon** L.—Tomato

1. **Lycopersicon esculentum** Mill. Tomato. May–Sept. Native to tropical America; occasionally escaped from cultivation but rarely persisting.

5. **Nicandra** Adans.—Apple-of-Peru

1. **Nicandra physalodes** (L.) Gaertn. Apple-of-Peru. July–Sept. Native to Peru; naturalized in disturbed areas; scattered in IL.

6. **Nicotiana** L.—Tobacco

1. Cauline leaves lanceolate to lance-ovate, sessile; inflorescence racemose; calyx lobes subulate, about as long as the calyx tube; corolla pale yellow, the tube up to 12 cm long .... 1. *N. longiflora*
1. Cauline leaves broadly ovate, petiolate; inflorescence paniculate; calyx lobes deltoid, about ¼ as long as the calyx tube; corolla greenish yellow, the tube up to 5 cm long .... 2. *N. rustica*

1. **Nicotiana longiflora** Cav. Long-flowered Tobacco. July–Sept. Native to S. America; occasionally planted but rarely escaped into disturbed areas; Cook and St. Clair cos.

2. **Nicotiana rustica** L. Wild Tobacco. Sept.–Oct. Native to Peru; adventive in disturbed areas; Cook, Menard, and Peoria cos.

7. **Petunia** Juss.—Petunia

1. Corolla white, salverform .... 1. *P. axillaris*
1. Corolla red, violet, or purple, funnelform.
  2. Corolla 5–9 cm long .... 2. *P. X hybrida*
  2. Corolla 3–4 cm long .... 3. *P. violacea*

1. **Petunia axillaris** (Lam.) BSP. White Petunia. July–Sept. Native to S. America; planted in gardens but rarely escaped and persistent from cultivation.

2. **Petunia X hybrida** Vilm. Garden Petunia. June–Aug. Horticultural plant; rarely escaped and persistent from cultivation.

3. **Petunia violacea** Lindl. Violet Petunia. June–Aug. Native to S. America; planted in gardens but infrequently escaped and persistent from cultivation. *Petunia X atkinsiana* D. Don ex Loud.

8. **Physalis** L.—Ground Cherry

1. Flowers white; calyx during fruiting red or scarlet .... 1. *P. alkekengi*
1. Flowers yellow; calyx during fruiting green or brownish.
 2. Stems glabrous or nearly so, never uniformly villous, hispid, hirsute, or glandular-hairy.
  3. Annuals without rhizomes; leaves mostly conspicuously dentate.
   4. Corolla darkened toward the center; peduncles up to 6 mm long; calyx often puberulent, with deltoid teeth; berries purple ..... 10. *P. philadelphica*
   4. Corolla without a darkened center; peduncles 10 mm long or longer; calyx usually glabrous, with acute to attenuate teeth; berries yellow.
    5. Calyx in fruit strongly 10-angled; peduncles 1–2 cm long .... 2. *P. angulata*
    5. Calyx in fruit obscurely 10-angled; peduncles nearly all over 2 cm long .... 9. *P. pendula*
  3. Perennials with rhizomes; leaves entire or merely repand-dentate.
   6. Leaves lanceolate to linear; berries yellow .... 7. *P. longifolia*
   6. Leaves ovate to ovate-oblong; berries red or purple.
    7. Leaves translucent; fruiting calyx pyramidal, to 6 cm long .... 8. *P. macrophysa*
    7. Leaves opaque; fruiting calyx ovoid, to 3 cm long.
     8. Leaves tapering to the symmetrical base, up to 6 cm long; peduncle up to 2 cm long; fruiting calyx sunken at base .... 13. *P. subglabrata*
     8. Leaves rounded at the asymmetrical base, up to 4 cm long; peduncle up to 1 cm long; fruiting calyx not sunken at base .... 14. *P. texana*
 2. Stems uniformly villous, hispid, hirsute, or glandular-hairy.
  9. Pubescence of stem glandular-viscid.
   10. Annuals without rhizomes; fruiting calyx abruptly acuminate.
    11. Fruiting calyx long-tapering, 3–4 cm long; reticulations between the lateral nerves obscure .... 3. *P. barbadensis*
    11. Fruiting calyx short-tapering, 2–3 cm long; reticulations between the lateral nerves prominent .... 4. *P. grisea*
   10. Perennials with rhizomes; fruiting calyx merely acute.
    12. Leaves more or less tapering to the base, most of them less than 6 cm long; berries red .... 15. *P. virginiana*
    12. Leaves more or less rounded or cordate at the base, many of them more than 6 cm long; berries yellow .... 5. *P. heterophylla*
  9. Pubescence of stem not glandular-viscid.
   13. Peduncles during flowering up to 5 mm long; annuals without rhizomes .... 11. *P. pubescens*
   13. Peduncles during flowering 1 cm long or longer; perennials with rhizomes.
    14. Stems hispid or hirsute; calyx in fruit not sunken at the base; leaves lanceolate to oblanceolate .... 6. *P. lanceolata*
    14. Stems villous; calyx in fruit shallowly or deeply sunken at the base; leaves mostly oblong to ovate-lanceolate to ovate.
     15. Leaves more or less rounded or cordate at the base .... 5. *P. heterophylla*
     15. Leaves more or less tapering to the base.
      16. Some or all of the hairs branched; calyx in fruit mostly 4 cm long or longer, shallowly sunken at base .... 12. *P. pumila*

16. All the hairs unbranched; calyx in fruit mostly less than 4 cm long, deeply sunken at base .... 15. *P. virginiana*

1. **Physalis alkekengi** L. Chinese Lantern. July–Sept. Native to Asia; cultivated in gardens but rarely escaped and seldom persisting.

2. **Physalis angulata** L. Ground Cherry. July–Sept. Native to the s. U.S.; adventive in disturbed areas; scattered in IL.

3. **Physalis barbadensis** Jacq. Ground Cherry. June–Sept. Native to tropical America; adventive in disturbed areas; Jackson Co.

4. **Physalis grisea** (Waterf.) M. Martinez. Ground Cherry. June–Oct. Disturbed habitats; scattered throughout the state. *Physalis pruinosa* L., misapplied.

5. **Physalis heterophylla** Nees. Three varieties occur in Illinois:

a. At least some of the hairs on the stem glandular .... 5a. *P. heterophylla* var. *heterophylla*
a. None of the hairs on the stem glandular.
  b. Leaves thick, dentate .... 5b. *P. heterophylla* var. *ambigua*
  b. Leaves thin, entire... 5c. *P. heterophylla* var. *nyctaginea*

5a. **Physalis heterophylla** Nees var. **heterophylla.** Ground Cherry. May–Sept. Woods, disturbed areas; occasional throughout the state.

5b. **Physalis heterophylla** Nees var. **ambigua** (Gray) Rydb. Ground Cherry. May–Sept. Disturbed areas; occasional throughout the state.

5c. **Physalis heterophylla** Nees var. **nyctaginea** (Dunal) Rydb. Ground Cherry. May–Sept. Disturbed areas; confined to the s. ½ of IL.

6. **Physalis lanceolata** Michx. Narrow-leaved Ground Cherry. June–Aug. Native to the w. U.S.; adventive in dry waste ground; scattered in IL, but not common. *Physalis hispida* (Waterf.) Cronq.

7. **Physalis longifolia** Nutt. Ground Cherry. June–Aug. Native to the w. U.S. and Mexico; adventive in disturbed areas; scattered in IL.

8. **Physalis macrophysa** Rydb. Large-fruited Ground Cherry. June–July. Low woods, rare; Champaign and Peoria cos.

9. **Physalis pendula** Rydb. Ground Cherry. July–Sept. Native to the sw. U.S.; occasionally adventive in disturbed areas; scattered in the s. cos., rare elsewhere.

10. **Physalis philadelphica** Lam. Tomatillo. July–Aug. Native to the sw. U.S. and Mexico; adventive in disturbed areas; scattered in IL. *Physalis ixocarpa* Hornem.

11. **Physalis pubescens** L. Annual Ground Cherry. June–Oct. Disturbed areas; occasional throughout the state.

12. **Physalis pumila** Nutt. Dwarf Ground Cherry. June–Aug. Dry hillside, very rare; Henry, McHenry, and Peoria cos.

13. **Physalis subglabrata** Mack. & Bush. Smooth Ground Cherry. June–Sept. Pastures, fields, disturbed woods; occasional throughout the state.

14. **Physalis texana** Rydb. Texas Ground Cherry. June–July. Sandy river banks, rare; St. Clair Co.

15. **Physalis virginiana** Mill. Ground Cherry. May–Aug. Woods, disturbed areas; occasional throughout the state.

9. **Solanum** L.—Nightshade

1. Plants prickly.
  2. Leaves deeply lobed, usually nearly to the midvein; calyx prickly.
    3. Flowers yellow; stem pubescence stellate................................ 8. *S. rostratum*
    3. Flowers violet to purple; stem pubescence of simple hairs .... 5. *S. heterodoxum*
  2. Leaves entire to wavy-toothed to coarsely toothed, never lobed nearly to the midvein; calyx without prickles.
    4. Stems hirsute, the stellate hairs sessile .... 1. *S. carolinense*
    4. Stems with short tomentum, the stellate hairs stipitate.
      5. Leaves linear-lanceolate to narrowly oblong, silvery gray; calyx lobes linear .... 4. *S. elaeagnifolium*
      5. Leaves ovate, green; calyx lobes ovate............................. 2. *S. dimidiatum*
1. Plants glabrous, pilose, puberulent, or stellate-pubescent, but not prickly.

6. Plants woody at base, climbing; berry red .... 3. *S. dulcamara*
6. Plants herbaceous, not climbing; berry green or black.
7. Leaves compound .... 10. *S. tuberosum*
7. Leaves simple, although sometimes deeply lobed.
8. Leaves deeply lobed; berry green... 9. *S. triflorum*
8. Leaves entire or shallowly toothed; berry black.
9. Plants without viscid glandular hairs; calyx scarcely enlarging in fruit .... 7. *S. ptychanthum*
9. Plants with viscid glandular hairs; calyx enlarging in fruit .... 6. *P. physalifolium*

1. **Solanum carolinense** L. Horse-nettle. June–Oct. Open woods and disturbed areas; in every co. White-flowered specimens may be known as f. *albiflorum* (Kuntze) Benke.

2. **Solanum dimidiatum** Raf. Torrey's Horse-nettle. July–Sept. Native s. and w. of IL; adventive in disturbed areas; Greene, Grundy, and Henry cos.

3. **Solanum dulcamara** L. Two varieties occur in Illinois:

a. Stems and leaves glabrous or short-pubescent ...... 3a. *S. dulcamara* var. *dulcamara*
a. Stems and leaves densely villous .... 3b. *S. dulcamara* var. *villosissimum*

3a. **Solanum dulcamara** L. var. **dulcamara.** Bittersweet Nightshade. June–Oct. Native to Europe; naturalized in thickets, marshes, low woods, along fences; common in the n. ½ of IL, occasional in the s. ½. Specimens with white flowers may be known as f. *albiflorum* House.

3b. **Solanum dulcamara** L. var. **villosissimum** Desv. Hairy Bittersweet Nightshade. June–Oct. Native to Europe; naturalized in thickets and along fences; scattered in IL.

4. **Solanum elaeagnifolium** Cav. Silvery Horse-nettle. July–Oct. Native to the w. U.S. and Mexico; adventive in disturbed areas, particularly along railroads; Adams, Cook, Crawford, and St. Clair cos.

5. **Solanum heterodoxum** Dunal var. **novomexicanum** Bartl. Nightshade. July–Oct. Native to the sw. U.S.; rarely adventive in IL; Crawford Co.

6. **Solanum physalifolium** Rusby. Hairy Nightshade. Aug.–Sept. Native to S. America; rarely adventive in disturbed soil; DeKalb, DuPage, and St. Clair cos. *Solanum sarachoides* Sendtn.

7. **Solanum ptychanthum** Dunal. Black Nightshade. May–Nov. Woods, along streams, fields, roadsides, along railroads; probably in every co. *Solanum americanum* Mill.

8. **Solanum rostratum** Dunal. Buffalo-bur. July–Oct. Native to the w. U.S.; naturalized in disturbed areas; occasional throughout the state. *Solanum cornutum* Lam.

9. **Solanum triflorum** Nutt. Cut-leaved Nightshade. June–Aug. Native to the w. U.S.; adventive in disturbed soil; occasional in the n. cos.

10. **Solanum tuberosum** L. Potato. June–Aug. Native to S. America; commonly planted but rarely persisting as an adventive.

148. STAPHYLEACEAE—BLADDERNUT FAMILY

1. **Staphylea** L.—Bladdernut

1. **Staphylea trifolia** L. Bladdernut. Apr.–May. Moist woods, common; in every co.

149. STERCULIACEAE—CHOCOLATE FAMILY

1. **Melochia** L.

1. **Melochia corchorifolia** L. Chocolate-weed. Aug.–Sept. Native to the s. U.S.; adventive along a highway; Randolph Co.

150. STYRACACEAE—STORAX FAMILY

1. Calyx and corolla 4-parted; ovary inferior; fruit an elongated, winged nut .... 1. *Halesia*
1. Calyx and corolla 5-parted; ovary superior; fruit a globose drupe .... 2. *Styrax*

1. **Halesia** Ellis—Silverbell

1. **Halesia carolina** L. Silverbell Tree. Apr.–May. Moist woods, very rare; Massac and Pulaski cos. *Halesia tetraptera* L., misapplied.

2. **Styrax** L.—Storax

1. Leaves glabrous or puberulent beneath; flowers 3–4 in a cluster .... 1. *S. americanum*
1. Leaves densely white-hairy beneath; flowers more than 4 in a raceme .... 2. *S. grandifolium*

1. **Styrax americanum** Lam. Storax. Apr.–May. Swampy woods, rare; confined to extreme s. IL; also Clay, Clinton, Hamilton, Kankakee, Lawrence, and Wayne cos.
2. **Styrax grandifolium** Ait. Big-leaf Snowbell Bush. May. Along a stream in woods, very rare; Alexander Co.

151. TAMARICACEAE—TAMARISK FAMILY

1. **Tamarix** L.—Tamarisk

1. **Tamarix parviflora** DC. Tamarisk. June–Aug. Native to Europe and Asia; planted as an ornamental but rarely escaped. *Tamarix gallica* L.

152. THYMELAEACEAE—THYMELAEA FAMILY

1. Shrubs; flowers appearing before the leaves .... 1. *Dirca*
1. Annual herbs; flowers appearing after the leaves .... 2. *Thymelaea*

1. **Dirca** L.—Leatherwood

1. **Dirca palustris** L. Leatherwood. Apr.–May. Rich, shaded woods; uncommon to rare throughout the state.

2. **Thymelaea** Tourn.

1. **Thymelaea passerina** (L.) Coss. & Germ. Thymelaea. June–Aug. Native to Europe; rarely adventive in IL; confined to the ne. cos. *Thymelaea annua* L.

153. TILIACEAE—BASSWOOD FAMILY

1. **Tilia** L.—Basswood

1. Some or all the leaves at maturity at least 10 cm long.
  2. Leaves glabrous beneath, except in the axils of the veins .... 1. *T. americana*
  2. Leaves pubescent beneath with stellate hairs.
    3. Leaves green beneath; peduncles and pedicels glabrous .... 1. *T. americana*
    3. Leaves white beneath; peduncles and pedicels pubescent .... 3. *T. heterophylla*
1. All leaves less than 6 cm long .... 2. *T. cordata*

1. **Tilia americana** L. Two varieties occur in Illinois:

a. Leaves glabrous beneath except in the axils of the veins .... 1a. *T. americana* var. *americana*
a. Leaves pubescent beneath with stellate hairs .... 1b. *T. americana* var. *neglecta*

1a. **Tilia americana** L. var. **americana**. Basswood. May–July. Rich woods; occasional to common throughout the state.
1b. **Tilia americana** L. var. **neglecta** (Spach) Fosberg. Basswood. May–July. Rich woods, rare; in some n. cos.
2. **Tilia cordata** Mill. Little-leaved Linden. May–July. Native to Europe; rarely escaped from cultivation; Coles and Jackson cos.
3. **Tilia heterophylla** Vent. White Basswood. May–July. Low rocky woods, rare; Hardin, Massac, Pope, and Pulaski cos.

154. ULMACEAE—ELM FAMILY

1. Leaves with a pair of strong lateral veins arising from the main vein at the very base of the blade; pith of branches chambered; sepals free; fruit a drupe .... 1. *Celtis*
1. Leaves with strong lateral veins arising from the main vein at a distance above the very base of the blade; pith of branches not chambered; sepals united; fruit a samara or nut.
  2. Fruit a wingless nut; at least some of the flowers unisexual, appearing with the leaves; leaves mostly singly toothed .... 2. *Planera*

2. Fruit a samara; all flowers perfect, appearing before the leaves; leaves mostly doubly toothed .... 3. *Ulmus*

1. **Celtis** L.—Hackberry

1. Leaves conspicuously several toothed.
2. Leaves harshly scabrous on the upper surface .... 2. *C. occidentalis*
2. Leaves smooth or nearly so on the upper surface.
3. Most of the leaves more than ½ as long as broad, very strongly asymmetrical at the base .... 2. *C. occidentalis*
3. Most of the leaves less than ½ as long as broad, only slightly asymmetrical at the base.
4. Drupe 8–11 mm long, dark purple or dark brown .... 2. *C. occidentalis*
4. Drupe 5–7 mm long, orange, pale brown, or red .... 1. *C. laevigata*
1. Leaves entire or only sparingly toothed.
5. Leaves harshly scabrous on the upper surface
6. Leaves mostly over ½ as broad as long .... 3. *C. tenuifolia*
6. Leaves mostly less than ½ as broad as long .... 1. *C. laevigata*
5. Leaves smooth or nearly so on the upper surface.
7. Leaves more than ½ as broad as long, acute to short-acuminate at the apex. .... 3. *C. tenuifolia*
7. Leaves less than ½ as broad as long, long-tapering at the apex .... 1. *C. laevigata*

1. **Celtis laevigata** Willd. Three varieties may be distinguished in Illinois:

a. Leaves more or less smooth on the upper surface; petioles essentially glabrous.
b. Leaves entire or nearly so .... 1a. *C. laevigata* var. *laevigata*
b. Leaves regularly toothed .... 1b. *C. laevigata* var. *smallii*
a. Leaves harshly scabrous on the upper surface; petioles pubescent .... 1c. *C. laevigata* var. *texana*

1a. **Celtis laevigata** Willd. var. **laevigata.** Sugarberry. Apr.–May. Low woods; occasional to common in the s. ½ of the state.

1b. **Celtis laevigata** Willd. var. **smallii** (Beadle) Sarg. Sugarberry. Apr.–May. Low woods; occasional in the extreme s. cos.

1c. **Celtis laevigata** Willd. var. **texana** Sarg. Dwarf Sugarberry. Apr.–May. Dry woods and cliffs; occasional in the s. ⅕ of the state.

2. **Celtis occidentalis** L. Three varieties occur in Illinois:

a. Leaves harshly scabrous on the upper surface .... 2a. *C. occidentalis* var. *occidentalis*
a. Leaves smooth or nearly so on the upper surface.
b. Most of the leaves less than ½ as broad as long .... 2b. *C. occidentalis* var. *canina*
b. Most of the leaves more than ½ as broad as long .... 2c. *C. occidentalis* var. *pumila*

2a. **Celtis occidentalis** L. var. **occidentalis.** Common Hackberry. Apr.–May. Low woods, wooded slopes; common throughout the state.

2b. **Celtis occidentalis** L. var. **canina** (Raf.) Sarg. Hackberry. Apr.–May. Low woods, wooded slopes; occasional throughout the state.

2c. **Celtis occidentalis** L. var. **pumila** (Pursh) Gray. Hackberry. Apr.–May. Low woods, wooded slopes, bluffs; occasional throughout the state.

3. **Celtis tenuifolia** Nutt. Two varieties occur in Illinois:

a. Leaves more or less smooth on the upper surface, membranous; petioles glabrous or nearly so .... 3a. *C. tenuifolia* var. *tenuifolia*
a. Leaves harshly scabrous on the upper surface, leathery; petioles densely pubescent .... 3b. *C. tenuifolia* var. *georgiana*

3a. **Celtis tenuifolia** Nutt. var. **tenuifolia.** Dwarf Hackberry. Apr.–May. Dry woods and cliffs, occasional; mostly in the s. ⅕ of the state and along the Mississippi River.

3b. **Celtis tenuifolia** Nutt. var. **georgiana** (Small) Fern. & Schub. Dwarf Hackberry. Apr.–May. Dry woods and cliffs, occasional; mostly in the s. ⅕ of the state.

2. **Planera** J. F. Gmel.—Water Elm

1. **Planera americana** J. F. Gmel. Water Elm; Planer Tree. Mar.–Apr. Swampy woods, rare; Alexander, Johnson, Massac, Pope, and Pulaski cos.

3. **Ulmus** L.—Elm

1. Upper surface of leaves scabrous.
 2. Winter buds rusty-pubescent; calyx with reddish hairs; samaras eciliate.
  3. Leaves harshly scabrous; body of samara pubescent; calyx shallowly lobed; bark roughened .... 7. *U. rubra*
  3. Leaves more or less scabrous, but not harshly so; only the central vein of the samaras pubescent; calyx deeply lobed; bark smooth .... 4. *U. glabra*
 2. Winter buds glabrous or pubescent, but not rusty-pubescfent; calyx without reddish hairs; samaras ciliate .... 2. *U. americana*
1. Upper surface of leaves not scabrous.
 4. None of the branches corky-winged.
  5. Leaves doubly serrate, usually strongly asymmetrical at the base.
   6. All or most of the leaves 4 cm long or longer; pubescence often in axillary tufts in the leaf axils beneath; branches not pendulous .... 2. *U. americana*
   6. Most or all of the leaves less than 4 cm long; pubescence never in axillary tufts beneath; lower branches often pendulous .... 3. *U. carpinifolia*
  5. Leaves mostly singly serrate, usually nearly symmetrical at the base .... 6. *U. pumila*
 4. At least some of the branches corky-winged.
  7. Buds glabrous or nearly so; leaves sessile or on petioles up to 2 mm long; samaras (excluding cilia) up to 5 mm broad .... 1. *U. alata*
  7. Buds downy-pubescent; leaves on petioles 3 mm long or longer; samaras (excluding cilia if present) 1.0–1.5 cm broad.
   8. Most of the mature leaves over 8 cm long, glabrous above; flowers pendulous from long pedicels; samaras ciliate .... 8. *U. thomasii*
   8. Most of the mature leaves less than 8 cm long, scabrous above; flowers sessile or nearly so; samaras eciliate .... 5. *U. procera*

1. **Ulmus alata** Michx. Winged Elm. Feb.–Mar. Wooded slopes, cliffs, or low woods; common in the s. ¼ of the state, apparently absent elsewhere.

2. **Ulmus americana** L. American Elm. Feb.–Apr. Rich woods, wooded floodplains, common; in every co.

3. **Ulmus carpinifolia** Gled. Smooth-leaved Elm. Mar.–Apr. Native to Europe, Asia, and Africa; escaped into a woods in Williamson Co. One of the trees is nearly 50 feet tall.

4. **Ulmus glabra** Huds. Scotch Elm. Mar.–May. Native to Europe and Asia; edge of woods; Williamson Co. *Ulmus campestris* L.

5. **Ulmus procera** Salisb. English Elm. Apr. Native to Europe; rarely escaped in IL; Jackson Co.

6. **Ulmus pumila** L. Siberian Elm. May. Native to Asia; escaped into disturbed areas; occasional to common throughout the state.

7. **Ulmus rubra** Muhl. Slippery Elm; Red Elm. Feb.–Apr. Woods, often in disturbed areas, common; in every co. Hybrids between this species and *U. pumila* in Illinois are fairly common and have been given the provisional name *U. X notha* by Wilhelm and Ware (Swink and Wilhelm, 1994). It differs from *U. rubra* by lacking hispid petioles and by having nearly symmetrical leaf bases. It differs from *U. pumila* by having buds scales white-ciliate and leaves commonly acuminate.

8. **Ulmus thomasii** Sarg. Rock Elm. Mar.–May. Rich woods, not common in the n. ½ of IL, absent elsewhere.

155. URTICACEAE—NETTLE FAMILY

1. Leaves opposite (a few alternate leaves may be present in *Boehmeria*).
 2. Leaves and stems glabrous, the stems translucent .... 4. *Pilea*

2. Leaves and stems pubescent, the stems opaque.
 3. Pistillate flowers with free sepals; stinging hairs often present..........5. *Urtica*
 3. Pistillate flowers with united sepals; stinging hairs never present..................
 ....1. *Boehmeria*
1. Leaves all alternate.
 4. Leaves serrate, more than 4 cm broad; stinging hairs present ....2. *Laportea*
 4. Leaves essentially entire, less than 3 cm broad; stinging hairs absent ....
 ....3. *Parietaria*

1. **Boehmeria** Jacq.—False Nettle

1. **Boehmeria cylindrica** (L.) Sw. Two varieties occur in Illinois:
a. Leaves smooth or slightly scabrous above.................1a. *B. cylindrica* var. *cylindrica*
a. Leaves harshly scabrous above..........................1b. *B. cylindrica* var. *drummondiana*

1a. **Boehmeria cylindrica** (L.) Sw. var. **cylindrica**. False Nettle. July–Oct. Moist woods, common; throughout the state.

1b. **Boehmeria cylindrica** (L.) Sw. var. **drummondiana** (Wedd.) Wedd. False Nettle. July–Oct. Edge of bogs and marshes; occasional throughout the state.

2. **Laportea** Gaud.—Wood Nettle

1. **Laportea canadensis** (L.) Wedd. Wood Nettle. June–Sept. Moist soil; common throughout the state.

3. **Parietaria** L.—Pellitory

1. **Parietaria pensylvanica** Muhl. Pellitory. May–Sept. In shade beneath cliffs and around buildings; disturbed dry soil; common throughout the state.

4. **Pilea** Lindl.—Clearweed

1. Achenes black, averaging 1.5 mm wide..1. *P. fontana*
1. Achenes green, averaging 1 mm wide..2. *P. pumila*

1. **Pilea fontana** (Lunell) Rydb. Clearweed. July–Sept. Moist soil; occasional and scattered in IL. (Including *P. opaca* [Lunell] Rydb.)

2. **Pilea pumila** (L.) Gray. Clearweed. July–Sept. Moist, usually shaded ground, common; probably in every co. (Including var. *deamii* [Lunell] Fern.)

5. **Urtica** L.—Nettle

1. Perennials often attaining a height of 1 m or more; inflorescences 2 cm long or longer ....2. *U. gracilis*
1. Annuals never attaining a height of 1 m; inflorescences less than 2 cm long.
 2. Leaves dentate, tapering to the base ....3. *U. urens*
 2. Leaves crenate, some of them cordate at the base ....1. *U. chamaedryoides*

1. **Urtica chamaedryoides** Pursh. Nettle. Apr.–June. Swampy woods, riverbanks, rare; Alexander, Jackson, Pope, and Union cos.

2. **Urtica gracilis** Ait. Stinging Nettle. June–Sept. Rich woods, moist ground, thickets; occasional to common in the n. ½ of the state, rare in the s. ½. *Urtica dioica* L., misapplied; *U. procera* Muhl.

3. **Urtica urens** L. Burning Nettle. June–Sept. Native to Europe and Asia; rarely escaped into waste areas; Champaign Co.

156. VALERIANACEAE—VALERIAN FAMILY

1. Cauline leaves, or some of them, pinnately compound; calyx divided into 5–15 setiform lobes..1. *Valeriana*
1. Cauline leaves entire or toothed, not pinnately compound; calyx lobes reduced to short teeth or absent..2. *Valerianella*

1. **Valeriana** L.—Valerian

1. Flowers unisexual, white; leaves parallel-veined; roots tuberous............. 1. *V. edulis*
1. Flowers perfect, pink or rose, rarely white; leaves pinnately veined; roots fibrous.
 2. All basal leaves pinnately compound; corolla 4–5 mm long ....2. *V. officinalis*
 2. All or most of the basal leaves simple; corolla 5–15 mm long.

3. Flowers 10–15 mm long; basal leaves cordate at base .... 3. *V. pauciflora*
3. Flowers 5–7 mm long; basal leaves tapering to base .... 4. *V. sitchensis*

1. **Valeriana edulis** Nutt. ssp. **ciliata** (Torr. & Gray) F. G. Mey. Valerian. May–June. Wet prairies, calcareous fens; occasional in the n. ⅕ of the state.

2. **Valeriana officinalis** L. Garden Heliotrope. June–July. Native to Europe; occasionally cultivated but infrequently escaped; Cook, DuPage, Kane, and Lake cos.

3. **Valeriana pauciflora** Michx. Pink Valerian. May–June. Mesic woods; occasional in the s. ½ of the state; also Pike, Scott, and Vermilion cos.

4. **Valeriana sitchensis** Bong. ssp. **uliginosa** (Torr. & Gray) F. G. Mey. Valerian. June–July. Wet soil, very rare; McHenry Co. *Valeriana uliginosa* Torr. & Gray.

2. **Valerianella** Mill.

1. Corolla 1.5–2.0 mm long; bracts and bractlets spinulose-ciliate.
2. Corolla lobes bluish; stamens included or shortly exserted .... 3. *V. locusta*
2. Corolla lobes white; stamens long-exserted .... 5. *V. radiata*
1. Corolla 3–5 mm long; bracts and bractlets eciliate except sometimes at the tip.
3. Fruits ovoid to ovoid-ellipsoid, longer than broad.
4. Fruits 3–4 mm long, 2.5 mm broad or broader.................. 1. *V. chenopodifolia*
4. Fruits up to 2.5 mm long, 0.7–1.2 mm broad .... 2. *V. intermedia*
3. Fruits orbicular or globose, as broad as long.
5. Fruits flattened, 3.0–3.5 mm long, about 3 mm broad .... 4. *V. patellaria*
5. Fruits not flattened, 2.0–2.5 mm long, about 2 mm broad....... 6. *V. umbilicata*

1. **Valerianella chenopodifolia** (Pursh) DC. Valerianella. May–June. Open habitat, very rare; Iroquois and Will cos.

2. **Valerianella intermedia** Dyal. Corn Salad. May–June. Wet ground, rare; Kankakee, LaSalle, and Will cos.

3. **Valerianella locusta** (L.) Betcke. Corn Salad. May. Native to Europe; rarely adventive in a disturbed area; LaSalle Co.

4. **Valerianella patellaria** (Sulliv.) Wood. Corn Salad. May–June. Wet ground, rare; LaSalle Co.

5. **Valerianella radiata** (L.) Dufr. Two varieties occur in Illinois:

a. Seed-bearing cavity of fruit broader than the combined width of the 2 empty cells; fruits 1.4–1.5 mm long, 0.7–0.8 mm broad .... 5a. *V. radiata* var. *radiata*
a. Seed-bearing cavity of fruit not as broad as the combined width of the 2 empty cells; fruits 1.7–1.8 mm long, about 1.2 mm broad .... 5b. *V. radiata* var. *missouriensis*

5a. **Valerianella radiata** (L.) Dufr. var. **radiata.** Corn Salad. Apr.–May. Wet fields, prairies, low woods; occasional to common in the s. ½ of the state; also Fulton Co.

5b. **Valerianella radiata** (L.) Dufr. var. **missouriensis** Dyal. Corn Salad. Apr.–May. Wet ground, rare; Jackson Co.

6. **Valerianella umbilicata** (Sulliv.) Wood. Corn Salad. May–June. Wet ground, rare; Grundy, Kankakee, LaSalle, and Will cos.

157. VERBENACEAE—VERVAIN FAMILY

1. Corolla strongly 2-lipped; calyx teeth 4; flowers in a dense, rounded head .... 2. *Phyla*
1. Corolla nearly regular; calyx lobes 5; flowers not in a dense, rounded head.
2. Corolla at least 1 cm long; calyx at least 8 mm long .... 1. *Glandularia*
2. Corolla up to 8 mm long; calyx up to 5 mm long .... 3. *Verbena*

1. **Glandularia** J. F. Gmel.

1. Corolla tube slightly longer than calyx; calyx lobes setaceous, elongated .... 1. *G. canadense*
1. Corolla tube about twice as long as calyx; calyx lobes deltoid, short .... 2. *G. peruviana*

1. **Glandularia canadensis** (L.) Nutt. Rose Verbena. Mar.–Oct. Rocky woods, edges of fields, prairies; occasional in the s. ⅔ of the state, rare or absent elsewhere. *Verbena canadensis* (L.) Britt.

2. **Glandularia peruviana** (L.) Small. Peruvian Vervain. June–Aug. Native to S. America; rarely escaped from cultivation; Kane and Kankakee cos. *Verbena peruviana* (L.) Britt.

2. **Phyla** Lour.—Fog-fruit

1. Leaves broadly rounded at apex .... 1. *P. cuneifolia*
1. Leaves acute to acuminate at apex .... 2. *P. lanceolata*

1. **Phyla cuneifolia** (Torr.) Greene. Hoary Fog-fruit. May–Sept. Native to the w. U.S.; adventive in a disturbed area; Menard Co.

2. **Phyla lanceolata** (Michx.) Greene. Fog-fruit. May–Sept. Wet soil, common; in every co. (Including var. *recognita* Fern. & Grisc.)

3. **Verbena** L.—Vervain

1. Some or all the cauline leaves auriculate-clasping .... 2. *V. bonariensis*
1. None of the cauline leaves auriculate-clasping.
  2. Flowers densely crowded into short-cylindric spikes arranged in compound cymes .... 8. *V. litoralis*
  2. Flowers not densely crowded, in slender spikes arranged in simple cymes.
    3. Stems often spreading or sprawling to ascending but not strictly erect; bracts slightly to much longer than the flowers.
      4. Stems more or less ascending; bracts 3–4 mm long .... 10. *V. X perriana*
      4. Stems spreading or sprawling; bracts 5 mm long or longer.
        5. Bracts 6 mm long or longer; inflorescence 1 cm broad or broader .... 3. *V. bracteata*
        5. Bracts about 5 mm long; inflorescence up to 1 cm broad .... 4. *V. X deamii*
    3. Stems erect; bracts not longer than the flowers.
      6. Leaves up to 2 cm broad.
        7. Flowers up to 6 mm broad; stem glabrous or sparsely pilose .... 12. *V. simplex*
        7. Flowers up to 10 mm broad; stem densely hairy.
          8. Leaves lanceolate .... 1. *V. X blanchardii*
          8. Leaves elliptic to elliptic-ovate .... 9. *V. X moechina*
      6. Leaves, or most of them, more than 2 cm broad.
        9. Leaves sessile or on petioles less than 5 mm long; spikes 1–3 per stem, 7–10 mm thick .... 13. *V. stricta*
        9. Leaves petiolate, the petioles at least 5 mm long; spikes usually more than 3 per stem, mostly less than 7 mm thick.
          10. Some or all the flowers overlapping, the calyx of 1 reaching beyond the base of the calyx of the flower above.
            11. Calyx 3–4 mm long; leaves sparsely pubescent .... 6. *V. hastata*
            11. Calyx 4–5 mm long; leaves softly pubescent .... 11. *V. X rydbergii*
          10. None or few of the flowers overlapping, the calyx of 1 at most reaching only to the base of the calyx of the flower above.
            12. Flowers blue .... 5. *V. X engelmannii*
            12. Flowers white.
              13. Stems hirtellous or nearly glabrous; each flower not reaching the base of the 1 immediately above .... 14. *V. urticifolia*
              13. Stems long-hairy; each flower reaching just to the base of the calyx of the flower immediately above .... 7. *V. X illicita*

1. **Verbena X blanchardii** Moldenke. Blanchard's Vervain. July–Aug. Dry fields; Cook, DuPage, and Winnebago cos. Reputed to be the hybrid between *V. hastata* L. and *V. simplex* Lehm.

2. **Verbena bonariensis** L. Clasping-leaved Vervain. Aug. Native to the tropics; rarely adventive in disturbed soil; Coles Co.

3. **Verbena bracteata** Lag. & Rodr. Creeping Vervain. Apr.–Oct. Disturbed soil; occasional to common throughout the state.

4. **Verbena X deamii** Moldenke. Deam's Vervain. July–Aug. Dry soil; scattered throughout the state. Reputed to be the hybrid between *V. bracteata* Lag. & Rodr. and *V. stricta* Vent.

5. **Verbena X engelmannii** Moldenke. Engelmann's Vervain. July–Sept. Woods, fields; scattered throughout the state. Reputed to be the hybrid between *V. hastata* L. and *V. urticifolia* L.

6. **Verbena hastata** L. Blue Vervain. June–Oct. Wet woods, wet prairies, wet disturbed areas, common; in every co.

7. **Verbena X illicita** Moldenke. Hybrid White Vervain. June–Sept. Low ground; scattered throughout the state. Reputed to be the hybrid between *V. stricta* Vent. and *V. urticifolia* L.

8. **Verbena litoralis** Kunth. Brazilian Vervain. June–July. Native to S. America; adventive along a railroad; Union Co. *Verbena brasiliensis* Vellozo.

9. **Verbena X moechina** Moldenke. Hybrid Vervain. June–July. Dry ground, particularly in pastures; Adams, Cook, Hardin, Peoria, and Winnebago cos. Reputed to be the hybrid between *V. simplex* Lehm. and *V. stricta* Vent.

10. **Verbena X perriana** Moldenke. Perry's Vervain. June–Aug. Dry soil; scattered throughout the state. Reputed to be the hybrid between *V. bracteata* Lag. & Rodr. and *V. urticifolia* L.

11. **Verbena X rydbergii** Moldenke. Rydberg's Vervain. June–Sept. Wet ground; scattered throughout the state. Reputed to be the hybrid between *V. hastata* L. and *V. stricta* Vent.

12. **Verbena simplex** Lehm. Narrow-leaved Vervain. May–Sept. Prairies, fields, disturbed soil; occasional and scattered throughout the state.

13. **Verbena stricta** Vent. Hoary Vervain. May–Sept. Prairies, pastures, fields, common; probably in every co. White-flowered forms may be known as f. *albiflora* Wadmond; rose-flowered forms may be known as f. *roseiflora* Benke.

14. **Verbena urticifolia** L. Two varieties occur in Illinois:

a. Leaves hirsute on the lower surface; nutlets about 2 mm long, corrugated on the back .... 14a. *V. urticifolia* var. *urticifolia*

a. Leaves velutinous on the lower surface; nutlets about 1.5 mm long, not corrugated on the back .... 14b. *V. urticifolia* var. *leiocarpa*

14a. **Verbena urticifolia** L. var. **urticifolia.** White Vervain. June–Oct. Fields, thickets, disturbed woods, common; in every co.

14b. **Verbena urticifolia** L. var. **leiocarpa** Perry & Fern. White Vervain. June–Oct. Low ground; Cook and Kane cos.

158. VIOLACEAE—VIOLET FAMILY

1. Stamens united; sepals attached at the base, not auriculate; flowers greenish .... 1. *Hybanthus*

1. Stamens free; sepals attached above the base, auriculate; flowers not greenish .... 2. *Viola*

1. **Hybanthus** Jacq.—Green Violet

1. **Hybanthus concolor** (T. F. Forst.) Spreng. Green Violet. Apr.–June. Rich woods; rather common in the s. cos., becoming rare and scattered northward.

2. **Viola** L.—Violet

1. Plants acaulescent, the leaves and peduncles seeming to arise from out of the ground.

  2. Flowers basically blue or white with deep purple veins.

    3. None of the leaves lobed.

      4. Leaves less than ½ as long as broad .... 22. *V. sagittata*

      4. Leaves more than ½ as long as broad.

        5. Style with downward turned beak, not particularly thickened at tip .... 13. *V. odorata*

        5. Style thickened at tip, not beaked nor turned downward.

          6. Spurred petal beardless within or nearly so.

            7. Petioles and leaves uniformly and densely villous .... 24. *V. sororia*

7. Petioles and leaves glabrous or only sporadically and sparsely pubescent.
8. Lateral petals with hairs swollen at the tip; leaf blades somewhat pubescent on the upper surface.. 6. *V. cucullata*
8. Lateral petals with hairs slender throughout; leaf blades entirely glabrous.
9. Leaves elongate-triangular, tapering to an elongated tip .... 11. *V. missouriensis*
9. Leaves cordate-ovate, rounded to short-pointed at the tip.
10. Petals blue or purple............................ 17. *V. pratincola*
10. Petals white with purple veins.................... 3. *V. bicolor*
6. Spurred petal bearded within.
11. Leaves and petioles glabrous (microscopically pubescent on upper surface of blade in *V. affinis*).
12. Sepals obtuse; peduncles longer than petioles .... 12. *V. nephrophylla*
12. Sepals acute; peduncles about equaling petioles .... 1. *V. affinis*
11. Leaves and petioles sparsely hirsute to pilose.
13. Leaves sparsely pubescent above, hirsutulous below; petioles sparsely hirsute .... 23. *V. septentrionalis*
13. Leaves uniformly puberulent on both surfaces; petioles pilose .... 7. *V. fimbriatula*
3. Some of the leaves hastately or palmately lobed.
14. All petals beardless.. 15. *V. pedata*
14. Lateral petals and sometimes the spurred petal bearded.
15. Leaves hastately lobed.
16. Sepals ciliate.. 7. *V. fimbriatula*
16. Sepals glabrous.. 22. *V. sagittata*
15. Leaves pinnately lobed.
17. Leaves cordate at base.
18. Early leaves unlobed; central lobe of lobed leaves undivided..... .... 14. *V. palmata*
18. All leaves lobed; central lobe several-divided.............................. .... 26. *V. subsinuata*
17. Leaves cuneate or truncate at base.
19. Leaves deeply divided into narrow segments; spurred petal villous within .... 16. *V. pedatifida*
19. Leaves divided into broad segments; spurred petal glabrous within .... 28. *V. viarum*
2. Flowers basically white.
20. Leaves deeply palmately cleft .... 15. *V. pedata*
20. Leaves unlobed.
21. Style with downward turned beak, not particularly thickened at tip .... 13. *V. odorata*
21. Style thickened at tip, not beaked nor turned downward.
22. Lateral petals bearded.
23. Peduncles and leaves glabrous or sparsely pubescent; seeds brown .... 3. *V. bicolor*
23. Peduncles and leaves finely pubescent; seeds buff .... 4. *V. blanda*
22. Lateral petals beardless.
24. Leaves cordate at base .... 10. *V. macloskeyi*
24. Leaves tapering to base.
25. Leaves linear, lanceolate, or oblanceolate............ 9. *V. lanceolata*
25. Leaves oblong to ovate .... 18. *V. primulifolia*
1. Plants with stems bearing alternate leaves and axillary flowers.
26. Stipules not foliaceous; tip of style slender or only slightly swollen.
27. Flowers blue .... 8. *V. labradorica*
27. Flowers white or yellow.

28. Flowers white.
29. Stipules entire.
30. Rhizomes short; leaves more or less glabrous on the lower surface .... 5. *V. canadensis*
30. Rhizomes long and slender; leaves pubescent on the lower surface ....21. *V. rugulosa*
29. Stipules toothed..25. *V. striata*
28. Flowers yellow ....10. *V. pubescens*
26. Stipules foliaceous; tip of style enlarged into a hollow, globose tip.
31. Base of stipules palmately divided; upper leaves entire or nearly so .... 20. *V. rafinesquii*
31. Base of stipules pinnately divided; upper leaves distinctly toothed.
32. Petals and sepals subequal, the petals pale yellow .... 2. *V. arvensis*
32. Petals 1 ½–3 times as long as the sepals, variously colored.
33. Flowers 2–3 cm across, variously colored............. 29. *V. X wittrockiana*
33. Flowers 1.5–2.0 cm across, purple ....27. *V. tricolor*

1. **Viola affinis** LeConte. Woodland Blue Violet. Apr.–June. Low woods, occasional; scattered throughout IL.

2. **Viola arvensis** Murr. Wild Pansy. Apr.–Aug. Native to Europe; rarely adventive in IL; Jackson, Cook, and DuPage cos.

3. **Viola bicolor** Pursh. Confederate Violet. Mar.–May. Woods, disturbed areas; occasional throughout IL. *Viola priceana* Pollard.

4. **Viola blanda** Willd. White Violet. May–June. Wooded slopes, rare; Cook, Jo Daviess, Kane, Lake, and McHenry cos. *Viola incognita* Brainerd.

5. **Viola canadensis** L. var. **corymbosum** Nutt. White Violet. Apr.–June. Rich woods, rare; Cook and Lake cos.

6. **Viola cucullata** Marsh. Marsh Blue Violet. Apr.–June. Moist soil in woods and marshes; occasional in the n. ½ of IL, apparently rare in the s. ½. *Viola obliqua* Hill.

7. **Viola fimbriatula** Smith. Sand Violet. Apr.–May. Dry woods, sandy soil, occasional; apparently restricted to the n. ½ of IL; also Richland Co.

8. **Viola labradorica** Schrank. Dog Violet. Apr.–June. Moist woods, rare; Cook, DeKalb, DuPage, Lake, McHenry, and Richland cos. *Viola conspersa* Reichenb. A hybrid between this species and *V striata,* called *Viola eclipsis* H. E. Ballard, is known from Richland Co.

9. **Viola lanceolata** L. Two subspecies occur in Illinois:

a. Leaves lanceolate, 3 ½–5 times as long as broad..9a. *V. lanceolata* ssp. *lanceolata*
a. Leaves linear, 6–15 times as long as broad.................... 9b. *V. lanceolata* ssp. *vittata*

9a. **Viola lanceolata** L. ssp. **lanceolata**. Lance-leaved Violet. May–June. Wet soil, occasional; confined to the n. ¾ of IL.

9b. **Viola lanceolata** L. ssp. **vittata** (Greene) Russell. Narrow-leaved Violet. May–June. Moist soil, rare; Cook, Kankakee, and Lake cos.

10. **Viola macloskeyi** Lloyd ssp. **pallens** (Banks) M. S. Baker. Smooth White Violet. Apr.–May. Moist soil, rare; confined to the n. ⅛ of IL. *Viola pallens* Banks.

11. **Viola missouriensis** Greene. Missouri Violet. Mar.–May. Low woods, occasional; scattered throughout the state.

12. **Viola nephrophylla** Greene. Northern Blue Violet. May–June. Usually moist soil; confined to the n. ½ of IL.

13. **Viola odorata** L. Sweet Violet. Apr.–May. Native to Europe; occasionally escaped from cultivation; scattered throughout IL.

14. **Viola palmata** L. Two varieties occur in Illinois:

a. Leaves with 3–5 usually shallow lobes...........................14a. *V. palmata* var. *palmata*
a. Leaves with 5–7 very deep lobes ....14b. *V. palmata* var. *dilatata*

14a. **Viola palmata** L. var. **palmata.** Cleft Violet. Apr.–June. Woods and wooded slopes; occasional in the s. ½ of IL, becoming rarer northward. *Viola triloba* Schwein.

14b. **Viola palmata** L. var. **dilatata** Ell. Cleft Violet. Apr.–June. Woods; occasional in the s. ⅔ of IL. *Viola triloba* Schwein. var. *dilatata* (Ell.) Brainerd.

15. **Viola pedata** L. Bird's-foot Violet. Apr.–June. Cherty slopes, borders of woods; occasional throughout IL. Plants with white petals may be known as f. *alba* (Thurb.) Britt. (Including var. *lineariloba* DC.)

16. **Viola pedatifida** G. Don. Prairie Violet. May–June. Prairies and dry woods; occasional in the n. ½ of IL, becoming rare southward. This species is known to hybridize with *V. sororia* to form *V. X bernardii* Greene.

17. **Viola pratincola** Greene. Common Blue Violet. Mar.–May. Woods, lawns, disturbed areas, common; probably in every co.

18. **Viola primulifolia** L. Primrose Violet. May–June. Moist soil, rare; Grundy, Iroquois, and Kankakee cos.

19. **Viola pubescens** Ait. Two varieties occur in Illinois:

a. Leaves and at least the upper ½ of the stem densely pubescent; basal leaves 1–2; each margin of the leaf with 18 or more teeth....... 19a. *V. pubescens* var. *pubescens*

a. Leaves and stems glabrous or puberulent; basal leaves often 3 or more; each margin of the leaf with up to 18 teeth .... 19b. *V. pubescens* var. *eriocarpa*

19a. **Viola pubescens** Ait. var. **pubescens.** Downy Yellow Violet. Apr.–May. Rich woods, rare; confined to the n. ½ of the state. Plants with glabrous capsules have been called var. *peckii* House.

19b. **Viola pubescens** Ait. var. **eriocarpa** (Schwein.) Russell. Smooth Yellow Violet. Apr.–May. Woods, common; probably in every co.

20. **Viola rafinesquii** Greene. Johnny-jump-up. Mar.–May. Native to Europe and Asia; commonly naturalized in disturbed areas in the s. ⅗ of IL, rare elsewhere.

21. **Viola rugulosa** Greene. Western White Violet. May–June. Wet woods, rare; Carroll, Jo Daviess, and Winnebago cos.

22. **Viola sagittata** Air. Arrow-leaved Violet. Mar.–June. Moist or dry woods; occasional throughout the state. *Viola emarginata*(Nutt.) LeConte.

23. **Viola septentrionalis** Greene. Northern Blue Violet. May–June. Woods, very rare; McHenry Co.

24. **Viola sororia** Willd. Woolly Blue Violet. Mar.–May. Woods, common; in every co.

25. **Viola striata** Ait. Cream Violet. Apr.–June. Moist soil in woods, meadows, and fields; common in the s. ½ of IL, becoming rare in the n. ½.

26. **Viola subsinuata** (Greene) Greene. Lobed Violet. Apr.–June. Prairies, savannas; scattered in IL. Some botanists believe this is a hybrid between *V. pedatifida* G. Don and *V. sororia* Willd.

27. **Viola tricolor** L. Wild Pansy. May–June. Native to Europe and Asia; rarely adventive in disturbed areas; scattered in IL.

28. **Viola viarum** Pollard. May. Blue Violet. Dry, gravelly hilltops, very rare; Adams and Peoria cos.

29. **Viola X wittrockiana** Gams. Pansy. May–Aug. Native to Europe; occasionally escaped from cultivation; Cook, Kendall, and McHenry cos. Reputed to be the hybrid between *V. tricolor* L. and *V. lutea* L.

159. VISCACEAE—MISTLETOE FAMILY

1. **Phoradendron** Nutt.—Mistletoe

1. **Phoradendron leucarpum** (Raf.) Rev. & M. C. Johnston. Mistletoe. Sept.–Oct. On various deciduous trees, mostly in low areas; occasional in the s. ⅙ of IL and extending northward along the Wabash River to Clark Co. *Phoradendron serotinum* (Raf.) M. C. Johnston.

160. VITACEAE—GRAPE FAMILY

1. Leaves compound.

2. Leaves palmately compound, with (3–) 5 leaflets .... 2. *Parthenocissus*

2. Leaves ternately or pinnately compound, with 11 or more leaflets........................ .... 1. *Ampelopsis*

1. Leaves simple.

3. Pith continuous, white; flowering and fruiting clusters broader than long .... 1. *Ampelopsis*

3. Pith interrupted at the nodes, brown; flowering and fruiting clusters longer than broad .... 3. *Vitis*

1. **Ampelopsis** Michx.

1. Leaves compound or variously divided.
  2. Leaves with some pubescence, doubly compound .... 1. *A. arborea*
  2. Leaves glabrous, once-divided or some of them undivided .... 2. *A. brevipedunculata*
1. Leaves simple, never divided .... 3. *A. cordata*

1. **Ampelopsis arborea** (L.) Koehne. Pepper-vine. June–Aug. Moist woods and thickets, rare; Alexander, Jackson, Pope, Pulaski, Randolph, and Union cos.

2. **Ampelopsis brevipedunculata** (Maxim.) Trautv. Turquoise-berry. June–Aug. Native to Asia; naturalized in disturbed areas; DuPage and Kane cos.

3. **Ampelopsis cordata** Michx. Raccoon Grape. May–July. Moist woods and thickets; common to occasional in the s. ⅔ of the state.

2. **Parthenocissus** Planch.—Virginia Creeper

1. Leaflets 3 .... 3. *P. tricuspidata*
1. Leaflets 5.
  2. Each tendril ending in an adhesive disk; leaves dull on the upper surface; berries 5–7 mm in diameter; petiolules shorter than the leaflets .... 2. *P. quinquefolia*
  2. Tendrils without adhesive disks; leaves shiny on the upper surface; berries 8–10 mm in diameter; petiolules longer than the leaflets .... 1. *P. inserta*

1. **Parthenocissus inserta** (Kern.) K. Fritsch. Woodbine. May–July. Woods, thickets; common in the n. ¼ of IL, occasional elsewhere. *Parthenocissus vitacea* (Knerr) A. Hitchc.

2. **Parthenocissus quinquefolia** (L.) Planch. Virginia Creeper. June–July. Woods, along fences, common; in every co.

3. **Parthenocissus tricuspidata** (Sieb. & Zucc.) Planch. Boston Ivy. June–Aug. Native to Asia; rarely escaped from cultivation.

3. **Vitis** L.—Grape

1. Leaves felted on the lower surface, covered with cobwebby hairs, or merely silvery-glaucous on the lower surface.
  2. Leaves felted on the lower surface but not cobwebby or merely silvery-glaucous .... 3. *V. labrusca*
  2. Leaves cobwebby on the lower surface nor merely silvery-glaucous.
    3. Seeds 5–8 mm long; branchlets not angular .... 1. *V. aestivalis*
    3. Seeds 4–5 mm long; branchlets angular .... 2. *V. cinerea*
1. Leaves glabrous on the lower surface, green, or with hairs merely in the vein axils.
  4. Leaves unlobed or shallowly 3-lobed .... 6. *V. vulpina*
  4. Leaves, or at least those on the fertile branches, sharply 3- or 5-lobed.
    5. Berries black; partitions in pith at the nodes 4–5 mm thick; leaves scarcely or not ciliate on the margins, very deeply lobed .... 4. *V. palmata*
    5. Berries bluish, glaucous; partitions in pith at the nodes up to 2 mm thick; leaves ciliate on the margins, usually lobed less than halfway to the middle .... 5. *V. riparia*

1. **Vitis aestivalis** Michx. Two varieties occur in Illinois:

a. Petioles tomentose or rusty-pubescent .... 1a. *V. aestivalis* var. *aestivalis*
a. Petioles glabrous or nearly so .... 1b. *V. aestivalis* var. *argentifolia*

1a. **Vitis aestivalis** Michx. var. **aestivalis.** Summer Grape. May–July. Rocky woods, bluff tops; occasional to common throughout the state. (Including *V. lincecumii* Buckl.)

1b. **Vitis aestivalis** Michx. var. **argentifolia** (Munson) Fern. Silver-leaved Grape. May–July. Rocky woods; occasional throughout the state.

2. **Vitis cinerea** (Engelm.) Engelm. Winter Grape. May–July. Low woods, thickets; occasional in the s. ¾ of the state, rare elsewhere.

3. **Vitis labrusca** L. Labruscan Grape. May–June. Cultivated vineyard species; rarely escaped into waste ground; Jackson Co. *Vitis labruscana* Bailey.

4. **Vitis palmata** M. Vahl. Catbird Grape. May–July. Swamps, low woods; confined to the s. ½ of the state.

5. **Vitis riparia** Michx. Three varieties occur in Illinois:

a. Petioles glabrous.
  b. Berries 8–12 mm in diameter .... 5a. *V. riparia* var. *riparia*
  b. Berries 6–7 mm in diameter .... 5b. *V. riparia* var. *praecox*
a. Petioles densely pubescent .... 5c. *V. riparia* var. *syrticola*

5a. **Vitis riparia** Michx. var. **riparia.** Riverbank Grape. May–June. Low woods; occasional to common throughout the state. *Vitis rupestris* Scheele, misapplied.

5b. **Vitis riparia** Michx. var. **praecox** Engelm. Riverbank Grape. Apr. Low woods; occasional in the sw. cos. of the state.

5c. **Vitis riparia** Michix. var. **syrticola** (Fern. & Wieg.) Fern. Riverbank Grape. May–June. Low woods, not common; scattered in IL.

6. **Vitis vulpina** L. Fox Grape; Frost Grape. May–June. Moist soil; occasional to common throughout the state.

161. ZYGOPHYLLACEAE—CALTROP FAMILY

1. Fruits without prickles .... 1. *Kallstroemia*
1. Fruits with 2–4 sharp prickles .... 2. *Tribulus*

1. **Kallstroemia** Scop.—Kallstroemia

1. **Kallstroemia parviflora** Norton. June–Sept. Kallstroemia. Native to the w. U.S.; adventive in disturbed soil; Cook, Macoupin, and St. Clair cos. *Kallstroemia intermedia* Rydb.

2. **Tribulus** L.—Caltrop

1. **Tribulus terrestris** L. Puncture-weed. June–Sept. Native to Europe; occasional in disturbed soil; scattered throughout IL.

162. ACORACEAE—SWEET FLAG FAMILY

1. **Acorus** L.—Sweet Flag

1. Midvein and lateral veins prominent .... 1. *A. americanus*
1. Midvein prominent, the lateral veins inconspicuous .... 2. *A. calamus*

1. **Acorus americanus** (Raf.) Raf. Sweet Flag. May–Aug. Marshes, low areas; apparently confined to the n. ¼ of IL; also Cass Co.

2. **Acorus calamus** L. Sweet Flag. May–Aug. Native to Eurasia. Marshes, low areas; occasional to common throughout the state.

163. AGAVACEAE—AGAVE FAMILY

1. Ovary inferior; flowers greenish yellow, up to 2.5 cm across; leaves without filaments along the margins .... 1. *Manfreda*
1. Ovary superior; flowers white, up to 4 cm across; leaves with filaments along the margins .... 2. *Yucca*

1. **Manfreda** Salisb.—Aloe

1. **Manfreda virginica** (L.) Rose. American Agave. May–July. Sandstone outcroppings, dry woods, occasional; mostly restricted to the s. ¼ of the state; also Jersey and St. Clair cos. *Agave virginica* L.; *Polianthes virginica* (L.) Shinners.

2. **Yucca** L.—Yucca

1. Perianth parts 3–5 cm long; inflorescence pubescent .... 1. *Y. flaccida*
1. Perianth parts 5–7 cm long; inflorescence glabrous .... 2. *Y. smalliana*

1. **Yucca flaccida** Haworth. Adam's Needle. May–July. Dry soil; adventive in s. IL. in a cemetery; Jackson Co.

2. **Yucca smalliana** Fern. Yucca; Adam's Needle. May–July. Native to the e. coast of the U.S.; escaped from cultivation along roads and in cemeteries; occasional throughout IL. *Yucca filamentosa* L., misapplied; *Y. filamentosa* L. var. *smalliana* (Fern.) Ahles.

164. ALISMATACEAE—WATER PLANTAIN FAMILY

1. Receptacle flat, bearing a single ring of pistils; stamens 6–9; flowers perfect .... 1. *Alisma*
1. Receptacle convex, bearing several rows of pistils; stamens 12–numerous (6–9 in *Echinodorus parvulus*); flowers unisexual or perfect.
   2. Achenes not winged; base of whorled inflorescence branches bearing 3 bracts and several bracteoles; flowers perfect; stamens never more than 21 .... 2. *Echinodorus*
   2. Achenes winged; base of whorled inflorescence branches bearing 3 bracts and no bracteoles; flowers unisexual; stamens usually more than 21...... 3. *Sagittaria*

1. **Alisma** L.—Water Plantain

1. Petals 1.0–2.5 mm long; flowers at most 3.5 mm broad; achenes 1.5–2.2 mm long .... 1. *A. subcordatum*
1. Petals 3.5–6.0 mm long; flowers at least 7 mm broad; achenes 2.1–3.0 mm long .... 2. *A. triviale*

1. **Alisma subcordatum** Raf. Small-flowered Water Plantain. July–Aug. Shallow water, marshes, ditches; common throughout the state. *Alisma plantago-aquatica* L. var. *parviflorum* (Pursh) Torr.

2. **Alisma triviale** Pursh. Water Plantain. July–Aug. Shallow water, marshes, ditches, not common; confined primarily to the n. ½ of the state; also Wabash Co. *Alisma plantago-aquatica* L. var. *americanum* Roem. & Schultes.

2. **Echinodorus** Rich.—Burhead

1. Plants erect, less than 10 cm tall; leaves linear to lanceolate; flowers at most only 6 mm across; stamens 6–9; achenes 10–20, beakless or nearly so......... 3. *E. tenellus*
1. Plants erect and more than 10 cm tall, or plants creeping or arching; leaves broadly ovate, rarely lanceolate; flowers at least 8 mm across; stamens 12–21; achenes more than 40, beaked.
   2. Scape creeping or arching; stamens 21; style shorter than the ovary; beak of achene incurved .... 2. *E. cordifolius*
   2. Scape erect; stamens 12; style longer than the ovary; beak of achene straight.... .... 1. *E. berteroi*

1. **Echinodorus berteroi** (Spreng.) Fassett. Burhead. July–Sept. Wet ditches, edges of swamps; scattered in the s. ¾ of the state; also Bureau and Carroll cos. *Echinodorus berteroi* (Spreng.) Fassett var. *lanceolatus* (Wats. & Coult.) Fassett; *Echinodorus rostratus* (Nutt.) Engelm. ex Gray.

2. **Echinodorus cordifolius** (L.) Griseb. Creeping Burhead. July–Sept. Wet ditches, edges of swamps; scattered in IL.

3. **Echinodorus tenellus** (Mart.) Buch. Small Burhead. Aug.–Sept. Wet ditches, rare; Cass, Mason, and St. Clair cos. *Echinodorus tenellus* (Mart.) Buchenau var. *parvulus* (Engelm.) Fassett; *E. parvulus* Engelm.

3. **Sagittaria** L.—Arrowhead

1. Pedicels of fruits recurved, rarely spreading; sepals of pistillate flowers mostly erect.
   2. Fruiting heads 1.2–2.0 cm in diameter; leaves hastate to sagittate .... 4. *S. calycina*
   2. Fruiting heads up to 1.2 cm in diameter; leaves linear-ovate to lance-elliptic .... 9. *S. platyphylla*
1. Pedicels of fruits spreading to ascending or absent; sepals of pistillate flowers mostly spreading to recurved.
   3. Filaments pubescent.
      4. Pistillate flowers sessile or nearly so .... 10. *S. rigida*
      4. Pistillate flowers pedicellate.
         5. Rhizomes present; stolons and corms absent .... 7. *S. graminea*
         5. Rhizomes absent; stolons and corms present .... 5. *S. cristata*
   3. Filaments glabrous.

6. Emersed leaves linear to ovate.
   7. Leaves lanceolate to ovate; emersed plants with erect to ascending petioles........ 1. *S. ambigua*
   7. Leaves linear to sagittate; emersed plants with recurved petioles........ 6. *S. cuneata*
6. Emersed leaves cordate, sagittate, or hastate.
   8. Bracts free or connate for less than ¼ their length.
      9. Petioles winged in cross-section; beak of achene strongly recurved........ 2. *S. australis*
      9. Petioles ridged in cross-section; beak of achene ascending, not recurved........ 3. *S. brevirostra*
   8. Bracts connate for at least ¼ their length.
      10. Beak of achene 1–2 mm long, horizontal........ 8. *S. latifolia*
      10. Beak of achene less than 1 mm long, erect or curved........ 6. *S. cuneata*

1. **Sagittaria ambigua** J. G. Sm. Arrowhead. May–Sept. Shallow water, ditches, borders of lakes and ponds, scattered but rare; primarily in the cent. cos.

2. **Sagittaria australis** (J. G. Sm.) Small. Southern Arrowhead. July–Sept. Springy woods, rare; Alexander, Pope, and Union cos.; also in a few nw. cos. *Sagittaria longirostra* (Micheli) J. G. Sm.; *S. engelmanniana* J. G. Sm. ssp. *longirostra* (J. G. Sm.) Bogin.

3. **Sagittaria brevirostra** Mack. & Bush. Short-beaked Arrowhead. June–Sept. Sloughs, shorelines, and shallow water; occasional throughout the state. *Sagittaria engelmanniana* J. G. Sm. ssp. *brevirostra* (Mack. & Bush) Bogin.

4. **Sagittaria calycina** Engelm. Arrowhead. June–Sept. Marshes, pond margins, shorelines, shallow water; locally throughout cent. and s. IL, rare elsewhere. (Including var. *maxima* Engelm.) *Sagittaria montevidensis* Cham. & Schlecht.; *Lophotocarpus calycinus* (Engelm.) J. G. Sm.; *Sagittaria montevidensis* Cham. & Schlecht. ssp. *calycina* (Engelm.) Bogin.

5. **Sagittaria cristata** Engelm. Arrowhead. June–Sept. Borders of lakes and ponds, rare; confined to the nw. cos.

6. **Sagittaria cuneata** Sheld. Arrowhead. June–Sept. Mud and water in sloughs and along waterways; confined to the n. ½ of the state; also Calhoun Co.

7. **Sagittaria graminea** Michx. Grass-leaved Arrowhead. May–Sept. Swamps, mud, sand, or shallow water; uncommon but scattered in the n. ½ of the state, rare in the s. ½.

8. **Sagittaria latifolia** Willd. Common Arrowhead. June–Oct. Swamps, sloughs, ponds, shorelines, shallow water, mud; occasional to common throughout the state. (Including f. *diversifolia* [Engelm.] B. L. Robins.)

9. **Sagittaria platyphylla** (Engelm.) J. G. Sm. Arrowhead. July–Sept. Streams and lakes, very rare; St. Clair Co. *Sagaittaria graminea* Michx. var. *platyphylla* Engelm.

10. **Sagittaria rigida** Pursh. Stiff Arrowhead. May–Oct. Shallow water, mud, sand, in swamps, margins of ponds, or along waterways; not common but scattered in IL.

165. ALLIACEAE—ONION FAMILY

1. Bulbs with a strong odor of garlic or onion........ 1. *Allium*
1. Bulbs without an odor of garlic or onion........ 2. *Nothoscordum*

1. **Allium** L.—Onion

1. Leaves absent at flowering time, some or all of them, when present, more than 1.5 cm broad.
   2. Petioles and leaf sheaths reddish; blades mostly 2.6–6.0 cm broad, elliptic........ 12. *A. tricoccum*
   2. Petioles and leaf sheaths greenish or whitish; blades mostly 0.8–2.0 cm broad, lanceolate........ 2. *A. burdickii*
1. Leaves present at flowering time, up to 1.5 cm broad.
   3. Leaves flat or channeled, not hollow.
      4. Leaves extending about ½ way up the stem; flowers whitish, greenish, or deep purple.
         5. Umbel producing bulblets; flowers whitish or greenish........ 9. *A. sativum*

5. Umbel not producing bulblets; flowers deep purple .... 1. *A. ampeloprasum*
4. Leaves basal or nearly so; some of the flowers usually pinkish.
6. Umbels producing bulblets .... 3. *A. canadense*
6. Umbels not producing bulblets.
7. Stems solid .... 8. *A. porrum*
7. Stems hollow.
8. Ovary and capsule crested near apex; outer bulb scales membranous; perianth parts shorter than the stamens at maturity.
9. Umbel nodding; leaves soft .... 5. *A. cernuum*
9. Umbel erect; leaves stiff .... 11. *A. stellatum*
8. Ovary and capsule without crests; outer bulb scales fibrous; perianth parts usually longer than or about equal to the stamens at maturity .... 7. *A. mutabile*
3. Leaves terete, hollow.
10. Stems inflated.
11. Pedicels much longer than the flowers .... 4. *A. cepa*
11. Pedicels about as long as or shorter than the flowers .... 6. *A. fistulosum*
10. Stems not inflated.
12. Leaves extending nearly to middle of stem; umbel bulblet-bearing .... 13. *A. vineale*
12. Leaves basal or nearly so; umbel not bulblet-bearing .... 10. *A. schoenoprasum*

1. **Allium ampeloprasum** L. var. **atroviolaceum** (Boiss.) Regel. Wild Leek. July–Sept. Native to Europe and Asia; rarely escaped along roads; Jackson, Pope, and Union cos.

2. **Allium burdickii** (Hanes) A. G. Jones. Burdick's Leek. June–July. Low, moist woods; occasional in the n. ½ of the state, rare elsewhere. *Allium tricoccum* Ait. var. *burdickii* Hanes.

3. **Allium canadense** L. Two varieties occur in Illinois:

a. Some or all the flowers replaced by bulblets .... 3a. *A. canadense* var. *canadense*
a. Bulblets absent .... 3b. *A. canadense* var. *lavendulare*

3a. **Allium canadense** L. var. **canadense** Wild Onion. May–July. Dry woods, prairies, disturbed soil, common; in every co.

3b. **Allium canadense** L. var. **lavendulare** (J. M. Bates) Ownbey & Aase. Wild Onion. May–July. Calcareous soil; n. IL.

4. **Allium cepa** L. Onion. May–Aug. Native to sw. Asia; seldom escaped from cultivation; scattered in IL.

5. **Allium cernuum** Roth. Nodding Onion. July–Sept. Wooded banks, not common; restricted to ne. IL.

6. **Allium fistulosum** L. Spring Onion. May–Aug. Native to Asia; rarely escaped from cultivation; Lake and Will cos.

7. **Allium mutabile** Michx. Wild Onion. May–July. Dry habitats; occasional throughout the state. *Allium canadense* L. var. *mobilense* (Regel) Ownbey.

8. **Allium porrum** L. Leek. May–July. Cultivated species rarely escaped from cultivation; Edwards, Johnson, and Union cos.

9. **Allium sativum** L. Garlic. June–Sept. Native to Europe and w. Asia; occasionally escaped from cultivation.

10. **Allium schoenoprasum** L. Two varieties occur in Illinois:

a. Bulbs several .... 10a. *A. schoenoprasum* var. *schoenoprasum*
a. Bulbs 1 or 2 .... 10b. *A. schoenoprasum* var. *sibiricum*

10a. **Allium schoenoprasum** L. var. **schoenoprasum.** Chives. June–Aug. Native to Europe; rarely escaped from cultivation; scattered in IL.

10b. **Allium schoenoprasum** L. var. **sibiricum** (L.) Hartm. Chives. June–Aug. Native to Europe and Asia; rarely escaped from cultivation; Kankakee Co.

11. **Allium stellatum** Ker. Cliff Onion. July–Sept. Hill prairies in calcareous areas; Jackson, McHenry, Monroe, Randolph, and Union cos.

12. **Allium tricoccum** L. Wild Leek; Wild Ramp. June–July. Moist, rich woods; occasional in the n. ½ of the state, rare elsewhere.

13. **Allium vineale** L. Field Garlic. May–Aug. Native to Europe; naturalized in waste ground, common; in every co.

2. **Nothoscorsum** Kunth—False Garlic

1. **Nothoscordum bivalve** (L.) Britt. False Garlic. Apr.–June; Sept.–Oct. Dry woods, bluffs, prairies; rather common in the s. ½ of the state; also DuPage Co.

166. AMARYLLIDACEAE—AMARYLLIS FAMILY

1. Flowers campanulate, with the white perianth segments tipped with green .... 2. *Leucojum*
1. Flowers not campanulate, the perianth parts not green-tipped.
  2. Flowers red, blooming when the leaves are absent .... 3. *Lycoris*
  2. Flowers yellow, white, or yellow-green, blooming when the leaves are present.
    3. Flowers recurved; perianth tube (not corona) up to 2 (–4) cm long; stamens included within the corona; ovules several per cell of ovary; bract 1 .... 4. *Narcissus*
    3. Flowers ascending; perianth tube (not corona) 5–10 cm long; stamens exserted from the corona; ovules 2 per cell of ovary; bracts 2–several .... 1. *Hymenocallis*

1. **Hymenocallis** Salisb.—Spider Lily

1. **Hymenocallis caroliniana** (L.) Herb. Spider Lily. July–Sept. Low woods, swamps; restricted to the s. ⅙ of the state; also Wabash Co. *Hymenocallis occidentalis* (LeConte) Kunth.

2. **Leucojum** L.—Snowflake

1. **Leucojum aestivum** L. Summer Snowflake. Apr.–May. Native to Europe; rarely escaped from cultivation; Pope Co.

3. **Lycoris** Herb.—Surprise Lily

1. **Lycoris radiata** Herb. Surprise Lily. June–July. Native to Asia; rarely escaped from cultivation; Jackson and Pulaski cos.

4. **Narcissus** L.—Daffodil; Narcissus

1. Flower solitary; leaves flat.
  2. Perianth white; corona white or yellow, cupular, less than ¼ as long as the perianth.
    3. Corona white with a red rim; flower solitary .... 3. *N. poeticus*
    3. Corona yellow; flowers several .... 2. *N. X medioluteus*
  2. Perianth yellow; corona yellow, tubular, about as long as the perianth .... 4. *N. pseudo-narcissus*
1. Flowers 2–20; leaves terete .... 1. *N. jonquilla*

1. **Narcissus jonqjuill** Jonquil. Mar. Native to Portugal and Spain; escaped from gardens; scattered in IL.

2. **Narcissus X medioluteus** Mill. Primrose Peerless. Apr.–May. Rarely escaped from cultivation; DuPage and Lake cos. Reputed to be the hybrid between *N. poeticus* L. and *N. tazetta* L.

3. **Narcissus poeticus** L. Poet's Narcissus. Apr.–May. Native to Europe; rarely escaped from cultivation along roads; scattered but not common in IL.

4. **Narcissus pseudo-narcissus** L. Daffodil. Apr.–May. Native to Europe; occasionally escaped from cultivation; scattered in IL.

167. ARACEAE—ARUM FAMILY

1. Free-floating aquatic plants, with leaves in rosettes .... 5. *Pistia*
1. Terrestrial plants or, if aquatic, not free-floating; leaves not in rosettes.
  2. Some or all the flowers perfect; spadix globose or short-cylindric; plants with rhizomes.
    3. Spadix globose; spathe more or less enclosing the spadix, not petaloid; perianth 4-parted; plants foul-smelling .... 6. *Symplocarpus*

3. Spadix short-cylindric; spathe broad, flat, petaloid, subtending but not enclosing the spadix; perianth absent; plants not foul-smelling .... 3. *Calla*
2. All flowers unisexual; spadix elongated; plants with corms or fleshy roots.
4. Staminodia absent; berries red; leaves compound .... 1. *Arisaema*
4. Staminodia present; berries brownish or greenish; leaves simple.
5. Plants rhizomatous; spathe tubular only at base; flowers confined to the lower ½ of the spadix .... 2. *Arum*
5. Plants with fleshy roots; spathe tubular at both ends, opening at the middle; flowers covering all or most of the spadix.................. 4. *Peltandra*

1. **Arisaema** Mart.—Jack-in-the-pulpit

1. Leaflets (3–) 5–13; spathe convolute below and above, open near the middle; spadix slender-tapering, exserted beyond the spathe .... 1. *A. dracontium*
1. Leaflets 3; spathe convolute below, open and arching above; spadix club-shaped, obtuse, covered by the arching spathe .... 2. *A. triphyllum*

1. **Arisaema dracontium** (L.) Schott. Green Dragon. May–June. Rich woods, common; probably in every co.

2. **Arisaema triphyllum** (L.) Schott. Two subspecies may be recognized in Illinois:

a. Lateral leaflets strongly asymmetrical; extended portion of spathe 4–7 cm wide; fruiting head 3–6 cm long .... 2a. *A. triphyllum* ssp. *triphyllum*
a. Lateral leaflets scarcely asymmetrical; extended portion of spathe to 3 cm wide; fruiting head to 2 cm long .... 2b. *A. triphyllum* ssp. *pusillum*

2a. **Arisaema triphyllum** (L.) Schott. ssp. **triphyllum**. Jack-in-the-pulpit. Apr.–May. Rich woods, common; in every co. There is considerable color variation in the spathes. *Arisaema atrorubens* (Ait.) Blume.

2b. **Arisaema triphyllum** (L.) Schott. ssp. **pusillum** (Peck) Huttleston. Small Jack-in-the-pulpit. Apr.–May. Rich woods, rare; Alexander and Jackson cos.

2. **Arum** L.—Arum

1. **Arum italicum** Mill. Arum. May. Native to w. Europe and n. Africa; rarely escaped from cultivation; Jackson Co.

3. **Calla** L.—Water Arum

1. **Calla palustris** L. Wild Calla. May–July. Wet soil, rare; Lake Co.

4. **Peltandra** Raf.—Arrow Arum

1. **Peltandra virginica** (L.) Schott. & Endl. Arrow Arum. May–June. Swamps, shallow water, wet ditches; occasional in the s. ⅘ of the state; also Cook, Kane, and Lee cos.

5. **Pistia** L.—Water Lettuce

1. **Pistia stratiotes** L. Water Lettuce. June–July. Native to the se. U.S.; rarely adventive in IL; Coles and DuPage cos.

6. **Symplocarpus** Salisb.—Skunk Cabbage

1. **Symplocarpus foetidus** (L.) Nutt. Skunk Cabbage. Feb.–Mar. Swamps, low areas; occasional in the n. ¾ of the state.

168. ASPARAGACEAE—ASPARAGUS FAMILY

1. **Asparagus** L.—Asparagus

1. **Asparagus officinalis** L. Asparagus. May–June. Native to Europe; commonly escaped from cultivation; probably in every co.

169. BUTOMACEAE—FLOWERING RUSH FAMILY

1. **Butomus** L.—Flowering Rush

1. **Butomus umbellatus** L. Flowering Rush. July–Sept. Native to Europe; rarely naturalized in IL; Cook, DuPage, and Rock Island cos.

170. COLCHICACEAE—COLCHICUM FAMILY

1. **Uvularia** L.—Bellwort

1. Leaves perfoliate, puberulent beneath; flowers bright yellow, the perianth parts nearly all over 25 mm long .... 1. *U. grandiflora*
1. Leaves sessile, not perfoliate, glabrous; flowers stramineous, the perianth parts nearly all less than 25 mm long .... 2. *U. sessilifolia*

1. **Uvularia grandiflora** Sm. Yellow Bellwort. Apr.–May. Rich woods; occasional to common throughout the state.

2. **Uvularia sessilifolia** L. Sessile-leaved Bellwort. Apr.–May. Rich woods, not common; confined to the s. ⅓ of the state.

171. COMMELINACEAE—DAY-FLOWER FAMILY

1. Calyx bilaterally symmetrical, with 2 of the sepals connate at the base; corolla bilaterally symmetrical; fertile and sterile stamens each 3 (rarely 3 plus 2), the filaments glabrous; ovary 3-celled, with 1 cell aborted; inflorescence cymose, from a spathe .... 1. *Commelina*
1. Calyx radially symmetrical; corolla radially symmetrical; fertile stamens 6, the filaments villous; ovary 3-celled, with all cells fertile; inflorescence appearing umbellate, from an elongated, leaflike bract .... 2. *Tradescantia*

1. **Commelina** L.—Day-flower

1. Plants rooting at the nodes; margins of spathes free; seeds black, rugose or reticulate.
  2. Plants annual; leaf sheaths glabrous at summit; anthers 6; seeds 4; petals 10–15 mm long, the lower petal white; seeds 3.5–4.0 mm long, rugose .... 1. *C. communis*
  2. Plants perennial; leaf sheaths ciliate at summit; anthers 5; seeds 5; petals 6–8 mm long, the lower petal blue; seeds 2.0–2.5 mm long, reticulate .... 2. *C. diffusa*
1. Plants not rooting at the nodes; margins of spathes united for about ⅓ the way up from the base; seeds reddish or brown, smooth or puberulent, not rugose or reticulate.
  3. Plants with thick, fleshy roots; lowest petal white, much smaller than the other 2; seeds about 3 mm long, brown .... 3. *C. erecta*
  3. Plants with rhizomes; lowest petal blue, only slightly smaller than the other 2; seeds about 6 mm long, reddish .... 4. *C. virginica*

1. **Commelina communis** L. Common Day-flower. June–Oct. Native to Asia; naturalized in moist waste ground; occasional to common throughout the state.

2. **Commelina diffusa** Burm. f. Small Day-flower. July–Oct. Moist woodlands and moist waste ground; occasional in the s. ⅔ of the state.

3. **Commelina erecta** L. Three varieties may be distinguished in Illinois:

a. Stems 0.6–1.2 m tall; leaves lanceolate to lance-ovate, some or all of them 2–4 cm wide .... 3a. *C. erecta* var. *erecta*
a. Stems up to 0.6 m tall; leaves linear to linear-lanceolate, up to 2 cm wide.
  b. Spathes 1.0–2.5 cm long; leaves 4–10 cm long........ 3b. *C. erecta* var. *angustifolia*
  b. Spathes 2.5–3.0 cm long; leaves 7–15 cm long .... 3c. *C. erecta* var. *deamiana*

3a. **Commelina erecta** L. var. **erecta.** Day-flower. June–Sept. Moist or dry sandy soil; occasional throughout the state.

3b. **Commelina erecta** L. var. **angustifolia** (Michx.) Fern. Narrow-leaved Day-flower. June–Sept. Dry, sandy soil, rare; confined to the s. ⅙ of IL; also Grundy Co.

3c. **Commelina erecta** L. var. **deamiana** Fern. Deam's Day-flower. June–Sept. Dry, sandy soil, rare; Cook and Kankakee cos.

4. **Commelina virginica** L. Virginia Day-flower. June–Sept. Low woods, not common; mostly restricted to the s. ⅓ of the state; also McDonough Co.

2. **Tradescantia** L.—Spiderwort

1. Stems flexuous; leaves usually 2–4 cm wide; cymes several, terminal and lateral; sepals 5–10 mm long .... 3. *T. subaspera*

1. Stems straight; leaves usually 0.5–1.5 cm wide; cyme usually solitary, terminal; sepals 8–15 mm long.
  2. Stems generally 4–10 dm tall; leaves and stems glabrous and glaucous; sepals glabrous or with a tuft of eglandular hairs at the tip .... 2. *T. ohiensis*
  2. Stems generally 0.5–4.0 dm tall; leaves and stems glabrous or pubescent, but not glaucous; sepals pubescent throughout with glandular or eglandular hairs.
    3. Pubescence of sepals eglandular..4. *T. virginiana*
    3. Pubescence of sepals glandular.. 1. *T. bracteata*

1. **Tradescantia bracteata** Small. Prairie Spiderwort. May–Aug. Prairies; scattered in the w. side of the state.

2. **Tradescantia ohiensis** Raf. Ohio Spiderwort. Apr.–Aug. Edges of woods, prairies, common; probably in every co.

3. **Tradescantia subaspera** Ker. Two varieties occur in Illinois:

a. Some or all the uppermost lateral cymes sessile..... 3a. *T. subaspera* var. *subaspera*
a. All of the lateral cymes pedunculate .... 3b. *T. subaspera* var. *montana*

3a. **Tradescantia subaspera** Ker var. **subaspera.** Broad-leaved Spiderwort. May–Aug. Woods; occasional in the s. ⅔ of the state, rare in the n. ⅓.

3b. **Tradescantia subaspera** Ker var. **montana** (Shuttlew.) Anders. & Woodson. Mountain Spiderwort. May–Aug. Rich woods, very rare; Jackson Co.

4. **Tradescantia virginiana** L. Virginia Spiderwort. Apr.–June. Woods, prairies; common in the s. ⅔ of the state, rare in the n. ⅓.

172. CYPERACEAE—SEDGE FAMILY

1. Flowers bisexual; achenes not enclosed in a saclike structure (perigynium) nor white in color (except sometimes in *Cyperus erythrorhizos*).
  2. Spikelets 1- or 2-flowered.
    3. Spikelets 1-flowered; plants with a sweet odor ....13. *Kyllinga*
    3. Spikelets 1- or 2-flowered; plants without a sweet odor.
      4. Achene crowned with a tubercle; bristles subtending lower flower ....15. *Rhynchospora*
      4. Achene without a tubercle; bristles absent.................................4. *Cladium*
  2. Spikelets 2- or more-flowered.
    5. Spikelets flattened; scales 2-ranked.
      6. Inflorescence terminal; bristles absent; achene without a tubercle............ .... 5. *Cyperus*
      6. Inflorescence axillary; bristles present; achene with a tubercle.................. ....6. *Dulichium*
    5. Spikelets not flattened; scales spirally arranged.
      7. Spikelet 1 per culm; leaf blades absent.....................................7. *Eleocharis*
      7. Spikelets more than 1 per culm; leaf blades usually present (if absent, the inflorescence lateral).
        8. Achene crowned by a tubercle .... 2. *Bulbostylis*
        8. Achene without a tubercle.
          9. Involucral bract 1, appearing like the bract of the spikelet and not appearing like the continuation of the culm ....19. *Trichophorum*
          9. Involucral bracts 1–several, if 1, then appearing like the continuation of the culm.
            10. Flowers and fruits subtended by bristles or scales.
              11. Flowers and fruits subtended only by scales.
                12. Flowers and fruits subtended by minute, rudimentary scales... 14. *Lipocarpha*
                12. Flowers and fruits subtended by papery scales................ .... 11. *Hemicarpha*
              11. Flowers and fruits subtended by capillary bristles (in addition, 3 scales present in *Fuirena*).
                13. Flowers and fruits subtended by 3 bristles and 3 scales... ....10. *Fuirena*
                13. Flowers and fruits subtended only by bristles.

14. Inflorescence lateral, the culm extending beyond the inflorescence .... 16. *Schoenoplectus*
14. Inflorescence terminal, the culm not extending beyond the inflorescence.
15. Bristles white at maturity, forming cottony heads .... 8. *Eriophorum*
15. Bristles reddish, brown, or tawny, not forming cottony heads.
16. Spikelets at least 1 cm long, the scales pubescent .... 1. *Bolboschoenus*
16. Spikelets up to 1 cm long, the scales glabrous .. .. 17. *Scirpus*
10. Flowers and fruits not subtended by bristles or scales.
17. Inflorescence lateral, the culm extending beyond the inflorescence.
18. Stigmas 3; fruits trigonous .... 12. *Isolepis*
18. Stigmas 2; fruits biconvex.
19. Achenes with vertical pebbles .... 10. *Hemicarpha*
19. Achenes cross-wrinkled .... 15. *Schoenoplectus*
17. Inflorescence terminal, the culm not extending beyond the inflorescence.
20. Leaves up to 4 mm broad .... 9. *Fimbristylis*
20. Leaves usually at least 5 mm broad .... 17. *Scirpus*
1. Flowers unisexual; achenes either enclosed in a saclike structure (perigynium) or white in color.
21. Flowers and achenes enclosed in a saclike structure (perigynium); achenes usually not white .... 3. *Carex*
21. Flowers and achenes not enclosed in a saclike structure (perigynium); achenes white .... 18. *Scleria*

1. **Bolboschoenus** (Asch.) Palla

1. Achene sharply trigonous, 4–5 mm long; bristles subtending flowers and fruits stout, persistent .... 1. *B. fluviatilis*
1. Achene obtusely trigonous or not trigonous, 2.5–4.0 mm long; bristles subtending flowers and fruits weak, deciduous .... 2. *B. maritimus*

1. **Bolboschoenus fluviatilis** (Torr.) Sojak. River Bulrush. July–Sept. Margins of streams and lakes, marshes; occasional in the n. ¾ of IL, rare elsewhere. *Scirpus fluviatilis* (Torr.) Gray.

2. **Bolboschoenus maritimus** (L.) Palla. Bayonet-grass. July–Sept. Shores and margins of ponds, rare; Cook, DuPage, Kankakee, Kendall, and LaSalle cos. *Scirpus maritimus* L.; *S. paludosus* A. Nels.; *S. maritimus* L. var. *paludosus* (A. Nels.) Kuk.; *Bolboschoenus paludosus* (A. Nels.) Soo.

2. **Bulbostylis** (Kunth) C. B. Clarke

1. **Bulbostylis capillaris** (L.) C. B. Clarke. Bulbostylis. July–Oct. Moist, sandysoil; occasional throughout the state. (Including var. *crebra* Fern. and var. *isopoda* Fern.)

3. **Carex** L.—Sedge

1. Hairs present on leaves, sheaths, and/or culms (this lead does not include plant parts that are merely scabrous or perigynia that may be papillose or granular .... Group 1
1. Leaves, sheaths, and/or culms all glabrous.
2. Perigynia pubescent .... Group 2
2. Perigynia glabrous.
3. Stigmas 3; achenes trigonous.
4. Spike 1, androgynous (with staminate flowers at tip, pistillate flowers below) .... 104. *C. leptalea*
4. Spikes more than 1 (sometimes only 1 in *C. squarrosa*), some of them entirely staminate or gynecandrous (with pistillate flowers at tip, stami-

nate flowers below) or androgynous in rare specimens of *C. sprengelii* and *C. retrorsa*).

5. Lowest scale of pistillate flower leaflike, 15–50 mm long (do not confuse this with bract that subtends the spike); pistillate spike with 2–4 perigynia .... Group 3

5. Lowest scale of pistillate flower less than 20 mm long, not leaflike; pistillate spikes usually with more than 4 perigynia.

6. Terminal spike entirely staminate.

7. Perigynia 10 mm long or longer .... Group 4

7. Perigynia up to 10 mm long, usually much shorter.

8. Staminate spikes more than 1 per culm .... Group 5

8. Staminate spike 1 per culm.

9. Leaves threadlike, 0.5–1.0 mm wide; perigynia 1.5–2.0 mm long .... 59. *C. eburnea*

9. Leaves filiform or wider, some or all of them more than 1 mm wide; perigynia usually more than 2 mm long.

10. Flowering culms hidden among the leaves; plants to 15 cm tall .... 176. *C. tonsa*

10. Flowering culms not hidden among the leaves; plants usually more than 15 cm tall.

11. Perigynium prominently 2-toothed at the tip or prolonged into a conspicuous bidentate beak .... Group 6

11. Perigynium ending abruptly at the tip, either without teeth or merely a small notch.

12. Most or all of the leaves, particularly on sterile shoots, 15 mm wide or wider .... Group 7

12. Most or all the leaves less than 15 mm wide.

13. At least the lowest pistillate spikes on flexuous or pendulous peduncles .... Group 8

13. All pistillate spikes ascending to erect.

14. Beak or tip of perigynium bent or curved to 1 side .... Group 9

14. Beak or tip of perigynium not bent or curved to 1 side .... Group 10

6. Terminal spike androgynous or gynecandrous .... Group 11

3. Stigmas 2; achenes lenticular.

15. Some or all the spikes pedunculate; staminate flowers usually on separate spikes .... Group 12

15. All spikes sessile, essentially alike, with staminate flowers either at the tip or at the base of each spike (exceptional specimens of *C. sterilis, C. bromoides,* and *C. praegracilis* may have a few all staminate or all pistillate flowers.)

16. Culms solitary or forming stoloniferous or rhizomatous colonies .... Group 13

16. Culms cespitose, not arising singly from extensive stolons and/or rhizomes.

17. Spikes androgynous .... Group 14

17. Spikes gynecandrous.

18. Perigynia plano-convex, with rounded margins or with narrowly rimmed margins .... Group 15

18. Perigynia more or less flat and scalelike, with winged margins .... Group 16

Group 1

Hairs present on leaves, sheaths, and/or culms.

1. Perigynia pubescent.

2. Terminal spike(s) entirely staminate.

3. Staminate spike 1, sessile or nearly so; perigynia 3.5–5.0 mm long, with teeth of the beak minute .... 88. *C. hirtifolia*

3. Staminate spikes 1–3, on long peduncles; perigynia 4.5–9.0 mm long, with teeth of the beak 1–2 mm long.

4. Leaves 2–4 mm wide; perigynia up to 35 per spke, 5–9 mm long; achenes 2.7–3.0 mm long .... 87. *C. hirta*

4. Leaves 3–8 mm wide; perigynia up to 70 per spike, 4.5–6.5 mm long; achenes 1.6–3.0 mm long .... 90. *C. houghtoniana*

2. Terminal spike gynecandrous (pistillate flowers at top, staminate flowers below.)

5. Spikes linear-cylindric; perigynia elliptic-ovoid to lance-ellipsoid, prominently nerved.

6. Perigynia elliptic-ovoid, 2.0–2.5 mm long, more or less compressed .... 189. *C. virescens*

6. Perigynia lance-ellipsoid, 3.5–5.0 mm long, angled .... 129. *C. oxylepis*

5. Spikes short-cylindric to subglobose; perigynia broadly ovoid, finely nerved .... 172. *C. swanii*

1. Perigynia glabrous.

7. At least the terminal spike staminate.

8. Staminate spikes 2–5; remainder of spikes usually androgynous; pistillate scales shorter than the perigynia; perigynia numerous per spike, lance-ovoid, 7–9 mm long, with a beak 1.5–3.0 mm long .... 14. *C. atherodes*

8. Staminate spike 1; lateral spikes entirely pistillate; pistillate scales as long as or longer than the perigynia; perigynia up to 40 per spike, elliptic-triangular to obovoid, up to 5 mm long, beakless or with a beak about 1 mm long.

9. Leaves and sheaths pubescent; leaves up to 3 mm wide; perigynia up to 40 per spike, elliptic-triangular, up to 3 mm long, beakless .... 130. *C. pallescens*

9. Only the sheaths pubescent; leaves 3–7 mm wide; perigynia up to 10 per spike, obovoid, 4–5 mm long, with beak about 1 mm long .... 89. *C. hitchcockiana*

7. Terminal spike gynecandrous.

10. Spikes narrowly cylindric, not clavate at the base, the lowest on long, flexuous or pendulous peduncles; perigynia ellipsoid to narrowly ovoid to oblong-ovoid, more or less trigonous, short-beaked.

11. Leaves up to 1.8 mm wide; terminal spike up to 2 mm thick; pistillate scales obtuse to acute; perigynia up to 20 per spike, ellipsoid .... 70. *C. formosa*

11. Leaves 3–8 mm wide; terminal spike more than 2 mm thick; pistillate scales awned; perigynia up to 40 per spike, narrowly ovoid to oblong-ovoid.

12. Pistillate spikes 5–7 mm thick; pistillate scales as long as or longer than the perigynia; perigynia oblong-ovoid, loose around the achene .... 50. *C. davisii*

12. Pistillate spikes 3–5 mm thick; pistillate scales shorter than the perigynia; perigynia narrowly ovoid, closely enveloping the achene .... 129. *C. oxylepis*

10. Spikes broadly cylindric, clavate at the base, on short, ascending peduncles; perigynia obovoid, rounded on all faces, or with a flat, inner face, pointed at the tip.

13. Perigynia trigonous in cross-section.

14. Leaves densely pubescent; pistillate scales acute to cuspidate .... 86. *C. hirsutella*

14. Leaves sparsely pubescent; pistillate scales acute to short-awned .... 38. *C. complanata*

13. Perigynia circular in cross-section.

15. Pistillate scales awned, longer than the perigynia; leaves densely pubescent; achenes 2.0–2.5 mm long .... 27. *C. bushii*

15. Pistillate scales acute to acuminate, shorter than the perigynia; leaves sparsely pubescent; achenes 1.5–2.2 mm long .... 31. *C. caroliniana*

Group 2
Leaves, stems, and sheaths all glabrous; perigynia pubescent.

1. Terminal spike androgynous (staminate flowers at the tip, pistillate flowers below) .... 132. *C. pedunculata*
1. Terminal spike gynecandrous (pistillate flowers at the tip, staminate flowers below) or entirely staminate.
  2. Terminal spike gynecandrous .... 31. *C. caroliniana*
  2. Terminal spike entirely staminate.
    3. Staminate spikes 2 or more per culm.
      4. Lower spikes sometimes androgynous; pistillate scales lanceolate to broadly lanceolate, slightly shorter than to longer than the perigynia; perigynia 2–5 mm long, the teeth of the beak up to 1 mm long.
        5. Culms smooth; leaves up to 2 mm wide, convolute, without a midvein; pistillate scales slightly shorter than the perigynia; perigynia obscurely nerved, 3–5 mm long, the teeth of the perigynia up to 0.6 mm long .... 100. *C. lasiocarpa*
        5. Culms usually somewhat scabrous; leaves 2–5 mm wide, flat, with a conspicuous midvein; pistillate scales as long as or longer than the perigynia; perigynia obviously nerved, 2.0–3.5 mm long, the teeth of the perigynia about 1 mm long .... 133. *C. pellita*
      4. Lower spikes almost always pistillate; pistillate scales broadly ovate, shorter than the perigynia; perigynia 4.2–10.0 mm long, the teeth of the beak 1.2–2.0 mm long.
        6. Perigynia strongly nerved, up to 40 per pistillate spike, 5–10 mm long .... 181. *C. trichocarpa*
        6. Perigynia obscurely nerved, up to 135 per pistillate spike, 4.2–6.4 mm long .... 172. *C. X subimpressa*
    3. Staminate spike 1 per culm.
      7. Perigynia 12 mm long or longer, crowded into a globose spike .... 79. *C. grayi*
      7. Perigynia up to 4.5 mm long, not forming a globose spike.
        8. Plants more than 60 cm tall; terminal staminate spike up to 4 cm long; perigynia 40–75 per spike .... 133. *C. pellita*
        8. Plants up to 60 cm tall, usually much shorter; terminal staminate spike up to 2 cm long; perigynia up to 25 per spike.
          9. Perigynia spongy at base; culms capillary to slender and flexuous; leaves pale green or glaucous.
            10. Culms and peduncles capillary; perigynia 1–8 per spike, 3.8–4.5 mm long .... 132. *C. pedunculata*
            10. Culms and peduncles slender and flexuous but not capillary; perigynia 10–25 per spike, 2.0–3.5 mm long .... 150. *C. richardsonii*
          9. Perigynia firm at base; culms neither capillary nor flexuous; leaves green to dark green.
            11. Some or all the fertile culms hidden among the leaf bases.
              12. Some fertile culms sometimes elevated above the leaves; plants with stolons; body of perigynium ellipsoid, yellow-green; beak of the perigynium about ⅓ as long as the body of the perigynium .... 123. *C. nigromarginata*
              12. All spikes hidden at base of plant; plants without stolons; body of perigynium orbicular, not yellow-green; beak of the perigynium about ½ as long as the body of the perigynium.
                13. Perigynia very sparsely pubescent .... 177. *C. tonsa*
                13. Perigynia uniformly pubescent.
                  14. Beak of the perigynium 2.5–4.0 mm long, about as long as the body of the perigynium; leaves bright green .... 152. *C. rugosperma*
                  14. Beak of the perigynium 0.8–1.7 mm long, ½ to ¾ as long as the body of the perigynium; leaves green.

15. Beak of perigynium about ¾ as long as the body; pistillate scales acuminate; perigynia 2.2–3.3 mm long .... 185. *C. umbellata*

15. Beak of perigynium ½–⅔ as long as the body; pistillate scales acute; perigynia 3.2–4.5 mm long .... 1. *C. abdita*

11. All fertile culms usually protruding above the leaves.

16. Perigynia ovoid to orbicular.

17. Beak of the perignium at least ⅔ the length of the body of the perigynium; perigynia yellow-green......................... .... 110. *C. lucorum*

17. Beak of the perigynium up to ¼ to ⅓ as long as the body of the perigynium; perigynia pale green or green.

18. Perigynia more or less terete in cross-section; leaves usually stiff .... 84. *C. heliophila*

18. Perigynia trigonous in cross-section; leaves usually soft.

19. Plants with slender stolons; leaves 1.5–3.0 mm wide .... 139. *C. pensylvanica*

19. Plants without stolons; leaves 1.8–5.0 mm wide .... 30. *C. communis*

16. Perigynia ellipsoid.

20. Plants with long stolons; beak of the perigynium ⅓ to ½ as long as the body of the perigynium .... 135. *C. physorhyncha*

20. Plants without stolons; beak of the perigynium up to ½ as long as the body of the perigynium

21. Culms and leaves often reclining, weak; staminate scales long-acuminate; leaves up to 1.5 mm wide .... 62. *C. emmonsii*

21. Culms and leaves more or less erect or ascending, not weak; staminate scales obtuse to short-acuminate; leaves often 1.5–5.0 mm wide.

22. Pistillate scales shorter than the perigynia, exposing the perigynia; achenes 2.0–2.5 mm long .... 131. *C. peckii*

22. Pistillate scales about as long as the perigynia, not exposing the perigynia; achenes 1.3–2.0 mm long.

23. Leaves 1.8–5.0 mm wide; staminate scales acute to short-acuminate; perigynia 1.2–1.6 mm wide... 36. *C. communis*

23. Leaves 1.0–2.5 mm wide; staminate scales obtuse to acute; perigynia 0.8–1.1 mm wide .... 5. *C. albicans*

Group 3

Leaves, sheaths, and/or culms all glabrous; perigynia glabrous; stigmas 3; achenes trigonous; perigynia less than 10 mm long; lowest scale of pistillate flower leaflike, 15–30 mm long; pistillate spikes with 2–4 perigynia.

1. Culms green at base; staminate part of spike often more than 6 mm long; staminate scales 1–2 mm long.

2. Perigynia orbicular, 2.3–2.5 mm wide; achenes 2.5 mm wide; stamnate scales truncate.. 95. *C. jamesii*

2. Perigynia mostly ellipsoid, about 1.5 mm wide; achenes 1.5–2.0 mm wide; staminate scales acute .... 192. *C. willdenowii*

1. Culms red-purple at base; staminate part of spike 3.3–6.0 mm long; staminate scales 2.0–3.2 mm long.. 176. *C. timida*

Group 4

Hairs absent on leaves, sheaths, culms, and perigynia; stigmas 3; achenes trigonous; terminal spike entirely staminate; perigynia 10 mm long or longer.

1. Perigynia not inflated, lanceoloid, up to 3.5 mm wide at its widest point .... 69. *C. folliculata*
1. Perigynia strongly inflated, subuloid to lanceoloid, usually 3.5 mm wide or wider at its widest point.
   2. Pistillate spikes globose or subglobose.
      3. Perigynia radiating in all directions in a spike; perigynia not shiny, cuneate at the base .... 79. *C. grayi*
      3. Perigynia radiating in all directions except retrorse; perigynia shiny, rounded at the base .... 94. *C. intumescens*
   2. Pistillate spikes thick-cylindric, longer than broad.
      4. All perigynia horizontally spreading; beak of perigynium 2–3 times longer than the body; achene truncate at summit, broader than long .... 73. *C. gigantea*
      4. Perigynia ascending; beak of perigynium somewhat shorter than or slightly longer than the body; achene narrowed at summit, longer than broad or about as long as broad.
         5. Pistillate spikes more than twice as long as broad; achenes about as long as broad, with conspicuously knobby angles and concave sides .... 111. *C. lupuliformis*
         5. Pistillate spikes up to twice as long as broad; achenes longer than broad, without conspicuous knobby angles and with nearly flat sides.
            6. Perigynia crowded, 13–20 mm long; pistillate scales long-acuminate to awned; leaves 5 mm wide or wider; culms several in a tuft........ 112. *C. lupulina*
            6. Perigynia loosely arranged, 10–14 mm long; pistillate scales obtuse to acute; leaves up to 6 mm wide; culms few or solitary, never in a tuft........ 108. *C. louisianica*

Group 5

Leaves, sheaths, and/or culms all glabrous; perigynia glabrous; stigmas 3; achenes trigonous; terminal spikes entirely staminate; perigynia up to 10 mm long, usually much shorter; staminate spikes more than 1 per culm.

1. Style disarticulating from mature achene; perigynium abruptly contracted to a beak as long as the body; lowest spikes often on pendulous peduncles........ 162. *C. sprengelii*
1. Style continuous and persistent on the mature achene; perigynium tapering to the beak, the beak usually somewhat shorter than the body of the perigynium; lowest spikes erect or ascending.
   2. Plants growing in large colonies from extensive rhizomes; perigynia thick and firm, not inflated.
      3. Culms red-purple at the base; leaves not glaucous nor blue-green; nerves of the perigynia conspicuous and slightly elevated.
         4. Teeth of the beak of the perigynium up to 1 mm long; most of the leaves 8 mm wide or wider........ 97. *C. lacustris*
         4. Teeth of the beak of the perigynium 1.0–2.2 mm long; leaves mostly 2–6 mm wide........ 98. *C. laeviconica*
      3. Culms brown at the base; leaves glaucous or blue-green; nerves of the perigynia finely impressed or absent.
         5. Leaves 8–15 mm wide, flat; perigynia up to 150 per spike, 6–8 mm long........ 91. *C. hyalinolepis*
         5. Leaves 1.5–3.2 mm wide, plicate; perigynia up to 30 per spike, 3.0–4.5 mm long........ 85. *C. heterostachya*
   2. Plants growing in dense clumps, usually without extensive rhizomes; perigynia thin and papery, inflated.

6. Perigynia horizontally spreading or even reflexed at maturity; sheaths loose .... 149. *C. retrorsa*
6. Perigynia spreading to ascending; sheaths tight.
   7. Culms spongy at the base; some or all of the leaves 7–12 mm wide........ 186. *C. utriculata*
   7. Culms firm at the base; leaves 2–7 mm wide.
      8. Achenes invaginated on 1 side; perigynia 7–10 mm long........ 183. *C. tuckermanii*
      8. Achenes not invaginated on 1 side; perigynia 5–8 mm long........ 188. *C. vesicaria*

Group 6

Leaves, sheaths, and/or culms glabrous; perigynia glabrous; stigmas 3; achenes trigonous; terminal spike entirely staminate; perigynia up to 10 mm long, usually much shorter; staminate spike 1 per culm; perigynia prominently 2-toothed at the tip or prolonged into a conspicuous, bidentate beak.

1. Pistillate spikes 2–4 mm thick.
   2. Pistillate spikes linear-cylindric, up to 8 cm long; lowest spikes on pendulous peduncles; perigynia 3.5–8.0 mm long, all ascending.
      3. Perigynia up to 45 per spike; pistillate scales usually with awns up to 1.5 mm long; perigynia 3–5 mm long, with several faint nerves........ 12. *C. arctata*
      3. Perigynia up to 25 per spike; pistillate scales obtuse to acute to cuspidate; perigynia 5.0–8.5 mm long, with 2 strong ribs and several faint nerves.
         4. Perigynia strongly overlapping, fusiform, gradually tapering to an elongated beak nearly ½ as long as the perigynia........ 68. *C. flexuosa*
         4. Perigynia barely overlapping, lance-ellipsoid, abruptly narrowed to an elongated bek about ⅓ as long as the perigynia........52. *C. debilis*
   2. Pistillate spikes short-cylindic, up to 3 cm long; lowest spikes ascending; perigynia 2–5 mm long, spreading to reflexed to ascending.
      5. Perigynia up to 30 per spike, spreading to reflexed, 2–3 mm long........ 190. *C. viridula*
      5. Perigynia up to 15 per spike, ascending, 3.5–5.0 mm loing........ 187. *C. vaginata*
1. Pistillate spikes usually at least 1 cm thick (sometimes slightly more slender in *C. sprengelii*).
   6. Some of the perigynia reflexed.
      7. Perigynia up to 35 per spike, 3.2–6.3 mm long, yellowish to bright yellow, the beak at least ½ as long as the body of the perigynium.
         8. Perigynia yellowish, 3.2–4.5 mm long, the beak glabrous, usually straight, the beak about as long as the body of the perigynium; leaves up to 4.5 mm wide........ 48. *C. cryptolepis*
         8. Perigynia bright yellow, 4.0–6.3 mm long, the beak scabrous, curved, the beak longer than the body of the perigynium; leaves up to 6 mm wide........ 67. *C. flava*
      7. Perigynia 50–100 per spike, 5–10 mm long, yellow-green, the beak up to ½ as long as the body of the perignyium.
         9. Lowest spikes on pendulous peduncles; perigynia 3.5–7.0 mm long, the beak 1–2 mm long; pistillate scales serrulate-awned.
            10. Teeth of the beak of the perigynium nearly straight, 0.7–1.2 mm long........ 145. *C. pseudocyperus*
            10. Teeth of the beak of the perigynium curved outward, 1.2–2.0 mm long........ 37. *C. comosa*
         9. Lowest spikes ascending; perigynia 7–10 mm long, the beak 2.5–3.5 mm long; pistillate scales acuminate........ 149. *C. retrorsa*
   6. All perigynia spreading to ascending, never reflexed.
      11. Culms filiform; leaves 1–3 mm wide; perigynia up to 15 per spike........ 127. *C. oligosperma*
      11. Culms broader than filiform; leaves 2.25–12.00 mm wide (if narrower, then plicate in *C. heterostachya*); perigynia 30–100 per spike.

12. Culms purple-red at the base; perigynia 30–100 per spike.
13. Pistillate scales much longer than the perigynia; spikes subtended by bracts many times their length; perigynia obconic, broadest above the middle.
14. Rhizomes short, not forming colonies; staminate scales 0.5–1.0 mm wide; beak of perigynium up to 2.5 mm long .... 71. *C. frankii*
14. Rhizomes long, forming colonies; staminate scales 1.0–1.5 mm wide; beak of perigynium 1.2–2.0 mm long .... 17. *C. aureolensis*
13. Pistillate scales not longer than the perigynia; spikes not subtended by exceptionally long bracts; perigynia ovoid, broadest near the base.
15. Lowest spikes on slender, often pendulous peduncles; perigynia 4–6 mm long, 1.5–2.0 mm broad, the beak 1.8–2.2 mm long.......................... 76. *C. hystericina*
15. Lowest spikes ascending; perigynia 5–9 mm long, 2–4 mm broad, the beak 3–4 mm long.
16. Leaves up to 7 mm wide; perigynia 3–4 mm wide, the beak about equaling the body .... 94 *C. lurida*
16. Leaves 2–4 mm wide; perigynia 2.0–2.5 mm wide, the beak longer than the body.. 14. *C. baileyi*
12. Culms brown at the base; perigynia 30–50 per spike.
17. Perigynia 5.0–6.5 mm long, the beak ½ as long as or equalling the body of the perigynium; leaves 2.5–7.0 mm wide, usually flat.
18. Beak of the perigynium as long as the body of the perigynium; leaves 2.5–4.0 mm wide; perigynia strongly nerved......................... 162. *C. sprengelii*
18. Beak of the perigynium about ½ as long as the body of the perigynium; leaves up to 7 mm wide; perigynia finely nerved......... 34. *C. cherokeensis*
17. Perigynia 3.0–4.5 mm long, the beak much shorter than the body; leaves 1.5–3.2 mm wide, often plicate .... 70. *C. heterostachya*

Group 7

Leaves, sheaths, and/or culms all glabrous; perigynia glabrous; stigmas 3; achenes trigonous; terminal spike entirely staminate; perigynia up to 10 mm long, usually much shorter; staminate spike 1 per culm; perigynia ending abruptly at the tip, either without teeth or with merely a small notch; most or all of the leaves, particularly on sterile shoots, 15 mm wide or wider.

1. Culms leafless, bearing only long-tubular purplish sheaths; staminate spike purple .... 138. *C. plantaginea*
1. Culms leafy; staminate spike brown to brown-purple.
2. Blade of uppermost leaf at most only 3 times longer than its sheath.
3. Basal leaves green; culms purple at base; perigynia 5.0–6.5 mm long .... 30. *C. careyana*
3. Basal leaves glaucous; culms brown at base; perigynia 2.5–4.5 mm long ...... 139. *C. platyphylla*
2. Blade of uppermost leaf many times longer than its sheath.
4. Uppermost bract much longer than the inflorescence; culms somewhat winged.
5. Leaves pale green; pistillate scales obtuse to acute; perigynia obovoid .... 7. *C. albursina*
5. Leaves dark green; pistillate scales acute to short-awned; perigynia oblongoid.. 96. *C. kraliania*
4. Uppermost bract about as long as or barely longer than the inflorescence; culms wingless .... 102. *C. laxiflora*

Group 8

Leaves, sheaths, and/or culms all glabrous; perigynia glabrous; stigmas 3; achenes trigonous; terminal spike entirely staminate; perigynia up to 10 mm long, usually much shorter; staminate spike 1 per culm; perigynia ending abruptly at the tip, either without teeth or with merely a small notch; most or all the leaves less than 15 mm wide; at least the lowest pistillate spikes on flexuous or pendulous peduncles.

1. Perigynia up to 15 per spike.
  2. Perigynia fusiform.
    3. Perigynia 3.5–4.5 mm long, curved at tip, with rounded angles .... 169. *C. styloflexa*
    3. Perigynia 2.5–3.3 mm long, straight at tip, with pointed angles .... 56. *C. digitalis*
  2. Perigynia ellipsoid to ovoid to oblongoid.
    4. Culms purple-red at base; leaves yellow-green.................... 75. *C. gracilescens*
    4. Culms brown at base; leaves pale green, bright green, or glaucous.
      5. Staminate spike sessile or nearly so; pistillate scales about as long as the perigynia; perigynia strongly curved at tip; angles of perigynia rounded .... 22. *C. blanda*
      5. Staminate spike pedunculate; pistillate scales shorter than the perigynia; perigynia straight at tip; angles of perigynia pointed.
        6. Some or all the leaves up to 12 mm wide; pistillate scales somewhat short-awned.
          7. Leaves bright green, harshly scabrous; leaves up to 8 mm wide..... .... 41. *C. copulata*
          7. Leaves green, slightly scabrous; leaves up to 12 mm wide .... 101. *C. laxiculmis*
        6. Leaves 2.0–5.3 mm wide; pistillate scales all acute.
          8. Perigynia 2.8–3.5 mm long, with 7–10 nerves; leaves 2.0–2.9 mm wide; peduncle of staminate spike 8–16 cm long, surpassing the uppermost bract.. 114. *C. macropoda*
          8. Perigynia 2.5–3.3 mm long, with 11–15 nerves; leaves 2.5–5.3 mm wide; peduncle of staminate spike 1–7 cm long, surpassed by the uppermost bract.. 56. *C. digitalis*
1. Perigynia 15–50 per spike.
  9. Leaves up to 2.5 mm wide; plants stoloniferous; perigynia glaucous-green .... 106. *C. limosa*
  9. Leaves 3–6 mm wide; plants not stoloniferous; perigynia green .... 142. *C. prasina*

Group 9

Leaves, sheaths, and/or culms all glabrous; perigynia glabrous; stigmas 3; achenes trigonous; terminal spike staminate; perigynia up to 10 mm long, usually much shorter; staminate spike 1 per culm; perigynia ending abruptly at the tip, either without teeth or with merely a small notch; most or all the leaves less than 15 mm wide; all pistillate spikes ascending to erect; beak or tip of perigynium bent or curved to 1 side.

1. Perigynia crowded in pistillate spikes, always overlapping.
  2. Perigynia rounded at the base, olive-green to brown; leaves more or less glaucous.
    3. Perigynia ellipsoid, not inflated, 2.3–3.0 mm long, 1.0–1.5 mm wide; cauline leaves up to 16 mm wide.. 82. *C. haleana*
    3. Perigynia obovoid, inflated, 2.5–4.0 mm long, 1.5–2.5 mm wide; cauline leaves up to 12 mm wide.. 7. *C. granularis*
  2. Perigynia tapering to the base, yellow-green to green to pale green; leaves pale green, green, or gray-green, but not glaucous.
    4. Perigynia witih 2–3 conspicuous nerves.............................. 105. *C. leptonervia*
    4. Perigynia with 4 or more faint or somewhat conspicuous nerves, or nerveless.
      5. Perigynia distichously arranged..................................... 137. *C. planispicata*
      5. Perigynia spirally arranged.
        6. Leaves gray-green, up to 7 mm wide; pistillate spikes 5–7 mm thick; pistillate scales cuspidate to short-awned; perigynia 3–5 mm long, turgid .... 114. *C. meadii*

6. Leaves green, up to 5 mm wide; pistillate spikes about 5 mm thick; pistillate scales obtuse to acute; perigynia 2.5–3.5 mm long, not turgid .... 174. *C. tetanica*
1. Perigynia loosely or remotely arranged in pistillate spikes, scarcely overlapping.
7. Leaves up to 4 mm wide; plants stoloniferous; pistillate scales with purple-brown margins.. 193. *C. woodii*
7. Leaves 5–12 mm wide; plants tufted; pistillate scales without purple-brown margins.
8. Pistillate spikes 4.5 mm thick; perigynia 4–5 mm long.............. 167. *C. striatula*
8. Pistillate spikes 3–4 mm thick; perigynia 3.0–4.5 mm long...... 102. *C. laxiflora*

Group 10

Leaves, sheaths, and/or culms all glabrous; perigynia glabrous; stigmas 3; achenes trigonous; terminal spike entirely staminate; perigynia up to 10 mm long, usually much shorter; staminate spike 1 per culm; perigynia ending abruptly at the tip, either without teeth or with merely a small notch; most or all the leaves less than 15 mm wide; all pistillate spikes ascending to erect; beak or tip of perigynium not bent or curved to 1 side.

1. Pistillate scales dark red-purple; perigynia more biconvex than trigonous .... 23. *C. buxbaumii*
1. Pistillate scales not dark red-purple; perigynia trigonous.
2. Staminate spike elevated above the pistillate spikes, on peduncles 2 cm long or longer.
3. Pistillate spikes loosely or remotely flowered.
4. Perigynia with rounded angles; leaves 6–20 mm wide, pale green.
5. Achenes 1.8–2.0 mm long; staminate scales usually cuspidate; pistillate spikes 3–4 mm thick; perigynia up to 15 per spike .... 140. *C. laxiflora*
5. Achenes 2.2–2.5 mm long; staminate scales obtuse; pistillate spikes 4–5 mm thick; perigynia up to 20 per spike..................... 167. *C. striatula*
4. Perigynia with pointed angles; leaves to 8 mm wide, green or dark green.
6. Peduncle of staminate spike up to 7 cm long, surpassed by the uppermost bract; perigynia usually fusiform, with 11–15 nerves; leaves 2.5–3.5 mm wide .... 56. *C. digitalis*
6. Peduncle of staminate spike 8–16 cm long, surpassing the uppermost bract; perigynia never fusiform, with 6–10 nerves; leaves 2.0–2.9 mm wide .... 114. *C. macropoda*
3. Pistillate spikes with crowded perigynia.
7. Perigynia distichously arranged...................................... 137. *C. planispicata*
7. Perigynia spirally arranged.
8. Leaves 1.5–3.5 mm wide; perigynia up to 45 per spike, strongly nerved.. 34. *C. crawei*
8. Leaves 3–8 mm wide; perigynia up to 27 per spike, finely nerved.
9. Staminate spike 0.5–1.5 mm thick; pistillate scales acute................. .... 2. *C. abscondita*
9. Staminate spikie 2.5–3.0 mm thick; pistillate scales awned.
10. Culms brownish at the base; leaves up to 5 mm wide; perigynia 3–4 mm long; achenes 2.0–2.2 mm long.............. 40. *C. conoidea*
10. Culms purple-red at the base; leaves 5–8 mm wide; perigynia 4.0–7.5 mm long; achenes 2.5–3.3 mm long.
11. Sheaths loose; pistillate spikes 3–6 mm thick; perigynia up to 12 per spike, oblongoid to obovoid, dull, the angles pointed .... 9. *C. amphibola*
11. Sheaths tight; pistillate spikes up to 10 mm thick; perigynia up to 27 per spike, ellipsoid, shiny, the angles rounded......... .... 42. *C. corrugata*
2. Staminate spike not strongly elevated above the pistillate spikes, sessile or on peduncles less than 2 cm long.
12. Perigynia up to 10 (–12) per spike; leaves green or glaucous, 2–8 mm wide.

13. Pistillate spikes loosely flowered, up to 1.5 cm long....... 126. *C. oligocarpa*
13. Pistillate spikes with crowded perigynia, (1.0–) 1.5–2.5 cm long.
14. Leaves more or less glaucous or strongly glaucous.
15. Pistillate scales up to ½ as long as the perigynia; perigynia 2.0–2.3 mm wide; plants more or less glaucous; plants usually of rather wet habitats .... 66. *C. flaccosperma*
15. Pistillate scales nearly as long as the perigynia; perigynia 1.5–1.8 mm wide; plants strongly glaucous; plants usually of rather dry habitats .... 74. *C. glaucodea*
14. Leaves green.
16. Perigynia distichously arranged........................... 137. *C. planispicata*
16. Perigynia spirally arranged.
17. Perigynia 3.5–4.7 mm long; one or more pistillate spikes usually near base of plant.
18. Sheaths loose; pistillate spikes 3–6 mm thick; perigynia up to 12 per spike, oblongoid to obovoid, dull, the angles pointed ....9. *C. amphibola*
18. Sheaths tight; pistillate spikes up to 10 mm thick; perigynhia up to 27 per spike, ellipsoid, shiny, the angles rounded......... ....42. *C. corrugata*
17. Perigynia 4.5–5.5 mm long; no pistillate spikes near base of plant .... 80. *C. grisea*
12. Perigynia 10–60 per spike; leaves more or less glaucous (except in *C. grisea*), up to 10 mm wide.
19. Pistillate scales ½ as long as the perigynia or nearly as long; leaves more or less glaucous, or strongly glaucous.
20. Perigynia 2.3–4.0 mm long, usually strongly nerved; sheaths tight; leaves more or less glaucous.
21. Perigynia ellipsoid, not inflated, 2.3–3.0 mm long, 1.0–1.5 mm wide; cauline leaves up to 16 mm wide .... 82. *C. haleana*
21. Perigynia obovoid, inflated, 2.5–4.0 mm long, 1.5–2.5 mm wide; cauline leaves up to 12 mm wide .... 77. *C. granularis*
20. Perigynia 3.5–6.0 mm long, finely impressed-nerved; sheaths loose; leaves strongly glaucous.
22. Pistillate scales up to ½ as long as the perigynia; perigynia 2.0–2.3 mm wide; plants more or less glaucous; plants of rather wet habitats .... 6. *C. flaccosperma*
22. Pistillate scales nearly as long as the perigynia; perigynia 1.5–1.8 mm wide; plants strongly glaucous; plants of usually dry habitats .... 74. *C. glaucodea*
19. Pistillate scales about as long as the perigynia or slightly longer; leaves green .... 80. *C. grisea*

Group 11
Leaves, sheaths, and/or culms all glabrous; perigynia glabrous; stigmas 3; achenes trigonous; terminal spike androgynous or gynecandrous.

1. Terminal spike androgynous.
2. Pistillate spikes up to 3.5 cm long, up to 1 cm thick, the lowest on pendulous peduncles; perigynia loosely arranged, 5.0–6.5 mm long, 1.8–2.0 mm wide, 2-ribbed but otherwise nerveless, the beak as long as the body; leaves 2.5–4.0 mm wide, shorter than the culms .... 162. *C. sprengelii*
2. Pistillate spikes up to 8 cm long, up to 2 cm thick, none of them on pendulous peduncles; perigynia crowded, 7–10 mm long, 2.5–3.5 mm wide, strongly nerved, the beak prominent but usually not ½ as long as the body; leaves mostly 4–10 mm wide, longer than the culms .... 149. *C. retrorsa*
1. Terminal spike gynecandrous.
3. Perigynia obconic; pistillate scales setaceous, much longer than the perigynia; spikes located about midway on plant, much surpassed by leaves and leaflike bracts.

- **4.** Rhizomes short, not forming colonies; staminate scales 0.5–1.0 mm wide; perigynia up to 100 per spike; beak of perigynium up to 2.5 mm long .... 71. *C. frankii*
- **4.** Rhizomes long, forming colonies; staminate scales 1.0–1.5 mm wide; perigynia up to 80 per spike; beak of prigynium 1.2–2.0 mm long .... 17. *C. aureolensis*

3. Perigynia narrowly lanceoloid to narrowly ellipsoid to ovoid to obovoid (sometimes obconic in *C. typhina*); pistillate scales lanceolate to ovate, shorter than to about as long as the perigynia; most spikes in upper part of plant or even surpassing the leaves and bracts.

- **5.** Pistillate spikes (or terminal spike if only 1 spike present) 10–22 mm thick.
  - **6.** Perigynia horizontally spreading, or the lowest ones reflexed; pistillate scales acute to acuminate to cuspidate .... 163. *C. squarrosa*
  - **6.** Perigynia ascending, none of them reflexed; pistillate scales usually obtuse, less commonly acute .... 184. *C. typhina*
- **5.** Pistillate spike up to 10 mm thick.
  - **7.** At least the lowest spikes pendulous or spreading on long, slender peduncles.
    - **8.** Perigynia narrowly lanceoloid or fusiform, 4.0–8.5 mm long.
      - **9.** Perigynia strongly overlapping, fusiform, 5.5–8.5 mm long, gradually tapering to an elongated beak nearly ½ as long as the perigynium; pistillate scales obtuse to acute, white or pale .... 52. *C. debilis*
      - **9.** Perigynia barely overlapping, narrowly lanceoloid, 4–7 mm long, abruptly narrowed to an elongated beak about ⅓ as long as the perigynium; pistillate acute to cuspidate, with reddish streaks .... 68. *C. flexuosa*
    - **8.** Perigynia narrowly ellipsoid to ovoid, 2–4 mm long.
      - **10.** Perigynia narrowly ellipsoid, beakless; leaves up to 8 mm wide .... 76. *C. gracillima*
      - **10.** Perigynia ovoid, with a beak up to 2 mm long; leaves 3–6 mm wide .... 142. *C. prasina*
  - **7.** All spikes sessile or on short, ascending peduncles.
    - **11.** Culms red-purple at the base; pistillate scales dark red-purple; perigynia more or less biconvex, glaucous-green; leaves glaucous below .... 28. *C. buxbaumii*
    - **11.** Culms usually brownish at the base; pistillate scales brown, red-brown, or greenish; perigynia trigonous, yellow, olive-green, or brown; leaves not glaucous.
      - **12.** Perigynia squarrose, none of them reflexed; leaves 4–9 mm wide; perigynia nerveless.
        - **13.** Pistillate spikes 4–6 mm thick; beak of perigynium absent or up to 0.5 mm long .... 157. *C. shortiana*
        - **13.** Pistillate spikes 7–8 mm thick; beak of perigynium 1.0–1.5 mm long .... 51. *C. X deamii*
      - **12.** Perigynia not squarrose, the lowest ones usually reflexed; leaves up to 4.5 mm wide; perigynia with a few nerves.
        - **14.** Pistillate spikes 6–10 mm thick; pistillate scales acute to acuminate, shorter than the perigynia; perigynia 3.2–4.5 mm long, narrowly ovoid, yellowish, the beak about as long as the body .... 48. *C. cryptolepis*
        - **14.** Pistillate spikes 2–3 mm thick; pistillate scales obtuse, about as long as the perigynia; perigynia 2–3 mm long, ovoid, green or yellow-green, the beak about ⅓ as long as the body .... 190. *C. viridula*

Group 12

Leaves, sheaths, and/or culms all glabrous; perigynia glabrous; stigmas 2; achenes lenticular; some or all the spikes pedunculate; staminate flowers usually on separate spikes.

1. Terminal spike gynecandrous.
2. Culms to 50 cm tall, slender; leaves 1.0–3.4 mm wide; pistillate scales obtuse to acute, shorter than to about as long as the perigynia; perigynia up to 30 per spike, yellow to orange to white-papillate; pistillate spikes up to 2 cm long.
3. Perigynia yellow to orange .... 16. *C. aurea*
3. Perigynia white-papillate .... 72. *C. garberi*
2. Culms to 1.5 m tall, stout; leaves 4–14 mm wide; pistillate scales long-awned, much longer than the perigynia; perigynia more than 30 per spike, green to pale brown to stramineous; pistillate spikes up to 12 cm long.
4. Lower sheaths smooth; perigynia 2.0–3.5 mm long, inflated; achenes often crimped along one margin .... 95. *C. crinita*
4. All sheaths scabrous; perigynia 3–4 mm long, scarcely inflated; achenes not crimped along one margin .... 81. *C. gynandra*
1. Terminal spike entirely staminate.
5. Culms to 50 cm tall, slender; leaves 1–3 mm wide; perigynia yellow to orange .... 16. *C. aurea*
5. Culms to 1.5 m tall, usually stout; leaves 3–14 mm wide; perigynia brown to olive-green to deep green.
6. Pistillate scales long-awned, longer than the perigynia.
7. Lower sheaths smooth; perigynia 2.0–3.5 mm long, inflated; achenes often crimped along one margin .... 95. *C. crinita*
7. All sheaths scabrous; perigynia 3–4 mm long, scarcely inflated; achenes not crimped along one margin .... 81. *C. gynandra*
6. Pistillate scales obtuse to acuminate, shorter than to barely longer than the perigynia.
8. Lowest leaf sheath with well-developed blades; perigynia broadest above the middle .... 11. *C. aquatilis*
8. Lowest leaf sheath bladeless; perigynia broadest at or below the middle.
9. Perigynia with a distinct beak 0.3–1.0 mm long; pistillate scales purple-black to purple-brown.
10. Beak of the perigynium twisted, 0.3–1.0 mm long; pistillate scales purple-black, shorter than the perigynia; leaves 3–5 mm wide .... 178. *C. torta*
10. Beak of perigynium straight or nearly so, 0.3–0.6 mm long; pistillate scales purple-brown, about as long as the perigynia; leaves up to 12 mm wide .... 122. *C. nebrascensis*
9. Perigynia beakless or the beak up to 0.3 mm long; pistillate scales red-brown, not purplish.
11. Ligule longer than the width of the blade, V-shaped; perigynia nerveless or faintly nerved.
12. Pistillate scales shorter than the perigynia; pistillate spikes to 10 cm long; perigynia more or less flat, 1.7–3.4 mm long; lowest sheath deep red or purple .... 168. *C. stricta*
12. Pistillate scales longer than the perigynia; pistillate spikes to 5 cm long; perigynia biconvex, 1.5–2.8 mm long; lowest sheath red-brown .... 83. *C. haydenii*
11. Ligule very short, not forming a V; perigynia distinctly nerved .... 63. *C. emoryi*

Group 13

Leaves, sheaths, and/or culms all glabrous; perigynia glabrous; stigmas 2; achenes lenticular; all spikes sessile, essentially alike, with staminate flowers either at the tip or at the base of each spike; culms solitary or forming stoloniferous or rhizomatous colonies.

1. Culms arising from axils of last year's dead leaves along prostrate or reclining stems; culms usually smooth; tip of perigynia smooth, not serrulate .... 35. *C. chordorrhiza*
1. Culms arising directly from the rhizomes; culms rough on the angles (except *C. duriuscula*); tip of perigynia serrulate.

2. Inflorescence headlike, with all spikes crowded; leaves filiform, up to 1.5 mm wide .... 58. *C. duriuscula*
2. Inflorescence elongated, often the lowest spikes not contiguous with the ones above; leaves 1–3 mm wide, not filiform.
   3. Perigynia biconvex; sheaths tight; inflorescence usually 5–10 cm long; rhizomes short.
      4. Sheaths copper-colored at the summit; perigynia 2.5–4.0 mm long, 1.2–1.3 mm wide, stramineous to brown .... 141. *C. prairea*
      4. Sheaths not copper-colored at the summit; perigynia 2–3 mm long, about 1 mm wide, dark brown to olive-black .... 55. *C. diandra*
   3. Perigynia plano-convex; sheaths loose and open (except in *C. socialis*); inflorescence to 6 cm long, often shorter; rhizomes elongated.
      5. Inflorescence 5–12 mm wide; perigynia narrowly winged, conspicuously nerved ventrally (except in a rare variety of *C. foenea*); pistillate scales usually red-brown.
         6. Rhizomes slender, cordlike, brown; inflorescence 1–4 cm long; perigynia 4.5–6.0 mm long, the beak 2–3 mm long .... 158. *C. siccata*
         6. Rhizomes thick, black; inflorescence 2.5–6.5 cm long; perigynia 2.0–4.6 mm long, the beak about 0.75 mm long .... 154. *C. sartwellii*
      5. Inflorescence 3–5 mm wide; perigynia wingless or nearly so, usually nerveless or obscurely nerved ventrally; pistillate scales pale brown to green and hyaline.
         7. Rhizomes stout; pistillate scales pale brown, longer than the perigynia; perigynia 8–12 per spike, ovoid-lanceoloid, not spongy at the base .... 140. *C. praegracilis*
         7. Rhizomes slender; pistillate scales green and hyaline, shorter than the perigynia; perigynia 1–9 per spike, narrowly lanceoloid, spongy at the base .... 159. *C. socialis*

Group 14

Leaves, sheaths, and/or culms glabrous; perigynia glabrous; stigmas 2; achenes lenticular; all spikes sessile, essentially alike, with staminate flowers at the tip of each spike; culms cespitose, not arising singly from extensive stolons and/or rhizomes; spikes androgynous.

1. Perigynia 1–3 per spike; culms capillary .... 57. *C. disperma*
1. Perigynia 4–many per spike (sometimes 1–3 perigynia per spike in *C. socialis*); culms not capillary.
   2. Culms soft, wing-angled, easily compressed (except in *C. oklahomensis*); sheaths loose and open.
      3. Culms firm, not easily compressed .... 124. C. *oklahomensis*
      3. Culms soft, spongy, easily compressed.
         4. Beak of perigynium ½ as long as to nearly as long as the body; sheaths with red-brown dots; spikes 8–12 per inflorescence.
            5. Sheaths septate-nodulose; pistillate scales green and hyaline; beak of perigynium about as long as the body .... 39. *C. conjuncta*
            5. Sheaths usually not septate-nodulose; pistillate scales brownish; beak of perigynium about ½ as long as the body .... 8. *C. alopecoidea*
         4. Beak of perigynium 1 ½–3 times longer than the body; sheaths without red-brown dots; spikes 10–25 per inflorescence.
            6. Pistillate scales shorter than the perigynia; perigynia yellow to brown; inflorescence up to 20 cm long, up to 6 cm thick; spikes 15–25 per inflorescence.
               7. Perigynia 3.5–6.0 mm long, strongly nerved on the inner face, the base tapering to the beak .... 165. *C. stipata*
               7. Perigynia 6–8 mm long, nerveless or faintly nerved on the inner face, the base abruptly enlarged below the beak .... 47. *C. crus-corvi*
            6. Pistillate scales about as long as the perigynia; perigynia greenish; inflorescence up to 6 cm long, up to 1.5 cm thick; spikes 10–15 per inflorescence .... 99. *C. laevivaginata*

2. Culms firm, at least not conspicuously wing-angled and not easily compressed; sheaths loose or tight.

8. Inflorescence consisting of 10 or more spikes, compound at the lower nodes.

9. Sheaths not septate-nodulose; leaves 1.0–3.1 mm wide; perigynia biconvex, 1.0–1.3 mm wide; pistillate scales acute to cuspidate.

10. Sheaths copper-colored at the summit; perigynia 2.5–4.0 mm long, 1.2–1.3 mm wide, stramineous to brown .... 141. *C. prairea*

10. Sheaths not copper-colored at the summit; perigynia 2–3 mm long, about 1 mm wide, dark brown to olive-black .... 55. *C. diandra*

9. Sheaths septate-nodulose; leaves 2–8 mm wide; perigynia plano-convex (biconvex in *C. decomposita*), 1.2–2.0 mm wide; pistillate scales awned or at least mucronate.

11. Perigynia plano-convex, green to stramineous to yellow-brown or golden brown, not spongy at the base; inflorescence up to 10 cm long, up to 1.5 cm thick.

12. Spikelets and perigynia greenish or stramineous at maturity.

13. Leaves longer than the culms.

14. Spikelets not subtended by setaceous bristles much longer than the perigynia, not appearing bristly; spikes usually up to 40 per inflorescence; perigynia lance-ovoid to ovoid, 1.2–1.8 mm wide .... 191. *C. vulpinoidea*

14. Spikelets subtended by setaeous bristles much longer than the perigynia, appearing bristly; spikes usually more than 40 per inflorescence; perigynia lanceoloid, 1.2–1.4 mm wide. .... 156. *C. setacea*

13. Leaves shorter than the culms.

15. Spikelets not subtended by setaceous bristles much longer than the perigynia, not appearing bristly; perigynia more or less orbicular, 3–4 mm long, 2.0–2.5 mm wide; achenes 1.5–2.0 mm long.. 65. *C. fissa*

15. Spikelets subtended by setaceous bristles much longer than the perigynia, appearing bristly; perigynia lanceoloid, 2.2–2.8 mm long, 1.2–1.4 mm wide; achenes 1.2–1.6 mm long .... 156. *C. setacea*

12. Spikelets and perigynia yellow-brown or golden brown at maturity.

16. Spikelets and perigynia golden brown; perigynia 2.2–2.6 mm long, 1.4–1.8 mm wide, obscurely nerved, the beak 1.2–1.6 mm long .... 23. *C. brachyglossa*

16. Spikelets and perigynia yellow-brown; perigynia 2.5–5.0 mm long, 1.7–3.0 mm wide, witih 2–5 nerves on the dorsal face, the beak 0.5–1.2 mm long.

17. Pistillate scales awn-tipped; perigynia 2.5–3.0 mm wide, the beak 0.8–1.2 mm long................................ 179. *C. triangularis*

17. Pistillate scales long-awned; perigynia 1.7–2.4 mm wide, the beak 0.5–0.7 mm long ....10. *C. annectens*

11. Perigynia biconvex, olive-black, spongy at the base; inflorescence up to 18 cm long, up to 4 cm thick.................................53. *C. decomposita*

8. Inflorescence consisting of up to 10 spikes (occasionally up to 15 in *C. gravida, C. lunelliana,* and *C. sparganioides*), not compound at the lower nodes.

18. Sheaths loose, with green mottling or green septations; leaves 4–10 mm wide (sometimes only 3 mm wide in *C. aggregata*).

19. Pistillate scales as long as or longer than the perigynia, awn-tipped.... .... 78. *C. gravida*

19. Pistillate scales ½ as long as the perigynia to nearly as long, acute (rarely awned in *C. aggregata*).

20. Inflorescence 5–15 cm long, the lowest spikes remote from the others; leaves up to 10 mm wide; perigynia 15–50 per spike............ .... 160. *C. sparganioides*

20. Inflorescence up to 5 cm long, the spikes all contiguous; leaves 3–8 mm wide; perigynia 5–20 per spike.
21. Leaves 3–5 mm wide; beak of the perigynium 1.0–1.5 mm long; summit of sheaths concave .... 3. *C. aggregata*
21. Leaves 4–8 mm wide; beak of the perigynium 0.5–1.3 mm long; summit of sheaths truncate.
22. Leaves dark green; bracts subulate or absent; pistillate scales acute; perigynia narrowly ovoid, 3.3–4.5 mm long, 1.5–2.5 mm wide, obscurely nerved dorsally .... 32. *C. cephaloidea*
22. Leaves bright green; bracts setaceous; pistillate scales acuminate to cuspidate; perigynia broadly ovoid to orbicular, 3.7–5.0 mm long, 3–5 mm wide, prominently nerved .... 110. *C. lunelliana*
18. Sheaths tight, without green mottling or green septations; leaves up to 4 mm wide (sometimes up to 5 mm wide in *C. cephalophora*).
23. All spikes crowded into a head; perigynia usually not spongy-thickened at the base.
24. Leaves 5–10 per culm; perigynia 2.7–4.5 mm long, 2–3 mm wide; ligule as wide as or wider than long; plants to 1 m tall.
25. Pistillate scales about equaling the perigynia; inflorescence up to twice as long as thick; ligule wider than long .... 116. *C. mesochorea*
25. Pistillate scales shorter than the perigynia; inflorescence 2–4 times longer than thick; ligule as wide as long.
26. Perigynia with conspicuous nerves on one or both faces, 3.0–4.5 mm long.
27. Perigynia 3.0–3.5 mm long, 2.0–2.5 mm wide; achenes 2.0–2.2 mm long .... 120. *C. muehlenbergii*
27. Perigynia 3.5–4.5 mm long, 2.5–3.0 mm wide; achenes 2.2–2.5 mm long .... 18. *C. austrina*
26. Perigynia nerveless on both faces, 2.7–3.1 mm long .... 136. *C. plana*
24. Leaves 3–5 per culm; perigynia 2.5–3.2 mm long, 1.2–2.0 mm wide; ligule longer than wide; plants to 75 cm tall.
28. Leaves 2–5 mm wide; perigynia broadest above the base, 1.5–2.0 mm wide .... 33. *C. cephalophora*
28. Leaves up to 3 mm wide; perigynia broadest at the base, 1.2–2.0 mm wide .... 103. *C. leavenworthii*
23. Spikes in an elongated, often interrupted, inflorescence; perigynia spongy thickened at the base (except in *C. austrina, C. muehlenbergii,* and *C. plana*).
29. Perigynia 2–3 mm wide; inflorescence up to 1 cm thick, often capitate above.
30. Perigynia not spongy-thickened at the base; leaves 5–10 per culm; beak of the perigynium up to 1 mm long.
31. Perigynia with conspicuous nerves on one or both faces, 3.0–4.5 mm long.
32. Perigynia 3.0–3.5 mm long, 2.0–2.5 mm wide; achenes 2.0–2.2 mm wide .... 120. *C. muehlenbergii*
32. Perigynia 3.5–4.5 mm long, 2.5–3.0 mm wide; achenes 2.2–2.5 mm wide .... 18. *C. austrina*
31. Perigynia nerveless on both faces, 2.8–3.1 mm long .... 136. *C. plana*
30. Perigynium spongy-thickened at the base; leaves 3–6 per culm; beak of the perigynium usually 1.0–1.8 mm long.
33. Bracts leaflike, up to 25 cm long; leaves 1.0–2.5 mm wide; perigynia 3.5–4.0 mm long, pale green; beak of the perigynium 1.0–1.2 mm long .... 13. *C. arkansana*

33. Bracts setaceous, up to 2 cm long; leaves 2–4 mm wide; perigynia 4.0–5.5 mm long, green-brown to black; beak of the perigynium 1.5–1.8 mm long .... 161. *C. spicata*

29. Perigynia 0.7–2.0 mm wide; inflorescence up to 5 mm thick, never capitate at the tip.

34. Beak of the perigynium smooth; bracts absent or only poorly developed; some perigynia reflexed.

35. Perigynia broadly ovoid, 1.3–2.0 mm wide, greenish brown at the base; leaves up to 2.5 mm wide ....148. *C. retroflexa*

35. Perigynia narrowly lanceoloid, 0.7–1.2 mm wide, pale green throughout; leaves up to 1.5 mm wide...........175. *C. texensis*

34. Beak of the perigynium serrulate; bracts usually present, setaceous, up to 15 cm long; perigynia spreading to ascending.

36. Stolons present, the plants colonial; perigynia narrowly lanceoloid... 159. *C. socialis*

36. Stolons absent, the plants not colonial; perigynia ovoid- lanceoloid.

37. Stigmas twisted; leaves up to 3 mm wide; beak of the perigynium 0.8–1.5 mm long.......................... 151. *C. rosea*

37. Stigmas more or less straight; leaves up to 1.9 mm wide; beak of the perigynium 0.5–1.0 mm long.... 146. *C. radiata*

Group 15

Leaves, sheaths, and/or culms all glabrous; perigynia glabrous; stigmas 2; achenes lenticular; all spikes sessile, essentially alike, with staminate flowers at the base of each spike; culms cespitose, not arising singly from extensive stolons and/or rhizomes; spikes gynecandrous; perigynia plano-convex, with rounded margins or with narrowly rimmed margins.

1. Perigynia with rounded margins, ascending.

2. Plants weak and often reclining; spikes 1–3 (–4) per culm; perigynia 1–15 per spike, oblongoid.. 182. *C. trisperma*

2. Plants firm or, if weak, scarcely reclining; spikes usually 4 or more per culm (rarely fewer in *C. bromoides* and *C. brunnescens*); perigynia 5 or more per spike, lanceoloid to ellipsoid to narrowly ovoid (rarely only 4 in *C. bromoides*).

3. All spikes overlapping; perigynia 3.5–4.5 mm long; beak of the perigynium 1.25–1.50 mm long.

4. Perigynia lanceoloid, nerved on both faces; most leaves up to 2 mm wide... 25. *C. bromoides*

4. Perigynia ovoid-lanceoloid, nerveless or faintly nerved only on the convex face; most leaves more than 2 mm wide.................... 54. *C. deweyana*

3. Some or all of the spikes separated from each other; perigynia 1.8–3.0 mm long; beak of the perigynium up to 1.2 mm long.

5. Leaves up to 2 mm wide, dark green; perigynia 5–10 per spike; ventral band of sheath red-brown dotted.................................26. *C. brunnescens*

5. Leaves 2–4 mm wide, glaucous to pale green; perigynia 10–30 per spike; ventral band of sheath not red-brown dotted.................... 29. *C. canescens*

1. Perigynia with narrowly rimmed margins, spreading to reflexed.

6. Terminal spike usually unisexual, either entirely staminate or entirely pistillate; perigynia castaneous to nearly black...164. *C. sterilis*

6. Terminal spike gynecandrous; perigynia green to dark brown.

7. Ventral face of the perigynium nerveless or nearly so; staminate part of the terminal spike proportionally longer than in the next 2 species....................... .... 93. *C. interior*

7. Ventral face of the perigynium usually with up to 10–12 nerves; staminate part of the terminal spike proportionally shorter than in *C. interior.*

8. Perigynia broadly ovoid to suborbicular, 2.0–3.5 mm long, 1.3–2.8 mm wide, the beak 0.5–1.2 mm long... 15. *C. atlantica*

8. Perigynia lanceoloid to ovoid, 2.5–4.5 mm long, 1–2 mm wide, the beak 1–2 mm long .... 60. *C. echinata*

Group 16

Leaves, sheaths, and/or culms all glabrous; perigynia glabrous; stigmas 2; achenes lenticular; all spikes sessile, essentially alike, with staminate flowers at the base of each spike; culms cespitose, not arising singly from extensive stolons and/or rhizomes; spikes gynecandrous; perigynia more or less flat and scalelike, with winged margins.

1. Perigynia 7–10 mm long, the beak 4.0–4.8 mm long; spikes 12–27 mm long .... 120. *C. muskingumensis*
1. Perigynia up to 7 mm long (–7.5 mm long in *C. bicknellii*), usually much shorter, the beak rarely as much as 4 mm long; spikes up to 12 (rarely–15) mm long.
   2. Tip of pistillate scales exceeding, or at least equaling, the tip of the perigynium.
      3. Tip of pistillate scales exceeding the tip of the perigynium; perigynia 4.5–6.5 mm long, 1.5–2.0 mm wide, not broadly winged .... 143. *C. praticola*
      3. Tip of pistillate scales as long as the tip of the perigynium but not exceeding it; perigynia 4–5 mm long, 2.5–4.0 mm wide, broadly winged .... 4. *C. alata*
   2. Tip of pistillate scales not reaching the tip of the perigynium.
      4. Perigynia up to 2 mm wide.
         5. Perigynia 0.8–1.1 mm wide, the wing very narrow to almost absent .... 44. *C. crawfordii*
         5. Perigynia 1.2 mm wide or wider, obviously winged.
            6. Perigynia widest at the middle.
               7. Perigynia evenly winged all the way to the base.
                  8. Perigynia appressed or ascending.
                     9. Perigynia lanceolate.
                        10. Perigynia usually nerved only on the dorsal face; sterile tufts of leaves usually not present; perigynia 3.8–5.5 mm long .... 156. *C. scoparia*
                        10. Perigynia nerved on both faces; sterile tufts of leaves present; perigynia 3.0–3.6 mm long .... 20. *C. bebbii*
                     9. Perigynia ovate to obovate to ovate-lanceolate to broadly elliptic.
                        11. Perigynia faintly nerved; wing of the perigynium not reaching the base of the beak of the perigynium; perigynia 1.2–1.5 mm wide .... 153. *C. sangamonensis*
                        11. Perigynia strongly nerved; wing of the perigynium reaching the base of the beak of the perigynium; perigynia 1.5–2.6 mm wide.
                           12. Perigynia ovate; pistillate scales usually awn-tipped .... 166. *C. straminea*
                           12. Perigynia obovate; pistillate scales acute to acuminate, not awn-tipped .... 107. *C. longii*
                  8. Perigynia spreading.
                     13. Wing of perigynium not reaching tip of perigynium; perigynia obovate; spikes usually tapering to tip .... 6. *C. albolutescens*
                     13. Wing of perigynium extending to top of perigynium; perigynia ovate; spikes usually subglobose, rounded at tip .... 124. *C. normalis*
               7. Wing of perigynium narrowed above the base and not extending to the base.
                  14. Perigynia lance-ovate to ovate, usually crimped on the shoulder, all widely spreading; spikes globose .... 46. *C. cristatella*
                  14. Perigynia lanceolate, not crimped on the shoulder, at least some of them usually ascending; spikes usually not globose.
                     15. All but the uppermost spikes remote from each other .... 144. *C. projecta*

15. Spikes crowded, with only 1 or 2 remote spikes, if any .... 180. *C. tribuloides*

6. Perigynia widest above or below the middle but not at the middle (at the middle in *C. missouriensis*).

16. Perigynia widest a little above the middle; wing of the perigynium not reaching the tip of the perigynium .... 6. *C. albolutescens*

16. Perigynia widest a little below the middle or nearly at the base; wing of the perigynium reaching the tip of the perigynium.

17. Perigynia lanceolate, the wing diminishing above the base of the perigynium .... 144. *C. projecta*

17. Perigynia ovate to orbicular, the wing extending all the way to the base of the perigynium.

18. All spikes except sometimes the lowest 1–2 crowded .... 124. *C. normalis*

18. Several of the lower spikes usually remote.

19. Sheaths papillose; pistillate scales ovate; beak of the perigynium 0.8–1.2 mm long .... 173. *C. tenera*

19. Sheaths glabrous; pistillate scales lanceolate to narrowly ovate; beak of the perigynium 1–2 mm long.

20. Perigynia 2.5–3.5 mm long .... 64. *C. festucacea*

20. Perigynia 3.5–4.5 mm long .... 61. *C. echinodes*

4. Some or all of the perigynia more than 2 mm wide.

21. Pistillate scales aristate to short-awned.

22. Perigynia ovate, 2.5–3.5 mm long, conspicuously nerved .... 166. *C. straminea*

22. Perigynia rhombic to suborbicular to ovate to elliptic, 4–7 mm long, nerveless or finely nerved.

23. Perigynia 4.5–7.0 mm long, widest at the middle, elliptic to ovate, the beak 2.0–2.8 mm long; achenes 1.2–1.5 mm wide .... 117. *C. missouriensis*

23. Perigynia 4–5 mm long, widest above the middle, rhombic to suborbicular, the beak 0.7–1.5 mm long; achenes 1 mm wide.

24. Wing of perigynium very broad, usually extending to the base of the perigynium; perigynia finely nerved on the ventral face, rounded at the base .... 4. *C. alata*

24. Wing of perigynium not particularly broad, diminishing before reaching the base of the perigynium; perigynia more or less nerveless, cuneate at the base .... 170. *C. suberecta*

21. Pistillate scales obtuse to acute to acuminate, never aristate or short-awned.

25. Perigynia broadest above the middle.

26. Wing of perigynium diminishing before reaching the base of the perigynium; perigynia more or less nerveless on both faces .... 170. *C. suberecta*

26. Wing of perigynium extending to the base of the perigynium; perigynia at least finely nerved on both faces.

27. Perigynia spreading, the wing not reaching the tip of the perigynium .... 6. *C. albolutescens*

27. Perigynia appressed to ascending, the wing reaching the tip of the perigynium.

28. Perigynia finely nerved, rhombic-orbicular .... 49. *C. cumulata*

28. Perigynia strongly nerved, obovate .... 107. *C. longii*

25. Perigynia broadest at or below the middle.

29. Perigynia less than 4 mm long.

30. Perigynia broadest below the middle .... 64. *C. festucacea*

30. Perigynia broadest at the middle.

31. Perigynia spreading.

32. Wing of the perigynium not reaching the tip of the perigynium .... 6. *C. albolutescens*

32. Wing of the perigynium reaching the tip of the perigynium .... 124. *C. normalis*
31. Perigynia appressed to ascending .... 89. *C. longii*
29. Some or all of the perigynia 4 mm long or longer.
33. Perigynia nerveless on the ventral face, orbicular.
34. Perigynia 2.5–3.5 mm wide .... 24. *C. brevior*
34. Perigynia 3.5–5.0 mm wide .... 147. *C. reniformis*
33. Perigynia nerved, at least finely so, on the ventral face, ovate to obovate to suborbicular.
35. Spikes globose to subglobose.
36. Sterile shoots present; perigynia 3.0–4.5 mm long, 1.5–2.1 mm wide .... 124. *C. normalis*
36. Sterile shoots absent; periynia 4.0–5.5 mm long, 2.0–3.5 mm wide.
37. Perigynia faintly nerved; beak of the perigynium 1.0–1.3 mm long; achenes elliptic to narrowly oblong, 1.2–1.3 mm wide .... 117. *C. molesta*
37. Perigynia conspicuously nerved; beak of the perigynium 1.3–1.8 mm long; achenes broadly ovate to orbicular, 1.4–1.8 mm wide .... 119. *C. molestiformis*
35. Spikes longer than broad.
38. Perigynia 4.5–7.5 mm long, 2.5–5.0 mm wide; beak of the perigynium 1.0–2.8 mm long.
39. Perigynia translucent, exposing the achene; beak of the perigynium 1.0–1.8 mm long; sheaths papillate .... 21. *C. bicknellii*
39. Perigynia opaque, concealing the achene; beak of the perigynium 1.8–2.8 mm long; sheaths glabrous.
40. Beak of the perigynium ½ as long as the body of the perigynium; pistillate scales acuminate .... 117. *C. missouriensis*
40. Beak of the perigynium ¼ as long as the body of the perigynium; pistilate scales obtuse .... 127. *C. opaca*
38. Perigynia 2.6–4.5 mm long, 1.5–2.7 mm wide; beak of the perigynium 0.5–1.0 mm long.
41. Perigynia spreading, the wing not reaching the tip of the perigynium .... 6. *C. albolutescens*
41. Perigynia appressed to ascending, the wing reaching the tip of the perigynium .... 89. *C. longii*

1. **Carex abdita** Bickn. Sedge. Apr.–May. Dry prairies, dry woods, sandy soil; throughout the state but more common in the n. cos. *Carex umbellata* Schk., misapplied.

2. **Carex abscondita** Mack. Thicket Sedge. May–June. Rich woods, very rare; Wabash Co. Specimen at the Missouri Botanical Garden.

3. **Carex aggregata** Mack. Clustered Sedge. May–July. Shaded woods, meadows; scattered in the n. ¾ of IL; also Alexander, Union, and Williamson cos. *Carex sparganioides* Muhl. var. *aggregata* (Mack.) Gl.

4. **Carex alata** Torr. & Gray. Broad-winged Sedge. May. Wet ground, rare; Jackson, Massac, Pope, and Wabash cos.

5. **Carex albicans** Willd. Sedge. Apr.–May. Dry woods, common; throughout the state. *Carex artitecta* Mack.

6. **Carex albolutescens** Schwein. Sedge. Apr.–May. Moist woods, very rare; Lee, Monroe, Pope, and Union cos.

7. **Carex albursina** Sheldon. White Bear Sedge. Apr.–May. Rich, wooded ravines, wooded slopes, often calcareous; occasional throughout the state.

8. **Carex alopecoidea** Tuckerm. Sedge. May–July. Wet meadows, moist fields, low woods, occasionally along roads; scattered in the n. ⅔ of IL; also Pope Co.

9. **Carex amphibola** Steud. Sedge. Apr.–June. Forests, wet meadows; occasional throughout the state but very rare in the ne. cos.

10. **Carex annectens** (Bickn.) Bickn. Yellow-brown Fox Sedge. May–June. Around ponds and lakes, fens, marshes, old fields, sometimes as an emergent aquatic; DuPage Co.

11. **Carex aquatilis** Wahl. var. **substricta** Kuk. Aquatic Sedge. Apr.–June. Marshes, wet meadows, wet ditches, along streams; occasional in the n. cos., much rarer elsewhere. *Carex substricta* (Kuk.) Mack.; *Carex aquatilis* Wahl. var. *altior* Fern.

12. **Carex arctata** Boott. Drooping Woodland Sedge. May–June. Mesic woods, very rare; Wabash Co.

13. **Carex arkansana** L. H. Bailey. Arkansas Sedge. May–June. Moist flatwoods, very rare; Douglas, Jackson, Saline, and Washington cos.

14. **Carex atherodes** Spreng. Sedge. May–June. Wet meadows, marshes; occasional in the n. ¼ of IL; also Champaign and St. Clair cos.

15. **Carex atlantica** L. H. Bailey. Two subspecies occur in Illinois:

a. Larger leaves more than 1.6 mm wide; inflorescence mostly more than 2 cm long .... 15a. *C. atlantica* ssp. *atlantica*

a. Larger leaves up to 1.6 mm wide; inflorescence up to 2 cm long .... 15b. *C. atlantica* ssp. *capillacea*

15a. **Carex atlantica** L. H. Bailey ssp. **atlantica.** Star Sedge. Apr.–May. Swampy woods, very rare; Cook, DuPage, and Pope cos. *Carex incomperta* Bickn.; *Carex atlantica* L. H. Bailey var. *incomperta* (Bickn.) F. J. Hermann.

15b. **Carex atlantica** L. H. Bailey ssp. **capillacea** (L. H. Bailey) Reznicek. Star Sedge. Apr.–May. Swampy woods, very rare; Pulaski Co. *Carex interior* L. H. Bailey var. *capillacea* L. H. Bailey; *Carex howei* Mack.

16. **Carex aurea** Nutt. Golden Sedge. May–June. Calcareous swales, sandy prairies, rare; Cook, Kane, Lake, Menard, and Washington cos.

17. **Carex aureolensis** Steud. Golden-fruit Sedge. May–Sept. Moist woods, wet ditches, not common; s. ¼ of IL.

18. **Carex austrina** (Small) Mack. Southern Sedge. May–June. Native to the s. U.S.; adventive along a railroad in Champaign Co.; also Jackson, Macon and Perry cos. *Carex muehlenbergii* Schk. var. *austrina* Small.

19. **Carex baileyi** Britt. Bailey's Sedge. May–Aug. Marshes, very rare; Jackson Co. Specimen at the Missouri Botanical Garden.

20. **Carex bebbii** Olney. Bebb's Sedge. May–June. Wet prairies, bogs, calcareous fens, marshes; occasional to frequent in the n. ½ of the state, uncommon in the s. ½.

21. **Carex bicknellii** Britt. Bicknell's Sedge. May–June. Dry prairies, old fields, dry slopes; occasional to common in the n. ¾ of the state, apparently absent elsewhere.

22. **Carex blanda** Dewey. Sedge. Apr.–June. Woods, meadows, mesic prairies, common; throughout the state.

23. **Carex brachyglossa** Mack. Larger Yellow Fox Sedge. May–June. Fields, disturbed low ground, mesic forests; scattered throughout the state. *Carex annectens* Bickn. var. *xanthocarpa* (Bickn.) Wieg.; *Carex vulpinioidea* Michx. var. *xanthocarpa* (Bickn.) Kuk.

24. **Carex brevior** (Dewey) Mack. Sedge. Apr.–June. Sandy prairies, dry woods, along railroads, often in disturbed areas; scattered to common in the n. ¾ of IL, uncommon elsewhere.

25. **Carex bromoides** Schk. Sedge. Apr.–May. Low woods, seep springs, swamps, prairie bogs; scattered throughout IL, but less common in the s. cos.

26. **Carex brunnescens** (Pers.) Poir. Brown Sedge. May. Alkaline bog, very rare; Lake Co.

27. **Carex bushii** Mack. Bush's Sedge. May–June. Usually dry woods, dry meadows, old fields; common in the s. ½ of IL, becoming less common northward; apparently adventive in Cook, DuPage, Lake and Will cos.

28. **Carex buxbaumii** Wahl. Buxbaum's Sedge. May. Marshes, wet prairies, swales, usually in calcareous areas; occasional in the n. ½ of IL; also Montgomery, Richland, Shelby, St. Clair, and Washington cos.

29. **Carex canescens** L. Two varieties may be distinguished in Illinois:

a. Spikes 6–12 mm long; perigynia 2.2–3.0 mm long .... 29a. *C. canescens* var. *disjuncta*
a. Spikes 4–7 mm long; perigynia up to 2.2 mm long .... 29b. *C. canescens* var. *subloliacea*

29a. **Carex canescens** L. var. **disjuncta** Fern. Silvery Sedge. Apr.–May. Sphagnum bogs, very rare; Lake Co.

29b. **Carex canescens** L. var. **subloliacea** Laestad. Silvery Sedge. Apr.–May. Sphagnum bogs, very rare; Lake Co.

30. **Carex careyana** Torr. Carey's Sedge. May–June. Rich woods; occasional in the s. ¼ of IL, not common in the e. cent. cos.; also Brown, Jo Daviess, Vermilion, and Will cos.

31. **Carex caroliniana** Schwein. Sedge. May–June. Wet meadows, wet woods; occasional in the s. ½ of IL; also Henry and McDonough cos.

32. **Carex cephaloidea** (Dewey) Dewey. Sedge. May–July. Rich woods, sometimes disturbed meadows; occasional in the n. ⅗ of IL, rare elsewhere.

33. **Carex cephalophora** Muhl. Capitate Sedge. Apr.–July. Woods, fields, lawns; common throughout the state.

34. **Carex cherokeensis** Schwein. Cherokee Sedge. June–July. Bottomland forests, rare; Alexander, Jackson, and Union cos.

35. **Carex chordorrhiza** Ehrh. Cordroot Sedge. Apr.–June. Sphagnum swamps, very rare; Lake and McHenry cos.

36. **Carex communis** L. H. Bailey. Sedge. May. Mesic woods, dry woods; scattered in IL, but not particularly common.

37. **Carex comosa** Boott. Porcupine Sedge. May–June. Swamps, boggy areas, wet ditches, pond margins; frequent in the n. ½ of IL, becoming less common southward.

38. **Carex complanata** Torr. & Hook. Hirsute Sedge. May–June. Dry woods, roadsides, rare; Lawrence and Williamson cos. Specimen at the Missouri Botanical Garden.

39. **Carex conjuncta** Boott. Soft Fox Sedge. May–June. Moist woods, swamps, wet prairies; occasional throughout the state.

40. **Carex conoidea** Schk. Apr.–June. Sedge. Wet meadows, wet prairies; occasional in the n. ½ of the state; also Massac Co.

41. **Carex copulata** (Bailey) Mack. Narrow-leaved Spreading Sedge. May–June. Mesic woods, wet woods, along streams; occasional throughout IL. *Carex digitalis* Willd. var. *copulata* Bailey.

42. **Carex corrugata** Fern Prune-fruit Sedge. May–June. Floodplain forests, not common; Jackson, Saline, Union, and White cos. *Carex amphibola* Steud. var. *globosa* Bailey.

43. **Carex crawei** Dewey. Crawe's Sedge. Apr.–May. Sandy flats, calcareous prairies, fens; occasional in the n. ½ of the state; also St. Clair Co.

44. **Carex crawfordii** Fern. Crawford's Sedge. May–June. Degraded marsh, very rare; Lake Co.

45. **Carex crinita** Lam. Two varieties may be distinguished in Illinois:

a. Lowest 2–6 spikes entirely pistillate; perigynia 2.0–3.5 mm long, often crimped on 1 side .... 45a. *C. crinita* var. *crinita*
a. Lowest 2–6 spikes pistillate, but some of them with a few staminate flowers at apex; perigynia 3–4 mm long, not crimped .... 45b. *C. crinita* var. *brevicrinis*

45a. **Carex crinita** Lam. var **crinita**. Fringed Sedge. May–Aug. Swampy woods, marshes; occasional throughout the state.

45b. **Carex crinita** Lam. var. **brevicrinis** Fern. Fringed Sedge. May–Aug. Swampy woods, marshes, rare; Alexander, Cook, Gallatin, Lake, and Pope cos.

46. **Carex cristatella** Britt. Two forms may be found in Illinois:

a. Spikelets all contiguous .... 46a. *C. cristatella* f. *cristatella*
a. Spikelets spread out in a moniliform inflorescence .... 46b. *C. cristatella* f. *catelliformis*

46a. **Carex cristatella** Britt. f. **cristatella.** Round-spikelet Sedge. May–July. Wet woods, marshes, swales, streambanks, ditches, meadows, bogs; throughout the state but more frequent in n. and w. cent. cos.

46b. **Carex cristatella** Britt. f. **catelliformis** (Farw.) Fern. Round-spikelet Sedge. May–July. Wet areas, occasional; scattered in IL.

47. **Carex crus-corvi** Shuttlw. Crowfoot Sedge. May–Aug. Swamps, wet woods, especially pin oak woods, upland swampy depressions; occasional in the s. ½ of the state, becoming less common northward.

48. **Carex cryptolepis** Mack. Yellow Sedge. May. Fens; confined to the extreme ne. cos. *Carex flava* L. var. *fertilis* Peck.

49. **Carex cumulata** (L. H. Bailey) Mack. Sedge. May. Mesophytic depression in black oak savannas, very rare; Iroquois and Kankakee cos.

50. **Carex davisii** Schwein. & Torr. Davis' Sedge. May–June. Moist woods, dry woods, wet ditches; occasional to common throughout the state.

51. **Carex X deamii** F. J. Herm. Deam's Sedge. May. Wet woods, alluvial woods, rare; Fayette, Macon, Pike, and Shelby cos. Reputed to be the hybrid between *C. shortiana* Dewey and *C. typhina* Michx.

52. **Carex debilis** Michx. White-edge Sedge. May–June. Dry woods, very rare; Carroll Co.

53. **Carex decomposita** Muhl. Cypress Knee Sedge. June–Aug. Usually on fallen logs or on swollen bases of trees in cypress swamps, rare; confined to the s. tip of the state.

54. **Carex deweyana** Schwein. Dewey's Sedge. May. Sandy oak woods, very rare; Winnebago Co. Not seen in IL since 1954.

55. **Carex diandra** Schrank. Sedge. May–July. Calcareous wet meadows, swamps, bogs, rare; confined to the ne. corner of the state and to a few cent. cos. near the Illinois River.

56. **Carex digitalis** Willd. Apr.–May. Sedge. Dry woods; occasional in the s. ⅙ of the state; scattered elsewhere.

57. **Carex disperma** Dewey. Soft-leaved Sedge. May–Aug. Tamarack and sphagnum swamps and bogs, very rare; Kane and Lake cos.

58. **Carex duriuscula** C. A. Mey. Needleleaf Sedge. May–July. Gravel bluff prairies, very rare; Carroll, Kane, and Winnebago cos. *Carex stenophylla* Wahl. var. *enervis* (C. A. Mey.) Kukenth.; *Carex eleocharis* Bailey.

59. **Carex eburnea** Boott. Sedge. Apr.–May. Calcareous ledges, wooded ravines; occasional in the n. ½ of IL; also Monroe and Union cos.

60. **Carex echinata** Murray. Sedge. May–June. Wet meadows, very rare; Pope Co. *Carex cephalantha* (L. H. Bailey) Bickn.

61. **Carex echinodes** (Fern.) P. E. Rothrock, Reznicek, & Hipp. Echinate Sedge. May–June. Bottomland forests; apparently confined to the n. ⅙ of IL. *Carex straminea* Willd. var. *echinodes* Fern.

62. **Carex emmonsii** Dewey. Emmons' Sedge. Apr. Dry woods, rare; Alexander, Cook, Fulton, Gallatin, Hardin, Iroquois, Kankakee, and Union cos. *Carex albicans* Willd. var. *emmonsii* (Dewey) Rettig.

63. **Carex emoryi** Dewey. Emory's Sedge. Apr.–June. Along streams, sedge meadows; occasional in the n. ⅔ of IL; also Johnson, Randolph, Union, and Williamson cos.

64. **Carex festucacea** Schk. Sedge. Apr.–June. Moist prairies, moist savannas, low woods; scattered in IL, but apparently more frequent in the n. ½ of the state.

65. **Carex fissa** Mack. Short-leaved Fox Sedge. May–June. Swamps, moist open ground, rare; Jackson and Union cos.

66. **Carex flaccosperma** Dewey. Sedge. May–June. Wet woods, swamps; mostly in the s. ¼ of the state; also Fayette, Macon, McDonough, Menard, and St. Clair cos.

67. **Carex flava** L. Yellow Sedge. June–July. Moist soil, very rare; Cook, Lake, DuPage, and McHenry cos.

68. **Carex flexuosa** Muhl. ex Willd. Rudge's Sedge. May–June. Mesic or dry woods, very rare; Hardin Co. *Carex debilis* Michx. var. *rudgei* Bailey.

69. **Carex folliculata** L. Sedge. May–June. Sedge. Swamps, very rare; reported from Cook Co. prior to 1926.

70. **Carex formosa** Dewey. Sedge. June. Moist savannas, very rare; Cook and Lake cos.

71. **Carex frankii** Kunth. Frank's Sedge. May–Sept. Moist woods, along streams, wet ditches; common to occasional throughout the state, except for the northernmost tier of cos.

72. **Carex garberi** Fern. Garber's Sedge. June. Sandy beaches near Lake Michigan, very rare; Cook and Lake cos.

73. **Carex gigantea** Rudge. Sedge. May–Sept. Wet woods, swampy woods, meadows, rare; Jackson, Johnson, Lawrence, Massac, Pulaski, and Union cos.

74. **Carex glaucodea** Tuckerm. Blue Sedge. May–July. Woods, occasional in the s. 3/5 of the state, absent elsewhere. *Carex flaccosperma* Dewey var. *glaucodea* (Tuckerm.) Kuk.

75. **Carex gracilescens** Steud. Sedge. Apr.–June. Woods; occasional throughout the state.

76. **Carex gracillima** Schwein. Graceful Sedge. Apr.–June. Moist or dry woods; occasional in the n. 1/4 of IL; also Alexander, Crawford, Jackson, and McDonough cos.

77. **Carex granularis** Muhl. Sedge. Apr.–June. Woods, old fields, fens, wet meadows; common throughout the state.

78. **Carex gravida** L. H. Bailey. Heavy Sedge. May–July. Dry prairies, old fields, disturbed meadows; scattered throughout the state.

79. **Carex grayi** Carey. Gray's Sedge. May–Sept. Wet woods, in floodplains, along streams, wooded swamps, common; throughout the state. (Including var. *hispidula* L. H. Bailey.)

80. **Carex grisea** Wahl. Sedge. May–June. Low woods, roadside ditches, common; throughout the state.

81. **Carex gynandra** Schwein. Fringed Look-alike Sedge. June–July. Swampy woods, edge of marsh, very rare; Alexander and Jackson cos. *Carex crinita* Lam. var. *gynandra* (Schwein.) Schwein. & Torr.

82. **Carex haleana** Olney. Hale's Sedge. May–June. Wet meadows; scattered throughout the state, but apparently more common in the n. 1/2 of the state. *Carex granularis* (Muhl.) var. *haleana* (Olney) Porter.

83. **Carex haydenii** Dewey. Hayden's Sedge. May–June. Sedge meadows, sandy wetlands; occasional in the n. 1/2 of the state; also Wabash Co.

84. **Carex heliophila** Mack. Plains Sedge. May. Prairies, plains, savannas, rare; Hardin, Jo Daviess, McHenry, and Rock Island cos. *Carex pensylvanica* Lam. ssp. *heliophila* (Mack.) W. A. Weber; *Carex inops* L. H. Bailey ssp. *heliophila* (Mack.) Crins.

85. **Carex heterostachya** Bunge. Sedge. Apr.–July. Native to Asia; adventive on a gravel bluff, very rare; Winnebago Co. *Carex X fulleri* Ahles.

86. **Carex hirsutella** Mack. Hairy-leaved Sedge. May–June. Dry woods, fields, dry meadows; common throughout the state except apparently absent from the nw. cos. *Carex complanata* Torr. & Hook. var. *hirsuta* (L. H. Bailey) Gl.

87. **Carex hirta** L. Hairy Sedge. June. Native to Europe; adventive in prairie restoration; disturbed soil; Cook, Ford, Iroquois, and Lake cos.

88. **Carex hirtifolia** Mack. Hairy Sedge. Mar.–May. Dry or mesic woods, common; throughout the state.

89. **Carex hitchcockiana** Dewey. Hitchcock's Sedge. May–June. Rich woods; scattered in the n. 1/2 of IL; also Fayette, Jackson, and Williamson cos.

90. **Carex houghtoniana** Torr. ex Dewey. Houghton's Sedge. July. Disturbed, sandy area (in IL); Whiteside Co.

91. **Carex hyalinolepis** Steud. Ditch Sedge. Apr.–July. Wet ditches, swamps; scattered in IL.

92. **Carex hystericina** Muhl. Porcupine Sedge. May–July. Swamps, calcareous fens, wet ditches; occasional to common in the n. 2/3 of the state, rare elsewhere.

93. **Carex interior** Bailey. Inland Sedge. Apr.–May. Bogs, wet meadows, moist prairies, wet woods, swamps; scattered throughout the state.

94. **Carex intumescens** Rudge. Sedge. May–Sept. Wet woods, swamps, marshes, bogs; scattered in IL, but not common. (Including var. *fernaldii* L. H. Bailey.)

95. **Carex jamesii** Schwein. James' Sedge. Apr.–May. Mesic woods, dry woods; common throughout the state.

96. **Carex kraliana** Naczi & Bryson. Kral's Sedge. May–June. Mesic woods, very rare; Jackson Co.

97. **Carex lacustris** Willd. Lake Sedge. May–Sept. Swampy woods, calcareous marshes, bogs, sometimes in standing water; throughout the state but infrequent in the s. cos.

98. **Carex laeviconica** Dewey. Sedge. June–July. Wet prairies, marshes; occasional in the n. ½ of the state, rare elsewhere, and apparently absent from the s. ¼ of IL.

99. **Carex laevivaginata** (Kuk.) Mack. Sedge. May–Aug. Swamps and wet woods, wooded seeps, fens, moist limestone barrens; scattered in IL, but not particularly common.

100. **Carex lasiocarpa** Ehrh. var. **americana** Fern. Hairy-fruited Sedge. May–June. Sphagnum bogs, sedge meadows, sometimes in shallow water; confined to the n. ⅓ of the state; also Champaign and Jefferson cos. Typical var. *lasiocarpa* is Eurasian.

101. **Carex laxiculmis** Schwein. Sedge. Apr.–May. Rich woods; occasional but scattered throughout the state.

102. **Carex laxiflora** Lam. Sedge. Apr.–May. Rich woods, not common; scattered in IL.

103. **Carex leavenworthii** Dewey. Leavenworth's Sedge. Apr.–June. Dry, open woods, either sandy or calcareous; scattered in IL.

104. **Carex leptalea** Wahl. Slender Sedge. May. Bogs, fens, wet meadows; known from several ne. cos.; also Fayette, Peoria, Tazewell, Vermilion, Washington, and Woodford cos.

105. **Carex leptonervia** (Fern.) Fern. Few-nerved Sedge. Apr.–May. Mesic woods, very rare; Pope Co. *Carex laxiflora* Lam. var. *leptonervia* Fern.

106. **Carex limosa** L. Sedge. June–July. Sphagnum bogs; confined to a few ne. cos.; also Peoria and Tazewell cos.

107. **Carex longii** Mack. Long's Sedge. Apr.–June. Flatwoods, mesic sand prairies, wet woods; scattered throughout the state.

108. **Carex louisianica** Bailey. Louisiana Sedge. May–Oct. Wet woods, wooded swamps, floodplains, meadows; mostly confined to extreme s. IL; also Wabash, Washington, and Wayne cos.

109. **Carex lucorum** Willd. Sedge. Apr. Woods, very rare; Pope Co. *Carex pensylvanica* Lam. var. *lucorum* (Willd.) Fern.

110. **Carex lunelliana** Mack. Lunell's Sedge. May–July. Prairies, old fields; scattered in IL, but apparently absent from the ne. cos. *Carex gravida* Bailey var. *lunelliana* (Mack.) F. J. Hermann.

111. **Carex lupuliformis** Sartw. Sedge. June–Oct. Wet woods, wooded swamps, marshes, meadows, roadside ditches; scattered throughout the state. *Carex eggertii* Bailey.

112. **Carex lupulina** Willd. Hop Sedge. June–Oct. Wet woods, wooded swamps, meadows, wet prairies, bogs, roadside ditches, common; throughout the state.

113. **Carex lurida** Wahl. Lurid Sedge. May–July. Swamps, wet woods, bogs, peaty fens, deep marshes, common; throughout the state.

114. **Carex macropoda** (Fern.) Mohlenbr. Southern Woodland Sedge. May–June. Dry or mesic woods, rare; Jackson, Johnson, and Pope cos. *Carex digitalis* Willd. var. *macropoda* Fern.

115. **Carex meadii** Dewey. Mead's Sedge. Apr.–June. Prairies, barrens, fens, meadows; scattered in IL, but not common in the s. ⅓ of the state. *Carex tetanica* Schk. var. *meadii* (Dewey) L. H. Bailey.

116. **Carex mesochorea** Mack. Midland Sedge. May–June. Dry woods, fields, rare; n. ⅔ of IL.

117. **Carex missouriensis** P. E. Rothrock & Reznicek. Missouri Sedge. May–June. Low areas in prairies; scattered in c. IL; also Grundy Co.

118. **Carex molesta** Mack. Sedge. Apr.–June. Old fields, moist prairies, swamps, wet depressions, ditches; scattered throughout the state.

119. **Carex molestiformis** Reznicek & P. E. Rothrock. Molesta-like Sedge. May–June. Bottomland forests, rare; Jackson and Union cos.

120. **Carex muehlenbergii** Schk. Muhlenberg's Sedge. May–July. Dry woods, scrubby black oak woods, old fields, sand prairies; occasional to common throughout the state.

121. **Carex muskingumensis** Schwein. Muskingum Sedge. Apr.–June. Low, swampy woods and floodplains of major streams, wooded depressions; occasional throughout the state.

122. **Carex nebrascensis** Dewey. Nebraska Sedge. May–June. Native to the w. U.S.; adventive along a railroad and alkaline edge of highway pavement; DuPage, Kane, and Will cos.

123. **Carex nigromarginata** Schwein. Sedge. Apr.–May. Woods, rare; confined to the s. ⅙ of IL; also Kankakee, Montgomery, and Wabash cos. *Carex lucorum* Willd. var. *nigromarginata* (Schwein.) Chapm.

124. **Carex normalis** Mack. Two forms occur in Illinois:

a. Most or all the spikes contiguous, congested .... 124a. *C. normalis* f. *normalis*
a. Some or all the spikes not contiguous, moniliform ..... 124b. *C. normalis* f. *perlonga*

124a. **Carex normalis** Mack f. **normalis.** Sedge. May–June. Seep springs, mesic woods, floodplains, streambanks, mesic savannas, marshes, pond borders, moist fields, ditches; common throughout the state.

124b. **Carex normalis** Mack. f. **perlonga** (Fern.) Fern. Sedge. May–June. Marsh, rare; Jackson Co.

125. **Carex oklahomensis** Mack. False Spongy Sedge. Oklahoma Sedge. Bottomland forest, marsh, very rare; Williamson Co. Specimen at the Missouri Botanical Garden.

126. **Carex oligocarpa** Schk. Sparse-fruited Sedge. May–June. Woods; scattered throughout the state.

127. **Carex oligosperma** Michx. Few-seeded Sedge. May–June. Bogs; confined to the ne. corner of IL.

128. **Carex opaca** (F. J. Hermann) P. E. Rothrock & Reznicek. Opaque Sedge. May–June. Wet prairies, ditches; Saline, St. Clair, and Washington cos. *Carex bicknellii* Britt. var. *opaca* F. J. Hermann.

129. **Carex oxylepis** Torr. & Hook. Two varieties occur in Illinois:

a. Perigynia glabrous .... 129a. *C. oxylepis* var. *oxylepis*
a. Perigynia pubescent... 129b. *C. oxylepis* var. *pubescens*

129a. **Carex oxylepis** Torr. & Hook. var. **oxylepis**. Sedge. Apr.–May. Swampy woods, rare; confined to the extreme s. tip of IL.

129b. **Carex oxylepis** Torr. & Hook. var. **pubescens** J. K. Underw. Sedge. Apr.–May. Swampy woods, very rare; Hardin Co.

130. **Carex pallescens** L. Pale Sedge. May–June. Rocky barrens, very rare; Fulton, Hancock, Johnson, McHenry, and Saline cos. *Carex pallescens* L. var. *neogaea* Fern.

131. **Carex peckii** Howe. Peck's Sedge. June. Dry woods, rare; McHenry Co.

132. **Carex pedunculata** Muhl. Long-stalked Hummock Sedge. Mar.–Apr. Dry, calcareous slopes, moist ravines, rare; Cook, Jo Daviess, Kane, Lake, McHenry, Ogle, and Winnebago cos.

133. **Carex pellita** Willd. Woolly Sedge. Apr.–June. Wet prairies, fens, marshes, sedge meadows, swamps; occasional to common throughout the state. *Carex lanuginosa* Michx.

134. **Carex pensylvanica** Lam. Pennsylvania Sedge. Apr.–May. Savannas, open woods; occasional to common in the n. ½ of the state, much rarer southward.

135. **Carex physorhyncha** Liebm. Sedge. Apr.–May. Rocky woods, particularly in chert; confined to the s. ¼ of the state; also Effingham Co. *Carex albicans* Willd. var. *australis* (L. H. Bailey) Rettig.

136. **Carex plana** Mack. Nerveless Sedge. May–July. Fields, woods, often in calcareous areas; scattered in IL; apparently more common than *Carex muehlenbergii. Carex muehlenbergii* Schk. ex Willd. var. *enervis* Boott.

137. **Carex planispicata** Naczi. Flat-spiked Sedge. May–June. Moist to wet woods, occasional; throughout the s. ⅔ of IL. *Carex amphibola* Steud. var. *rigida* (Bailey) Fern.

138. **Carex plantaginea** Lam. Plantain-leaved Sedge. Mar.–Apr. Rich woods, very rare; Cook, Johnson, and LaSalle cos.

139. **Carex platyphylla** Carey. Broad-leaved Sedge. May. Rich woods, very rare; Mason, McHenry, Saline, and St. Clair cos.

140. **Carex praegracilis** Boott. Sedge. Apr.–Aug. Low prairies, roadsides, particularly where salt has been applied during the winter months, dry sterile soil, low areas of medians and drainage swales; occasional in the n. ⅕ of IL; also Champaign and Christian cos.

141. **Carex prairea** Dewey. Sedge. May–July. Bogs, fens, drainage swales, floating sedge mats, wet meadows, wet prairies, swamps; occasional in ne. IL and in cos. along the Illinois River.

142. **Carex prasina** Wahl. Sedge. May–June. Rich woods; scattered in the nw. ¼ of IL; also Fayette, Johnson, Lawrence, and Vermilion cos.

143. **Carex praticola** Rydb. Sedge. May. Native n. and w. of IL; adventive in a cemetery in Cook Co.

144. **Carex projecta** Mack. Sedge. May–June. Riparian terraces, moist woods, swampy woods; scattered throughout the state.

145. **Carex pseudocyperus** L. Cyperus-like Sedge. June. Along a river (in IL); Fulton Co.

146. **Carex radiata** (Wahl.) Small. Star Sedge. Apr.–June. Woods, both mesic and dry, disturbed areas; scattered throughout the state. *Carex rosea* Schk., misapplied.

147. **Carex reniformis** (Bailey) Small. Sedge. May. Wet ground, very rare; Massac Co.

148. **Carex retroflexa** Muhl. Sedge. Apr.–June. Low wooded ridges, bluffs, dry woods; occasional to common in the s. ½ of IL, scattered in the n. ½.

149. **Carex retrorsa** Schwein. Sedge. May–July. Occasional in the n. ½ of IL; also Coles, Richland, and Union cos.

150. **Carex richardsonii** R. Br. Hummock Sedge. Apr.–May. Rocky areas, hill prairies, sand dunes; occasional in the n. ½ of the state.

151. **Carex rosea** Schk. Star Sedge. Apr.–May. Mesic woods; occasional throughout the state. *Carex convoluta* Mack.

152. **Carex rugosperma** Mack. Wrinkle-fruit Sedge. Apr.–June. Dry, rocky woods, very rare; Pope Co. *Carex tonsa* (Fern.) Bickn. var. *rugosperma* (Mack.) Crins.

153. **Carex sangamonensis** (Clokey) Mohlenbr. Sangamon Sedge. May–June. Rich soil, rare; Macon and Sangamon cos. Type from Macon Co.

154. **Carex sartwellii** Dewey. Sartwell's Sedge. Apr.–June. Low wet prairies and calcareous meadows, creek and river bottoms, marshes, dunes, peaty swamps, open cold bogs; occasional in the n. ½ of IL; also Randolph, St. Clair, and Washington cos.

155. **Carex scoparia** Schk. Two varieties occur in Illinois:

a. Spikes congested .... 155a. *C. scoparia* var. *scoparia*
a. Spikes remote .... 155b. *C. scoparia* var. *moniliformis*

155a. **Carex scoparia** Schk. var. **scoparia.** Sedge. May–June. Wet open woods, wet prairies, wet meadows, seeps, calcareous fens; common throughout the state.

155b. **Carex scoparia** Schk. var. **moniliformis** Tuckerm. Sedge. May–June. Wet meadows, marshes; scattered in IL.

156. **Carex setacea** Dewey. Bristly Sedge. July–Aug. Wet ground, not common; Calhoun, Jackson, Lake, and Lee cos.

157. **Carex shortiana** Dewey. Short's Sedge. May–June. Mesic woods, bottomlands, at the head of ravines, wet meadows; common in the s. ½ of the state, occasional to rare elsewhere.

158. **Carex siccata** Dewey. Two varieties occur in Illinois:

a. Perigynia nerved on both faces, tapering gradually to the beak .... 158a. *C. siccata* var. *siccata*
a. Perigynia nerveless on the ventral face, tapering abruptly to the beak .... 158b. *C. siccata* var. *enervis*

158a. **Carex siccata** Dewey var. **siccata**. Sedge. Apr.–May. Prairie or sandy soil, sometimes sandy woods and roadsides; mostly confined to the n. ⅙ of the state; also Kankakee, Menard, and Peoria cos. *Carex foenea* Willd., misapplied.

158b. **Carex siccata** Dewey var. **enervis** Mohlenbr. Sedge. Sandy plains, very rare; exact locality unknown; known only from the type collection. *Carex foenea* Dewey var. *enervis* Mohlenbr. & Evans.

159. **Carex socialis** Mohlenbr. & Schwegm. Colonial Sedge. Apr.–June. Wet woods, often in floodplains, not common; confined to the s. ⅙ of IL; also Washington Co.

160. **Carex sparganioides** Muhl. Sedge. May–July. Dry or moist woods; scattered throughout the state.

161. **Carex spicata** Huds. Spicate Sedge. May. Native to Europe and Asia; rarely adventive in disturbed soil; DeKalb, DuPage, and Winnebago cos.

162. **Carex sprengelii** Dewey. Long-beaked Sedge. May. Moist woods, wooded terraces; occasional to common in the n. ½ of IL; also Washington Co.

163. **Carex squarrosa** L. Squarrose Sedge. June–Sept. Low woods, swamps, along streams, wet meadows; scattered throughout the state but more common in the s. cos.

164. **Carex sterilis** Willd. Sedge. Apr.–May. Wet meadows, fens, marly seeps; confined to the n. ½ of IL; also Coles, St. Clair, and Washington cos.

165. **Carex stipata** Muhl. Spongy Sedge. Apr.–Aug. Wet meadows, marshes, swamps, fens, wet ditches, bogs; occasional to common throughout the state.

166. **Carex straminea** Willd. Sedge. June–July. Wet savannas; also along a railroad; rare in IL; n. ⅔ of IL.

167. **Carex striatula** Michx. Sedge. Apr.–May. Rich woods, rare; confined to the s. 3 tiers of cos. Specimen at the Missouri Botanical Garden.

168. **Carex stricta** Lam. Tussock Sedge. Apr.–June. Sedge meadows, fens, marshes; occasional to common in the n. ¾ of IL, apparently absent elsewhere.

169. **Carex styloflexa** Buckl. Sedge. Apr.–May. Rich woods, low woods, rare; confined to the s. ⅙ of IL. Specimen at the Missouri Botanical Garden.

170. **Carex suberecta** (Olney) Britt. Sedge. May–June. Marshes, fens, wet prairies, wet meadows; occasional in the n. ⅔ of IL, apparently absent in the s. ⅓.

171. **Carex X subimpressa** Clokey. Sedge. June. Marshes, very rare; DuPage, Macon, Montgomery, and St. Clair cos. This is the reputed hybrid between *C. hyalinolepis* Steud. and *C. pellita* Willd.

172. **Carex swanii** (Fern.) Mack. Swan's Sedge. May–June. Savannas, dry woods, swamp forests; scattered throughout the state but not particularly common.

173. **Carex tenera** Dewey. Remote Sedge. May–June. Floodplain woods, wet meadows, mesic prairies, swampy depressions, wet ditches; occasional to common throughout the state.

174. **Carex tetanica** Schk. Sedge. Apr.–June. Wet prairies, fens, wet meadows; occasional in the n. ½ of the state, rare elsewhere.

175. **Carex texensis** (G. S. Torr.) Bailey. Texas Sedge. Apr.–June. Disturbed soil, particularly in lawns and in cemeteries; rare and scattered in IL. *Carex retroflexa* Muhl. var. *texensis* (G. S. Torr.) Fern.

176. **Carex timida** Naczi & B. A. Ford. James' Look-alike Sedge. May–June. Moist or dry woods; scattered in s ½ of IL.

177. **Carex tonsa** (Fern.) Bickn. Shaved Sedge. Apr.–May. Sand dunes, sand prairies, rocky woods, not common; known from cos. in extreme nw. IL; also Lake, Mason, LaSalle, and Pope cos.

178. **Carex torta** Boott. Twisted Sedge. Apr.–June. Along and in streams; occasional in the s. ¼ of IL; also Whiteside Co.

179. **Carex triangularis** Boeck. Eastern Fox Sedge. June. Wet ditch, very rare; Massac Co.

180. **Carex tribuloides** Vahl. Sedge. May–June. Wet woods, swamps, wet ditches, peaty marshes, swales, wet prairies, wet meadows, peaty fens, oxbows, shores of lakes and ponds; common throughout the state.

181. **Carex trichocarpa** Muhl. Sedge. June–Aug. Calcareous meadows, sloughs, seeps, marshes; occasional to frequent in the n. ½ of the state; also Washington Co.

182. **Carex trisperma** Dewey. Three-fruited Sedge. June–July. Tamarack swamps, sphagnum bogs, very rare; Lake Co.

183. **Carex tuckermanii** Boott. Tuckerman's Sedge. Apr.–June. Upland depressions in wet savannas, rare; mostly confined to the ne. corner of IL; also Hancock, Henderson, and Winnebago cos.

184. **Carex typhina** Michx. Sedge. June–Sept. Bottomland woods, swamps, wet meadows; occasional in the s. ½ of the state, usually infrequent elsewhere.

185. **Carex umbellata** Schk. Sedge. Apr. Dry, rocky woods, very rare; Jackson, Pope, and Randolph cos.

186. **Carex utriculata** Boott. Beaked Sedge. May–June. Marshes, bogs, sometimes in standing water, not common; confined to ne. IL; also Fulton and St. Clair cos. *Carex rostrata* Michx., misapplied.

187. **Carex vaginata** Tausch. Sheathed Sedge. May. Along a road (in IL); Lake Co.

188. **Carex vesicaria** L. var. **monile** (Tuckerm.) Fern. Sedge. May–Aug. Upland swamps and depressions, wet meadows; occasional in the n. ½ of the state, extending southward to Lawrence, St. Clair, Wabash, and Washington cos. *Carex monile* Tuckerm.

189. **Carex virescens** Muhl. Hairy-fruited Sedge. May–June. Dry woods, not common; apparently confined to the s. ½ of the state; also Fulton and Vermilion cos.

190. **Carex viridula** Michx. Sedge. May–Sept. Calcareous pond shores, pannes, seeps, fens, flat gravelly prairies, not common; confined to the extreme ne. cos. of IL.

191. **Carex vulpinoidea** Michx. Fox Sedge. May–Aug. Swamps, wet meadows, low areas, moist open ground, common; in every co.

192. **Carex willdenowii** Schk. Willdenow's Sedge. Apr.–May. Rocky woods; not common; confined to the s. ⅛ of IL; also Iroquois Co.

193. **Carex woodii** Dewey. Wood's Sedge. Apr.–May. Mesic woods; rare in the n. ¼ of IL. *Carex tetanica* Schk. var. *woodii* (Dewey) L. H. Bailey.

4. **Cladium** P. Browne—Twig Rush

1. **Cladium mariscoides** (Muhl.) Torr. Twig Rush. Aug.–Oct. Swamps, bogs, not common; restricted to the extreme ne. corner of IL.

5. **Cyperus** L.—Flatsedge

1. Stigmas 2; achenes lenticular.
  2. Achenes black, nearly as broad as long, with horizontal wrinkles; scales straw-colored .... 10. *C. flavescens*
  2. Achenes drab or gray, longer than broad, without horizontal wrinkles; scales usually suffused with purple.
    3. Styles divided nearly to base, persistent and conspicuously exserted to 4 mm from scales .... 4. *C. diandrus*
    3. Styles divided to about the middle, falling away early, hidden by the scales or exserted to 2 mm from the scales.
      4. Scales closely appressed, strongly suffused with purple (straw-colored in *C. bipartitus* f. *elutus*) .... 2. *C. bipartitus*
      4. Scales with tips somewhat spreading, the spikelets appearing serrate, straw-colored or purple on the margins .... 9. *C. filicinus*
1. Stigmas 3; achenes trigonous.
  5. Scales with strongly recurved tips .... 21. *C. squarrosus*
  5. Scales with tips either appressed or only slightly spreading.
    6. Clusters of spikelets spherical or globose, with spikelets radiating in all directions.
      7. Scales appressed but with the tips shortly recurved; spikes to 8 mm across.
        8. Achenes linear, 1.0–1.4 mm long; perennial plants at least 35 cm tall at maturity .... 18. *C. pseudovegetus*
        8. Achenes ellipsoid to oblongoid, 0.5–1.0 mm long; annual plants never more than 35 cm tall .... 1. *C. acuminatus*
      7. Scales appressed to spreading, their tips straight; some of the heads over 1 cm across.
        9. Scales appressed; spikelets 2- to 3-flowered .... 5. *C. echinatus*
        9. Scales spreading; spikelets 5- to several-flowered .... 11. *C. grayoides*
    6. Cluster of spikelets hemispherical, cylindrical, ellipsoid, or lanceoloid, but not spherical or globose.
      10. Spikelets arising from a common point on the axis.
        11. Annual with soft bases; achenes tan, broadest above the middle .... 3. *C. compressus*

11. Perennial with short, firm rhizomes; achenes light brown, dark brown, or black, broadest at the middle.
12. Scales obtuse or the uppermost acute, without a mucro, 5- to 9-nerved; achenes black .... 15. *C. lupulinus*
12. Scales terminated by a mucro up to 1.5 mm long, 9- to 15-nerved; achenes light or dark brown.
13. Achenes 1.2–1.8 mm long, dark brown; scales 2.0–2.5 mm long. .... 12. *C. houghtonii*
13. Achenes 2.5–3.3 mm long, light brown; scales 2.5–4.0 mm long.
14. Culms harshly scabrous .... 20. *C. schweinitzii*
14. Culms smooth or nearly so .... 16. *C. X mesochorus*
10. Spikelets arising from either side of an elongated axis.
15. Scales 1.0–1.5 mm long; achenes 0.9–1.0 mm long.
16. Achenes white; spikelets about 1 mm broad; scales closely appressed; roots usually reddish .... 7. *C. erythrorhizos*
16. Achenes brown or black; spikelets about 1.5 mm broad; scales somewhat remote from each other; roots not reddish .... 13. *C. iria*
15. Scales 1.5–4.5 mm long; achenes 1.0–2.8 mm long.
17. Scales remote, the tip of one just reaching the base of the one above it, giving the spikelet a zigzag appearance .... 6. *C. engelmannii*
17. Scales approximate and overlapping.
18. Some or all the mature spikelets reflexed.
19. Spikelets 8–15 mm long; scales 4–5 mm long; achenes about 2.5 mm long .... 14. *C. lancastriensis*
19. Spikelets 3–5 mm long; scales 2.0–2.5 mm long; achenes about 1.5 mm long .... 19. *C. retrorsus*
18. All except sometimes the lowest pair of spikelets spreading or ascending.
20. Rhizomes scaly and usually ending in a tuber; scales at the tips of the spikelets slightly spreading, giving the spikelet a serrated margin and an obtuse apex .... 8. *C. esculentus*
20. Rhizomes absent or merely becoming hard and cormlike; scales at the tips of the spikelets appressed, giving the spikelet a smooth margin and a pointed apex.
21. Plants annual, without rhizomes; scales ferruginous or golden brown, 1.7–3.0 mm long; achenes obovoid-oblongoid .... 17. *C. odoratus*
21. Plants perennial, with hardened bases; scales straw-colored, 3.5–5.0 mm long; achenes linear .... 22. *C. strigosus*

1. **Cyperus acuminatus** Torr. & Hook. Pointed Flatsedge. June–Oct. Wet ground; scattered throughout IL.

2. **Cyperus bipartitus** Torr. Two forms occur in Illinois:

a. Scales strongly suffused with red-brown or purple .... 2a. *C. bipartitus* f. *bipartitus*
a. Scales stramineous throughout except for the greenish midrib .... 2b. *C. bipartitus* f. *elutus*

2a. **Cyperus bipartitus** Torr. f. **bipartitus.** Flatsedge. July–Oct. Wet ground along banks and shores; throughout the state but uncommon in the s. ⅓ of IL. *Cyperus rivularis* Kunth. (Including *C. rivularis* Kunth f. *elongatus* Boeckl.)

2b. **Cyperus bipartitus** Torr. f. **elutus** (C. B. Clarke) Mohlenbr. Flatsedge. July–Oct. Wet ground; scattered in IL, but rare. *Cyperus rivularis* Kunth. f. *elutus* C. B. Clarke.

3. **Cyperus compressus** L. Compressed Flatsedge. Sept. Native to the s. U.S. and tropical America; adventive in a roadside ditch; Lawrence and Washington cos.

4. **Cyperus diandrus** Torr. Flatsedge. June–Oct. Wet ground along banks and shores, not common; scattered in IL.

5. **Cyperus echinatus** (L.) A. Wood. Round-headed Flatsedge. June–Sept. Dry

sandy woods, old fields; common in the s. ⅓ of the state, scattered elsewhere. *Cyperus ovularis* (Michx.) Torr. (Including *C. ovularis* [Michx.] Torr. var. *robustus* Brigg., var. *sphaericus* Boeckl., and var. *wolfii* [Wood] Kuk.)

6. **Cyperus engelmannii** Steud. Engelmann's Flatsedge. July–Oct. Wet ground; rare but scattered in IL

7. **Cyperus erythrorhizos** Muhl. Red-rooted Flatsedge. Aug.–Oct. Moist, often sandy soil, occasional; throughout the state.

8. **Cyperus esculentus** L. Two varieties may be distinguished in Illinois:

a. Spikelets to 15 mm long, 1.5–3.0 mm broad .... 8a. *C. esculentus* var. *esculentus*
a. Spikelets more than 15 mm long, more than 3 mm broad .... 8b. *C. esculentus* var. *leptostachyus*

8a. **Cyperus esculentus** L. var. **esculentus.** Yellow Nut Sedge. June–Oct. Moist, frequently disturbed, soil, common; in every co.

8b. **Cyperus esculentus** L. var. **leptostachyus** Boeckl. Yellow Nut Sedge. June–Oct. Moist, often disturbed, soil; scattered in IL.

9. **Cyperus filicinus** Vahl. Flatsedge. July–Sept. Native to the s. U.S.; adventive in a wet roadside ditch; Jackson Co.

10. **Cyperus flavescens** L. Flatsedge. July–Oct. Wet open soil; scattered in IL. (Including var. *poiformis* [Pursh] Fern.)

11. **Cyperus grayoides** Mohlenbr. Sand Prairie Flatsedge. Aug. Sand prairies and blowouts, rare; scattered in the nw. ¼ of IL.

12. **Cyperus houghtonii** Torr. Houghton's Flatsedge. June–Sept. Sandy soil, rare; Boone, Cook, Grundy, Kane, Kankakee, Lake, Mason, Whiteside, and Will cos.

13. **Cyperus iria** L. Rice-field Flatsedge. July–Oct. Native to Europe and Asia; adventive in wet meadows, very rare; Alexander, Coles, Pulaski, and Union cos.

14. **Cyperus lancastriensis** Porter. Lancaster Flatsedge. Aug.–Oct. Moist, sandy woods, rare; Massac, Perry, Pope, and Pulaski cos.

15. **Cyperus lupulinus** (Spreng.) Marcks. Two varieties occur in Illinois:

a. Plants with a central, sessile glomerule of spikelets and 3–6 (–8) well-developed rays; achenes (1.7–) 1.8–2.2 mm long .... 15a. *C. lupulinus* var. *lupulinus*
a. Plants with only a central, sessile glomerule of spikelets or occasionally with 1–2 rays; achenes 1.2–1.8 (–2.0) mm long .... 15b. *C. lupulinus* var. *macilentus*

15a. **Cyperus lupulinus** (Spreng.) Marcks var. **lupulinus.** Flatsedge. May–Oct. Rocky or sandy soil; common in the n. ½ of the state, less common in the s. ½. *Cyperus filiculmis* Vahl, misapplied.

15b. **Cyperus lupulinus** (Spreng.) Marcks var. **macilentus** (Fern.) Marcks. Flatsedge. May–Oct. Sandy soil; common in the n. ½ of the state, rare southward.

16. **Cyperus X mesochorus** Geise. Hybrid Flatsedge. July–Oct. Sandy areas, often where disturbed; occasional and scattered in the n. ½ of IL, rarer southward. Reputed to be the hybrid between *C. lupulinus* (Spreng.) Marcks var. *lupulinus* and *C. schweinitzii* Torr.

17. **Cyperus odoratus** L. Rusty Flatsedge. Aug.–Oct. Rich, moist soil; scattered throughout IL. *Cyperus ferruginescens* Boeckl.

18. **Cyperus pseudovegetus** Steud. Flatsedge. June–Oct. Moist soil; occasional in the s. ⅓ of IL.

19. **Cyperus retrorsus** Chapm. Reflexed Flatsedge. June–July. Native to the s. U.S.; disturbed cultivated soil; Jackson Co.

20. **Cyperus schweinitzii** Torr. Schweinitz' Flatsedge. June–Sept. Sandy soil; occasional in the n. ½ of IL; adventive in Jackson, Madison, and St. Clair cos.

21. **Cyperus squarrosus** L. Flatsedge. May–Oct. Moist soil; scattered throughout the state. *Cyperus aristatus* Rottb.; *C. inflexus* Muhl.

22. **Cyperus strigosus** L. Straw-colored Flatsedge. June–Oct. Wet ground, common; in every co. (Including f. *robustior* Kunth and var. *stenolepis* [Torr.] Kuk.)

6. **Dulichium** Rich.—Three-way Sedge

1. **Dulichium arundinaceum** (L.) Britt. Three-way Sedge. July–Oct. Swamps and low ground in woods; throughout the state but not common; rare in the s. cos.

7. **Eleocharis** R. Br.—Spikerush

1. Spikelet about the same thickness as the culm; plants robust, often 1 m tall or taller.
   2. Culms terete, septate by cross partitions; bristles shorter than the achene, or lacking .... 6. *E. equisetoides*
   2. Culms 4-sided, sharply angled, not septate; bristles slightly longer than the achene .... 17. *E. quadrangulata*
1. Spikelet usually conspicuously thicker than the culm; plants usually less than 1 m tall (except *E. rostellata* and *E. macrostachya*).
   3. Style 2-cleft; achenes obovoid-lenticular (rarely sometimes trigonous in *E. olivacea*).
      4. Plants perennial, with rhizomes.
         5. Achenes at most 1 mm long, the small, conical tubercle ⅛–⅙ as long .... 12. *E. olivacea*
         5. Achenes averaging somewhat longer than 1 mm, the conical tubercle ¼–½ as long.
            6. Basal scale 1, suborbicular, completely encircling the culm; culms averaging 0.6 mm wide .... 7. *E. erythropoda*
            6. Basal scales 1–3, never completely encircling the culm; culms averaging 1.3 mm wide.
               7. Basal sheaths of mature culms with prominent V-shaped sinuses; culms sometimes soft and inflated, up to 75 cm tall .... 14. *E. palustris*
               7. Basal sheaths of mature culms truncate to slightly oblique at apex; culms usually rigid, often more than 75 cm tall .... 10. *E. macrostachya*
      4. Plants annual, with fibrous roots.
         8. Achenes yellow to deep brown; plants usually 15–50 cm tall.
            9. Tubercle ½–⅔ the width of the achene .... 13. *E. ovata*
            9. Tubercle more than ⅔ the width of the achene.
               10. Tubercle up to ¼ the height of the achene .... 5. *E. engelmannii*
               10. Tubercle ¼–½ the height of the achene .... 11. *E. obtusa*
         8. Achenes lustrous black or purple; plants usually less than 15 cm tall .... 8. *E. geniculata*
   3. Style 3-cleft; achenes usually trigonous.
      11. Tubercle conspicuously differentiated from the achene.
         12. Achenes with 8–19 longitudinal ribs.
            13. Scales 1.5–2.2 mm long; culms 0.1–0.5 mm wide, capillary.
               14. Perennial with slender rhizomes; scales obtuse to acute, not recurved; achenes 0.7–1.1 mm long, with 30–60 trabeculae .... 1. *E. acicularis*
               14. Annual; scales acute to acuminate, recurved; achenes 0.60–0.75 mm long, with 20–30 trabeculae .... 2. *E. bella*
            13. Scales 2.5–3.0 mm long; culms about 1 mm wide, often inrolled .... 20. *E. wolfii*
         12. Achenes pitted or with papillae, but not ribbed.
            15. Tubercle subulate; achene about 1.5 mm long (including the tubercle); annual .... 9. *E. intermedia*
            15. Tubercle not subulate; achene 1.0–1.2 mm long (including the tubercle); perennial.
               16. Culm with 5 vascular bundles and conspicuously (4- or) 5-angled; mature achene usually olivaceous and warty .... 19. *E. verrucosa*
               16. Culm with 6–14 vascular bundles and appearing low-angular or compressed; mature achene usually yellow, brown, or orange-brown, reticulate or slightly warty.
                  17. Culm with 6–8 vascular bundles, appearing low-angular; achenes slightly warty or reticulate .... 4. *E. elliptica*
                  17. Culm with 9–14 vascular bundles, compressed; achenes reticulate .... 3. *E. compressa*

11. Tubercle confluent with top of achene, long-conical.
18. Achenes up to 1.5 mm long; scales up to 2.5 mm long .... 15. *E. parvula*
18. Achenes 1.8–3.0 mm long; scales 3–8 mm long.
19. Culms flat, to 2 mm wide; scales elliptic, obtuse; plants tufted.. 18. *E. rostellata*
19. Culms usually 3-angled, less than 1 mm wide; scales lanceolate, acute; plants with stoloniferous rhizomes..................... 16. *E. pauciflora*

1. **Eleocharis acicularis** (L.) Roem. & Schultes. Three varieties occur in Illinois:

a. Culms cylindrical, 3- to 4-angled, up to 0.3 mm in diameter; tubercle as high as wide.
b. Culms up to 15 cm long; spikelets to 4 mm long...1a. *E. acicularis* var. *acicularis*
b. Culms up to 60 cm long; spikelets up to 8 mm long ....1b. *E. acicularis* var. *gracilescens*
a. Culms more or less compressed, with 6–12 ribs, 0.3–0.5 mm in diameter; tubercle as high as wide.. 1c. *E. acicularis* var. *porcata*

1a. **Eleocharis acicularis** (L.) Roem. & Schultes var. **acicularis** Needle Spikerush. July–Oct. Low, wet ground; occasional throughout the state.

1b. **Eleocharis acicularis** (L.) Roem. & Schultes var. **gracilescens** Svenson. Needle Spikerush. July–Oct. Usually in standing water; St. Clair Co.

1c. **Eleocharis acicularis** (L.) Roem. & Schultes var. **porcata** S. G. Smith. July–Aug. Wet ground, rare; one historical collection, according to Rothrock.

2. **Eleocharis bella** (Piper) Svenson. Pretty Spikerush. June–Aug. Alluvial banks of Illinois River, very rare; Peoria Co. Not seen since 1901. *Eleocharis acicularis* (L.) Roem. & Schultes var. *bella* Piper.

3. **Eleocharis compressa** Sull. Flat-stemmed Spikerush. May–July. Low areas; scattered throughout the state. *Eleocharis elliptica* Kunth var. *compressa* (Sull.) Drap. & Mohlenbr.

4. **Eleocharis elliptica** Kunth. Spikerush. May–July. Low areas; occasional throughout the state.

5. **Eleocharis engelmannii** Steud. Engelmann's Spikerush. May–Oct. Wet ground; occasional throughout the state. *Eleocharis engelmannii* Steud. var. *detonsa* Gray; *Eleocharis obtusa* (Willd.) Schultes var. *engelmannii* (Steud.) Gilly; *Eleocharis obtusa* (Willd.) Schultes var. *detonsa* (Gray) Drap. & Mohlenbr.; *Eleocharis ovata* (Roth) Roem. & Schultes var. *detonsa* (Gray) Mohlenbr.

6. **Eleocharis equisetoides** (Ell.) Torr. Horsetail Spikerush. July–Oct. Wet ground or standing water, very rare; Cass, Cook, and Lake cos.

7. **Eleocharis erythropoda** Steud. Red-based Spikerush. June–Sept. Wet soil; common in the n. ½ of the state, less common elsewhere. *Eleocharis calva* (Gray) Torr.

8. **Eleocharis geniculata** (L.) Roem. & Schultes. Jointed Spikerush. June–Sept. Wet sand, very rare; Cook Co. *Eleocharis caribaea* (Rottb.) S. F. Blake.

9. **Eleocharis intermedia** (Muhl.) Schultes. Spikerush. July–Oct. Riverbanks and swampy areas; fairly common along the Illinois River, occasional elsewhere in the n. ½ of IL.

10. **Eleocharis macrostachya** Britt. Large Spikerush. July–Oct. Edge of swamps, sloughs, not common; DuPage, Lake, and McHenry cos. *Eleocharis palustris* (L.) Roem. & Schultes, misapplied.

11. **Eleocharis obtusa** (Roth) Schultes. Blunt Spikerush. May–Oct. Wet ground; common throughout the state. *Eleocharis ovata* (Roth) Roem. & Schultes var. *obtusa* (Willd.) Kuk.

12. **Eleocharis olivacea** Torr. Spikerush. July–Sept. Wet sands, very rare; Cook, Lake, and Mason cos.

13. **Eleocharis ovata** (Roth) Roem. & Schultes. Spikerush. May–Oct. Wet ground; scattered but uncommon in IL.

14. **Eleocharis palustris** (L.) Roem. & Schultes. Marsh Spikerush. June–Sept. Edge of swamps, sloughs, ponds, streams; occasional throughout the state. *Eleocharis smallii* Britt.

15. **Eleocharis parvula** (Roem. & Schultes) Link. Small Spikerush. July–Sept. Wet soil, very rare; Coles and Effingham cos.

16. **Eleocharis pauciflora** (Lightf.) Link. Few-flowered Spikerush. July–Oct. Wet areas; restricted to the extreme ne. cos.

17. **Eleocharis quadrangulata** (Michx.) Roem. & Schultes. Square-stemmed Spikerush. June–Oct. Shallow water in ponds and lakes, not common; primarily in the s. ¼ of the state. (Including var. *crassior* Fern.)

18. **Eleocharis rostellata** (Torr.) Torr. Spikerush. July–Sept. Marshy, calcareous soils, rare; Cook, Kendall, Lake, McHenry, Wabash, and Will cos.

19. **Eleocharis verrucosa** (Svenson) Harms. Warty Spikerush. May–Sept. Low, wet ground, moist crevices on dry bluffs; occasional throughout the state. *Eleocharis tenuis* Schultes var. *verrucosa* (Svenson) Svenson.

20. **Eleocharis wolfii** Gray. Wolf's Spikerush. May–July. Wet ground; rare and scattered in IL.

8. **Eriophorum** L.—Cotton Sedge

1. Leaves up to 1.5 mm wide, channeled throughout; involucral bract 1.
  2. Culms glabrous throughout; sheath of leaf longer than blade; achene 1.5–2.0 mm long ..... 2. *E. gracile*
  2. Culms scabrous above; sheath of leaf shorter than blade; achene 2.5–3.1 mm long ..... 3. *E. tenellum*
1. Leaves 1.5–8.0 mm wide, channeled only above the middle; involucral bracts 2–5.
  3. Stamen 1; scales several-nerved ..... 4. *E. virginicum*
  3. Stamens 3; scales 1-nerved.
    4. Peduncles glabrous; anthers 2.5–5.0 mm long ..... 1. *E. angustifolium*
    4. Peduncles puberulent; anthers 1.0–1.3 mm long ..... 5. *E. viridi-carinatum*

1. **Eriophorum angustifolium** Honck. Cotton Sedge. June–Aug. Calcareous fens; scattered in the n. ⅓ of IL. (Including var. *majus* Schultz.)

2. **Eriophorum gracile** Koch. Cotton Sedge. May–July. Bogs, very rare; McHenry and Peoria cos.

3. **Eriophorum tenellum** Nutt. Cotton Sedge. June–Sept. Bogs, rare; Cass, Lake, and McHenry cos.

4. **Eriophorum virginicum** L. Cotton Sedge. Aug.–Oct. Bogs, rare; Lake and McHenry cos. (Including f. *album* [Gray] Wieg.)

5. **Eriophorum viridi-carinatum** (Engelm.) Fern. Cotton Sedge. May–Aug. Swamps and bogs, rare; Lake, Rock Island, and Winnebago cos.

9. **Fimbristylis** Vahl—Fimbry

1. Style branches 3; achene trigonous ..... 2. *F. autumnalis*
1. Style branches 2; achene lenticular.
  2. Spikelets in a close head; achene 0.4–0.5 mm long; cespitose annuals ..... 4. *F. vahlii*
  2. Spikelets on elongated rays, cymose; achene 1.0–1.8 mm long; tufted perennials or annuals with hardened bases.
    3. Outer scales puberulent; achene 1.3–1.7 mm long, reticulate ..... 3. *F. pubescens*
    3. All scales glabrous; achene 1.0–1.3 mm long, with horizontal and transverse ribs, or sometimes irregularly tuberculate ..... 1. *F. annua*

1. **Fimbristylis annua** (All.) Roem. & Schultes. Fimbry. June–Oct. Moist soil, rare; Alexander, Johnson, and Massac cos. *Fimbristylis baldwiniana* (Schultes) Torr.

2. **Fimbristylis autumnalis** (L.) Roem. & Schultes. Fimbry. June–Oct. Moist or dry, sandy soil; occasional throughout the state. (Including var. *mucronulata* [Michx.] Fern.)

3. **Fimbristylis puberula** (Michx.) Vahl var. **drummondii** (Boeckl.) Ward. Fimbry. Moist or dry soil, rare; restricted to the n. ½ of the state; also St. Clair Co. *Fimbristylis drummondii* Boeckl.; *F. caroliniana* (Lam.) Fern.

4. **Fimbristylis vahlii** (Lam.) Link. Vahl's Fimbry. July–Oct. Moist soil, Cass Co; shore of lake, Aalexander Co.

10. **Fuirena** Rottb.—Umbrella Grass

1. **Fuirena scirpoidea** Michx. Umbrella Grass. July. Along edge of lake, very rare; Hamilton Co. *Fuirena pumila* Michx., misapplied.

11. **Hemicarpha** Nees

1. Second scale of spikelet up to 0.2 mm long, or absent .... 3. *H. micrantha*
1. Scond scale of spikelet 0.5 mm long or longer.
2. First scale of spikelet widest at the middle; second scale of spikelet veinless, or with obscure veins .... 1. *H. aristulata*
2. First scale of spikelet widest above the middle; second scale of spikelet with 2–4 conspicuous veins.. 2. *H. drummondii*

1. **Hemicarpha aristulata** (Cov.) Smyth. Hemicarpha. July–Aug. Wet soil, very rare; Alexander Co. *Lipocarpha aristulata* (Cov.) G. C. Tucker.

2. **Hemicarpha drummondii** Nees. Hemicarpha. Aug.–Oct. Wet, sandy soil, very rare; Cook, Iroquois, and Lake cos. *Scirpus micranthus* Vahl var. *drummondii* (Nees) Mohlenbr.; *Lipocarpha drummondii* (Nees) G. C. Tucker.

3. **Hemicarpha micrantha** (Vahl) Pax. Hemicarpha. Aug.–Oct. Sandy banks; scattered throughout the state. *Scirpus micranthus* Vahl; *Lipocarpha micrantha* (Vahl) G. C. Tucker.

12. **Isolepis** R. Br.

1. Achenes 1.0–1.5 mm long, 0.7–1.0 mm wide; scales usually attached to the achenes; leaf sheaths green or brown... 1. *I. carinatus*
1. Achenes 0.8–1.0 mm long, 0.5–0.7 mm wide; scales free from the achenes; leaf sheaths usually reddish ....2. *I. cernua*

1. **Isolepis carinatus** Hook. & Arn. Dwarf Bulrush. May–June. Pastures, fields, rare; confined to the s. tip of the state. *Scirpus koilolepis* (Steud.) Gl.

2. **Isolepis cernua** (Vahl) Roem. & Schultes. Nodding Dwarf Bulrush. May–June. Native to the w. U.S., S. America, Eurasia; Africa; Australia. Wet ground, rare: Cook Co. *Scirpus cernuus* Vahl.

13. **Kyllinga** Rottb.

1. **Kyllinga pumila** Michx. Sweet-scented Flatsedge. July–Sept. Moist, open soil; restricted to the s. ½ of the state; also Adams and Crawford cos. *Cyperus densicaespitosus* Mattf. & Kuk.; *C. tenuifolius* (Steud.) Dandy.

14. **Lipocarpha** R. Br.

1. **Lipocarpha maculata** (Michx.) Torr. Lipocarpha. June–Sept. Around a pond, very rare; Cass Co.

15. **Rhynchospora** Vahl—Beaked Rush

1. Leaves 6–20 mm broad; achene 4.7–5.5 mm long, 2.5–3.0 mm broad; tubercle 12–20 mm long .... 4. *R. corniculata*
1. Leaves 0.3–6.0 (–7.0) mm broad; achene 1.2–2.6 mm long, 0.8–1.6 mm broad; tubercle 0.3–1.6 mm long.
2. Bristles antrorsely pubescent, or entirely lacking; tubercle 0.3–0.6 mm long, broadly deltoid ....5. *R. globularis*
2. Bristles retrorsely pubescent; tubercle 0.6–1.8 mm long, conical.
3. Spikelets lanceoloid; bristles 3.6–4.3 mm long; leaves capillary, at most 0.4 mm broad..2. *R. capillacea*
3. Spikelets ovoid to lance-ovoid; bristles 2.0–3.5 mm long (occasionally to 4.0 mm long in *R. glomerata*); leaves 0.5–7.0 mm broad.
4. Achene emarginate or nearly so, slightly rugulose .... 1. *R. alba*
4. Achene with a conspicuous margin, smooth.
5. Achene 1.2–1.4 mm broad, with broad shoulders; bristles 3–4 mm long .... 6. *R. glomerata*
5. Achene 0.8–1.2 mm broad, pyriform; bristles 2.0–2.8 mm long .... 3. *R. capitellata*

1. **Rhynchospora alba** (L.) Vahl. White Beaked Rush. July–Sept. Sphagnum bogs, rare; Cook, Lake, McHenry, and Peoria cos.

2. **Rhynchospora capillacea** Torr. Beaked Rush. July–Sept. Bogs and marshes; rare in the ne. corner of the state; also St. Clair Co. (Including f. *leviseta* [E. J. Hill] Fern.)

3. **Rhynchospora capitellata** (Michx.) Vahl. Beaked Rush. July–Oct. Moist soil, not common in the ne. corner of the state; absent elsewhere except for Lee, Massac, Pope, and Williamson cos.

4. **Rhynchospora corniculata** (Lam.) Gray. Great Beaked Rush. June–Sept. Swamps, low roadside ditches; mostly restricted to the s. tip of IL; also Hamilton Co. (Including var. *interior* Fern.)

5. **Rhynchospora globularis** (Chapm.) Small. Beaked Rush. June–July. Moist, sandy soil; very rare; Cook and Kankakee cos.

6. **Rhynchospora glomerata** (L.) Vahl. Beaked Rush. July–Oct. Moist, often sandy soil, rare; Cook, Johnson, Kankakee, and Pope cos.

16. **Schoenoplectus** (Reichenb.) Palla

1. Culms 3-angled.
- **2.** Annuals without rhizomes; lowest leaves all reduced to bladeless sheaths .... 7. *S. mucronatus*
- **2.** Perennials, often with rhizomes; usually at least 1 or more leaves at base of culm with a blade.
  - **3.** Leaves with blades usually more than 3 in number at base of culm; scales acute at tip, not notched, several-nerved .... 14. *S. torreyi*
  - **3.** Leaves with blades usually 1–3 in number at base of culm; scales notched at tip, 1-nerved.
    - **4.** Upper leaves shorter than to equaling the leaf sheath; apex of floral scales with a notch 0.1–0.4 mm deep .... 2. *S. americanus*
    - **4.** Upper leaves usually much longer than the leaf sheah; apex of floral scales with a notch at least 0.5 mm deep.
      - **5.** Perianth bristles shorter than to as long as the achenes; plants up to 01.5 m tall .... 8. *S. pungens*
      - **5.** Perianth bristles longer than the achenes; plants often 2 m tall or taller .... 4. *S. deltarum*

1. Culms terete or sometimes more or less flattened, not 3-angled.
- **6.** Some or all of the spikelets pedicellate.
  - **7.** Achenes trigonous; all spikelets pedicellate .... 6. *S. heterochaetus*
  - **7.** Achenes biconvex; at least some of the spikelets sessile.
    - **8.** Basal sheaths markedly fibrillose on the margins; achenes subtended by 2–4 bristles .... 3. *S. californicus*
    - **8.** Basal sheaths smooth along the margins or with only a few fibers; achenes subtended by (4–) 6 bristles.
      - **9.** Scales viscid, purplish-dotted, mucronulate; stems dark green .... 1. *S. acutus*
      - **9.** Scales not viscid, not purplish-dotted, acute; stems light green .... 13. *S. tabernaemontani*
- **6.** All spikelets sessile or nearly so.
  - **10.** Perennial with rhizomes; achenes 2.2–3.8 mm long.
    - **11.** Scales viscid, purplish-dotted; achenes 2.2–2.6 mm long .... 1. *S. acutus*
    - **11.** Scales not viscid, not purplish-dotted; achenes 2.5–3.8 mm long .... 12. *S. subterminalis*
  - **10.** Tufted annual; achenes 1.3–2.0 mm long.
    - **12.** Achenes smooth .... 11. *S. smithii*
    - **12.** Achenes transversely corrugated or pitted.
      - **13.** Achenes transversely corrugated; scales awned.
        - **14.** Achenes 2-sided; stigmas 2 .... 5. *S. hallii*
        - **14.** Achenes trigonous; stigmas 3 .... 10. *S. saximontanus*
      - **13.** Achenes pitted; scales obtuse .... 9. *S. purshianus*

1. **Schoenoplectus acutus** (Muhl.) A. Love & D. Love. Two forms occur in Illinois:

a. Spikelets on divergent rays to 5 cm long.................................. 1a. *S. acutus* f. *acutus*
a. Spikelets all sessile, forming a dense glomerule............... 1b. *S. acutus* f. *congestus*

1a. **Schoenoplectus acutus** (Muhl.) A. Love & D. Love f. **acutus.** Hard-stem Bulrush. Aug.–Sept. Shallow water; occasional throughout the state. *Scirpus acutus* Muhl. *Schoenoplectus acutus* hybridizes with *S. heterochaetus,* forming *Schoenoplectus X oblongus* (T. Koyama) Sojak.

1b. **Schoenoplectus acutus** (Muhl.) A. Love & D. Love f. **congestus** (Farw.) Mohlenbr. Hard-stem Bulrush. Aug.–Sept. Shallow water, rare; Lake and Williamson cos. *Scirpus acutus* Muhl. f. *congestus* (Farw.) Fern.

2. **Schoenoplectus americanus** (Pers.) Volkart ex Schinz & R. Keller. Olney's Bulrush. July–Aug. Around ponds, rare; c. cos. *Scirpus americanus* Pers.

3. **Schoenoplectus californicus** (C. A. Mey.) Sojak. Giant Bulrush. July–Oct. Native s. and w. of IL; adventive around a pond; Pope Co. *Scirpus californicus* (C. A. Mey.) Steud.

4. **Schoenoplectus deltarum** (Schuyler) Sojak. Delta Bulrush. July–Oct. Wet roadside ditch, very rare; Williamson Co. *Scirpus deltarum* Schuyler.

5. **Schoenoplectus hallii** (Gray) S. G. Sm. Hall's Bulrush. Aug.–Oct. Shores of ponds, rare; Alexander, Cass, Kankakee, Mason, Menard, and Morgan cos. *Scirpus hallii* Gray.

6. **Schoenoplectus heterochaetus** (Chase) Sojak. Bulrush. July–Sept. Shores and swamps; scattered throughout the state but not common. *Scirpus hetereochaetus* Chase.

7. **Schoenoplectus mucronatus** (L.) Palla. Mucronate Bulrush. July–Sept. Native to Europe and Asia; adventive in wet ground, rare; Alexander, Jasper, and Mason cos. *Scirpus mucronatus* L.

8. **Schoenoplectus pungens** (Vahl) Palla. Three-square. June–Sept. Along shores, in marshes; scattered throughout the state. *Scirpus americanus* Pers., misapplied; *Scirpus pungens* Vahl.

9. **Schoenoplectus purshianus** (Fern.) M. T. Strong. Pursh's Bulrush. Aug.–Oct. Shores of lakes, rare; confined to the s. ⅔ of IL; also Kankakee Co. *Scirpus purshianus* Fern.; *S. purshianus* f. *williamsii* (Fern.) Fern.; *S. smithii* Gray var. *williamsii* (Fern.) Beetle.

10. **Schoenoplectus saximontanus** (Fern.) J. Raynal. Rocky Mountain Bulrush. July–Aug. Wet ground; very rare. *Scirpus saximontanus* Fern.

11. **Schoenoplectus smithii** (Gray) Sojak. Smith's Bulrush. July–Oct. Wet soil; scattered in IL. *Scirpus smithii* Gray.

12. **Schoenoplectus subterminalis** (Torr.) Sojak. Bulrush. July–Oct. Shallow water, very rare; Cook and Lake cos. *Scirpus subterminalis* Torr.

13. **Schoenoplectus tabernaemontani** (K. C. Gmel.) Palla. Soft-stem Bulrush. June–Sept. Swamps, marshes, shallow water; common in the n. ½ of the state, less common in the s. ½. *Scirpus validus* Vahl; *S. validus* Vahl var. *creber* Fern.; *Scirpus tabernaemontani* K. C. Gmel.

14. **Schoenoplectus torreyi** (Olney) Palla. Torrey's Bulrush. July–Oct. Shores of ponds, very rare; Lee, St. Clair, and Winnebago cos. *Scirpus torreyi* Olney.

17. **Scirpus** L.—Bulrush

1. Stolons present (sometimes absent in *S. polyphyllus*); bristles concealed by the scales at maturity, retrorsely barbed.
  2. Culms with 10–20 leaves; scales reddish brown.......................... 9. *S. polyphyllus*
  2. Culms with 2–10 leaves; scales brown or black.
    3. Lowest sheaths reddish; bristles pubescent throughout...... 6. *S. microcarpus*
    3. All sheaths greenish; bristles (if present) glabrous at base.
      4. Bristles 0–3, shorter than the achenes................................ 4. *S. georgianus*
      4. Bristles usually 5 or 6, shorter than to slightly longer than the achenes.
        5. Lower leaf blades and sheaths usually nodulose-septate; scales mostly brownish; longer bristles frequently longer than the achenes .... 2. *S. atrovirens*

5. Lower leaf blades and sheaths usually not nodulose-septate; scales mostly blackish; longer bristles usually shorter than or about equaling the achenes....5. *S. hattorianus*

1. Stolons absent; bristles usually exceeding the scales at maturity (except in *S. pendulus*), smooth or sparsely pubescent.
   6. Bristles at maturity equaling or shorter than the scales; scales usually with prominent green midveins....8. *S. pendulus*
   6. Bristles at maturity much longer than the scales; scales usually with inconspicuous midveins.
      7. Spikelets and involucels reddish brown.
         8. All spikelets sessile or subsessile....3. *S. cyperinus*
         8. Lateral spikelets pedicellate....10. *S. rubricosus*
      7. Spikelets and involucels drab, pale brown, or blackish.
         9. All spikelets sessile or subsessile....3. *S. cyperinus*
         9. Lateral spikelets pedicellate.
            10. Spikelets and involucels drab or pale brown....7. *S. pedicellatus*
            10. Spikelets and involucels blackish....1. *S. atrocinctus*

1. **Scirpus atrocinctus** Fern. Blackish Bulrush. June–July. Swamps and boggy areas, very rare; Lake Co.

2. **Scirpus atrovirens** Willd. Dark Green Bulrush. June–Aug. Wet soil; common throughout the state; in every co. An unnamed hybrid between this species and *Scirpus georgianus* Harper has been found in Effingham Co.

3. **Scirpus cyperinus** (L.) Kunth. Two varieties occur in Illinois:

a. Spikelets and involucels reddish brown....3a. *S. cyperinus* var. *cyperinus*
a. Spikelets and involucels drab....3b. *S. cyperinus* var. *pelius*

3a. **Scirpus cyperinus** (L.) Kunth var. **cyperinus** Wool Grass. Aug.–Oct. Swamps and marshes; common throughout the state.

3b. **Scirpus cyperinus** (L.) Kunth var. **pelius** Fern. Pale Wool Grass. Aug.–Oct. Swamps, marshes; Cook and Lake cos.

4. **Scirpus georgianus** Harper. Dark Green Bulrush. June–Sept. Moist soil; occasional throughout the state. *Scirpus atrovirens* Willd. var. *georgianus* (Harper) Fern.

5. **Scirpus hattorianus** Mak. Bulrush. June–Aug. Ditches, very rare; Carroll, Cook, DuPage, Kankakee, and Kendall cos.

6. **Scirpus microcarpus** Presl. Small-fruited Bulrush. Aug.–Sept. Marshes and swamps, very rare; Lake Co. *Scirpus rubrotinctus* Fern.

7. **Scirpus pedicellatus** Fern. Bulrush. July–Aug. Wet soil, very rare; Cook, DuPage, and Pope cos.

8. **Scirpus pendulus** Muhl. Nodding Bulrush. June–Aug. Low woods and along streams, common; probably in every co. *Scirpus lineatus* Michx., misapplied.

9. **Scirpus polyphyllus** Vahl. Many-leaved Bulrush. July–Sept. Low woods; rare and scattered in IL.

10. **Scirpus rubricosus** Fern. Pedicellate Wool Grass. July–Oct. Swamps; s. ⅓ of IL; also Stephenson Co. *Scirpus eriophorum* Michx., *nomen illeg.; Scirpus cyperinus* (L.) Kunth var. *rubricosus* (Fern.) Gilly.

18. **Scleria** Berg.—Nut Rush

1. Achenes smooth; perennials.
   2. Tubercles of hypogynium none; achenes 2.0–2.5 mm long; culms firm, to 1 m tall....4. *S. triglomerata*
   2. Tubercles of hypogynium 8–9; achenes 2.5–3.5 mm long; culms soft, to 70 cm tall....2. *S. oligantha*
1. Achenes papillose, pitted, or wrinkled; annuals (except *S. pauciflora*).
   3. Achenes papillose; tubercles of hypogynium 6 or 9; perennial....3. *S. pauciflora*
   3. Achenes pitted or wrinkled; tubercles of hypogynium none; annual.
      4. Achenes pitted; hypogynium 3-lobed; inflorescence closely paniculate; achenes 1.8–2.2 mm long....1. *S. muehlenbergii*
      4. Achenes wrinkled; hypogynium absent; inflorescence verticillate; achenes 1.5 mm long....5. *S. verticillata*

1. **Scleria muehlenbergii** Steud. Muhlenberg's Nut Rush. Aug.–Oct. Sandy shore of pond, very rare; Cass and Lee cos. *Scleria reticulata* Michx., misapplied; *S. reticulata* Michx. var. *pubescens* Britt.

2. **Scleria oligantha** Michx. Nut Rush. June–Sept. Moist woods; scattered in IL.

3. **Scleria pauciflora** Muhl. Two varieties occur in Illinois:

a. Culms and leaves glabrous or sparsely hirtellous ....3a. *S. pauciflora* var. *pauciflora*
a. Culms and leaves densely pilose..............................3b. *S. pauciflora* var. *caroliniana*

3a. **Scleria pauciflora** Muhl. var. **pauciflora.** Few-flowered Nut Rush. June–Sept. Dry soil, in woods, on bluffs; occasional in the s. tip of the state, also Iroquois, Kankakee, and Will cos.

3b. **Scleria pauciflora** Muhl. var. **caroliniana** (Willd.) Wood. Few-flowered Nut Rush. June–Sept. Dry soil, rare; Lee and Pope cos.

4. **Scleria triglomerata** Michx. Nut Rush. June–Sept. Moist or dry woods, fields; occasional in the n. ½ of the state; also Effingham and Pope cos.

5. **Scleria verticillata** Muhl. Whorled Nut Rush. July–Sept. Bogs and marshes, rare; scattered in the n. ½ of the state.

19. **Trichophorum** Pers.

1. Culms smooth, terete; bristles longer than the achene .... 1. *T. cespitosum*
1. Culms scabrous, 3-angled; bristles about as long as the achene ....
.... 2. *T. verecundum*

1. **Trichophorum cespitosum** (L.) Hartm. var. **callosum** (Bigel.) Mohlenbr. Tufted Bulrush. June–Aug. Boggy situations, rare; Lake and McHenry cos. *Scirpus cespitosus* L. var. *callosus* Bigel.

2. **Trichophorum planifolium** (Sprengl.) Palla. Bashful Bulrush. Apr. Cherty slopes, very rare; Alexander and Union cos. *Scirpus verecundus* Fern.; *Trichophorum verecundum* (Fern.) Mohlenbr.

173. DIOSCOREACEAE—YAM FAMILY

1. **Dioscorea** L.—Yam

1. Leaves usually with basal lobes; plants usually bearing tubers in the axils of the leaves during late summer and autumn .... 1. *D. polystachys*
1. Leaves without basal lobes; plants not bearing tubers in the axils of the leaves during late summer and autumn.
  2. All leaves (except sometimes the lowermost) alternate; capsule 1.5–2.5 cm long; seeds (including wing) 7–14 mm broad; petiole essentially glabrous at point of attachment of blade ....3. *D. villosa*
  2. Lowest leaves whorled, becoming opposite or alternate on the upper part of the stem; capsule 2.5–3.0 cm long; seeds (including wing) 15–18 mm broad; petiole puberulent at point of attachment of blade .... 2. *D. quaternata*

1. **Dioscorea polystachys** Turcz. Chinese Yam. June–Aug. Native to Asia; adventive in disturbed areas; scattered in s. IL; also Piatt Co. *Dioscorea batatas* Dcne.; *Dioscorea oppositifolia* L.

2. **Dioscorea quaternata** (Walt.) J. F. Gmel. Two varieties occur in Illinois:

a. Stems and leaves green ....2a. *D. quaternata* var. *quaternata*
a. Stems and leaves more or less glaucous.......................2b. *D. quaternata* var. *glauca*

2a. **Dioscorea quaternata** (Walt.) J. F. Gmel. var. **quaternata.** Wild Yam. June–July. Dry or moist woods, occasional; confined to the s. ¼ of the state; also Vermilion Co.

2b. **Dioscorea quaternata** (Walt.) J. F. Gmel. var. **glauca** (Muhl.) Fern. Wild Yam. June–July. Dry or moist woods, occasional; confined to the s. ⅓ of the state.

3. **Dioscorea villosa** L. Two forms occur in Illinois:

a. Leaves with some pubescence ....3a. *D. villosa* f. *villosa*
a. Leaves glabrous .... 3b. *D. villosa* f. *glabrifolia*

3a. **Dioscorea villosa** L. f. villosa. Wild Yam. June–July. Dry or moist woods, common; in every co.

3b. **Dioscorea villosa** L. f. **glabriflora** (Bartlett) Fern. Wild Yam. June–July. Dry or moist woods, occasional; scattered in IL.

174. HEMEROCALLACEAE—DAY-LILY FAMILY

1. **Hemerocallis** L.—Day-Lily

1. Flowers orange .... 1. *H. fulva*
1. Flowers yellow .... 2. *H. lilio-asphodelus*

1. **Hemerocallis fulva** (L.) L. Orange Day Lily. June–Aug. Native to Europe and Asia; adventive along roads and in waste ground, common; probably in every co.
2. **Hemerocallis lilio-asphodelus** L. Yellow Day Lily. June–Aug. Native to Asia; rarely escaped into disturbed soil. *Hemerocallis flava* L.

175. HYACINTHACEAE—HYACINTH FAMILY

1. Leaves cauline, elliptic to ovate, not grasslike .... 2. *Chionodoxa*
1. Leaves basal, usually grasslike.
  2. Perianth parts white with a ventral green stripe .... 4. *Ornithogalum*
  2. Perianth parts lavender, purple, or blue (rarely white in *Camassia*).
    3. Perianth parts united nearly to tip; racemes densely flowered .... 3. *Muscari*
    3. Perianth parts essentially free; racemes less densely flowered.
      4. Each perianth segment with 3 veins .... 1. *Camassia*
      4. Each perianth segment with 1 vein .... 5. *Scilla*

1. **Camassia** Lindl.—Wild Hyacinth

1. Scape with 3–24 persistent bracts; inflorescence at anthesis 2–3 (–3.5) cm broad; capsule longer than broad; perianth parts deep lavender to pale purple; plants beginning to flower in May .... 1. *C. angusta*
1. Scape with 0–2 (–3) deciduous bracts; inflorescence at anthesis 3–5 cm broad; capsule as broad as long; perianth parts white to pale blue to pale lilac; plants beginning to flower in early Apr. .... 2. *C. scilloides*

1. **Camassia angusta** (Engelm. & Gray) Blankinship. Wild Hyacinth. May–July. Prairies and woods, very rare; Macon and Peoria cos.
2. **Camassia scilloides** (Raf.) Cory. Wild Hyacinth. Apr.–June. Moist prairies and woods; occasional throughout the state.

2. **Chionodoxa** Boiss.—Glory-of-the-snow

1. **Chionodoxa luciliae** Boiss. Glory-of-the-snow. Apr. Native to Asia; rarely escaped from cultivation; Coles Co. *Chionodoxa forbesii* Baker.

3. **Muscari** Mill.—Grape Hyacinth

1. Leaves terete, not more than 3 mm broad .... 4. *M. racemosum*
1. Leaves flat, most or all of them more than 3 mm broad.
  2. Raceme up to 6 cm long; perianth to 6 mm long .... 3. *M. comosum*
  2. Raceme at least 9 cm long; perianth 9–11 mm long.
    3. All flowers fertile, purple or blue; perianth 3.5–5.0 mm long .... 2. *M. botryoides*
    3. Fertile flowers deep violet with white teeth; sterile flowers pale blue; perianth 5–6 mm long .... 1. *M. armeniacum*

1. **Muscari armeniacum** Leicht. Heavenly Blue. Apr.–May. Native to sw. Asia; rarely escaped from cultivation; Piatt Co.
2. **Muscari botryoides** (L.) Mill. Grape Hyacinth. Apr. Native to Europe; escaped to fields, particularly grassy places; occasional throughout the state.
3. **Muscari comosum** (L.) Mill. Grape Hyacinth. Apr. Native to s. Europe; rarely escaped from cultivation; Jackson and Madison cos.
4. **Muscari racemosum** (L.) Mill. Blue Bottles. Apr.–May. Native to Europe; occasionally escaped into disturbed soil; scattered in IL. *Muscari atlanticum* Boiss. & Reut.

4. **Ornithogalum** L.—Star-of-Bethlehem

1. Perianth segments 2.5 cm long or longer; flowers in racemes .... 1. *O. nutans*

1. Perianth segments up to 2 cm long; flowers more or less appearing in umbels .... 2. *O. umbellatum*

1. **Ornithogalum nutans** L. Star-of-Bethlehem. Apr.–May. Native to Europe; rarely adventive in IL; Bond Co.

2. **Ornithogalum umbellatum** L. Star-of-Bethlehem. Apr.–June. Native to Europe; adventive in fields, open woods, lawns, and along roads; common throughout the state.

5. **Scilla** L.—Squill

1. **Scilla sibirica** Andr. Squill. Apr.–May. Native to Europe and Asia; rarely spreading from cultivation; scattered in the n. ⅔of IL.

176. HYDROCHARITACEAE—FROG'S-BIT FAMILY

1. Leaves borne along the stem, never more than 5 mm broad.
  2. Leaves in whorls of 4 or 6, generally more than 2 cm long; petals 9–12 mm long .... 1. *Egeria*
  2. Leaves opposite or in whorls of 3, generally less than 2 cm long; petals up to 5 mm long .... 2. *Elodea*
1. Leaves all basal, 5 mm or more broad.
  3. Leaves ovate to orbicular, often with a layer of spongy cells beneath; fertile stamens 6–12, the filaments united into a column .... 3. *Limnobium*
  3. Leaves narrow, elongate, ribbonlike, without a layer of spongy cells beneath; fertile stamens 2, the filaments free .... 4. *Vallisneria*

1. **Egeria** Planch.—Giant Waterweed

1. **Egeria densa** Planch. Giant Waterweed. July–Sept. Native to S. America; infrequently found as an adventive, usually in mine ponds; DuPage, Edwards, Franklin, Jefferson, Lake, and Williamson cos. *Elodea densa* (Planch.) Caspary; *Anacharis densa* (Planch.) Vict.

2. **Elodea** Michx.—Waterweed

1. Staminate flowers pedicellate, not liberated at maturity, with petals 3.5–6.0 mm long; leaves 1–5 mm broad .... 1. *E. canadensis*
1. Staminate flowers sessile, liberated at maturity, with petals 1.5–2.0 mm long; leaves usually 0.5–1.5 mm broad .... 2. *E. nuttallii*

1. **Elodea canadensis** L. C. Rich. Waterweed; Elodea. July–Sept. Quiet water; occasional in the n. ½ of the state, rare in the s. ½. *Anacharis canadensis* (Michx.) Rich.

2. **Elodea nuttallii** (Planch.) St. John. Waterweed; Elodea. July–Sept. Quiet water; occasional in the n. cos., not common in the s. cos. *Anacharis nuttallii* Planch.

3. **Limnobium** Rich.—Frog's-bit

1. **Limnobium spongia** (Bosc) Steud. Sponge Plant. July–Aug. Swamps, rare; Alexander, Jackson, Johnson, Massac, Pope, Pulaski, and Union cos.

4. **Vallisneria** L.—Eelgrass

1. **Vallisneria americana** Michx. Eelgrass; Water Celery. June–Sept. Quiet water; restricted to the n. ½ of the state; also Alexander Co.

177. HYPOXIDACEAE—YELLOW STAR-GRASS FAMILY

1. **Hypoxis** L.—Yellow Star-grass

1. **Hypoxis hirsuta** (L.) Coville. Yellow Star-grass. Apr.–June. Dry woods, prairies, field, sandstone outcroppings, calcareous fens; occasional throughout the state.

178. IRIDACEAE—IRIS FAMILY

1. Sepals somewhat recurved; petals spreading or erect; styles petal-like; stamens concealed by the arching styles .... 3. *Iris*
1. Sepals and petals similar, spreading or tubular; styles not petal-like; stamens not concealed.

2. Perianth tubular, forming a short funnel, scarlet; upper 3 perianth segments longer than the lower 3 .... 2. *Gladiolus*
2. Perianth spreading and orange or rotate and blue or white; all perianth segments essentially equal.
   3. Flowers orange, with purplish spots, at least 3 cm broad; filaments distinct; flowers bracteate, the bracts soon withering; capsule at least 2 cm long; seeds attached to a central column, resembling a blackberry; plants rhizomatous .... 1. *Belamcanda*
   3. Flowers blue or white, less than 2 cm broad; filaments united to apex; flowers borne from 1 or 2 persistent spathes; capsule up to 6 mm long; seeds not attached to a central column; plants with fibrous roots .... 4. *Sisyrinchium*

1. **Belamcanda** Adans.—Blackberry Lily

1. **Belamcanda chinensis** (L.) DC. Blackberry Lily. July–Aug. Native to Asia; adventive along roads; scattered throughout the state. *Iris domestica* (L.) Goldblatt & Mabberley.

2. **Gladiolus** L.—Gladiolus

1. Upper petals spreading, not forming a hood .... 1. *G. X colvillea*
1. Upper petals arching, forming a hood .... 2. *G. X ganvadensis*

1. **Gladiolus X colvillei** Sweet. Scarlet Gladiolus. July–Aug. Escaped from cultivation along roads; Johnson and Massac cos. Reputed to be a hybrid between *G. tristis* Salisb. and *G. cardinalis* Curt.

2. **Gladiolus X ganvadensis** Van Houtte. Garden Gladiolus. July–Aug. Escaped from cultivation; Kankakee Co. Reputed to be a hybrid between *G. dalenii* Van Geel and *G. oppositiflorus* Herb.

3. **Iris** L.—Iris

1. Upper surface of sepals with a beard of hairs.
   2. Stems very short or none, bearing 1 flower; leaves to 15 cm long, to 7 mm broad; sepals to 17 mm broad .... 7. *I. pumila*
   2. Stems to nearly 1 m tall, bearing several flowers; leaves to 100 cm long, to 3 cm broad; sepals over 20 mm broad.
      3. Flowers blue .... 5. *I. X germanica*
      3. Flowers yellow .... 3. *I. flavescens*
1. Sepals without a beard of hairs.
   4. Rhizomes stout, at least 1 cm thick; leaves 40–100 cm long; flowering stem usually branched, 20–100 cm long; spathes subtending flower unequal in length; flowers several per stem; sepals 1- to 2-ridged or without ridges; capsule obtusely 3-angled or 6-angled, 3–9 cm long; plants often of more aquatic situations.
      5. Ovary and capsule 6-angled; capsule indehiscent; leaves rather soft.
         6. Flowering stem 20–40 cm tall; flowers dark blue; capsule 3–5 cm long; leaves 15–30 mm broad; seeds more or less globose .... 1. *I. brevicaulis*
         6. Flowering stem at least 50 cm tall; flowers copper-colored, rarely yellow; capsule 4.5–7.5 cm long; leaves 10–15 mm broad; seeds flattened .... 4. *I. fulva*
      5. Ovary and capsule 3-angled (occasionally 6-angled in the yellow-flowered *I. pseudacorus*); capsule dehiscent; leaves firm.
         7. Flowers usually some shade of blue, violet, or lavender; sepals not 2-ridged on upper surface; perianth tube constricted above the ovary.
            8. Leaves up to 8 mm wide .... 9. *I. sibirica*
            8. Most or all the leaves 10 mm wide or wider .... 8. *I. shrevei*
         7. Flowers basically yellow; sepals 2-ridged on upper surface; perianth not constricted above the ovary .... 6. *I. pseudacorus*
   4. Rhizomes slender, less than 1 cm thick; leaves (at flowering time) to 25 cm long, at maturity as long as 40 cm; flowering stem unbranched, to 4.5 cm long; spathes subtending flower nearly equal in length; flowers 1–2 per stem; sepals

sharply 3-ridged above; capsule sharply 3-angled, about 1 cm long; plants of rich woodlands, usually near streams .... 2. *I. cristata*

1. **Iris brevicaulis** Raf. Blue Iris. May–July. Marshes and wet prairies; scattered in the s. ⅔ of the state.

2. **Iris cristata** Soland. Dwarf Crested Iris. Apr.–May. Lowland woods, usually along streams; restricted to the s. tip of IL.

3. **Iris flavescens** DC. Yellow Iris. Apr.–June. Origin uncertain; escaped from cultivation; rarely adventive in IL; DuPage, Grundy, Henry, Kankakee, and Lake cos.

4. **Iris fulva** Ker. Swamp Red Iris; Copper Iris. May–June. Swamps, usually in shallow water; Alexander, Hamilton, Jackson, Johnson, Massac, Pulaski, and Union cos. Yellow-flowered plants occur in Johnson Co. They have been named forma *fulvaurea* Small.

5. **Iris X germanica** L. Bearded Iris. Apr.–June. Origin uncertain; escaped from cultivation into disturbed soil; Jackson, Lake, McLean, Piatt, and Pope cos.

6. **Iris pseudacorus** L. Yellow Iris. June–Aug. Native to Europe; adventive in disturbed soil; scattered in IL.

7. **Iris pumila** L. Dwarf Iris. Mar.–Apr. Native to Europe and Asia; escaped into waste ground; DuPage, Mason, and Schuyler cos.

8. **Iris shrevei** Small. Wild Blue Iris; Blue Flag. May–June. Wet situations, common; probably in every co. *Iris virginica* L. var. *shrevei* (Small) E. Anders.

9. **Iris sibirica** L. Siberian Iris. June–July. Native to Europe; rarely escaped from cultivation; Cass and DuPage cos.

4. **Sisyrinchium** L.—Blue-eyed Grass

1. Spathes long-pedunculate from the axils of leaflike bracts; capsules dark brown.
  2. Leaves bright green; stems broadly winged, 3–5 mm wide; capsules 4–6 mm long; plants becoming black upon drying .... 2. *S. angustifolium*
  2. Leaves pale green or glaucous; stems narrowly winged, 1–3 mm wide; capsules 3.0–4.5 mm long; plants generally not becoming black upon drying.......... ....3. *S. atlanticum*
1. Spathes sessile, terminal; capsules pale brown, stramineous, or greenish (occasionally becoming dark brown in *S. montanum*).
  3. Spathes 2 .... 1. *S. albidum*
  3. Spathe 1.
    4. Capsules 2–4 mm long; margins of outer bract free to base or united for less than 2 mm.
      5. Stems winged; margins of outer bract free to base; flowers pale blue or white..4. *S. campestre*
      5. Stems marginate; margins of outer bract united for 1–2 mm above base; flowers bright violet...6. *S. mucronatum*
    4. Capsules 4–6 mm long; margins of outer bract united for 2–6 mm above base ....5. *S. montanum*

1. **Sisyrinchium albidum** Raf. Blue-eyed Grass. Apr.–June. Open woods, fields, prairies, common; probably in every co.

2. **Sisyrinchium angustifolium** Mill. Blue-eyed Grass. May–June. Low woods, moist prairies, common; probably in every co. *Sisyrinchium graminoides* Bickn.

3. **Sisyrinchium atlanticum** Bickn. Blue-eyed Grass. May–June. Wet prairies, rare; DuPage, Iroquois, Kankakee, Macoupin, Pope, and Union cos.

4. **Sisyrinchium campestre** Bickn. Blue-eyed Grass. May–June. Prairies, particularly in sandy soil; mostly in the n. ½ of the state; also Hardin and Madison cos.

5. **Sisyrinchium montanum** Greene. Two varieties occur in Illinois:

a. Plants pale green or glaucous, drying pale; capsule whitish, greenish, or stramineous, even at maturity .... 5a. *S. montanum* var. *montanum*
a. Plants bright green, drying blackish; capsule greenish or pale brown at first, becoming dark at maturity .... 5b. *S. mucronatum* var. *crebrum*

5a. **Sisyrinchium montanum** Greene var. **montanum**. Blue-eyed Grass. May–June. Sandy prairies, rare; Cook, DuPage, Lake, and Winnebago cos.

5b. **Sisyrinchium montanum** Greene var. **crebrum** Fern. Blue-eyed Grass. May–June. Fields, very rare; Kankakee Co.

6. **Sisyrinchium mucronatum** Michx. Blue-eyed Grass. May–June. Sandy prairies, rare; scattered in the n. ½ of IL.

179. JUNCACEAE—RUSH FAMILY

1. Plants glabrous; capsule several-seeded .... 1. *Juncus*
1. Plants pubescent; capsule 3-seeded .... 2. *Luzula*

1. **Juncus** L.—Rush

Technical Key

1. Leaf sheaths usually without blades, apiculate or mucronate; inflorescence appearing lateral on the stems.
    2. Stems densely cespitose; stamens 3; anthers 0.5–0.8 mm long; seeds 0.5 mm long; capsules beakless .... 15. *J. effusus*
    2. Stems single at intervals from elongated rhizomes; stamens 6; anthers 1.5–2.0 mm long; seeds 1 mm long; capsules with beaks 0.5–1.0 mm long .... 4. *J. arcticus*
1. Leaf sheaths with definite blades; inflorescence terminal.
    3. Flowers in heads, not prophyllate (i.e., not with bracteoles).
        4. Leaves flat, not terete nor cross-septate; anthers purplish brown.
            5. Stems solitary, approximately 3.5 cm apart on conspicuous, scaly rhizomes, 4.9–8.8 (–10.2) dm tall; leaves 2.0–6.5 mm wide; heads (13–) 20–135 .... 6. *J. biflorus*
            5. Stems cespitose, 0.45–5.90 dm tall; leaves 1.0–2.5 (–2.9) mm wide; heads 3–28 (–32) .... 19. *J. marginatus*
        4. Leaves terete and cross-septate, or flat and cross-septate in *J. validus;* anthers yellow.
            6. Leaves flat .... 27. *J. validus*
            6. Leaves terete.
                7. Seeds fusiform, 0.7–1.9 mm long, caudate.
                    8. Seeds 1.2–1.9 mm long, the tails comprising (⅓–) ½–⅝ the total length of the seeds; heads 5- to 50-flowered; stamens 3 .... 10. *J. canadensis*
                    8. Seeds 0.7–1.2 mm long, the tails comprising ¼–⅖ the total length of the seeds; heads 2- to 5- (to 10-) flowered; stamens 3 or 6.
                        9. Sepals obtuse to subacute .... 8. *J. brachycephalus*
                        9. Sepals acute to acuminate .... 24. *J. subcaudatus*
                7. Seeds ellipsoid to oblongoid to ovoid, 0.4–0.6 mm long, apiculate.
                    10. Stamens 6.
                        11. Involucral leaves shorter than the inflorescences; heads hemispherical or turbinate, 2–7 mm wide, 2- to 9-flowered; sepals 1.9–3.0 mm long, acute to acuminate to obtuse; capsules oblongoid or ellipsoid, acute to obtuse.
                            12. Petals mostly shorter and blunter than the sepals; capsules obtuse and abruptly short-pointed .... 2. *J. alpinoarticulatus*
                            12. Petals as long as or a little longer than the sepals; capsules more attenuate at tip .... 5. *J. articulatus*
                        11. Involucral leaves usually exceeding the inflorescences; heads spherical or hemispherical, 8–15 mm wide, 9- to 90-flowered; sepals 2.5–5.0 mm long, subulate; capsules lanceoloid, subulate.
                            13. Sepals 2.5–4.0 mm long; petals equaling to exceeding the sepals by 0.8 mm; heads 8–11 (–12) mm across; stems to 6 dm tall .... 21. *J. nodosus*
                            13. Sepals 4–5 mm long; petals 1 mm shorter than to nearly equaling the sepals; heads 10–15 mm across; stems to 10.7 dm tall .... 26. *J. torreyi*

10. Stamens 3.
14. Capsules linear-lanceoloid, acute, exceeding the sepals by at least 1.5 mm (usually by 2.0 mm)....................13. *J. diffusissimus*
14. Capsules ellipsoid or oblongoid, obtuse to subulate, shorter than to exceeding the sepals by 1 mm.
15. Capsules oblongoid, subulate, exceeding the sepals by 0.75–1.00 mm; perianth segments subulate.............................
....22. *J. scirpoides*
15. Capsules ellipsoid, acute to obtuse, shorter than to exceeding the sepals by 0.75 mm; perianth segments subulate to acuminate.
16. Perianth segments acuminate; petals 0.7 mm shorter than to equaling the sepals; capsules shorter than the petals to exceeding the sepals by 0.75 mm.
17. Heads turbinate, 3–5 mm across.
18. Heads 150–280; leaves 1.5–4.5 mm wide.................
....20. *J. nodatus*
18. Heads 3–35; leaves 0.5–1.0 mm wide.......................
.... 12. *J. debilis*
17. Heads hemispherical, 5–10 mm across.........................
....1. *J. acuminatus*
16. Perianth segments subulate; petals distinctly (approximately 1 mm) shorter than the sepals; capsules usually 0.5 mm shorter than the petals.............. 7. *J. brachycarpus*
3. Flowers borne singly on the inflorescence branches, not in heads, prophyllate (i.e., with bracteoles).
19. Annual; auricle at top of sheath absent; sepals 4–7 mm long; inflorescence comprising ¼–⅘ the total height of the plants .... 9. *J. bufonius*
19. Perennial; auricle at top of sheath present; sepals 2.3–6.0 mm long; inflorescence comprising less than ½ the total height of the plants.
20. Sepals obtuse; anthers 1 mm long, 3 times longer than the filaments.
21. Capsues longer than the perianth................................11. *J. compressus*
21. Capsules equaling the perianth... 16. *J. gerardii*
20. Sepals acuminate, subulate, or aristate; anthers shorter than to as long as the filaments.
22. Leaves terete, at least distally; capsules usually exceeding the perianth; seeds 0.5–1.3 mm long.
23. Leaves involute near the summit of the sheath, becoming closed and terete above; inflorescences 1.5–6.5 (–8.0) cm long; petals acute or obtuse; capsules exceeding the sepals by (0.75–) 1.0–1.6 mm; seeds 0.5–0.6 mm long, apiculate at both ends ....
.... 17. *J. greenei*
23. Leaves terete throughout; inflorescences (1.0–) 2.0–3.5 cm long; petals acuminate or aristate; capsules slightly shorter than to exceeding the sepals by 1 mm; seeds 1.0–1.3 mm long, caudate at both ends ....28. *J. vaseyi*
22. Leaves flat or involute; capsules shorter than to exceeding the perianth by 0.1 mm; seeds 0.3–0.5 mm long.
24. Petals 0.2 mm shorter than to exceeding the sepals by 0.5 mm; tips of the inflorescence branches incurved; leaves to 13 cm long
.... 23. *J. secundus*
24. Petals 1 mm shorter than to equaling the sepals; tips of the inflorescence branches not incurved; leaves to 30 cm long.
25. Auricles friable, not firm or rigid, scarious, hyaline, and lanceolate, white or brownish, prolonged 1.0–4.5 (–5.0) mm beyond point of insertion; perianth segments spreading.
26. Capsules less than ¾ the length of sepals, the flowers more or less interrupted in the inflorescence; plants at least usually 70 cm tall...3. *J. anthelatus*

26. Capsules ¾ or more the length of the sepals, the flowers congested in the inflorescence; plants usually less than 70 cm tall .... 25. *J. tenuis*

25. Auricles firm at apex or rigid, cartilaginous or membranous, occasionally prolonged to 2 mm beyond point of insertion; perianth segments spreading or appressed.

27. Auricles cartilaginous, opaque, rigid, often slightly flaring, and obtuse, yellow or orange-brown, less than 1 mm long and usually 0.75 mm prolonged beyond point of insertion; perianth segments spreading; bracteoles obtuse to acute .... 14. *J. dudleyi*

27. Auricles membranous, hyaline, not cartilaginous or rigid, usually firm at apex, pale or brown, very slightly prolonged to exserted 2 mm beyond point of insertion; perianth segments appressed; bracteoles acuminate to aristate................ .... 18. *J. interior*

Nontechnical Key

1. Leaves absent; flowers appearing laterally.

2. Stems without longitudinal ribs; stems in rows from a rhizome....... 4. *J. arcticus*

2. Stems with longitudinal ribs; stems cespitose.................................. 15. *J. effusus*

1. Leaves present; flowers appearing terminally.

3. Flowers prophyllate, subtended by 2 small opposite bracteoles.

4. Annuals; inflorescence ⅓ or more the total height of the plant...................... .... 9. *J. bufonius*

4. Perennials; inflorescence less than ½ the total height of the plant.

5. 2–4 leaves present on the stem.

6. Capsules longer than the perianth............................... 11. *J. compressus*

6. Capsules as long as the perianth .... 16. *J. gerardii*

5. Cauline leaves absent; all leaves basal.

7. Leaves terete for some or all of their length.

8. Sepals 3.5–4.5 mm long; capsules 4.0–5.5 mm long......28. *J. vaseyi*

8. Sepals 2.3–3.5 mm long; capsules 3–4 mm long.......... 17. *J. greenei*

7. Leaves flat or involute.

9. Flowers conspicuously on 1 side of branches .... 23. *J. secundus*

9. Flowers not on 1 side of branches.

10. Auricles scarious, projecting 1–3 mm beyond base of blade.

11. Capsules less than ¾ length of the sepals; flowers more or less interrupted in the inflorescence; plants usually at least 70 cm tall ....3. *J. anthelatus*

11. Capsules ¾ or more the length of the sepals; flowers congested in the inflorescence; plants usually less than 70 cm tall .... 25. *J. tenuis*

10. Auricles membranous or cartilaginous, not prolonged beyond base of blade.

12. Auricles membranous, green or pale .... 18. *J. interior*

12. Auricles cartilaginous, yellow .... 14. *J. dudleyi*

3. Flowers not prophyllate, not subtended by 2 small, opposite bracteoles.

13. Leaves flat.

14. Leaves not cross-septate.

15. Plants with 15–20 heads, each head 6–8 mm across; blades 1–3 mm wide .... 19. *J. marginatus*

15. Plants with 20 or more heads, each head 4–6 mm across; blades 4–6 mm wide.. 6. *J. biflorus*

14. Leaves cross-septate..27. *J. validus*

13. Leaves terete, usually cross-septate.

16. Capsules much longer than the perianth...................... 13. *J. diffusissimus*

16. Capsules shorter than, equaling, or barely longer than the perianth.

17. Heads spherical or hemispherical.
 18. Heads hemispherical.
  19. Heads 6–10 mm in diameter.
   20. Seeds 0.3–0.4 mm long; flowers often more than 10 per head .... 1. *J. acuminatus*
   20. Seeds 0.7–1.2 mm long; flowers usually 5 (–10) per head .... 24. *J. subcaudatus*
  19. Heads 10–20 mm in diameter .... 10. *J. canadensis*
 18. Heads spherical.
  21. Heads up to 10 mm in diameter.
   22. Sepals 3.5–6.0 mm long; heads 30–100 per plant........ 7. *J. brachycarpus*
   22. Sepals 2.0–3.2 mm long; heads 20–60 per plant........ 22. *J. scirpoides*
  21. Heads 10–20 mm in diameter.
   23. Leaves 0.5–1.5 mm wide; heads 5–25 per plant........ 21. *J. nodosus*
   23. Some leaves at least 2 mm wide; heads 25–100 per plant .... 26. *J. torreyi*
17. Heads turbinate.
 24. Heads 35–200 per plant.
  25. Perianth about ⅔ as long as the capsules, obtuse to acute .... 8. *J. brachycephalus*
  25. Perianth nearly as long as the capsules, acuminate........ 20. *J. nodatus*
 24. Heads 3–35 per plant.
  26. Leaves more than 1 mm wide.
   27. Heads 6–8 mm across; capsules short-apiculate .... 2. *J. alpinoarticulatus*
   27. Heads 8–10 mm across; capsules long-apiculate .... 5. *J. articulatus*
  26. Leaves 0.5–1.0 mm wide........ 12. *J. debilis*

1. **Juncus acuminatus** Michx. Pointed Rush. July–Sept. Wet, low ground of ditches, ponds; occasional to common throughout the state.

2. **Juncus alpinoarticulatus** Chaix. Two subspecies occur in Illinois:

a. 1–several of the flowers elevated above the others on slightly elongated pedicels .... 2a. *J. alpinoarticulatus* ssp. *alpinoarticulatus*
a. Flowers sessile or equally short-pedicellate .... 2b. *J. alpinoarticulatus* ssp. *fuscescens*

2a. **Juncus alpinoarticulatus** Chaix ssp. **alpinoarticulatus.** Alpine Rush. July–Aug. Wet sandy shores and marshes, rare; confined to extreme ne. IL; also Ogle Co. *Juncus alpinus* Vill.; *J. alpinus* Vill. var. *rariflorus* Hartm.; *J. alpinoarticulatus* Chaix ssp. *americanus* Hamet-Ahti. A hybrid between this species and *J. articulatus* L. is known from Cook Co. It may be called *J. X alpiniformis* Fern.

2b. **Juncus alpinoarticulatus** Chaix ssp. **fuscescens** (Fern.) Hamet-Ahti. Alpine Rush. July–Aug. Wet sandy shores and marshes, rare; confined to extreme ne. IL. *Juncus alpinus* Vill. var. *fuscescens* Fern.

3. **Juncus anthelatus** (Wieg.) R. E. Brooks. Path Rush. June–Oct. Moist or dry soil, often in disturbed areas; scattered in IL. *Juncus tenuis* Willd. var. *anthelatus* Wieg.

4. **Juncus arctatus** L. var. **balticus** (Willd.) Trautv. Baltic Rush. June–Sept. Wet sandy shores, swamps, not common; confined to the n. ¼ of the state.

5. **Juncus articulatus** L. Northern Rush. July–Aug. Along a railroad, rare; Cook, DuPage, and Lake cos.

6. **Juncus biflorus** Ell. Two forms occur in Illinois:

a. Heads predominantly 2- to 4- (to 5-) flowered........ 6a. *J. biflorus* f. *biflorus*
a. Heads predominantly 6- to 10-flowered .... 6b. *J. biflorus* f. *andinus*

6a. **Juncus biflorus** Ell. f. **biflorus.** Two-flowered Rush. May–Sept. Wet ground; occasional to common in the s. ½ of the state; also Cook, Kankakee, Lee, and Winnebago cos.

6b. **Juncus biflorus** Ell. f. **andinus** Fern. & Grisc. Rush. May–Sept. Wet ground; scattered in the s. ½ of the state.

7. **Juncus brachycarpus** Engelm. Short-fruited Rush. July–Aug. Low, wet ground of fields, prairies, ditches, roadsides; occasional throughout the state, except for the nw. cos.

8. **Juncus brachycephalus** (Engelm.) Buch. Short-headed Rush. July–Sept. Marshes, wet ground along bodies of water; local in the n. ½ of the state; also Jackson, Johnson, and Montgomery cos.

9. **Juncus bufonius** L. Two varieties occur in Illinois:

a. Terminal flowers and flowers on the inflorescence branches borne singly, separated .... 9a. *J. bufonius* var. *bufonius*

a. Many of the terminal flowers and flowers on the inflorescence branches closely aggregated into 2- and 4-flowered fascicles .... 9b. *J. bufonius* var. *congestus*

9a. **Juncus bufonius** L. var. **bufonius.** Toad Rush. June–Aug. Wet roadsides, fields; local and scattered in IL.

9b. **Juncus bufonius** L. var. **congestus** Wahl. Toad Rush. June–Aug. Wet roadsides, very rare; scattered in s. IL.

10. **Juncus canadensis** J. Gay. Canada Rush. July–Oct. Wet ground of bogs, swamps, meadows; occasional in the n. ½ of the state, becoming less common in the s. ½. (Including f. *conglobatus* Fern.)

11. **Juncus compressus** L. Black Grass. July–Aug. Native to Europe; adventive in disturbed marshy area; Cook, DuPage, and Lake cos.

12. **Juncus debilis** Gray. Slender Rush. May–June. Wet ground, very rare; Jackson Co.

13. **Juncus diffusissimus** Buckl. Long-fruited Rush. July–Aug. Wet ground, ditches; local in the s. ⅓ of the state.

14. **Juncus dudleyi** Wieg. Dudley's Rush. June–Oct. Moist ground; occasional throughout the state. *Juncus tenuis* Willd. var. *dudleyi* (Wieg.) F. J. Herm.

15. **Juncus effusus** L. var. **solutus** Fern. & Wieg. Soft Rush. June–Sept. Wet ground of shores, swamps, and ditches; common in the s. ½ of the state, occasional in the n. ½.

16. **Juncus gerardii** Loisel. Black Grass. July–Aug. Native to Europe; adventive in disturbed wet soil; Cook and St. Clair cos.

17. **Juncus greenei** Oakes & Tuckerm. Greene's Rush. July–Sept. Wet, usually sandy ground of meadows, prairies, and fields; local in a few n. cos.

18. **Juncus interior** Wieg. Inland Rush. June–Oct. Wet ground of fields, prairies, ditches, roadsides; common throughout the state.

19. **Juncus marginatus** Rostk. Grass-leaved Rush. June–Sept. Wet, often sandy ground of pond borders, ditches, fields; occasional throughout the state.

20. **Juncus nodatus** Coville. Stout Rush. July–Sept. Wet ground, ditches, edges of water bodies; occasional in the s. ½ of the state; also Hancock Co.

21. **Juncus nodosus** L. Knotted Rush. July–Aug. Wet, often sandy ground of marshes, swamps, fields, and shores; mostly confined to the n. ½ of the state; also St. Clair, Washington, and Williamson cos.

22. **Juncus scirpoides** Lam. Sedgelike Rush. July–Aug. Wet and often sandy ground, rare; Alexander, Cass, Cook, Lawrence, Mason, Menard, and Will cos.

23. **Juncus secundus** Beauv. One-sided Rush. July–Sept. Open sandstone ledges and outcroppings; confined to the s. ½ of the state, except for Hancock Co.

24. **Juncus subcaudatus** (Engelm.) Coville & S. F. Blake. Rush. July–Oct. Along streams, very rare; Jackson Co.

25. **Juncus tenuis** Willd. Path Rush. June–Oct. Moist or dry ground, often in disturbed areas, common; in every co.

26. **Juncus torreyi** Coville. Torrey's Rush. July–Aug. Wet ground; common throughout the state. (Including var. *globularis* Farw.)

27. **Juncus validus** Coville. Stout Rush. May–June. Wet ground, very rare; Alexander Co.

28. **Juncus vaseyi** Engelm. Vasey's Rush. July–Sept. Wet meadows, bogs, shores, very rare; Cook, McHenry, and Winnebago cos.

2. **Luzula** DC.—Wood Rush

1. Flower solitary (rarely paired) at the tips of the inflorescence rays; seeds more than 2 mm long, including the strongly curved caruncle .... 1. *L. acuminata*
1. Flowers crowded in glomerulate spikes; seeds 1.2–2.0 mm long, including the conical caruncle.
  2. Plants producing small, white bulblets at base .... 2. *L. bulbosa*
  2. Plants not producing small, white bulblets at base.
    3. Rays of umbel erect to ascending .... 4. *L. multiflora*
    3. Rays of umbel horizontally spreading to reflexed .... 3. *L. echinata*

1. **Luzula acuminata** Raf. Wood Rush. May. Open woods, rare; Jo Daviess, LaSalle, and Ogle cos.
2. **Luzula bulbosa** (A. W. Wood) Smyth. Bulbous Wood Rush. Apr.–May. Woods; occasional throughout the state. *Luzula campestris* (L.) DC. var. *bulbosa* A. W. Wood.
3. **Luzula echinata** (Small) F. J. Herm. Wood Rush. Apr.–May. Woods and cliffs; occasional to common in the s. ⅕ of the state, rare elsewhere. (Including var. *mesochorea* F. J. Herm.) *Luzula multiflora* (Retz.) Lejuene var. *echinata* (Small) Mohlenbr.; *L. campestris* (L.) DC. var. *echinata* (Small) Fern. & Wieg.
4. **Luzula multiflora** (Retz.) Lejeune. Wood Rush. Apr.–May. Woods and shaded cliffs; common in the s. cos., less common northward. *Luzula campestris* (L.) DC. var. *multiflora* (Retz.) Selak.

180. JUNCAGINACEAE—ARROW-GRASS FAMILY

1. **Triglochin** L.—Arrow-grass

1. Carpels usually 6, the axis between them slender .... 1. *T. maritima*
1. Carpels 3, the axis between them broadly 3-winged .... 2. *T. palustris*

1. **Triglochin maritima** L. Arrow-grass. May–Aug. Sandy shores, swamps, wet ditches, rare; known only from the extreme ne. cos.; also Peoria and Tazewell cos.
2. **Triglochin palustris** L. Arrow-grass. May–July. Low areas, rare; known only from the ne. cos.; also Peoria and Tazewell cos.

181. LEMNACEAE—DUCKWEED FAMILY

1. Rootlets 1–several per plant; reproductive pouches lateral, 2 per plant.
  2. Rootlet 1 per plant .... 1. *Lemna*
  2. Rootlets (1–) 2 or more per plant .... 2. *Spirodela*
1. Rootlets none; reproductive pouch basal, 1 per plant.
  3. Frond thick, globose or ellipsoid .... 3. *Wolffia*
  3. Frond thin, linear .... 4. *Wolffiella*

1. **Lemna** L.—Duckweed

1. Fronds spatulate with long, persistent stipes, often submerged in compact masses; rootlet frequently absent .... 8. *L. trisulca*
1. Fronds orbicular to elliptic, floating; rootlet present.
  2. Fronds producing rounded, rootless turions or specialized overwintering structures late in the year; fronds with a distinct line of papillae near apex .... 9. *L. turionifera*
  2. Fronds not producing rounded, rootless turions late in the year; fronds without papillae or with indistinct lines of papillae.
    3. Fronds 3- or 5-nerved.
      4. Rootlet sheaths cylindrical, unwinged.
        5. Lower surface of frond flattened or only weakly convex, apex rounded to acute; frond usually 3-nerved .... 3. *L. minor*
        5. Lower surface of frond moderately to strongly convex, apex broadly rounded or often obtuse; frond 3- or 5-nerved .... 2. *L. gibba*
      4. Rootlet sheaths winged.
        6. Fronds with a single apical papilla; seeds with 8–26 strong ribs .... 1. *L. aequinoctialis*

6. Fronds with 2–3 papillae; seeds with 35 or more ribs
7. Fronds firm, weakly to strongly asymmetrical, thick, the nerves often inconspicuous .... 6. *L. perpusilla*
7. Fronds membranous, symmetrical or nearly so, membranous, the 3 nerves prominent ....7. *L. trinervis*
3. Fronds 1-nerved, or the nerves obscure.
8. Fronds 1-nerved, the lower surface green.
9. Fronds ovate-elliptic to elliptic, weakly to moderately asymmetrical, often floating in compact masses ....10. *L. valdiviana*
9. Fronds obovate to orbicular, symmetrical, usually solitary .... 4. *L. minuta*
8. Fronds obscurely nerved, the lower surface frequently reddish purple ....5. *L. obscura*

1. **Lemna aequinoctialis** Welw. Duckweed. Standing water; Cook, Henry, Jackson, Jo Daviess, Union, and Williamson cos.

2. **Lemna gibba** L. Duckweed. Standing water; DuPage, Henry, Knox, Mason, and Will cos.

3. **Lemna minor** L. Duckweed. Standing water; common throughout the state.

4. **Lemna minuta** HBK. Duckweed. Standing water; Carroll, Cook, DeKalb, DuPage, Madison, and Will cos. *Lemna minima* Phil.; *L. miniuscula* Hertel.

5. **Lemna obscura** (Austin) Daubs. Duckweed. Standing water; scattered throughout the state.

6. **Lemna perpusilla** Torr. Duckweed. Standing water; scattered throughout the state but not common.

7. **Lemna trinervis** (Austin) Small. Duckweed. Standing water; scattered throughout the state but not common.

8. **Lemna trisulca** L. Star Duckweed. Standing water; local throughout the state, but rare in the s. cos.

9. **Lemna turionifera** Landolt. Duckweed. Standing water; scattered throughout the state.

10. **Lemna valdiviana** Phil. Duckweed. Standing water; mostly in the s. ⅓ of the state; also Cook and Lake cos.

2. **Spirodela** Schleiden—Greater Duckweed

1. Fronds orbicular, 5- to 8-nerved; rootlets 2–10 ....1. *S. polyrhiza*
1. Fronds obovate to reniform, obscurely 3- (or 5-) nerved; rootlets (1–) 2–5................ ....2. *S. punctata*

1. **Spirodela polyrhiza** (L.) Schleiden. Greater Duckweed. Standing water; common throughout the state.

2. **Spirodela punctata** (Mey.) C. H. Thompson. Greater Duckweed. Standing water; Alexander, Johnson, Lawrence, and Union cos. *Spirodela oligorhiza* (Kurtz) Hegelm.; *Landoltia punctata* (Mey.) Les & D. J. Crawford.

3. **Wolffia** Schleiden—Water Meal

1. Plants about as deep as wide or not as deep as wide, boat-shaped.
2. Plants 1 ½–2 times longer than wide, subacute at apex, without a papilla ....1. *W. borealis*
2. Plants 1–1 ½ times longer than wide, obtuse at apex, with a prominent central papilla ....2. *W. brasiliensis*
1. Plants deeper than wide, globose to ovoid.
3. Plants up to 1.3 times longer than wide, 0.4–1.2 mm wide ....3. *W. columbiana*
3. Plants 1.3–2.0 times longer than wide, 0.3–0.5 mm wide................. 4. *W. globosa*

1. **Wolffia borealis** (Engelm.) Landolt. Water Meal. Standing water; scattered throughout the state but not common. *Wolffia punctata* Griseb.

2. **Wolffia brasiliensis** Weddell. Water Meal. Standing water; scattered throughout the state but apparently not common. *Wolffia papulifera* C. H. Thompson.

3. **Wolffia columbiana** Karst. Water Meal. Standing water; throughout the state.

4. **Wolffia globosa** (Roxb.) Hartog & Plas. Water Meal. Standing water; scattered throughout the state.

4. **Wolffiella** Hegelm.—Mud Midget

1. **Wolffiella gladiata** (Hegelm.) Hegelm. Mud Midget. Standing water; scattered throughout the state but apparently not common. *Wolffiella floridana* (J. D. Sm.) C. H. Thompson.

182. LILIACEAE—LILY FAMILY

1. Flowers borne in umbels; leaves basal .... 1. *Clintonia*
1. Flowers borne variously, but not in umbels; leaves basal or cauline.
  2. Leaves cauline; flowers at least 5 cm long .... 4. *Lilium*
  2. Leaves basal; flowers less than 5 cm long.
    3. Leaves 1–few, coriaceous, sometimes speckled; flower solitary.
      4. Leaves speckled; flower nodding .... 2. *Erythronium*
      4. Leaves not speckled; flower erect .... 5. *Tulipa*
    3. Leaves several, not coriaceous, usually not speckled; flowers several in racemes .... 3. *Hosta*

1. **Clintonia** L.—Bluebead Lily

1. **Clintonia borealis** (Ait.) Raf. Bluebead Lily. May. Moist soil, very rare; Cook Co.

2. **Erythronium** L.—Dog-tooth Violet

1. Flowers yellow; stigmas united .... 2. *E. americanum*
1. Flowers white; stigmas free.
  2. Perianth parts ascending or spreading .... 1. *E. albidum*
  2. Perianth parts reflexed .... 3. *E. mesochoreum*

1. **Erythronium albidum** Nutt. White Dog-tooth Violet; White Trout Lily. Apr.–May. Woods and occasionally fields; common throughout the state.

2. **Erythronium americanum** Ker. Yellow Dog-tooth Violet; Yellow Trout Lily. Apr.–May. Moist woods, frequently on shaded bluffs; occasional in the ne., e. cent., and s. cos.; also Calhoun Co.

3. **Erythronium mesochoreum** Knerr. White Dog-tooth Violet; White Trout Lily. Apr.–May. Prairies, rare; Macoupin, Morgan, and Pike cos.

3. **Hosta** Tratt—Plantain Lily

1. **Hosta lancifolia** (Thunb.) Engl. Plantain Lily. Aug. Native to e. Asia; rarely escaped into disturbed areas; Carroll, Cook, and DuPage cos. *Hosta japonica* (Thunb.) Voss.

4. **Lilium** L.—Lily

1. All the leaves alternate, the upper with bulblets in the axils; stems scabrous .... 1. *L. lancifolium*
1. Some of the leaves borne in whorls; bulblets absent; stems glabrous.
  2. Flowers erect; perianth parts with claws; only the uppermost group of leaves in a whorl; plants less than 1 m tall .... 3. *L. philadelphicum*
  2. Flowers nodding; perianth parts without claws; several whorls of leaves borne on the stem; plants generally over 1 m tall.
    3. Bulbs yellow; margins and nerves of leaves roughened; midvein of outer perianth parts rounded on the back; anthers 8–15 mm long; filament attached 1–2 mm from end of anther .... 2. *L. michiganense*
    3. Bulbs white; margins and nerves of leaves smooth; midvein of outer perianth parts sharply ridged on the back; anthers 17–25 mm long; filament attached 4–8 mm from end of anther .... 4. *L. superbum*

1. **Lilium lancifolium** Thunb. Tiger Lily. July–Aug. Native to e. Asia; escaped from cultivation into disturbed soil; scattered in IL. *Lilium tigrinum* L.

2. **Lilium michiganense** Farw. Turk's-cap Lily. May–July. Moist woods and prairies; rather common throughout the state.

3. **Lilium philadelphicum** L. var. **andinum** (Nutt.) Ker. Wood Lily. June–July. Prairies, occasional; mostly restricted to the upper ½ of the state; also Coles and Macoupin cos.

4. **Lilium superbum** L. Superb Lily. July. Low, moist woods, rare; Jackson, Pope, and Williamson cos.

5. **Tulipa** L.—Tulip

1. **Tulipa sylvestris** L. Tulip. Apr.–May. Native to the Caucasus; rarely escaped from gardens; Kane Co.

183. MARANTHACEAE—ARROWROOT FAMILY

1. **Thalia** L.—Fireflag

1. **Thalia dealbata** Roscoe. Fireflag. July–Oct. Wet roadside ditch, very rare; Alexander Co.

184. MELANTHIACEAE—MELANTHIUM FAMILY

1. Axis of inflorescence glabrous.
  2. Plants dioecious; capsules loculicidal .... 1. *Chamaelirium*
  2. Plants monoecious; capsules septicidal.
    3. Leaves mostly basal; perianth parts with a large gland just below the middle .... 5. *Zigadenus*
    3. Leaves cauline; perianth parts eglandular .... 3. *Stenanthium*
1. Axis of inflorescence pubescent.
  4. Perianth parts with a pair of glands near base; leaves broadly linear .... 2. *Melanthium*
  4. Perianth parts eglandular; leaves broadly oval .... 4. *Veratrum*

1. **Chamaelirium** Willd.—Devil's Bit

1. **Chamaelirium luteum** (L.) Gray. Fairy Wand. May–June. Low, wooded hillsides, rare; Hardin, Massac, and Pope cos.

2. **Melanthium** L.—Bunch-flower

1. **Melanthium virginicum** L. Bunch-flower. June–July. Meadows, wet prairies; rare in the w. and sw. cos., absent elsewhere.

3. **Stenanthium** Gray—Grass-leaved Lily

1. **Stenanthium gramineum** (Ker) Morong. Grass-leaved Lily. June–Aug. Moist woods and along streams, not common; confined to the s. ½ of the state. (Including var. *robustum* [S. Wats.] Fern.)

4. **Veratrum** L.—False Hellebore

1. **Veratrum woodii** Robbins ex Wood. False Hellebore. July–Aug. Rich, moist woods, rare; apparently confined to a few cent. cos. *Melanthium woodii* (Robins ex Wood) Bodkin.

5. **Zigadenus** Michx.—Camass

1. **Zigadenus elegans** Pursh. White Camass. July–Sept. Limestone cliffs or low areas near rivers, rare; Jo Daviess, Kane, and Kankakee cos.

185. NAJADACEAE—NAIAS FAMILY

1. **Najas** L.—Naiad

1. Leaves seemingly entire or serrulate, not spinulose along the lower side of the midvein; plants monoecious; achenes 1.5–3.5 mm long.
  2. Leaves linear, 0.5–2.0 mm wide, with slightly dilated bases; marginal spinules microscopic.
    3. Achenes lustrous, smooth but with 30–50 rows of obscure, minute, often square areolae; style and 2 stigmas 0.8–1.6 mm long .... 1. *N. flexilis*
    3. Achenes dull, pitted, with 16–24 rows of distinct, large, hexagonal or rectangular areolae; style and 2–3 stigmas less than 1 mm long .... 3. *N. guadalupensis*
  2. Leaves filiform, 0.1–0.3 mm wide, with abruptly dilated bases; marginal spinules macroscopic although minute.
    4. Achenes with roughened appearance, with 22–40 rows of longer-than-broad, rectangular areolae; style and 2–3 stigmas 0.8–1.2 mm long .... 2. *N. gracillima*

4. Achenes with ribbed appearance, with 10–18 rows of broader-than-long, rectangular areolae; style and 2 stigmas 1.0–1.4 mm long .... 5. *N. minor*

1. Leaves coarsely toothed, spinulose along the lower side of the midvein; plants dioecious; achenes 4.0–7.5 mm long .... 4. *N. marina*

1. **Najas flexilis** (Willd.) Rostk. & Schmidt. Naiad. July–Sept. Shallow water; occasional throughout the state.

2. **Najas gracillima** (A. Br.) Magnus. Naiad. July–Sept. Pools, ponds, and lakes with sandy and muddy substrata, rare; scattered in the s. ⅓ of IL; also Ford Co.

3. **Najas guadalupensis** (Spreng.) Magnus. Two varieties occur in Illinois:

a. Leaves with 50–100 teeth per side; stems 0.2–1.0 mm thick .... 3a. *N. guadalupensis* var. *guadalupensis*

a. Leaves with 20–40 teeth per side; stems 1–2 mm thick .... 3b. *N. guadalupensis* var. *olivacea*

3a. **Najas guadalupensis** (Spreng.) Magnus var. **guadalupensis.** Naiad. July–Oct. Near shores of ponds and shallow lakes; occasional to uncommon throughout the state.

3b. **Najas guadalupensis** (Spreng.) Magnus var. **olivacea** (C. Rosend. & Butters) Haynes. Naiad. July–Oct. Edge of ponds and lakes, rare; scattered in IL. *Najas olivacea* C. Rosend. & Butters.

4. **Najas marina** L. Naiad. July–Oct. In water, very rare; DuPage, Kane, Lake, and McHenry cos.

5. **Najas minor** All. Naiad. July–Oct. Native to Europe and Asia; shallow water along lake shores; occasional to common in the s. ⅔ of IL; also Carroll, Cook, and DuPage cos.

186. NARTHECIACEAE—NARTHECIUM FAMILY

1. Leaves evergreen, linear and grasslike; perianth pink, not scaly; flowers 4–6 mm across .... 2. *Liriope*

1. Leaves deciduous, lanceolate, not grasslike; perianth whitish, scaly; flowers 3–4 mm across .... 1. *Aletris*

1. **Aletris** L.—Colic Root

1. **Aletris farinosa** L. Colic Root. June–Aug. Moist, sandy prairies, sandy flats; restricted to the n. ⅓ of the state.

2. **Liriope** Lour.—Lily-turf

1. **Liriope spicatum** Lour. Grass-leaved Lily; Lily-turf. Aug.–Oct. Native to China and Japan; rarely escaped from cultivation; Jackson Co.

187. ORCHIDACEAE—ORCHID FAMILY

1. Plants without green leaves, at least at flowering time.
   2. Green leaves never produced; stem brown or purple, with colored sheaths; rhizomes coralline.
      3. Leaves without longitudinal ridges; flowers at most 1 cm long, brown, yellow-green, or white spotted with purple .... 5. *Corallorhiza*
      3. Leaves with 5–6 longitudinal ridges; flowers 1.7–2.3 cm long, madder-purple .... 10. *Hexalectris*
   2. Green leaves produced, but these usually absent at flowering time; stem green; plants with fleshy roots or with corms or tubers connected in a series.
      4. Flower deep crimson-pink, more than 2.5 cm long, solitary .... 2. *Arethusa*
      4. Flowers not deep crimson-pink, usually less than 2.5 cm long, 2–several.
         5. Leaves more than 1; plants with fleshy roots but without connected tubers; flowers whitish .... 16. *Spiranthes*
         5. Leaf 1; plants with connected tubers; flowers yellow, brown, or madder-purple.
            6. Leaves green on both sides; flowers not spurred .... 1. *Aplectrum*
            6. Leaves purple beneath; flowers spurred .... 17. *Tipularia*
1. Plants with green leaves at flowering time.
   7. Leaves whorled .... 11. *Isotria*
   7. Leaves alternate or basal.

8. Lip inflated, saclike, at least 18 mm long; anthers 2................. 6. *Cypripedium*
8. Lip flat or, if inflated, less than 18 mm long; anther 1.
9. Leaves 1 or 2, basal (leaves 1–2 and cauline in *Malaxis*).
10. Flowers totally pink, resupinate; leaf generally 1, linear to linear-lanceolate ....3. *Calopogon*
10. Flowers greenish yellow, madder-purple, or white and pink, not resupinate; leaves generally 2, lanceolate to elliptic to ovate to orbicular.
11. Flowers spurred, usually fragrant; anther persistent, difficult to detach.
12. Flowers white and pink; sepals and petals united to form a hooded structure (galea) behind the column .... 8. *Galearis*
12. Flowers greenish yellow; sepals and petals not forming a galea .... 14. *Platanthera*
11. Flowers spurless, not fragrant; anther easily and soon detachable .... 12. *Liparis*
9. Leaves more than 2, basal or cauline or, if only 1–2, then the leaves cauline.
13. Leaves 1–2, cauline.. 13. *Malaxis*
13. Leaves 3 or more, basal, cauline, or both.
14. All leaves basal.
15. Leaves green throughout; lip not saclike............... 16. *Spiranthes*
15. Leaves conspicuously marked with white; lip saclike ....9. *Goodyera*
14. At least 1 or more cauline leaves present (basal leaves also may be present).
16. Flowers spurred; anther persistent, not easily detached.
17. Spur of flower at least 4 mm long, as long as or longer than the lip; gland at base of pollen masses not surrounded by a membrane... 14. *Platanthera*
17. Spur of flower 1–3 mm long, up to ½ as long as the lip; gland at base of pollen masses surrounded by a membrane ....4. *Coeloglossum*
16. Flowers not spurred; anther easily detached.
18. Leaves both basal and cauline; flowers 1–2, terminal or numerous, sessile, the inflorescence spicate.
19. Cauline leaf 1; sepals and petals free; flowers 1–2, pink, at least 15 mm long .... 15. *Pogonia*
19. Cauline leaves 2 or more; upper sepal united with petals; flowers numerous, white, at most 10 mm long....... ..... 16. *Spiranthes*
18. Leaves all cauline; flowers numerous, pedicellate, the inflorescence racemose.
20. Flowers white (rarely pink), nodding, 1.0–1.5 cm long, the lip not saclike; sepals and petals lanceolate; leaves weak, oval..18. *Triphora*
20. Flowers green and purple, ascending, well over 1.5 cm long, the lip saclike at the base; sepals and petals ovate; leaves firm, lanceolate to ovate ....7. *Epipactis*

1. **Aplectrum** (Nutt.) Torr.—Putty-root Orchid

1. **Aplectrum hyemale** (Willd.) Nutt. Putty-root Orchid; Adam-and-Eve. May–June. Rich woods, occasional; scattered throughout the state.

2. **Arethusa** L.—Dragon's-mouth Orchid

1. **Arethusa bulbosa** L. Dragon's-mouth Orchid. May. Bog, very rare; Cook Co.

3. **Calopogon** L. C. Rich.—Grass Pink Orchid

1. Flowering stem as long as or shorter than the leaf; perianth pale pink; club-shaped hairs near tip of lip pink...1. *C. oklahomensis*

1. Flowering stem much longer than the leaf; perianth bright pink; club-shaped hairs near tip of lip white, yellow, or orange .... 2. *C. tuberosus*

1. **Calopogon oklahomensis** D. H. Goldman. Oklahoma Grass Pink Orchid. May–July. Prairies, rare; n. ½ of the state.

2. **Calopogon tuberosus** (L.) BSP. Grass Pink Orchid. May–July. Wet prairies, not common; confined to the n. ½ of the state; also Jackson and Macoupin cos. *Calopogon pulchellus* (Salisb.) R. Br.

4. **Coeloglossum** Hartm.—Frog Orchid

1. **Coeloglossum viride** (L.) Hartm. Long-bracted Orchid; Frog Orchid. May–June. Rich woods, rare; confined to the n. ½ of the state. *Habenaria viridis* (L.) R. Br. var. *bracteata* (Muhl.) Gray.

5. **Corallorhiza** Chat.—Coral-root Orchid

1. Lip with 2 lateral lobes or teeth; plants flowering generally from mid-May to mid-Aug.
  2. Stems yellowish; sepals and petals yellow-green, rarely spotted with purple, 4–5 mm long, 1-nerved; spur absent; lip 4–5 mm long, with 2 short lateral lobes; capsule greenish, 6–10 mm long; rhizome white........................ 3. *C. trifida*
  2. Stems purplish; sepals and petals white with purple spots, 6–8 mm long, 3-nerved; spur present; lip 6–8 mm long, with 2 well-developed lateral teeth or lobes; capsule brownish, 10–20 mm long; rhizome brown.............. 1. *C. maculata*
1. Lip entire or erose, not lobed nor toothed; plants flowering generally from late Mar. to mid-May or from mid-Aug. to Oct.
  3. Stems purple; sepals and petals greenish yellow, spotted with purple, linear-lanceolate, 6–8 mm long; lip 5–6 mm long, notched at apex, long-clawed; capsule 8–12 mm long; plants flowering in early spring............... 4. *C. wisteriana*
  3. Stems purple or brown below, greenish above; sepals and petals purplish green to purple, oblong, 3–5 mm long; lip 3–4 mm long, undulate at apex but not notched, short-clawed or clawless; capsule 5–8 mm long; plants flowering in autumn .... 2. *C. odontorhiza*

1. **Corallorhiza maculata** (Raf.) Raf. Two varieties occur in Illinois:

1. Middle lob of lip not expanded; floral bracts up to 1 mm long.................................... .... 1a. *C. maculata* var. *maculata*
1. Middle lobe of lip expanded; floral bracts 1.0–2.8 mm long.. .... 1b. *C. maculata* var. *occidentalis*

1a. **Corallorhiza maculata** (Raf.) Raf. var. **maculata** Spotted Coral-root Orchid. June–Aug. Woods rich in humus, rare; confined to the n. ¼ of the state.

1b. **Corallorhiza maculata** (Raf.) Raf. var. **occidentalis** (Lindl.) Ames. Western Spotted Coral-root Orchid. Apr.–July. Woods, rare; w. IL.

2. **Corallorhiza odontorhiza** (Willd.) Nutt. Fall Coral-root Orchid. Aug.–Oct. Woods; scattered throughout the state, but not common.

3. **Corallorhiza trifida** Chat. Pale Coral-root Orchid. May. Moist woods, very rare; Cook Co.

4. **Corallorhiza wisteriana** Conrad. Spring Coral-root Orchid. Mar.–May. Rich woods; scattered in the s. ½ of the state; also LaSalle Co.

6. **Cypripedium** L.—Lady's-slipper Orchid

1. Lip cleft down the middle, pink; leaves 2, basal .... 1. *C. acaule*
1. Lip not cleft, yellow, white, or with pink or purple streaks; leaves more than 2, cauline.
  2. Sepals and petals acuminate, yellow, greenish yellow, or streaked with purple or brown.
    3. Lip yellow.
      4. Lateral petals purple-brown; staminodium ovate............. 5. *C. parviflorum*
      4. Lateral petals yellow-green striped with brown; staminodium oblong .... 4. *C. X favillianum*

3. Lip white, occasionally marked with purple streaks.
5. Lip white; staminodium narrowly triangular .... 2. *C. X andrewsii*
5. Lip white streaked with purple; staminodium oblong .... 3. *C. candidum*
2. Sepals and petals obtuse, white .... 6. *C. reginae*

1. **Cypripedium acaule** Ait. White Lady's-slipper Orchid. May–July. Bogs and acid woods, very rare; Cook, Lake, McHenry, Ogle, and Stephenson cos.

2. **Cypripedium X andrewsii** A. M. Fuller. Andrews' Lady's-slipper Orchid. June. Low, springy areas, rare; Cook and McHenry cos. Reputed to be the hybrid between *C. candidum* Willd. and *C. parviflorum* Salisb. var. *makasin* (Farw.) Sheviak.

3. **Cypripedium candidum** Willd. White Lady's-slipper Orchid. May–June. Bogs, springy areas, rare; restricted to the n. ½ of the state.

4. **Cypripediuim X favillianum** J. T. Curtis. Faville's Lady's-slipper Orchid. June. Shaded ground, rare; Cook, Henderson, and Woodford cos. Reputed to be the hybrid between *C. candidum* Willd. and *C. parviflorum* Salisb. var. *pubescens* (Willd.) O. W. Knight.

5. **Cypripedium parviflorum** Salisb. Three varieties occur in Illinois:

a. Lower surface of distalmost bract glabrous or sparsely pubescent when young; lip petal 15–29 mm long; plants strongly aromatic .... 5b. *C. parviflorum* var. *makasin*
a. Lower surface of distalmost bract densely silvery-pubescent when young; lip petal 20–54 mm long; plants moderately aromatic.
b. Lip petal 22–34 mm long; sepals and petals densely dark red-brown spotted...... .... 5a. *C. parviflorum* var. *parviflorum*
b. Lip petals to 54 mm long; sepals and petals spotted but not densely so .... 5c. *C. parviflorum* var. *pubescens*

5a. **Cypripedium parviflorum** Salisb. var. **parviflorum** Spotted Yellow Lady's-slipper Orchid. Apr.–June. Mesic forests, very rare; s. IL. *Cypripedium calceolus* L. var. *parviflorum* (Salisb.) Fern.

5b. **Cypripedium parviflorum** Salisb. var. **makasin** (Farw.) Sheviak. Small-flowered Yellow Lady's-slipper Orchid. May–Aug. Fens, wet prairies, wet meadows, very rare; n. ½ of IL.

5c. **Cypripedium parviflorum** Salisb. var. **pubescens** (Willd.) O. W. Knight. Large Yellow Lady's-slipper Orchid. Apr.–Aug. Mesic forests, meadows, prairies, fens, not common; scattered throughout the state. *Cypripedium calceolus* L. var. *pubescens* (Willd.) Correll.

6. **Cypripedium reginae** Walt. Showy Lady's-slipper Orchid. June–July. Bogs, low springy areas, rare; restricted to the n. ½ of the state.

7. **Epipactis** Sw.—Helleborine

1. **Epipactis helleborine** (L.) Crantz. Helleborine. July–Aug. Native to Europe; adventive in disturbed woods; ne. cos.; also Coles Co. *Epipactis latifolia* (L.) Crantz.

8. **Galearis** Raf.—Orchis

1. **Galearis spectabilis** (L.) Raf. Showy Orchis. Apr.–June. Rich woods, not common; scattered throughout the state. *Orchis spectabilis* L.

9. **Goodyera** R. Br.—Rattlesnake-plantain Orchid

1. **Goodyera pubescens** (Willd.) R. Br. Rattlesnake-plantain Orchid. June–Aug. Rich, moist woods, not common; scattered throughout the state.

10. **Hexalectris** Raf.—Crested Coral-root Orchid

1. **Hexalectris spicata** (Walt.) Barnh. Crested Coral-root Orchid. June–Sept. Dry woods, limestone ledges, rare; Hardin, Jackson, Monroe, Pope, and Randolph cos.

11. **Isotria** Raf.—Whorled Pogonia

1. Sepals 1.5–2.5 cm long; petals 1.2–1.5 cm long; capsule up to 2.5 cm long, on a peduncle up to 2 cm long .... 1. *I. medeoloides*
1. Sepals 3.5–6.0 cm long; petals 1.5–2.5 cm long; capsule 2.5–3.5 cm long, on a peduncle at least 2.5 cm long .... 2. *I. verticillata*

1. **Isotria medeoloides** (Pursh) Raf. Small Whorled Pogonia. May. Wooded slope of sandstone cliff, very rare; Randolph Co.

2. **Isotria verticillata** (Willd.) Raf. Whorled Pogonia. May–June. Lowland woods, very rare; Pope Co.

12. **Liparis** Rich.—Twayblade Orchid

1. Pedicels 5–10 mm long, as long as the flowers; sepals greenish white, 10–12 mm long; lateral petals 10–12 mm long, madder-purple; lip 8–12 mm long, madder-purple, flat .... 1. *L. liliifolia*
1. Pedicels 4–5 mm long, usually a little shorter than the flowers; sepals yellow-green, 4–6 mm long; lateral petals 4–6 mm long, yellowish; lip 4–6 mm long, yellow-green, turned up slightly along the margins .... 2. *L. loeselii*

1. **Liparis liliifolia** (L.) Rich. Twayblade Orchid. May–July. Rich woods, occasional; scattered throughout the state.

2. **Liparis loeselii** (L.) Rich. Lesser Twayblade Orchid. June–July. Low woods, bogs, calcareous fens, rare; restricted to the n. ½ of the state; also Crawford Co.

13. **Malaxis** Sw.—Adder's-mouth Orchid

1. Pedicels 1–3 mm long; lip entire, deflexed; upper sepal 2.0–2.5 mm long .... 1. *M. brachypoda*
1. Pedicels 4–9 mm long; lip bilobed, ascending; upper sepal 1.2–1.6 mm long .... 2. *M. unifolia*

1. **Malaxis brachypoda** (Gray) Fern. Adder's-mouth Orchid. July. Bogs, very rare; Kane Co. *Malaxis monophylla* (L.) Sw.; *M. monophylla* (L.) Sw. var. *brachypoda* (Gray) F. Morris.

2. **Malaxis unifolia** Michx. Adder's-mouth Orchid. June–Aug. Dry or moist woods, rare; Hancock, Henderson, and Menard cos.

14. **Platanthera** L. C. Rich.—Fringed Orchid

1. Leaves 2, basal, suborbicular; lip elongated, entire, 8–20 mm long, the spur 13–40 mm long.
  2. Lip linear, 12–20 mm long; flowers greenish white, with pedicels 5–13 mm long; stem with bracts .... 12. *P. orbiculata*
  2. Lip triangular-elongated, 8–12 mm long; flowers yellow-green, sessile; stem without bracts .... 9. *P. hookeri*
1. Leaves 1–several, cauline, linear to lanceolate to oval; lip lobed, toothed, erose, or fringed, if entire and elongated, then the lip 4–8 mm long and the spur 2–12 mm long.
  3. Lip entire, erose, 2- to 3-toothed, or very shallowly 3-lobed with the lobes entire; spur 2–12 mm long; flowers sessile.
    4. Leaf 1; inflorescence 2–6 cm long; all bracts shorter than the flowers; spur 8–12 mm long; roots slender .... 4. *P. clavellata*
    4. Leaves 2–several; inflorescence 5–30 cm long; at least the lower bracts longer than the flowers; spur 2–8 mm long; rootstocks tuberous.
      5. Lip erose, with a tubercle borne near the summit .... 7. *P. flava*
      5. Lip entire, without a tubercle.
        6. Flowers greenish white, faintly odorous; lip gradually broadened at the base .... 1. *P. aquilonis*
        6. Flowers creamy white, with a spicy fragrance; lip abruptly broadened at the base .... 5. *P. dilatata*
  3. Lip fringed or deeply 3-lobed and toothed; spur 12 mm long or longer; flowers pedicellate.
    7. Lip simple, fringed.
      8. Flowers orange; leaves lanceolate, some or all over 2 cm broad .... 3. *P. ciliaris*
      8. Flowers white; leaves linear to linear-lanceolate, less than 2 cm broad .... 2. *P. blephariglottis*
    7. Lip deeply 3-lobed, the lobes fringed, long-toothed, or erose.

9. Flowers yellow-green or white; lobes of the lip fringed.
  10. Flowers yellow-green or greenish white; sepals 4.5–7.0 mm long; spur 14–20 mm log .... 10. *P. lacera*
  10. Flowers white; sepals 7–12 mm long; spur 20–48 mm long .... 11. *P. leucophaea*
9. Flowers reddish purple; lobes of the lip long-toothed or erose.
  11. Lip shallowly erose, 13–22 mm long, the terminal lobe notched .... 13. *P. peramoena*
  11. Lip long-toothed, 9–13 mm long, the terminal lobe not notched.
    12. Lip up to 1.3 cm wide; raceme compact and densely flowered .... 6. *P. fissa*
    12. Lip more than 1.3 cm wide; raceme laxly flowered .... 8. *P. grandiflora*

1. **Platanthera aquilonis** Sheviak. Green Orchid. June–Aug. Swampy areas, rare; n. ½ of IL. *Platanthera hyperborea* (L.) Lindl. var. *huronensis* (Nutt.) Luer; *Habeneria hyperborea* (L.) R. Br. var. *huronensis* (Nutt.) Farw.

2. **Platanthera blephariglottis** (Willd.) Lindl. White Fringed Orchid. June–July. Low areas, very rare; Macon Co. *Habenaria blephariglottis* (Willd.) Hook.

3. **Platanthera ciliaris** (L.) Lindl. Yellow Fringed Orchid. June–Aug. Low ground, very rare; Cook and Union cos.; also reported from Kankakee Co. *Habenaria ciliaris* (L.) R. Br.

4. **Platanthera clavellata** (Michx.) Luer. Wood Orchid. July–Aug. Moist, shaded areas, rare; Cass, Cook, Kankakee, Lake, Lee, Pope, and Will cos. *Habenaria clavellata* (Michx.) Spreng.

5. **Platanthera dilatata** (Pursh) Lindl. White Orchis. June–July. Springy habitats, rare; Carroll, Kankakee, and McHenry cos. *Habenaria dilatata* (Pursh) Hook.

6. **Platanthera fissa** (Muhl. ex Willd.) Linidl. Purple Fringed Orchid. June–Aug. Low woods, bogs, rare; Cook, Lake, and Winnebago cos. *Platanthera psycodes* (L.) Lindl.; *Habenaria psycodes* (L.) Spreng.

7. **Platanthera flava** (L.) Lindl. Two varieties occur in Illinois:

a. Bracts of the lowest flowers as long as the flowers; inflorescence lax; lip suborbicular .... 7a. *P. flava* var. *flava*
a. Bracts of the lowest flowers surpassing the flowers; inflorescence crowded; lip oblong .... 7b. *P. flava* var. *herbiola*

7a. **Platanthera flava** (L.) Lindl. var. **flava**. Tubercled Orchid. June–Aug. Low, shaded areas, rare; Johnson, Massac, and Wabash cos. *Habenaria flava* (L.) R. Br.

7b. **Platanthera flava** (L.) Lindl. var. **herbiola** (R. Br.) Luer. Tubercled Orchid. June–July. Moist woods, rare; restricted to the n. ½ of the state; also St. Clair and Wabash cos. *Habenaria flava* (L.) R. Br. var. *herbiola* (R. Br.) Ames & Correll.

8. **Platanthera grandiflora** (Bigel.) Lindl. Purple Fringed Orchid. June–Aug. Low woods, very rare; Lake Co. *Habenaria grandiflora* Bigel.; *H. psycodes* (L.) Lindl. var. *grandiflora* (Bigel.) Gray.

9. **Platanthera hookeri** (Torr.) Lindl. Hooker's Orchid. June–July. Moist woods, rare; Cook, Hancock, and Lake cos. *Habenaria hookeri* Torr.

10. **Platanthera lacera** (Michx.) G. Don. Green Fringed Orchid. June–July. Swamps, rare but scattered in IL. *Habenaria lacera* (Michx.) Lodd.

11. **Platanthera leucophaea** (Nutt.) Lindl. White Fringed Orchid. June–July. Swampy areas, not common; scattered in the n. ⅔ of the state. *Habenaria leucophaea* (Nutt.) Gray.

12. **Platanthera orbiculata** (Pursh) Lindl. Round-leaved Orchid. July–Aug. Moist woods, very rare; Cook and Kane cos. *Habenaria orbiculata* (Pursh) Torr.

13. **Platanthera peramoena** (Gray) Gray. Purple Fringeless Orchid. June–Sept. Low woods, wet prairies; occasional in the s. ⅖ of the state. *Habenaria peramoena* Gray.

15. **Pogonia** Juss.—Pogonia

1. **Pogonia ophioglossoides** (L.) Ker-Gawler. Snake-mouth Orchid. June–Aug. Low ground, rare; restricted to the n. ¼ of the state.

16. **Spiranthes** Rich.—Ladies' Tresses

1. Flowers produced in 2–3 rows, the spikes usually crowded; rachis pubescent (see also *S. vernalis* under second number 1 of couplet).
  2. Lip conspicuously constricted below the summit; lateral sepals united at base .... 7. *S. romanzoffiana*
  2. Lip not constricted below the summit; lateral sepals free.
    3. Lip yellow; flowers produced from May to July .... 4. *S. lucida*
    3. Lip white or cream; flowers produced from Aug. to Oct.
      4. Sepals and petals 4–5 mm long; lip 4–5 mm long, with 2 basal, conspicuous, incurved callosities .... 6. *S. ovalis*
      4. Sepals and petals 7–12 mm long; lip 7–12 mm long, with 2 basal, rather obscure (sometimes prominent in *S. cernua*), rounded callosities.
        5. Leaves usually persisting, at least through anthesis; flowers white, not strongly scented .... 1. *S. cernua*
        5. Leaves usually absent during anthesis; flowers cream, strongly scented .... 5. *S. magnicamporum*

1. Flowers produced in a single row, often appearing secund, the spikes laxly flowered; rachis glabrous (pubescent in *S. vernalis*).
  6. Rachis pubescent; sepals, petals, and the lip 5–10 mm long, the lip yellowish; lowest cauline leaves usually resembling the basal leaves, present at anthesis .... 9. *S. vernalis*
  6. Rachis glabrous; sepals, petals, and the lip 3–4 mm long, the lip white or white with a green area; all leaves basal, withered or absent at anthesis.
    7. Tuberous roots several; lip white with a green central area; largest leaves 2–4 cm long.
      8. Leaves present but withered at anthesis, thin; spike little spiraling; summit of lip green with a broad white border; plants flowering from mid-June to early Sept., averaging Aug. 5 .... 3. *S. lacera*
      8. Leaves absent at anthesis, thick; spike strongly spiraling; summit of lip green with a narrow white border; plants flowering from late July to early Oct., averaging Sept. 2 .... 2. *S. gracilis*
    7. Tuberous root 1 or occasionally 2 or 3; lip white throughout; largest leaves at most 2 cm long .... 8. *S. tuberosa*

1. **Spiranthes cernua** (L.) Rich. Nodding Ladies' Tresses. Aug.–Oct. Moist or dry situations in usually open areas; occasional throughout the state.

2. **Spiranthes gracilis** (Bigel.) Beck. Slender Ladies' Tresses. July–Oct. Dry, open woods, rare; restricted to the s. ½ of the state.

3. **Spiranthes lacera** (Raf.) Raf. Slender Ladies' Tresses. June–Sept. Mostly dry woods, rare; scattered in IL.

4. **Spiranthes lucida** (H. H. Eaton) Ames. Yellow-lipped Ladies' Tresses. May–June. Wet situations, rare; Cook, Hancock, Kane, Lake, Will, and Woodford cos.

5. **Spiranthes magnicamporum** Sheviak. Prairie Ladies' Tresses. Aug.–Oct. Dry prairies; scattered in most of the state except the southernmost cos.

6. **Spiranthes ovalis** Lindl. var. **erostellata** Catling. Ladies' Tresses. Sept.–Oct. Rich woods, not common; scattered throughout the state.

7. **Spiranthes romanzoffiana** Cham. Hooded Ladies' Tresses. July–Sept. Open habitats, rare; Coles, Cook, McHenry, and Peoria cos.

8. **Spiranthes tuberosa** Raf. Little Ladies' Tresses. July–Oct. Dry, open woods; confined to the s. ⅕ of the state.

9. **Spiranthes vernalis** Engelm. & Gray. Spring Ladies' Tresses. July–Aug. Rich woods, prairies, rare; Effingham, Madison, Massac, Menard, Pope, St. Clair, Union, Wabash, and Williamson cos.

17. **Tipularia** Nutt.—Crane-fly Orchid

1. **Tipularia discolor** (Pursh) Nutt. Crane-fly Orchid. July–Aug. Rich woods, rare; confined to the s. ⅙ of IL; also Effingham Co.

18. **Triphora** Nutt.—Nodding Pogonia

1. **Triphora trianthophora** (Sw.) Rydb. Nodding Pogonia. Aug.–Oct. Rich woods, not common; scattered throughout the state.

188. POACEAE—GRASS FAMILY

1. Culms woody .... 13. *Arundinaria*
1. Culms herbaceous.
 2. Spikelets enclosed by a spiny bur .... 25. *Cenchrus*
 2. Spikelets not enclosed by a spiny bur.
  3. Spikelets with 1 or more perfect florets.
   4. Inflorescence solitary, racemose, paniculate, or spicate, but not digitate.
    5. Inflorescence spicate or spikelike, with 1 spike per culm .... Group 1
    5. Inflorescence solitary, racemose or paniculate, but not composed of single spikes.
     6. Each spikelet with 2 or more perfect florets.
      7. Some part of the spikelets awned .... Group 2
      7. Spikelets without awns .... Group 3
     6. Each spikelet with 1 perfect floret (sterile or staminate lemmas may be present, in addition).
      8. Some part of the spikelets awned .... Group 4
      8. Spikelets without awns .... Group 5
   4. Inflorescence digitate (the spikes and racemes radiating from near the same point) .... Group 6
  3. Spikelets unisexual (i.e., either all staminate or all pistillate); some spikelets perfect in *Rottboellia* .... Group 7

Group 1

Inflorescence spicate or spikelike, with 1 spike per culm; spikelets with 1 or more perfect florets.

1. Spikelets cylindrical, borne at swollen rachis joints, the entire spikelet falling at maturity; each glume with 1 awn and 1 tooth .... 1. *Aegilops*
1. Spikelets not as above; rachis joints not swollen; glumes awned or awnless, but not with 1 awn and 1 tooth.
 2. Spikelets borne edgewise to the rachis; inner glume absent except in the terminal spikelet .... 61. *Lolium*
 2. Spikelets borne flatwise to the rachis; glumes present on all spikelets.
  3. Each spikelet subtended by and usually surpassed by 1 or more sterile bristles (not to be confused with awns).
   4. Bristles persistent on the rachis after spikelets fall .... 85. *Setaria*
   4. Bristles falling as the spikelets fall .... 72. *Pennisetum*
  3. Each spikelet not subtended by bristles.
   5. Each spikelet with 2 or more perfect florets (*Anthoxanthum* and *Phalaris* have 3 lemmas, but 2 of them are sterile).
    6. At least some part of the spikelets awned.
     7. Upper spikelets paired, the lowermost solitary .... 42. *Elyhordeum*
     7. Spikelets either all paired, all borne in 3s, or all solitary.
      8. Spikelets either all paired or all borne in 3s.
       9. Spikelets all borne in 3s.
        10. Spike erect or ascending .... 54. *Hordeum*
        10. Spike nodding .... 42. *Elyhordeum*
       9. Spikelets all paired .... 43. *Elymus*
      8. Spikelet solitary.
       11. Glumes 3-nerved; annuals .... 96. *Triticum*
       11. Glumes usually 1-nerved; annuals or perennials.
        12. Spikelets more than 3 cm long; annuals .... 84. *Secale*
        12. Spikelets up to 3 cm long; perennials.
         13. Plants without rhizomes, cespitose; spikelets disarticulating above the glumes.
          14. Awns up to 1 mm long .... 56. *Koeleria*
          14. Awns more than 1 mm long.
           15. Lemmas 5–7 mm long; glumes 2–5 mm long; spikelets pectinately arranged .... 2. *Agropyron*
           15. Lemmas 8–25 mm long; glumes 8–18 mm long; spikelets not pectinately arranged .... 43. *Elymus*

13. Plants with rhizomes, forming colonies; spikelets disarticulating below the glumes .... 44. *Elytrigia*

6. Spikelets awnless throughout.

16. Annuals .... 96. *Triticum*

16. Penennials.

17. Spikelets paired..60. *Leymus*

17. Spikelet solitary.

18. Glumes and lemmas obtuse or truncate at apex.................... .... 44. *Elytrigia*

18. Glumes acute at apex.

19. Plants without rhizomes, cespitose; spikelets disarticulating above the glumes.

20. Glumes 2.5–5.0 mm long.

21. Blades 3–10 mm broad; lemmas densely pubescent on the nerves....................................93. *Tridens*

21. Blades 1–3 mm broad; lemmas merely scabrous ... . 56. *Koeleria*

20. Glumes 8–18 mm long ....43. *Elymus*

19. Plants with rhizomes, forming colonies; spikelets disarticulating below the glumes.................... 44. *Elytrigia*

5. Each spikelet with a single perfect floret (1–2 sterile lemmas present in addition in *Anthoxanthum* and *Phalaris;* 1 staminate lemma present in addition in *Holcus*).

22. Upper spikelets paired, the lowermost solitary .... 42. *Elyhordeum*

22. Spikelets either all paired, all borne in 3s, or all solitary.

23. Spikelets all paired or all borne in 3s.

24. Spikelets all paired...43. *Elymus*

24. Spikelets all borne in 3s .... 55. *Hordeum*

23. Spikelet solitary.

25. Some part of the spikelet awned.

26. Lemmas awned; glumes awned .... 74. *Phleum*

26. Lemmas awned; glumes usually awnless.

27. Lemmas awned from the middle...........23. *Calamagrostis*

27. Lemmas awned from the tip.

28. Lemma 1 per spikelet, perfect.

29. Glumes united near the base ....5. *Alopecurus*

29. Glumes free at the base.............. 66. *Muhlenbergia*

28. Lemmas 2–3 per spikelet, but only 1 perfect.

30. Spikelets 5–10 mm long, each with 1 perfect floret and 2 empty lemmas ....8. *Anthoxanthum*

30. Spikelets 3.5–5.0 mm long, each with 1 perfect floret and 1 staminate floret....................53. *Holcus*

25. Spikelets not awned.

31. First glume absent; second glume enclosing the membranous lemma... 102. *Zoysia*

31. Both glumes present, neither enclosing the lemma.

32. Glumes 9–15 mm long; lemma 7–14 mm long.................. .... 6. *Ammophila*

32. Glumes to 7 (–10 in *Phalaris*) mm long; lemmas to 7 mm long.

33. Each spikelet with 1 perfect floret and (1–) 2 empty lemmas...73. *Phalaris*

33. Each spikelet with 1 perfect floret only.

34. Lemma 3-nerved .... 66. *Muhlenbergia*

34. Lemma 1-nerved.

35. Spikes broad, ¼ to nearly as broad as long ....30. *Crypsis*

35. Spikes slender, ⅕ or less as broad as long......... .... 90. *Sporobolus*

Group 2

Inflorescence solitary, racemose, or paniculate, but not spicate nor digitate; spikelets with 2 or more perfect florets; some part of the spikelet awned.

1. Lemmas 2-toothed at the apex.
  2. Awn of lemma arising between the teeth.
    3. Lemmas 5- to 9-nerved.
      4. Callus of lemmas bearded .... 81. *Schizachne*
      4. Callus of lemmas not bearded.
        5. Glumes much shorter than the entire spikelet .... 21. *Bromus*
        5. Glumes equaling or longer than the uppermost floret ....34. *Danthonia*
    3. Lemmas 3-nerved.
      6. Panicles 3–5 (–8) cm long; spikelets 2- to 5-flowered .... 94. *Triplasis*
      6. Panicles 10–20 cm long; spikelets 6- to 12-flowered .... 58. *Leptochloa*
  2. Awn of lemma arising near the middle or base of the lemma.
    7. Glumes 17–30 mm long; awn of lemmas up to 40 mm long .... 15. *Avena*
    7. Glumes 2.5–5.0 mm long; awn of lemmas 2.5–6.0 mm long.
      8. Awn arising from near the middle of the lemma; lemmas 3-nerved .... 4. *Aira*
      8. Awn arising from near the base of the lemma; lemmas 5-nerved .... 35. *Deschampsia*
1. Lemmas acute or obtuse at the apex, not 2-toothed.
  9. Lemmas 3-nerved .... 58. *Leptochloa*
  9. Lemmas 5-nerved (all the nerves sometimes obscure in *Festuca*).
    10. Blades involute, about 1 mm in diameter, if flat, less than 2 mm broad.
      11. Plants annual; stamen 1 .... 98. *Vulpia*
      11. Plants perennial; stamens 3 .... 48. *Festuca*
    10. Blades flat, 2–8 mm broad.
      12. Lemmas glabrous; spikelets not crowded in 1-sided panicles .... 48. *Festuca*
      12. Lemmas ciliate along the keel; spikelets crowded in 1-sided panicles .... 32. *Dactylis*

Group 3

Inflorescence solitary, racemose, or paniculate, but not spicate nor digitate; spikelets with 2 or more perfect florets; spikelets awnless.

1. Lemmas distinctly 2-toothed at the apex.
  2. Perennial; blades to 3 mm broad; panicle branches erect or spreading; spikelets 5- to 12-flowered; glumes 1 cm long .... 21. *Bromus*
  2. Annual; blades 5–15 mm broad; panicle branches lax; spikelets 2-flowered; glumes 1.5–2.5 cm long .... 15. *Avena*
1. Lemmas acute to obtuse at the apex, not 2-toothed.
  3. Glumes at least 15 mm long, as long as the spikelets .... 15. *Avena*
  3. Glumes less than 10 mm long, shorter than the spikelets.
    4. Rachilla with long silky hairs, the hairs longer than the spikelets; culms to 4 m tall .... 75. *Phragmites*
    4. Rachilla without long silky hairs longer than the spikelets; culms to 1.5 m tall.
      5. Lemmas 3-nerved.
        6. Lemmas 6–10 mm long; grain beaked .... 36. *Diarrhena*
        6. Lemmas 1.5–5.0 mm long; grain not beaked.
          7. Lemmas glabrous .... 45. *Eragrostis*
          7. Lemmas pubescent.
            8. Lemmas densely hairy at base, frequently with a tuft of hairs.
              9. Lemmas villous at the base, but without a tuft of cobwebby hairs.
                10. Lemmas retuse or obtuse, 3.5–4.0 mm long ....93. *Tridens*
                10. Lemmas acute and mucronate, 4.5 mm long .... 78. *Redfieldia*

9. Lemmas with a tuft of cobwebby hairs at the base, pubescent on the nerves .... 76. *Poa*
8. Lemmas pubescent only on the nerves.
11. Lemmas keeled .... 45. *Eragrostis*
11. Lemmas rounded on the back.
12. Lemmas 1.0–2.5 mm long; spikelets 1–5 mm long .... 58. *Leptochloa*
12. Lemmas 4.0 mm long; spikelets 5–8 mm long .... 93. *Tridens*
5. Lemmas 5- to many-nerved or apparently nerveless, with only the midnerve conspicuous.
13. Spikelets crowded and overlapping on 1 side of a zigzag rachis .... 83. *Sclerochloa*
13. Spikelets not overlapping on 1 side of a zigzag rachis.
14. Lemmas apparently nerveless.
15. Spikelets disarticulating above the glumes.
16. Glumes 2.0–4.5 mm long; plants of moist or dry woods .... 48. *Festuca*
16. Glumes up to 2 mm long; plants of waste ground .... 77. *Puccinellia*
15. Spikelets disarticulating below the glumes .... 89. *Sphenopholis*
14. Lemmas obviously nerved.
17. Lemmas 4–10 mm long.
18. Lemmas as broad as long; inflorescence with up to 8 spikelets .... 20. *Briza*
18. Lemmas longer than broad; inflorescence with more than 8 spikelets.
19. Lemmas with 9 or more nerves, 4–10 mm long.
20. Spikelets compressed, 6- to 18-flowered .... 26. *Chasmanthium*
20. Spikelets not compressed, 2- to 3-flowered .... 62. *Melica*
19. Lemmas 5- to 7-nerved, 7 (–8) mm long.
21. Lemmas obscurely 7-nerved; spikelets 10–20 mm long .... 49. *Glyceria*
21. Lemmas 5-nerved; spikelets less than 10 mm long.
22. Spikelets not crowded in 1-sided panicles, not compressed .... 48. *Festuca*
22. Spikelets crowded in 1-sided panicles, compressed .... 32. *Dactylis*
17. Lemmas 1.5–5.0 mm long.
23. Lemmas distinctly keeled .... 76. *Poa*
23. Lemmas rounded on the back.
24. Nerves of lemma parallel to the summit.
25. Sheaths closed; lodicules united .... 49. *Glyceria*
25. Sheaths open; lodicules free from each other .... 91. *Torreyochloa*
24. Nerves of lemma converging toward the summit.
26. Lemmas glabrous.
27. Plants annual; stamen 1 .... 98. *Vulpia*
27. Plants perennial; stamens 3 .... 48. *Festuca*
26. Lemmas pubescent, at least on the nerves or keel or at the base .... 76. *Poa*

Group 4

Inflorescence solitary, racemose, or paniculate, but not composed of single spikes; each spikelet with 1 perfect floret (sterile or staminate lemmas may be present, in addition); some part of the spikelet awned.

1. Spikelets borne singly (i.e., not paired).
2. Lemmas 3-awned.

3. Spikelets borne on 1 side of a long, arching raceme; lemma rounded on the back; spikelets with 1 perfect lemma and 1–2 sterile ones .... 18. *Bouteloua*
3. Spikelets borne in a more or less erect inflorescence, not 1-sided; lemma inrolled around the palea; no sterile lemma present .... 10. *Aristida*
2. Lemmas 1-awned or awnless.
4. First glume reduced to a sheath, united with the lowest, swollen joint of the rachilla .... 47. *Eriochloa*
4. First glume not reduced to a sheath and not united with the rachillar joint.
5. Lemma awnless; glumes awned.
6. Plants over 1 m tall; lemmas 7–10 mm long .... 88. *Spartina*
6. Plants less than 1 m tall; lemmas 2–4 mm long......... 66. *Muhlenbergia*
5. Lemma awned; glumes awnless or awned.
7. Spikelets arranged in 4 or more crowded ranks, each spikelet composed of 1 fertile and 1 sterile floret .... 40. *Echinochloa*
7. Spikelets not arranged in 4 or more crowded ranks, each spikelet composed of 1 fertile floret (each spikelet composed of 1 fertile floret and 1 staminate floret in *Arrhenatherum* or 1 sterile floret sometimes in *Gymnopogon*).
8. Blades 1–3 mm wide.
9. First glume less than 1 mm long .... 66. *Muhlenbergia*
9. First glume at least 1.5 mm long.
10. At least 1 awn of lemma 2 cm long or longer.
11. Tufted annuals from a cluster of fibrous roots .... 10. *Aristida*
11. Cespitose or stout perennials.
12. Sheaths villous on the margins and at the summit; ligule less than 1 mm long; glumes 5–11 mm long; lemma 4.5–6.0 mm long, pubescent throughout, the awn 2–4 cm long .... 67. *Nassella*
12. Sheaths more or less glabrous; ligule (at least of the upper leaves) 3–6 mm long; glumes 15–40 mm long; lemma 9–25 mm long, pubescent at base, becoming glabrate above, the awn 10–20 cm long .... 51. *Heterostipa*
10. Awn of lemma up to 2 cm long.
13. Lemmas 2.2–4.5 mm long.
14. Glumes 5-nerved; lemmas indurated ......69. *Oryzopsis*
14. Glumes 1-nerved; lemmas not indurated .... 66. *Muhlenbergia*
13. Lemmas 0.5–2.2 mm long.
15. Lemmas firm; glumes unequal......................... 9. *Apera*
15. Lemmas not firm; glumes equal .... 3. *Agrostis*
8. Blades 3 mm wide or wider.
16. Plants 2 m tall or taller; all leaves 2 cm broad or broader........... .... 14. *Arundo*
16. Plants usually less than 2 m tall; none or only a few of the leaves as much as 2 cm broad.
17. Second glume 5- to 7-nerved.
18. Awns straight or curved, not twisted near base; second glume 7-nerved .... 69. *Oryzopsis*
18. Awns twisted near base; second glume 5-nerved.
19. Sheaths villous on the margins and at the summit; ligule less than 1 mm long; glumes 5–11 mm long; lemma 4.5–6.0 mm long, pubescent throughout, the awn 2–4 cm long .... 67. *Nasella*
19. Sheaths more or less glabrous; ligule (at least of the upper leaves) 3–6 mm long; glumes 15–40 mm long; lemma 9–25 mm long, pubescent at base, becoming glabrate above, the awn 10–20 cm long .... 51. *Heterostipa*

17. Second glume 1- to 3-nerved.
20. Lemma (excluding awns) 5–10 mm long.
21. First glume less than 1 mm long...... 19. *Brachyelytrum*
21. First glume 2.5–8.0 mm long.
22. Awn 10–20 mm long; spikelet (excluding awns) 7–10 mm long; lemma 5- to 7-nerved .... 11. *Arrhenatherum*
22. Awn up to 1.5 mm long; spikelet (excluding awns) 2.5–6.5 mm long; lemma 3-nerved .... 28. *Cinna*
20. Lemma (excluding awns) 1.5–5.0 mm long.
23. Spikelets remote along 1 side of a slender rachis, forming very slender unilateral spikes .... 50. *Gymnopogon*
23. Spikelets in contracted or open panicles.
24. Lemma with a tuft of hairs at the base on the callus, awned from near the middle............................. .... 23. *Calamagrostis*
24. Lemma glabrous or pubescent but without a large tuft of hairs on the callus, awned from the tip.
25. Plants usually at least 1 m tall; spikelets disarticulating below the glumes .... 28. *Cinna*
25. Plants up to 1 m tall, usually smaller; spikelets disarticulating above the glumes .... 66. *Muhlenbergia*

1. Spikelets borne in pairs.
26. Both spikelets pedicellate, the pedicels unequal in length .... 65. *Miscanthus*
26. 1 spikelet sessile, the other pedicellate (or represented merely by a pedicel).
27. Pedicellate spikelet represented only by a pedicel................. 86. *Sorghastrum*
27. Pedicellate spikelet present or at least represented by a lemma.
28. Both spikelets of the pair with perfect florets.
29. Robust perennials; racemes very numerous, in large dense panicles... .... 46. *Erianthus*
29. Matted annual; racemes 1–few, not in dense panicles........................... .... 63. *Microstegium*
28. Only the sessile spikelet of the pair perfect.
30. Decumbent annual; blades cordate .... 12. *Arthraxon*
30. Upright perennials; blades not cordate.
31. Inflorescence paniculate... 87. *Sorghum*
31. Inflorescence racemose or nearly spicate.
32. Flowering culms much branched into many short leafy branchlets terminated by 1–6 racemes.
33. Racemes 2 or more from the sheaths............. 7. *Andropogon*
33. Raceme solitary at the tip of the peduncle............................. .... 82. *Schizachyrium*
32. Flowering culms unbranched .... 17. *Bothriochloa*

Group 5

Inflorescence solitary, racemose or paniculate, but not spicate; each spikelet with 1 perfect floret (sterile or staminate lemmas may be present, in addition); no part of the spikelet awned.

1. Spikelets borne in pairs.
2. 1 spikelet of the pair sessile, the other pedicellate.
3. Both spikelets fertile.. 63. *Microstegium*
3. Only 1 of the spikelets fertile, the other represented only by a lemma........... .... 12. *Arthraxon*
2. Both spikelets either sessile or pedicellate.
4. First glume as long as or longer than the lemmas; plants 2.5–4.0 m tall......... .... 65. *Miscanthus*
4. First glume absent or up to 0.5 mm long, much shorter than the lemmas; plants up to 1.5 m tall.

5. Spikelets with long, tawny hairs longer than the spikelets .... 92. *Trichachne*
5. Spikelets without long, tawny hairs exceeding the spikelets .... 71. *Paspalum*

1. Spikelet solitary (i.e., not borne in pairs).
   6. First glume reduced to a sheath and united with the lowest, swollen joint of the rachilla........ 47. *Eriochloa*
   6. First glume absent, reduced, or normal, neither sheathlike nor united with a swollen rachillar joint.
      7. Spikelets plano-convex........ 71. *Paspalum*
      7. Spikelets not plano-convex.
         8. Plants with stiff, papillose-based hairs; spikelets partially embedded in axis of inflorescence .... 79. *Rottboellia*
         8. Plants variously pubescent or glabrous, but without stiff, papillose-based hairs; spikelets not embedded in axis of inflorescence.
            9. Both glumes absent; nodes bearded .... 57. *Leersia*
            9. Both glumes present, although the first often much reduced or, if absent, the plants not producing seeds; nodes not bearded (except in some *Dichanthelium*).
               10. First glume absent; plants with creeping rhizomes, rarely producing seeds........ 102. *Zoysia*
               10. First glume present, although occasionally strongly reduced; rhizome present or absent; plants producing seeds.
                   11. First glume up to ½ (–⅔ in a few species of *Panicum* and *Dichanthelium*), as long as the second glume, the second glume usually identical to the lemma in size and shape.
                       12. Each floret subtended by 1 or more bristles .... 85. *Setaria*
                       12. Florets not subtended by bristles.
                           13. Spikelet solitary at the end of long, often capillary, pedicels.
                               14. Fertile lemma leathery........ 59. *Leptoloma*
                               14. Fertile lemma indurated.
                                   15. Lemma with cross wrinkles........ 97. *Urochloa*
                                   15. Lemma without cross wrinkles.
                                       16. Plants with horizontal stolons........ 54. *Hopia*
                                       16. Plants without horizontal stolons.
                                           17. Annuals or rhizomatous perennials (or with a stout caudex in some species); winter rosettes absent; all panicles uniform, bearing spikelets in summer and autumn.
                                               18. Panicle branches one-sided; spikelets usually on pedicels less than 1 mm long .... 29. *Coleataenia*
                                               18. Panicle branches not one-sided; spikelets on pedicels 2 mm or more long........ 70. *Panicum*
                                           17. Perennials without rhizomes; winter rosettes present; autumnal panicles smaller than vernal panicles........ 37. *Dichanthelium*
                           13. Spikelets grouped in 2 or more ranks.
                               19. Inflorescence a dense panicle, often dark purple-brown .... 40. *Echinochloa*
                               19. Inflorescence racemose or paniculate with remote, ascending racemes, usually not dark purple-brown.
                                   20. Racemes over 2 cm long........ 71. *Paspalum*
                                   20. Racemes 1–2 (–3) cm long........ 40. *Echinochloa*
                   11. First glume nearly as long as the second glume, not conspicuously different in size, or reduced to a cuplike structure.
                       21. Glumes reduced to a cuplike structure........ 68. *Oryza*
                       21. Glumes not reduced to a cuplike structure.
                           22. Glumes 3-nerved.

23. Spikelets 4.5–8.0 mm long; glumes 4–6 mm long; blades 2–5 mm broad .... 52. *Hierochloe*
23. Spikelets 2.0–3.5 mm long; glumes 2–3 mm long; some or all the blades over 5 mm broad.
24. Lemmas 5-nerved; spikelet with 1 perfect and 1 sterile lemma; blades to 8 mm broad .... 16. *Beckmannia*
24. Lemmas nerveless; spikelet with 1 perfect lemma only; blades to 20 mm broad .... 64. *Milium*
22. Glumes 1-nerved.
25. Lemma with a conspicuous tuft of hairs on the callus; spikelets 6–7 mm long.......................... 24. *Calamovilfa*
25. Lemma glabrous or pubescent but without a large tuft of hairs on the callus; spikelets 1–6 mm long.
26. Lemma 3- to 5-nerved, the nerves sometimes obscure.
27. First glume longer than the lemma .... 3. *Agrostis*
27. First glume shorter than the lemma.
28. Spikelets appressed on 2 sides of a triangular rachis.............. 80. *Schedonnardus*
28. Spikelets not confined to 2 sides of the rachis .... 66. *Muhlenbergia*
26. Lemma 1-nerved............................. 90. *Sporobolus*

Group 6
Inflorescence digitate, the spikes and racemes radiating from near the same point.

1. Some part of the spikelet awned.
2. Spikelets borne in pairs, 1 sessile and perfect, the other pedicellate and staminate .... 7. *Andropogon*
2. Spikelet borne singly.
3. Spikelets with 3–5 perfect florets; second glume and lemmas awned .... 33. *Dactyloctenium*
3. Spikelets with 1 perfect floret (also 1–2 empty lemmas present); only the fertile lemma awned .... 27. *Chloris*
1. Spikelet awnless.
4. Spikelets with 3–6 perfect florets.. 41. *Eleusine*
4. Spikelets with 1 perfect floret.
5. First glume 1.0–1.5 mm long; second glume 1-nerved; no sterile lemmas present.. 31. *Cynodon*
5. First glume absent or up to 0.8 mm long; second glume 5-nerved; 1 of the lemmas sterile .... 38. *Digitaria*

Group 7
Spikelets unisexual (i.e., either all staminate or all pistillate).

1. Plants to 40 cm tall, dioecious; staminate spikelets 3- to 75-flowered.
2. Lemmas with a tuft of cobwebby hairs at base .... 76. *Poa*
2. Lemmas without a tuft of cobwebby hairs at base.
3. Both staminate and pistillate spikelets 20- to 75-flowered .... 45. *Eragrostis*
3. Staminate spikelets (2- or) 3- to 15-flowered; pistillate spikelets 1- to 9-flowered.
4. Staminate spikelets (2- or) 3-flowered; pistillate spikelets 1-flowered .... 22. *Buchloe*
4. Staminate spikelets 8- to 15-flowered; pistillate spikelets 7- to 9-flowered .... 39. *Distichlis*
1. Plants 1–4 m tall, monoecious; staminate spikelets 1- to 2-flowered.
5. Some spikelets perfect, some unisexual; plants with stiff, papillose-based hairs .... 79. *Rottboellia*

5. All spikelets unisexual; plants without stiff, papillose-based hairs.
6. Staminate spikelets 2-flowered; glumes membranous or coriaceous.
7. Annual; staminate and pistillate spikelets in different inflorescences; pistillate spikelets borne in pairs .... 99. *Zea*
7. Perennial; staminate and pistillate spikelets in the same inflorescence; pistillate spikelet solitary .... 95. *Tripsacum*
6. Staminate spikelets 1-flowered; glumes none.
8. Pistillate spikelets confined to the uppermost erect branches of the inflorescence, the staminate spikelets confined to the lower spreading branches; margin of leaf more or less smooth .... 100. *Zizania*
8. Pistillate and staminate spikelets on the same branches of the inflorescence; margin of leaf harsh and cutting .... 101. *Zizaniopsis*

1. **Aegilops** Host—Goat Grass

1. **Aegilops cylindrica** Host. Jointed Goat Grass. June–Sept. Native to Europe; adventive in waste ground; scattered throughout the state. *Triticum cylindricum* (Host) Ces.

2. **Agropyron** Gaertn.—Wheat Grass

(See also Elymus and Elytrigia.)

1. Blades involute, at least when dry; spikelets 5–7 mm long .... 1. *A. cristatum*
1. Blades flat; spikelets 8–12 mm long .... 2. *A. desertorum*

1. **Agropyron cristatum** (L.) Gaertn. Crested Wheat Grass. June–Aug. Native to Russia; adventive in IL; Fulton and Jo Daviess cos.

2. **Agropyron desertorum** (Fisch.) Schult. Wheat Grass. June–Aug. Native to Russia; introduced in IL; Champaign, Cook, Fulton, and Jo Daviess cos.

3. **Agrostis** L.—Bent Grass

1. Annuals; lemma sharply nerved, with a flexuous awn to 10 mm long .... 2. *A. elliottiana*
1. Perennials; lemma obscurely nerved, awnless (rarely a very short awn present in a variety of *A. scabra*).
2. Tufted perennials without rhizomes; palea absent or minute and nerveless.
3. Blades 1–2 mm broad; spikelets 1.2–2.0 mm long; lemma 0.5–1.0 mm long .... 4. *A. hyemalis*
3. Blades 2–6 mm broad; spikelets 2–3 mm long; lemma 1.3–2.0 mm long.
4. Panicle branches harshly scabrous, bearing florets only near the tip .... 6. *A. scabra*
4. Panicle branches glabrous or nearly so, bearing florets from near the middle to the tip .... 5. *A. perennans*
2. Rhizomatous or stoloniferous perennials; palea at least ½ as long as the lemma, 2-nerved.
5. Some of the panicle branches bearing florets to the base; ligule 2–6 mm long.
6. Blades 5–10 mm broad; rhizomes present; culms erect; panicle purple, the branches spreading .... 3. *A. gigantea*
6. Blades 1–5 mm broad; stolons present; culms decumbent; panicle straw-colored, the branches ascending .... 7. *A. stolonifera*
5. None of the panicle branches bearing florets to the base; ligule less than 2 mm long .... 1. *A. capillaris*

1. **Agrostis capillaris** L. Rhode Island Bent. June–Sept. Native to Europe; escaped from lawns into adjacent areas; scattered in some n. cos; also Perry Co. *Agrostis tenuis* Sibth.

2. **Agrostis elliottiana** Schult. Awned Bent Grass. May–July. Dry soil, particularly on bluffs; restricted to the s. ½ of IL.

3. **Agrostis gigantea** Roth. Red Top. June–Sept. Fields, common; in every co. *Agrostis alba* L., misapplied.

4. **Agrostis hyemalis** (Walt.) BSP. Tickle Grass. Mar.–June. Woods and fields, common; probably in every co.

5. **Agrostis perennans** (Walt.) Tuckerm. Upland Bent Grass. June–Sept. Dry

woods; common in the s. ½ of the state, becoming increasingly less common northward. (Including var. *aestivalis* Vasey.)

6. **Agrostis scabra** Willd. Two forms occur in Illinois:

a. Lemmas awnless .... 6a. *A. scabra* f. *scabra*
a. Lemmas awned .... 6b. *A. scabra* f. *tuckermanii*

6a. **Agrostis scabra** Willd. f. **scabra.** Tickle Grass. June–Sept. Moist or dry, usually open, soil; occasional throughout the state.

6b. **Agrostis scabra** Willd. f. **tuckermanii** Fern. Tickle Grass. June–Sept. Moist ground, rare; Cook Co.

7. **Agrostis stolonifera** L. var. **palustris** (Huds.) Farw. Creeping Bent Grass. June–Sept. Wet ground; occasional throughout the state. *Agrostis alba* L. var. *palustris* (Huds.) Pers.

4. **Aira** L.—Hairgrass

1. **Aira caryophyllaea** L. Slender Hairgrass. May–June. Native to Europe; adventive in waste ground; Piatt and Pope cos.

5. **Alopecurus** L.—Foxtail

1. Spikelets 4.0–5.5 mm long; awn of lemma exserted 4–5 mm beyond lemma .... 4. *A. pratensis*
1. Spikelets 2.0–3.5 mm long; awn of lemma exserted 1–3 mm beyond lemma.
  2. Awn of lemma straight, exserted about 1 mm beyond lemma .... 1. *A. aequalis*
  2. Awn of lemma bent or twisted, exserted 1–3 mm beyond lemma.
    3. Spikelets 2.0–2.5 mm long; annual .... 2. *A. carolinianus*
    3. Spikelets 2.5–3.5 mm long; perennial .... 3. *A. geniculatus*

1. **Alopecurus aequalis** Sobol. Foxtail. May–July. Wet ground; occasionally in shallow water, not common; scattered throughout the state.

2. **Alopecurus carolinianus** Walt. Common Foxtail. Apr.–June. Moist ground; occasional throughout the state.

3. **Alopecurus geniculatus** L. Marsh Foxtail. May–June. Native to Europe and Asia; sparingly adventive in the U.S.; Cook, DuPage, and McHenry cos.

4. **Alopecurus pratensis** L. Meadow Foxtail. June. Native to Europe and Asia; along railroads and other disturbed areas; scattered throughout the state.

6. **Ammophila** Host—Beach Grass

1. **Ammophila breviligulata** Fern. Beach Grass. July–Sept. Sand dunes, very rare; Cook and Lake cos.

7. **Andropogon** L.—Beardgrass

1. Pedicellate spikelet staminate, with well developed glumes.
  2. Plants glaucous; awn of sessile spikelet less than 7 mm long .... 4. *A. hallii*
  2. Plants not glaucous; awn of sessile spikelet 7 mm long or longer .... 1. *A. gerardii*
1. Pedicellate spikelet underdeveloped, with at least 1 of the glumes obsolete.
  3. Sessile spikelet 5–6 mm long, concealed by silvery hairs; stamens 3 .... 5. *A. ternarius*
  3. Sessile spikelet up to 5 mm long, not concealed by silvery hairs; stamen 1.
    4. Upper sheaths greatly inflated and rust-colored, 6–12 mm long; culms villous at the upper nodes; racemes nearly entirely enclosed by the sheath, the peduncles more than 10 mm long; awn twisted or curved near base .... 3. *A. gyrans*
    4. Upper sheaths slightly inflated or, if greatly inflated, not rust-colored, 2–6 cm long; culms more or less glabrous; racemes enclosed only at their base, the peduncles 2–10 mm long; awn straight or nearly so.
      5. Spikes slender, up to 2 cm broad, the branches simple .... 6. *A. virginicus*
      5. Spikes stout, 2–6 cm broad, the branches forking .... 2. *A. glomeratus*

1. **Andropogon gerardii** Vitman. Big Bluestem. July–Sept. Prairies; occasional to common throughout the state; in every co.

2. **Andropogon glomeratus** (Walt.) BSP. Bushy Broom Sedge. Aug.–Oct. Native to the s. U.S.; rarely adventive in wet disturbed soil; Williamson Co.

3. **Andropogon gyrans** Ashe. Elliott's Broom Sedge. Sept.–Oct. Fields, open woods; restricted to the s. ¼ of the state. *Andropogon elliottii* Chapm.

4. **Andropogon hallii** Hack. Sand Bluestem. July–Sept. Native to the w. U.S.; adventive in a dry habitat, rare; Bureau Co.

5. **Andropogon ternarius** Michx. Silver Broom Sedge. July–Oct. Roadsides, rare; Gallatin, Pope, and Saline cos.

6. **Andropogon virginicus** L. Broom Sedge. Aug.–Oct. Fields, open woods; occasional or common in the s. ⅔ of IL, less common in the n. ⅓.

8. **Anthoxanthum** L.—Vernal Grass

1. Spikelets whitish green, 5–7 mm long; glumes glabrous; awns of empty lemmas long-exserted; annuals to 35 cm tall .... 1. *A. aristatum*
1. Spikelets brownish green, 8–10 mm long; glumes pubescent; awns of empty lemmas included or barely exserted; perennials to nearly 1 m tall .... 2. *A. odoratum*

1. **Anthoxanthum aristatum** Boiss. Annual Sweet Grass. Aug. Native to Europe; waste ground; Rock Island Co. *Anthoxanthum puellii* Lecoq. & Lamotte.

2. **Anthoxanthum odoratum** L. Sweet Vernal Grass. June. Native to Europe; adventive in fields; Cook, DuPage, Grundy, Jackson, and Lake cos.

9. **Apera** Adans.—Windgrass

1. **Apera interrupta** (L.) P. Beauv. Italian Windgrass. June. Native to Europe and Asia; adventive in an IL lawn; DuPage and Macon cos. *Agrostis interrupta* L.

10. **Aristida** L.—Three-awn

1. Leaf sheaths and nodes woolly .... 6. *A. lanosa*
1. Leaf sheaths and nodes glabrous or pubescent, but not woolly.
  2. Awns of lemma twisted and united into a basal column 8–15 mm long; glumes 20–30 mm long .... 11. *A. tuberculosa*
  2. Awns of lemma usually not twisted but sometimes coiled, free to base or united into a basal column up to 3 mm long (except *A. oligantha* and *A. ramosissima*); glumes up to 20 (–32) mm long.
    3. Central awn of most lemmas at least 20 mm long.
      4. Lateral awns of lemmas at least 12 mm long.
        5. Awns of lemmas united at base into a column 1–3 mm long, the awns articulated with the summit of the lemmas; sheaths villous on the margins .... 3. *A. desmantha*
        5. Awns of lemmas free to base, the awns not articulated with the summit of the lemmas; sheaths glabrous or nearly so.
          6. All awns of lemmas 35–70 mm long; glumes 12–32 mm long; lemmas 12–20 mm long .... 8. *A. oligantha*
          6. Central awn of lemmas 15–33 mm long, the lateral awns 12–24 mm long; glumes 6–14 mm long; lemmas 5.5–8.5 mm long.
            7. Perennial; first glume longer than the second .... 9. *A. purpurascens*
            7. Annual; first glume equal to second or shorter .... 5. *A. intermedia*
      4. Lateral awns of lemmas up to 6 mm long or even absent .... 10. *A. ramosissima*
    3. Central awn of lemmas up to 20 mm long, usually considerably shorter.
  8. Lateral awns 12–20 mm long; central awn curved or divergent at base, not coiled.
    9. Central awn of lemmas conspicuously bent near base; glumes obscurely 3-nerved .... 7. *A. longespica*
    9. Central awn of lemmas not conspicuously bent near base; glumes sharply 1-nerved .... 5. *A. intermedia*
  8. Lateral awns up to 12 mm long; central awn coiled at base (or bent conspicuously near base in *A. longespica*).

10. Glumes essentially equal, 4–10 mm long; lemmas 4.0–8.5 mm long; inflorescence usually reduced to a raceme.
    11. Central awn conspicuously bent at base, not coiled, 6.5–20.0 mm long; lateral awns 3–15 mm long; glumes obscurely 3-nerved .... 7. *A. longespica*
    11. Central awn loosely coiled at base, 3–10 mm long; lateral awns 0.7–3.3 mm long; glumes 1-nerved .... 4. *A. dichotoma*
10. Second glume longer than first glume by at least 2 mm, 7–15 mm long; lemmas 7.5–10.5 mm long; inflorescence a slender panicle.
    12. Lateral awns of lemmas 5–12 mm long; central awn 10–19 mm long......... .... 1. *A. basiramea*
    12. Lateral awns of lemmas 2–4 mm long; central awn 7.0–12.5 mm long....... .... 2. *A. curtissii*

1. **Aristida basiramea** Engelm. Three-awn. Aug.–Oct. Dry, sandy soil, not common; confined to the n. ⅔ of the state; also Gallatin and Jackson cos.

2. **Aristida curtissii** (Gray) Nash. Three-awn. Aug.–Oct. Dry soil, very rare; Massac and Ogle cos. *Aristida dichotoma* Michx. var. *curtissii* Gray.

3. **Aristida desmantha** Trin. & Rupr. Three-awn. Aug.–Sept. Sandy soil, very rare; Cass, Mason, Menard, Morgan, and Tazewell cos.

4. **Aristida dichotoma** Michx. Three-awn. Aug.–Oct. Dry fields and along highways; occasional in the s. ½ of the state, less common in the n. ½.

5. **Aristida intermedia** Scribn. & Ball. Two varieties may be recognized in Illinois:

a. Glumes nearly equal in length; central awn of lemmas longer than the lateral ones .... 5a. *A. intermedia* var. *intermedia*
a. Second glume longer than the first glume; awns of lemmas equal............ .... 5b. *A. intermedia* var. *necopina*

5a. **Aristida intermedia** Scribn. & Ball var. **intermedia.** Three-awn. Sept.–Oct. Sandy soil, rare; Cass, DuPage, Grundy, Henry, Lake, LaSalle, Lee, McHenry, and Lee cos.

5b. **Aristida intermedia** Scribn. & Ball var. **necopina** (Shinners) Mohlenbr. Three-awn. Sept.–Oct. Sandy soil, very rare; Lee Co. *Aristida necopina* Shinners.

6. **Aristida lanosa** Muhl. ex Ell. Woolly Three-awn. July–Aug. Dry field, very rare; Winnebago Co.

7. **Aristida longespica** Poir. Two varieties may be recognized in Illinois:

a. Lateral awns 3–4 mm long; central awn 6.5–13.0 mm long; glumes 4–6 mm long.... .... 7a. *A. longespica* var. *longespica*
a. Lateral awns 4–15 mm long; central awn 10–20 mm long; glumes 5–9 mm long...... .... 7b. *A. longespica* var. *geniculata*

7a. **Aristida longespica** Poir. var. **longespica.** Three-awn. Aug.–Oct. Sandy soil, particularly in fields and along highways; occasional throughout the state.

7b. **Aristida longespica** Poir. var. **geniculata** (Raf.) Fern. Three-awn. Aug.–Sept. Sandy soil, rare; Brown, Carroll, Henry, Iroquois, Jackson, Kankakee, Lake, Lee, and Ogle cos.

8. **Aristida oligantha** Michx. Three-awn. Aug.–Oct. Dry soil of fields and woods, along railroads; common throughout the state.

9. **Aristida purpurascens** Poir. Arrowfeather. Aug.–Sept. Sandy soil; occasional throughout the state.

10. **Aristida ramosissima** Engelm. Two forms occur in Illinois:

a. Lateral awns 0.5–0.6 mm long .... 10a. *A. ramosissima* f. *ramosissima*
a. Lateral awns absent or less than 0.5 mm long...... 10b. *A. ramosissima* f. *uniaristata*

10a. **Aristida ramosissima** Engelm. f. **ramosissima.** Slender Three-awn. Aug.–Sept. Dry soil of fields; occasional in the s. ½ of the state, absent elsewhere.

10b. **Aristida ramosissima** Engelm. f. **uniaristata** (Gray) Mohlenbr. Slender Three-awn. Aug.–Sept. Dry soil of fields, rare; Marion and St. Clair cos.

11. **Aristida tuberculosa** Nutt. Needle Grass. Aug.–Oct. Sandy soil; occasional in the n. ½ of the state; also Union Co.

11. **Arrhenatherum** Beauv.—Oat Grass

1. **Arrhenatherum elatius** (L.) J. S. & C. Presl. Tall Oat Grass. Native to Europe; occasionally escaped into waste ground; widely scattered in IL.

12. **Arthraxon** P. Beauv.—Mat Grass

1. **Arthraxon hispidus** (Thunb.) Makino. Arthraxon. Sept.–Oct. Native to Asia and the s. Pacific; rarely adventive in disturbed soil; Massac Co.

13. **Arundinaria** Michx.—Cane

1. **Arundinaria gigantea** (Walt.) Muhl. Giant Cane. Apr.–May. Low ground but occasionally migrating upslope; confined to the s. ½ of IL; also Greene Co.

14. **Arundo** L.—Giant Reed

1. **Arundo donax** L. Giant Reed. Sept.–Oct. Native along the Mediterranean Sea; adventive in Alexander, Pope, and Pulaski cos.

15. **Avena** L.—Oats

1. Spikelets 3-flowered; lemmas pubescent, with a bent awn .... 1. *A. fatua*
1. Spikelets 2-flowered; lemmas glabrous, awnless or with a staight awn .... 2. *A. sativa*

1. **Avena fatua** L. Wild Oats. May–Sept. Native to Europe; rarely escaped into waste ground.
2. **Avena sativa** L. Oats. May–Aug. Native to Europe and Asia; occasionally escaped into waste ground. *Avena fatua* L. var. *sativa* (L.) Haussk.

16. **Beckmannia** Host—Slough Grass

1. **Beckmannia syzigachne** (Steud.) Fern. American Slough Grass. July–Sept. Wet ground, very rare; Cook, Lake, and McDonough cos.

17. **Bothriochloa** Kuntze—Beardgrass

1. Panicles reddish at maturity; hairs not obscuring the spikelets................1. *B. bladhii*
1. Panicles silvery at maturity; hairs more or less obscuring the spikelets.................... 2. *B. laguroides*

1. **Bothriochloa bladhii** (Retz.) S. T. Blake. Australian Bluestem. Aug.–Oct. Native to Africa and Asia; rarely adventive in disturbed soil; Champaign Co.
2. **Bothriochloa laguroides** (DC.) Herter. Silver Beardgrass. Aug.–Oct. Native to the s. and w. of IL; adventive in waste ground; Alexander, Clark, DuPage, Grundy, Jackson, Johnson, Sangamon, and Union cos. (Including ssp. *torreyana* [Steud.] Allred & Gould.) *Bothriochloa saccharoides* (Swartz) Rydb.

18. **Bouteloua** Lag.—Grama Grass

1. Spikes 10–50, spreading or nodding, to 2 cm long, falling entire at maturity; spikelets 7–10 mm long ....1. *B. curtipendula*
1. Spikes 1–3 (–6), straight or curved backward, 2–5 cm long, the florets falling from the glumes; spikelets 5–6 mm long.
  2. Rachis of inflorescence not projecting beyond last spikelet; second glume sparsely pilose-papillose on keel; empty lemma long-villous at base; blades glabrous or scabrous..................2. *B. gracilis*
  2. Rachis of inflorescence projecting 2–5 mm beyond last spikelet; second glume papillose-hirsute on keel; empty lemma glabrous at base; blades papillose-pubescent..................3. *B. hirsuta*

1. **Bouteloua curtipendula** (Michx.) Torr. Side-oats Grama. July–Sept. Prairies, dry hills; occasional in the n. ½ of the state; rather common on the bluffs bordering the Mississippi River from Jo Daviess to Alexander cos.
2. **Bouteloua gracilis** (HBK.) Lag. Blue Grama. July–Sept. Sand flats, very rare; Carroll, Champaign, DuPage, Henry, Jo Daviess, Lake, and Morgan cos.; also adventive at a strip mine in Williamson Co.

3. **Bouteloua hirsuta** Lag. Grama Grass. July–Sept. Prairies; mostly in the nw. ¼ of IL; also Kane, Kankakee, and Tazewell cos.

19. **Brachyelytrum** P. Beauv.—False Brome

1. **Brachyelytrum erectum** (Roth) P. Beauv. False Brome. May–Aug. Moist or occasionally dry woods; occasional throughout the state.

20. **Briza** L.—Quaking Grass

1. Spikelets 1–2 cm long, 1.0–1.5 cm broad, 10- to 20-flowered, stramineous to pale brown .... 1. *B. maxima*
1. Spikelets 3–5 mm long, 3–6 mm wide, 3- to 6-flowered, whitish.......... 2. *B. minor*

1. **Briza maxima** L. Big Quaking Grass. June–July. Native to Europe; rarely adventive in waste ground; Coles and St. Clair cos.
2. **Briza minor** L. Little Quaking Grass. May–June. Native to Europe; rarely adventive in disturbed soil; Jackson Co.

21. **Bromus** L.—Brome Grass

1. Some or all the awns more than 12 mm long; teeth of lemmas 2–5 mm long.
  2. Lemmas 16–35 mm long, scabrous or puberulent on the back; first glume 8–25 mm long; second glume 13–35 mm long; awns 20–65 mm long.
    3. Lemmas 16–21 mm long; first glume 8–12 mm long; second glume 13–18 mm long; awns 20–30 mm long.......... 20. *B. sterilis*
    3. Lemmas 20–35 mm long; first glume 15–25 mm long; second glume 20–35 mm long; awns 30–65 mm long.......... 7. *B. diandrus*
  2. Lemmas 10–12 mm long, villous throughout on the back, becoming hispidulous at the summit, rarely entirely glabrous; first glume 4–7 mm long; second glume 8–10 mm long; awns (10–) 12–15 mm long.......... 21. *B. tectorum*
1. All of most of the awns 1–12 mm long; teeth of lemmas usually less than 2 mm long (occasionally up to 3 mm long in *B. squarrosus*).
  4. First glume 3- to 7-nerved (1-nerved in *B. nottowayanus*); second glume 5- to 9-nerved; annuals or perennials.
    5. Perennials from rhizomes; blades 6–13 mm broad (occasionally only 5 mm broad in *B. kalmii*).
      6. Glumes glabrous or nearly so; lemmas sharply keeled.
        7. Inflorescence erect; awns 4–6 mm long; sheaths spreading to retrorsely pilose.......... 14. *B. marginatus*
        7. Inflorescence spreading or drooping; awns, or most of them, 7–12 mm long; sheaths glabrous or nearly so except for the ciliate summit.......... 3. *B. carinatus*
      6. Glumes pilose or densely appressed pubescent; lemmas not strongly keeled.
        8. Awns 5–8 mm long; cauline leaves 6–8 in number.......... 15. *B. nottowayanus*
        8. Awns 1–3 mm long; cauline leaves 3–5 (–6) in number.......... 12. *B. kalmii*
    5. Annuals from fibrous roots; blades 2–6 mm broad (to 8 mm broad in *B. secalinus*).
      9. Lemmas strongly keeled, 12–15 mm long.......... 4. *B. catharticus*
      9. Lemmas rounded on back, up to 12 mm long.
        10. Awns to 6 mm long or absent.
          11. Blades softly villous; lemmas 9–11 (–12) mm long; awns to 1 mm long or absent.......... 2. *B. briziformis*
          11. Blades harsh-pubescent above or glabrous; lemmas 5–8 mm long; awns 1–6 mm long, rarely absent.......... 18. *B. secalinus*
        10. Most or all the awns more than 6 mm long.
          12. Inflorescence erect or ascending.
            13. Lemmas plicate, conspicuously nerved; inflorescence compact. .... 9. *B. hordeaceus*
            13. Lemmas not plicate, faintly nerved; inflorescence open.

14. Lower lemmas 7–9 mm long; branches of inflorescence solitary or paired, usually shorter than the spikelets; anthers 2.0–2.5 mm long .... 17. *B. racemosus*
14. Lower lemmas 9–11 mm long; branches of inflorescence 2–5, usually much longer than the spikelets; anthers 1.5–2.0 mm long .... 6. *B. commutatus*
12. Inflorescence spreading or drooping.
15. Awns straight or nearly so.
16. Lemmas all nearly the same length; anthers 4 mm long .... 1. *B. arvensis*
16. Lowest lemmas longer than the upper; anthers 2.0–2.5 mm long .... 17. *B. racemosus*
15. Awns flexuous or twisted.
17. Lemmas with a hyaline margin 1 mm broad; rachilla not exposed at maturity .... 19. *B. squarrosus*
17. Lemmas without a broad hyaline margin; rachilla exposed at maturity .... 11. *B. japonicus*
4. First glume 1-nerved; second glume 3- or 5-nerved; perennials. (*Bromus nottowayanus* has the first glume 1-nerved, the second glume 5- or 7-nerved.)
18. Awns absent or up to 2 mm long; blades and sheaths glabrous .... 10. *B. inermis*
18. Awns 2–8 mm long; blades and sheaths (particularly the lower) pubescent, rarely glabrous.
19. Inflorescence narrow, erect; blades 2–3 mm broad .... 8. *B. erectus*
19. Inflorescence spreading or drooping; blades (3–) 4–17 mm broad.
20. Leaves 10–20 per culm, the blades auriculate at base .... 13. *B. latiglumis*
20. Leaves 5–8 per culm, the blades not auriculate at base.
21. Lemmas pubescent throughout on the back or glabrous .... 16. *B. pubescens*
21. Lemmas pubescent only on the margins in the lower ½–¾ of the lemma .... 5. *B. ciliatus*

1. **Bromus arvensis** L. Chess. May–July. Native to Europe; adventive in waste ground; not common but scattered in IL.

2. **Bromus briziformis** Fisch. & Mey. Rattlesnake Chess. May–July. Native to Europe; adventive in waste ground, rare; Richland and Washington cos.

3. **Bromus carinatus** Hook. California Brome. June–July. Native to the w. U.S.; rarely adventive in disturbed soil; DuPage and Kane cos.

4. **Bromus catharticus** Vahl. Rescue Grass. May–July. Native to tropical America; adventive in disturbed soil; Champaign Co. *Bromus unioloides* Kunth; *B. willdenovii* Kunth.

5. **Bromus ciliatus** L. Canada Brome Grass. June–Sept. Open woods; common throughout the state. (Including var. *intonsus* Fern.)

6. **Bromus commutatus** Schrad. Hairy Chess. May–Aug. Native to Europe; adventive in waste ground; common throughout the state.

7. **Bromus diandrus** Roth ssp. **rigidus** (Roth) Lainz. Ripgut Grass. May–June. Native to Europe; rarely adventive in IL; St. Clair Co. *Bromus rigidus* Roth.

8. **Bromus erectus** Huds. Erect Brome Grass. July. Native to Europe; adventive in waste ground; St. Clair Co.

9. **Bromus hordeaceus** L. Two subspecies occur in Illinois:

a. Lemmas 8–11 mm long, densely hairy .... 9a. *B. hordeaceus* ssp. *hordeaceus*
a. Lemmas 7–8 mm long, nearly glabrous .... 9b. *B. hordeaceus* spp. *pseudothominei*

9a. **Bromus hordeaceus** L. ssp. **hordeaceus.** Soft Chess. May–July. Native to Europe; adventive in disturbed soil; not common; Cook, DuPage, Effingham, Jackson, Lawrence, and Pulaski cos. *Bromus mollis* L.

9b. **Bromus hordeaceus** L. ssp. **pseudothominei** (P. M. Sm.) H. Scholz. Soft Chess. May–July. Native to Europe; adventive in disturbed soil; DuPage Co. *Bromus X pseudothominei* P. M. Sm.

10. **Bromus inermis** Leyss. Awnless Brome Grass; Smooth Brome. May–July. Native to Europe; roadsides, fields, waste ground, common; in every co. (Including f. *aristatus* [Schur] Fern.)

11. **Bromus japonicus** Thunb. Japanese Chess. May–July. Native to Europe and Asia; adventive in waste ground; common throughout the state.

12. **Bromus kalmii** Gray. Kalm's Brome Grass. June–July. Dry woods, sandy soil, not common; confined to the n. ½ of the state; also Calhoun Co.

13. **Bromus latiglumis** (Shear) Hitchc. Nodding Brome. June–Sept. Moist, open woods, occasional in the n. ½ of the state, rare in the s. ½. (Including f. *incanus* [Shear] Fern.) *Bromus purgans* L., misapplied.

14. **Bromus marginatus** Nees. Brome Grass. May–Aug. Native to the w. U.S.; adventive in waste ground; Cook and Kane cos.

15. **Bromus nottowayanus** Fern. Brome Grass. June–Aug. Moist, wooded ravines; Adams, Brown, Cook, Jackson, Peoria, Stark, and Woodford cos.

16. **Bromus pubescens** Muhl. Canada Brome Grass. June–Aug. Moist, open woods; common throughout the state.

17. **Bromus racemosus** L. Chess. May–Aug. Native to Europe; adventive in waste ground; occasional throughout the state.

18. **Bromus secalinus** L. Smooth Chess. May–Aug. Native to tropical America; adventive in waste ground; common; probably in every co.

19. **Bromus squarrosus** L. Nodding Brome; One-way Brome. Native to Europe and Asia; adventive in waste ground; ne. cos.

20. **Bromus sterilis** L. Brome Grass. May–June. Native to Europe; adventive in waste ground; Champaign, Cook, Jackson, McDonough, and McHenry cos.

21. **Bromus tectorum** L. Downy Chess. Apr.–July. Native to Europe; naturalized in waste areas, common; in every co.

22. **Buchloe** Engelm.—Buffalo Grass

1. **Buchloe dactyloides** (Nutt.) Engelm. Buffalo Grass. Aug.–Sept. Prairies, rare; confined to the n. ⅔ of IL. *Bouteloua dectyloides* Nutt.

23. **Calamagrostis** Adans.—Reed Grass

1. Callus hairs of lemma exceeding the lemma.. 2. *C. epigeios*
1. Callus hairs of lemma ½ as long as to equaling the lemma.
  2. Callus hairs of lemmas ½ as long as the lemma; awn of lemma abruptly twisted or bent... 4. *C. insperata*
  2. Callus hairs of lemma equaling the lemma; awn of lemma straight.
    3. Blades 4–8 mm broad; panicle open, more or less nodding; glumes spreading in fruit; lemmas translucent at tip.......................... 1. *C. canadensis*
    3. Blades 2–4 mm broad; panicle contracted, erect; glumes connivent at tip in fruit; lemmas opaque throughout.
      4. Blades usually flat, scabrous; glumes opaque throughout.......................... .... 3. *C. inexpansa*
      4. Blades involute, smooth; glumes translucent at tip.................. 5. *C. stricta*

1. **Calamagrostis canadensis** (Michx.) P. Beauv. Two varieties have been found in Illinois:

a. Panicle loosely flowered; spikelets 2.8–3.8 mm long; glumes distinctly exceeding the lemma, acute to acuminate...............................1a. *C. canadensis* var. *canadensis*
a. Panicle densely flowered; spikelets 2.2–2.8 mm long; glumes nearly equaled by the lemma, obtuse to acute.................................. 1b. *C. canadensis* var. *macouniana*

1a. **Calamagrostis canadensis** (Michx.) P. Beauv. var. **canadensis.** Bluejoint Grass. June–July. Moist soil; occasional in the n. ½ of the state, rare in the s. ½, absent from the extreme s. cos.

1b. **Calamagrostis canadensis** (Michx.) P. Beauv. var. **macouniana** (Vasey) Stebbins. Bluejoint Grass. Aug. Wet ditch, very rare; Henry Co.

2. **Calamagrostis epigeios** (L.) Roth. Feathertop. June–July. Native to Europe; adventive in strip mine, rare; Randolph Co.

3. **Calamagrostis inexpansa** Gray. Northern Reed Grass. June–July. Wet ground, rare; confined to the n. ⅓ of IL. (Including var. *brevior* [Vasey] Stebbins.)

4. **Calamagrostis insperata** Swallen. Ofer Hollow Reed Grass. June–Aug. Sandstone ledges, rare; Pope Co.

5. **Calamagrostis stricta** (Timm.) Koeler. Reed Grass. June–Aug. Native n. of IL; rarely adventive in disturbed soil; n. IL. *Calamagrostis neglecta* (Ehrh.) Gaertn., Mey., & Scherb.

24. **Calamovilfa** (Gray) Hack.—Sand Reed

1. **Calamovilfa longifolia** (Hook.) Scribn. Sand Reed. July–Sept. Sandy areas; scattered in the n. ½ of the state; also St. Clair Co.

25. **Cenchrus** L.—Sand Bur

1. Inner bristles in fascicles of 45–75 .... 1. *C. longispinus*
1. Inner bristles in fascicles of 8–50 .... 2. *C. spinifex*

1. **Cenchrus longispinus** (Hack.) Fern. Sand Bur. July–Sept. Sandy soil; occasional throughout the state. *Cenchrus pauciflorus* Benth.

2. **Cenchrus spinifex** Cav. Common Sand Bur. July–Sept. Fields, disturbed sandy areas; scattered in IL; *Cenchrus tribuloides* L., misapplied.

26. **Chasmanthium** Link—Inland Oats

1. **Chasmanthium latifolium** (Michx.) Yates. Inland Oats. July–Oct. Moist soil; common in the s. ½ of IL, rare in the n. ½. *Uniola latifolia* Michx.

27. **Chloris** Swartz—Windmill Grass

1. Lowest lemma with conspicuous spreading marginal hairs .... 1. *C. gayana*
1. Lowest lemma with glabrous, scabrous, or short-pubescent marginal hairs.
  2. Leaves 1–3 mm wide; panicle branches whorled .... 2. *C. verticillata*
  2. Leaves 3–15 mm wide; panicle branches digitate .... 3. *C. virgata*

1. **Chloris gayana** Kunth. Finger Grass. June–Sept. Native to Africa; adventive in a strip mine; Perry Co.

2. **Chloris verticillata** Nutt. Windmill Grass. June–Sept. Native to the w. U.S.; adventive in lawns and other disturbed areas; scattered in IL.

3. **Chloris virgata** Sw. Feather Windmill Grass. July–Aug. Disturbed areas, very rare; Champaign and Union cos.

28. **Cinna** L.—Wood Reed

1. Panicle gray-green, the branches mostly ascending; spikelets 4.0–6.5 mm long; second glume 3-nerved; awn less than 0.5 mm long .... 1. *C. arundinacea*
1. Panicle green, the branches mostly spreading; spikelets 2.5–4.0 mm long; second glume 1-nerved; awn 0.5–1.0 mm long .... 2. *C. latifolia*

1. **Cinna arundinacea** L. Stout Wood Reed. July–Sept. Moist woods, damp soil; occasional to common throughout the state.

2. **Cinna latifolia** (Trev.) Griseb. Drooping Wood Reed. July–Sept. Moist woods, rare; Cook, DeKalb, Kane, Lake, and Winnebago cos.

29. **Coleataenia** Griseb.—One-sided Panicum

1. Spikelets pubescent .... 4. *C. rigidula*
1. Spikelets glabrous.
  2. Spikelets up to 2.5 mm long.
    3. Spikelets less than 2 mm long .... 4. *C. rigidula*
    3. Spikelets 2.0–2.5 mm long.
      4. Ligules ciliate, 2–3 mm long .... 3. *C. longifolia*
      4. Ligules membranous, up to 1 mm long.
        5. Spikelets 2.5–2.8 mm long; grain short-stipitate .... 5. *C. stipata*
        5. Spikelets 2.0–2.5 mm long; grain not stipitate.
          6. Panicle branches spreading to ascending; spikelets up to 2.2 mm long .... 1. *C. rigidula*

6. Panicle branches erect; spikelets 2.2–2.5 mm long ... 2. *C. condensa*
2. Spikelets 3 mm long or longer .... 1. *C. anceps*

1. **Coleataenia anceps** (Michx.) Soreng. Beaked Panicum. June–Oct. Moist soil, woods, prairies, roadside ditches, stream banks; commonin the s. ⅓ of the state, rare to absent elsewhere. *Panicum anceps* Michx.

2. **Coleataenia condensa** (Nash) Mohlenbr., comb. nov. (basionym: *Panicum condensum* Nash). Munro Grass. June–Oct. Moist soil, rare; Jackson, Johnson, and Massac cos. *Panicum rigidulum* Bosc var. *condensum* (Nash) Mohlenbr.; *Panicum agrostoides* Spreng. var. *condensum* (Nash) Fern.

3. **Coleataenia longifolia** (Torr.) Soreng. Long-leaved Panicum. June–Oct. Rocky ledges in wooded ravine, very rare; Monroe Co. *Panicum longifolium* Torr.; *Panicum rigidulum* Bosc var. *pubescens* (Vasey) Lelong.

4. **Coleataenia rigidula** (Bosc ex Nees) LeBlond. Munro Grass. June–Oct. Moist soil in prairies, along ponds or creeks, low woods; occasional in the s. ⅓ of the state, less common elsewhere. *Panicum rigidulum* Bosc ex Nees; *Panicum agrostoides* Spreng.

5. **Coleataenia stipitata** (Nash) LeBlond. Stalked Panicum. June–Oct. Moist soil, very rare; Johnson Co. *Panicum stipitatum* Nash; *Panicum agrostoides* Spreng. var. *elongatum* Scribn.

30. **Crypsis** Ait.—Crypsis

1. **Crypsis schoenoides** (L.) Lam. Crypsis. Aug.–Sept. Native to Europe; introduced into disturbed areas; Cook, Grundy, Kendall, Monroe, St. Clair, and Will cos. *Heleochloa schoenoides* (L.) Host.

31. **Cynodon** Rich.—Bermuda Grass

1. **Cynodon dactylon** (L.) Pers. Bermuda Grass. July–Oct. Native to Europe; adventive in lawns and waste ground; occasional in the s. ½ of the state, much less common in the n. ½.

32. **Dactylis** L.—Orchard Grass

1. **Dactylis glomerata** L. Orchard Grass. May–July. Native to Europe; adventive in waste ground; in every co.

33. **Dactyloctenium** Willd.—Crowfoot Grass

1. **Dactyloctenium aegyptium** (L.) Beauv. Crowfoot Grass. Aug. Native to the Old World; adventive in waste ground; St. Clair Co.

34. **Danthonia** Lam. & DC.—Poverty Oat Grass

1. **Danthonia spicata** (L.) Roem. & Schultes. Curly Grass; Poverty Oat Grass. May–Aug. Dry woods and bluffs; occasional to common throughout the state.

35. **Deschampsia** P. Beauv.—Hairgrass

1. Lemmas glabrous on the nerves; awn of lemmas straight; blades flat or plicate, up to 5 mm wide .... 1. *D. caespitosa*
1. Lemmas roughened on the nerves; awn of lemmas geniculate; leaves involute, 1–2 mm wide .... 2. *D. flexuosa*

1. **Deschampsia caespitosa** (L.) P. Beauv. var. **glauca** (Hartm.) Lindm. f. Tufted Hairgrass. June–July. Along creeks and in swamps, rare; confined to extreme ne. IL.

2. **Deschampsia flexuosa** (L.) Trin. Hairgrass. June–July. Dry soil, very rare; Cook Co.

36. **Diarrhena** P. Beauv.—Beakgrass

1. Leaf sheaths pubescent; lemmas 7–11 mm long, ovate, acuminate .... 1. *D. americana*
1. Leaf sheaths glabrous; lemmas 5–7 mm long, obovate, cuspidate....... 2. *D. obovata*

1. **Diarrhena americana** P. Beauv. American Beakgrass. June–Sept. Rich woods, rare; Alexander, Hardin, Jackson, Pope, and Union cos.

2. **Diarrhena obovata** (Gl.) Brandenburg. Beakgrass. June–Sept. Low, shaded woods, moist ledges, base of limestone cliffs; throughout the state. *Diarrhena americana* P. Beauv. var. *obovata* Gl.

37. **Dichanthelium** (Hitchc. & Chase) Gould—Rosette Grass

1. Spikelets glabrous.
  2. Spikelets 1.0–1.9 mm long.
    3. Spikelets up to 0.8 mm broad; all nodes always densely bearded .... 22. *D. microcarpon*
    3. Spikelets 2.0–2.9 mm broad; only the lower nodes usually bearded .... 10. *D. dichotomum*
  2. Spikelets 2 mm long or longer.
    4. Spikelets 2.0–2.9 mm long.
      5. Spikelets 1.2–1.7 mm broad.
        6. Blades 3–6 mm broad; spikelets 2.1–2.4 mm long, 1.2–1.3 mm broad .... 39. *D. werneri*
        6. Blades 6–12 mm broad; spikelets 2.9–3.0 mm long, 1.6–1.7 mm broad .... 11. *D. helleri*
      5. Spikelets up to 1 mm broad.
        7. At least the lower nodes bearded; sheaths softly pubescent .... 10. *D. dichotomum*
        7. Nodes glabrous or sparsely pilose; sheaths glabrous, ciliate, pilose, or appressed-pubescent but not softly pubescent.
          8. Spikelets 2.2–2.5 mm long; sheaths with pale glandular spots .... 41. *D. yadkinense*
          8. Spikelets 2.0–2.2 mm long; sheaths not glandular-spotted .... 10. *D. dichotomum*
    4. Spikelets 3 mm long or longer.
      9. Spikelets obtuse.
        10. Culms appressed-pubescent; first glume sparsely hirsute .... 24. *D. oligosanthes*
        10. Culms glabrous or spreading-pubescent; first glume glabrous .... 35. *D. scribnerianum*
      9. Spikelets acute .... 9. *D. depauperatum*
1. Spikelets pubescent.
  11. Spikelets 1.0–1.9 mm long.
    12. Some or all the sheaths papillose-pilose; spikelets obtuse; blades pubescent throughout on the upper surface (except in some specimens of *D. acuminatum* and *D. lindheimeri*).
      13. Culms pilose with horizontally spreading hairs .... 29. *D. praecocius*
      13. Culms variously pubescent but the hairs not horizontally spreading.
        14. Upper surface of blades glabrous except for long hairs at the base; autumnal form matted.
          15. Axis of panicle branches long-pilose .... 1. *D. acuminatum*
          15. Axis of panicle branches glabrous .... 17. *D. lindheimeri*
        14. Upper surface of blades pilose or appressed-pubescent or velvety; autumnal form erect or spreading (becoming matted in *D. meridionale* and *D. auburne*).
          16. Spikelets 1.3–1.5 mm long; grain 1.2–1.3 mm long.
            17. Plants velvety-pubescent, including the axis of the panicle .... 4. *D. auburne*
            17. Plants not velvety-pubescent, the axis of the panicle long-pilose, puberulent, or glabrous.
              18. Panicle branches (at least the lower) drooping, the axes long-pilose .... 12. *D. implicatum*
              18. Panicle branches ascending, the axes glabrous or puberulent.
                19. Vernal blades 1.5–4.0 cm long; plants greenish yellow; autumnal form nearly erect .... 21. *D. meridionale*

19. Vernal blades 4.5–7.0 cm loing; plants grayish; autumnal form spreading to ascending, but eventually forming mats.. *D. albemarlense*

16. Spikelets 1.6–1.9 mm long; grain 1.5–1.6 mm long.

20. Upper surface of leaves pilose, the hairs 3–5 mm long; first glume acute or acuminate, about ½ as long as the spikelet ....37. *D. subvillosum*

20. Upper surface of leaves short-pubescent, the hairs less than 3 mm long; first glume obtuse and truncate, ¼–⅓ as long as the spikelet..1. *D. acuminatum*

12. None of the sheaths papillose-pilose; spikelets obtuse or acute; upper surface of blades glabrous except near the base (rarely puberulent throughout in *D. laxiflorum*).

21. Ligule 3–5 mm long ....1. *D. acuminatum*

21. Ligule less than 2 mm long.

22. Grain 1.7–1.8 mm long.

23. Panicle branches ascending, viscid; sheaths viscid; autumnal culms leafy to base .... 23. *D. nitidum*

23. Panicle branches spreading, not viscid; sheaths not viscid; autumnal culms essentially leafless below middle.......................... .... 10. *D. dichotomum*

22. Grain 1.3–1.5 mm long.

24. Sheaths retrorsely pilose... 15. *D. laxiflorum*

24. Sheaths glabrous or ciliate or ascending-pilose.

25. Sheaths ascending-pilose, with long, soft hairs intermingled with short, crisp ones; leaves to 2.5–8.0 mm broad.

26. Spikelets 1.5–2.0 mm long; leaves 2.5–4.5 mm broad............ .... 28. *D. portoricense*

26. Spikelets 1.8–2.5 mm long; leaves 3.5–8.0 mm wide.............. .... 32. *D. sabulorum*

25. Sheaths glabrous or ciliate, without 2 types of hairs; blades 7–25 mm broad.

27. Spikelets ellipsoid, about 0.7 mm broad; nodes bearded....... ....22. *D. microcarpon*

27. Spikelets obovoid-spherical, 1.0–1.3 mm broad; nodes appressed-puberulent or glabrous.

28. Panicle nearly as broad as long; culms spreading; nodes appressed-puberulent.......................36. *D. sphaerocarpon*

28. Panicle ¼–½ as broad as long; culms erect; nodes more or less glabrous.......................................27. *D. polyanthes*

11. Spikelets 2 mm long or longer.

29. Spikelets 2.0–2.9 mm long.

30. Nodes bearded but with a sticky ring immediately beneath them............. ....34. *D. scoparium*

30. Nodes beardless or, if with a beard, then without a sticky ring.

31. Spikelets up to 2.5 mm long.

32. Ligule 2–5 mm long; sheaths papillose-pilose.

33. Pubescence of culms horizontally spreading; autumnal form freely branched; ligule 4–5 mm long............38. *D. villosissimum*

33. Pubescence of culms appressed or ascending; autumnal form sparsely branched; ligule 2–3 mm long.

34. Pubescence of blades and sheaths silky; upper surface of blades pubescent along both margins ....30. *D. pseudopubescens*

34. Pubescence of blades and sheaths short and stiff; upper surface of blades more or less glabrous........33. *D. scoparioides*

32. Ligule 1 mm long or less; sheaths not papillose-pilose (except *D. linearifolium*).

35. Spikelets 1.2–1.5 mm broad (occasionally narrower in *D. mattamuskeetense*); grain 2.0–2.1 mm long, 1.1–1.2 mm broad.

36. Sheaths papillose-pilose; spikelets 1.3–1.5 mm broad .... 18. *D. linearifolium*
36. Sheaths puberulent, velvety, or glabrous; spikelets 1.2–1.3 mm broad.
37. Lowermost nodes bearded; lower sheaths velvety-pubescent .... 20. D. mattamuskeetense
37. Nodes without a beard; lower sheaths glabrous or puberulent but rarely velvety.
38. Sheaths puberulent; culms crisp-puberulent; spikelets at least 2.4 mm long .... 3. *D. ashei*
38. Sheaths glabrous; culms sparsely pilose on the nodes; spikelets 2.2–2.4 mm long ...... 39. *D. werneri*
35. Spikelets 0.9–1.1 mm broad; grain 1.5–1.9 mm long, 0.9–1.0 mm broad.
39. Sheaths retrorsely pilose; spikelets papillose-pilose; grain 1.5 mm long .... 15. *D. laxiflorum*
39. Sheaths puberulent, appressed-puberulent, or merely ciliate; spikelets not papillose; grain 1.7–1.9 mm long.
40. Upper sheaths viscid-spotted; nodes with reflexed hairs .... 23. *D. nitidum*
40. Upper sheaths not viscid; nodes glabrous or sparsely pilose.
41. Leaves tapering to base, not subcordate or broadly rounded .... 10. *D. dichotomum*
41. Leaves subcordate or broadly rounded at base or sometimes even clasping.
42. Sheaths glabrous or sparsely pilose .... 5. *D. boreale*
42. Sheaths, at least the lower ones, copiously pilose or villous .... 25. *D. ovale*
31. Spikelets 2.5–2.9 mm long.
43. Sheaths with papillose hairs.
44. Blades 12–30 mm wide .... 7. *D. clandestinum*
44. Blades 2–12 wide.
45. Blades glabrous or scabrous on the upper surface.
46. Blades 6–12 mm wide.
47. Culms appressed-pubescent; first glume sparsely hirsute .... 24. *D. oligosanthes*
47. Culms glabrous or spreading-pubescent; first glume glabrous .... 35. *D. scribnerianum*
46. Blades 2–5 mm wide.
48. Spikelets 1.6–1.7 mm broad; grain 2.4 mm long, 1.5–1.6 mm broad .... 26. *D. perlongum*
48. Spikelets 1.3–1.5 mm broad; grain 2.0–2.1 mm long, 1.2 mm broad .... 18. *D. linearifolium*
45. Blades hirsute or velvety on the upper surface.
49. Blades long-hirsute on both surfaces; grain 2.4–2.5 mm long .... 40. *D. wilcoxianum*
49. Blades velvety on both surfaces; grain 2.2 mm long .... 19. *D. malacophyllum*
43. Sheaths without papillose hairs.
50. Lowermost nodes bearded; lower sheaths velvety .... 20. *D. mattamuskeetense*
50. None of the nodes bearded; sheaths glabrous or puberulent but not velvety.
51. Spikelets 2.6–2.8 mm long, obtuse to subacute; blades firm, more or less cordate at base .... 8. *D. commutatum*
51. Spikelets more than 2.8 mm long, abruptly short-pointed; blades thin, narrowed or rounded at base .... 13. *D. joori*

29. Spikelets 3 mm long or longer.
52. Grain 2.1–2.5 mm long.
53. Some or all the blades at least 12 mm wide.
54. Culms and usually the sheaths papillose-hispid; spikelets subacute to acute .... 7. *D. clandestinum*
54. Culms and sheaths glabrous; spikelets abruptly short-pointed .... 13. *D. joori*
53. Blades 2–12 mm wide.
55. Blades glabrous above, glabrous or sparsely pilose beneath.
56. Blades 6–12 mm wide.
57. Culms appressed-pubescent; first glume sparsely hirsute .... 24. *D. oligosanthes*
57. Culms glabrous or spreading-pubescent; first glume glabrous .... 35. *D. scribnerianum*
56. Blades 2–5 mm wide.
58. Spikelets acute, the second glume and sterile lemma beaked .... 9. *D. depauperatum*
58. Spikelets obtuse, the second glume and sterile lemma not beaked .... 26. *D. perlongum*
55. Blades long-hirsute or velvety above and beneath.
59. Blades long-hirsute on both surfaces; grain 2.4–2.5 mm long .... 40. *D. wilcoxianum*
59. Blades velvety on both surfaces; grain 2.2 mm long .... 19. *D. malacophyllum*
52. Grain 2.8–3.5 mm long.
60. Blades papillose-pubescent on both surfaces; first glume a little more than ½ as long as the spikelet .... 16. *D. leibergii*
60. Blades not papillose-pubescent; first glume up to ½ as long as the spikelet.
61. Pubescence of sheath papillose.
62. Nodes not retrorsely bearded.
63. Culms appressed-pubescent; first glume sparsely hirsute .... 24. *D. oligosanthes*
63. Culms glabrous or spreading-pubescent; first glume glabrous .... 35. *D. scribnerianum*
62. Nodes retrorsely bearded.
64. Ligule 3–4 mm long .... 31. *D. ravenelii*
64. Ligule 1 mm long .... 6. *D. boscii*
61. Pubescence of sheath not papillose.
65. Nodes glabrous or sparsely pilose; spikelets 3.0–3.7 mm long .... 14. *D. latifolium*
65. Nodes retrorsely bearded; spikelets 3.8–5.0 mm long .... 6. *D. boscii*

1. **Dichanthelium acuminatum** (Sw.) Gould & Clark. Two varieties occur in Illinois:

a. Axis of panicle branches with some long-spreading hairs .... 1a. *D. acuminatum* var. *fasciculatum*
a. Axis of panicle branches either glabrous or with appressed hairs .... 1b. *D. acuminatum* var. *septentrionale*

1a. **Dichanthelium acuminatum** (Sw.) Gould & Clark var. **fasciculatum** (Torr.) Freckm. Panic Grass. May–Oct. Low, moist, open situations, swampy soil, moist depressions of sandstone cliffs, sandy soil; common throughout the state. *Panicum lanuginosum* Ell.; *P. dichotomum* L. var. *fasciculatum* Torr.; *P. huachucae* Ashe; *P. tennesseense* Ashe; *P. lanuginosum* Ell. var. *huachucae* (Ashe) Hitchc.; *P. lanuginosum* Ell. var. *fasciculatum* (Torr.) Fern.; *P. lanuginosum* Ell. var. *tennesseense* (Ashe) Gl.

1b. **Dichanthelium acuminatum** (Sw.) Gould & Clark var. **septentrionale** (Fern.) Mohlenbr. Panic Grass. May–Oct. Moist soil; occasional throughout the state. *Panicum lindheimeri* (Nash) Gould var. *septentrionale* Fern.; *P. lanuginosum* Ell. var. *septentrionale* (Fern.) Fern.

2. **Dichanthelium albemarlense** (Nash) Mohlenbr., comb. nov. (basionym: *Panicum albemarlense* Ashe). Gray Panic Grass. May–Sept. Margins of a wet peaty meadow, very rare; Kankakee Co. *Panicum albemarlense* Nash; *P. meridionale* Ashe var. *albemarlense* (Ashe) Fern.

3. **Dichanthelium ashei** (T. G. Pearson in Ashe) Mohlenbr., comb. nov. (basionym: *Panicum ashei* T. G. Pearson in Ashe). Ashe's Panic Grass. June–Oct. Dry woods, rare; Jackson Co. *Panicum ashei* T. G. Pearson in Ashe; *Panicum commutatum* Schult. var. *ashei* (T. G. Pearson in Ashe) Fern.; *Dichanthelium commutatum* (Schult.) Gould var. *ashei* Mohlenbr.; *D. commutatum* (Schult.) Gould ssp. *ashei* (T. G. Pearson in Ashe) Freckm. & Lelong. Specimen in Missouri Botanical Garden herbarium.

4. **Dichanthelium auburne** (Ashe) Mohlenbr. Red-brown Panicum. May–June; Oct. Sandy soil, very rare; Cook and Will cos. *Panicum auburne* Ashe.

5. **Dichanthelium boreale** (Nash) Freckm. Northern Panic Grass. June–Oct. Moist sand, very rare; Cook, Lake, St. Clair, and Will cos. *Panicum boreale* Nash.

6. **Dichanthelium boscii** (Poir.) Gould & Clark. Two varieties occur in Illinois:

a. Leaves, sheaths, and culms glabrous or sparsely pubescent .... 6a. *D. boscii* var. *boscii*
a. Leaves, sheaths, and culms softly pubescent .... 6b. *D. boscii* var. *molle*

6a. **Dichanthelium boscii** (Poir.) Gould & Clark var. **boscii.** Large-fruited Panic Grass. June–Oct. Woods; common in the s. ½ of the state, rare or absent elsewhere. *Panicum boscii* Poir.

6b. **Dichantheliuim boscii** (Poir.) Gould & Clark var. **molle** (Vasey) Mohlenbr. Large-fruited Panic Grass. June–Oct. Woods; occasional in the s. ¼ of the state. *Panicum boscii* Poir. var. *molle* (Vasey) Hitchc. & Chase.

7. **Dichanthelium clandestinum** (L.) Gould. Deer-tongue Grass. June–Oct. Moist, often sandy, soil; common in the s. ½ of the state, becoming less frequent northward. *Panicum clandestinum* L.

8. **Dichanthelium commutatum** (Schult.) Gould. Broad-leaved Panic Grass. June–Oct. Mostly dry woods; occasional in the s. ⅓ of the state; also Clark and Coles cos. *Panicum commutatum* Schult.

9. **Dichanthelium depauperatum** (Muhl.) Gould. Panic Grass. May–Sept. Dry, open woods, prairies, disturbed areas; occasional throughout the state. *Panicum depauperatum* Muhl.

10. **Dichanthelium dichotomum** (L.) Gould. Two varieties occur in Illinois:

a. Nodes glabrous or sparsely pubescent, not bearded; grain slightly exserted .... 10a. *D. dichotomum* var. *dichotomum*
a. At least the lowermost nodes bearded; grain included .... 10b. *D. dichotomum* var. *barbulatum*

10a. **Dichanthelium dichotomum** (L.) Gould var. **dichotomum.** Panic Grass. May–Oct. Dry soil, usually in woodlands; rather common in the s. ¼ of the state, occasional to rare elsewhere. *Panicum dichotomum* L.

10b. **Dichanthelium dichotomum** (L.) Gould var. **barbulatum** (Michx.) Mohlenbr. Panic Grass. May–Oct. Dry, usually rocky, woods; scattered in IL. *Panicum barbulatum* Michx.; *P. dichotomum* L. var. *barbulatum* (Michx.) Wood.

11. **Dichanthelium helleri** (Nash) Mohlenbr., comb. nov. (basionym: *Panicum helleri* Nash). Heller's Panic Grass. May–Sept. Limestone ledges, rare; Jackson and Randolph cos. *Panicum helleri* Nash; *P. oligosanthes* Schult. var. *helleri* (Nash) Fern.; *Dichanthelium oligosanthes* (Schult.) Gould var. *helleri* (Nash) Mohlenbr.

12. **Dichanthelium implicatum** (Scribn.) Kerguelen. Panic Grass. May–Oct. Moist soil; occasional throughout the state. *Panicum implicatum* Scribn.; *P. lanuginosum* L. var. *implicatum* (Scribn.) Fern.; *Dichanthelium acuminatum* (Sw.) Gould & Clark var. *implicatum* (Scribn.) Gould & Clark.

13. **Dichanthelium joori** (Vasey) Mohlenbr. Joor's Panic Grass. June–Oct. Low swampy woods, very rare; Johnson and Union cos. *Panicum joori* Vasey; *P. commutatum* Schult. var. *joori* (Vasey) Fern.; *Dichanthelium commutatum* (Schult.) Gould ssp. *joorii* (Vasey) Freckm. & Lelong.

14. **Dichanthelium latifolium** (L.) Gould & Clark. Broad-leaved Panic Grass. June–Oct. Dry, rocky woods; occasional throughout the state. *Panicum latifolium* L.

15. **Dichanthelium laxiflorum** (Lam.) Gould. Panic Grass. May–Sept. Woods; common in the s. tip of the state, uncommon elsewhere. *Panicum laxiflorum* Lam.

16. **Dichanthelium leibergii** (Vasey) Freckm. Leiberg's Panic Grass. June–Oct. Dry soil, mostly in sandstone woods; occasional in the n. ½ of the state, rare in the s. ½. *Panicum leibergii* Vasey.

17. **Dichanthelium lindheimeri** (Nash) Gould. Lindheimer's Panic Grass. May–Sept. Moist, sandy soil; occasional throughout the state. *Panicum lindheimeri* Nash; *P. lanuginosum* Ell. var. *lindheimeri* (Nash) Fern.; *Dichanthelium acuminatum* (Sw.) Gould var. *lindheimeri* (Nash) Gould & Clark.

18. **Dichanthelium linearifolium** (Scribn.) Gould. Narrow-leaved Panic Grass. May–Sept. Dry woods; occasional throughout the state. *Panicum linearifolium* Scribn.

19. **Dichanthelium malacophyllum** (Nash) Gould. Panic Grass. June–Oct. Limestone bluffs, dry woods, rare; confined to the s. ⅙ of IL. *Panicum malacophyllum* Nash.

20. **Dichanthelium mattamuskeetense** (Ashe) Mohlenbr. Panic Grass. May–Sept. Along a levee, very rare; Massac Co. *Panicum mattamuskeetense* Ashe; *P. dichotomum* L. var. *mattamuskeetense* (Ashe) Lelong.

21. **Dichanthelium meridionale** (Ashe) Freckm. Panic Grass. May–Sept. Sandy woods, peaty meadows; scattered in IL. *Panicum meridionale* Ashe.

22. **Dichanthelium microcarpon** (Muhl.) Mohlenbr. Panic Grass. May–Oct. Wet ground, often in woods; occasional in the s. ⅓ of the state; also Fulton and Peoria cos. *Panicum microcarpon* Muhl.; *P. dichotomum* L. var. *ramulosum* (Torr.) Lelong.

23. **Dichanthelium nitidum** (Lam.) Mohlenbr. Shiny Panic Grass. June–Sept. Xeric limestone bluff top, very rare; Jackson Co. *Panicum nitidum* Lam.

24. **Dichanthelium oligosanthes** (Schult.) Gould. Panic Grass. June–Oct. Sandy soil, mostly in woods; occasional throughout the state. *Panicum oligosanthes* Schult.

25. **Dichanthelium ovale** (Ell.) Gould & Clark var. **addisonii** (Nash) Gould & Clark. Addison's Panicum. May–June. Sandy soil, very rare; Iroquois and LaSalle cos. *Panicum addisonii* Nash; *P. commonsonianum* Ashe; *P. commonsonianum* Ashe ssp. *addisonii* (Nash) W. Stone.

26. **Dichanthelium perlongum** (Nash) Freckm. Panic Grass. May–Sept. Dry soil, particularly in prairies and upland woods; occasional in the n. ½ of IL, rare in the s. ½. *Panicum perlongum* Nash.

27. **Dichanthelium polyanthes** (Schult.) Mohlenbr. Panic Grass. June–Oct. Woods and woodland openings; occasional in the s. ⅓ of the state; also Peoria Co. *Panicum polyanthes* Schult.; *P. sphaerocarpon* Ell. var. *polyanthes* (Schult.) Sherif; *Dichanthelium sphaerocarpon* (Ell.) Gould var. *polyanthes* (Schult.) Gould. (Including *Dichanthelium sphaerocarpon* [Ell.] Gould. var. *isophyllum* [Scribn.] Gould & Clark.)

28. **Dichanthelium portoricense** (Desv.) B. F. Hansen & Wunderlin. Panic Grass. June–Oct. Sandy woods; rare in the n. ¼ of IL. *Dichanthelium columbianum* (Scribn.) Freckm.; *D. sabulorum* (Lam.) Gould & Clark var. *thinium* (Hitchc. & Chase) Gould & Clark; *Panicum columbianum* Scribn.; *P. tsugetorum* Nash.

29. **Dichanthelium praecocius** (Hitchc. & Chase) Mohlenbr. Panic Grass. May–Sept. Dry soil, often in prairies; occasional throughout the state, becoming rare in the extreme s. cos. *Panicum praecocius* Hitchc. & Chase; *Dichanthelium villosissimum* (Nash) Freckm. var. *praecocius* (Hitchc. & Chase) Freckm.; *Dichanthelium ovale* (Ell.) Gould & Clark ssp. *praecocius* (Hitchc. & Chase) Freckm. & Lelong.

30. **Dichanthelium pseudopubescens** (Nash) Mohlenbr., comb. nov. (basionym: *Panicum pseudopubescens* Nash). Hairy Panic Grass. May–Sept. sandy soil; occasional in the n. ⅔ of IL. *Panicum pseudopubescens* Nash; *P. villosissimum* Nash var. *pseudopubescens* (Nash) Fern.; *Dichanthelium ovale* (Ell.) Gould & Clark ssp. *pseudopubescens* (Nash) Freckm. & Lelong.

31. **Dichanthelium ravenelii** (Scribn. & Merrill) Gould. Ravenel's Panicum. June–Oct. Cherty ravines, sandstone ledges, rare; Hardin, Pope, and Union cos. *Panicum ravenelii* Scribn. & Merrill.

32. **Dichanthelium sabulorum** (Lam.) Gould & C. A. Clark var. **patulum** (Scribn. & Merr.) Gould & C. A. Clark. Rosette Grass. May–Sept. Woods, sandstone ledges, very rare; n. ½ of IL.

33. **Dichanthelium scoparioides** (Ashe) Mohlenbr. Panic Grass. June–Oct. Dry

fields, very rare; Lake Co. *Panicum scoparioides* Ashe; *P. villosissimum* Nash var. *scoparioides* (Ashe) Fern.

34. **Dichanthelium scoparium** (Lam.) Gould. Panic Grass. June–Oct. Moist fields, roadsides, not common; Gallatin, Johnson, Pope, Saline, White, and Williamson cos. *Panicum scoparium* Lam.

35. **Dichnthelium scribnerianum** (Nash) Gould. Scribner's Panic Grass. June–Oct. Dry, sandy soil, often in prairies; occasional throughout the state. *Panicum scribnerianum* Nash; *P. oligosanthes* Schult. var. *scribnerianum* (Nash) Fern.; *Dichanthelium oligosanthes* (Schult.) Gould var. *scribnerianum* (Nash) Gould.

36. **Dichanthelium sphaerocarpon** (Ell.) Gould. Panic Grass. June–Oct. Sandy soil; occasional in the s. ½ of the state. *Panicum sphaerocarpon* Ell.

37. **Dichanthelium subvillosum** (Ashe) Mohlenbr. Panic Grass. May–June. Sandy soil, rare; Lake and Mason cos. *Panicum subvillosum* Ashe.

38. **Dichanthelium villosissimum** (Nash) Freckm. Hairy Panic Grass. May–Sept. Sandy soil, often in woods; occasional throughout the state. *Panicum villosissimum* Nash.

39. **Dichanthelium werneri** (Scribn.) Mohlenbr., comb. nov. (basionym: *Panicum werneri* Scribn.). Werner's Panic Grass. May–Sept. Wooded slopes, very rare; LaSalle Co. *Panicum werneri* Scribn.; *P. linearifolium* Scribn. var. *werneri* (Scribn.) Fern.; *Dichanthelium linearifolium* (Scribn.) Gould) var. *werneri* (Scribn.) Mohlenbr.

40. **Dichanthelium wilcoxianum** (Vasey) Freckm. Wilcox's Panicum. June–Sept. Dry soil, mostly in prairies; confined to the extreme nw. cos.; also Kankakee, Pope, and Scott cos. *Panicum wilcoxianum* Vasey.

41. **Dichanthelium yadkinense** (Ashe) Mohlenbr. Panic Grass. May–Sept. Moist ground, rare; restricted to a few s. cos. *Panicum yadkinense* Ashe; *P. dichotomum* L. var. *yadkinense* (Ashe) Lelong; *Dichanthelium dichotomum* (L.) Gould. ssp. *yadkinense* (Ashe) Freckm. & Lelong.

38. **Digitaria** Heist.—Finger Grass

1. Culms rooting at the lower nodes, decumbent at the base; rachis broadly winged, about 1 mm broad.
  2. Sheaths (at least the lower) papillose-pilose; blades pilose to scabrous; spikelets 2.5–3.5 mm long; second glume about ½ as long as spikelet, usually 1.2–1.6 mm long; fertile lemma greenish brown.
    3. Spikelets 2.5–3.0 mm long; sterile lemma appressed-pubescent .... 4. *D. sanguinalis*
    3. Spikelets 3.0–3.5 mm long; sterile lemma with cilia to 1.5 mm long .... 1. *D. ciliaris*
  2. Sheaths glabrous; blades glabrous; spikelets 1.7–2.2 mm long; second glume about as long as spikelet, 1.7–2.2 mm long; fertile lemma dark brown to blackish .... 3. *D. ischaemum*
1. Culms erect, not rooting at the lower nodes; rachis narrowly winged, less than 1 mm broad.
  4. Racemes less than 10 cm long; spikelets 1.5–1.7 (–2.0) mm long; second glume and sterile lemma more or less glabrous to short-pubescent, 1.5–1.7 (–2.0) mm long .... 2. *D. filiformis*
  4. Racemes over 10 cm long; spikelets 2.0–2.5 mm long; second glume and sterile lemma long-pubescent, 2.0–2.5 mm long .... 5. *D. villosa*

1. **Digitaria ciliaris** (Retz.) Koeler. Southern Crabgrass. June–Oct. Native to Europe and Asia; rarely adventive in IL; s. ⅔ of the state. *Digitaria sanguinalis* (L.) Scop. var. *ciliaris* (Retz.) Parl.

2. **Digitaria filiformis** (L.) Koel. Slender Crabgrass. Aug.–Sept. Sandy soil; scattered in IL.

3. **Digitaria ischaemum** (Schreb.) Schreb. Smooth Crabgrass. July–Oct. Native to Europe; adventive in waste ground; occasional throughout the state.

4. **Digitaria sanguinalis** (L.) Scop. Common Crabgrass. June–Aug. Native to Europe and Asia; adventive in waste ground and lawns; in every co.

5. **Digitaria villosa** (Walt.) Pers. Hairy Finger Grass. July–Sept. Sandy soil, very rare; Cass, Jackson, Mason, and Saline cos.

39. **Distichlis** Raf.—Salt Grass

1. **Distichlis spicata** (L.) Greene var. **stricta** (Torr.) Scribn. Salt Grass. July–Oct. Native to w. N. America; adventive along railroads and in saline waste areas; scattered in the n. ½ of the state. *Distichlis stricta* (Torr.) Rydb.

40. **Echinochloa** P. Beauv.—Barnyard Grass

1. Second glume with an awn 2–10 mm long; sheaths papillose-hirsute (glabrous in f. *laevigata*); grain about 3 times longer than broad ....5. *E. walteri*
1. Second glume awnless or with an awn less than 2 mm long; sheaths glabrous or scabrous; grain at most about twice as long as broad, usually shorter.
  2. Racemes of panicle slender, distant; grain 2.0–2.5 mm long; blades 2–6 mm broad .... 1. *E. colonum*
  2. Racemes of panicle broader, more crowded; grain 2.5–3.5 mm long; blades 5–25 mm broad.
    3. Fertile lemma with a weak, easily broken tip, with a ring of minute setae just below the tip; rachis and axis of panicle setose.
      4. Panicle green or purple; sterile lemma short-awned; pubescence of second glume and sterile lemma hispidulous or setose.............2. *E. crus-galli*
      4. Panicle dark brownish purple; sterile lemma awnless; pubescence of second glume and sterile lemma appressed ....3. *E. frumentacea*
    3. Fertile lemma without a ring of setae just below the firm tip; rachis and axis of panicle not setose ....4. *E. muricata*

1. **Echinochloa colonum** (L.) Link. Jungle Rice. Aug.–Oct. Native to Europe and Asia; adventive in waste ground; scattered in IL.

2. **Echinochloa crus-galli** (L.) P. Beauv. Barnyard Grass. Aug.–Oct. Native to Europe and Asia; adventive in disturbed soil; occasional to common throughout IL.

3. **Echinochloa frumentacea** (Roxb.) Link. Billion Dollar Grass. Aug.–Oct. Native to Asia; occasionally escaped from cultivation; scattered in IL. *Echinochloa crus-galli* (L.) P. Beauv. var. *frumentacea* (Roxb.) W. Wight.

4. **Echinochloa muricata** (Michx.) Fern. Three varieties occur in Illinois:

a. Spikelets and grains 3.5–4.5 mm long; sterile lemma with an awn over 3 mm long; anthers 0.7–0.9 mm long; panicle more or less open, the branches spreading ....4a. *E. muricata* var. *muricata*
a. Spikelets and grains 2.5–3.5 mm long; sterile lemma awnless or with an awn less than 3 mm long; anthers 0.3–0.7 mm long; panicle contracted, the branches ascending.
  b. Panicle purple; second glume and sterile lemma papillose-hispid ....4b. *E. muricata* var. *microstachya*
  b. Panicle green; second glume and sterile lemma appressed-pubescent, not papillose ....4c. *E. muricata* var. *wiegandii*

4a. **Echinochloa muricata** (Michx.) Fern. var. **muricata.** Wild Millet. Aug.–Oct. Low ground; common throughout the state; in every co. *Echinochloa pungens* (Poir.) Rydb.

4b. **Echinochloa muricata** (Michx.) Fern. var. **microstachya** Wieg. Wild Millet. Aug.–Oct. Low ground; scattered in IL. *Echinochloa microstachya* (Wieg.) Rydb.; *E. pungens* (Poir.) Rydb. var. *microstachya* (Wieg.) Fern. & Grisc.

4c. **Echinochloa muricata** (Michx.) Fern. var. **wiegandii** (Fassett) Mohlenbr. Wild Millet. Aug.–Oct. Low ground; common throughout the state; in every co. *Echinochloa pungens* (Poir.) Rydb. var. *wiegandii* Fassett.

5. **Echinochloa walteri** (Pursh) Heller. Two forms occur in Illinois:

a. Sheaths (at least the lower) papillose-hirsute ....5a. *E. walteri* f. *walteri*
a. Sheaths glabrous ....5b. *E. walteri* f. *laevigata*

5a. **Echinochloa walteri** (Pursh) Heller f. **walteri**. Southern Wild Millet. Aug.–Oct. Low ground, swamps, rarely in standing water; occasional and scattered throughout IL.

5b. **Echinochloa walteri** (Pursh) Heller f. **laevigata** Wieg. Southern Wild Millet. Aug.–Oct. Swamps, very rare; Union Co.

41. **Eleusine** Gaertn.—Goose Grass

1. Panicle with 4 or more branches, the branches 3.0–3.5 mm wide .... 1. *E. indica*
1. Panicle with 1–3 branches, the branches 5–15 mm wide .... 2. *E. tristachya*

1. **Eleusine indica** (L.) Gaertn. Goose Grass; Water Grass. June–Oct. Native to Europe and Asia; adventive in disturbed areas and lawns, common; probably in every co.

2. **Eleusine tristachya** (Lam.) Lam. Three-spike Goose Grass. June–Sept. Native to tropical America; adventive in disturbed soil; Winnebago Co.

42. **Elyhordeum** Mansf.—Hybrid Barley

1. Upper spikelets paired, the lowermost solitary .... 1. *E. X macounii*
1. Spikelets all borne in 3s .... 2. *E. X montanense*

1. **Elyhordium X macounii** (Vasey) Backworth & D. R. Dewey. Macoun's Wild Rye. June–July. Native to w. N. America; adventive in disturbed soil; Cook Co. This is reputed to be the hybrid between *Elymus trachycaulus* (Link) Gould and *Hordeum jubatum* L. *Elymus macounii* Vasey; *Agrohordeum X macounii* (Vasey) Lepage.

2. **Elyhordeum X montanense** (Scribn.) Bowden. Hybrid Barley. June–July. Prairies and roadsides, not common; Marshall, Peoria, and Stark cos. This is reputed to be the hybrid between *Elymus virginicus* L. and *Hordeum jubatum* L. *Hordeum X montanense* Scribn.

43. **Elymus** L.—Wild Rye

1. Each spikelet with 2 or more perfect florets.
  2. At least some of the spikelets awned.
    3. Spikelets either all paired or all borne in 3s.
      4. Glumes well over 40 mm long; axis of inflorescence breaking apart at maturity .... 10. *E. longifolius*
      4. Glumes up to 40 mm long; axis of inflorescence not breaking apart at maturity.
        5. Glumes reduced to unequal, filiform bristles up to 15 mm long, or absent.
          6. Spikelets strongly ascending to erect; lemmas curved or bent, pubescent .... 3. *E. diversiglumis*
          6. Spikelets widely spreading; lemmas straight or nearly so, glabrous (except for var. *bigeloviana*) .... 8. *E. hystrix*
        5. Glumes subequal in length, not reduced to filiform bristles.
          7. Spikelets widely spreading, somewhat separate ..... 4. *E. X ebingeri*
          7. Spikelets ascending, crowded.
            8. Glumes soft throughout, 3- or 5-nerved throughout .... 6. *E. glaucus*
            8. Glumes firm, at least at base, 3- or 5-nerved only above the base.
              9. Base of glumes firm for only about 1 mm; paleas 8.5–12.0 (–14.0) mm long; awns usually strongly curved or twisted at maturity .... 1. *E. canadensis*
              9. Base of glumes firm for more than 1 mm; paleas up to 8.5 mm long; awns straight or occasionally slightly curved at maturity.
                10. Glumes 0.5–0.8 mm wide, hardened 1–3 mm adaxially.
                  11. Lemmas glabrous or hispidulous; paleas 7.0–8.5 mm long, the apices bidentate; sheaths glabrous; spikelets more or less glaucous .... 13. *E. riparius*
                  11. Lemmas villous-hirsute; paleas 5.5–6.5 mm long, the apices obtuse; sheaths villous-hirsute; spikelets not glaucous .... 15. *E. villosus*
                10. Glumes (0.8–) 0.9–2.3 mm wide, swollen on at least ½ or all of the adaxial surface.
                  12. Some or all the spikes enclosed at the base by the sheaths.

13. Spikes glaucous; lemmas hispid or hirsute .... 6. *E. hirsuticaulis*
13. Spikes not glaucous; lemmas glabrous or scabrous .... 16. *E. virginicus*
12. None of the spikes enclosed at the base by the sheaths.
14. Awn of glumes up to 3 mm long, or absent; awn of lemmas 1–3 mm long .... 2. *E. curvatus*
14. Awn of glumes 3–25 mm long; awn of lemmas 8–35 mm long.
15. Awn of glumes 3–10 mm long; awn of lemmas 8–20 mm long; auricles at base of leaves pale brown.
16. Lemmas hispid to hirsute .... 6. *E. hirsuticaulis*
16. Lemmas glabrous or scabrous ....9. *E. jejunus*
15. Awn of glumes 15–25 mm long; awn of lemmas 20–35 mm long; auricles at base of leaves purple-brown.
17. Auricles at base of leaves 2–3 mm long; plants flowering from mid-May to mid-June .... 11. *E. macgregorii*
17. Auricles at base of leaves 0–2 mm long; plants flowering from mid-June to mid-July .... 5. *E. glabriflorus*
3. Spikelet solitary.
18. Glumes 8–12 mm long .... 14. *E. trachycaulus*
18. Glumes 12–18 mm long .... 12. *E. pauciflorus*
2. Spikelets awnless throughout.
19. Glumes 8–12 mm long .... 14. *E. trachycaulus*
19. Glumes 12–18 mm long .... 12. *E. pauciflorus*
1. Each spikelet with a single perfect floret .... 15. *E. villosus*

1. **Elymus canadensis** L. Three interdrading varieties occur in Illinois:

a. Internodes 4–7 mm long, strongly glaucous; lemmas hirsute .... 1a. *E. canadensis* var. *canadensis*
a. Internodes 2–3 mm long, not strongly glaucous or not glaucous; lemmas glabrous or hirsute.
b. Lemmas glabrous or scabrous, the awn 20–30 mm long; spikes 6–20 mm long .... 1b. *E. canadensis* var. *brachystachys*
b. Lemmas glabrous or hirsute, the awn 30–40 mm long; spikes 15–30 cm long .... 1c. *E. canadensis* var. *robustus*

1a. **Elymus canadensis** L. var. **canadensis** Nodding Wild Rye. June–Sept. Woods, roadsides, dry prairies, common; in every co. (Including var. *crescendus* [Ramaley] Bush.)

1b. **Elymus canadensis** L. var. **brachystachys** (Scribn. & Ball) Farw. Short-spike Wild Rye. June–Sept. Woods, prairies; scattered in the state.

1c. **Elymus canadensis** L. var. **robustus** (Scribn. & J. G. Sm.) Mack. & Bush. Large Wild Rye. June–Sept. Woods, prairies; scattered in IL.

2. **Elymus curvatus** Piper. Short-awned Wild Rye. June–Aug. Wet fields, bottomland forests, disturbed soil; not common in the n. ½ of the state. *Elymus submuticus* (Hook.) Smythe; *E. virginicus* L. var. *submuticus* Hook.

3. **Elymus diversiglumis** Scribn. & C. R. Ball. Unequal-glumed Wild Rye. June–Sept. Rocky woods, very rare; Union Co.

4. **Elymus X ebingeri** G. C. Tucker. Ebinger's Wild Rye. June–Sept. Dry soil, not common. This is reputed to be the hybrid between *E. virginicus* L. and *E. hystrix* L.

5. **Elymus glabriflorus** (Vasey ex L. H. Dewey) Scribn. & Ball. Southeastern Wild

Rye. June–July. Moist or damp woods, roadsides, old fields; Jackson and Knox cos. *Elymus virginicus* L. var. *glabriflorus* Vasey ex L. H. Dewey.

6. **Elymus glaucus** Buckl. Blue Wild Rye. June–Oct. Dry woods, very rare; Union Co.

7. **Elymus hirsutiglumis** Scribn. Hairy-glumed Wild Rye. June–Aug. Fields, woods, low ground; scattered in IL. *Elymus virginicus* L. var. *intermedius* (Vasey ex Gray) Bush; *E. intermedius* Scribn., *nomen illeg.*

8. **Elymus hystrix** L. Two varieties are known from Illinois:

a. Lemmas glabrous .... 8a. *E. hystrix* var. *hystrix*
a. Lemmas pubescent .... 8b. *E. hystrix* var. *bigeloviana*

8a. **Elymus hystrix** L. var. **hystrix.** Bottlebrush Grass. June–Aug. Woods, common; in every co. *Hystrix patula* Moench.

8b. **Elymus hystrix** L. var. **bigeloviana** (Fern.) Bowden. Bottlebrush Grass. June–Aug. Woods; occasional throughout the state. *Elymus hystrix* L. f. *bigeloviana* (Fern.) Dore; *Hystrix patula* Moench var. *bigeloviana* (Fern.) Deam; *H. patula* Moench f. *bigeloviana* (Fern.) Gl.

9. **Elymus jejunus** (Ramaley) Rydb. Spring Wild Rye. May–June. Rocky woods, disturbed soil; scattered in IL. *Elymus virginicus* L. var. *jejunus* (Ramaley) Bush.

10. **Elymus longifolius** (J. G. Sm.) Gould. Squirrel-tail Grass. July–Sept. Native to Europe; adventive along a railroad; Mason Co. *Sitanion hystrix* (Nutt.) J. G. Sm.; *Sitanion longifolium* J. G. Sm.; *Elymus elymoides* (Raf.) Swezey.

11. **Elymus macgregorii** R. Brooks & J. J. P. Campb. Early Wild Rye. May–June. Deep woods; scattered in a few w. cos.

12. **Elymus pauciflorus** (Link) Gould ssp. **subsecundus** (Link) Gould. Bearded Wheat Grass. June–Aug. Woods, fields, rare; confined to a few n. cos.; also Piatt Co. *Agropyron trachycaulum* (Link) Malte var. *unilaterale* (Cassidy) Malte.

13. **Elymus riparius** Wieg. Riverbank Rye. July–Sept. Woods; scattered throughout the state.

14. **Elymus trachycaulus** (Link) Gould. Slender Wheat Grass. June–Aug. Fields, disturbed areas, along railroads, rare; known from a few n. cos. *Agropyron trachycaulum* (Link) Malte.

15. **Elymus villosus** Muhl. Two forms occur in Illinois:

a. Glumes and lemmas hispid to hirsute .... 15a. *E. villosus* f. *villosus*
a. Glumes scabrous; lemmas glabrous or scabrous .... 15b. *E. villosus* f. *arkansanus*

15a. **Elymus villosus** Muhl. f. **villosus.** Hairy Wild Rye. June–Sept. Woods, common; in every co.

15b. **Elymus villosus** Muhl. f. **arkansanus** (Scribn. & Ball) Fern. Hairy Wild Rye. June–Sept. Woods, rare; DuPage, Henry, Jackson, and Stark cos.

16. **Elymus virginicus** L. Virginia Wild Rye. June–Sept. Fields, woods, low ground; common throughout the state; in every co.

44. **Elytrigia** Desv.—Wheat Grass

1. At least some of the spikelets awned.
  2. Blades flat, 5–10 mm broad; sheaths pubescent .... 3. *E. repens*
  2. Blades involute, at least when dry, 1–5 mm broad; sheaths glabrous.
    3. Lemmas glabrous, scabrous, or pubescent only at base .... 4. *E. smithii*
    3. Lemmas short-pilose throughout .... 1. *E. dasystachya*
1. Spikelets awnless throughout.
  4. Glumes and lemmas obtuse or truncate at apex .... 2. *E. elongata*
  4. Glumes and sometimes the lemmas acute at apex.
    5. Blades flat, 5–10 mm broad; sheaths pubescent .... 3. *E. repens*
    5. Blades involute, at least when dry, 2–5 mm broad; sheaths glabrous.
      6. Lemmas glabrous, scabrous, or pubescent only at the base .... 4. *E. smithii*
      6. Lemmas short-pilose throughout .... 1. *E. dasystachya*

1. **Elytrigia dasystachya** (Hook.) A. & D. Love. Western Wheat Grass. June–Aug. Native to the w. U.S.; adventive along railroads; Cook, DuPage, and Will cos. *Agropy-*

ron smithii Rydb. var. *molle* (Scribn. & Smith) Jones; *A. dasystachyum* (Hook.) Scribn.; *Elymus lanceolatus* (Scribn. & J. G. Sm.) Gould.

2. **Elytrigia elongata** (Host) Nevski. Tall Wheat Grass. July–Aug. Native to Europe; adventive in disturbed saline soils; Cook and DuPage cos. *Triticum elongatum* Host; *Elytrigia pontica* (Podp.) Holub; *Agropyron elongatum* (Host) P. Beauv.; *Thinopyrum ponticum* (Podp.) Z. W. Lu & R. C. Wang.

3. **Elytrigia repens** (L.) Desv. Two forms occur in Illinois:

a. Lemmas awnless .... 3a. *E. repens* f. *repens*
a. Lemmas with an awn up to 10 mm long .... 3b. *E. repens* f. *aristata*

3a. **Elytrigia repens** (L.) Desv. f. **repens.** Quack Grass. June–July. Native to Europe and Asia; adventive in fields and waste ground; common in the n. ¾ of the state, rare in the s. ¼. *Agropyron repens* (L.) P. Beauv.; *Elymus repens* (L.) Gould.

3b. **Elytrigia repens** (L.) Desv. f. **aristata** (Schum.) Mohlenbr. Awned Quack Grass. June–July. Native to Europe and Asia; rarely adventive in waste ground; Cook Co. *Agropyron repens* (L.) P. Beauv. f. *aristatum* (Schum.) Holmb.

4. **Elytrigia smithii** (Rydb.) Nevski. Western Wheat Grass. June–Aug. Native to the w. U.S.; adventive along railroads; occasional in the n. ⅔ of the state, rare in the s. ⅓. *Agropyron smithii* Rydb.; *Pascopyrum smithii* (Rydb.) A. Love.

45. **Eragrostis** P. Beauv.—Love Grass

1. Plants forming mats, the culms rooting at the nodes.
2. Sheaths more or less glabrous; inflorescence 2–8 cm long, the peduncle glabrous; lemmas 1.5–2.0 mm long, acute, glabrous; anthers 0.2–0.5 mm long; plants monoecious .... 7. *E. hypnoides*
2. Sheaths pubescent; inflorescence 10–25 cm long, the peduncle villous; lemmas 2–4 mm long, acuminate, sparsely villous along the nerves; anthers 1.5–2.0 mm long; plants dioecious .... 13. *E. reptans*
1. Plants erect or ascending, not forming mats.
3. All lemmas 2.5–3.5 mm long; perennials .... 15. *E. trichodes*
3. Lemmas, or most of them, 1.0–2.5 mm long; annuals (except *E. spectabilis, E. curvula,* and *E. hirsuta*).
4. Leaf blades with small, rounded, wartlike projections (glands) along the margins; lemmas often glandular along the keel.
5. Spikelets 2.2–3.0 mm broad; lemmas 2.0–2.5 mm long, glandular along the keel; anthers 0.5 mm long .... 2. *E. cilianensis*
5. Spikelets 1.5–2.0 mm broad; lemmas 1.5–2.0 mm long, glandular or eglandular along the keel; anthers 0.2 mm long .... 10. *E. minor*
4. Leaf blades and lemmas without wartlike projections (glands).
6. Inflorescence very narrow, dense, the branches stiffly arching.
7. Spikelets gray-green, 8–10 mm long; blades to 3 (–4) mm broad, involute; perennials .... 3. *E. curvula*
7. Spikelets green, 2–3 mm long; blades 3–5 mm broad, flat; annuals .... 8. *E. japonica*
6. Inflorescence open, the branches ascending to spreading.
8. Perennials with short rhizomes or, if rhizomes absent, the remains of the previous year's leaves evident.
9. Short rhizomes present; spikelets purplish, 3–8 mm long, 7- to 12-flowered; lateral nerves of lemmas rather conspicuous .... 14. *E. spectabilis*
9. Rhizomes absent; spikelets usually not purplish, 2–4 mm long, 2- to 6-flowered; lateral nerves of lemmas faint .... 3. *E. curvula*
8. Tufted annuals (perennials in *E. hirsuta*), without rhizomes.
10. Lemmas 1.8–2.5 mm long; second glume 1.6–2.0 mm long.
11. Spikelets 8- to 12-flowered; pubescence of lower sheaths not papillose-based .... 9. *E. mexicana*
11. Spikelets 2- to 3-flowered; pubescence of lower sheaths papillose-based .... 6. *E. hirsuta*
10. Lemmas 1.0–1.8 mm long; second glume 0.5–1.6 mm long.

12. Lemmas conspicuously 3-nerved; spikelets 4–8 mm long.
  13. Panicle 15–25 (–30) cm long .... 11. *E. pectinacea*
  13. Panicle 30–50 cm long .... 4. *E. diffusa*
12. Lemmas obscurely 3-nerved; spikelets 1.5–4.0 mm long.
  14. Panicle at least ⅔ as long as the entire plant; first glume 1.0–1.5 mm long .... 1. *E. capillaris*
  14. Panicle not more than ½ as long as the entire plant; first glume 0.5–1.2 mm long.
    15. Panicle broadest near base; second glume ½–¾ as long as the lowest lemma .... 12. *E. pilosa*
    15. Panicle broadest near middle; second glume as long as the lowest lemma .... 5. *E. frankii*

1. **Eragrostis capillaris** (L.) Nees. Lace Grass. July–Sept. Dry, rocky soil, particularly in woodlands; occasional throughout the state.

2. **Eragrostis cilianensis** (All.) Vign. Stinking Love Grass. May–Oct. Native to Europe and Asia; introduced in fields and waste ground; in every co. *Eragrostis megastachya* (Koel.) Link.

3. **Eragrostis curvula** (Schrad.) Nees. Weeping Love Grass. Aug.–Oct. Native to Africa; introduced into waste ground; Morgan Co.

4. **Eragrostis diffusa** Buckl. Western Love Grass. July–Oct. Native to the w. U.S.; adventive along a road; Menard Co.

5. **Eragrostis frankii** C. A. Meyer. Two varieties occur in Illinois:

a. Spikelets 2- to 5-flowered, 2–3 mm long .... 5a. *E. frankii* var. *frankii*
a. Spikelets 6- to 7-flowered, 3–4 mm long .... 5b. *E. frankii* var. *brevipes*

5a. **Eragrostis frankii** C. A. Meyer var. **frankii.** Sandbar Love Grass. July–Oct. Sandy soil, occasional; scattered throughout the state.

5b. **Eragrostis frankii** C. A. Meyer var. **brevipes** Fassett. Sandbar Love Grass. July–Oct. Sandy soil, rare; Henderson Co.

6. **Eragrostis hirsuta** (Michx.) Nees. Bigtop Love Grass. Aug.–Sept. Dry soil, rare; Massac Co.

7. **Eragrostis hypnoides** (Lam.) BSP. Pony Grass. July–Oct. Wet ground, usually in sandy or muddy areas; occasional throughout the state.

8. **Eragrostis japonica** L. Pond Love Grass. Aug.–Oct. Disturbed soil along banks of Ohio River; Massac Co. *Eragrostis glomerata* (Walt.) L. H. Dewey.

9. **Eragrostis mexicana** (Hornem.) Link. Love Grass. July–Oct. Native to the sw. U.S.; adventive along a railroad; St. Clair Co. *Eragrostis neomexicana* Vasey.

10. **Eragrostis minor** Host. Love Grass. June–Oct. Native to Europe; adventive in disturbed habitats; probably in every co. *Eragrostis poaeoides* P. Beauv.

11. **Eragrostis pectinacea** (Michx.) Nees. Tufted Love Grass. July–Oct. Disturbed soil; in every co.

12. **Eragrostis pilosa** (L.) P. Beauv. India Love Grass. July–Sept. Native to Europe; adventive in waste ground; restricted to a few s. cos.; also Douglas and McDonough cos.

13. **Eragrostis reptans** (Michx.) Nees. Pony Grass. July–Sept. Sandy soil; scattered throughout the state. *Neeragrostis reptans* (Michx.) Nicora.

14. **Eragrostis spectabilis** (Pursh) Steud. Tumble-grass; Purple Love Grass. June–Oct. Sandy soil; in every co. (Including var. *sparsihirsuta* Farw.)

15. **Eragrostis trichodes** (Nutt.) Wood. Two varieties occur in Illinois:

a. Inflorescence purplish; spikelets 3- to 6-flowered, 4–7 mm long .... 15a. *E. trichodes* var. *trichodes*
a. Inflorescence yellowish; spikelets 8- to 15-flowered, 8–12 mm long .... 15b. *E. trichodes* var. *pilifera*

15a. **Eragrostis trichodes** (Nutt.) Wood var. **trichodes** Sand Thread Love Grass. July–Oct. Open, sandy areas; scattered in IL.

15b. **Eragrostis trichodes** (Nutt.) Wood var. **pilifera** (Scheele) Fern. Sand Thread Love Grass. July–Oct. Sand prairies, not common.

46. **Erianthus** Michx.—Plume Grass

1. Awn of fertile lemma up to 6 mm long .... 4. *E. ravennae*
1. Awn of fertile lemma 10 mm long or longer.
   2. Flowering stems glabrous below the inflorescence.
      3. Leaves 10–25 mm broad; glumes 5–6 mm long .... 2. *E. brevibarbis*
      3. Leaves 1–10 (–12) mm broad; glumes 7–10 mm long .... 5. *E. strictus*
   2. Flowering stems sericeous below the inflorescence.
      4. Awn of fertile lemma spirally twisted; inflorescence silvery white .... 1. *E. alopecuroides*
      4. Awn of fertile lemma straight or slightly curved but not spirally twisted; inflorescence light brown to even purplish .... 3. *E. giganteus*

1. **Erianthus alopecuroides** (L.) Ell. Plume Grass. Sept.–Oct. Open woods; occasional in the s. tip of the state. *Saccharum alopecuroides* (L.) Nutt.

2. **Erianthus brevibarbis** Michx. Brown Plume Grass. Aug.–Oct. Dry hills, very rare; sw. IL. *Saccharum brevibarbis* (Michx.) Nutt.

3. **Erianthus giganteus** (Walt.) P. Beauv. Sugarcane Plume Grass. Sept.–Oct. Wet roadside ditch, rare; Massac Co. *Saccharum giganteum* (Walt.) Pers.

4. **Erianthus ravennae** (L.) P. Beauv. Ravenna Grass. Aug.–Oct. Native to Europe; escaped along roads; scattered in IL. *Saccharum ravennae* (L.) P. Beauv.

5. **Erianthus strictus** Baldwin. Narrow Plume Grass. July–Oct. Wet roadside ditch, rare; Alexander Co. *Saccharum strictum* (Baldwin) Nutt.; *S. baldwinii* Spreng.

47. **Eriochloa** HBK.—Cup Grass

1. Pedicels and rachis villous; spikelets about 5 mm long .... 3. *E. villosa*
1. Pedicels and rachis short-pilose; spikelets 3.5–5.0 mm long.
   2. Grain 2.0–2.5 mm long, with an awn to 1 mm long; blades pubescent, 3–7 mm broad .... 2. *E. contracta*
   2. Grain 3 mm long, apiculate; blades glabrous, 5–10 mm broad ..... 1. *E. acuminata*

1. **Eriochloa acuminata** (J. Presl) Kunth. Cup Grass. July–Oct. Native to the w. U.S.; adventive along a levee; Union Co. *Eriochloa gracilis* (Fourn.) Hitchc.; *E. lemmonii* Vasey & Scribn. var. *gracilis* (Fourn.) Gould.

2. **Eriochloa contracta** A. Hitchc. Prairie Cup Grass. June–Sept. Native to the w. U.S.; moist soil; scattered in the s. ½ of IL and the ne. cos.

3. **Eriochloa villosa** (Thunb.) Kunth. Cup Grass. Aug. Native to Asia; adventive in waste ground; scattered in the n. ½ of IL; also Fayette Co.

48. **Festuca** L.—Fescue

1. Leaf blades involute or plicate, 0.4–1.2 mm broad.
   2. First glume 1–2 mm long; second glume 1.8–3.0 mm long; lemmas 2.5–3.5 mm long; awns absent, or up to 0.5 mm long .... 2. *F. filiformis*
   2. First glume 2.5–4.5 mm long; second glume 3.5–5.5 mm long; lemmas 4–7 mm long; awns 1.0–3.5 mm long.
      3. Lowest sheaths whitish, not becoming fibrous; lemmas essentially nerveless .... 7. *F. trachyphylla*
      3. Lowest sheaths brown or reddish, becoming fibrous; lemmas 3- to 5-nerved .... 5. *F. rubra*
1. Leaf blades flat, 3–11 mm broad.
   4. Lemmas 5.5–10.0 mm long.
      5. Spikelets 6- to 11-flowered; lemmas 5.5–8.0 mm long .... 4. *F. pratensis*
      5. Spikelets 4- to 5-flowered; lemmas 7–10 mm long .... 1. *F. arundinacea*
   4. Lemmas 3.3–5.2 mm long.
      6. Inflorescence spreading at maturity; spikelets to 4 mm broad; lemmas acute or subacute .... 6. *F. subverticillata*
      6. Inflorescence ascending at maturity; spikelets about 5 mm broad; lemmas obtuse .... 3. *F. paradoxa*

1. **Festuca arundinacea** Schreb. Tall Fescue. May–Aug. Native to Europe; adventive in waste ground; scattered in IL. *Festuca elatior* L. var. *arundinacea* (Schreb.) Wimmer; *Schedonorus arundinacea* (Schreb.) Dumort.

2. **Festuca filiformis** Pourret. Slender Fescue. May–July. Native to Europe; adventive in disturbed soil; scattered in IL, usually around metropolitan areas. *Festuca capillata* Lam.; *F. tenuifolia* Sibth.

3. **Festuca paradoxa** Desv. Woodland Fescue. May–July. Dry or moist woods; occasional in the s. ¾ of the state, absent elsewhere except for Iroquois, Jo Daviess, and Will cos.

4. **Festuca pratensis** Huds. Meadow Fescue. May–Aug. Native to Europe; adventive in disturbed soil, common; in every co. *Schedonorus pratensis* (Huds.) Holub.

5. **Festuca rubra** L. Red Fescue. June–July. Disturbed soil; infrequent throughout the state.

6. **Festuca subverticillata** (Pers.) E. B. Alexeev. Nodding Fescue. May–July. Moist woods, occasional; scattered throughout the state. *Festuca obtusa* Biehler.

7. **Festuca trachyphylla** (Hack.) Krajina. Sheep Fescue. May–June. Native to Europe; adventive in disturbed areas, not common; scattered throughout the state. *Festuca duriuscula* L.; *F. ovina* L. ssp. *trachyphylla* Hack.; *F. ovina* L. var. *duriuscula* (L.) Koch.

49. **Glyceria** R. Br.—Manna Grass

1. Spikelets at least 10 mm long; sheaths compressed.
  2. Some or all of the spikelets on pedicels 3 mm long or longer; culms usually 1.5–5.0 mm thick; lemmas usually scabrous.
    3. Culms 1.5–2.5 mm thick; leaves 4–8 mm wide; lemmas 3.5–6.0 mm long; upper glume 2.5–5.0 mm long .... 4. *G. declinata*
    3. Culms 3–5 mm thick; leaves 2–5 mm wide; lemmas 3–4 mm long; upper glume 2–3 mm long .... 2. *G. borealis*
  2. Spikelets on pedicels up to 3 mm long, or pedicels absent; culms usually 5–12 mm thick; lemmas hirtellous.
    4. Leaves 4–12 mm wide; lemmas obscurely nerved, 3.5–5.5 mm long; upper glume 3.2–5.0 mm long; anthers 1–2 mm long .... 6. *G. septentrionalis*
    4. Leaves 10–18 mm broad; lemmas sharply nerved, 2.5–3.0 mm long; anthers less than 1 mm long .... 1. *G. arkansana*
1. Spikelets 2–8 mm long; sheaths terete or subterete.
  5. Lemmas 3–4 mm long, obscurely nerved; spikelets 3–4 mm broad .... 3. *G. canadensis*
  5. Lemmas 1.5–2.8 mm long, sharply nerved; spikelets 2.0–2.5 mm broad.
    6. Inflorescence 5–20 cm long; spikelets 2.0–4.5 mm long; first glume 0.5–1.0 mm long; second glume 0.8–1.3 mm long; lemmas 1.5–2.0 mm long .... 7. *G. striata*
    6. Inflorescence 20–40 cm long; spikelets 5–6 mm long; first glume 1.2–2.0 mm long; second glume 1.5–2.5 mm long; lemmas 2.0–2.7 mm long .... 5. *G. grandis*

1. **Glyceria arkansana** Fern. Southern Manna Grass. May–June. Shallow water, swamps, low woods, very rare; Jackson and Union cos. *Glyceria septentrionalis* Hitchc. var. *arkansana* (Fern.) Steyerm. & Kucera.

2. **Glyceria borealis** (Nash) Batchelder. Northern Manna Grass. June–July. Shallow water, rare; Cook, Jo Daviess, Lake, and Stephenson cos.

3. **Glyceria canadensis** (Michx.) Trin. Rattlesnake Manna Grass. June–Sept. Wet ground, very rare; Cook, Peoria, and Tazewell cos.

4. **Glyceria declinata** Breb. Low Glyceria. June–July. Native to Europe; adventive in wet disturbed soil; DuPage Co.

5. **Glyceria grandis** S. Wats. American Manna Grass. June–Aug. Wet ground, not common; restricted to the extreme n. cos.

6. **Glyceria septentrionalis** Hitchc. Manna Grass. May–Aug. Shallow water, marshy soil, swampy meadows, occasional; scattered throughout the state.

7. **Glyceria striata** (Lam.) Hitchc. Two varieties occur in Illinois:

a. Spikelets green; uppermost branches of the panicle more or less nodding; lemmas with a minutely scarious apex .... 7a. *G. striata* var. *striata*
a. Spikelets purple; uppermost branches of the panicle ascending; lemmas with a broadly scarious apex .... 7b. *G. striata* var. *stricta*

7a. **Glyceria striata** (Lam.) Hitchc. var. **striata.** Fowl Manna Grass. May–Aug. Moist soil, common; in every co.

7b. **Glyceria striata** (Lam.) Hitchc. var. **stricta** (Scribn.) Fern. Fowl Manna Grass. May–Aug. Wet ground; occasional in the n. ⅓ of the state, absent elsewhere.

50. **Gymnopogon** P. Beauv.—Beard Grass

1. **Gymnopogon ambiguus** (Michx.) BSP. Beard Grass. July–Oct. Sandy or gravelly soil in open areas, very rare; Pope Co.

51. **Heterostipa** (Elias) Barkworth—Needle Grass

1. Glumes 15–28 mm long; lemma 9–13 mm long, the flexuous but obscurely geniculate awn 10–15 cm long; ligule of upper leaves 3–4 mm long .... 1. *H. comata*
1. Glumes 28–42 mm long; lemma 16–25 mm long, the twice geniculate awn 12–20 cm long; ligule of upper leaves 4–6 mm long .... 2. *H. spartea*

1. **Heterostipa comata** (Trin. & Rupr.) Barkworth. Needle Grass. July–Aug. Dry or loamy soil, usually in prairies, rare; Cook, Kane, Lake, and Winnebago cos. *Stipa comata* Trin. & Rupr.

2. **Heterostipa spartea** (Trin.) Barkworth. Porcupine Grass. May–June. Sandy soil, particularly in prairies; occasional in the n. ⅔ of IL, nearly absent in the s. ⅓. *Stipa spartea* Trin.

52. **Hierochloe** R. Br.—Sweet Grass

1. **Hierochloe odorata** (L.) P. Beauv. Sweet Grass. May–June. Meadows; occasional in the n. ¼ of the state; also St. Clair Co. *Anthoxanthum hirtum* (Schrank) Schouten & Veldkamp.

53. **Holcus** L.—Velvet Grass

1. **Holcus lanatus** L. Velvet Grass. June–July. Native to Europe; occasionally escaped into waste ground; widely scattered in IL.

54. **Hopia** Zuolaga & Morrone.

1. **Hopia obtusa** (Kunth) Zuloaga & Morrone. Vine Mesquite. July. Native to the sw. U.S.; adventive in disturbed soil; Pike Co. *Panicum obtusum* Kunth.

55. **Hordeum** L.—Barley

1. Each spikelet with 2 or more perfect florets .... 5. *H. vulgare*
1. Each spikelet with a single perfect floret.
  2. Most or all of the awns 3 cm long or longer .... 2. *H. jubatum*
  2. None of the awns 3 cm long.
    3. Spikes 1.5 cm broad or broader; stems branching from the base .... 3. *H. marinum*
    3. Spikes less than 1.5 cm broad; stems unbranched from the base.
      4. Annual; 4 glumes of each group of 3 spikelets setiform, the other 2 glumes dilated at base; awn of lemma of central spikelet 8–15 mm long .... 4. *H. pusillum*
      4. Perennial; all glumes setiform; awn of lemma of central spikelet 5–8 mm long .... 1. *H. brachyantherum*

1. **Hordeum brachyantherum** Nevski. Meadow Barley. May–July. Native to the w. U.S.; infrequently adventive in IL; Cook, DuPage, and Jackson cos.

2. **Hordeum jubatum** L. Squirrel-tail Grass. June–Sept. Fields, roadsides; common in the n. ⅔ of IL, occasional elsewhere.

3. **Hordeum marinum** Huds. Mediterranean Barley. June–July. Native to Europe; rarely adventive in disturbed soil; DuPage Co. *Hordeum geniculatum* All.; *H. marinum* Huds. ssp. *gussoneanum* (Parl.) Thell.

4. **Hordum pusillum** Nutt. Little Barley. May–July. Fields, disturbed soil; common in the s. ¾ of the state, rare in the n. ¼, where it may be introduced.

5. **Hordeum vulgare** L. Two varieties occur in Illinois:

a. Lemmas awned .... 5a. *H. vulgare* var. *vulgare*
a. Lemmas 3-lobed at apex, awnless .... 5b. *H. vulgare* var. *trifurcatum*

5a. **Hordeum vulgare** L. var. **vulgare.** Common Barley. May–July. Native to Asia; occasionally escaped along roads; throughout the state.

5b. **Hordeum vulgare** L. var. **trifurcatum** (Schindl.) Alefeld. Pearl Barley. May–July. Native to Asia; occasionally escaped along roads; scattered in IL.

56. **Koeleria** Pers.—June Grass

1. **Koeleria macrantha** (Ledeb.) Spreng. June Grass. Prairies, sandy black oak woods; occasional throughout the state. *Koeleria cristata* (L.) Pers.; *K. pyramidata* (Lam.) P. Beauv.

57. **Leersia** Sw.—Cut Grass

1. Spikelets broadly rounded, 3–4 mm broad, over ½ as wide as broad .... 1. *L. lenticularis*
1. Spikelets oblongoid, 1–2 mm broad, less than ½ as wide as long.
  2. Sheaths conspicuously retrorse-scabrous; blades spinulose on the margins; lowest panicle branches whorled; stamens 3 .... 2. *L. oryzoides*
  2. Sheaths glabrous or scaberulous; blades scaberulous; lowest panicle branch solitary; stamens 1–2 .... 3. *L. virginica*

1. **Leersia lenticularis** Michx. Catchfly Grass. Aug.–Oct. Low woods, swamps, marshes; occasional throughout the state except for the extreme n. cos.

2. **Leersia oryzoides** (L.) Swartz. Rice Cut Grass. Aug.–Oct. Low, moist soil; occasional to common throughout the state.

3. **Leersia virginica** Willd. White Grass. July–Sept. Moist woods; common throughout the state. (Including var. *ovata* [Poir.] Fern.)

58. **Leptochloa** P. Beauv.—Sprangletop

1. Spikelets 2- to 4-flowered; lemmas 0.7–1.5 mm long; sheaths papillose-pilose; ligules 1–2 mm long.
  2. Lemmas 1.0–1.5 mm long; grain 0.7–0.9 mm long; glumes acute .... 4. *L. filiformis*
  2. Lemmas 0.7–1.0 mm long; grain 0.4–0.5 mm long; glumes aristate .... 2. *L. attenuata*
1. Spikelets 5- to 10-flowered; lemmas 2–8 mm long; sheaths glabrous or nearly so; ligules 3–7 mm long.
  3. Blades becoming involute, 1–5 mm broad.
    4. Lemmas 5–8 mm long, purplish to lead-colored .... 1. *L. acuminata*
    4. Lemmas up to 5 mm long, pale green.
      5. Lemma with an awn 0.5–1.0 mm long .... 3. *L. fascicularis*
      5. Lemma mucronate, without an awn .... 6. *L. uninervia*
  3. Blades flat, 5–10 mm broad .... 5. *L. panicoides*

1. **Leptochloa acuminata** (Nash) Mohlenbr. Salt Meadow Grass. July–Sept. Native to the w. U.S.; adventive along railroads and in disturbed saline habitats in the n. ¼ of IL. *Leptochloa fascicularis* (Lam.) Gray var. *acuminata* (Nash) Gl.; *Diplachne acuminata* Nash.

2. **Leptochloa attenuata** (Nutt.) Steud. Sprangletop. July–Sept. Sandy shores; occasional in the s. ½ of the state. *Leptochloa panicea* (Retz.) Ohwi ssp. *mucronata* (Michx.) Nowack.

3. **Leptochloa fascicularis** (Lam.) Gray. Salt Meadow Grass. July–Sept. Sandy soil; scattered in the s. ⅗ of the state; also Cook, Iroquois, Kane, and LaSalle cos. *Diplachne fascicularis* (Lam.) P. Beauv.; *Leptochloa fusca* L. Kunth ssp. *fascicularis* (Lam.) N. Snow.

4. **Leptochloa filiformis** (Lam.) P. Beauv. Red Sprangletop. July–Sept. Low, sandy soil, particularly along rivers; occasional in the s. ⅖ of the state. *Leptochloa panicea* (Retz.) Ohwi ssp. *brachiana* (Steud.) N. Snow.

5. **Leptochloa panicoides** (Presl) Hitchc. Salt Meadow Grass. July–Sept. Low areas, rare; Alexander, Calhoun, and Pike cos. *Diplachne panicoides* Presl; *D. halei* Nash.

6. **Leptochloa uninervia** (Presl) Hitchc. & Chase. One-nerved Sprangletop. July–Sept. Disturbed soil, rare; McDonough Co. *Leptochloa fusca* (L.) Kunth ssp. *uninervia* (Presl) N. Snow.

59. **Leptoloma** Chase—Fall Witch Grass

1. **Leptoloma cognatum** (Schult.) Chase. Fall Witch Grass. July–Sept. Sandy soil; occasional throughout the state. *Digitaria cognata* (Schultes) Pilger.

60. **Leymus** Hochst.—Beach Rye

1. **Leymus arenarius** L. Beach Rye. July–Oct. Sandy shores of Lake Michigan, rare; Cook, Kane, and Lake cos. *Elymus arenarius* L.; *E. arenarius* L. var. *villosus* Mey.; *E. mollis* Trin.

61. **Lolium** L.—Rye Grass

1. Glume as long as or longer than the spikelet, 15–20 mm long....... 3. *L. temulentum*
1. Glume shorter than the spikelet, 4–12 mm long.
  2. Lemmas, or at least the uppermost, awned; spikelets 10- to 20-flowered; annual .... 1. *L. multiflorum*
  2. Lemmas awnless; spikelets 6- to 10-flowered; perennial .... 2. *L. perenne*

1. **Lolium multiflorum** Lam. Italian Rye Grass. May–Sept. Native to Europe; adventive in waste ground, fields; occasional throughout the state. *Lolium perenne* L. var. *multiflorum* (Lam.) Parn.; *L. perenne* L. var. *aristatum* Willd.
2. **Lolium perenne** L. English Rye Grass. June–Sept. Native to Europe; adventive in waste ground, fields, lawns, common; in every co.
3. **Lolium temulentum** L. Darnel. June–July. Native to Europe; adventive in disturbed soil, fields, not common; scattered in IL.

62. **Melica** L.—Melic Grass

1. Cauline leaves 3–4, 2–5 mm broad; sheaths scabrous; glumes nearly equal in length; first glume oblong, at least twice as long as broad; fertile lemmas usually 2 .... 1. *M. mutica*
1. Cauline leaves 5–8, 5–12 mm broad; sheaths glabrous; glumes unequal in length; first glume ovate, less than twice as long as broad; fertile lemmas usually 3 .... 2. *M. nitens*

1. **Melica mutica** Walt. Two-flowered Melic Grass. May–June. Flat woods, rare; scattered throughout the state.
2. **Melica nitens** (Scribn.) Nutt. Three-flowered Melic Grass. May–July. Rocky woods, prairies; occasional throughout the state, but uncommon in the n. cos.

63. **Microstegium** Nees—Natal Grass

1. **Microstegium vimineum** (Trin.) A. Camus. Natal Grass. Sept.–Oct. Native to Asia; adventive in disturbed soil; occasional in s. IL; also Will Co. *Eulalia viminea* (Trin.) Kuntze.

64. **Milium** L.—Millet Grass

1. **Milium effusum** L. Millet Grass. June–Aug. Moist woods, rare; Cook, Kane, and Tazewell cos.

65. **Miscanthus** Anderss.—Plume Grass

1. Fertile lemma awnless; blades 10–18 mm broad.......................... 1. *M. sacchariflorus*
1. Fertile lemma awned; blades up to 10 mm broad .... 2. *M. sinensis*

1. **Miscanthus sacchariflorus** (Maxim.) Hack. Plume Grass. Aug.–Oct. Native to Asia; adventive in disturbed soil; scattered in the n. ⅓ of IL.
2. **Miscanthus sinensis** Anderss. Eulalia. Sept.–Oct. Native to Asia; escaped along roads; scattered in IL.

66. **Muhlenbergia** Schreb.—Satin Grass; Muhly Grass

1. Panicle diffuse, open, at least 5 cm across; spikelets on pedicels longer than the lemmas.

2. Rhizomes present; panicle 5–20 cm long; spikelets about 1.5 mm long; lemma glabrous, 1.3–1.7 mm long, awnless.. 1. *M. asperifolia*

2. Rhizomes lacking; panicle 20–45 cm long; spikelets (excluding awns) 3.0–4.5 mm long; lemma scabrous, 3.0–4.5 mm long, with an awn 5–20 mm long.... ....3. *M. capillaris*

1. Panicle narrower, contracted to less than 2 cm thick; spikelets on pedicels shorter than or as long as the lemmas.

3. Plants densely tufted from firm bases or decumbent and rooting at the lower nodes, without rhizomes.

4. Lower portion of culms decumbent, rooting at the nodes; blades (1–) 2–4 mm broad, flat; glumes 0.1–0.5 (–1.7) mm long; lemma awned.

5. Glumes obtuse, the first 0.1–0.2 mm long, the second 0.1–0.3 mm long; awn of lemma 1.5–4.0 mm long.. 11. *M. schreberi*

5. Glumes acute to aristate, the first 0.5 mm long, the second 1.0–1.5 (–1.7) mm long; awn of lemma 0.5–1.5 mm long ....4. *M. X curtisetosa*

4. Culms stiffly erect, tufted from firm bases; blades 1–2 mm broad, flat or involute; glumes 1.7–2.8 mm long; lemma acuminate, awnless......................... .... 5. *M. cuspidata*

3. Plants rhizomatous, not rooting at the lower nodes.

6. Internodes glabrous or minutely scabrous but not puberulent or pilose (rarely pubescent at summit of internodes in *M. racemosa*).

7. Panicle to 4 mm broad; glumes ovate-lanceolate, 1.3–2.0 (–2.5) mm long.

8. Body of lemma 1.7–2.3 mm long, awnless or with an awn 1–2 (–4) mm long...12. *M. sobolifera*

8. Body of lemma 2.5–3.3 mm long, with an awn 2–7 mm long .... 2. *M. bushii*

7. Panicle 5–15 mm broad; glumes lance-subulate, 1.6–8.0 mm long.

9. Spikelets 2–4 mm long; glumes 2–3 mm long.................6. *M. frondosa*

9. Spikelets 4–8 mm long; glumes 4–8 mm long .... 10. *M. racemosa*

6. Internodes puberulent or pilose.

10. Lemma glabrous at base.. 7. *M. glabrifloris*

10. Lemma pilose at base.

11. Glumes (including awns) (3.2–) 4.5–8.0 mm long......... 8. *M. glomerata*

11. Glumes 1.5–3.5 mm long.

12. Glumes ovate-lanceolate; anthers 1.0–1.5 mm long; mature grain 2.0–2.3 mm long...14. *M. tenuiflora*

12. Glumes linear-lanceolate; anthers 0.3–0.8 mm long; mature grain 1.3–1.8 mm long.

13. Glumes silvery or whitish; ligules 1.0–2.5 mm long; anthers 0.5–0.8 mm long.. 13. *M. sylvatica*

13. Glumes green or purplish; ligules 0.5–1.0 mm long; anthers 0.3–0.5 mm long.. 9. *M. mexicana*

1. **Muhlenbergia asperifolia** (Nees & Meyen) Parodi. Scratch Grass. June–Sept. Sandy soil; scattered in the n. ½ of the state; also St. Clair Co.

2. **Muhlenbergia bushii** Pohl. Nodding Muhly. Aug.–Oct. Low woods, not common; confined to the s. ⅔ of IL, except for Kankakee and LaSalle cos.

3. **Muhlenbergia capillaris** (Lam.) Trin. Hair-awn Muhly. Sept.–Oct. Sandy woods; occasional in the s. ½ of the state.

4. **Muhlenbergia X curtisetosa** (Scribn.) Bush. Hybrid Muhly. July–Oct. Woods, rare; Champaign, Fulton, Jackson, and Peoria cos. This is reputed to be the hybrid between *M. frondosa* (Poir.) Fern. and *M. schreberi* J. F. Gmel.

5. **Muhlenbergia cuspidata** (Torr.) Rydb. Plains Muhly. Aug.–Oct. Gravelly soil, not common; scattered in IL.

6. **Muhlenbergia frondosa** (Poir.) Fern. Two forms occur in Illinois:

a. Lemma awnless.. 6a. *M. frondosa* f. *frondosa*

a. Lemma with an awn 4–11 mm long.............................6b. *M. frondosa* f. *commutata*

6a. **Muhlenbergia frondosa** (Poir.) Fern. f. **frondosa**. Satin Grass. Aug.–Sept. Moist woods, roadsides, fields, common; throughout the state.

6b. **Muhlenbergia frondosa** (Poir.) Fern. f. **commutata** (Scribn.) Fern. Satin Grass. Aug.–Sept. Moist woods; occasional throughout the state.

7. **Muhlenbergia glabrifloris** Scribn. Inland Muhly. July–Oct. Moist woods; occasional in the cent. and s. cos.; also Will Co. *Muhlenbergia glabriflora* Scribn.

8. **Muhlenbergia glomerata** (Willd.) Trin. Spike Muhly. Aug.–Sept. Dry or wet ground, rare; scattered in the n. ¼ of IL; also St. Clair Co.

9. **Muhlenbergia mexicana** (L.) Trin. Two varieties occur in Illinois:

a. Lemma awnless .... 9a. *M. mexicana* var. *mexicana*
a. Lemma with an awn to 9 mm long .... 9b. *M. mexicana* var. *filiformis*

9a. **Muhlenbergia mexicana** (L.) Trin. var. **mexicana.** Wirestem Muhly. Aug.–Oct. Moist soil, usually in woods; occasional throughout the state.

9b. **Muhlenbergia mexicana** (L.) Trin. var. **filiformis** (Torr.) Vasey. Wirestem Muhly. Aug.–Oct. Moist soil, usually in woods, very rare; Lake Co. *Muhlenbergia mexicana* (L.) Trin. f. *ambigua* (Torr.) Fern.

10. **Muhlenbergia racemosa** (Michx.) BSP. Marsh Muhly. Aug.–Sept. Moist soil; occasional throughout the state.

11. **Muhlenbergia schreberi** J. F. Gmel. Nimble Will. July–Oct. Disturbed soil, woods; common throughout the state.

12. **Muhlenbergia sobolifera** (Muhl. ex Willd.) Trin. Two forms occur in Illinois:

a. Lemma awnless .... 12a. *M. sobolifera* f. *sobolifera*
a. Lemma with an awn up to 3 mm long .... 12b. *M. sobolifera* f. *setigera*

12a. **Muhlenbergia sobolifera** (Muhl. ex Willd.) Trin. f. **sobolifera.** Rock Muhly. July–Oct. Dry rocky woods; occasional throughout the state, but rare in the ne. cos.

12b. **Muhlenbergia sobolifera** (Muhl. ex Willd.) Trin. f. **setigera** (Scribn.) Deam. Rock Muhly. July–Oct. Dry woods, rare; Wabash Co.

13. **Muhlenbergia sylvatica** (Torr.) Torr. Woodland Muhly. Aug.–Oct. Rich, often rocky woods; occasional throughout the state.

14. **Muhlenbergia tenuiflora** (Willd.) BSP. Slender Satin Grass. July–Oct. Rocky woods, moist bluffs; occasional throughout the state.

67. **Nassella** E. Desv.—Feather Grass

1. Collar of leaf sheath glabrous; leaves 0.2–0.6 mm wide; panicle 8–25 cm long; glumes aristate; florets 1.5–2.5 mm long, 0.7–0.9 mm wide; caryopsis 1.2 mm long .... 1. *N. trichotoma*
1. Collar of leaf sheath hispidulous; leaves 3–7 mm wide; panicle 3–7 mm long; glumes apiculate; florets 3.35–5.50 mm long, 1.0–1.2 mm wide; caryopsis 3.5 mm long .... 2. *N. viridula*

1. **Nassella trichotoma** (Nees) Hack. ex Arechev. Serrated Tussockgrass. June–July. Native to S. America; disturbed soil; mostly in the cent. cos.; noxious weed, apparently extirpated from IL on purpose.

2. **Nassella viridula** (Trin.) Barkworth. Feather Grass. June–July. Edge of woods and along railroad track near pond, rare; Kane and McHenry cos. *Stipa viridula* Trin.

68. **Oryza** L.—Rice

1. **Oryza sativa** L. Rice. Aug. Native in warm regions of the world; rarely escaped from cultivation; Union Co.

69. **Oryzopsis** Michx.—Rice Grass

1. Blades flat, 5–15 mm broad; spikelets (excluding the awn) 6–9 mm long; glumes acute to acuminate, 7–9 mm long, conspicuously 7-nerved; lemma 5.5–8.5 mm long, with the awn 5–25 mm long.
  2. Upper leaves longer than lower leaves; lemma dark brown to blackish, the awn 12–25 mm long .... 3. *O. racemosa*
  2. Upper leaves shorter than lower leaves; lemma pale green to yellowish, the awn 5–10 mm long .... 1. *O. asperifolia*
1. Blades involute, 1–2 mm broad; spikelets (excluding the awn) 3–4 mm long; glumes obtuse, 3.5–4.0 mm long, obscurely 5-nerved; lemma 3.5–4.0 mm long, with the awn 1–2 mm long .... 2. *O. pungens*

1. **Oryzopsis asperifolia** Michx. Rice Grass. Apr.–July. Dry woods, very rare; Cook and Lake cos.

2. **Oryzopsis pungens** (Torr.) Hitchc. Rice Grass. Apr.–June. Dry soil, very rare; Menard Co. *Piptatherum pungens* (Torr.) Barkworth.

3. **Oryzopsis racemosa** (J. E. Smith) Ricker. Rice Grass. July–Sept. Rich, rocky woods, rare; confined to the n. ½ of IL. *Piptatherum racemosum* (J. E. Smith) Barkworth.

70. **Panicum** L.—Panic Grass

(See also Coleataenia, Dichanthelium, and Hopia.)

1. First glume up to ¼ as long as the spikelet .... 2. *P. dichotomiflorum*
1. First glume at least ⅓ as long as the spikelet.
  2. Spikelets up to 2.9 mm long.
    3. Some or all the sheaths papillose-pilose; spikelets not warty.
      4. Panicle at least ½ the length of the plant; spikelets acuminate .... 1. *P. capillare*
      4. Panicle up to ⅓ the length of the entire plant; spikelets acute.
        5. Fruits stramineous; blades 6–10 mm broad; spikelets 0.9–1.0 mm broad; culms stout, to 100 cm tall .... 4. *P. gattingeri*
        5. Fruits nigrescent; blades 2–8 mm broad; spikelets 0.7 mm broad; culms slender, to 50 cm tall .... 6. *P. philadelphicum*
    3. None of the sheaths papillose-pilose; spikelets warty .... 7. *P. verrucosum*
  2. Spikelets 3 mm long or longer.
    6. First glume at least ½ as long as the spikelet; grain 2.4–3.0 mm long.
      7. Sheaths papillose-hispid; panicle more or less nodding..... 5. *P. miliaceum*
      7. Sheaths ciliate or villous at the throat but not papillose-hispid; panicle ascending or spreading .... 8. *P. virgatum*
    6. First glume up to ½ as long as the spikelet; grain 1.7–2.3 mm long.
      8. Axis of panicle glabrous; grain 2 mm long .... 3. *P. flexile*
      8. Axis of panicle short-pilose; grain 1.7–1.8 mm long .... 1. *P. capillare*

1. **Panicum capillare** L. Two varieties occur in Illinois:

a. Lowest branches of panicle included at the base; spikelets mostly 2.0–2.5 mm long; grain about 1.5 mm long .... 1a. *P. capillare* var. *capillare*
a. Panicle long-exserted; spikelets mostly 2.5–4.0 mm long; grain 1.6–1.7 mm long .... 1b. *P. capillare* var. *occidentale*

1a. **Panicum capillare** L. var. **capillare.** Witch Grass. July–Oct. Fields, waste ground; common throughout the state. (Including var. *agreste* Gatt.)

1b. **Panicum capillare** L. var. **occidentale** Rydb. Witch Grass. July–Oct. Open ground, rare; Cook, Edgar, Henderson, Lake, Pike, and Whiteside cos.

2. **Panicum dichotomiflorum** Michx. Three varieties occur in Illinois:

a. Spikelets 2.4–3.5 mm long; at least some of the leaves over 5 mm broad.
  b. Culms mostly upright, not geniculate, the nodes not swollen; sheaths not inflated .... 2a. *P. dichotomiflorum* var. *dichotomiflorum*
  b. Culms mostly spreading, geniculate, some of the nodes swollen; sheaths inflated .... 2b. *P. dichotomiflorum* var. *geniculatum*
a. Spikelets 1.7–2.3 mm long; blades up to 5 (–8) mm broad .... 2c. *P. dichotomiflorum* var. *puritanorum*

2a. **Panicum dichotomiflorum** Michx. var. **dichotomiflorum.** Fall Panicum. June–Oct. Fields, waste ground, damp soil; common throughout the state.

2b. **Panicum dichotomiflorum** Michx. var. **geniculatum** (Muhl.) Fern. Fall Panicum. June–Oct. Low, disturbed soil, fields; scattered throughout the state.

2c. **Panicum dichotomiflorum** Michx. var. **puritanorum** Svenson. Fall Panicum. June–Sept. Along streams, very rare; Cook Co.

3. **Panicum flexile** (Gatt.) Scribn. Slender Panicum. July–Oct. Moist sandy soil; occasional throughout the state.

4. **Panicum gattingeri** Nash. Gattinger's Panicum. July–Oct. Dry or moist open ground, disturbed soil; scattered throughout the state; apparently absent from the

n. 3 tiers of cos. *Panicum capillare* L. var. *campestre* Gatt.; *Panicum philadelphicum* Trin. ssp. *gattingeri* (Nash) Freckman & Lelong.

5. **Panicum miliaceum** L. Broomcorn Millet. June–Oct. Native to Europe and Asia; escaped from cultivation into disturbed ground; scattered in IL. (Including var. *ruderale* Fern.)

6. **Panicum philadelphicum** Trin. Panicum. July–Oct. Dry, usually sandy soil; locally scattered in IL.

7. **Panicum verrucosum** Muhl. Warty Panicum. June–Oct. Low roadside ditch, very rare; Alexander Co.

8. **Panicum virgatum** L. Switch Grass. June–Oct. Fields, prairies, rocky stream beds, woods, disturbed soil; common throughout the state.

71. **Paspalum** L.—Bead Grass

1. Rachis foliaceous, the margins folded over and clasping the spikelets at their bases.
- **2.** Racemes of each inflorescence 1–5; rachis shorter than the rows of spikelets .... 3. *P. dissectum*
- **2.** Racemes of each inflorescence 5–50, usually more than 10; rachis longer than the rows of spikelets .... 5. *P. fluitans*

1. Rachis firm, narrow or broad, the margins not folded over the rows of spikelets.
- **3.** Spikelet solitary .... 6. *P. laeve*
- **3.** Spikelets borne in pairs along the rachis.
  - **4.** Spikelets surrounded by soft white hairs often longer than the spikelets.
    - **5.** Racemes up to 8 per plant .... 2. *P. dilatatum*
    - **5.** Racemes 9–15 per plant .... 10. *P. urvillei*
  - **4.** Spikelets not surrounded by soft white hairs.
    - **6.** All spikelets 3 mm long or longer.
      - **7.** Spikelets 3.5 mm long or longer; foliage distinctly bluish .... 4. *P. floridanum*
      - **7.** Spikelets 3.0–3.4 mm long; foliage green.
        - **8.** Some or all the spikelets arranged in 4 rows .... 8. *P. pubiflorum*
        - **8.** All spikelets arranged in 2 rows .... 7. *P. praecox*
    - **6.** Some or all the spikelets less than 3 mm long.
      - **9.** Spikelets pubescent .... 1. *P. bushii*
      - **9.** Spikelets glabrous.
        - **10.** Spikelets up to 2.5 mm long; sterile lemma 3-nerved .... 9. *P. setaceum*
        - **10.** Spikelets 2.5 mm long or longer; sterile lemma 5-nerved .... 7. *P. praecox*

1. **Paspalum bushii** Nash. Hairy Bead Grass. June–Sept. Fields, edge of woods, bluffs; infrequent but scattered throughout the state.

2. **Paspalum dilatatum** Poir. Dallis Grass. July–Aug. Native to the s. U.S.; adventive in disturbed soil; Jackson Co.

3. **Paspalum dissectum** (L.) L. Mudbank Paspalum. June–Sept. Moist soil, edges of shallow swamps, very rare; Perry, Pulaski, St. Clair, and Williamson cos.

4. **Paspalum floridanum** Michx. Giant Bead Grass; Blue Bead Grass. June–Sept. Low, moist, sandy soil; occasional in the s. ⅓ of the state.

5. **Paspalum fluitans** (Ell.) Kunth. Swamp Bead Grass. June–Sept. Floating in shallow standing water, low ground in swampy woods; occasional in the s. ⅔ of the state.

6. **Paspalum laeve** Michx. Two varieties occur in Illinois:

a. Spikelets up to 2.5 mm broad, a little longer than broad .... 6a. *P. laeve* var. *laeve*
a. Spikelets at least 2.8 mm broad, about as broad as long .... 6b. *P. laeve* var. *circulare*

6a. **Paspalum laeve** Michx. var. **laeve** Field Bead Grass. June–Sept. Moist soil of roadside ditches, meadows, and stream borders; common in the s. ½ of the state, rare elsewhere.

6b. **Paspalum laeve** Michx. var. **circulare** (Nash) Stone. June–Sept. Moist soil; occasional throughout IL. *Paspalum circulare* Nash; *P. tenue* Darby.

7. **Paspalum praecox** Walt. Bead Grass. June–Sept. Wet, roadside ditch, rare; Pulaski Co. *Paspalus lentiferum* Lam.

8. **Paspalum pubiflorum** Rupr. var. **glabrum** (Vasey) Vasey. Bead Grass. June–Sept. Moist soil in ditches, along roads, and along streams; scattered in the s. ½ of the state, rare elsewhere.

9. **Paspalum setaceum** Michx. Four varieties occur in Illinois:

a. Blades glabrous except for a few long hairs at the base .... 9b. *P. setaceum* var. *ciliatifolium*
a. Blades pubescent throughout.
  b. Spikelets 1.4–1.9 mm long, 1.0–1.5 mm broad........9a. *P. setaceum* var. *setaceum*
  b. Spikelets 1.8–2.4 mm long, 1.5–2.4 mm broad.
    c. Midvein of lemma of sterile floret present; hairs on the blades all the same length..........9c. *P. setaceum* var. *muhlenbergii*
    c. Midvein of lemma of sterile floret absent; hairs on the blades of 2 different lengths.......... 9d. *P. setaceum* var. *stramineum*

9a. **Paspalum setaceum** Michx. var. **setaceum.** Slender Bead Grass. June–Sept. Moist or dry soils, rare; confined to a few s. cos.

9b. **Paspalum setaceum** Michx. var. **ciliatifolium** (Michx.) Vasey. Slender Bead Grass. June–Sept. Moist or dry soils, occasionally in woods; common to occasional throughout the state. *Paspalum ciliatifolium* Michx.

9c. **Paspalum setaceum** Michx. var. **muhlenbergii** (Nash) D. J. Banks. Slender Bead Grass. June–Sept. Moist or dry soil; occasional throughout the state. *Paspalum muhlenbergii* Nash; *P. pubescens* Muhl. var. *muhlenbergii* (Nash) House; *P. ciliatifolium* Michx. var. *muhlenbergii* (Nash) Fern.

9d. **Paspalum setaceum** Michx. var. **stramineum** (Nash) D. J. Banks. Slender Bead Grass. June–Sept. Usually sandy soil; occasional in the s. ⅔ of IL. *Paspalum stramineum* Nash; *P. ciliatifolium* Michx. var. *stramineum* (Nash) Fern.

10. **Paspalum urvillei** Steud. Vasey Grass. June–Aug. Native to S. America; adventive in an old field; Jackson Co.

72. **Pennisetum** Rich.—Fountain Grass

1. Perennial from short, stout rhizomes; culms strongly compressed, densely villous throughout; inflorescence soft, to 15 cm long; bristles of spikelets up to 20 mm long; fertile lemma chartaceous.......... 1. *P. alopecuroides*
1. Annual; culms terete, with densely hairy nodes; inflorescence stiff, to 75 cm long; bristles of spikelets up to 6 mm long except for one much longer bristle; fertile lemma coriaceous.......... 2. *P. glaucum*

1. **Pennisetum alopecuroides** (L.) Spreng. Fountain Grass. Aug.–Oct. Native to Asia; rarely escaped from cultivation; Crawford Co.

2. **Pennisetum glaucum** (L.) R. Br. Pearl Millet; Italian Millet. Sept.–Oct. Native to Europe and Asia; rarely adventive along a roadside; Williamson Co. *Pennisetum americanum* (L.) Leeke.

73. **Phalaris** L.—Canary Grass

1. Keel of glumes wingless, the glumes 4.5–6.5 mm long; sterile florets 1–2 mm long; perennial from creeping rhizomes.......... 1. *P. arundinacea*
1. Keel of glumes broadly winged, the glumes 7–10 mm long; sterile florets 2.5–4.5 mm long; annual.......... 2. *P. canariensis*

1. **Phalaris arundinacea** L. Reed Canary Grass. May–July. Meadows, moist areas; occasional to common throughout the state. (Including f. *picta* [L.] Asch. & Graebn.)

2. **Phalaris canariensis** L. Canary Grass. June–July. Native to Europe; adventive in disturbed ground; occasional throughout the state.

74. **Phleum** L.—Timothy

1. **Phleum pratense** L. Timothy. June–Aug. Native to Europe; adventive in disturbed soil, fields, very common; in every co.

75. **Phragmites** Trin.—Reed

1. **Phragmites australis** (Cav.) Trin. Two subspecies occur in Illinois:

a. Culms green .... 1a. *P. australis* ssp. *australis*
a. Culms reddish .... 1b. *P. australis* ssp. *americana*

1a. **Phragmites australis** (Cav.) Trin. ssp. **australis** Reed. July–Sept. Native to the Old World; disturbed areas; adventive in IL and very invasive; throughout the state.

1b. **Phragmites australis** (Cav.) Trin. ssp. **americana** Saltonstall, P. M. Peterson, & Soreng. American Reed. July–Sept. Usually wet areas; scattered in IL.

76. **Poa** L.—Blue Grass

1. Plants dioecious; pistillate spikelets woolly; staminate spikelets glabrous or nearly so .... 4. *P. arachnifera*
1. Plants monoecious; spikelets perfect, variously pubescent or glabrous.
  2. Some spikelets transformed into bulblets .... 7. *P. bulbosa*
  2. Spikelets all normal.
    3. Lemmas without a tuft of cobwebby hairs at base.
      4. Tufted annual to about 30 cm tall, sometimes rooting at the lower nodes; lemmas elliptic to ovate .... 3. *P. annua*
      4. Tufted perennial to 75 cm tall, not rooting at the lower nodes; lemmas oblong .... 6. *P. autumnalis*
    3. Lemmas with a tuft of cobwebby hairs at base.
      5. Nerves and keel of the lemmas glabrous (except for the cobwebby tuft) .... 11. *P. languida*
      5. Keel and sometimes the nerves of the lemmas pubescent.
        6. Keel of the lemma pubescent, the nerves glabrous.
          7. Culms beneath the panicle and the sheaths scabrous; lemmas sharply nerved; ligule of upper leaves 4–8 mm long .... 17. *P. trivialis*
          7. Culms beneath the panicle and the sheaths usually glabrous; lemmas obscurely nerved; ligule of upper leaves about 1 mm long .... 1. *P. alsodes*
        6. Keel and at least some of the nerves of the lemma pubescent.
          8. Marginal nerves of the lemmas pubescent, the intermediate nerves glabrous.
            9. Plants with rhizomes; lemmas with 5 prominent nerves.
              10. Lemmas at least 3.6 mm long .... 5. *P. arida*
              10. Lemmas up to 3.5 mm long.
                11. Basal leaves flat, at least as broad as the culm; culm compressed at base, 2–3 mm thick at base; glumes broadly lanceolate, straight .... 15. *P. pratensis*
                11. Basal leaves involute or filiform, narrower than the culm; culm terete at base, 1–2 mm thick at base; glumes narrowly lanceolate, arching .... 2. *P. angustifolia*
            9. Plants without rhizomes; lemmas with 3 prominent nerves and 2 obscure nerves.
              12. Culms very weak, solitary or in small tufts; sheaths scabrous; lowest branches of the panicle mostly paired .... 13. *P. paludigena*
              12. Culms more firm, usually densely tufted; sheaths usually glabrous; lowest branches of the panicle in clusters of 3–5.
                13. Ligule 0.5–1.0 mm long; anthers 1.2–1.6 mm long .... 12. *P. nemoralis*
                13. Ligule 2–5 mm long; anthers up to 1 mm long .... 14. *P. palustris*
          8. All nerves of the lemmas pubescent.
            14. First glume 2.5–3.5 mm long; lemmas 3.5–4.5 mm long; blades 1–2 mm broad .... 18. *P. wolfii*

14. First glume 1.5–2.5 (–3.0) mm long; lemmas 1.5–3.5 mm long; blades 1–5 mm broad.
15. Tufted perennial; inflorescence often 10 cm long or longer.
16. Lemmas 3-nerved .... 10. *P. interior*
16. Lemmas 5-nerved .... 16. *P. sylvestris*
15. Tufted annual or rhizomatous perennial; inflorescence usually less than 10 cm long.
17. Tufted annual to 30 cm tall; culms terete; anthers 0.1–0.2 mm long... 8. *P. chapmaniana*
17. Rhizomatous perennial to 70 cm tall; culms compressed; anthers about 1 mm long .... 9. *P. compressa*

1. **Poa alsodes** Gray. Woodland Blue Grass. May–June. Moist woods, rare; Jackson, Lake, Pope, and St. Clair cos.

2. **Poa angustifolia** L. Narrow-leaved Blue Grass. June–Sept. Woods and clearings, rare; Jackson, Pope, Union, and Williamson cos.

3. **Poa annua** L. Annual Blue Grass. May–Oct. Native to Europe and Asia; mostly in moist waste ground, common; in every co.

4. **Poa arachnifera** Torr. Texas Blue Grass. June–Aug. Native to the sw. U.S.; disturbed soil, very rare; Winnebago Co.

5. **Poa arida** Vasey. Plains Blue Grass. May. Native to the w. U.S.; adventive along roadsides; Cook, DuPage, Grundy, Kane, Kendall, Washington, and Will cos.

6. **Poa autumnalis** Muhl. Blue Grass. Mar.–June. Moist woods, very rare; Jackson and Pope cos.

7. **Poa bulbosa** L. Bulbous Blue Grass. May–July. Native to Europe; adventive in disturbed soil; scattered in IL.

8. **Poa chapmaniana** Scribn. Chapman's Blue Grass. Apr.–Sept. Fields and waste ground; occasional in the s. ⅔ of the state, rare elsewhere.

9. **Poa compressa** L. Canada Blue Grass. May–Aug. Native to Europe and Asia; dry soil, common; in every co.

10. **Poa interior** Rydb. Inland Blue Grass. May–June. Along a sandstone ledge, very rare; Randolph Co. Specimen at Missouri Botanical Garden.

11. **Poa languida** Hitchc. Woodland Blue Grass. June–Sept. Moist woods, rare; Cook, Jo Daviess, Lake, and LaSalle cos. *Poa saltuensis* Fern. & Wieg. var. *languida* (Hitchc.) A. Haynes.

12. **Poa nemoralis** L. Woodland Blue Grass. June–Aug. Native to Europe; adventive in open woods and disturbed soil, rare; n. ⅔ of IL.

13. **Poa paludigena** Fern. & Wieg. Marsh Blue Grass. June–July. Bogs, rare; Kane Co.

14. **Poa palustris** L. Fowl Blue Grass. June–Sept. Wet soil; occasional in the n. ½ of the state, rare in the s. ½.

15. **Poa pratensis** L. Kentucky Blue Grass. Apr.–Sept. Native to Europe and Asia; adventive in fields, woods, disturbed soil, very common; in every co.

16. **Poa sylvestris** Gray. Woodland Blue Grass. May–July. Moist woods; occasional throughout the state.

17. **Poa trivialis** L. Meadow Blue Grass. May–Aug. Native to Europe; adventive in disturbed soil, rare; Cook, DuPage, Jo Daviess, McHenry, Stark, and Will cos.

18. **Poa wolfii** Scribn. Meadow Blue Grass. Apr.–June. Meadows and woods, rare; confined to the n. ½ of IL; also Brown Co.

77. **Puccinellia** Parl.—Alkali Grass

1. **Puccinellia distans** (Jacq.) Parl. Alkali Grass. June–Oct. Native to Europe; adventive in disturbed, often saline, soil; occasional in the n. ¼ of the state.

78. **Redfieldia** Vasey—Blowout Grass

1. **Redfieldia flexuosa** (Thurb.) Vasey. Blowout Grass. July–Oct. Native to the w. U.S.; adventive in disturbed soil; Hancock Co.

79. **Rottboellia** L. f.—Itchgrass

1. **Rottboellia cochinchinensis** (Lour.) Clayton. Itchgrass. July–Oct. Native to Asia; adventive along a road; Jersey Co.

80. **Schedonnardus** Steud.—Tumble Grass

1. **Schedonnardus paniculatus** (Nutt.) Trel. Tumble Grass. July–Sept. Salt licks, roadsides, very rare; Hancock and Saline cos.

81. **Schizachne Hack.**—False Melic Grass

1. **Schizachne purpurascens** (Torr.) Swallen. False Melic Grass. June. Moist wooded slope, very rare; Jo Daviess Co.

82. **Schizachyrium** Nees—Little Bluestem

1. Plants branching from the lower nodes; collar of leaf elongate and narrow; sand and sand dunes .... 1. *S. littorale*
1. Plants not branching from the lower nodes; collar of leaf neither elongate nor narrower; plants not of sand dunes .... 2. *S. scoparium*

1. **Schizachyrium littorale** (Nash) E. P. Bickn. Seaside Bluestem. Aug.–Oct. Sandy soil, very rare; Lake Co. *Schizachyrium scoparium* (Michx.) Nash var. *littorale* E. P. Bickn.
2. **Schizachyrium scoparium** (Michx.) Nash. Little Bluestem. Aug.–Oct. Prairies, fields, open woods; occasional to common throughout the state. *Andropogon scoparius* Michx.

83. **Sclerochloa** P. Beauv.—Hard Grass

1. **Sclerochloa dura** (L.) P. Beauv. Hard Grass. May–July. Native to Europe; adventive in disturbed soil; scattered in IL.

84. **Secale** L.—Rye

1. **Secale cereale** L. Rye. May–July. Native to Europe and Asia; occasionally adventive in disturbed soil; scattered in IL.

85. **Setaria** P. Beauv.—Foxtail Grass

1. Perennial from short, knotty rhizomes; second glume 1.0–1.5 mm long .... 3. *S. parviflora*
1. Annuals from fibrous roots; second glume 1.8–2.7 mm long.
  2. Spike with retrorsely scabrous bristles .... 5. *S. verticillata*
  2. Spike with antrorsely scabrous bristles.
    3. Each spikelet subtended by 5–20 bristles; blades more or less twisted .... 4. *S. pumila*
    3. Each spikelet subtended by 1–3 bristles; blades usually not twisted.
      4. Spike strongly arching; blades strigose above, puberulent below .... 1. *S. faberi*
      4. Spike erect or scarcely arching; blades glabrous (sometimes ciliate).
        5. Spike usually not green, often lobed or interrupted; first glume short-acuminate, 1.2–1.5 mm long; fertile lemma smooth .... 2. *S. italica*
        5. Spike usually green, unlobed (except in var. *major*); first glume acute, 0.7–1.0 mm long; fertile lemma rugose or rugulose .... 6. *S. viridis*

1. **Setaria faberi** F. Herrm. Giant Foxtail. July–Oct. Native to Asia; adventive in disturbed soil, very common; in every co.
2. **Setaria italica** (L.) P. Beauv. Italian Millet. July–Sept. Native to Europe and Asia; adventive in disturbed soil; scattered in IL.
3. **Setaria parviflora** (Poir.) Kerg. Perennial Foxtail. July–Sept. Fields and roadsides, not common; scattered in IL. *Setaria geniculata* (Lam.) P. Beauv.
4. **Setaria pumila** (Poir.) Roem. & Schultes. Yellow Foxtail. June–Sept. Disturbed habitats, very common; in every co. *Setaria glauca* (L.) P. Beauv.; *Setaria lutescens* (Weigel) Hubb. Schwegman has collected a very robust specimen in Massac Co., which has culms over 1 m tall and spikes about 20 cm long. Although this plant may be ssp. *pallidefusca* (Schumach.) B. K. Simon, this subspecies is said to have reddish bristles, while Schwegman's specimen has yellow bristles. Further study is needed for this interesting discovery.
5. **Setaria verticillata** (L.) P. Beauv. Bristly Foxtail. July–Sept. Native to Europe and Asia; adventive in waste ground; occasional throughout the state.

6. **Setaria viridis** (L.) P. Beauv. Two varieties occur in Illinois:

a. Culms to 75 cm tall; blades to 12 mm broad; panicle unlobed .... 6a. *S. viridis* var. *viridis*
a. Culms usually at least 1.5 m tall; blades to 25 mm broad; panicle appearing lobed near base .... 6b. *S. viridis* var. *major*

6a. **Setaria viridis** (L.) P. Beauv. var. **viridis.** Green Foxtail. June–Sept. Native to Europe and Asia; adventive in waste ground, very common; in every co.

6b. **Setaria viridis** (L.) P. Beauv. var. **major** (Gaudin) Pospichal. Giant Green Foxtail. June–Sept. Native to Europe; adventive in cultivated fields; occasional in IL.

86. **Sorghastrum** Nash—Indian Grass

1. **Sorghastrum nutans** (L.) Nash. Indian Grass. Aug.–Oct. Prairies, fields, open woods; common throughout the state; in every co.

87. **Sorghum** Moench—Sorghum

1. Perennials from rhizomes; awn of fertile lemma twisted near base.
  2. Culms up to 2 m tall; blades 10–20 mm broad; awn of fertile lemma 10–15 mm long .... 3. *S. halepense*
  2. Culms more than 2 m tall; some or all of the blades 25–50 mm broad; awn of fertile lemma usually less than 10 mm long .... 1. *S. almum*
1. Annuals; awn of fertile lemma usually not twisted near base.
  3. Sessile spikelet 4.5–5.5 mm long, shorter than the pedicellate spikelet; awn of fertile lemma falling away early .... 2. *S. bicolor*
  3. Sessile spikelet 6–7 mm long, as long as the pedicellate spikelet; awn of fertile lemma persistent .... 4. *S. sudanense*

1. **Sorghum almum** L. Sorghum Grass. July–Oct. Native to Argentina; adventive along roads and edges of fields; Jackson, Pulaski, and Williamson cos.

2. **Sorghum bicolor** (L.) Moench. Sorghum. Aug.–Oct. Native to Asia; occasionally escaped from cultivation; occasional throughout the state. *Sorghum vulgare* Pers. Several cultivated varieties occasionally are adventive, including var. *drummondii* (Nees) Mohlenbr., var. *caffrorum* (Retz.) Mohlenbr., and var. *saccharatum* (L.) Mohlenbr.

3. **Sorghum halepense** (L.) Pers. Johnson Grass. June–Oct. Native to Europe and Asia; adventive in disturbed soil, very common; in every co.

4. **Sorghum sudanense** (Piper) Stapf. Sudan Grass. Sept.–Oct. Native to Africa; adventive along a highway; Jackson Co.

88. **Spartina** Schreb.—Cord Grass

1. **Spartina pectinata** Link. Cord Grass; Slough Grass. June–Sept. Wet prairies, marshes; occasional throughout the state although more common in the n. cos.

89. **Sphenopholis** Scribn.—Wedge Grass

1. First glume narrowly oblong, at least 0.5 mm broad; second lemma scabrous; anthers 1.0–1.5 mm long .... 2. *S. nitida*
1. First glume subulate, less than 0.5 mm broad; second lemma smooth to scabrous; anthers less than 1 mm long.
  2. Panicle dense and spikelike; second glume firm, rounded or truncate at the apex .... 3. *S. obtusata*
  2. Panicle loose; second glume scarious, acute to apiculate at the apex .... 1. *S. intermedia*

1. **Sphenopholis intermedia** (Rydb.) Rydb. Wedge Grass. May–July. Moist woods, moist prairies, occasional; throughout the state. *Sphenopholis obtusata* (Michx.) Scribn. var. *major* (Torr.) Erdman.

2. **Sphenopholis nitida** (Biehler) Scribn. Shiny Wedge Grass. May–July. Dry woods, prairies, not common; scattered throughout IL.

3. **Sphenopholis obtusata** (Michx.) Scribn. Wedge Grass. May–July. Woods, prairies; occasional throughout the state.

90. **Sporobolus** R. Br.—Dropseed

1. Panicle more or less open, spreading, over 10 cm long (except in most specimens of *S. pyramidatus*); perennials.
  2. Spikelets 3.5–6.0 mm long; glumes subulate, acuminate .... 4. *S. heterolepis*
  2. Spikelets 1.5–2.8 mm long; glumes narrowly lanceolate, acute.
    3. Sheaths densely villous, at least near the summit; panicles more than 10 cm long .... 3. *S. cryptandrus*
    3. Sheaths glabrous or sometimes merely ciliate; panicles up to 10 (–15) cm long .... 8. *S. pyramidatus*
1. Panicle contracted, spikelike, less than 10 cm long (except in *S. compositus*); perennials or annuals.
  4. First and second glumes considerably unequal in length.
    5. Spikelets 4 mm long or longer.
      6. Lemma glabrous; blades pilose on upper surface near base .... 2. *S. compositus*
      6. Lemma sparsely villous; blades scabrous on margin and tip but not pilose .... 1. *S. clandestinus*
    5. Spikelets up to 3 mm long.
      7. Second glume at least 1.5 mm long, lanceolate; spikelets never black .... 3. *S. cryptandrus*
      7. Second glume up to 1.3 mm long, elliptic-ovate; spikelets often blackish .... 5. *S. indicus*
  4. First and second glumes nearly the same length.
    8. Spikelets 2–3 mm long; first glume 1.5–2.5 mm long; second glume 1.7–2.7 mm long; lemma 2–3 mm long .... 6. *S. neglectus*
    8. Spikelets 3.5–6.5 mm long; first glume 2.8–4.0 mm long; second glume 3.0–4.5 mm long; lemma 3–5 mm long.
      9. Lower sheaths papillose-pilose; lemma glabrous .... 7. *S. ozarkanus*
      9. Lower sheaths long-ciliate, not papillose-pilose; lemma appressed-pubescent .... 9. *S. vaginiflorus*

1. **Sporobolus clandestinus** (Biehler) Hitchc. Hidden Dropseed. Aug.–Oct. Sandy soil, not common; scattered in IL.

2. **Sporobolus compositus** (Poir.) Merr. Dropseed. Sept.–Oct. Dry, often sandy soil; occasional throughout the state. *Sporobolus asper* (Michx.) Kunth.

3. **Sporobolus cryptandrus** (Torr.) Gray. Sand Dropseed. Aug.–Oct. Sandy soil; occasional in the n. ⅓ of the state and in the w. cos., rare elsewhere.

4. **Sporobolus heterolepis** (Gray) Gray. Prairie Dropseed. Aug.–Sept. Dry soil, often in prairies; occasional in the n. ½ of the state, rare in the s. ½.

5. **Sporobolus indicus** (L.) R. Br. Smut Grass. July–Oct. Native to Asia; rarely adventive along roadsides; Alexander Co.

6. **Sporobolus neglectus** Nash. Sheathed Dropseed. Sept.–Oct. Dry soil; occasional throughout the state.

7. **Sporobolus ozarkanus** Fern. Ozark Dropseed. Sept.–Oct. Dry soil, very rare; Clay Co. *Sprorobolus vaginiflorus* (Torr.) A. Wood var. *ozarkana* (Fern.) Shinners.

8. **Sporobolus pyramidatus** (Lam.) Hitchc. Whorled Dropseed. Aug.–Oct. Native to the w. U.S.; adventive in disturbed soil; Alexander, Jackson, and Williamson cos.

9. **Sporobolus vaginiflorus** (Torr.) A. Wood. Poverty Grass. Sept.–Oct. Dry soil; occasional throughout the state. (Including var. *inaequalis* Fern.)

91. **Torreyochloa** Church—Swamp Manna Grass

1. **Torreyochloa pallida** (Torr.) Church. Swamp Manna Grass. May–Aug. Swamps, very rare; Jackson, Montgomery, and Union cos. *Puccinellia pallida* (Torr.) Clausen.

92. **Trichachne** Nees—Sour Grass

1. **Trichachne insularis** (L.) Nees. Sour Grass. July–Sept. Native to the se. U.S.; adventive in a roadside ditch; Williamson Co. *Digitaria insularis* (L.) Fedde.

93. **Tridens** Roem. & Schultes—Purple-top

1. Panicle loose, open; glumes oblong to ovate; lemmas 3.5–4.0 mm long .... 2. *T. flavus*

1. Panicle contracted, spikelike; glumes linear-lanceolate; lemmas 2.5–3.0 mm long.
  2. Leaves strongly folded and involute; veins of lemmas glabrous, or pubescent only at base .... 1. *T. albescens*
  2. Leaves flat or slightly folded; veins of lemmas pubescent to above the middle .... 3. *T. strictus*

1. **Tridens albescens** (Vasey) Wooton & Standl. White Tridens. June–July. Native to the w. U.S.; adventive near a soccer field, Williamson Co. *Triodia albescens* Vasey.

2. **Tridens flavus** (L.) Hitchc. Two forms occur in Illinois:

a. Spikelets yellow at maturity .... 2a. *T. flavus* f. *flavus*
a. Spikelets purplish at maturity .... 2b. *T. flavus* f. *cupreus*

2a. **Tridens flavus** (L.) Hitchc. f. **flavus.** Purple-top. June–Sept. Fields, roadsides, edge of woods; occasional throughout the state. *Triodia flava* (L.) Smyth.

2b. **Tridens flavus** (L.) Hitchc. f. **cupreus** (Jacq.) Fosb. Purple-top. June–Sept. Fields, roadsides, edge of woods, very common; in every co.

3. **Tridens strictus** (Nutt.) Nash. Spicate Purple-top. July–Oct. Disturbed soil, fields, not common; scattered throughout the state. *Triodia stricta* (Nutt.) Benth.

94. **Triplasis** P. Beauv.—Sand Grass

1. **Triplasis purpurea** (Walt.) Chapm. Sand Grass. Aug.–Oct. Sandy soil; occasional in the n. ¾ of the state, absent in the s. ¼.

95. **Tripsacum** L.—Gama Grass

1. **Tripsacum dactyloides** (L.) L. Gama Grass. May–Sept. Low ground; occasional in the s. ⅔ of the state; also DuPage Co.

96. **Triticum** L.—Wheat

1. **Triticum aestivum** L. Wheat. June–Aug. Native to the Old World; sporadic in fields and along roadsides. Several cultivated varieties may be found, including Bearded Wheat, with long awns on the lemmas.

97. **Urochloa** P. Beauv.—Signal Grass

1. Spikelet solitary .... 1. *U. platyphylla*
1. Spikelets in pairs .... 2. *U. ramosa*

1. **Urochloa platyphylla** (Nash) R. D. Webster. Broadleaf Signal Grass. Aug.–Nov. Native to the se. U.S. and the w. Indies; adventive along roadsides; Alexander Co.

2. **Urochloa ramosa** (L.) T. Q. Nguyen. Browntop Millet. July–Aug. Native to Africa and Asia; adventive in disturbed soil; Jackson Co.

98. **Vulpia** K. C. Gmel.—Annual Fescue

1. Awns of lemmas, if present, up to 5.5 mm long .... 4. *V. octoflora*
1. Some of the awns of the lemmas 1 cm long or longer.
  2. Second glume at least twice as long as the first glume; first glume less than 3 mm long .... 3. *V. myuros*
  2. Second glume not twice as long as the first glume; first glume more than 3 mm long.
    3. Lemmas glabrous or scabrous .... 1. *V. bromoides*
    3. Lemmas pubescent .... 2. *V. elliotea*

1. **Vulpia bromoides** (L.) S. F. Gray. Bromelike Fescue. June–July. Native to Europe; adventive in disturbed soil, rare; Massac and Pope cos. *Festuca bromoides* L.

2. **Vulpia elliotea** (Raf.) Fern. Sand Fescue. Apr.–May. Dry sandy prairie, very rare; Alexander Co. *Vulpia sciurea* (Nutt.) Henrard; *Festuca sciurea* Nutt.

3. **Vulpia myuros** (L.) K. C. Gmel. Foxtail Fescue. June–July. Native to Europe; adventive in waste ground, rare; Jackson, Johnson, Massac, Perry, and Williamson cos. *Festuca myuros* L.

4. **Vulpia octoflora** (Walt.) Rydb. Three varieties occur in Illinois:

a. Inflorescence appearing racemose; awns 3.5–5.5 mm long .... 4a. *V. octoflora* var. *octoflora*

a. Inflorescence loosely or densely spicate; awns up to 3 mm long or absent.
  b. Inflorescence densely spicate; lower glume 1.5–3.0 mm long; awns absent or up to 2 mm long .... 4b. *V. octoflora* var. *glauca*
  b. Inflorescence loosely spicate; lower glume 2.5–4.0 mm long; awns 1–3 mm long .... 4c. *V. octoflora* var. *tenella*

4a. **Vulpia octoflora** (Walt.) Rydb. var. **octoflora.** Six-weeks Fescue. May–July. Dry soil, often on bluff tops; scattered throughout the state. *Festuca octoflora* Walt.

4b. **Vulpia octoflora** (Walt.) Rydb. var. **glauca** (Nutt.) Fern. Six-weeks Fescue. May–July. Dry soil, not common; scattered in IL.

4c. **Vulpia octoflora** (Walt.) Rydb. var. **tenella** (Willd.) Fern. Six-weeks Fescue. May–July. Dry soil; occasional throughout the state.

99. **Zea** L.—Corn

1. **Zea mays** L. Corn. June–Sept. Occasionally escaped from cultivation along roads but seldom persistent.

100. **Zizania** L.—Wild Rice

1. Pistillate lemma scabrous, slenderly nerved; aborted spikelets up to 1 mm broad, subulate, gradually tapering into the awn .... 1. *Z. aquatica*
1. Pistillate lemma glabrous, broadly nerved; aborted spikelets 1.5–2.0 mm broad, linear, abruptly tapering into the awn .... 2. *Z. interior*

1. **Zizania aquatica** L. Wild Rice. June–Sept. Shallow water, not common; scattered in IL, except for the s. cos. *Zizania palustris* L.

2. **Zizania interior** (Fassett) Rydb. Wild Rice. June–Sept. Shallow water, very rare; Cook, Lake, and Union cos. *Zizania aquatica* L. var. *interior* Fassett; *Z. palustris* L. var. *interior* (Fassett) Dore.

101. **Zizaniopsis** Doell. & Aschers.

1. **Zizaniopsis miliacea** (Michx.) Doell. & Aschers. Southern Wild Rice. July–Oct. Edge of lake, very rare; Montgomery Co.

102. **Zoysia** Willd.—Zoysia

1. **Zoysia japonica** Steud. Zoysia. June–Sept. Native to Asia; often planted but seldom escaped from cultivation; Jackson Co.

189. PONTEDERIACEAE—PICKERELWEED FAMILY

1. Inflorescence with 50 or more flowers; fruit a 1-seeded utricle .... 3. *Pontederia*
1. Inflorescence with 1–30 flowers; fruit a capsule with 10–many seeds.
  2. Leaves linear; inflorescence 1-flowered .... 4. *Zosterella*
  2. Leaves lanceolate to ovate to reniform; inflorescence 1- to several-flowered.
    3. Petiole swollen; perianth lobes more than 2 cm long; stamens 6 .... 1. *Eichhornia*
    3. Petiole not swollen; perianth lobes less than 2 cm long; stamens 3 .... 2. *Heteranthera*

1. **Eichhornia** Kunth—Water Hyacinth

1. **Eichhornia crassipes** (Mart.) Solms-Laub. Water Hyacinth. May–June. Native to the tropics; adventive in shallow water; DuPage, Hardin, Kendall, Massac, and Will cos.

2. **Heteranthera** Ruiz & Pavon—Mud Plantain

1. Stems short, erect, with the leaves clustered at the tip; leaves truncate to cuneate .... 1. *H. limosa*
1. Stems creeping, with scattered leaves; leaves cordate.
  2. Flower solitary; leaves longer than broad .... 4. *H. rotundifolia*
  2. Flowers 1–3 in a cluster; leaves as broad as long or broader.
    3. Flowering spathes more or less sessile; perianth purple .... 2. *H. multiflora*
    3. Flowering spathes on short stalks; perianth white .... 3. *H. reniformis*

1. **Heteranthera limosa** (Sw.) Willd. Mud Plantain. June–July. Muddy shores, shallow water; scattered in IL.

2. **Heteranthera multiflora** (Griseb.) C. N. Horn. Mud Plantain. Roadside ditch, rare; Hardin Co.

3. **Heteranthera reniformis** Ruiz & Pavon. Mud Plantain. July–Aug. Muddy shores; shallow water; restricted to the s. ⅓ of IL.

4. **Heteranthera rotundifolia** (Kunth) Griseb. Mud Plantain. June–Oct. Muddy shores, rare; Alexander Co.

3. **Pontederia** L.—Pickerelweed

1. **Pontederia cordata** L. Pickerelweed. May–Sept. Wet areas, sometimes in standing water; occasional in most of IL. except for the e. cent. cos. (Including var. *angustifolia* [Pursh] Torr.)

4. **Zosterella** Small—Water Star Grass

1. **Zosterella dubia** (Jacq.) Small. Water Star Grass. July–Aug. Shallow water, muddy shores, not common; confined to the n. ¾ of IL. *Heteranthera dubia* (Jacq.) MacM.

190. POTAMOGETONACEAE—PONDWEED FAMILY

1. Leaves septate, 0.3–0.5 (–2.0) mm broad; peduncles flexible .... 2. *Stuckenia*
1. Leaves not septate, some or all over 0.5 mm broad; peduncles rigid .... 1. *Potamogeton*

1. **Potamogeton** L.—Pondweed

1. Leaves uniform.
  2. Stipules adnate to the leaf base; leaves auriculate at base............20. *P. robbinsii*
  2. Stipules free from the leaf base (or sometimes adnate in *P. diversifolius*); leaves rounded, cordate, or tapering at base, never auriculate.
    3. Leaves sharply serrulate, crisped along the margin; fruits 5–6 mm long.................. 3. *P. crispus*
    3. Leaves entire or minutely denticulate, usually flat; fruits 1.6–5.0 mm long.
      4. Leaves 5–30 mm broad, rounded or cordate at base
        5. Leaves entire; fruits 4–5 mm long, dorsally keeled .....15. *P. praelongus*
        5. Leaves minutely denticulate; fruits 2.5–3.5 mm long, not keeled .... 19. *P. richardsonii*
      4. Leaves up to 5 mm broad, tapering to base.
        6. Leaves with 15–35 nerves; spikes with 6–11 whorls of flowers; fruits quadrate, 4–5 mm long ....24. *P. zosteriformis*
        6. Leaves with 3–14 nerves; spikes with 1–5 whorls of flowers; fruits obovoid to ovoid, 1.9–3.0 mm long.
          7. Stipules connate, forming cylinders.
            8. Spikes subcapitate, 2–5 mm long; fruits keeled; sepaloid connectives 0.5–1.0 mm long; leaves glandless at base.................. 7. *P. foliosus*
            8. Spikes elongate, more or less interrupted, 6–15 mm long; fruits rounded on back; sepaloid connectives 1.2–1.5 mm long; leaves biglandular at base.
              9. Leaves 5- to 7-nerved, thin, translucent.................. 8. *P. friesii*
              9. Leaves 3- to 18-nerved, firm, opaque.
                10. Stipules chartaceous, white, becoming fibrous.
                  11. Stems compressed; leaves with (5–) 9–14 nerves, cuspidate to aristate; fruits 2.6–3.1 mm long, 1.7–2.3 mm wide ....11. *P. X haynesii*
                  11. Stems not compressed; leaves with 3–5 (–7) nerves, acute; fruits 1.9–2.1 mm long, 1.3–1.8 mm wide.................. 22. *P. strictifolius*
                10. Stipules membranous, olive, not becoming fibrous.................. 17. *P. pusillus*
          7. Stipules free ....2. *P. berchtoldii*
1. Leaves of two kinds, the floating usually shorter and broader than the submersed, or floating leaves sometimes absent.

12. Leaves entire.
13. Submersed leaves up to 2 mm broad, 1- to 5-nerved.
14. Spikes uniform; stipules free from base of leaf; fruits 1.6–5.0 mm long.
15. Submersed leaves 0.1–0.5 mm broad, 1- to 3-nerved .... 23. *P. vaseyi*
15. Submersed leaves 0.5–2.0 mm broad, 3- to 5-nerved .... 13. *P. natans*
14. Spikes of 2 or more kinds, those in the axils of the submersed leaves subgloboid, those in the axils of the floating leaves elongate; stipules adnate to base of leaf; fruits 1.0–1.5 mm long.
16. Fruits beaked, with sharp-tipped keels; floating leaves obtuse to acute .... 5. *P. diversifolius*
16. Fruits beakless, without sharp-tipped keels; floating leaves acute to acuminate .... 3. *P. bicupulatus*
13. Submersed leaves 5–75 mm broad, 7- to 21-nerved.
17. Stems compressed; submersed leaves linear, 5–10 mm broad; fruit capped with a minute, toothlike beak .... 6. *P. epihydrus*
17. Stems terete; submersed leaves lanceolate to ovate, 10–75 mm broad; fruit capped with a prominent beak.
18. Fruiting spike 4 cm long or longer; stipules of submersed leaves subpersistent; fruit tapering to base; some nerves of floating leaves more prominent than other nerves .... 1. *P. amplifolius*
18. Fruiting spike 2.0–3.5 cm long; stipules of submersed leaves falling away early; fruit rounded at base; all nerves of floating leaves uniform .... 16. *P. pulcher*
12. Leaves minutely denticulate.
19. Fruits (including beak) 3.5–4.3 mm long .... 14. *P. nodosus*
19. Fruits (including beak) 1.7–3.5 mm long.
20. Stems simple or once-branched; stipules strongly keeled.
21. Floating leaves 2.0–6.5 cm broad; fruits normally developed .... 12. *P. illinoensis*
21. Floating leaves 0.5–1.0 cm broad; fruits not maturing .... 10. *P. X hagstromii*
20. Stems repeatedly branched; stipules faintly keeled.
22. Floating leaves elliptic to ovate, the petioles usually as long as or longer than the blades; stipules obtuse.
23. Submersed leaves acute at apex; fruit strongly 3-keeled, well developed .... 9. *P. gramineus*
23. Submersed leaves cuspidate at apex; fruit obscurely 3-keeled, poorly developed .... 21. *P. X spathuliformis*
22. Floating leaves oblong- to linear-lanceolate, the petioles shorter than the blades; stipules acuminate .... 18. *P. X rectifolius*

1. **Potamogeton amplifolius** Tuckerm. Broad-leaved Pondweed. June–Oct. Lakes and roadside ditches, rare; Cook, DuPage, Lake, and McHenry cos.

2. **Potamogeton berchtoldii** Fieber. Berchtold's Pondweed. June–Sept. Ponds, lakes, and streams; occasional in the n. cos., apparently absent in the s. cos. *Potamogeton pusillus* L. ssp. *tenuissimus* (Mertens & W. D. J. Koch) R. R. Haynes & Hellquist.

3. **Potamogeton bicupulatus** Fern. Slender-leaved Pondweed. June–Sept. Acid waters, very rare; Wabash Co.

4. **Potamogeton crispus** L. Curly Pondweed. May–Sept. Native to Europe; naturalized in muddy to calcareous ponds, lakes, and streams; occasional throughout the state.

5. **Potamogeton diversifolius** Raf. Water-thread Pondweed. May–Oct. Quiet waters; occasional throughout the state.

6. **Potamogeton epihydrus** Raf. Ribbon-leaved Pondweed. June–Sept. Quiet ponds and lakes, rare; Fulton, Hancock, and Lake cos.

7. **Potamogeton foliosus** Raf. Two varieties may be distinguished in Illinois:

a. Leaves dark green; fruits olivaceous .... 7a. *P. foliosus* var. *foliosus*
a. Leaves bright green; fruits green .... 7b. *P. foliosus* var. *macellus*

7a. **Potamogeton foliosus** Raf. var. **foliosus.** Leafy Pondweed. June–Sept. Ponds, lakes, rivers, and streams; occasional throughout the state.

7b. **Potamogeton foliosus** Raf. var. **macellus** Fern. Leafy Pondweed. June–Sept. Ponds, lakes; occasional throughout the state but not as common as the typical variety.

8. **Potamogeton friesii** Rupr. Fries' Pondweed. June–Sept. Ponds and streams, rare; Cook, Lake, McHenry, and Shelby cos.

9. **Potamogeton gramineus** L. Grass-leaved Pondweed. June–Sept. Lakes, ponds, and streams, rare; Cook, Lake, McHenry, and Wabash cos.

10. **Potamogeton X hagstromii** Benn. Hagstrom's Pondweed. July–Sept. Lakes, very rare; Cook, Lake, and McHenry cos. Reputed to be the hybrid between *P. gramineus* L. and *P. richardsonii* (Benn.) Rydb.

11. **Potamogeton X haynesii** Hellq. & Crow. Haynes' Pondweed. June–Sept. Lakes, very rare; Lake Co. This is the hybrid between *P. strictifolius* Benn. and *P. zosteriformis* Fern.

12. **Potamogeton illinoensis** Morong. Illinois Pondweed. June–Oct. Ponds, lakes, and streams; occasional in the n. ½ of the state, rare in the s. ½.

13. **Potamogeton natans** L. Floating-leaved Pondweed. June–Sept. Lakes and streams; occasional throughout the state except for the extreme s. cos.

14. **Potamogeton nodosus** Poir. Long-leaved Pondweed. May–Oct. Ponds and streams; occasional throughout the state.

15. **Potamogeton praelongus** Wulfen. White-stemmed Pondweed. June–Sept. Lakes and rivers, rare; Cook, Lake, and McHenry cos.

16. **Potamogeton pulcher** Tuckerm. Spotted Pondweed. June–Sept. Shallow water, rare; Jackson, Kane, Mason, and Menard cos.

17. **Potamogeton pusillus** L. Two varieties may be distinguished in Illinois:

a. Primary leaves 1–3 mm wide .... 17a. *P. pusillus* var. *pusillus*
a. Primary leaves 0.3–1.0 mm wide .... 17b. *P. pusillus* var. *minor*

17a. **Potamogeton pusillus** L. var. **pusillus.** Little Pondweed. June–Sept. Ponds, lakes, and streams; occasional throughout the state, but rarer in the s. cos.

17b. **Potamogeton pusillus** L. var. **minor** (Biv.) Fern. & Schub. Little Pondweed. June–Sept. Ponds and lakes; occasional in the n. ½ of IL.

18. **Potamogeton X rectifolius** Benn. Hybrid Pondweed. June–Sept. Ditches, in 1–3 feet of water; Cook Co. Reputed to be the hybrid between *P. nodosus* Poir. and *P. richardsonii* (Benn.) Rydb.

19. **Potamogeton richardsonii** (Benn.) Rydb. Richardson's Pondweed. June–Sept. Lakes and rivers, rare; Cook, Kankakee, Lake, and McHenry cos.

20. **Potamogeton robbinsii** Oakes. Robbins' Pondweed. June–Sept. Stagnant water, very rare; Cook, Lake, and McHenry cos.

21. **Potamogeton X spathuliformis** (Robbins) Morong. Pondweed. June–Sept. Ponds and streams, very rare; Wabash Co. Reputed to be the hybrid between *P. gramineus* L. and *P. illinoensis* Morong.

22. **Potemogeton strictifolius** Benn. Pondweed. June–Sept. Lakes, very rare; Cook and Lake cos.

23. **Potamogeton vaseyi** Robbins. Vasey's Pondweed. June–Sept. Lakes and ponds, rare; McHenry Co.

24. **Potamogeton zosteriformis** Fern. Flat-stemmed Pondweed. June–Sept. Ponds, lakes, and streams, not common; confined to the n. ½ of the state.

2. **Stuckenia** Borner

1. **Stuckenia pectinata** (L.) Borner. Sago Pondweed; Fennel-leaved Pondweed. May–Sept. Calcareous water; occasional throughout the state except the southernmost cos. *Potamogeton pectinatus* L.; *Coleogeton pectinatus* (L.) Les & R. R. Haynes.

191. RUPPIACEAE—DITCH GRASS FAMILY

1. **Ruppia** L.—Ditch Grass

1. **Ruppia cirrhosa** (Petagna) Grande. Ditch Grass. July–Oct. Saline waters, rare; Henry, Lake, and Vermilion cos. *Ruppia maritima* L. var. *rostrata* Agardh., misapplied.

192. RUSCACEAE—RUSCUS FAMILY

1. Leaves several, not cordate.
  2. Flowers axillary from the cauline leaves .... 3. *Polygonatum*
  2. Flowers terminal .... 4. *Smilacina*
1. Leaves 1–3, cordate or not cordate.
  3. Leaves cordate; perianth parts 4; stamens 4 .... 2. *Maianthemum*
  3. Leaves not cordate; perianth parts 6; stamens 6................................ 1. *Convallaria*

1. **Convallaria** L.—Lily-of-the-valley

1. **Convallaria majalis** L. Lily-of-the-valley. May–June. Native to Europe; occasionally escaped from cultivation.

2. **Maianthemum** Weber—False Lily-of-the-valley

1. **Maianthemum canadense** Desf. Two varieties occur in Illinois:

a. Leaves glabrous beneath; transverse veins of leaf conspicuous in transmitted light .... 1a. *M. canadense* var. *canadense*
a. Leaves pubescent beneath; transverse veins of leaf obscure in transmitted light..... .... 1b. *M. canadense* var. *interius*

1a. **Maianthemum canadense** Desf. var. **canadense.** False Lily-of-the-valley. May–June. Moist woods, rare; Cook Co.

1b. **Maianthemum canadense** Desv. var. **interius** Fern. False Lily-of-the-valley. May–June. Moist woods, occasional; restricted to the n. ¼ of the state.

3. **Polygonatum** Mill.—Solomon's-seal

1. Leaves pilose on the nerves beneath .... 3. *P. pubescens*
1. Leaves glabrous beneath.
  2. Leaves sessile, the largest ones with less than 100 nerves; perianth 10–17 mm long, the lobes 3–4 mm long .... 1. *P. biflorum*
  2. Leaves more or less clasping or sheathing at the base, the largest ones with over 100 nerves; perianth 17–20 mm long, the lobes 5–7 mm long .... 2. *P. commutatum*

1. **Polygonatum biflorum** (Walt.) Ell. Small Solomon's-seal. Apr.–June. Dry woods, sandstone cliffs; restricted to the s. tip of the state.

2. **Polygonatum commutatum** (Schult.) A. Dietr. Solomon's-seal. Apr.–June. Woods, common; in every co.

3. **Polygonatum pubescens** (Willd.) Pursh. Small Solomon's-seal. May–June. Moist, shaded woods; restricted to a few extreme n. cos.

4. **Smilacina** Desf.—False Solomon's-seal

1. Flowers paniculate; stamens longer than the perianth parts; perianth parts to 3 mm long; berry red .... 1. *S. racemosa*
1. Flowers racemose; stamens shorter than or equaling the perianth parts; perianth parts 4–6 mm long; berry black or greenish black .... 2. *S. stellata*

1. **Smilacina racemosa** (L.) Desf. False Solomon's-seal. Apr.–June. Rich, moist woods, common; in every co. *Maianthemum racemosum* (L.) Link.

2. **Smilacina stellata** (L.) Desf. Starry Solomon's-seal. May–June. Moist woods, prairies; occasional in the n. ⅗ of the state; also Crawford, St. Clair, and Wabash cos. *Maianthemum stellatum* (L.) Link.

193. SCHEUCHZERIACEAE—SCHEUCHZERIA FAMILY

1. **Scheuchzeria** L.—Arrow-grass

1. **Scheuchzeria palustris** L. var. **americana** Fern. Arrow-grass. May–July. Bogs; Fulton, Lake, McHenry, and Menard cos.

194. SMILACACEAE—GREENBRIER FAMILY

1. **Smilax** L.—Greenbrier

1. Stems woody, with few to many prickles, rarely without prickles; ovule 1 per cell of ovary.

2. Leaves pale beneath, usually glaucous ....3. *S. glauca*
2. Leaves green beneath, never glaucous.
   3. Stems flexuous, with stout spines or spines sometimes absent; leaves subcoriaceous to coriaceous.
      4. Leaves thick-margined, coriaceous, often blotched with white, often lobed at base; midvein on lower leaf surface usually prickly; peduncles much longer than the subtending petioles; berries not glaucous ....1. *S. bona-nox*
      4. Leaves thin-margined, subcoriaceous, not blotched with white, not lobed at base; midvein on lower leaf surface not prickly; peduncles about as long as the subtending petioles; berries glaucous.............7. *S. rotundifolia*
   3. Stems not flexuous, with weak spines or spines sometimes absent; leaves membranous..8. *S. tamnoides*

1. Stems herbaceous, without prickles; ovules 2 per cell of ovary.
   5. Stems climbing or twining, with numerous tendrils all along the stem; peduncles borne from the axils of developed leaves.
      6. Blades light green, not shiny beneath; petiole generally short; berries blue... ....5. *S. lasioneuron*
      6. Blades dark green, shiny beneath; petiole long; berries black ....6. *S. pulverulenta*
   5. Stems erect, with few or no tendrils near the apex; peduncles borne from bladeless sheaths.
      7. Basal leaves narrowly ovate or elliptic, the base mostly truncate to subcordate; petiole usually equal to or longer than the blade ....4. *S. illinoensis*
      7. Basal leaves broadly ovate, the base cordate; petiole generally equal to or shorter than the blade..2. *S. ecirrhata*

1. **Smilax bona-nox** L. Two varieties occur in Illinois:

a. Leaves spinulose on the margins, with patches of white coloration.......................... .... 1a. *S. bona-nox* var. *bona-nox*

a. Leaves entire or with very weak marginal spinules, green throughout...................... .... 1b. *S. bona-nox* var. *hederaefolia*

1a. **Smilax bona-nox** L. var. **bona-nox**. Catbrier; Greenbrier. May–June. Dry woods and fields, on bluffs, thickets, common; confined to the s. ¼ of the state.

1b. **Smilax bona-nox** L. var. **hederaefolia** (Beyrich) Fern. Catbrier; Greenbrier. May–June. Woods and fields, not common; restricted to the s. ¼ of the state.

2. **Smilax ecirrhata** Kunth. Carrion Flower. May–June. Moist woods; occasional in the n. ½ of IL, becoming uncommon in the s. cos.

3. **Smilax glauca** Walt. Two varieties may be distinguished in Illinois:

a. Leaves minutely pubescent beneath .... 3a. *S. glauca* var. *glauca*

a. Leaves glabrous beneath.. 3b. *S. glauca* var. *leurophylla*

3a. **Smilax glauca** Walt. var. **glauca**. Glaucous Catbrier; Glaucous Greenbrier. May–June. Dry woods, edges of fields and bluffs; common in the s. ⅓ of IL, absent elsewhere.

3b. **Smilax glauca** Walt. var. **leurophylla** Blake. Glaucous Catbrier; Glaucous Greenbrier. May–June. Edges of swamps; Jackson, Johnson, Pope, and Union cos.

4. **Smilax illinoensis** Mangaly. Carrion Flower. May–June. Thickets, often near roads; occasional throughout the state.

5. **Smilax lasioneuron** Hook. Carrion Flower. Apr.–June. Moist woods, edge of fields; in every co.

6. **Smilax pulverulenta** Michx. Carrion Flower. Apr.–June. Moist or dry woods, edges of fields; occasional in the s. ⅖ of the state, rare elsewhere.

7. **Smilax rotundifolia** L. Catbrier; Greenbrier. Apr.–May. Dry woods, edge of fields; common in the s. ⅓ of the state, absent elsewhere.

8. **Smilax tamnoides** L. var. **hispida** (Muhl.) Fern. Catbrier; Greenbrier. May–June. Usually moist woods; probably in every co. *Smilax hispida* Muhl.

195. SPARGANIACEAE—BUR-REED FAMILY

1. **Sparganium** L.—Bur-reed

1. Plants floating or suberect; stems to 30 cm long; staminate head 1; pistillate heads about 1 cm in diameter, the lowest short-pedunculate; beak of achene 0.5–1.5 mm long .... 5. *S. natans*
1. Plants erect; stems (5–) 39–120 cm tall; staminate heads 3–25; pistillate heads 1.5–3.5 cm in diameter, all sessile; beak of achene 2–6 mm long.
  2. Central axis bearing only staminate heads; achene not stipitate, obpyramidal, truncate at the summit; stigmas 2 .... 4. *S. eurycarpum*
  2. Central axis bearing 1–4 pistillate heads; achene stipitate, fusiform, tapering to the summit; stigma 1.
    3. At least 1 of the pistillate heads borne above the subtending bract (supra-axillary); inflorescence simple; achene usually greenish brown, even at maturity .... 3. *S. emersum*
    3. All heads axillary in the subtending bract; inflorescence usually with 1 or more branches; achene pale or dark brown.
      4. Leaves stiff, strongly keeled; branches of inflorescence without any pistillate heads; pistillate heads 2.5–3.5 cm in diameter; body of achene 5–7 mm long, shiny, pale brown .... 2. *S. androcladum*
      4. Leaves soft, usually not strongly keeled; branches of inflorescence with 1–3 pistillate heads; pistillate heads 1.2–2.5 cm in diameter; body of achene 3–5 mm long, dull, dark brown .... 1. *S. americanum*

1. **Sparganium americanum** Nutt. Bur-reed. June–July. Shallow water; restricted to the n. ¼ of the state.

2. **Sparganium androcladum** (Engelm.) Morong. Bur-reed. June–July. Scattered throughout the state but not common.

3. **Sparganium emersum** Rehmann. Green-fruited Bur-reed. June–Aug. Shallow water, rare; n. ⅓ of IL; also Union Co. *Sparganium chlorocarpum* Rydb.; *S. angustifolium* Michx., misapplied.

4. **Sparganium eurycarpum** Engelm. Bur-reed. June–Aug. Shallow water; occasional throughout the state but less common in the s. cos.

5. **Sparganium natans** L. Least Bur-reed. July–Aug. Shallow water, very rare; McHenry Co. *Sparganium minimum* (Hartm.) Fried.

196. THISMIACEAE—THISMIA FAMILY

1. **Thismia** Griff.—Thismia

1. **Thismia americana** N. E. Pfeiff. Thismia. Aug. Wet prairies; undoubtedly extinct. Cook Co. Not seen since 1916.

197. TOFIELDIACEAE—TOFIELDIA FAMILY

1. **Triantha** (Nutt.) Baker

1. **Triantha glutinosa** Michx. False Asphodel. June–July. Bogs, rare; restricted to the extreme ne. cos. *Tofieldia glutinosa* (Michx.) Pers.

198. TRILLIACEAE—TRILLIUM FAMILY

1. Flowers in umbels; leaves in two whorls below the flowers .... 1. *Medeola*
1. Flower solitary; leaves in 1 whorl below the flower .... 2. *Trillium*

1. **Medeola** L.—Cucumber-root

1. **Medeola virginiana** L. Indian Cucumber-root. May–June. Rich, moist woods, rare; Cook and LaSalle cos.

2. **Trillium** L.—Trillium; Wake Robin

1. Flowers sessile, basically maroon or green (rarely yellow).
  2. Leaves petiolate; petals abruptly tapering to distinct claws; sepals reflexed .... 7. *T. recurvatum*
  2. Leaves sessile; petals scarcely or gradually tapering, usually without distinct claws; sepals spreading to erect, not reflexed.

3. Stamens ½ as long as petals; connectives 1–3 mm long.
  4. Stems and veins on lower leaf surface glabrous; anthers 5–6 times longer than the filaments; pollen yellow .... 8. *T. sessile*
  4. Some of the stems and veins on lower leaf surface minutely scabrous; anthers 4 times longer than the filaments; pollen olive to brownish .... 9. *T. viride*
3. Stamens ⅓ as long as petals; connectives less than 1 mm long .... 2. *T. cuneatum*

1. Flowers pedunculate, white or occasionally pink or purple, never green nor yellow.
  5. Ovary and fruit 3-angled .... 6. *T. nivale*
  5. Ovary and fruit 6-angled.
    6. Ovary purple (in IL specimens) .... 3. *T. erectum*
    6. Ovary white.
      7. Petals strongly recurved, 1.5–2.5 cm long; peduncle 10–40 mm long; anthers 4–7 mm long .... 1. *T. cernuum*
      7. Petals spreading, 2–6 cm long; peduncle 25–100 mm long; anthers (6–) 7–15 mm long.
        8. Stigmas erect; anthers slightly longer than the filaments .... 5. *T. grandiflorum*
        8. Stigmas spreading; anthers about twice as long as the filaments......... .... 4. *T. flexipes*

1. **Trillium cernuum** L. var. **macranthum** Eames & Wieg. Nodding Trillium. May–June. Moist woods, very rare; Cook, Lake, and McHenry cos.

2. **Trillium cuneatum** Raf. Trillium. Apr. Rich, moist woods, very rare; Jackson Co.

3. **Trillium erectum** L. Purple Trillium. May. Rich woods, very rare; Jo Daviess, Lake, and McHenry cos. White-flowered plants may be called var. *alba* (Michx.) Pursh.

4. **Trillium flexipes** Raf. White Trillium. Apr.–May. Rich woods; occasional throughout the state. *Trillium gleasonii* Fern.

5. **Trillium grandiflorum** (Michx.) Salisb. Large White Trillium. Apr.–May. Rich, moist woods; occasional in the n. ½ of the state, rare elsewhere.

6. **Trillium nivale** Riddell. Snow Trillium. Mar.–Apr. Rich woods, occasional; restricted to the n. ⅗ of the state.

7. **Trillium recurvatum** Beck. Three forms occur in Illinois:

a. Petals purple .... 7a. *T. recurvatum* f. *recurvatum*
a. Petals yellow.
  b. Stamens purple .... 7b. *T. recurvatum* f. *luteum*
  b. Stamens yellow .... 7c. *T. recurvatum* f. *shayii*

7a. **Trillium recurvatum** Beck f. **recurvatum.** Purple Trillium; Purple Wake Robin. Mar.–May. Rich, moist woods, common; in every co.

7b. **Trillium recurvatum** Beck f. **luteum** Clute. Purple Trillium; Purple Wake Robin. Mar.–May. Rich, moist woods; occasional throughout the state.

7c. **Trillium recurvatum** Beck f. **shayii** Palmer & Steyerm. Shay's Trillium. Apr.–May. Rich woods, very rare; Jackson Co.

8. **Trillium sessile** L. Two forms occur in Illinois:

a. Petals purple .... 8a. *T. sessile* f. *sessile*
a. Petals greenish yellow .... 8b. *T. sessile* f. *viridiflorum*

8a. **Trillium sessile** L. f. **sessile.** Sessile Trillium; Sessile Wake Robin. Mar.–Apr. Rich woods, occasional; scattered throughout the state except the w. cos.

8b. **Trillium sessile** L. f. **viridiflorum** Beyer. Sessile Trillium. Mar.–Apr. Rich woods; scattered in IL.

9. **Trillium viride** Beck. Green Trillium. Apr.–May. Rich woods, prairies, not common; Adams, Jackson, Macoupin, Perry, Pike, Union, and Williamson cos.

199. TYPHACEAE—CAT-TAIL FAMILY

1. **Typha** L.—Cat-tail

1. Staminate and pistillate spikes usually contiguous; pollen grains in groups of 4; stigma spatulate; pistillate flowers not subtended by bracts .... 4. *T. latifolia*

1. Staminate and pistillate spikes usually separated; pollen grains borne singly; stigma linear; pistillate flowers subtended by narrow, scalelike bracts.
2. Pistillate spikes green at maturity; pistillate flowers abortive .... 3. *T. X glauca*
2. Pistillate spikes brown at maturity; pistillate flowers fertile.
3. Pistillate spikes 1–2 cm in diameter, dark brown; leaves usually less than 8 .... 1. *T. angustifolia*
3. Pistillate spikes 2–3 cm in diameter, light brown; leaves more than 8 .... 2. *T. domingensis*

1. **Typha angustifolia** L. Narrow-leaved Cat-tail. June–Oct. Wet situations, occasional to common; scattered in IL.

2. **Typha domingensis** Pers. Southern Cat-tail. June–Oct. Native s. and w. of IL; power plant cooling pond, very rare; Will Co.

3. **Typha X glauca** Godr. Hybrid Cat-tail. June–Oct. Wet ground, rare; known from the n. ¼ of IL. This is the hybrid between *T. angustifolia* L. and *T. latifolia* L.

4. **Typha latifolia** L. Common Cat-tail. June–Oct. Low areas, very common; in every co.

200. XYRIDACEAE—YELLOW-EYED GRASS FAMILY

1. **Xyris** L.—Yellow-eyed Grass

1. Stems not twisted, from a nonbulbous base; lateral sepals 4–6 mm long, the keel toothed but not ciliate .... 1. *X. jupicai*
1. Stems twisted, from a bulbous base; lateral sepals 3–4 mm long, the keel ciliate .... 2. *X. torta*

1. **Xyris jupicai** L. C. Rich. Yellow-eyed Grass. July–Aug. Native s. of IL; low, sandy areas, very rare; McHenry Co. *Xyris caroliniana* Walt.

2. **Xyris torta** Sm. Twisted Yellow-eyed Grass. July–Aug. Low, sandy areas; restricted to the n. ½ of the state.

201. ZANNICHELLIACEAE—HORNED PONDWEED FAMILY

1. **Zannichellia** L.—Horned Pondweed

1. **Zannichellia palustris** L. var. **major** (Boenn) W. D. J. Koch. Horned Pondweed. June–Oct. Ditches, spring-fed streams, ponds, lakes; occasional in the n. ½ of IL; also Jackson and St. Clair cos.

Summary of the Taxa Treated in This Book

Family	*Genera*	*Species*	*Hybrids*	*Lesser Taxa*	*Non-Natives*
1. Aspleniaceae	1	7	6	3	
2. Azollaceae	1	2			
3. Blechnaceae	1	2			
4. Dennstaedtiaceae	2	2		1	
5. Dryopteridaceae	8	21	5	1	
6. Equisetaceae	1	9	4		
7. Hymenophyllaceae	2	2			
8. Isoetaceae	1	3			
9. Lycopodiaceae	5	9	2	1	
10. Marsileaceae	1	1			1
11. Onocleaceae	2	2			
12. Ophioglossaceae	4	11		1	
13. Osmundaceae	1	3			
14. Polypodiaceae	2	2			
15. Pteridaceae	6	8			1
16. Selaginellaceae	1	3			
17. Thelypteridaceae	3	5			1
18. Cupressaceae	2	4		2	
19. Ginkgoaceae	1	1			1
20. Pinaceae	3	16			12
21. Taxaceae	1	1			
22. Taxodiaceae	2	2			1
23. Acanthaceae	3	7		3	
24. Aceraceae	1	12	1	7	5
25. Adoxaceae	1	1			
26. Aizoaceae	1	1			1
27. Amaranthaceae	8	22		2	14
28. Anacardiaceae	3	9		4	2
29. Annonaceae	1	1			
30. Apiaceae	39	59		7	24
31. Apocynaceae	4	7	1	3	3
32. Aquifoliaceae	2	4			
33. Araliaceae	4	9			4
34. Aristolochiaceae	3	2		2	2
35. Asclepiadaceae	4	24		1	2
36. Asteraceae	119	387	20	73	129
37. Balsaminaceae	1	3			1
38. Berberidaceae	4	6		3	2
39. Betulaceae	2	10	2	3	5
40. Bignoniaceae	3	4			1
41. Boraginaceae	17	35		1	23
42. Brassicaceae	43	96		10	59
43. Buddlejaceae	2	2			1
44. Buxaceae	1	1			1

Family	Genera	Species	Hybrids	Lesser Taxa	Non-Natives
45. Cabombaceae	2	2			
46. Cactaceae	2	4			
47. Caesalpiniaceae	5	10		2	1
48. Callitrichaceae	1	4			
49. Calycanthaceae	1	1			1
50. Campanulaceae	4	17	1	5	4
51. Cannabinaceae	2	3		3	3
52. Caprifoliaceae	7	39	7	9	23
53. Caryophyllaceae	21	66		4	45
54. Celastraceae	2	12			8
55. Ceratophyllaceae	1	2			
56. Chenopodiaceae	11	44			30
57. Cistaceae	3	9		2	
58. Cleomaceae	4	7		1	4
59. Convolvulaceae	7	17		3	10
60. Cornaceae	1	12		1	2
61. Corylaceae	3	4		2	
62. Crassulaceae	3	12	1	1	10
63. Cucurbitaceae	9	11		1	8
64. Cuscutaceae	1	10		2	
65. Dipsacaceae	2	3			3
66. Droseraceae	1	2			
67. Ebenaceae	1	1		3	
68. Elaeagnaceae	2	4			3
69. Elatinaceae	2	3			
70. Ericaceae	10	19		3	1
71. Euphorbiaceae	9	42		5	15
72. Fabaceae	43	133	10	16	65
73. Fagaceae	3	26	21	3	4
74. Fumariaceae	4	12		1	3
75. Gentianaceae	8	15	3	1	2
76. Geraniaceae	2	12			8
77. Grossulariaceae	1	8			3
78. Haloragidaceae	2	7		1	1
79. Hamamelidaceae	2	2			
80. Heliotropaceae	1	4			3
81. Hippocastanaceae	1	5	1	1	2
82. Hippuridaceae	1	1			
83. Hydrangeaceae	1	2			
84. Hydrophyllaceae	5	12			1
85. Hypericaceae	3	24		1	1
86. Iteaceae	1	1			
87. Juglandaceae	2	12	3	5	
88. Lamiaceae	38	99	8	12	55
89. Lardizabalaceae	1	1			1
90. Lauraceae	2	2		2	
91. Lentibulariaceae	1	8			
92. Limnanthaceae	1	1			
93. Linaceae	1	8		1	2
94. Loasaceae	1	3			2
95. Loganiaceae	1	1			
96. Lythraceae	7	12		1	4
97. Magnoliaceae	2	3			1
98. Malvaceae	12	26		1	17
99. Melastomaceae	1	2			
100. Menispermaceae	3	3			
101. Menyanthaceae	2	2			1

Family	*Genera*	*Species*	*Hybrids*	*Lesser Taxa*	*Non-Natives*
102. Mimosaceae	4	4			2
103. Molluginaceae	1	1			1
104. Monotropaceae	1	2			
105. Moraceae	4	7			5
106. Myricaceae	2	2			1
107. Nelumbonaceae	1	2			2
108. Nyctaginaceae	1	5			4
109. Nymphaeaceae	2	4			
110. Nyssaceae	1	3		1	
111. Oleaceae	6	15	2		9
112. Onagraceae	6	41	1	2	11
113. Orobanchaceae	3	7			1
114. Oxalidaceae	1	5			1
115. Papaveraceae	7	11		2	9
116. Parnassiaceae	1	1			
117. Passifloraceae	1	2			
118. Paulowniaceae	1	1			1
119. Pedaliaceae	1	1			1
120. Penthoraceae	1	1			
121. Philadelphaceae	2	6			5
122. Phrymaceae	1	1			
123. Phytolaccaceae	1	1			
124. Plantaginaceae	1	12		1	5
125. Platanaceae	1	1			
126. Polemoniaceae	6	14		4	6
127. Polygalaceae	1	8		2	
128. Polygonaceae	11	63	1	4	30
129. Portulacaceae	3	6			2
130. Primulaceae	10	23	2	1	6
131. Pyrolaceae	3	5			
132. Ranunculaceae	20	69		11	20
133. Resedaceae	1	3			3
134. Rhamnaceae	4	12		4	6
135. Rosaceae	32	230	9	12	77
136. Rubiaceae	9	35		4	6
137. Rutaceae	5	5		1	3
138. Salicaceae	2	33	13	7	12
139. Santalaceae	1	1			
140. Sapindaceae	2	2			2
141. Sapotaceae	1	2			
142. Sarraceniaceae	1	1			
143. Saururaceae	1	1			
144. Saxifragaceae	4	8		2	
145. Scrophulariaceae	31	85		8	28
146. Simaroubaceae	1	1			1
147. Solanaceae	9	38	1	4	28
148. Staphyleaceae	1	1			
149. Sterculiaceae	1	1			1
150. Styracaceae	2	3			
151. Tamaricaceae	1	1			1
152. Thymelaeaceae	2	2			1
153. Tiliaceae	1	3		1	1
154. Ulmaceae	3	12		5	4
155. Urticaceae	5	8		1	1
156. Valerianaceae	1	10		1	2
157. Verbenaceae	3	11	7	1	4
158. Violaceae	2	30	3	1	5

Family	*Genera*	*Species*	*Hybrids*	*Lesser Taxa*	*Non-Natives*
159. Viscaceae	1	1			
160. Vitaceae	3	12		3	3
161. Zygophyllaceae	2	2			2
162. Acoraceae	2	2			
163. Agavaceae	2	3			2
164. Alismataceae	3	15			
165. Alliaceae	2	17		1	10
166. Amaryllidaceae	4	5		1	5
167. Araceae	6	7		1	2
168. Asparagaceae	1	1			1
169. Butomaceae	1	1			1
170. Colchicaceae	1	2			
171. Commelinaceae	2	8		3	1
172. Cyperaceae	19	291	4	15	14
173. Dioscoreaceae	1	3		2	1
174. Hemerocallaceae	1	2			2
175. Hyacinthaceae	5	10			8
176. Hydrocharitaceae	4	5			1
177. Hypoxidaceae	1	1			
178. Iridaceae	4	15	3	3	8
179. Juncaceae	2	32	1	3	2
180. Juncaginaceae	1	2			
181. Lemnaceae	4	17			
182. Liliaceae	5	10			3
183. Maranthaceae	1	1			
184. Melanthiaceae	5	5			
185. Najadaceae	1	5		1	1
186. Nartheciaceae	2	2			1
187. Orchidaceae	19	48	3	3	1
188. Poaceae	102	356	5	45	131
189. Pontederiaceae	4	7			1
190. Potamogetonaceae	2	21	4	2	1
191. Ruppiaceae	1	1			
192. Ruscaceae	4	7		1	1
193. Scheuchzeriaceae	1	1			
194. Smilacaceae	1	8		2	
195. Sparganiaceae	1	5			
196. Thismiaceae	1	1			
197. Tofieldiaceae	1	1			
198. Trilliaceae	2	10		4	
199. Typhaceae	1	3	1		1
200. Xyridaceae	1	2			1
201. Zannichelliaceae	1	1			
Totals	1,046	3,442	156	391	1,156

New Combinations in This Book

Eurybia chasei (G. N. Jones) Mohlenbr. (basionym: *Aster chasei* G. N. Jones)
Packera paupercula (Michx.) A. Love and D. Love var. **balsamitae** (Muhl. ex Willd.) Mohlenbr. (basionym: *Senecio balsamitae* Muhl. ex Willd.)
Symphyotrichum concinnum (Willd.) Mohlenbr. (basionym: *Aster concinnus* Willd.)
Symphyotrichum oolentangiense (Riddell) G. L. Nesom var. **laevicaule** (Fern.) Mohlenbr. (basionym: *Aster oolentangiense* Riddell var. *laevicaule* Fern.)
Symphyotrichum patentissimum (Lindl. ex DC.) Mohlenbr. (basionym: *Aster patentissimus* Lindl. ex DC.)
Boechera dentata (Raf.) Al-Shehbaz var. **phalacrocarpa** (M. Hopkins) Mohlenbr. (basionym: *Arabis dentata* Raf. var. *phalacrocarpa* M. Hopkins)
Campanulastrum americanum (L.) Small var. **illinoense** (Fresen.) Mohlenbr. (basionym: *Campanula illinoensis* Fresen.)
Cornus sericea L. var. **baileyi** (Coult. & Evans) Mohlenbr. (basionym: *Cornus baileyi* Coult. & Evans)
Stachys pilosa Nutt. var. **homotricha** (Fern.) Mohlenbr. (basionym: *Stachys palustris* (L.) var. *homotricha* Fern.)
Persicaria hydropiperoides (Michx.) small var. **bushiana** (Stanford) Mohlenbr. (basionym: *Polygonum hydropiperoides* Michx. var bushianum Stanford)
Persicaria pensylvanica (L.) Small var. **laevigata** (Fern.) Mohlenbr. (basionym: *Polygonum pensylvanicum* L. var. *laevigata* Fern.)
Tracaulon arifolium (L.) Raf. var. **pubescens** (Keller) Mohlenbr. (basionym: *Polygonum sagittatum* L. var. *pubescens* Keller)
Frangula alnus Muhl. var. **angustifolia** (Loud.) Mohlenbr. (basionym: *Rhamnus frangula* L. var. *angustifolia* Loud.)
Frangula caroliniana (Walt.) Gray var. **mollis** (Fern.) Mohlenbr. (basionym: *Rhamnus caroliniana* Walt. var. *mollis* Fern.)
Micranthes forbesii (Vasey) Mohlenbr. (basionym: *Saxifraga forbesii* Vasey)
Coleataenia condensa (Nash) Mohlenbr. (basionym: *Panicum condensum* Nash)
Dichanthelium albemarlense (Nash) Mohlenbr. (basionym: *Panicum albemarlense* Nash)

Dichanthelium ashei (Pearson in Ashe) Mohlenbr. (basionym: *Panicum ashei* T. G. Pearson in Ashe)
Dichanthelium helleri (Nash) Mohlenbr. (basionym: *Panicum helleri* Nash)
Dichanthelium pseudopubescens (Nash) Mohlenbr. (basionym: *Panicum pseudopubescens* Nash)
Dichanthelium werneri (Scribn.) Mohlenbr. (basionym: *Panicum werneri* Scribn.)

Additional Taxa

The following additions to the flora were encountered too late for inclusion in the text.

TAXODIACEAE
Metasequoia Hu & W. C. Cheng

Metasequoia glyptostroboides Hu & W. C. Cheng. Dawn Redwood. This ornamental species from China has been reported as an adventive in a flower bed in Coles Co.

ASTERACEAE
Ageratum L.

Ageratum conyzoides L. Ageratum. Native to S. America; rarely escaped from cultivation; Jackson Co. When I collected this plant, I called it *Conoclinium coelestinum,* which it looks very much like. On closer examination while preparing the Asteraceae volumes in my Illustrated Flora of Illinois series, I realized it was actually an *Ageratum.* The following key distinguishes the two genera:

1. Pappus of 25–30 barbellate bristles .... Conoclinium
1. Pappus of aristate scales .... Ageratum

Packera A. Love & D. Love

Three varieties of this species are now known from Illinois, in addition to the typical variety. The following key separates the four varieties:

1. Rhizomes and stolons absent; plants reproducing asexually by adventitious rosettes arising from the roots .... 1d. *P. paupercula* var. *savannarum*
1. Rhizomes or stolons present; plants not reproducing asexually by adventitious rosettes arising from theroots.
2. Plants with slender stolons .... 1c. *P. paupercula* var. *pseudotomentosa*
2. Plants with sturdy, thick rhizomes.
3. Plants with up to 8 heads; basal leaves up to 4 cm long ....
.... 1a. *P. paupercula* var. *paupercula*
3. Plants with more than 8 heads; basal leaves 4 cm long or longer....................
.... 1b. *P. paupercula* var. *balsamitae*

1a. **Packera paupercula** (Michx.) A. Love & D. Love var. **paupercula.** Northern Ragwort. May–June. Wet prairies, moist sand flats, sedge meadows, open woods; common in the n. 1/2 of IL, rare in the s. 1/2. *Senecio paupercula* Michx.

1b. **Packera paupercula** (Michx.) A. Love & D. Love var. **balsamitae** (Muhl. ex Willd.) Mohlenbr., comb. nov. (basionym: *Senecio balsamitae* Muhl. ex Willd.). Balsam Groundsel; Balsam Ragwort. May–June. Wet prairies, moist sand flats; rare in the n. tier of counties. *Senecio aurea* L. var. *balsamitae* (Muhl. ex Willd.) Torr. & Gray.

1c. **Packera paupercula** (Michx.) A. Love & D. Love var. **pseudotomentosa** (Mack. & Bush) R. Kowal. Northern Ragwort. May–June. Rocky woods, glades; known from Champaign, Grundy, Henry, and Winnebago cos. *Senecio pseudotomentosa* Mack. & Bush.

1d. **Packera paupercula** (Michx.) A. Love & D. Love var. **savannarum** R. Kowal. Savanna Ragwort. May–June. Savannas, mesic prairies; known from Mason and Peoria cos.

Youngia Cass.

Youngia japonica (L.) DC. Youngia. May–June. Native to Asia; adventive in a parking lot; Massac Co. *Prenanthes japonica* L.

CAPRIFOLIACEAE
Sambucus L.

Sambucus nigra L. var. **laciniata** L. Cleft-leaved Black Elderberry. May–June. Horticultural variant. DuPage Co.: edge of Crowley Marsh, Morton Arboretum. This species is very different looking from typical var. *nigra* because of its deeply cleft leaflets.

CARYOPHYLLACEAE
Montia L.

This genus differs from *Portulaca* by its superior ovary, from *Claytonia* by having several pairs of opposite leaves, and from *Phemeranthus* by lacking a basal rosette of leaves.

Montia linearis (Dougl. ex Hook.) Greene. Montia. April. Native of the w. U.S.; adventive in fallow cropland; Clay Co.

FAGACEAE
Quercus L.

Quercus nigra L. Water Oak. Mar.–Apr. Native to the s. U.S. Adventive near a railroad near the headquarters of the Shawnee National Forest, Harrisburg, Saline Co.

LYTHRACEAE
Lythrum L.

An additional species of *Lythrum* has been found in Illinois.

Lythrum hyssopifolium L. Hyssop-leaved Loosestrife. June–Aug. Wet ground; McHenry Co. This species is similar to *L. alatum* because of its axillary flowers but differs by its annual habit and its narrower leaves that taper to the base. Its stamens are also not exserted above the corolla.

NELUMBONACEAE
Nelumbo L.

A second species of *Nelumbo* has been found in Illinois. The following key separates the two species known from Illinois:

1. Petals cream-colored; seeds spherical .... *N. lutea*
1. Petals pink or white; seeds oblong .... *N. nucifera*

Nelumbo nucifera Gaertn. Sacred Lotus. May–Aug. Native to Asia. A large colony of this handsome species has been found in Lake Vermilion, Vermilion Co.

POLYGONACEAE
Persicaria Mill.

An adventive *Persicaria* and a new variety have been found in Jackson Co. The adventive is *P. posumbu,* which is similar to the adventive *P. longiseta,* but differs by characters seen in the following key:

1. Plants to 60 cm tall; leaves without a lunate blotch on the upper surface; sepals 2.2–2.8 mm long; achenes 1.6–2.3 mm long; stamens 5 .... *P. longiseta*
1. Plants to 150 cm tall; leaves with a lunate blotch on the upper surface; sepals 2.0–2.5 mm long; achenes 2. 0–2.5 mm long; stamens 5–8 .... *P. posumbu*

Persicaria posumbu (Buch.-Ham. ex D. Don) H. Gross. June–Nov. Native to Asia; in addition to the Illinois location, this species has been reported from Pennsylvania. Jackson Co.; adventive and seemingly invasive south of Carbondale; apparently introduced in mulch. The new variety is as follows:

Persicaria hydropiperoides (Michx.) Small var. **bushiana** (Stanford) Mohlenbr., comb. nov. (basionym: *Polygonum hydropiperoides* Michx. var. *bushianum* Stanford). Jackson Co.: edge of a rocky stream in rich, moist woods. This variety differs from var. *hydropiperoides* by its much longer ocreae and ocreolae.

ROSACEAE

Cotoneaster Medic.

Cotoneaster apiculatus Rehder & E. H. Wilson. Cranberry Cotoneaster. June–July. Escaped from cultivation; DuPage Co.

Cotoneaster magnificus J. Fryer & B. Hylmo. Cotoneaster. June–July. DuPage Co: along a road and at the edge of a woodland.

SOLANACEAE

Petunia L.

Petunia parviflora Juss. Small-flowered Petunia. May–June. DuPage Co: mudflats and banks of small creeks.

CYPERACEAE

Cyperus L.

Cyperus polystachyos Rottb. Flatsedge. June–July. Wet ground, very rare; Williamson Co., collected by D. Ketzner in 2010. This is one of five species of Cyperus in Illinois that has 2 stigmas and a lenticular achene. It differs from *C. flavescens* by having a tan or brownish achene and from *C. rivularis* and *C. diandrus* by lacking purple markings on its scales. It is most similar to *C. filicinus,* differing by its narrower spikelets (1.2–2.0 mm wide) and shorter achenes (0.8–1.0 mm long).

POACEAE

Aristida L.

Aristida purpurea Nutt. Purple Three-awn. July–Sept. Dry soil, rare; Jersey and Lee cos.

Glyceria L.

Another species and a hybrid are reported from Illinois. A new key to the taxa of *Glyceria* in Illinois is provided.

1. Spikelets at least 10 mm long; sheaths compressed.
  2. Some or all the spikelets on pedicels 1.7–5.0 mm long; culms usually 1.5–5.0 mm thick; lemmas usually scabrous.
    3. All panicle branches strongly ascending; upper glumes 2–5 mm long; lemmas 5- or 7-nerved.
      4. Culms 1.5–2.5 mm thick; leaves 4–8 mm wide; lemmas 3.5–6.0 mm long; upper glume 2.5–5.0 mm long .... *G. declinata*
      4. Culms 3–5 mm thick; leaves 2–5 mm wide; lemmas 3–4 mm long; upper glume2–3 mm long .... *G. borealis*
    3. Upper panicle branches strongly ascending; lower panicle branches often spreading; upper glumes about 6 mm long; lemmas 11-nerved .... *G. X pedicellata*
  2. Spikelets on pedicels up to 1.7 mm long, or pedicels absent; culms usually 5–12 mm thick; lemmas hirtellous.

5. Leaves 4–12 mm wide; lemmas obscurely nerved, 3.5–5.5 mm long; anthers 1–2 mm long .... *G. septentrionalis*
5. Leaves 10–18 mm wide; lemmas sharply nerved, 2.5–3.0 mm long; anthers less than 1 mm .... *G. arkansana*

1. Spikelets 2–8 mm long; sheaths terete or subterete.
   6. Lemmas 3–4 mm long, obscurely nerved; spikelets 3–4 mm broad .... *G. canadensis*
   6. Lemmas 1.5–2.7 mm long, sharply nerved; spikelets 2.0–2.5 mm long.
      7. First glume 0.5–1.0 mm long; second glume 0.8–1.3 mm long; lemmas 1.5–2.0 mm long.......... *G. striata*
      7. First glume 1.2–2.4 mm long; second glume 1.5–3.0 mm long; lemmas 1.9–2.8 mm long.
         8. Panicle wide-spreading, 12 cm wide or wider; midvein of glumes extending to apex; first glume up to 2.3 mm long; apex of lemmas flat; culms 8 mm thick or thicker; ligules 1–5 mm long; anthers 3 .... *G. grandis*
         8. Panicle narrow, up to 1.5 cm wide; midvein of glumes not extending to apex; first glume 2.3–2.4 mm long; apex of lemmas prow-shaped; culms up to 5 mm thick; ligules less than 1 mm long; anthers 2 *G. melicaria*

Glyceria melicaria (Michx.) F.T. Hubb. Melic Manna Grass. June–Aug. Wet woods; reported from two w. cent. cos. in Illinois. (*Flora of North America,* vol. 24, 2007.)

Glyceria x pedicellata F. Towns. Hybrid Manna Grass. June–July. Reported from Cook Co. This is reputed to be the hybrid between *G. fluitans* (L.) R. Br. and *G. notata* Chevill. (Mapped in online version of *Manual of the Grasses of the United States,* Utah State University Herbarium.)

Miscanthus L.

Miscanthus sinensis Anderss. var. **gracillimus** Hitchc. Maiden Grass. Native to Europe. Williamson Co.: escaped from cultivation and rampant in an old field. This variety differs from var. *sinensis* by its leaves that are up to 5 mm wide and deeply canaliculate. Typical var. *sinensis* has leaves usually 8–12 mm wide and shallowly canaliculate.

Muhlenbergia Schreb.

Muhlenbergia schreberi J. F. Gmel. var. **palustris** (Scribn.) Scribn. ex B. L. Rob. This variety has been found in a disturbed area at the edge of a woods in Jackson Co. It appears to be intermediate between *M. schreberi* and *M. curtisetosa.* Its second glume is about 1 mm long; the second glume in *M. schreberi* is up to 0.5 mm long; the second glume of *M. curtisetosa* is 2–3 mm long.

Phyllostachys Carr.

Phyllostachys aurea Carr. ex Riv. & Riv. Golden Bamboo. May–June. Native to Asia; running rampant in a woods 1 mile s. of Carbondale, Jackson Co.

Last-minute additions from DuPage County unless otherwise noted: *Aesculus parviflora* Walt. (also Lake Co.); *Allium aflatunense* B. Fedtsch.; *Allium giganteum* Regel (also Will Co.); *Allium tuberosum* Rottl. ex Spreng.; *Anemone blanda* Schott and Kotschy; *Aristolochia clemtitis* L.; *Asarum europaeum* L.; *Helianthus X intermedius* R. W. Long; *Malus astracanica* (Dums.-Cours.) DC. (also Will Co.); *Malus floribunda* Van Houtte (also Will Co.) *Malus X purpurea* (Barbier) Rehd.; *Pulmonaria officianalis* L. (also Lake Co.); *Pulmonaria saccharata* Mill.; *Viburnum carlesii* Hemsl. Reported by Scott Kobel.

Excluded Species

DRYOPTERIDACEAE

Gymnocarpium x brittonianum (Sarvela) Pryer & Haufler. I can find no evidence that this hybrid has been found in IL.

ASTERACEAE

Hieracium venosum L. The report of this species from IL in *Flora of North America* (2006) is based on a misidentification of *Hieracium gronovii.* (Strothers, personal communication, 2013.)

CAPRIFOLIACEAE

Viburnum nudum L. Possumhaw; Raisin Tree. McAtee (1956) attributed this small tree to Jackson and Williamson cos., but I have not been able to find any specimens to verify this, although there is a good possibility that this southern species occurs in a swampy woods in s. IL.

FABACEAE

Centrosema virginianum (L.) Benth. Butterfly-pea. Personal information from a former student of mine, Dr. Paul Fantz, reveals that there is no evidence of this species in IL.

LAMIACEAE

Agastachye foeniculum (Pursh) Kuntze. The collection of this non-native species from Menard Co. is from a garden. It apparently does not occur as an adventive in IL.

POLYGONACEAE

Eriogonum annuum Nutt. Reported from IL by Horton in 1952, but I cannot find any specimens to substantiate this.

SCROPHULARIACEAE

Veronica agrestis L. The record for this species is actually *V. biloba* L.

VITACEAE

Vitis rupestris Scheele. My reports of this grape in the past are based on misidentifications. As far as I know, there are no known authentic specimens of this species from IL.

CYPERACEAE

Carex tincta (Fern.) Fern. Although IL is listed for this species in *Flora of North America,* personal communication with Anton Reznicek indicates that the reference is in error.

POTAMOGETONACEAE

Stuckenia vaginata (Turcz.) Holub. The specimen on which this report is based is actually *S. pectinata.*

SMILACINACEAE

Smilax herbacea L. My previous reports of this species from Jackson Co. are based on a misidentification. This herbaceous carrion flower apparently does not occur in IL.

Glossary

Acaulescent. Seemingly without aerial stems.

Achene. A type of one-seeded, dry, indehiscent fruit with the seed coat not attached to the mature ovary wall.

Acicular. Tapering to a needlelike point.

Actinomorphic. Having radial symmetry; regular, in reference to a flower.

Acuminate. Gradually tapering to a point.

Acute. Sharply tapering to a point.

Adaxial. Toward the axis; when referring to a leaf, the upper surface.

Alternate. Said of the arrangement of leaves where the leaves are not across from each other.

Annual. Living for a single year.

Anther. The terminal, pollen-bearing part of a stamen.

Anthesis. Flowering time.

Antrorse. Projecting forward.

Apex. The tip.

Apical. Pertaining to the apex.

Apiculate. Abruptly short-pointed at the tip.

Appressed. Lying flat against the surface.

Arachnoid. Cobwebby.

Areole. A small area between leaf veins.

Aristate. Bearing an awn.

Aromatic. Having an odor.

Array. The arrangement of flowers in an inflorescence.

Attenuate. Gradually becoming narrowed.

Auricle. An earlike lobe.

Auriculate. Bearing an earlike process.

Awn. A bristle usually terminating a structure.

Axil. The junction of two structures.

Axillary. Borne from an axil.

Barbellate. Possessing small downward-pointing hairs.

Basal. Located at the bottom of the plant.

Beaked. Possessing a small, terminal, pointed projection.

Bidentate. Having two teeth.

Biennial. Living for two years.

Bifid. Two-cleft.

Bifurcate. Forked.

Bipinnate. Divided once into distinct segments, with each segment in turn divided into distinct segments.

Bipinnatifid. Divided partway to the center, with each lobe again divided partway to the center.

Bisexual. Referring to a flower that contains both stamens and pistils.

Biternate. Divided into three segments two times.

Bract. An accessory structure at the base of many flowers, usually appearing leaflike.

Bracteate. Bearing one or more bracts.

Bracteole. A secondary bract.

Bractlet. A small bract.

Bristle. A stiff hair or hairlike growth; a seta.

Bulblet. A small bulb.

Bulbous. Swollen at the base.

Calcareous. Living in an environment that is basic and not acidic.

Calyx. The outermost segments of the perianth of a flower, composed of sepals.

Campanulate. Bell-shaped.

Canescent. Grayish-hairy.

Capillary. Threadlike.

Capitate. Forming a head.

Cartilaginous. Firm but flexible.

Caudate. With a tail-like appendage.

Caudex. The woody base of a perennial plant.

Cauline. Belonging to a stem.

Cespitose. Growing in tufts.

Chaff. A scale.

Chaffy. Covered with scales.

Chartaceous. Papery.

Cilia. Marginal hairs.

Ciliate. Bearing cilia.

Ciliolate. Bearing small cilia.

Coherent. Referring to like parts that have grown together.

Colonial. Said of plants growing together.
Columnar. Forming an elongated structure.
Compressed. Flattened.
Concave. Curved on the inner surface; opposed to convex.
Connivent. Growing together.
Convex. Curved on the outer surface; opposed to concave.
Convolute. Rolled lengthwise.
Cordate. Heart-shaped.
Coriaceous. Leathery.
Corm. An underground, vertical stem with scaly leaves, differing from a bulb by lacking fleshy leaves.
Corolla. The segments of a flower just within the calyx, composed of petals.
Corrugated. Folded or wrinkled.
Corymb. A type of inflorescence where the pedicellate flowers are arranged along an elongated axis but with the flowers all attaining about the same height.
Corymbiform. Shaped like a corymb.
Crateriform. With an indented center, like a crater.
Crenate. With round teeth.
Crenulate. With small, round teeth.
Crest. A ridge.
Crisped. Curled.
Culm. The stem that terminates in an inflorescence.
Cuneate. Wedge-shaped; tapering to the base.
Cuspidate. Terminating in a very short point.
Cyme. A type of broad and flattened inflorescence in which the central flowers bloom first.
Cymose. Bearing a cyme.
Cypsela. A dry, indehiscent fruit derived from an inferior ovary in the Asteraceae.

Deciduous. Falling away.
Decumbent. Lying flat but with the tip ascending.
Decurrent. Adnate to the petiole or stem and then extending beyond the point of attachment.
Deflexed. Turned downward.
Dehiscent. Splitting at maturity.
Dentate. With sharp teeth, the tips of which project outward.
Denticulate. With small, sharp teeth, the tips of which project outward.
Depressed. With a slight indention.
Diffuse. Loosely spreading.
Dioecious. With staminate flowers on one plant, pistillate flowers on another.
Disarticulate. To come apart; to become disjointed.
Disc. The central part of a flowering head in the Asteraceae.
Divergent. Spreading apart.
Double series. Occurring in two rows.
Drupe. A type of fruit in which the seed is surrounded by a hard, dry covering that, in turn, is surrounded by fleshy material.

Ebracteate. Without bracts.
Echinate. Spiny.
Eciliate. Without cilia.
Eglandular. Without glands.
Ellipsoid. Referring to a solid object that is broadest at the middle, gradually tapering to both ends.
Elliptic. Broadest at the middle, gradually tapering to both ends.
Emarginate. Having a shallow notch at the extremity.
Entire. Referring to the edge of a leaf that is without teeth.
Epaleate. Said of a receptacle that has no scales attached to it.
Epunctate. Without dots.
Erose. With an irregularly notched margin.

Falcate. Sickle-shaped.
Fascicle. Cluster.
Ferruginous. Rust-colored.
Fetid. Foul-smelling.
Fibrous. Referring to roots borne in tufts.
Filament. The part of the stamen supporting the anther.
Filiform. Threadlike.
Fimbriate. Fringed.
Flaccid. Weak; flabby.
Flexible. Readily bent without breaking.
Flexuous. Zigzag.
Floret. A small flower.
Fusiform. Tapering to a narrow tip at both ends, broadened at the middle.

Geniculate. Bent.
Glabrate. Becoming smooth.
Glabrous. Without pubescence or hairs.
Gland. An enlarged, spherical body functioning as a secretory organ.
Glandular. Bearing glands.
Glaucous. With a whitish covering that can be rubbed off.
Globose. Round; globular.
Glomerule. A small, compact cluster.
Glutinous. Covered with a sticky secretion.

Hastate. Spear-shaped; said of a leaf that is triangular with spreading basal lobes.
Hemispherical. Referring to a solid object that is half-circular.
Hirsute. Bearing stiff hairs.
Hirsutulous. Bearing minute, stiff hairs.
Hirtellous. Finely hirsute.
Hispid. Bearing rigid hairs.
Hispidulous. Bearing minute, rigid hairs.
Hoary. Grayish-white, usually referring to pubescence.
Hyaline. Transparent.

Imbricate. Overlapping.
Inferior. Referring to the position of the ovary when it is surrounded by the adnate portion of the floral tube or is embedded in the receptacle.
Inflexed. Turned inward.
Inflorescence. A cluster of flowers.
Internode. The area between two adjacent nodes.
Involucre. A circle of bracts that subtends a flower cluster.
Involute. Rolled inward.

Keel. A ridgelike process.

Laciniate. Divided into narrow, pointed divisions.
Lamina. The blade of a leaf.
Lanceolate. Lance-shaped; broadest near base, gradually tapering to the narrower apex.
Lanceoloid. Referring to a solid object that is broadest near base, gradually tapering to the narrower apex.
Latex. Milky juice.
Leaflet. An individual unit of a compound leaf.
Lenticular. Lens-shaped.
Ligulate. Bearing ray flowers in a head in the Asteraceae.
Linear. Elongated and uniform in width throughout.
Lobulate. With small lobes.
Locular. Referring to the locule, or cavity of the ovary or the anther.
Locule. The cavity of an ovary or an anther.
Loment. A fruit divided into one-seeded segments by transverse partitions.
Lustrous. Shiny.
Lyrate. Pinnatifid, with the terminal lobe much larger than the lower ones.

Membranous. Very thin.
Mesic. Moist.
Midvein. The central vein of a leaf.
Monoecious. Bearing both sexes in separate flowers on the same plant.
Mucro. A short, abrupt tip.
Mucronate. Possessing a short, abrupt tip.
Mucronulate. Possessing a very short, abrupt tip.

Nerved. Having veins.
Neutral. Said of a flowering head in the Asteraceae that is neither pistillate, staminate, or fertile.
Nigrescent. Blackish.
Node. The place on the stem from which leaves and branchlets arise.

Obconic. An upside-down cone.
Oblanceolate. Reverse lance-shaped; broadest at apex, gradually tapering to narrow base.
Oblong. Broadest at the middle and tapering to both ends but broader than elliptic.
Oblongoid. Referring to a solid object that in side view is nearly the same width throughout.
Obovoid. Referring to a solid object that is broadly rounded at the apex, becoming narrowed below.
Obtuse. Rounded at the apex.
Opaque. Not allowing the passage of light; not see-through.
Opposite. Said of a leaf arrangement where the leaves are across from each other.
Orbicular. Round, in three dimensions.
Oval. Broadly elliptic.
Ovary. The lower swollen part of the pistil that produces the ovules.
Ovate. Broadly rounded at base, becoming narrowed above; broader than lanceolate.
Ovoid. Referring to a solid object that is broadly rounded at the base, becoming narrowed above.

Palea. The scale attached to the receptacle in the Asteraceae.
Paleate. A receptacle in the Asteraceae that bears scales.
Palmate. Divided radiately, like the fingers of a hand.
Panicle. A type of inflorescence composed of several racemes.
Paniculate. Said of flowers borne in a panicle.
Pannose. Having the texture of felt.
Papilla. A small wart.
Papillate. Bearing small warts, or papillae.

Papillose. Bearing pimplelike processes.
Pappus. The modified calyx in the Asteraceae.
Pectinate. Pinnatifid into close, narrow segments; comblike.
Pedicel. The stalk of a flower of an inflorescence.
Pedicellate. Bearing a pedicel.
Peduncle. The stalk of an inflorescence.
Peduncled. Provided with a peduncle.
Peltate. Attached away from the margin, in reference to a leaf.
Perennial. Living more than two years.
Perfect. Bearing both stamens and pistils in the same flower.
Perfoliate. Referring to a leaf that appears to have the stem pass through it.
Perianth. Those parts of a flower including both the calyx and corolla.
Petiolate. Bearing a petiole, or leafstalk.
Petiole. The stalk of a leaf.
Petiolulate. Bearing a petiolule, or leaflet-stalk.
Petiolule. The stalk of a leaflet.
Phyllary. A bract subtending the involucre in the Asteraceae.
Pilose. Bearing soft hairs.
Pinna. A primary division of a compound blade.
Pinnate. Divided once into distinct segments.
Pinnatifid. Said of a simple leaf or leaf-part that is cleft or lobed only partway to its axis.
Pistil. The ovule-producing organ of a flower normally composed of an ovary, a style, and a stigma.
Pistillate. Bearing pistils but not stamens.
Plicate. Folded.
Plumose. Bearing fine hairs, like the plume of a feather.
Prismatic. Referring to a structure whose faces are parallel to the same axis.
Procumbent. Lying on the ground.
Prostrate. Lying flat on the ground.
Puberulent. Bearing minute hairs.
Pubescent. Bearing some kind of hairs.
Punctate. Dotted.
Pustular. With a swollen base.
Pyramidal. Shaped like a pyramid.

Raceme. A type of inflorescence where pedicellate flowers are arranged along an elongated axis.
Racemose. Having the flowers borne in a raceme.
Receptacle. The part of the flower to which the perianth, stamens, and pistils are usually attached.
Reflexed. Turned downward.
Repent. Creeping.
Resinous. Producing a sticky secretion, or resin.
Reticulate. Resembling a network.
Retrorse. Pointing downward.
Retuse. Shallowly notched at a rounded apex.
Revolute. Rolled under from the margin.
Rhizomatous. Having rhizomes.
Rhizome. An underground horizontal stem bearing nodes, buds, and roots.
Rhombic. Becoming quadrangular.
Ribbed. With strong veins.
Rosette. A cluster of leaves in a circular arrangement at the base of a plant.
Rotate. Flat and circular.
Rufescent. Reddish brown.
Rufous. Red-brown.
Rugose. Wrinkled.
Rugulose. With small wrinkles.

Sagittate. Shaped like an arrowhead.
Salverform. Referring to a tubular corolla that abruptly expands into a flat limb.
Scabrous. Rough to the touch.
Scale. A small outgrowth.
Scape. A leafless stalk bearing a flower or inflorescence.
Scarious. Thin and membranous.
Scurfy. Bearing scaly particles.
Secund. Borne on one side.
Septate. With dividing walls.
Sericeous. Silky; bearing soft, appressed hairs.
Serrate. With teeth that project forward.
Serrulate. With very small teeth that project forward.
Sessile. Without a stalk.
Seta. Bristle.
Setaceous. Bearing bristles, or setae.
Setiform. Bristle-shaped.
Setose. Bearing setae.
Silicle. A short silique.
Silique. An elongated capsule with a central partition separating the valves.
Sinuate. Wavy along the margins.
Sinus. The cleft between two lobes or teeth.
Sordid. Dark-looking; having the color of mud.
Spatulate. Oblong but with the basal end elongated.
Spicate. Bearing a spike.
Spike. A type of inflorescence where sessile flowers are arranged along an elongated axis.
Spikelet. A small spike.
Spinescent. Becoming spiny.
Spinose. Bearing spines.

Spinule. A small spine.
Spinulose. Bearing small spines.
Stamen. The pollen-producing organ of a flower composed of a filament and an anther.
Staminate. Bearing stamens but not pistils.
Stellate. Star-shaped.
Stipe. A stalk.
Stipitate. Possessing a stipe.
Stipular. Pertaining to a stipule.
Stipule. A leaflike or scaly structure found at the point of attachment of a leaf to the stem.
Stolon. A slender, horizontal stem on the surface of the ground.
Stoloniferous. Bearing stolons.
Stramineous. Straw-colored.
Striate. Marked with grooves.
Strigillose. Bearing short, appressed, straight hairs.
Strigose. Bearing appressed, straight hairs.
Strigulose. Bearing short, appressed, straight hairs.
Style. The elongated part of the pistil between the ovary and the stigma.
Subacute. Nearly pointed at the tip.
Subcordate. Nearly heart-shaped.
Subcuneate. Nearly wedge-shaped.
Suborbicular. Nearly spherical.
Subsessile. Nearly sessile.
Subshrubby. Referring to a plant that is almost shrubby.
Subulate. With a very short, narrow point.
Succulent. Fleshy.
Suffused. Spread throughout; flushed.
Superior. Referring to the position of the ovary when the free floral parts arise below the ovary.

Tawny. Light brown or sand colored.
Taxon (*pl.* **taxa**). A category in classification, such as genus, species, variety, etc.
Terete. Round in cross-section.
Ternate. Divided three times.
Terrestrial. Growing on land.
Testa. The seed coat.
Thyrse. An arrangement of flowers.
Thyrsoid. Having the flowers arranged in a thyrse.
Tomentose. Pubescent with matted wool.
Tomentulose. Finely pubescent with matted wool.
Tomentum. Woolly hair.
Torulose. With small contractions.
Translucent. Partly transparent.
Trifoliolate. Divided into three leaflets.
Trigonous. Triangular in cross-section.
Truncate. Abruptly cut across.
Tuber. An underground fleshy stem formed as a storage organ at the end of a rhizome.
Tubular. In the form of a tube.
Turbinate. Shaped like a turban.

Undulate. Wavy.
Unisexual. Bearing either stamens or pistils in one flower but not both.
Urceolate. Urn-shaped.

Verrucose. Warty.
Verticil. A whorl.
Verticillate. Whorled.
Villous. Bearing long, soft, slender, unmatted hairs.
Virgate. Wandlike.
Viscid. Sticky.

Whorl. An arrangement of three or more structures at a point on the stem.

Zygomorphic. Bilaterally symmetrical.

Index to Families and Genera

Names in roman type are accepted names, while those in italics are synonyms and are not considered valid. Page numbers in italics indicate invalid references.

Index to Common Names

Robert H. Mohlenbrock taught botany at Southern Illinois University Carbondale for thirty-four years. After his retirement in 1990, he joined Biotic Consultants as a senior scientist teaching wetland identification classes. Among his hundreds of publications are those in the series Illustrated Flora of Illinois. Since 1984, he has written the monthly column This Land for *Natural History* magazine.